Sustainable Development Perspectives in Earth Observation

Earth Observation

Sustainable Development Perspectives in Earth Observation

Edited by

Mukunda Dev Behera
Centre for Ocean, River, Atmosphere and Land Sciences (CORAL), Indian Institute of Technology Kharagpur, Kharagpur, West Bengal, India

Swadhin K. Behera
Applications Laboratory, Research Institute for Value Added Information Generation, Japan Agency for Marine-Earth Science and Technology, Yokohama, Japan

Saroj Kanta Barik
Centre for Advanced Studies in Botany, North-Eastern Hill University, Shillong, India

Mrutyunjaya Mohapatra
Indian Meteorological Department, Government of India, New Delhi, India

Trilochan Mohapatra
Protection of Plant Varieties and Farmers Rights Authority, Ministry of Agriculture & Farmers Welfare, Government of India, New Delhi, India

Series Editor

George P. Petropoulos

ELSEVIER

Elsevier
Radarweg 29, PO Box 211, 1000 AE Amsterdam, Netherlands
125 London Wall, London EC2Y 5AS, United Kingdom
50 Hampshire Street, 5th Floor, Cambridge, MA 02139, United States

ISBN: 978-0-443-14072-3

For Information on all Elsevier publications
visit our website at https://www.elsevier.com/books-and-journals

Publisher: Candice Janco
Acquisitions Editor: Jenette McClain
Editorial Project Manager: Rupinder Heron
Production Project Manager: Paul Prasad Chandramohan
Cover Designer: Mark Rogers

Typeset by MPS Limited, Chennai, India

Contents

Section B
Forest and land cover systems and climate change

10. Remote sensing of biodiversity and climate change: adaptation and mitigation prospective

Mukunda Dev Behera, S. Mahanand, P. Tripathi, M.P. Menachery, Soumit K. Behera, Saroj Kanta Barik and Mrutyunjaya Mohapatra

11. Linking meteorological and phenological observations in forests

A.J. Prakash, Mukunda Dev Behera, M. Mukhopadhyay, C.P. Singh, B.K. Bhattacharya, C. Jeganathan, R.K. Chaturvedi and Mrutyunjaya Mohapatra

20. An overview of atmospheric carbon dioxide and global warming interlinks

Chiranjit Das, Prabir K. Patra and Ravi Kumar Kunchala

21. Advances in Earth Observation Satellites for global air quality monitoring

Jayanarayanan Kuttippurath and Vikas Kumar Patel

22. Assessment of nitrogen dioxide concentrations in nonattainment cities during COVID-19 lockdown using GIS technique

Aniket Raut, Prakhar Misra, Nirmal Kumar, Hitesh Supe and Ram Avtar

23. Forecasting CO_2 emissions over India using deep learning models

Pringale Kumar Das, Mukunda Dev Behera, Buddolla Jagadish and Pulakesh Das

24. Role of ocean and climate in terrestrial ecosystems

Swadhin K. Behera, Roma Varghese, J.V. Ratnam, Takeshi Doi and Mukunda Dev Behera

List of contributors

Dibyendu Adhikari Plant Ecology and Climate Change Science, CSIR-National Botanical Research Institute, Lucknow, Uttar Pradesh, India

F. Ahmad World Agroforestry (CIFOR-ICRAF), New Delhi, India

Roniki Anjaneyulu Department of Civil Engineering, National Institute of Technology (NIT), Rourkela, Odisha, India

K. Athira Department of Civil Engineering, Indian Institute of Technology Palakkad, Palakkad, Kerala, India

Ram Avtar Faculty of Environmental Earth Science, Hokkaido University, Sapporo, Japan

Chandrasekaran B. Regional Remote Sensing Centre-South, National Remote Sensing Centre (NRSC), Indian Space Research Organisation (ISRO), ISITE Campus, Bengaluru, Karnataka, India

Abigith Baby Deutsche Gesellschaft für Internationale Zusammenarbeit (GIZ), New Delhi, India

Saroj Kanta Barik Centre for Advanced Studies in Botany, North-Eastern Hill University, Shillong, India

Mukunda Dev Behera Centre for Ocean, River, Atmosphere and Land Sciences (CORAL), Indian Institute of Technology Kharagpur, Kharagpur, West Bengal, India

Soumit K. Behera CSIR-National Botanical Research Institute, Lucknow, Uttar Pradesh, India

Swadhin K. Behera Applications Laboratory, Research Institute for Value Added Information Generation, Japan Agency for Marine-Earth Science and Technology, Yokohama, Japan

B.K. Bhattacharya Space Application Center (ISRO-SAC), Ahmedabad, Gujarat, India

C.M. Biradar World Agroforestry (CIFOR-ICRAF), New Delhi, India

Anima Biswal Agricultural Sciences and Applications Group, National Remote Sensing Centre, Indian Space Research Organization, Hyderabad, Telangana, India

S.K. Biswal Odisha State Wildlife Department, Baripada, Odisha, India

Abhishek Chakraborty Agricultural Sciences and Applications Group, National Remote Sensing Centre, Indian Space Research Organization, Hyderabad, Telangana, India

R.K. Chaturvedi Department of Humanities and Social Sciences, BITS Pilani, K K Birla Goa Campus, Vasco da Gama, Goa, India

Prakash Chauhan National Remote Sensing Centre, Indian Space Research Organization, Hyderabad, Telangana, India

Sai Prabala Swetha Chittella Centre for Ocean, River, Atmosphere and Land Sciences (CORAL), Indian Institute of Technology Kharagpur, Kharagpur, West Bengal, India

Karun Kumar Choudhary Agricultural Sciences and Applications Group, National Remote Sensing Centre, Indian Space Research Organization, Hyderabad, Telangana, India

R. Choudhary World Agroforestry (CIFOR-ICRAF), New Delhi, India

V.M. Chowdary Agricultural Sciences and Applications Group, National Remote Sensing Centre, Indian Space Research Organization, Hyderabad, Telangana, India

Chiranjit Das Centre for Atmospheric Sciences, Indian Institute of Technology Delhi, New Delhi, NCT of Delhi, India

Pinaki Das Department of Geography, Savitribai Phule Pune University, Pune, Maharashtra, India

Pringale Kumar Das Centre for Ocean, River, Atmosphere and Land Sciences (CORAL), Indian Institute of Technology Kharagpur, Kharagpur, West Bengal, India

Pulakesh Das School of Forest Resources, University of Maine, Orono, ME, United States

Pranab Deb Centre for Ocean, River, Atmosphere and Land Sciences (CORAL), Indian Institute of Technology Kharagpur, Kharagpur, West Bengal, India

Pramit Kumar Deb Burman Centre for Climate Change Research, Indian Institute of Tropical Meteorology, Ministry of Earth Sciences, Pune,

Maharashtra, India; Department of Atmospheric and Space Sciences, Savitribai Phule Pune University, Pune, Maharashtra, India

S.K. Dhyani World Agroforestry (CIFOR-ICRAF), New Delhi, India

A. Dogra World Agroforestry (CIFOR-ICRAF), New Delhi, India

Takeshi Doi Applications Laboratory, Research Institute for Value Added Information Generation, Japan Agency for Marine-Earth Science and Technology, Yokohama, Japan

Sanjiban Dutta Centre for Ocean, River, Atmosphere and Land Sciences (CORAL), Indian Institute of Technology Kharagpur, Kharagpur, West Bengal, India

P.C. Gogineni Odisha State Wildlife Department, Baripada, Odisha, India

S. Gowda Odisha State Wildlife Department, Baripada, Odisha, India

R. Hebbar Regional Remote Sensing Centre-South, National Remote Sensing Centre (NRSC), Indian Space Research Organisation (ISRO), ISITE Campus, Bengaluru, Karnataka, India

Buddolla Jagadish Centre for Ocean, River, Atmosphere and Land Sciences (CORAL), Indian Institute of Technology Kharagpur, Kharagpur, West Bengal, India

C. Jeganathan Department of Remote Sensing, Birla Institute of Technology Mesra, Ranchi, Jharkhand, India

V.S. Jeyakanthan Deltaic Regional Centre, National Institute of Hydrology, Kakinada, Andhra Pradesh, India

C.S. Jha Agricultural Sciences and Applications Group, National Remote Sensing Centre, Indian Space Research Organization, Hyderabad, Telangana, India

R. Joshi Sikkim Regional Centre, G.B. Pant National Institute of Himalayan Environment, Gangtok, Sikkim, India

Nagajothi K. Regional Remote Sensing Centre-South, National Remote Sensing Centre (NRSC), Indian Space Research Organisation (ISRO), ISITE Campus, Bengaluru, Karnataka, India

A. Kashyap Centre for Ocean, River, Atmosphere and Land Sciences (CORAL), Indian Institute of Technology Kharagpur, Kharagpur, West Bengal, India

Stephanie Katsir Deutsche Gesellschaft für Internationale Zusammenarbeit (GIZ), New Delhi, India

Nirmal Kumar Graduate School of Environmental Science, Hokkaido University, Sapporo, Japan

S. Kumar Centre for Ocean, River, Atmosphere and Land Sciences (CORAL), Indian Institute of Technology Kharagpur, Kharagpur, West Bengal, India

Sourav Kumar Department of Geoinformatics, School of Natural Resources Management, Central University of Jharkhand, Ranchi, India

Mamta Kumari Agricultural Sciences and Applications Group, National Remote Sensing Centre, Indian Space Research Organization, Hyderabad, Telangana, India

Ravi Kumar Kunchala Centre for Atmospheric Sciences, Indian Institute of Technology Delhi, New Delhi, NCT of Delhi, India

A. Kurein Centre for Ocean, River, Atmosphere and Land Sciences (CORAL), Indian Institute of Technology Kharagpur, Kharagpur, West Bengal, India

Jayanarayanan Kuttippurath Centre for Ocean, River, Atmosphere and Land Sciences (CORAL), Indian Institute of Technology Kharagpur, Kharagpur, West Bengal, India

B. Laxman Agricultural Sciences and Applications Group, National Remote Sensing Centre, Indian Space Research Organization, Hyderabad, Telangana, India

Nagashree M.K. Regional Remote Sensing Centre-South, National Remote Sensing Centre (NRSC), Indian Space Research Organisation (ISRO), ISITE Campus, Bengaluru, Karnataka, India

S. Mahanand Centre for Ocean, River, Atmosphere and Land Sciences (CORAL), Indian Institute of Technology Kharagpur, Kharagpur, West Bengal, India

M.P. Menachery Centre for Ocean, River, Atmosphere and Land Sciences (CORAL), Indian Institute of Technology Kharagpur, Kharagpur, West Bengal, India

Prakhar Misra Department of Civil Engineering, Indian Institute of Technology, Roorkee, Uttarakhand, India

Mrutyunjaya Mohapatra Indian Meteorological Department, Government of India, New Delhi, India

Trilochan Mohapatra Protection of Plant Varieties and Farmers Rights Authority, Ministry of Agriculture & Farmers Welfare, Government of India, New Delhi, India

Kishore Mote J. Coffee Board, Ministry of Commerce, Government of India, Bengaluru, Karnataka, India

Sujoy Mudi Centre for Ocean, River, Atmosphere and Land Sciences (CORAL), Indian Institute of

Technology Kharagpur, Kharagpur, West Bengal, India

D.S. Mukharib Coffee Board, Ministry of Commerce, Government of India, Bengaluru, Karnataka, India

M. Mukhopadhyay Centre for Ocean, River, Atmosphere and Land Sciences (CORAL), Indian Institute of Technology Kharagpur, Kharagpur, West Bengal, India

Vinod P.V. Regional Remote Sensing Centre-South, National Remote Sensing Centre (NRSC), Indian Space Research Organisation (ISRO), ISITE Campus, Bengaluru, Karnataka, India

Varun Pandey Agricultural Sciences and Applications Group, National Remote Sensing Centre, Indian Space Research Organization, Hyderabad, Telangana, India

Bikash Ranjan Parida Department of Geoinformatics, School of Natural Resources Management, Central University of Jharkhand, Ranchi, India

Vikas Kumar Patel Centre for Ocean, River, Atmosphere and Land Sciences (CORAL), Indian Institute of Technology Kharagpur, Kharagpur, West Bengal, India

Prabir K. Patra Japan Agency for Marine-Earth Science and Technology (JAMSTEC), Yokohama, Kanagawa, Japan; Research Institute for Humanity and Nature, Kyoto, Kyoto Prefecture, Japan

A.J. Prakash Centre for Ocean, River, Atmosphere and Land Sciences (CORAL), Indian Institute of Technology Kharagpur, Kharagpur, West Bengal, India

A. Raj Centre for Ocean, River, Atmosphere and Land Sciences (CORAL), Indian Institute of Technology Kharagpur, Kharagpur, West Bengal, India

Parichay S. Raju Agricultural Sciences and Applications Group, National Remote Sensing Centre, Indian Space Research Organization, Hyderabad, Telangana, India

R. Venkata Ramana Deltaic Regional Centre, National Institute of Hydrology, Kakinada, Andhra Pradesh, India

Y.R. Satyaji Rao Deltaic Regional Centre, National Institute of Hydrology, Kakinada, Andhra Pradesh, India

J.V. Ratnam Applications Laboratory, Research Institute for Value Added Information Generation, Japan Agency for Marine-Earth Science and Technology, Yokohama, Japan

Aniket Raut Graduate School of Environmental Science, Hokkaido University, Sapporo, Japan

A.H. Rizvi World Agroforestry (CIFOR-ICRAF), New Delhi, India

J. Rizvi World Agroforestry (CIFOR-ICRAF), New Delhi, India

Rama Subramoniam S. Regional Remote Sensing Centre-South, National Remote Sensing Centre (NRSC), Indian Space Research Organisation (ISRO), ISITE Campus, Bengaluru, Karnataka, India

Bhavana Sahay Agricultural Sciences and Applications Group, National Remote Sensing Centre, Indian Space Research Organization, Hyderabad, Telangana, India

M.R. Sahu Centre for Ocean, River, Atmosphere and Land Sciences (CORAL), Indian Institute of Technology Kharagpur, Kharagpur, West Bengal, India

Sudam C. Sahu Department of Botany, Maharaja Sriram Chandra Bhanja Deo University, Baripada, Odisha, India

Chinju Saju Department of Civil Engineering, Indian Institute of Technology Palakkad, Palakkad, Kerala, India

Dhrubajyoti Samanta Earth Observatory of Singapore, Nanyang Technological University, Singapore

Arnab Sen Centre for Ocean, River, Atmosphere and Land Sciences (CORAL), Indian Institute of Technology Kharagpur, Kharagpur, West Bengal, India

Divya Shah Food and Agriculture Organisation of United Nations, New Delhi, India

A. Singh World Agroforestry (CIFOR-ICRAF), New Delhi, India

C.P. Singh Space Application Center (ISRO-SAC), Ahmedabad, Gujarat, India

R.K. Singh World Agroforestry (CIFOR-ICRAF), New Delhi, India; Centre for Ocean, River, Atmosphere and Land Sciences (CORAL), Indian Institute of Technology Kharagpur, Kharagpur, West Bengal, India

Sarmistha Singh Department of Civil Engineering, Indian Institute of Technology Palakkad, Palakkad, Kerala, India; ESSENCE, Indian Institute of Technology Palakkad, Palakkad, Kerala, India

R.B. Sinha Food and Agriculture Organisation of United Nations, New Delhi, India

K. Sreenivas Agricultural Sciences and Applications Group, National Remote Sensing Centre, Indian Space Research Organization, Hyderabad, Telangana, India

P. Srikanth Agricultural Sciences and Applications Group, National Remote Sensing Centre, Indian Space Research Organization, Hyderabad, Telangana, India

S.K. Srivastav Regional Centres, NRSC, Indian Space Research Organisation (ISRO), New Delhi, India

Ishita S. Srivastava School of Environmental Science and Engineering, Indian Institute of Technology Kharagpur, Kharagpur, West Bengal, India

Hitesh Supe Graduate School of Environmental Science, Hokkaido University, Sapporo, Japan

Ratnakar Swain Department of Civil Engineering, National Institute of Technology (NIT), Rourkela, Odisha, India

Sabyasachi Swain Department of Civil Engineering, Malaviya National Institute of Technology, Jaipur, Rajasthan, India

P. Tripathi International Centre for Integrated Mountain Development, Kathmandu, Nepal

Shivam Trivedi Regional Remote Sensing Centre-South, National Remote Sensing Centre (NRSC), Indian Space Research Organisation (ISRO), ISITE Campus, Bengaluru, Karnataka, India

Roma Varghese Centre for Ocean, River, Atmosphere and Land Sciences (CORAL), Indian Institute of Technology Kharagpur, Kharagpur, West Bengal, India; Indian Institute of Tropical Meteorology, Pune, Maharashtra, India

Swapnil Vyas Department of Geography, Savitribai Phule Pune University, Pune, Maharashtra, India

K.G. Yadhukrishna Centre for Ocean, River, Atmosphere and Land Sciences (CORAL), Indian Institute of Technology Kharagpur, Kharagpur, West Bengal, India

About the editors

Mukunda Dev Behera is a renowned scientist in Earth observation of ecological climatology, with over 25 years of experience. Currently, an Associate Professor at the Centre for Ocean, River, Atmosphere, and Land Sciences (CORAL) at the Indian Institute of Technology Kharagpur, he leads the Spatial Analysis and Modelling Laboratory. He previously worked as a scientist with the Indian Space Research Organization (ISRO). He has made outstanding contributions to the fields of forest informatics, land and water resource management, and nature conservation by theorizing, modeling, and conducting innovative experiments and field-based measurements with satellite remote sensing. He has developed innovative protocols and new methods for estimating a range of vegetation photosynthetic variables using a suite of satellite data products and modeling. With over 160 peer-reviewed publications and 5 co-edited books, he also serves on the editorial boards of several esteemed environmental journals and contributed to the IPCC AR-6 of WG-II.

Swadhin K. Behera is a Distinguished Scientist in climate variability and predictability research, with over 26 years of experience at JAMSTEC, Japan. Through the integration of Earth observation data and numerical modeling, his work focuses on climate analysis, prediction, and applications, particularly the role of the ocean in driving tropical and extratropical climate variations. He is also actively involved in the UN Ocean Decade's Digital Twins of the Ocean program, contributing to the development of ocean-based solutions to global challenges. He currently heads the Application Laboratory and is also an Adjunct Professor at the University of Tokyo. He has published over 200 peer-reviewed journal articles in highly cited international journals and has edited several books, including *Tropical and Extratropical Air-Sea Interactions* (Elsevier).

Saroj Kanta Barik is a Professor at the Department of Botany, North-Eastern Hill University, Shillong, India. He served as the Director of the Council of Scientific and Industrial Research (CSIR)-National Botanical Research Institute (NBRI), Lucknow, and a Director of the CSIR-Indian Institute of Toxicology Research (IITR), Lucknow. For the last three and half decades, he has been active in diverse research areas such as bioprospection and biodiversity conservation, ecological modeling, community, population and molecular ecology, and impact of climate change on species and ecosystems. His group has been successful in using satellite data and global climate models to predict the distributional range and impact of climate change on several forest species. He has published 150 research papers in scientific journals of international repute, 12 books, and 40 technical and research reports.

Mrutyunjaya Mohapatra is the Director General of Meteorology in the India Meteorological Department (IMD), Government of India. He is also the Permanent Representative of India and member of the Executive Council for the World Meteorological Organization (WMO), and Vice-President. He has significantly contributed to the improvement in cyclone warning services by IMD and led the upgradation of cyclone warning services in a holistic manner using Numerical Weather Prediction (NWP). He has received appreciations from the WMO and Indian Ministry of Earth Sciences for providing timely and accurate tropical cyclone advisories to Myanmar during cyclone, Nargis in 2008. He has published more than 125 research papers and 15 Books.

Trilochan Mohapatra is an eminent Agricultural Biotechnologist and Geneticist of international repute and is currently serving as a Chairperson of the Protection of Plant Varieties and Farmers' Rights Authority, Government of India (GoI). Earlier, he served as a GoI Secretary in the Department of Agricultural Research and Education, and a Director General of the Indian Council of Agricultural Research, Ministry of Agriculture and Farmers' Welfare. He also served as the Director-cum-Vice Chancellor of the Indian Agricultural Research Institute (IARI), New Delhi, and worked as the Director of the National Rice Research Institute, Cuttack, India. He worked at the National Research Centre on Plant Biotechnology, IARI, New Delhi as a Scientist for over 30 years. He has published more than 200 research papers in national and international peer-reviewed journals in the fields of improvement and functional genomics of agricultural crops.

Preface

In an era of unprecedented environmental change and looming resource scarcity, Earth observation has become essential to support analysis and prediction that drive mitigation and adaptation efforts in our collective quest for sustainability. As nations strive to achieve the United Nations Sustainable Development Goals (SDGs), innovative tools and methodologies that improve our understanding of environmental, social, and economic dynamics are essential. The increasing frequency of extreme weather events, shifting climate patterns, and the ecosystem degradation require a scientific approach that is both innovative and comprehensive. *Sustainable Development Perspectives in Earth Observation* was conceived in response to this need, providing a broad yet detailed exploration of how Earth observation technologies and methods can be used to address some of the most pressing environmental challenges of our time.

This book brings together expert contributions from a range of disciplines by providing insights into terrestrial, oceanic, and atmospheric processes. It addresses the complex interplay among these spheres, especially in the context of climate change, and explores how these interconnections influence global and regional ecosystems. With a focus on agricultural research, climate variation, ocean—atmosphere dynamics, forest resource management, and biogeophysical processes, the volume serves as a comprehensive resource for understanding the critical role of Earth observation in promoting sustainable development. For example, several chapters discuss how advances in Earth observation technologies are transforming agricultural practices, including crop forecasting, agroforestry, floriculture, and crop insurance, thereby enabling more precise and resilient responses to climate variability. In parallel, this book discusses geospatial insights that are critical for effective water management, particularly in dealing with hydroclimatic extremes and developing adaptive strategies for both terrestrial and coastal ecosystems. In this context, this book also delves into the dynamics of ocean—atmosphere interactions, highlighting phenomena such as El Niño, the Indian Ocean Dipole, and tropical cyclones that affect hydrological cycle and ecosystem. By offering insights into tools and techniques that facilitate informed decision-making, this work aims to contribute to more resilient agricultural practices, better natural resource management, and improved climate adaptation strategies. These discussions provide a solid scientific foundation for understanding climate variability while also underscoring the urgent need for more effective climate mitigation strategies. With a focus on linking theory to practice, this book offers practical solutions for professionals working to enhance ecosystem productivity, manage water resources, and preserve biodiversity in an era of climate uncertainty. As the global community works to achieve the SDGs, the knowledge and tools derived from Earth observation technologies will be critical to developing policies that promote environmental sustainability, economic resilience, and social equity.

This book is intended for a broad readership, scientists, academics, researchers, students, and professionals in the fields of climate science, atmospheric and oceanic science, agriculture, forestry, and environmental science. As the global community continues to confront the realities of environmental change, the knowledge shared within these pages serves as both a guide and a call to action for managers and policymakers in both the public and private sectors. They will find value in the scientific advancements discussed here valuable for informing and updating policies aimed at achieving SDGs. It is our hope that this volume will empower readers with the tools and understanding necessary to develop sustainable solutions and to engage meaningfully with the challenges that lie ahead. We extend our deepest gratitude to all the contributors, whose expertise and commitment have made this book a vital resource for those dedicated to safeguarding our planet's future.

Chapter 1

Inter-linkage among climate, terrestrial and marine systems: sustainable development perspectives

Mukunda Dev Behera[1], Swadhin K. Behera[2], Saroj Kanta Barik[3], Mrutyunjaya Mohapatra[4] and Trilochan Mohapatra[5]

[1]Centre for Ocean, River, Atmosphere and Land Sciences (CORAL), Indian Institute of Technology Kharagpur, Kharagpur, West Bengal, India, [2]Applications Laboratory, Research Institute for Value Added Information Generation, Japan Agency for Marine-Earth Science and Technology, Yokohama, Japan, [3]Centre for Advanced Studies in Botany, North-Eastern Hill University, Shillong, India, [4]Indian Meteorological Department, Government of India, New Delhi, India, [5]Protection of Plant Varieties and Farmers Rights Authority, Ministry of Agriculture & Farmers Welfare, Government of India, New Delhi, India

1.1 Agriculture, water, and climate change

The studies on how geospatial technologies are transforming agriculture reveal the critical role of these tools in addressing modern challenges in this field. Chowdary et al. (2025) provide a thorough examination of geospatial technologies, including remote sensing, GIS, and GPS, which have become indispensable for precise crop assessment and management, especially in the face of global issues like population growth and food security. Remote sensing, with its ability to assess crop health, monitor growth, and estimate yield across large areas, is central to this evident technological shift. By providing real-time data on vegetation, soil conditions, and crop health through indices like NDVI, LSWI, and so on, remote sensing enables decision-making, which is crucial for sustainable agricultural practices. Their research demonstrates the accuracy and effectiveness that these technologies provide, which facilitate the improved decision-making procedures for inventory control and agricultural production modeling. Optimizing agricultural operations is highly relevant, particularly in light of the growing uncertainty around climate change. In addition, Singh, Behera, et al. (2025) and Singh, Saju, and Athira (2025) explore agroforestry as a sustainable system of land use. The important ecological and economic advantages of agroforestry, which contribute to enhanced biodiversity, carbon sequestration, and climate adaptation, are highlighted. They also stress the value of citizen science and geoinformatics in advancing agroforestry techniques, which are in line with the larger objectives of sustainable development. Barik, Behera, and Adhikari (2025), who demonstrate for the first time the complex link between floral fragnance and changing climatic conditions using ecological niche models, emphasize the significance of floriculture in the context of climate change. Their study on the distribution of aromatic plant species, such as tuberose and jasmine, provides sustainable methods for growing floricultural crops to get the best fragrance in the face of climate change. The study also offers a useful foundation for comprehending the complex interactions between plants and their environment.

Crop insurance is another important aspect of agricultural sustainability that Kumar and Parida (2025) investigate using geospatial technologies. The potential of remote sensing-based models to inform crop insurance systems, alleviate climate risks, and help farming communities for their improved livelihood is highlighted by their work on crop damage assessment during flood occurrences in Bihar. Such an approach is essential not only for calculating premiums and payouts but also for strengthening the resilience capability of farmers in the face of climate change. Based on the most recent climate model simulations, Das, Patra, and Kunchala (2025), Das, Behera, Jagadish, and Das (2025), Das, Behera, Barik, and Mudi (2025), and Das, Das, Vyas, and Deb Burman (2025) explore the critical role of evapotranspiration (ET) and precipitation in India's water cycle, emphasizing their response to climate change. Using multi-model ensemble simulations from the Coupled Model Intercomparison Project (CMIP 6), this study provides significant insight into how climate variables are projected to change by the end of the century. With SSP5−8.5 showing the highest

Sustainable Development Perspectives in Earth Observation. DOI: https://doi.org/10.1016/B978-0-443-14072-3.00024-1

increases, the findings suggest that the water budget will be strained in some regions but may see surplus in others, such as the Indian Himalayan region and Western Ghats. This research calls for nuanced, adaptive water management strategies to address the uneven impacts of climate change across India.

The focus is widened to include hydroclimatic extremes by Singh, Behera, et al. (2025), Singh, Saju, et al. (2025), who investigate the intricate relationships between ocean-atmospheric phenomena and hydroclimatic extremes, such as floods, droughts, and heatwaves. Their research highlights the crucial role of large-scale climatic oscillations like ENSO, PDO, and NAO in shaping the frequency and intensity of these extreme events. As climate change amplifies these oscillations, it leads to more frequent and severe occurrences of compound droughts and heatwaves, posing significant risks to both ecosystems and human society. Despite progress in understanding these dynamics, the study emphasizes the need for region-specific research and deeper exploration into the nonlinear interactions between different oscillations. Swain, Rao, Jeyakanthan, and Ramana (2025) provide a paradigm for combining traditional ecological knowledge (TEK) with contemporary scientific methods as they address how TEK might improve the ecorestoration efforts of coastal ecosystems. This all-encompassing method of managing water resources using geospatial technologies provides important insights for tackling the problems that the coastal ecosystems are facing as a result of climate change. Anjaneyulu and Swain (2025) provide a comprehensive overview of dam break analysis (DBA) techniques and their evolution over the past two decades. The authors highlight the key models used for DBA, such as HEC-RAS, DAMBRK, FLDWAV, and MIKE11, emphasizing their technical capabilities and practical applications in breach modeling. They also discuss the role of remote sensing in post-dam break assessments, particularly for monitoring surface water changes.

1.2 Forest and land cover systems and climate change

In the face of climate change, the intricate interplay between biodiversity and ecosystem health becomes increasingly critical. Behera, Mahanand, et al. (2025), Behera, Yadhukrishna, et al. (2025), Behera, Varghese, Ratnam, Doi, and Behera (2025) emphasize that biodiversity, particularly plant diversity, serves as a cornerstone for ecological processes essential for climate regulation, hydrological cycles, and soil fertility. As climate change escalates the threat of species extinction, robust research is imperative to understand plant diversity's geographic patterns and responses to various environmental drivers. Utilizing high-resolution remote sensing data, the accuracy level of biodiversity monitoring can be enhanced, and effective conservation strategies can be formulated. They not only identify essential biodiversity variables derived from remote sensing data but also highlight the necessity of integrating global collaborations with local actions. The authors opine that by focusing on sustainable land management and nature-based solutions, the twin challenges of biodiversity loss and climate change can be addressed while achieving the Sustainable Development Goals for climate action and life on land. Similarly, Jaya Prakash et al. (2025) explore the intricate connections between meteorological factors and forest phenology through the lens of Earth Observation data. Their findings reveal how climate variables such as temperature, precipitation, net solar radiation, and soil moisture profoundly influence the timing of key biological events in forest ecosystems, affecting their overall health and productivity. The increased net solar radiation correlates with enhanced photosynthetic activity, as indicated by the Normalized Difference Vegetation Index (NDVI) and Enhanced Vegetation Index (EVI), highlighting the critical role of solar energy in fostering vegetative growth. However, they also caution that excessive temperature and water stress from drought can jeopardize forest vitality, emphasizing the delicate balance needed for optimal ecosystem health. By integrating advanced remote sensing technologies with robust statistical analyzes, this study underscores the necessity for adoption of such technologies in ongoing monitoring of forest phenology that helps forest conservation planning.

In light of the ongoing conversion of forests to croplands, there is a critical need for systematic monitoring of shifting cultivation areas, particularly in Northeast India. Utilizing advanced satellite image processing and a Random Forest machine learning algorithm, Das, Patra, et al. (2025), Das, Behera, Jagadish, et al. (2025), Das, Behera, Barik, et al. (2025), and Das, Das, et al. (2025) effectively mapped deforested and burned areas across Manipur, Meghalaya, and Nagaland using cloud-free Landsat imagery from 2016 to 2018. The findings reveal an alarming trend of forest loss. While the study showcases the potential of machine learning for precise land cover mapping, it also highlights the challenges posed by the rugged terrain and the limitations of existing satellite data. The authors advocate to incorporate higher-resolution satellite data and innovative AI techniques in future research to enhance the accuracy level of landuse and land cover monitoring efforts. In parallel, Kurein et al. (2025) explore the role of phenology in enhancing vegetation type mapping within the Similipal Biosphere Reserve, utilizing the capabilities of Sentinel-2 satellite imagery. Here they highlight the limitations of traditional field-based methods, advocating for the integration of high-temporal-frequency multispectral data to overcome these challenges. The authors also applied a random forest algorithm to time-series imagery to track temporal vegetation dynamics to discern seasonal patterns and improve classification outcomes. They found that incorporating multi-seasonal data not only improves model performance but also enhances the understanding of complex ecosystems.

For protecting and rehabilitating degraded soils, Deutsche Gesellschaft für Internationale Zusammenarbeit (GIZ) GmbH and the National Bank for Agriculture and Rural Development (NABARD) implemented the Soil Protection and Rehabilitation for Food Security project in India (ProSoil). Baby, Das, and Katsir (2025) investigate the impact of water development interventions on agricultural resilience in the Bhokardan region of Maharashtra, where ProSoil interventions started in 2015. Through comprehensive monitoring and evaluation, the authors assessed the effectiveness of various water conservation initiatives, including ponds, bunds, and irrigation systems, implemented. By utilizing satellite-derived data, the study reveals a positive trend in crop health and cropping intensity, particularly during the Zaid season, where an impressive increase of 167.92% was observed in the ProSoil intervention areas compared to the earlier periods. Notably, while neighboring regions experienced deteriorating crop health during severe drought years, the ProSoil area demonstrated resilience, underscoring the importance of the water harvesting structures in enhancing agricultural productivity. They emphasize the critical role of sustainable agricultural practices in mitigating drought impacts and highlight the potential for similar interventions in other regions to foster resilience against climate variability. Trivedi et al. (2025) explore the integration of Earth observation and artificial intelligence in assessing agroforestry land use systems, emphasizing their potential for sustainable agricultural development. The importance of creating a detailed geospatial database, which not only maps various agroforestry systems but also aids in prioritizing areas for agroforestry intensification. By aligning with India's National Agroforestry Policy, the findings advocate for the widespread adoption of agroforestry practices to enhance farmers' livelihoods, support food security, and contribute to climate change mitigation. They recommend ongoing development of regional databases, identification of key tree species, and effective policy implementation to maximize the benefits of agroforestry, particularly for smallholder farmers, thereby bridging the gap between scientific research and practical application in sustainable land management. Subramoniam et al. (2025) present a comprehensive assessment of India's coffee agro-ecosystem. They highlight the importance of Earth observation technologies in understanding the impacts of climate change on coffee cultivation. They identify approximately 1.20 lakh ha of suitable land for coffee expansion in various districts, demonstrating the potential for cultivating both Arabica and Robusta varieties in climate-sensitive areas. By integrating long-term climatic and terrain parameters, this research not only maps potential areas under different existing coffee types but also provides valuable insights for future cultivation strategies. As climate change poses significant threats to coffee production, there is a need for innovative approaches to enhance the resilience and sustainability of the coffee sector.

Like agroforestry systems, high altitude wetlands are highly sensitive to climate change and demand effective conservation strategies. Considering this, Behera, Mahanand, et al. (2025), Behera, Yadhukrishna, et al. (2025), and Behera, Varghese, et al. (2025) explore the seasonal dynamics of surface water and water quality in high-altitude wetlands of Sikkim, utilizing advanced remote sensing techniques and indices. The research reveals significant fluctuation in water levels and turbidity across seasons, with summer temperature and evaporation contributing to lower water indices, while monsoon rains lead to increased water levels and higher turbidity. Notably, the findings underscore the wetlands' vulnerability to climate change, with winter conditions presenting a complex scenario of fluctuating water quality. They emphasize the necessity for continuous monitoring and adaptive management strategies to ensure the health of these critical ecosystems.

1.3 Ocean-atmosphere dynamics and climate change

Since the 20th century, global mean sea level has been rising at an accelerated pace. Deb, Chittella, Sen, and Dutta (2025) dealt with the complexities of the Antarctic climate system, highlighting significant changes driven by climate dynamics. Notably, the Antarctic Peninsula shows the highest melt potential, which quantifies the frequency and intensity of summer temperature exceeding a critical melt threshold, revealing vulnerabilities in various ice shelves. They also identify intricate patterns of circulation influenced by global climate phenomena, such as El Niño, which are reshaping the dynamics of ice melt and potentially contributing to broader sea level rise. Samanta (2025) emphasizes the substantial regional variability in sea level changes, which can differ by as much as 20% from global averages due to various physical processes. As climate change continues to exert its influence, Asian coastal areas, especially Japan's east coast, are identified as particularly vulnerable, facing the potential for extreme sea level rise of up to 10 m along the China coast. The author stresses the urgent need for effective adaptation and mitigation strategies to address these challenges, noting that the impacts on coastal populations, infrastructure, and ecosystems are already being felt and will likely intensify in the coming decades. Like that of coastal areas, polar areas like Antarctica, primarily covered by a vast ice sheet, face challenges from climate change that are affecting the timing and extent of sea ice in the region.

The increasing concentration of atmospheric carbon dioxide (CO_2) is one of the most significant drivers of global warming and climate change. As the primary anthropogenic greenhouse gas, CO_2 has been responsible for a substantial rise in global temperatures over the past century. Das, Patra, et al. (2025), Das, Behera, Jagadish, et al. (2025), Das, Behera, Barik, et al. (2025), Das, Das, et al. (2025), explore the critical relationship between atmospheric CO_2 and global warming, focusing on the growing impact of CO_2 as the dominant anthropogenic greenhouse gas. Here, using in situ and space-based remote sensing, the global rise in surface air temperature, which is closely linked to increasing CO_2 levels, is monitored. The balance between carbon sources (such as fossil fuel emissions and land-use changes) and sinks (land and ocean) is examined, highlighting the lag in the natural carbon sequestration process relative to emissions. The concept of the "remaining carbon budget" emerges as a pivotal tool for policymakers in managing climate action, emphasizing the need for direct emission reductions and careful implementation of carbon dioxide removal (CDR) strategies. They also explored future carbon sink projections under various warming scenarios, emphasizing the declining efficiency of natural sinks by the latter half of the century and calling for urgent mitigation efforts to address the climate crisis. In connection with this, Kuttippurath and Patel (2025) highlight the remarkable advancements in Earth Observation Satellites (EOS) for global air quality monitoring over recent decades on a global scale basis, especially for tracking air pollution. While ground-based measurements have traditionally been used to monitor air quality, their limited geographical coverage, especially in rural and developing regions, poses challenges for real-time pollution monitoring. They go through how different satellites have transformed air quality monitoring by providing high-resolution, comprehensive data on pollutants like ammonia, carbon monoxide, and aerosols. The authors also touch on the impact of recent missions like TEMPO (Tropospheric Emissions: Monitoring Pollution) and GEMS (Geostationary Environment Monitoring Spectrometer), which offer hourly data across vast areas of the Northern Hemisphere, enhancing the ability to identify pollution sources and improve public health strategies. The authors stress that further advances in EOS are necessary to enhance real-time air quality assessments and inform more effective environmental policies.

Raut, Misra, Kumar, Supe, and Avtar (2025) examine the nitrogen dioxide (NO_2) concentrations in non-attainment cities in Maharashtra during the COVID-19 lockdown, utilizing satellite-based remote sensing and GIS techniques. They highlight that while cities like Akola and Jalgaon saw a reduction in NO_2 levels, other cities, such as Nagpur, experienced a substantial increase in pollution. The analysis emphasizes the role of factors such as power plants, industries, and population density in shaping these trends. The findings underscore that even stringent lockdown measures were insufficient to bring about a uniform reduction in pollution across all cities, pointing to the need for more comprehensive and sustainable air quality management strategies. Furthermore, this study advocates for continued monitoring of multiple pollutants beyond NO_2, contributing to the National Clean Air Programme (NCAP) and SDGs. Das, Patra, et al. (2025), Das, Behera, Jagadish, et al. (2025), Das, Behera, Barik, et al. (2025), and Das, Das, et al. (2025) explore the rising CO_2 emissions in India, focusing on both urban and national scales using deep learning models and satellite data from 2009 to 2019. They compare recurrent neural network (RNN) and long short-term memory (LSTM) algorithms for time-series analysis of CO_2 fluxes, finding that LSTM performs better, projecting a steady increase in emissions through 2030, except for a slight reduction in Bangalore. The study also employs regression models to analyze CO_2 emissions from fossil fuels, socio-economic factors like GDP and population, and renewable energy use, with Ridge and Multi-Layer Perceptron models showing higher emission predictions compared to linear and LASSO regressions. The authors underscore the importance of accurate emission forecasts for policymaking, highlighting the urgent need for proactive measures to mitigate the impacts of climate change and reduce greenhouse gas emissions across India's major urban areas.

Behera, Mahanand, et al. (2025), Behera, Yadhukrishna, et al. (2025), and Behera, Varghese, et al. (2025) delve into the pivotal role of oceanic processes in regulating global climate conditions and explore their repercussions on terrestrial ecosystems. The study focused on the response of global vegetation to large-scale climate phenomena such as the Indian Ocean Dipole (IOD), El Niño, and El Niño Modoki through atmospheric teleconnections. The findings reveal the spatio-temporal variability associated with the global NDVI anomalies in response to each of the climate variability patterns. A profound impact of these coupled ocean-atmospheric processes was observed over Asia, Africa, and some high-latitude areas, especially during the boreal summer. The study highlights the complex and varied impacts of these tropical climate modes, with IOD showing seasonal lag responses and El Niño Modoki having a notable negative influence on ecosystems in East and Southeast Asia. This research underscores the importance of monitoring and predicting oceanic-climate patterns for effective climate adaptation and mitigation, especially as tropical climate variability strongly affects mid- and high-latitude ecosystems.

Monitoring air quality in industrial regions is critical for understanding the impacts of industrialization and for assessing the effectiveness of air pollution control initiatives, such as India's National Clean Air Program (NCAP). Kumar, Das, and Behera (2025) investigate a marked rise in Aerosol Optical Depth (AOD) over the Kalinga Nagar Industrial area in Odisha, India, using MODIS satellite data from 2001 to 2021. The findings reveal a significant increase in AOD levels, particularly after 2007, which correlates with the establishment of the industrial cluster. AOD levels exceeded 0.9 by 2021, with the highest concentrations observed during the summer months of May, June, and July, followed by a decrease during the monsoon. The study also utilized MERRA-2 model data, confirming that dry dust deposition was highest in the summer, while wet dust deposition increased with rising humidity during the monsoon. The research underscores the growing air pollution in the region, stressing the value of remote sensing data for effective air quality monitoring, especially in areas lacking ground-based stations.

1.4 Conclusion

The integration of Earth observation technologies into sustainable development practices offers transformative solutions to the complex challenges posed by climate change and environmental degradation. The diverse studies presented in this volume collectively underscore the importance of leveraging remote sensing and geospatial analytics for effective decision-making in agriculture, water management, and ecosystem conservation. By understanding the interconnectedness of ecological processes and human activities, we can develop more resilient strategies in line with the sustainable development goals. As we face an increasingly uncertain future, the knowledge gained from these studies will not only inform policy and practice but also inspire innovative approaches to protecting our planet's natural resources and ensuring a sustainable legacy for future generations. The commitment to use Earth observation technologies as tools for sustainability is essential in our collective journey towards a more equitable and resilient world.

References

Anjaneyulu, R., & Swain, R. (2025). *Comprehensive insights into dam break analysis: A mini review. Earth observation: Sustainable development perspectives* (1st ed.).

Baby, A., Das, P., & Katsir, S. (2025). *Satellite-based monitoring of agriculture resilience succeeding water development interventions. Earth observation: Sustainable development perspectives* (1st ed.).

Barik, S. K., Behera, M. D., & Adhikari, D. (2025). *Sustaining floriculture and floral fragrance in a changing climate. Earth observation: Sustainable development perspectives* (1st ed.).

Behera, M. D., Mahanand, S., Tripathi, P., Menachery, M. P., Sahu, M., Behera, S., ... Mohapatra, M. (2025). *Biodiversity, climate change and remote sensing: Adaptation and mitigation perspective. Earth observation: Sustainable development perspectives* (1st ed.).

Behera, M. D., Yadhukrishna, K. G., Raj, A., Srivastava, I., Das, P., Kashyap, A., & Joshi, R. (2025). *Spatio temporal assessment of surface dynamics of high-altitude wetlands using earth-observation datasets. Earth observation: Sustainable development perspectives* (1st ed.).

Behera, S. K., Varghese, R., Ratnam, J. V., Doi, T., & Behera, M. D. (2025). *Role of ocean and climate in terrestrial ecosystems. Earth observation: Sustainable development perspectives* (1st ed.).

Chowdary, V. M., Chakraborthy, A., Sahay, B., Choudhary, K. K., Biswal, A., Srikanth, P., ... Chauhan, P. (2025). *Geospatial applications for crop assessment. Earth observation: Sustainable development perspectives* (1st ed.).

Das, C., Patra, P. K., & Kunchala, R. K. (2025). *An overview of atmospheric carbon dioxide and global warming interlinks. Earth observation: Sustainable development perspectives* (1st ed.).

Das, P. K., Behera, M. D., Jagadish, B., & Das, P. (2025). *Forecasting CO_2 emissions over India using deep learning models. Earth observation: Sustainable development perspectives* (1st ed.).

Das, P., Behera, M. D., Barik, S. K., & Mudi, S. (2025). *Earth observation for monitoring of shifting cultivation. Earth observation: Sustainable development perspectives* (1st ed.).

Das, P., Das, P., Vyas, S., & Deb Burman, P. K. (2025). *Terrestrial water cycle in future climate over India. Earth observation: Sustainable development perspectives* (1st ed.).

Deb, P., Chittella, S., Sen, A., & Dutta, S. (2025). *The Antarctic climate system: Recent changes and their drivers. Earth observation: Sustainable development perspectives* (1st ed.).

Kumar, S., & Parida, B. R. (2025). *Crop insurance using geo-information: A strategy for climate change mitigation. Earth observation: Sustainable development perspectives* (1st ed.).

Kumar, S., Das, P., & Behera, M. D. (2025). *Significant rise in aerosol concentration in past two decades over an industrial area. Earth observation: Sustainable development perspectives* (1st ed.).

Kurein, A., Sahu, M., Behera, M. D., Prakash, A. J., Mukhopadhyay, M., Behera, S. K., ... Sahu, S. C. (2025). *Role of phenology as discriminant in vegetation type mapping. Earth observation: Sustainable development perspectives* (1st ed.).

Kuttippurath, J., & Patel, V. K. (2025). *Advances in earth observation satellites for global air quality monitoring. Earth observation: Sustainable development perspectives* (1st ed.).

Prakash, A. J., Behera, M. D., Mukhopadhyay, M., Singh, C. P., Bhattacharya, B. K., Jeganathan, C., ... Mohapatra, M. (2025). *Linking meteorological and phenological observations in forests. Earth observation: Sustainable development perspectives* (1st ed.).

Raut, A., Misra, P., Kumar, N., Supe, H., & Avtar, R. (2025). *Assessment of nitrogen dioxide concentrations in non-attainment cities during COVID-19 lockdown using GIS technique. Earth observation: Sustainable development perspectives* (1st ed.).

Samanta, D. (2025). *Sea level change in Asia. Earth observation: Sustainable development perspectives* (1st ed.).

Singh, R. K., Behera, M. D., Biradar, C. M., Dhyani, S. K., Prakash, A. J., Das, P., ... Rizvi, J. (2025). *Agroforestry for sustainable livelihoods: Role of geoinformatics, ICT and citizen science. Earth observation: Sustainable development perspectives* (1st ed.).

Singh, S., Saju, C., & Athira, K. (2025). *Impacts of ocean-atmospheric phenomena on hydroclimate extremes. Earth observation: Sustainable development perspectives* (1st ed.).

Subramoniam, R., Nagashree, M. K., Trivedi, S., Mukharib, D. S., Mote, K., Hebbar, R., ... Chauhan, P. (2025). *Earth observation-based mapping and site suitability analysis of coffee agro-ecosystem in the face of climate change. Earth observation: Sustainable development perspectives* (1st ed.).

Swain, S., Rao, Y. R. S., Jeyakanthan, V. S., & Ramana, R. V. (2025). *Geospatial technology for coastal water resources management. Earth observation: Sustainable development perspectives* (1st ed.).

Trivedi, S., Vinod, P. V., Subramoniam, R., Chandrasekaran, B., Nagashree, M. K., Nagajothi, K., ... Chauhan, P. (2025). *Assessment of agroforestry land use systems for sustainable agriculture development: A geospatial perspective using AI. Earth observation: Sustainable development perspectives* (1st ed.).

Section A

Agriculture, water and climate change

Chapter 2

Geospatial applications for crop assessment

V.M. Chowdary[1], Abhishek Chakraborty[1], Bhavana Sahay[1], Karun Kumar Choudhary[1], Anima Biswal[1], P. Srikanth[1], Mamta Kumari[1], B. Laxman[1], Varun Pandey[1], Parichay S. Raju[1], K. Sreenivas[1] and Prakash Chauhan[2]

[1]*Agricultural Sciences and Applications Group, National Remote Sensing Centre, Indian Space Research Organization, Hyderabad, Telangana, India,* [2]*National Remote Sensing Centre, Indian Space Research Organization, Hyderabad, Telangana, India*

Abbreviations

NDVI	Normalized difference vegetation index
NDWI	Normalized difference water index
LSWI	Land surface wetness index
SAR	Synthetic aperture radar
BS	Backscatter
RTM	Radiative transfer model
RS	Remote sensing
GIS	Geographic information systems
GPS	Global positioning systems
IT	Information technology
NASA	National Aeronautics and Space Administration
ESA	European Space Agency
JAXA	Japan Aerospace Exploration Agency
CNSA	China National Space Administration
ISRO	Indian Space Research Organization
LAI	Leaf area index.
AGB	Above ground biomass
fAPAR	Fraction of absorbed photosynthetically active radiation
NPP	Net primary productivity
CCC	Canopy chlorophyll content
SIF	Solar-induced chlorophyll fluorescence
VHI	Vegetation health index
SPI	Standardized precipitation index
SPEI	Standardized precipitation evaporation index
VCI	Vegetation condition index
TCI	Temperature condition index

2.1 Introduction

For approximately 2.5 million years, the sustenance of the human species revolved around hunting animals and gathering plants, resulting in a minimal ecological footprint. Around 10,000 years ago, a pivotal shift occurred as humans commenced the domestication of select animal and plant species, giving rise to agrarian societies characterized by heightened food security. This transition marked the onset of population expansion and escalated agricultural

Sustainable Development Perspectives in Earth Observation. DOI: https://doi.org/10.1016/B978-0-443-14072-3.00017-4

cultivation. Over the course of history, humans have adeptly navigated this cyclical paradigm, underscoring their remarkable adaptability. However, the contemporary landscape presents limitations on the horizontal expansion of agricultural activities. Consequently, the focus has gravitated toward vertical expansion within the confines of constrained resources.

Forecasts unveiled by the United Nations (UN) project a surge in the global population from 6.8 billion to 9.1 billion by 2050. This demographic projection heralds a staggering 60% upsurge in the demand for agricultural produce (Alexandratos & Bruinsma 2012). The Food and Agriculture Organization of the United Nations (FAO, 2017) reported that there is an urgent need to produce 56% more food (as compared to 2010, under the growth scenario of "business as usual") to feed the burgeoning population of the world by 2050 (Trendov, Varas, & Zeng, 2019). Factors such as fragmented land tenure, soil degradation, diminished soil health, receding crop productivity, and the metamorphosis of global climatic patterns impose formidable impediments on agricultural progress. Confronted with the imminent challenge of nourishing a burgeoning world populace, it is imperative to reverse the waning trajectory of overall crop productivity, arrest the degradation of natural resources, and bolster agrarian incomes via the prism of sustainable resource management strategies.

The assimilation of emergent technologies such as remote sensing (RS), geographic information systems (GIS), global positioning systems (GPS), and information technology (IT) portends a seismic advancement in future agricultural productivity (Hakkim, Joseph, Gokul, & Mufeedha, 2016; Mitran, Meena, & Chakraborty, 2020). These technologies confer the capacity for continuous monitoring and evaluation of natural resource dynamics. Coalescing under the rubric of geospatial technology, this domain encompasses a spectrum of tools engineered for the collection, storage, analysis, interpretation, and dissemination of spatial data. This frugal methodology encompasses the acquisition of real-time satellite imagery via RS, the assimilation and orchestration of data through GIS, the provisioning of location-based services and geo-referencing via GPS, and the orchestration of web-enabled services facilitated by IT.

The cadence of RS innovations begets detailed datasets conducive to the inventorying, cartography, and vigilant tracking of crops, land, and water resources at an expansive scale (Gerhards, Schlerf, Mallick, & Udelhoven, 2019). A symbiotic union of satellite RS with GIS and mobile application-driven positional insights has emerged as a potent conduit for nurturing sustainable agricultural development. This synergy optimizes the judicious allocation of resource inputs, curtails production costs, and mitigates biotic and abiotic vulnerabilities. Beyond these merits, this technology furnishes pivotal "decision support scenarios," serving as linchpins for assiduous monitoring of the holistic well-being of the agricultural domain and facilitating informed decision-making. A pantheon of applications underpins the scope of geospatial technologies within the agricultural landscape, including crop inventory and monitoring (Ghazaryan et al., 2018; Heupel, Spengler, & Itzerott, 2018; Schmedtmann & Campagnolo, 2015), crop biophysical and biochemical parameter retrieval (Meena et al., 2018; Peñuelas, Gamon, Fredeen, Merino, & Field, 1994), crop growth simulation modeling, and the quantification of crop yields (Ban, Ahn, & Lee, 2019; Huang et al., 2019; Phung, Lam-Dao, Nguyen-Huy, Le-Toan, & Apan, 2020), etc.

Medium- and coarse-resolution RS datasets bequeath comprehensive, synoptic coverage of crops and soil resources at a regional or even continental scale. Conversely, fine-resolution satellite data proffers granularity for applications tailored to micro-level or farm-specific domains, encompassing meticulous oversight of fertilizer and pesticide administration, the employment of variable rate technology, assessment of crop insurance viability, and adjudication of crop damage incidents, etc. RS data sources spanning the optical, microwave, thermal, and hyperspectral spectrum emerge as potent instruments for unraveling the tapestry of crop and soil characteristics across a gamut of spatial and temporal scales. Research endeavors underscore the prowess of RS technology in furnishing spatiotemporally diverse insights for the agricultural sector (Mulla, 2013; Pareeth, Karimi, Shafiei, & De Fraiture, 2019; Phung et al., 2020; Rotairo, Durante, Lapitan, & Rao, 2019). A plenitude of satellite RS resources, including MODIS, Landsat, Sentinel, Resourcesat-2, Cartosat-1, Cartosat-2, Planet, and QuickBird, presently avail themselves to researchers entrusted with the stewardship of natural resources. The cumulative efforts of space agencies such as NASA, ESA, JAXA, CNSA, ISRO, and their counterparts have engendered an astronomical augmentation of spaceborne RS capabilities. It bears noting that nations spanning the Asia-Pacific, South Asia, North America, and Europe corridors are pioneering the establishment of Agricultural Market Information Systems, harnessing the prowess of geospatial tools to amalgamate socioeconomic and crop data, thus engendering an all-encompassing apparatus for agriculture management. An overview of geospatial solutions for crop assessment that include driving variables and services that are likely to be accrued is presented in Fig. 2.1.

In summation, geospatial technology has transcended its ancillary role to emerge as an integral constituent within the architecture of agriculture management systems. Its efficacy is eloquently substantiated through its capacity to engender resource optimization, amplify productivity, and propel the course of future advancements.

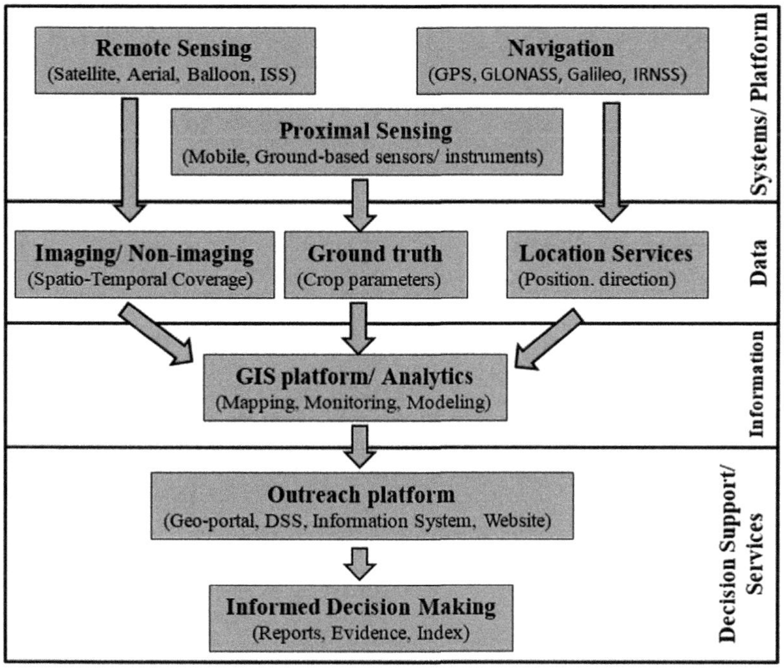

FIGURE 2.1 Overview of geospatial solutions for crop assessment.

2.1.1 Role of geospatial technologies in crop assessment

Geospatial technology significantly contribute to operational agriculture monitoring systems by providing a timely, synoptic, cost efficient and repetitive information over large areas (Justice et al., 2002) thus generating information on crop acreage (Bhattacharya et al., 2024; Galford et al., 2008; Goyal et al., 2022; Srikanth et al., 2024; Yao, Feng, & Zhang, 2014) and yield (Bernardes, Moreira, Adami, Giarolla, & Rudorff, 2012; Meroni, Marinho, Sghaier, Verstrate, & Leo, 2013; Mulianga, Bégué, Simoes, & Todoroff, 2013; Rembold, Atzberger, Savin, & Rojas, 2013; Mohit et al., 2022), crop phenology (Sakamoto et al 2005, Shen et al 2013) and stress condition (Gu, Brown, Verdin, & Wardlow, 2007; Mirik et al., 2014) in a standardized and regular manner at the (sub)regional to the national scale. These estimates and information should be provided during the growing season(s) in near real time and updated periodically through the season until the crop harvest. Thus, satellite remote sensing has become one of the potential techniques used for local, regional, and global crop monitoring since the 1970s (MacDonald & Hall, 1980).

Several remote sensing-based multispectral vegetation indices and matrices have been developed that are correlated with crop biophysical, morphological, and physiological attributes. Such indices include the normalized difference vegetation index (NDVI), leaf area index (LAI), crop canopy cover, above ground biomass (AGB), fraction of absorbed photosynthetically active radiation (fAPAR), net primary productivity (NPP), canopy chlorophyll content (CCC), solar-induced chlorophyll fluorescence (SIF), vegetation health index (VHI) and growth stages, etc. (Mohammed et al., 2019; Zhang, Wu, Yu, Zou, & Zheng, 2014). These remote sensing-based indices are used to assess crop growth status and crop stress condition during the crop growing season. Similarly, several indices are developed to analyze diseases and pest-induced stress on crops at local or regional scales (Zhang et al., 2019). The crop stress due to drought is assessed through several climate and vegetation-based indices such as the standardized precipitation index (SPI), standardized precipitation evaporation index (SPEI), vegetation condition index (VCI), temperature condition index (TCI), etc. Some of the popular regional and global crop monitoring systems are presented in Table 2.1.

2.2 Spectral response of crop/vegetation

Spectral reflectance in the context of remote sensing pertains to the measurement of how vegetation interacts with various wavelengths of light, either by reflecting or absorbing them. This spectral response offers invaluable insights into the health, composition, and physiological attributes of plant life. Multiple factors influence spectral reflectance, encompassing both the intrinsic characteristics of vegetation and external factors related to incident light conditions. The significant driving parameters and dynamics that exert influence over the spectral reflectance of vegetation are expounded

TABLE 2.1 Regional and global crop monitoring systems.

Crop monitoring parameters	Agro-climatic indicators	Remote sensing indicators	Coverage	System & reference
Crop type, crop stress and drought	Precipitation profile and departure from average;	NDVI and VCI	Global	ASIS Van Hoolst et al. (2016) https://www.fao.org/giews/
Crop type and area, crop stress and drought, crop yield	Precipitation, Temperature, PAR and potential biomass anomalies over 15 years and profiles	NDVI and VCI	Global	CropWatch Wu et al. (2015) http://cropwatch.cn
Crop type, crop stress and drought,	SPI1, SPI3, and GWSI	NDVI	Global	ASAP Anomaly hot Spots of Agricultural Production Rembold et al. (2019)
Crop type and area, crop stress and drought, crop yield	SPEI, ASI and PET	VCI, VPI, and NDVI	Europe & neighbors	MARS Monitoring Agricultural Resources López-Lozano & Baruth, (2019) https://mars.jrc.ec.europa.eu/asap/
Crop water production functions	Agro Climate for Crop Explorer	NDVI	Global	Crop Explorer https://ipad.fas.usda.gov/cropexplorer/
Crop type and area, crop yield	Precipitation and Temperature departures from normal	NDVI and VCI	USA	NASS, VegScape https://nassgeodata.gmu.edu/VegScape
Drought	Rainfall assumptions: average and accumulative	NDVI	30 countries	FEWS-NET; Famine Early Warning Systems Network https://fews.net/
Crop type, crop stress and drought,	Precipitation, Temperature and Evapotranspiration departures from normal		Global	GLAM Global Agricultural monitoring Becker-Reshef et al. (2010) https://glam.nasaharvest.org
Crop type, and drought,	Anomalies of Precipitation and Temperature sums		Global	Crop Monitor Becker-Reshef et al. (2020)) https://cropmonitor.org/
Crop type and area, crop yield	Anomaly map	VI	India	FASAL Forecasting Agricultural output using space, Agrometeorological and Land based observations Parihar & Oza, (2006)
Crop type and area, crop stress and drought, crop yield	Maps and profiles of cumulated Precipitation	NDVI and VCI	Russia	VEGA Tolpin et al. (2019) http://vega.geoglam.ru//?lang = eng
Crop type and area, crop stress and drought, crop yield	Agro-climatic models	NDVI	Canada	CALMS Chipanshi et al. (2015)

Abbreviations: *ASI*, Agricultural stress index; *NDVI*, normalized difference vegetation index; *SPI*, standardized precipitation index; *SPEI*, standardized precipitation-evapotranspiration index; *GWSI*, global water requirement satisfaction index; *PET*, potential evapotranspiration; *VCI*, vegetation condition index; *VHI*, vegetation health index; *VPI*, vegetation productivity indicator.
Source: Modified from Wu, B., Zhang, M., Zeng, H., Tian, F., Potgieter, A.B., Qin, X., ... Boken, V. (2023). Challenges and opportunities in remote sensing-based crop monitoring: A review. *National Science Review*, *10*(4), pp. nwac290. https://doi.org/10.1093/nsr/nwac290.

upon below, drawing upon the research contributions of Gitelson, Merzlyak, and Lichtenthaler (1996), Roy (1989), and Ripple (1986):

(i) **Chlorophyll content:** The quantity of chlorophyll, the primary pigment responsible for light absorption during photosynthesis, exerts a pronounced impact on spectral reflectance. Chlorophyll predominantly absorbs light in the blue and red regions of the spectrum, resulting in heightened absorption and diminished reflectance, especially within these spectral bands.

(ii) **Leaf structure:** The internal composition of leaves plays a pivotal role in shaping the reflectance spectrum. Variables such as leaf thickness, size, shape, orientation, and internal structure modulate the path length of light within leaves, thereby influencing the reflectance spectrum.
(iii) **Leaf pigments:** In addition to chlorophyll, other pigments like carotenoids and anthocyanins contribute significantly to the spectral reflectance of vegetation. These pigments absorb light at distinct wavelengths, thereby introducing variations in reflectance patterns.
(iv) **Leaf water content:** Water within plant tissues interacts with incident light, impacting reflectance properties. Specifically, the hydroxyl bonds interact with the Shortwave Infrared Region (SWIR) of the electromagnetic spectrum. Consequently, heightened moisture content results in augmented absorption within the SWIR region, and conversely, reduced moisture content leads to diminished absorption.
(v) **Leaf dry matter content:** The concentration of dry matter, encompassing components such as cellulose, lignin, and other structural elements, exerts a discernible influence on spectral reflectance. Elevated dry matter content typically corresponds to amplified reflectance within the visible and nearinfrared regions (NIRs).
(vi) **Canopy structure:** The arrangement of leaves, branches, and various elements within a vegetation canopy introduces complexities into reflectance patterns. Canopy architecture, leaf orientation, and LAI collectively determine the extent and direction of light interactions with vegetation, thus affecting the overall reflectance profile.
(vii) **Leaf senescence and stress:** Alterations in vegetation due to senescence (aging) or stress factors, such as nutrient deficiencies, diseases, or drought, can induce changes in spectral reflectance. These changes may be attributed to decreased chlorophyll content, modifications in leaf structure, or underlying physiological processes.

It is crucial to recognize that the dynamics of spectral reflectance in crops exhibit temporal variations owing to seasonal fluctuations, growth stages, and prevailing environmental conditions. Remote sensing techniques, leveraging satellite or airborne sensors, are adept at capturing these dynamic shifts over extensive geographical areas. The monitoring of crop spectral reflectance furnishes invaluable data for the study of plant health, productivity, ecological processes, and its manifold applications in domains such as agriculture, forestry, and environmental monitoring.

2.2.1 Crop discrimination using spectral observations

Spectral observations encompass the meticulous examination of the distinctive reflectance patterns exhibited by diverse crop types across the electromagnetic spectrum. Each crop category manifests its unique spectral signatures due to inherent disparities in biochemical and biophysical characteristics. These spectral response characteristics serve as indicators of how crops interact with incident light at varying wavelengths, thereby facilitating the differentiation among different crop varieties.

Fig. 2.2, showcases a false color composite (FCC) image wherein the different crops (black gram, turmeric, sugarcane, and banana) are distinctly discernible. This distinctiveness arises from tonal and textural disparities within the image, allowing for the effective discrimination of these crops.

To illustrate this phenomenon, Fig. 2.3 portrays the typical spectral response curves, often referred to as spectral signatures, of three distinct crops: black gram, turmeric, and sugarcane. This depiction spans the wavelength range from 350 to 2500 nm, providing a comprehensive view of the crops' spectral behavior.

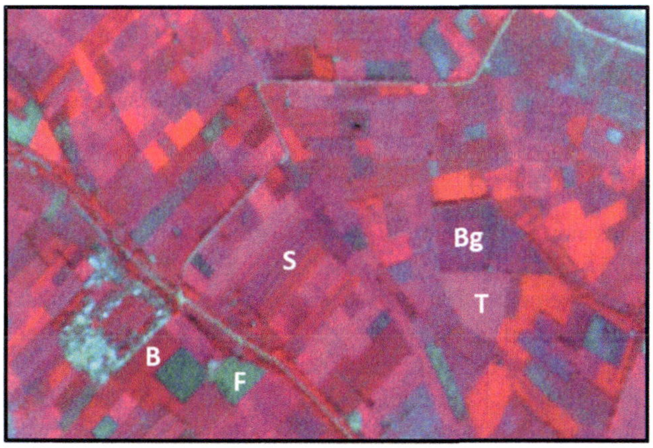

FIGURE 2.2 Tonal variation of the different crops as seen in Resourcesat-2 LISS IV FCC. *B*, Banana; *Bg*, black gram; *F*, fallow; *S*, sugarcane; *T*, turmeric.

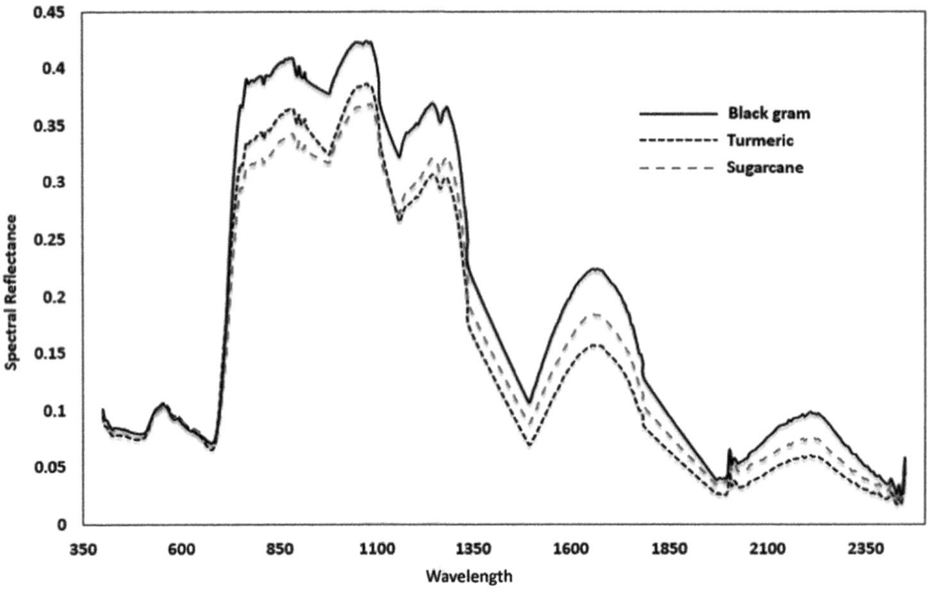

FIGURE 2.3 Spectral signature of three different crops.

2.2.2 Vegetation indices for crop monitoring

A vegetation index (VI) represents a mathematical transformation of spectral image bands, designed to accentuate the significance of vegetation properties while facilitating spatial and temporal comparisons pertaining to photosynthetic activity and canopy structure. These indices are shaped by the interaction between solar radiation and crops. Within this realm, a plethora of indices exist, many of which leverage the inverse correlation between red and near-infrared reflectance, a phenomenon associated with thriving green vegetation.

Since the 1960s, scientists have envisioned the utilization of satellite remote sensing techniques for monitoring vegetation dynamics on Earth's surface, encompassing critical attributes such as LAI, percentage of green cover, chlorophyll content, green biomass, and absorbed photosynthetically active radiation (APAR). Lyon, Yuan, Lunetta, and Elvidge (1998) undertook a classification of VIs based on their computational methodologies, categorizing them as subtraction, division, or rational transform indices. This categorization serves to evaluate the effectiveness of various VIs for vegetation monitoring. With the evolution of hyperspectral remote sensing capabilities, specific VIs tailored for hyperspectral data have emerged, including narrow band vegetation indices, which complement traditional multi-spectral VIs. These indices have demonstrated their worth in a multitude of applications, notably: (1) climate trend analysis; (2) remote estimation of soil water content; (3) drought monitoring; (4) crop irrigation scheduling and management; (5) monitoring evaporation and plant transpiration; (6) assessing changes in biodiversity; (7) vegetation classification; and (8) detection and quantification of crop diseases.

Several commonly utilized optical vegetation indices include:

Ratio vegetation index (RVI): The RVI is calculated as the ratio between the red and near-infrared bands of multispectral images, as proposed by Pearson and Miller in Pearson and Miller (1972). It is a straightforward index, yet its interpretability and comparability across time are limited due to its unbounded range.

$$\mathrm{RVI} = \mathrm{Red}/\mathrm{NIR} \tag{2.1}$$

NDVI: NDVI is a widely employed index for monitoring vegetation dynamics and quantifying biomass. It is computed as follows, following the formulation by Rouse et al. (1974):

$$\mathrm{NDVI} = (\mathrm{NIR} - \mathrm{Red})/(\mathrm{NIR} + \mathrm{Red}) \tag{2.2}$$

NDVI typically falls within the range of -1 to +1 due to its normalization. This index exhibits sensitivity to the presence of green vegetation, as noted by Sellers in Sellers (1985). Consequently, it finds extensive use in regional and global applications aimed at assessing vegetation conditions, as demonstrated by the works of Prince and Tucker (1986), Townshend and Justice (1986), Collet and Abednego (1987), and Deblonde and Cihlar (1993). NDVI's effectiveness as a descriptor of vegetation variations remains remarkable, even in the face of atmospheric effects (Kaufman, 1984), radiometric degradation in the red and near-infrared bands (Holben, Kaufman, & Kendall, 1990), and sensor calibration degradation. NDVI's success can be attributed to the normalization it provides, mitigating the impact of sensor

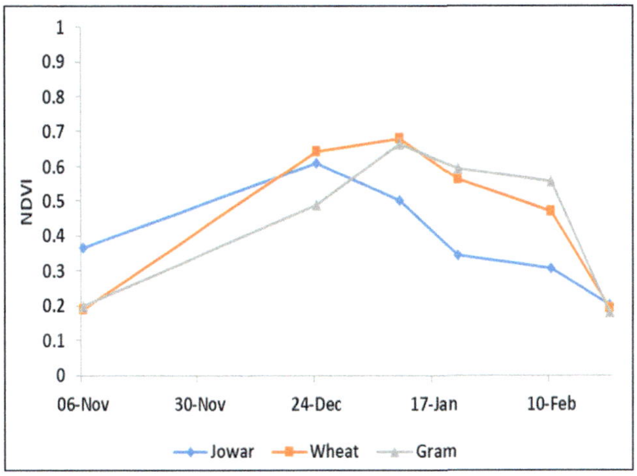

FIGURE 2.4 Temporal normalized difference vegetation index profile discriminating different crops.

calibration degradation, accounting for approximately 6% of the overall index value (Holben et al., 1990; Kaufman & Holben, 1993). Research conducted by Holben and Fraser (1984) suggests that NDVI is particularly sensitive to viewing and illumination geometry, especially for low-density vegetation covers.

Fig. 2.4 presents the temporal NDVI profiles of wheat crops, jowar and gram, captured during their growth stages from November to February. These temporal profiles simplify discrimination and classification tasks, revealing distinct crop phenology stages through the analysis of vegetation index temporal profiles.

Soil-adjusted vegetation index (SAVI): SAVI is a valuable tool employed to mitigate the influence of soil brightness in areas with low vegetative cover. Proposed by Huete in 1988 as a refinement of the NDVI, this index is unique due to its simple model that effectively accounts for soil-vegetation interactions. SAVI is defined as follows:

$$\text{SAVI} = \big((\text{NIR} - R)/(\text{NIR} + R + L)\big) \ * \ (1 + L) \tag{2.3}$$

Here, 'L' represents the soil adjustment factor. Through a simplified radiative transfer model, Huete (1988) established that optimal adjustment can be achieved when 'L' is set to 0.5. This value minimizes the secondary backscattering effect of canopy-transmitted soil background reflected radiation. When 'L' is set to zero ($L = 0$), SAVI is equivalent to NDVI.

Enhanced vegetation index (EVI): EVI is an optimized vegetation index designed to enhance sensitivity in high biomass regions and improve vegetation monitoring by minimizing the influence of the atmosphere and decoupling the canopy background signal. EVI is computed using the following equation:

$$\text{EVI} = G \times \lfloor(\text{NIR} - \text{Red})/(\text{NIR} + C1 \times \text{Red} - C2 \times \text{Blue} + L)\rfloor \tag{2.4}$$

Key components include:

- 'NIR,' 'red,' and 'blue' are surface reflectances that have undergone atmospheric correction, including partial corrections for effects like Rayleigh scattering and ozone absorption.
- 'L' represents the canopy background adjustment, accounting for non-linear, differential radiant transfer of NIR and red through a canopy.
- '$C1$' and '$C2$' are coefficients of the aerosol resistance term, which uses the blue band to correct for aerosol influences in the Red band.
- 'G' is a gain factor.
- The coefficients used in the MODIS-EVI algorithm are $L = 1$, $C1 = 6$, $C2 = 7.5$, and $G = 2.5$.

While NDVI is sensitive to chlorophyll content, EVI is more responsive to variations in canopy structure, including LAI, canopy type, plant physiognomy, and canopy architecture. Both indices complement each other in global vegetation studies and enhance the detection of vegetation changes and the extraction of canopy biophysical parameters. Another notable distinction is their response to the presence of snow: NDVI decreases, whereas EVI increases (Huete et al., 2002).

Normalized difference water index (NDWI): NDWI is a satellite-derived index that utilizes the near-infrared (NIR) and short-wave infrared (SWIR) channels, introduced by Gao in 1996.

$$NDWI = (NIR - SWIR)/(NIR + SWIR) \quad (2.5)$$

The SWIR reflectance captures variations in both vegetation water content and the spongy mesophyll structure within vegetation canopies. In contrast, NIR reflectance is influenced by leaf internal structure and dry matter content but not by water content. By combining NIR with SWIR, variations induced by leaf internal structure and dry matter content are eliminated, leading to improved accuracy in estimating vegetation water content (Ceccato, Flasse, Tarantola, Jacquemoud, & Grégoire, 2001). The spectral reflectance in the SWIR region is significantly influenced by the amount of water within the internal leaf structure, resulting in a negative relationship between SWIR reflectance and leaf water content (Tucker, 1980).

NDWI has demonstrated utility in drought monitoring and early warning in various studies (e.g., Ceccato, Flasse, & Gregoire, 2002; Gu et al., 2007). It is calculated using NIR and SWIR reflectance, making it sensitive to changes in liquid water content and the spongy mesophyll of vegetation canopies (Ceccato et al., 2001; Gao, 1996).

2.3 Satellite based crop inventory

Crop inventory, a widely embraced and meticulously validated application of satellite remote sensing, has emerged as an indispensable tool for the efficient, timely, and cost-effective mapping, monitoring, and management of agricultural resources. The realm of remote sensing technology has witnessed its profound evolution within the agricultural domain, driven by rigorous experimentation and extensive field studies. These applications have been meticulously conceived, scientifically grounded, and systematically executed through collaborative efforts encompassing various central government ministries, state governments, research organizations, academic institutions, and industry stakeholders.

In India, pioneering experiments, notably the detection of coconut root-wilt disease employing thermal aerial photography, marked the inception of this journey. The triumphant outcomes of these initiatives subsequently catalyzed the formulation and execution of a plethora of systematic experiments. At the core of remote sensing-based crop identification and discrimination lies the fundamental premise that each crop possesses a distinct spectral signature, an outcome of its unique architectural characteristics, phenological stages, sowing windows, and more.

The foundation of remote sensing in agriculture rests upon the diverse regions of the electromagnetic (EM) spectrum and the manner in which vegetation interacts with each spectral component. Materials such as water, soil, vegetation, buildings, and roads exhibit differential reflectance patterns in the visible and infrared spectra. The interpretation of optical images necessitates a comprehensive understanding of the spectral reflectance signatures attributed to various surface features, whether natural or man-made. Key spectral bands employed in vegetation studies encompass the visible, near-infrared, short-wave infrared, and thermal regions of the EM spectrum. These distinct EM regions are sensitive to various plant parameters, including pigmentation, leaf structure, and water content, all of which influence the spectral response of vegetation. The specific responses emanating from these parameters are pivotal in determining the crop type, growth stage, and overall health.

However, optical sensors are susceptible to the constraint of cloud cover, rendering them ineffective, particularly during the monsoon season. In contrast, sensors operating within the microwave region of the EM spectrum possess the invaluable capacity to penetrate cloud cover, rendering them suitable for mapping regions frequently shrouded in clouds. The interaction of microwaves with vegetation is contingent upon factors such as moisture content and dielectric properties. These properties, influenced by crop species, moisture content, and structural composition, introduce variations in the dielectric characteristics across different crops. The backscatter information obtained from crops through microwave sensors provides critical insights, including data on biomass and crop health.

In sum, satellite remote sensing has emerged as a transformative force in agriculture, offering comprehensive and versatile solutions for crop inventory, monitoring, and management. This technology bridges the gap between Earth observation and agricultural sustainability, empowering stakeholders to make data-informed decisions and optimize resource utilization in this vital sector.

2.3.1 Crop classification techniques

Crop classification techniques encompass a suite of methods and strategies employed to categorize and distinguish various crop types or vegetation from data obtained through remote sensing or alternative sources. These techniques hold pivotal significance in numerous domains, including agriculture, environmental monitoring, land use planning, and crop yield estimation. The choice of a specific crop classification technique hinges on several factors, including data accessibility, computational resources, accuracy prerequisites, and the distinctive characteristics of the study area. Here, we provide a concise overview of some commonly used crop classification techniques (Jensen, 1996; Lillesand, Kiefer, & Chipman, 2015):

Supervised classification: This technique involves training a classification algorithm using labeled training data, where each pixel is assigned a specific class label corresponding to a particular crop type. Prominent supervised classification algorithms include the maximum likelihood classifier (MLC), support vector machines (SVM), random forests, artificial neural networks (ANN), among others.

Unsupervised classification: Unsupervised classification does not necessitate labeled training data. Instead, it groups pixels into clusters based on their spectral attributes. Techniques such as K-means clustering and Gaussian Mixture Models (GMM) are often applied for unsupervised crop classification.

Object-based image analysis (OBIA): OBIA transforms remote sensing imagery by segmenting it into meaningful objects, moving beyond the traditional pixel-based approach. These objects are delineated based on a multitude of characteristics, encompassing shape, texture, spectral properties, and more. Subsequently, rule-based or machine learning methodologies are employed to classify these objects. OBIA techniques are particularly effective when applied to high-resolution satellite or aerial imagery, where individual objects are more discernible and contextual information greatly enhances accuracy. For instance, Fig. 2.5 shows a high-resolution optical image (LISS-IV and Cartosat-1 PAN merged) where different horticulture crops can be visually distinguished based on their tone, canopy structure, ground cover, leaf density, age, etc. Further, data fusion techniques combine information from multiple sources, such as optical and radar data, to improve crop classification accuracy. Fusion of different data types can provide complementary information and enhance the discrimination of crop types.

Decison tree classifier: Decision trees are hierarchical structures that partition data based on various spectral and spatial attributes. Each internal node represents a test on a specific attribute, while each leaf node corresponds to a class label. Decision trees offer interpretability and are widely favored for crop classification.

Deep learning: Deep learning techniques, notably convolutional neural networks (CNNs), have demonstrated remarkable potential in crop classification tasks. CNNs can autonomously discern intricate spectral and spatial features from remote sensing data, attaining exceptional accuracy in classification.

Transfer learning: Transfer learning entails utilizing pre-trained models trained on extensive datasets to extract features from remote sensing data. These models are then fine-tuned on smaller labeled datasets specific to crop classification tasks. Transfer learning proves valuable in scenarios with limited labeled data availability.

FIGURE 2.5 High resolution satellite image showing different horticulture crops.

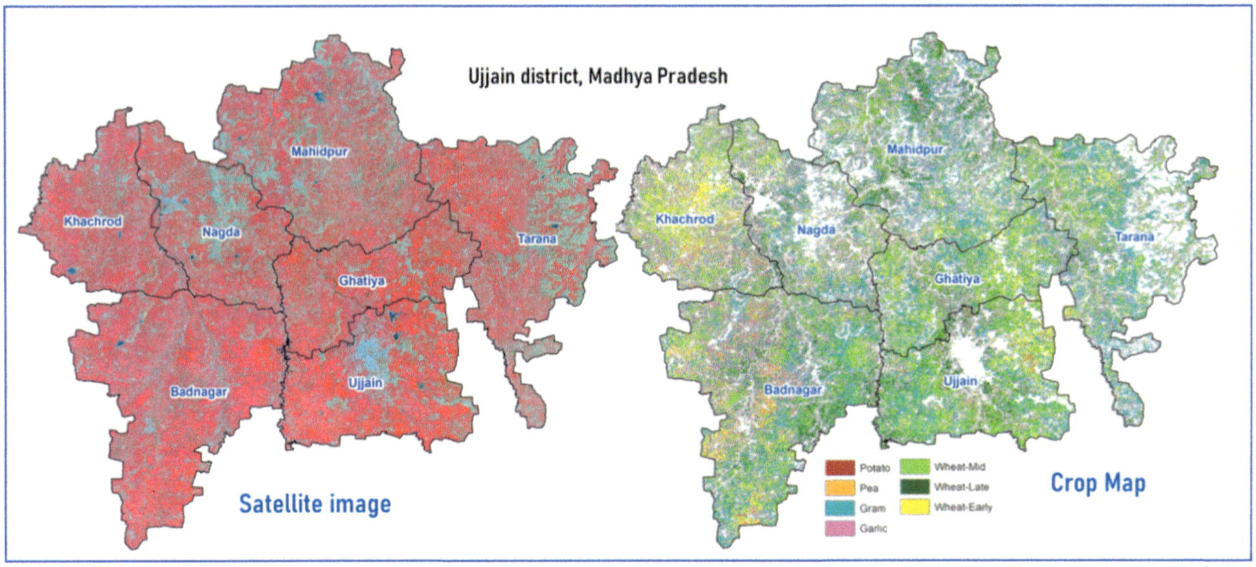

FIGURE 2.6 Satellite-based field crop discrimination with multispectral sensors. Map lines delineate study areas and do not necessarily depict accepted national boundaries. *MP AgriGIS Project, NRSC.*

These techniques collectively constitute a dynamic toolkit for crop classification, each with its own distinct advantages and suitability for specific applications. As technology continues to advance, the synergy between these techniques, data fusion approaches, and emerging technologies promises to enhance our ability to monitor and manage agricultural resources with unprecedented precision and insight.

Temporal satellite data spanning the period from crop sowing and emergence to its growth, maturity, and senescence can be of great help for crop inventory and crop monitoring. The temporal response of a crop pixel during crop growth period can be mapped to carry out crop-specific mapping. Fig. 2.6 shows the multispectral Sentinel-2 satellite image of Ujjain district of Madhya Pradesh and corresponding classified map where different field crops have been discriminated based on sowing time, spectral response, and crop stage. In the cloud-covered monsoon season, temporal SAR data is useful in area estimation of crops. The different transplanting periods of rice are delineated based on the backscatter response of temporal SAR data for Krishna district of Andhra Pradesh, as shown in Fig. 2.7.

2.3.2 Crop classification accuracy: precision in identifying crop types

Crop classification accuracy denotes the degree of correctness and precision in the identification and categorization of various crop types within a given dataset or remote sensing imagery. The accuracy of crop classification is subject to variation based on several factors, including data quality and resolution, the employed classification method, and the expertize of individuals involved. Notable factors influencing classification accuracy encompass the presence of mixed pixels (pixels containing multiple crop types), temporal fluctuations in crop appearance, and staggered crop sowings. Consequently, the presence of a representative and accurate ground truth dataset is pivotal for the evaluation and enhancement of crop classification accuracy.

Various metrics are employed to assess the accuracy of crop classification, with the error matrix serving as a central tool. The error matrix facilitates the computation of key parameters such as:

(i) **Overall accuracy:** This metric gauges the percentage of correctly classified pixels or areas relative to the total number of pixels or areas within the dataset.
(ii) **Error of omission and commission:** These metrics quantify the number of misclassified pixels, distinguishing between instances where a pixel is erroneously included in a class (commission error) and when it is incorrectly excluded from a class (omission error).
(iii) **User's accuracy:** The user's accuracy reflects the probability that a classified pixel or area genuinely belongs to a specific class, thereby quantifying the reliability of the classification from the user's perspective.
(iv) **Producer's accuracy:** The producer's accuracy assesses the probability that a pixel or area of a specific class is correctly classified as such, offering insights into the effectiveness of classification from the producer's standpoint.

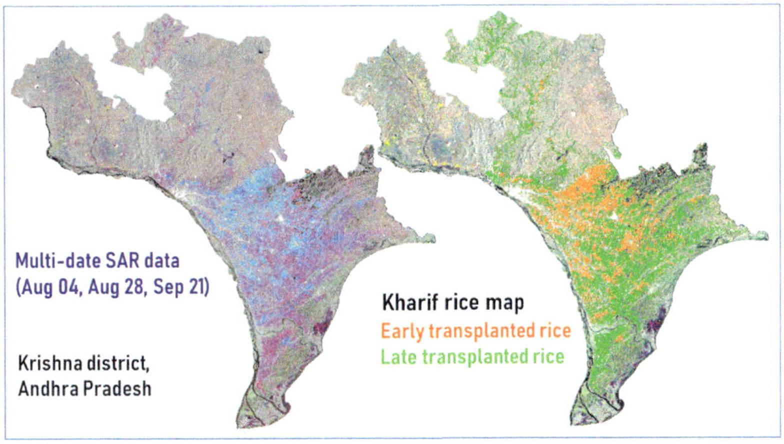

FIGURE 2.7 Kharif rice area estimation with microwave sensor. Map lines delineate study areas and do not necessarily depict accepted national boundaries. *FASAL Project, NRSC.*

(v) **Kappa coefficient:** The Kappa coefficient serves as a robust measure evaluating the classification's performance in contrast to random assignment. It ranges from −1 to 1, with a value of 0 indicating performance no better than random classification. Negative values signify a classification significantly worse than random, while a value approaching 1 signifies a classification significantly superior to random.

In essence, the accuracy of crop classification is not merely a binary measure but a multifaceted evaluation encompassing various dimensions of correctness and reliability. These metrics provide a comprehensive assessment of the precision and efficacy of crop classification endeavors, serving as indispensable tools in the pursuit of data-driven agricultural insights.

2.4 Crop parameter retrieval from satellite data

The estimation of crop biophysical and biochemical parameters stands as a pivotal cornerstone in diverse agricultural applications, ranging from crop health monitoring and stress detection to natural resource management, yield estimation, and hydrological modeling. These parameters furnish invaluable revelations concerning canopy function and structure, broadly categorized into two primary groups: biophysical parameters and biochemical parameters.

Biophysical parameters: This category encompasses a spectrum of variables, including:

(i) **LAI (unitless):** LAI offers insights into the extent of foliage cover within the crop canopy, serving as a pivotal metric for understanding canopy structure and biomass accumulation.
(ii) **Crop height (m):** Crop height, though labor-intensive to measure, provides tangible indicators of crop growth progression, facilitating effective management practices.
(iii) **Biomass (kg m^{-2}):** Biomass quantifies the total mass of vegetation in a given area, a critical parameter for assessing crop productivity and resource allocation.
(iv) **Stomatal conductance (mmol s^{-1}):** Stomatal conductance delineates the extent to which plants regulate gas exchange, offering crucial insights into crop water use and stress responses.

Biochemical parameters: This category encompasses parameters reflective of the chemical composition and health status of crops, including:

(i) **Chlorophyll content (μg cm^{-2}):** Chlorophyll content serves as a proxy for leaf nitrogen content and provides critical data on photosynthetic efficiency, thus indicating nitrogen deficiency and overall crop health.
(ii) **Nitrogen content (%N):** Nitrogen content, expressed as a percentage, is fundamental for understanding the nutrient status of crops and plays a vital role in fertilization management.
(iii) **Leaf water content (%):** Leaf water content quantifies the moisture content within leaves, offering crucial information about plant hydration and drought stress.
(iv) **Salinity (mmol):** Salinity levels are indicative of soil health and its suitability for crop cultivation, with high salinity levels posing potential threats to crop vitality.

These biophysical and biochemical parameters collectively serve as sensitive indicators of crop stress, enabling farmers and researchers to proactively detect and address issues affecting crop health and productivity. Their application is instrumental in advancing sustainable agriculture, optimizing resource utilization, and ensuring food security in the face of evolving environmental challenges.

2.4.1 Modeling approaches to retrieve crop parameter

Remote sensing techniques have been utilized to map crop biophysical and biochemical parameters across extensive areas at frequent intervals. There are different approaches for retrieving these parameters using satellite data, which can be broadly categorized into three groups: empirical approaches, semiempirical approaches, and physical process-based modeling approaches.

Empirical approaches establish statistical relationships between satellite-derived variables and ground-measured crop parameters. For practical applications, vegetation indices (VIs) or shape indices (e.g., red edge) are the most commonly used satellite-derived variables. VIs are straightforward to compute using optical satellite data and exhibit distinct differences between cropland and other land covers. Some studies have employed active synthetic aperture radar (SAR) data with cross-polarization (HV or VH), which canal leviate the cloud effect during the monsoon season. Several studies have demonstrated the use of NDVI and VH (SAR variable) for crop parameter retrieval using statistical modeling. However, empirical techniques have limitations as they rely on observed data and may not be reliable across different crop types, locations, time periods, or parameters that were not considered during the model development. Consequently, they are less feasible for regular mapping over large areas.

In contrast, physical-process-based modeling approaches describe the physiological processes of plants and crop canopies. These approaches employ radiative transfer models (RTMs) to simulate the spectral reflectance of the canopy and then invert the RTMs to obtain the desired parameters (Campos-Taberner et al., 2016; Féret, Gitelson, Noble, & Jacquemoud, 2017). Though these approaches are data-intensive and require information about crop architecture, sun illumination, and viewing geometries, soil backgrounds, however, offer the advantage of being applicable across various operational scenarios. Most crop models combine process-based and empirical models, resulting in semi-empirical models.

2.4.1.1 Empirical and semi-empirical modeling of crop parameter retrieval

Remote sensing can be utilized to determine various crop parameters through empirical and semi-empirical modeling. It enables the estimation of crucial crop growth parameters such as biomass, height, and LAI, which characterize the photosynthetic strength. LAI, among multiple crop growth parameters, plays a vital role in quantifying the exchange of mass and energy between vegetation and the atmosphere. It describes key physiological behaviors of plant canopies, including photosynthesis and transpiration, and is also employed for estimating crop yield. Optical and microwave satellite data are extensively employed in estimating LAI using empirical and semi-empirical modeling (Kira, Nguy-Robertson, Arkebauer, Linker, & Gitelson, 2016; Pu & Cheng, 2015). In optical remote sensing, vegetation indices (VIs) serve as the simplest satellite-derived variables widely used for LAI retrieval (Baret & Guyot, 1991; Delegido, Vergara, Verrelst, Gandía, & Moreno, 2011; Haboudane, Miller, Pattey, Zarco-Tejada, & Strachan, 2004). CLAIR (Clevers, Büker, Van Leeuwen, & Bouman, 1994) is an example of a semi-empirical model used in the optical region. The VI in the optical domain is widely used due to its computational ease. However, the quality of optical data is often compromised by cloud cover, posing challenges in estimating LAI for monsoon-season crops like rice. In the microwave region, SAR backscatter demonstrates the ability to mitigate cloud contamination. Moreover, SAR exhibits sensitivity to various target properties, such as dielectric and geometric properties, making it highly suitable for LAI estimation. Previous studies have investigated the satellite-based microwave data for the estimation of crop growth variables, including LAI, plant height, and vegetation water content, which exhibit sensitivity to radar backscattering (Mandal, Bhattacharya, & Rao, 2021; Steele-Dunne et al., 2017). With advancements in SAR imaging systems, including dual-polarimetry, full-polarimetry, and recently compact-polarimetry, the observation dimensions are enhanced, enabling quantitative retrieval of crop parameters. The dual-polarimetric temporal co-polarized ratio (HH/VV) has frequently been used to detect and monitor rice growth using a rule-based change detection technique (Lopez-Sanchez, Cloude, & Ballester-Berman, 2011). C-band Sentinel-1 data in dual-polarization were employed to monitor wheat crop growth using breakpoints and local extrema (Schlund & Erasmi, 2020) and to retrieve LAI, biomass, and height using machine learning approaches (Kumar et al., 2018; Mandal et al., 2020). As there is a loss of coherence under reflection symmetry conditions in dual-polarimetric SAR, fully polarimetric (FP) SAR data are exploited to derive growth parameters relevant to the vegetation volume scattering mechanism. Numerous studies have demonstrated the potential of full-polarimetric SAR data for retrieving growth parameters such as crop phenology (Canisius et al., 2018), biomass

(Hosseini & McNairn, 2017), LAI (Kumar, Kumari, & Saha, 2013; Mandal et al., 2021; Ratha et al., 2019), among others. The water cloud model and MIMICS are semi-empirical models widely used in the microwave region of the electromagnetic spectrum to retrieve crop biophysical parameters.

2.4.1.2 Canopy reflectance model

Canopy reflectance models analyze the radiation processes within crop canopies based on the equation of radiative transfer. Remote sensing can relate reflectance measurements and growth parameters using canopy reflectance models, based on an understanding of crop growth parameters, the environment, and their association with reflectance. The bio-physio-chemical properties of the canopy alter solar radiation, which is then reflected and intercepted by space-borne sensors. The following subsystems contribute to the relationship between canopy reflectance and satellite data (Goel 1988):

(i) Radiation source: Spectral intensity, source zenith angle, and source azimuth angle.
(ii) Atmosphere: Absorption and scattering properties of aerosol particles, water vapor, and ozone; environmental parameters such as temperature, relative humidity, wind speed, and precipitation.
(iii) Vegetation: Optical parameters (reflectance and transmittance) and structural parameters (geometrical shapes and positions) of vegetation components (leaves, stalks, stems, etc.), planting geometry.
(iv) Soil background: Reflectance and absorption, surface roughness, surface texture, bulk density, and moisture profile.
(v) Detector: Spectral sensitivity, aperture, calibration, view zenith angle, and view azimuth angle.

Canopy reflectance models can be categorized into four groups based on their approach and complexity level (Berger et al., 2018):

(i) Geometrical models: Canopies are represented by translucent geometric shapes. These models are suitable for characterizing sparse canopy patches with minimal multiple scattering.
(ii) Turbid medium models: Canopies are considered as horizontally uniform, plane-parallel layers with absorbing and scattering particles. These models are suitable for denser canopies, assuming that the canopy height exceeds that of vegetation elements.
(iii) Hybrid models: These models combine geometric and turbid medium approaches, offering a complex yet flexible modeling framework. They are suitable for characterizing moderate crop canopies.
(iv) Monte-Carlo ray tracing models: These models consider the radiation regime in vegetated canopies most accurately but are computationally intensive.

Furthermore, these models require expert knowledge for parameterization and inverse schemes compared to empirical and semi-empirical modeling approaches.

Comprehensive reviews of canopy reflectance models (CRMs) have been conducted, as evidenced by studies (Goel, 1988). One of the most widely used models is a turbid medium model called PROSAIL, which combines the PROSPECT leaf model (Jacquemoud & Baret, 1990) with the scattering by arbitrary inclined leaves (SAIL) canopy model (Verhoef, 1985). This coupling requires simulated leaf reflectance/transmittance from PROSPECT as a prerequisite for the SAIL model (Fig. 2.8).

2.4.2 Inversion of canopy reflectance model

PROSAIL can operate in two modes: forward mode and backward mode. In the forward mode, canopy reflectance is computed using various input parameters to examine the sensitivity of canopy reflectance to different factors. In backward modeling, crop growth variables are retrieved from remotely sensed data through model inversion. Model inversion aims to obtain the input parameters (crop growth parameters) using the output of the model, which is the remotely sensed reflectance from the crop canopy. The goal is to find the best match between the remotely sensed and model-simulated spectra.

Earlier, inversion techniques such as iterative numerical optimization and Markov chain Monte Carlo methods (Jacquemoud, Baret, Andrieu, Danson, & Jaggard, 1995) were employed to invert the model. However, due to limitations such as convergence to local minima or computational intensity, these techniques have been replaced by other techniques. One computationally less intensive inversion technique is the look-up table (LUT) approach, which is commonly used for the PROSAIL model (Weiss, Baret, Myneni, Pragnère, & Knyazikhin, 2000). However, the LUT approach also has limitations, such as uncertainty in solutions for various ill-posed problems or the requirement of

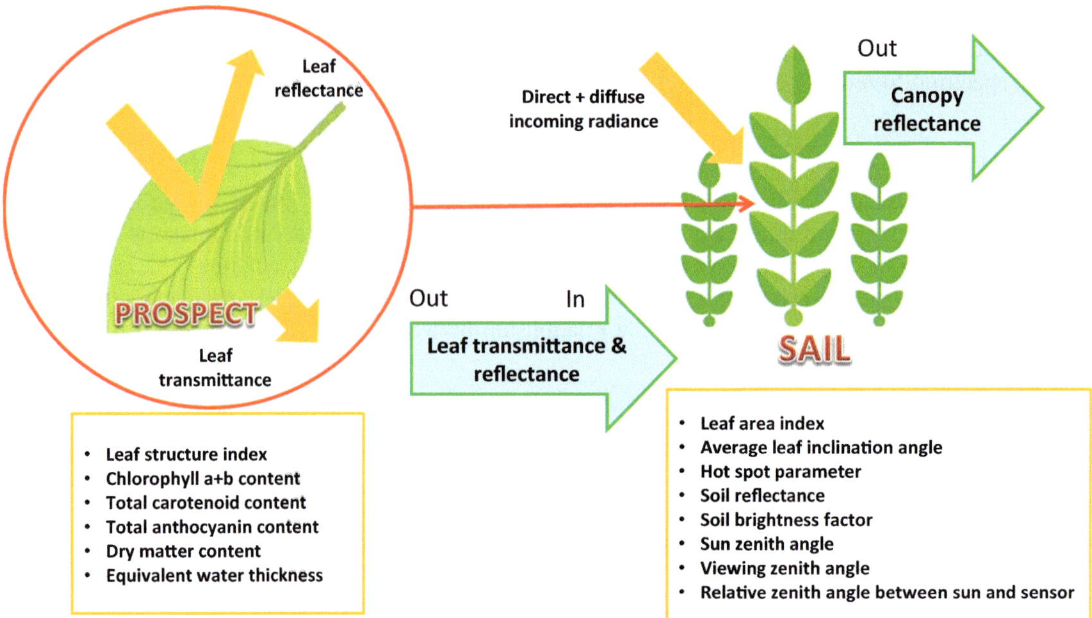

FIGURE 2.8 Parameters and processes occurring in PROSAIL (PROSPECT + SAIL) model. *Adapted from Berger, K., Atzberger, C., Danner, M., D'Urso, G., Mauser, W., Vuolo, F., & Hank, T. (2018). Evaluation of the PROSAIL model capabilities for future hyperspectral model environments: A review study.* Remote Sensing, *10(1), 85. https://doi.org/10.3390/rs10010085.*

relatively long computation times for large areas or hyperspectral data. In recent years, machine learning regression algorithms (MLRA) have emerged as alternatives to traditional inversion techniques. MLRA techniques, such as the genetic algorithm, kernel-based algorithm of Gaussian process regression, artificial neural networks, and support vector machine regression, have become the popular inversion (Durbha, King, & Younan, 2007; Schlerf & Atzberger, 2006). Though these are very efficient computationally, they share the common limitation of the ill-posed nature of model inversion; which is observed with different combinations of canopy parameters corresponding to almost similar spectra.

2.5 Crop phenology and cropping intensity

Satellite-based remote sensing has ushered in a revolutionary era in our capacity to monitor and analyze diverse facets of Earth's surface, with a particular impact on agricultural systems. Among the critical dimensions of agriculture, crop phenology, which delves into the study of plant growth and developmental stages, stands as a linchpin in monitoring and managing agricultural processes. Satellite imagery emerges as a valuable resource for tracking crop phenology on a large scale, unraveling the intricacies of planting, growth, and harvest cycles. In this section, we delve into the utility of satellite-based remote sensing techniques for monitoring crop phenology and explore its manifold applications in agriculture.

Three distinct methodologies exist for assessing vegetation phenological events: in situ observations, bioclimatic models, and remote sensing techniques, as documented in scholarly works (Fisher, Richardson, & Mustard, 2007; Schaber & Badeck, 2003). In situ observations, though grounded in accuracy, are marred by high costs, time intensiveness, susceptibility to operator biases, and limitations in extending coverage over expansive geographical areas. Bioclimatic models, while theoretically sound, necessitate precise and comprehensive data regarding vegetation and climatic conditions specific to forest or crop species, along with localized calibration, making them less feasible for large-scale implementation. It is within this context that satellite remote sensing emerges as a pragmatic alternative, offering routine and sometimes daily monitoring of Earth's entire land surface. Satellites, with their ability to capture data spanning entire ecosystems or regions, offer a unique perspective of Earth's surface, enabling the detection of broad-scale phenological changes that would be arduous, if not impossible, to discern through field surveys alone. Additionally, remote sensing data can be standardized and calibrated across sensors, rendering continuous insights over prolonged periods. This cost-effective approach has positioned satellite remote sensing as an invaluable tool for phenology studies, featuring excellent temporal repeatability across vast and otherwise inaccessible regions.

Extensive research endeavors have been undertaken to extract crop phenology metrics and detect substantial changes utilizing remote sensing techniques (Chakraborty, Seshasai, & Dadhwal, 2014; Fang & Liang, 2003; Heumann, Seaquist, Eklundh, & Jönsson, 2017; Kumari, Chakraborty, Chakravarathi, & Roy, 2023; Piao, Fang, Zhou, Ciais, & Zhu, 2006). Crop phenology, a prominent focus, is often scrutinized through the temporal dynamics of the NDVI. Derived from satellite imagery and available since the inception of Earth observation satellites (EOS), NDVI computation hinges on the comparison of near-infrared and red band reflectance values. This index furnishes insights into the photosynthetic capacity of plant canopies and the extent of vegetation cover at specific locations. Nevertheless, it is noteworthy that coarse-resolution sensors alone may not directly capture specific phenological events. Instead, researchers derive various phenology proxies, including the commencement and conclusion of the growing season (SGS and EGS), the duration of the growing season (LGS), the seasonal amplitude of NDVI (AMP), and the seasonally integrated NDVI (SiNDVI), among others. Time series NDVI analyzes have been instrumental in detecting positive trends in growing season length across diverse regions, thereby enriching our understanding of vegetation dynamics and their responses to environmental factors.

In sum, satellite-based remote sensing stands as a transformative force in the realm of crop phenology monitoring, equipping us with the tools and insights to navigate the intricacies of agricultural systems on a global scale. Its applications extend beyond scientific inquiry, encompassing agricultural management, resource allocation, and bolstering our collective efforts towards sustainable and resilient food production systems.

2.5.1 Time series normalized difference vegetation index and crop phenology metrics

Phenology, which encapsulates vegetation responses to seasonal climatic variations encompassing changes in irradiance, temperature, and rainfall, holds paramount importance as a land surface parameter in atmospheric and climate models. However, when scrutinizing satellite-derived time-series data, the ostensibly observed seasonal patterns may be subject to influence from various cyclic and noncyclic factors. These recurrent phenomena are aptly labeled "seasonality." Furthermore, it's worth noting that seasonality parameters derived from satellite-based time-series data often contend with elevated levels of noise, which can introduce uncertainty into the measurements.

To address these challenges and engender more robust measurements, a valuable approach involves the utilization of fitting functions. This strategy serves a dual purpose: mitigating noise and bolstering the stability of derived measurements. The practical implementation of fitting functions is depicted in Fig. 2.9, which illustrates a multi-year NDVI time series (depicted by the blue line). In this representation, actual NDVI values may appear erratic and frequently exhibit gaps due to noise and cloud contamination. The application of a fitting function, visualized by the red line in Fig. 2.9, effectively envelops the NDVI time series, rendering it continuous and primed for subsequent analyzes.

Fitting functions thus emerge as a pivotal tool in the realm of phenological analysis, enabling researchers to extract meaningful insights from satellite-derived time-series data amidst the complexities introduced by seasonality and noise. This approach, by promoting data continuity and enhancing measurement stability, contributes to the precision and

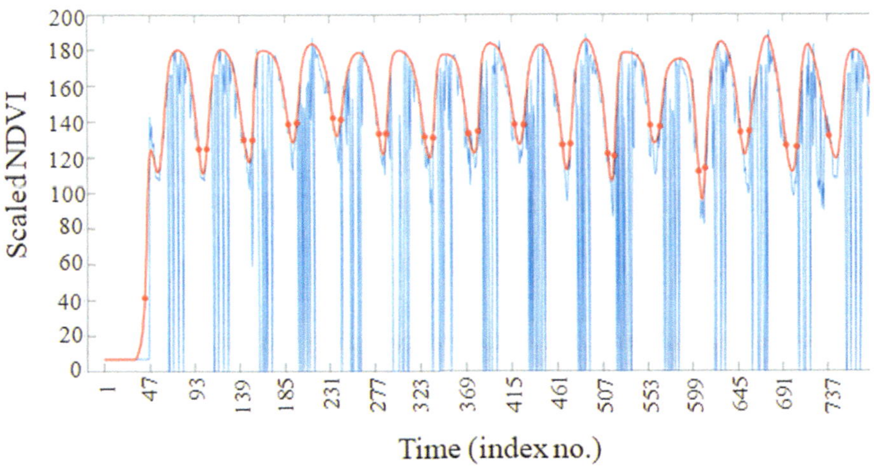

FIGURE 2.9 Scaled normalized difference vegetation index time series of multiple years (8-days composite, 2001–2014) showing crop phenology; blue line actual NDVI values, red line fitted NDVI values. *Adapted from Chakraborty, A., Seshasai, M.V.R., Reddy, C.S., & Dadhwal, V.K. (2018). Persistent negative changes in seasonal greenness over different forest types of India using MODIS time series NDVI data (2001–2014).* Ecological Indicators, 85, *887–903. https://doi.org/10.1016/j.ecolind.2017.11.032.*

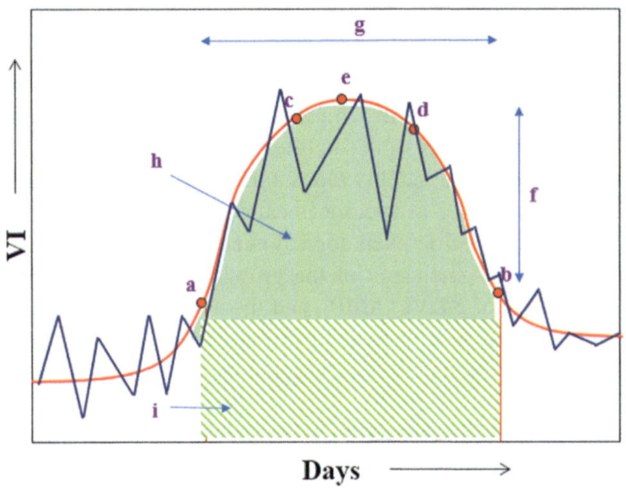

FIGURE 2.10 The seasonality parameters generated using normalized difference vegetation index time series; blue line original vegetation index (VI, such as NDVI) and red line smoothened VI. *Adapted from Eklundh, L., & Jönsson, P. (2017). Lund University TIMESAT 3.3 with seasonal trend decomposition and parallel processing —- Software Manual (92). Lund University, https://web.nateko.lu.se/timesat/docs/TIMESAT33_SoftwareManual.pdf. (Accessed 09.10.24).*

reliability of phenological assessments, thereby advancing our understanding of how vegetation responds to the ever-changing rhythms of the natural world.

The NDVI time series, as depicted in Fig. 2.10, serves as a rich source from which numerous crop phenology metrics and their derivatives can be extracted. These metrics provide valuable insights into the temporal dynamics of vegetation growth and offer a comprehensive perspective on the agricultural landscape. Here are some well-known phenological metrics derived from the NDVI time series:

(i) Start of the season (a): This represents the time at which the left edge of the NDVI curve has increased to a user-defined level, often expressed as a certain fraction of the seasonal amplitude. It is measured from the left minimum level.

(ii) End of the season (b): This metric corresponds to the time when the right edge of the NDVI curve has decreased to a user-defined level, again typically a fraction of the seasonal amplitude. It is measured from the right minimum level.

(iii) Length of the season (g): This metric quantifies the duration of the growing season, spanning from the start to the end of the season.

(iv) Base level: The base level is calculated as the average of the left and right minimum values of the NDVI curve.

(v) Mid of the season (c and d): These metrics represent the mean values of the times at which the left edge has increased to the 80% level (c) and the right edge has decreased to the 80% level (d), respectively.

(vi) Seasonal maximum value (e): This metric identifies the time at which the NDVI curve reaches its seasonal maximum value, which may occur between the midpoints c and d.

(vii) Seasonal amplitude (f): The seasonal amplitude is computed as the difference between the maximum value and the base level of the NDVI curve.

(viii) Small seasonal integral (h): This metric corresponds to the integral of the function describing the season from the start to the end, excluding the base value (i).

(ix) Large seasonal integral (h + i): The large seasonal integral encompasses the integral of the function describing the season from the start to the end, including the base value (i).

These phenology metrics, along with their derivatives, offer a comprehensive framework for characterizing crop growth patterns and monitoring seasonal changes. They serve as valuable tools for agricultural management, yield estimation, and the assessment of crop health and productivity, contributing to informed decision-making in the realm of agriculture and environmental management.

2.5.2 Time-series normalized difference vegetation index data and cropping season

India has three cropping seasons: *Kharif*, *Rabi*, and *Zaid*. Kharif crops are grown with the onset of the monsoon in different parts of the country, and these are harvested in September–October. Important crops grown during this season are paddy, maize, jowar, bajra, tur, moong, urad, cotton, jute, groundnut, and soyabean. Some of the most important rice-growing regions are Assam, West Bengal, the coastal regions of Orissa, Andhra Pradesh, Tamil Nadu, Kerala, and Maharashtra, particularly the Konkan coast along with Uttar Pradesh, Bihar, Punjab, and Haryana.

Rabi crops are sown in winter from October to December and harvested in summer from April to June. Some of the important *Rabi* crops are wheat, barley, peas, gram, and mustard. Though these crops are grown in large parts of India, states from the north and north-western parts such as Punjab, Haryana, Himachal Pradesh, Jammu and Kashmir, Uttarakhand, and Uttar Pradesh are important for the production of wheat and other *Rabi* crops. Availability of precipitation during the winter months due to the western temperate cyclones helps in the success of these crops. However, the success of the green revolution in Punjab, Haryana, western Uttar Pradesh, and parts of Rajasthan has also been an important factor in the growth of the above-mentioned *Rabi* crops.

There is a short season during the summer months, that is, between the *Rabi* and the Kharif seasons, known as the Zaid season. Some of the crops produced during Zaid are watermelon, muskmelon, cucumber vegetables, and fodder crops. The temporal NDVI profile has the potential to assess the crop season. Hence, 16-day composite MODIS NDVI values of November 2018 to December 2019 were stacked, and the temporal variation of the NDVI across the time frame over different parts of India was extracted/plotted as shown in Fig. 2.11. The NDVI value was found to be low at the beginning of the crop season during sowing. As the crop season progresses, NDVI increases with the crop growth, and both attain their peak. The NDVI further decreased as the crop matured and dried up. Hence, the temporal NDVI attained a bell-shaped curve for a crop season. If the patterns of temporal NDVI are observed over the year, a uni-modal bell-shaped curve would be found for a single cropped area. Whereas the double-cropped area would have bimodal curve and the triple-cropped area would have a trimodal curve.

Representative patterns of temporal NDVI are presented in Fig. 2.11. Over Punjab and Haryana, a bimodal curve was observed representing a double-cropped area (one crop in rabi and another in kharif). On the other hand, places over Odisha, Jharkhand, and Chhattisgarh were mainly single-cropped *(Kharif)*, so a bell-shaped uni-modal curve is found during the *kharif* season, that is, July—November. A representative pattern of triple-cropped area is seen over

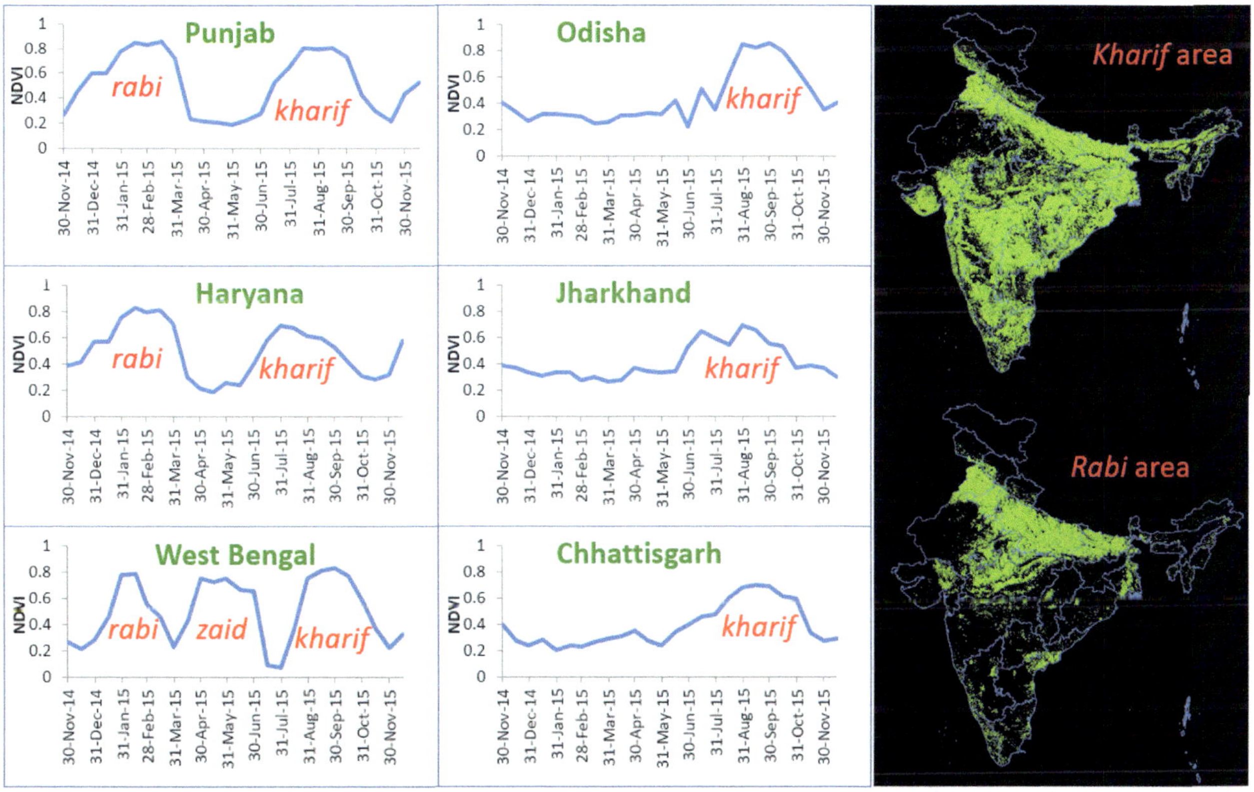

FIGURE 2.11 Temporal normalized difference vegetation index patterns over different cropping season across India. Mapping of *Kharif* and *Rabi* crop season at national level. Map lines delineate study areas and do not necessarily depict accepted national boundaries.

West Bengal, comprising *Rabi-Zaid-Kharif*. Using this NDVI pattern, we can also map the Kharif and Rabi areas over India, and the same is presented in Fig. 2.11.

2.5.3 Cropping intensity

Cropping intensity refers to the number of crops planted on a specific agricultural field within a year (Raut, Sitaula, Aune, & Bajracharya, 2011). It provides valuable information on the utilization of croplands and serves as a measure for crop intensification to enhance total production from the same agricultural land. Understanding cropping intensity is essential for crop yield modeling, assessing the impacts of climate change on yield and soil quality (Becker & Johnson, 2001; Challinor, Parkes, & Ramirez-Villegas, 2015; Sherrod, Peterson, Westfall, & Ahuja, 2003). Therefore, generating cropping intensity information for a region is crucial for improved agricultural management and crop production, such as monitoring irrigation interventions or the planting of short-duration crops in rainfed conditions.

Cropping intensity also influences the interactions between croplands and other physical components of the Earth system. Remote sensing techniques can assist in estimating cropping intensity by mapping crop and fallow lands during different agricultural seasons within a year. Croplands exhibit high spatial and temporal variability, varying from year to year. For example, in drought years, a smaller area is occupied by crops compared to wet years. Thus, satellite remote sensing offers a synoptic view of cropland identification over large areas and provides estimates of cropping intensity.

Approaches for estimating cropping intensity can be broadly categorized into two parts: (1) generation of crop maps in different seasons and integrating them, and (2) employing a phenology-based approach to determine the number of crops on the same field. In the first approach, the crop season within an agricultural year is identified, and crop-fallow maps are generated for each season. These maps are then integrated into various categories, such as triple crop, double crop, single crop, and fallow land, based on combinations like crop–crop–crop, crop–fallow–crop, crop–-crop–fallow, fallow–crop–crop, crop–-fallow–fallow, fallow–fallow–crop, fallow–crop, fallow, and fallow–fallow–fallow. In the second approach, vegetation indices are commonly used to generate temporal profiles, which are then smoothed to remove noise. The number of crop-fallow patterns on the same land is categorized. Cropping intensity can be computed using the multiple cropping index (MCI). MCI is calculated by dividing the sum of the areas cropped and harvested in a single year by the total cultivated area, multiplied by 100 (Dalrymple, 1971). The formula for computing MCI is provided below:

$$\mathrm{MCI} = \frac{\sum_{i=1}^{n} a_i}{A} \times 100 \tag{2.6}$$

where n = total number of crops, a_i = area planted and harvested in a particular season within a year, and A = total cultivated land area available. An MCI value of 300 indicates the sequential cultivation of three crops within a year, representing a triple crop scenario. An MCI value of 200 corresponds to the cultivation of two crops within the same year, referred to as a double crop. An MCI value of 100 indicates that the land has been cropped once in a year, representing a single crop. Finally, an MCI value of 0 signifies that the land has not been cropped within a year, indicating a fallow condition.

2.6 Crop yield modeling

Remote sensing has revolutionized the techniques of collecting information from crop systems, starting from microclimatic conditions and crop health to crop phonology using satellites, drones, and other remote sensing platforms. Remote sensing is a potential tool for crop yield estimation as it provides synoptic, periodical, and accurate crop information in a nondestructive, near-real-time, cost effective, and unbiased manner compared to conventional methods of crop yield estimation. Remote sensing-based crop yield estimation is categorized into 3 broad major approaches, namely (1) statistical approach (2) semi-physical method, (3) crop growth simulation model, and (4) ML-based approach.

2.6.1 Statistical approach

Crop yield depends on soil nutrition levels, fertilizer application rates, crop health, crop stress, and agro-meteorological parameters, viz., temperature, humidity, rainfall, and other factors. Further, each particular crop needs specific growing weather conditions and agronomic practices. Thus, several factors, including weather vagaries, possess a serious threat to agricultural crop production in India and also are noteworthy risks to the economy. Therefore, prognosticating crop yield at different scales is a challenging task for every nation. Statistical models are the most commonly used tools to forecast the crop yield, whereas statistical forecasting models for predicting dynamic behavior of crop yield should be able to take advantage not only of historical data of crop yield but also the impact of various driving forces of the external environment. The statistical methods like linear regression and time-series models predict crop yield efficiently and precisely.

The most investigated statistical yield forecasting models are regression models (Birthal, Khan, Negi, & Agarwal, 2014; Dkhar, Feroze, Singh, & Ray, 2017; Kumar et al., 2011) that offer flexibility in incorporating relevant auxiliary information generally through linear modeling framework; such covariates play an important role in explaining the variability of the response variable. A good amount of covariates with significant correlation is therefore crucial for accuracy of such kind models, but incorporating a large number of covariates becomes limited in terms of parsimony. The yield of any crop is mostly affected by technological change and weather variability. It can be assumed that the technological factors may increase yield smoothly through time, and weather is also having volatile effect on yield; therefore, time (in year) or some functions of time can be used to study the overall effect of technology and weather on crop yield. Time-series forecasting models based on historic data may be a suitable alternative for between- or within year crop yield prediction (Banakara, Sharma, Sahoo, Dubey, & Chowdary, 2023; Chattopadhyay et al., 2011). Particularly, when the past data are used as auxiliary, they may take care of the cumulative effect of existing technological factors, weather and all other relevant covariates.

The statistical models that are used for yield prediction are given by

2.6.1.1 *Linear regression model*

The linear regression method tries to establish a linear relationship between the response variable and predictor(s). Let Y be the response variable that depends on n independent variables $X_1, \ldots, X_n$. Then, the linear regression equation can be mathematically expressed as:

$$Y = b_0 + b_1X_1 + b_2X_2 + \ldots\ldots\ldots\ldots\ldots + b_nX_n + e \tag{2.7}$$

where bi's ($i = 1, \ldots, m$) are unknown parameters (coefficient) and e is the error term, usually assumed to be identically and independently distributed as N (0, σ^2). A regression equation involving one auxiliary is called a simple regression model, whereas a multiple regression equation involves more than one covariate. Regression models include simple or multiple linear regression models, where one variable or multiple variables are used, respectively. The multiple linear regression model uses different methods to enter independent variables into the analysis. Stepwise, all-variable enter method, backward, and forward methods are some of the techniques used in the multiple regression approach while handling multiple variables. Shastry, Sanjay, and Bhanusree (2017) estimated wheat, cotton, and maize yields using multiple linear regression techniques.

2.6.1.2 *Time series model*

Time-series analysis is a statistical technique used for studying data points collected sequentially over time. It is a powerful method for understanding and predicting future values based on historical observations. One fundamental characteristic of time series data is that future values are dependent on past values in a stochastic (random) manner. The primary goal of time-series analysis is to identify models that capture the underlying dynamics of the data, allowing for accurate forecasting of future trends and patterns.

One widely used class of models for capturing linear dynamic structures in time-series data is the autoregressive moving average (ARMA) models. ARMA models provide a framework for linear forecasting and are particularly useful when the data exhibit stationary behavior. Stationarity is a crucial property of time series, indicating that statistical characteristics (such as mean and variance) remain constant over time. Achieving stationarity is essential for making reliable statistical inferences.

However, real-world time series often contain time trends and cyclic patterns that cannot be adequately represented by stationary ARMA models alone. To address this issue, a common approach is to "de-trend" and "de-seasonalize" the data by taking differences between consecutive observations. The resulting differential data is then amenable to modeling with a stationary ARMA component. This combined process of differencing and modeling is known as an autoregressive integrated moving average (ARIMA) model.

Key components of an ARIMA model:

- **Autoregressive (AR) component:** This part of the model captures the linear relationship between the current value and past values of the time series. The order of the autoregressive component, denoted as '*p*,' represents how many past observations are used to predict the current value.
- **Integrated (I) component:** The integrated component refers to the number of differencing operations needed to make the data stationary. If no differencing is required, it is denoted as '$I = 0$.' If one difference is needed, it is '$I = 1$,' and so on.
- **Moving average (MA) component:** The moving average component models the relationship between the current value and past white noise (random) errors. The order of the moving average component, denoted as '*q*,' represents how many past errors are considered in the model.

ARIMA models provide a flexible framework for modeling a wide range of time series data, from purely stationary to non-stationary with trends and seasonality. By adjusting the values of '*p*,' '*I*,' and '*q*,' analysts can tailor ARIMA models to suit the specific characteristics of the data.

In summary, time-series analysis, particularly through ARIMA models, is a valuable tool for understanding and forecasting sequential data. It allows analysts to capture both linear and non-linear patterns in time series data, making it applicable in various fields, including finance, economics, and, as mentioned in the previous responses, agriculture for crop yield forecasting.

The AR model is given by

$$v(B)X_t = \beta_1 X_t + \beta_2 X_{t-1} + \beta_3 X_{t-2} + \ldots\ldots\ldots + \beta_m X_{t-m+1} + \text{et} \tag{2.8}$$

where m is the number of regressor in the model and {et} is assumed to follow standard white noise process with $N(0,\sigma^2)$ and β's are the coefficients, X_ts are time series data points. Fuzzy logic-based crop yield (Bang, Bishnoi, Chauhan, Dixit, & Chawla, 2019) was predicted using time series based ARMA, SARIMA models.

2.6.1.3 Partial least square regression

Partial least squares (PLS) regression is a powerful statistical method used for modeling the relationships between two datasets, typically a predictor matrix (*X*) and a response matrix (*Y*). It is a versatile technique that finds applications in various fields, including crop yield estimation and prediction. PLS regression is particularly useful when dealing with datasets with a high number of predictor variables, multicollinearity (correlation among predictor variables), and a limited number of observations.

Key characteristics and concepts of PLS regression:

(i) **Bilinear factor model:** PLS belongs to the bilinear factor model family of methods. It focuses on identifying linear combinations of variables (latent variables or factors) in both X and Y spaces that maximize the covariance between the two spaces.
(ii) **Covariance modeling:** PLS aims to capture the covariance structure between X and Y by identifying latent variables that explain the maximum variance in both datasets simultaneously. This makes it a useful tool for modeling complex relationships when traditional linear regression methods might fail.
(iii) **Dimension reduction:** PLS achieves dimension reduction by projecting the original variables in both X and Y spaces onto a smaller set of latent variables. These latent variables are linear combinations of the original variables and serve as predictors in the regression model.
(iv) **Multicollinearity handling:** PLS is robust against multicollinearity, which is a common issue in datasets where predictor variables are highly correlated. Instead of relying on strict orthogonality (as in principal component analysis), PLS focuses on finding directions of maximum covariance, even when predictor variables are correlated.

In summary, PLS regression is a versatile and powerful technique for modeling complex relationships between datasets, making it suitable for situations with high-dimensional, correlated data and limited observations. Its ability to handle multicollinearity and reduce the dimensionality of data while preserving important relationships has made it a valuable tool in various scientific and industrial applications, including crop yield estimation.

The PLS regression model is given by the equation

$$Y = T_h C'_h + E_h = XW_h * C'_h + E_h = XW_h(P'_h W_h) - 1 \;\; C'_h + E_h \tag{2.9}$$

where Y is the matrix of the dependent variables and X is the matrix of the explanatory variables. T_h, C'_h, $W*_h$, W_h, and P_h, are the matrices generated by the PLS algorithm, and E_h is the matrix of the residuals. Rice yield in East China (Guo, Xiang, Li, Ma, & Du, 2021) was predicted by the partial least square regression technique, and R^2 was obtained up to 0.614.

2.6.2 Crop growth simulation model

The crop growth model simulates crop yield by integrating the effects of several biophysical and environmental factors such as temperature, solar radiation, rainfall, humidity, water and nutrient availability, crop stresses due to environmental factors, as well as diseases and pests. Besides simulating crop yield, these models are used to explore the cause-and-effect relationship of yield with factors like crop management practices and climate change scenarios on a field as well as a regional scale. Crop models are important tools in agriculture as they help farmers, researchers, and policymakers to take informed decisions on several aspects, starting from crop management with optimization of input resources, seasonal crop yield estimation supporting market intelligence and crop insurance services, adaptability of crops to climate change, etc. Several crop models are available worldwide that are developed with specified objectives and modules having their own strengths and limitations. Most commonly used crop growth models include DSSAT (Decision Support System for Agrotechnology Transfer), Jones et al. (2003); APSIM (Agricultural Production Systems Simulator), Keating et al. (2003); STICS (SimulateurmulTIdisciplinaire pour les Cultures Standard), Brisson et al. (2002); CERES (Crop Environment Resource Synthesis), Jones, Kiniry, and Dyke (1986); WOFOST (World Food Studies), Van Diepen, Wolf, Van Keulen, and Rappoldt (1989); AquaCrop, Steduto, Hsiao, Raes, and Fereres (2009); CropSyst, Stöckle, Donatelli, and Nelson (2003). With rapid development in remote sensing technology, integration of remote sensing-derived biophysical parameters like LAI, chlorophyll content, canopy wetness, etc., with crop growth simulation models is becoming increasingly important. Different types of remote sensing data, such as multispectral, hyperspectral, and microwavedata acquired in different spectral regions of the electromagnetic spectrum at different multiple spatial and temporal resolutions, are being used to retrieve crop biophysical parameters. These parameters are assimilated into the crop model using different techniques such as the Bayesian method, Kalman filtering and ensemble Kalman filtering, ANN, etc. Several studies have demonstrated that the prediction accuracy of crop models increases significantly by assimilating the crop LAI and other biophysical parameters into the crop model. Use of a particular assimilation technique depends on the specific need of the crop model and the type of parameter retrieved from the remote sensing data.

2.6.3 Radiation/light use efficiency approach

Radiation/light use efficiency approach (RUE) models, which are production efficiency models, are widely employed for quantifying gross primary productivity (GPP) and NPP using remotely sensed data. These models are based on the radiation use efficiency concept originally proposed by Monteith (1972). According to this concept, when croplands are adequately watered and fertilized, their productivity demonstrates a linear relationship with the amount of fraction of absorbed photosynthetically active radiation (FAPAR) by the canopies.

However, under real-world environmental conditions, the potential efficiency of optical energy utilization is influenced by various factors such as water availability, temperature, and other environmental stressors. Consequently, GPP (or NPP) can be estimated using the LUE approach by multiplying APAR with the maximum LUE and accounting for environmental stresses. The LUE methodology has emerged as a crucial technique for regional or global mapping of GPP and NPP, as highlighted in studies by Potter et al. (1993) and Running et al. (2004).

Mathematically, Yield based on RUE model can be expressed as:

$$\text{Yield} = HI * \left(\sum_{sowing}^{Harvest} (APAR * \text{€} * Environmental\ stress(f)) \right. \tag{2.10}$$

$$\text{APAR} = PAR * fAPAR$$

Where, APAR = Absorbed PAR, PAR − Photosynthetically Active Radiation, '€' is the potential light use efficiency, and 'f' is the limitation imposed on light use efficiency by various environmental stresses

There are several RUE model for estimating the GPP or yield of crop, these models varies in terms of parameterization of 'f,' that is, the stress factor. Different environmental factors are considered by different models. The major current RUE models are summarized in Table 2.2:

Remote sensing data plays a crucial role in the RUE model by providing valuable information regarding vegetation type, growth status, and environmental conditions. FAPAR is derived either through spectral vegetation index relationships, as demonstrated by studies like Goward and Huemmrich (1992), or using FAPAR products derived from radiative transfer models such as MODIS FAPAR and Sentinel-3 FAPAR, as discussed by Zhao et al. (2005). Several studies Asrar, Fuchs, Kanemasu, and Hatfield (1984) and Fensholt, Sandholt, and Rasmussen (2004) have established a linear

TABLE 2.2 Different radiation/light use efficiency approach models for estimating gross primary productivity of vegetation.

Model	Scalers used	Reference
CASA Model	Air temperature, soil moisture	Potter et al. (1993)
GLO-PEM	Air temperature, soil moisture, VPD	Prince and Goward (1995)
MODIS-GPP	Air temperature, VPD	Running, Nemani, Glassy, and Thornton (1999)
VPM	Air temperature, radiation, water, phenology	Xiao, Zhang, Hollinger, Aber, and Moore (2005)
EC-LUE	Air temperature, water	Yuan et al. (2010)
3-PG	No limiting factor	Landsberg and Waring (1997)
C-Fix	No limiting factor	Veroustraete, Sabbe, and Eerens (2002)

relationship between the NDVI and the FAPAR, which in turn is linked to productivity through the vegetation's capacity for light absorption during photosynthesis. However, this relationship may vary depending on factors such as canopy type, structure, soil conditions, and sun-view orientation and can saturate at higher NDVI values.

RUE models are quite successful for estimating regional and global vegetation productivity. However, at local scales, the optimal energy utilization rate models still exhibit several shortcomings that restrict their application in vegetation productivity estimation. Most light-use efficiency models directly simulate GPP instead of NPP. Assumes the autotrophic respiration constant over the crop growth period.

2.6.4 Machine learning approach

Machine learning, a subset of artificial intelligence, empowers computers to learn from data without explicit programming. Machine learning models are non-parametric, enabling them to capture the complex and non-linear relationships between yield and the influencing features. By detecting patterns and correlations in the data, machine learning models can develop predictive models. During the training phase, historical data is utilized to determine the parameters of the predictive model, which incorporates multiple features. Machine learning has seen increasing use in research to enhance crop yield accuracy by incorporating satellite-derived vegetation indices, meteorological data, hydrological variables, edaphic factors, and more.

In the literature, two prominent machine learning approaches are commonly followed: decision tree-based models like random forest and deep neural network-based models. Random forest (RF) is a supervised ensemble learning model that combines the outputs of multiple decision trees to generate the final prediction. Each decision tree is trained independently and grows in random subspaces of the data using the classification and regression tree (CART) methodology. Deep neural networks (DNN) leverage the universal approximation theorem, which states that a fully connected feedforward neural network with a finite number of layers and neurons can approximate any relationship between input and output variables. This principle enables DNNs to model complex relationships between variables and make accurate predictions (Fig. 2.12).

The workflow of machine learning encompasses several essential steps, including pre-processing, feature design, data splitting into training and validation sets, selection of a suitable machine learning algorithm, training, optimization, evaluation, and testing. To ensure accurate and reliable results, datasets from diverse sources such as satellite-derived data, weather information, soil data, and yield records must undergo homogenization and alignment to achieve consistent spatial and temporal resolutions. Additionally, it is crucial for the training, validation, and testing datasets to be explicitly defined and mutually consistent to enable effective model training and assessment.

Since crop yield is the result of a complex interaction between weather, edaphic, management factors, and genotypes, datasets/features must represent these factors and should have physical meaning in terms of their impact on crop growth and development. Spectral data over the crop is the integrated manifestation of the effect of all the above factors, and hence, satellite data can be used as a feature in the machine learning models. Table 2.3 represents various datasets commonly used in the machine learning models.

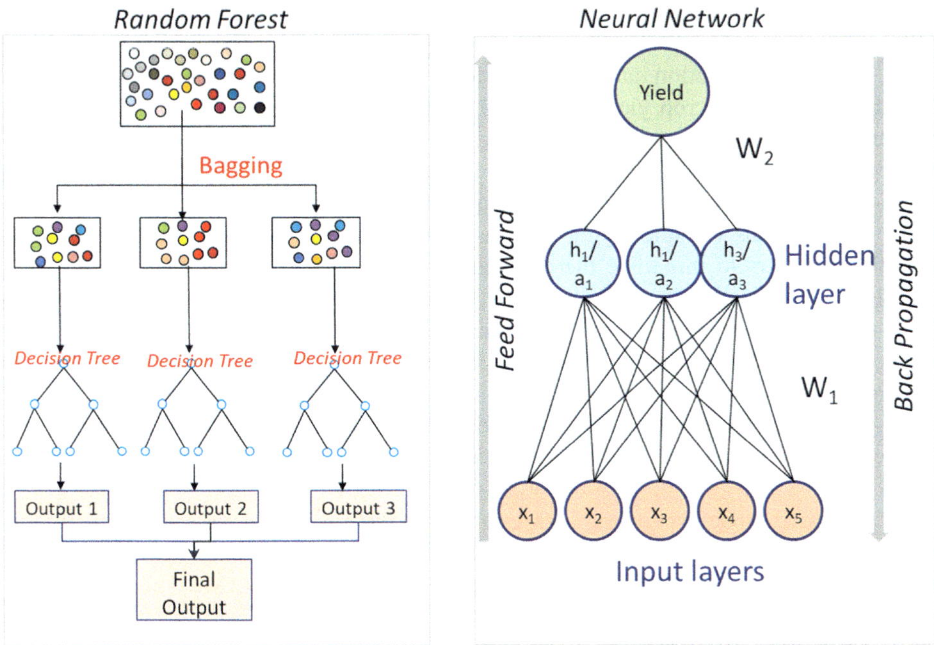

FIGURE 2.12 Schematic flowchart representing machine learning models.

TABLE 2.3 Commonly used datasets for machine learning based crop yield estimation.

Category	Features	Source
Satellite based	Reflectance bands	Resourcesat-2, Sentinel-2, Landsat-8, MODIS,
	Vegetation Indices-Greenness (NDVI, EVI, Red edge index)	Resourcesat-2, Sentinel-2, Landsat-8, MODIS
	Vegetation Indices-wetness (NDWI, LSWI, etc)	Resourcesat-2, Sentinel-2, Landsat-8, MODIS
	Radar backscatter (VH, VV, RVI etc)	EOS-4, Sentinel-1
Meteorological	Rainfall, Rainy Days	IMD gridded, CHIRPS, any other gridded/reanalysis data
	Dry-spell/Wet-spell	IMD gridded, CHIRPS, any other gridded/reanalysis data
	Temperature	IMD gridded, any other gridded/reanalysis data
	Growing degree days	IMD gridded, CHIRPS, any other gridded/reanalysis data
	Heat wave/cold wave	IMD gridded, CHIRPS, any other gridded/reanalysis data
Bio-physical	FAPAR	PROVA/Sentinel-3, MODIS, Sentinel-2
	LAI	PROVA/Sentinel-3, MODIS, Sentinel-2
Edaphic	Soil (texture, depth, AWC, etc)	NBSSLUP (1: 250K, 1:50K)
	DEM	CartoDEM, SRTM
	Soil moisture	SMAP, AMSR-E

Feature selection, training, and optimization are critical steps in developing accurate machine learning models for crop yield estimation. These steps ensure that the model effectively captures the relationships between input features and crop yield, leading to reliable predictions. Feature selection involves identifying the most relevant and informative features from the available data. It aims to reduce the dimensionality of the input space and eliminate redundant or irrelevant features that may introduce noise or overfitting. Proper feature selection enhances model performance and interpretability. Various techniques, such as correlation analysis, statistical tests, and domain knowledge, can be employed to select the most influential features for crop yield estimation. Various features derived from satellite and weather data during crop growing periods (start of the season to end of the season) are generally considered during feature selection.

Training the machine learning model involves feeding the selected features and corresponding crop yield data into the model. During this process, the model learns the underlying patterns and relationships in the data. The training phase optimizes the model's internal parameters to minimize the difference between predicted and actual crop yield values. Regularization criteria are followed to check the overfitting of the model. Optimization involves fine-tuning the model's hyperparameters to achieve the best performance. Hyperparameters are parameters that are not learned during training but are set prior to training, such as learning rate, regularization parameters, or the number of hidden layers in a neural network. Optimization techniques like grid search, random search, or Bayesian optimization can be employed to find the optimal combination of hyperparameters that maximize the model's accuracy and generalization capabilities. Statistical criteria need to be considered for accuracy, such as the mathematical expectation of mean absolute of forecast error MAE, RMSE, or NRMSE.

For model training and validation, stratified random sampling approaches from different yield classes are mostly followed, and models are trained with 60%–70% of the data and validated with 30%–40% of the data. High-quality training yield data are essential for accurate and reliable crop yield estimation using machine learning (ML) models.

2.7 Geospatial support to crop insurance

Indian agriculture is exposed to various risks of crop losses due to natural calamities like drought, flood, extreme weather events, etc., along with biotic stresses like pests and diseases. Crop insurance is a mechanism to distribute these risks among multiple stakeholders and helps farmers get financial support in the event of complete or partial damage to the crop. In India, crop insurance has been in effect since 1972, and several modifications to the scheme were introduced to bring the premium lower, hasten the claim settlement, and bring more transparency in the process (Tiwari, Chand, & Anjum, 2020). The latest crop insurance scheme in India is the Pradhan Mantri Fasal Bima Yojna (PMFBY), which is based on an area-yield approach and extends coverage for various risks to standing crops, prevented/failed sowing, localized damages, post-harvest losses, etc. The PMFBY guidelines have made it mandatory to incorporate technologies into crop insurance (PMFBY, 2020). These technologies include remote sensing, geospatial tools, mobile devices, and drones, which have been recognized as valuable resources for enhancing crop insurance support. There are multiple merits in using geospatial and remote sensing data for crop insurance, which includes unbiased, objective, and independent data to crosscheck and supplement other field information inputs, reducing fraud, moral hazard, and adverse selection, enabling faster and cheaper payouts to the insured, removing area discrepancy in coverage, etc.

Geospatial technology has significant role in different areas of crop insurance value chain. Some of these areas are enlisted in Table 2.4.

2.7.1 Rate making

Crop insurance pricing, or rate-making, is a procedure to determine the price of the insurance contract, that is, premium amount to be charged for a crop in a given district. It is a two-step process; in the first step, state is divided into a group of districts based on the available risk information; in the second step, crop-specific premiums are calculated based on the risk exposure and business opportunities. Including historical crop risk maps generated using satellite data, for example, flood maps, drought maps, etc., can improve the clustering mechanism. Flood risk information can be derived from the flood hazard atlas generated by the Disaster Management Support Group of NRSC using time series flood layers. Likewise, multiple-year drought maps can be used to generate drought vulnerability maps of the area. Land-related risks like degraded land, wasteland, soils, etc. can be incorporated from the existing thematic layers generated through satellite remote sensing technology. All these risk maps can be synthesized into a composite risk index to be used for clustering. Racine and Ker (2006) have documented various non-parametric techniques for rating crop insurance.

TABLE 2.4 Geospatial Technology intervention in crop insurance value chain.

Steps in crop insurance	Geospatial technology interventions	References
Rate making pricing, risk based clustering of districts for insurance pricing	Agriculture risk info maps, flood maps, drought maps, vulnerability maps, etc.	Benami et al. (2021) and Ker and Goodwin (2000)
Enrollment and expansion	High resolution satellite images, georeferenced cadastral maps, etc.	Aggarwal et al. (2016) and Sarvia, De Petris, and Borgogno Mondino (2020)
Crop monitoring prevented sowing risk, mid-season adversity	Moderate resolution satellite data, weather data, mobile apps, data analytics, etc.	Brock Porth et al. (2020) and Sawant, Mohite, Sakkan, and Pappula (2019)
Crop damages natural calamities, abnormal events	Optical and SAR satellite data, weather data, mobile apps, etc.	De Leeuw et al. (2014) and Tapia-Silva, Itzerott, Foerster, Kuhlmann, and Kreibich (2011)
Localized crop damages hailstorms, inundation	High resolution satellite data, georeferenced cadastral maps, weather data, mobile apps, etc.	Chakraborty, Srikanth, Murthy, Rao, and Chowdhury (2021), Choudhary, Chakraborty, Murthy, and Poddar (2022) and Silleos, Perakis, and Petsanis (2002)
Crop yield measurements and estimation smart sampling (CCE plot selection) technology based crop yield estimation	Moderate resolution satellite data, weather data, mobile apps, crop models, data analytics, etc.	Paulraj and Easwaran (2020) and Sawasawa (2013)
Post harvest damage	Weather data, mobile apps, etc.	Jha, Vishwakarma, Ahmad, Rai, and Dixit (2015)
Timely settlement of claims	Cadastral database, IT tools, etc.	Aggarwal et al. (2016)

2.7.2 Crop mapping and monitoring

The principal of crop monitoring for crop insurance is based on the local scale and more frequent information product generation. Crop area discrepancies, crop assessment, and mid-season adversities are the few important aspects in the crop insurance value chain where GPS-enabled mobile and satellite technology has a direct role to play. Mobile apps enable the most efficient way of field data collection, and such data are useful for crop monitoring. Mobile app-based field data also complements satellite-based crop mapping and crop assessment. Mobile apps are very effective for assessing in-season crop management information and assessing localized damages. The availability of comprehensive crop surveillance data through mobile apps allows insurance companies and agricultural departments to assess risks accurately. Fig. 2.13 shows the sowing variability of wheat captured through the mobile app.

Satellite-based mapping of total crop area as well as area under major crops permits area discrepancy analysis or prevented/failed sowing risk assessment. An accurate crop map is the basic information product required for developing technology-based solutions to improve crop insurance. Satellite sensors capture spectral information across different wavelengths, enabling the discrimination of various crops based on their unique spectral signatures. Medium-resolution multispectral satellite data like Resourcesat-2, Sentinel-2, Landsat-8/9, etc. provide valuable data for crop classification and mapping. Fig. 2.14 shows multiple-year crop maps and insurance unit-wise area statistics using temporal Sentinal-2 data.

Availability of cloud-free satellite data during rainy seasons is a limitation that can be overcome by integrating data from multiple satellites and microwave data (Fig. 2.15). Rice and jute crop mapping using microwave radar (SAR) data is already operational in India. Transplanted rice with its water background at the time of transplantation can be delineated with higher accuracies ($>90\%$) than upland rice. SAR-based mapping for other crops like soybean, cotton, jowar, pear-millet, etc. is also reported in the recent literature, providing accuracies of $80\%-85\%$.

The current availability of satellite data sets, derived indices, improved weather data sets, and geospatial tools and techniques presents significant potential benefits for enhancing crop monitoring within insurance units. Fig. 2.16 lists

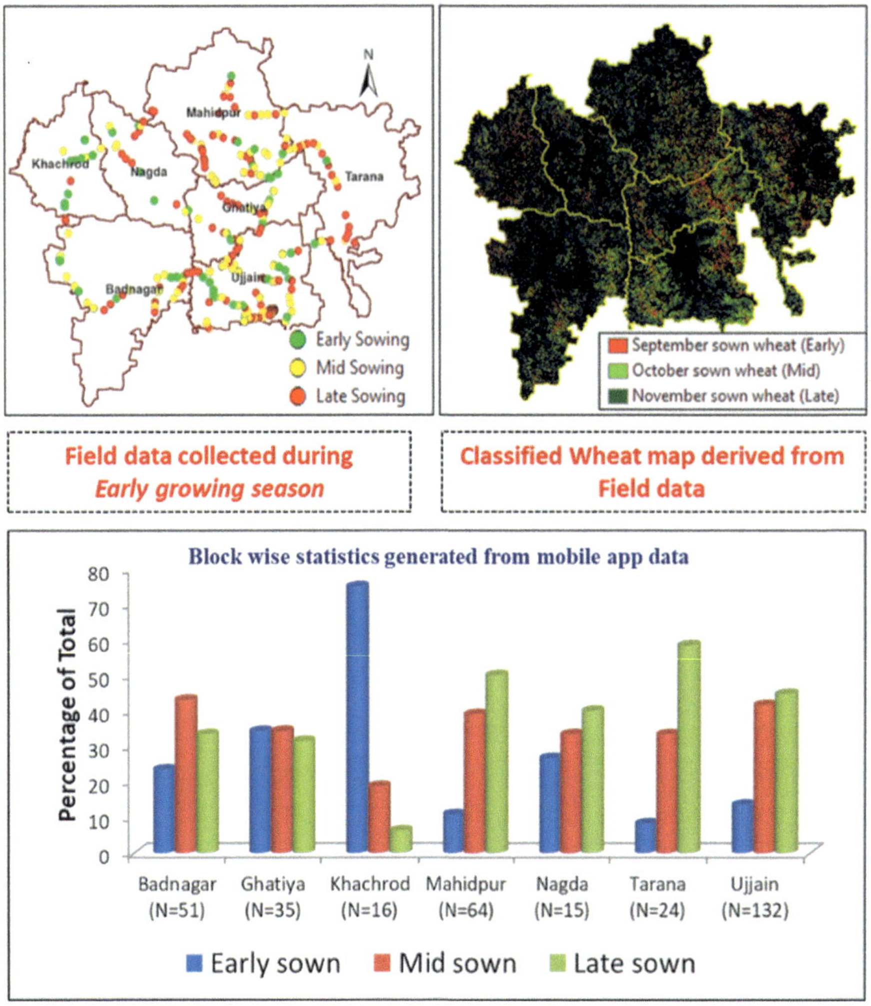

FIGURE 2.13 Mobile apps for crop surveillance and derived value-added information. Map lines delineate study areas and do not necessarily depict accepted national boundaries.

some of the geospatial datasets and their strengths in capturing crop growth, risks, and productivity of the crop. These advancements offer a promising opportunity to introduce transparency and objectivity into the entire value chain of crop insurance. Geospatial approaches for capturing mid-season adversaries like flood, drought, and damages caused by extreme weather events are covered.

2.7.3 Smart sampling for yield measurements

Yield measurement in insurance units is being carried out using crop cutting experiments (CCE), that is, cutting the crop in a specified area and measuring its yield. Currently, four CCE plots for a given crop in each insurance unit and season are considered for yield measurement and average yield estimation. Random selection of CCE plots may result in a skewed representation of crop condition, leading to a biased estimate of the average yield. Moreover, these random plots fail to capture in-season crop risk, like flood, drought, etc. as the sampling fields are identified at the beginning of the crop season. Satellite-based crop proxy can be effectively utilized to improve the distribution of these CCE sample points through incorporating crop health indicators derived from satellites. Smart sampling, or intelligent sampling, is a stratified random sampling method where a crop yield proxy variable is used to stratify the block or village into different strata of high to low yield. CCE locations are then selected randomly and proportionately from each stratum. The smart sampling approach offers numerous benefits. Firstly, by selecting samples according to the current season, they become highly representative of the crop situation throughout the entire insurance unit. This ensures a comprehensive understanding of the overall conditions. Additionally, by sharing CCE locations with primary workers at the last moment, the approach minimizes human bias and eliminates adverse selection. Furthermore, it enables the optimization of CCE numbers through stratified random sampling. This means that either the number of CCEs can be reduced while

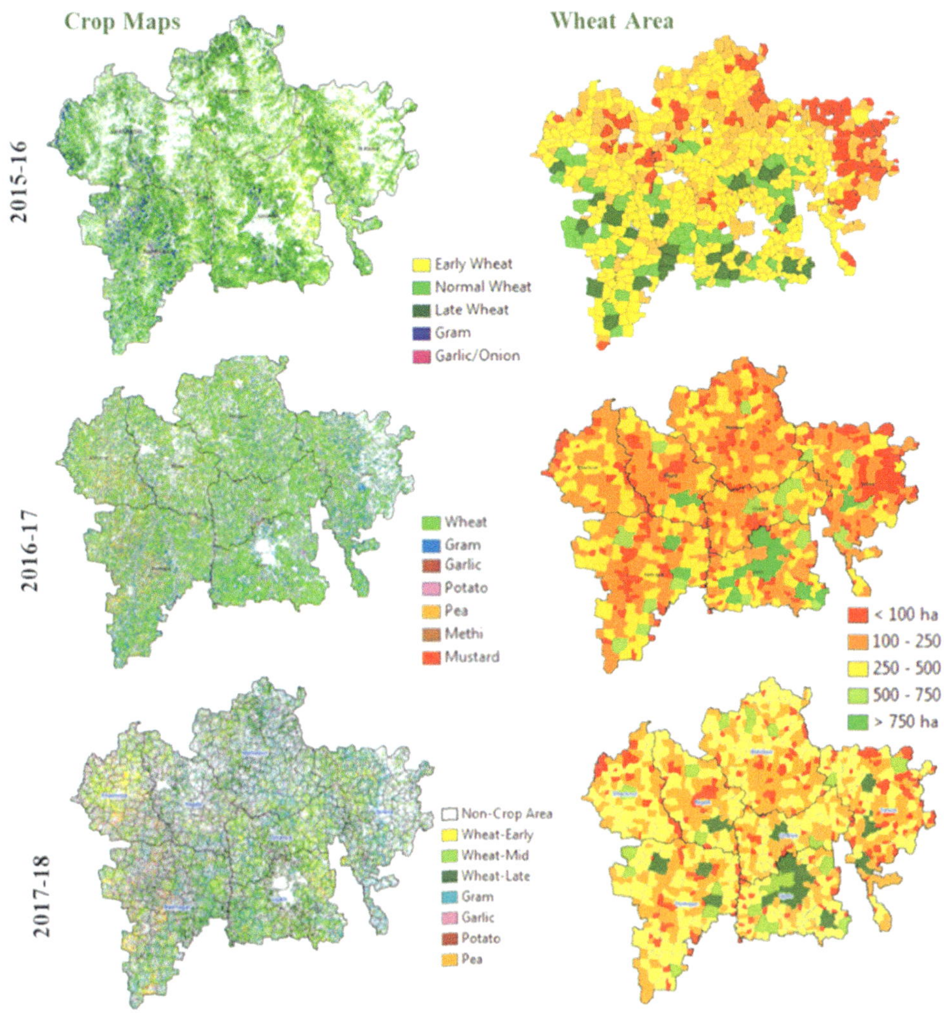

FIGURE 2.14 Satellite based crop maps and wheat area estimation in insurance units of Ujjain, Madhya Pradesh. Map lines delineate study areas and do not necessarily depict accepted national boundaries.

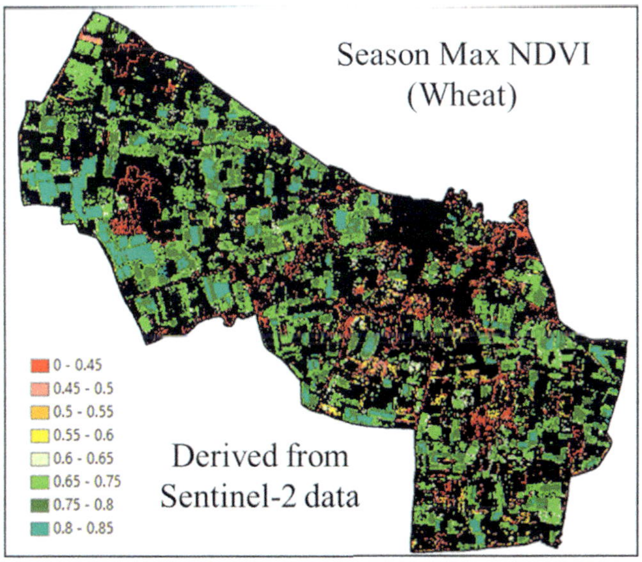

Bichrod-Sentinel-2		
Classes	NDVI Season max_Hecatres	NDVI Season max_Percentage
0.0-.45	34.83	11.07
0.45-0.50	13.69	4.35
0.50-0.55	10.72	3.41
0.55-0.60	8.10	2.57
0.60-0.65	22.57	7.17
0.60-0.70	53.54	17.01
0.70-0.75	82.81	26.31
0.75-0.80	88.43	28.10

Season maximum NDVI representing the variability in the greenness of the wheat crop in an insurance unit.

FIGURE 2.15 Crop monitoring using normalized difference vegetation index in an Insurance unit in Ujjain, Madhya Pradesh. Map lines delineate study areas and do not necessarily depict accepted national boundaries.

		Rainfall	Air Temperature	Evapo-transpiration	Soil Moisture	Vegetation indices (Optical)	Radar Backscatter	Land surface Temperature
Crop Status	Cropped Area					***	**	**
	Crop Type					***	**	*
	Phenology			*		***	**	**
	Canopy structure					*	**	
Crop Stress	Agril Drought	**	*	***	**	**	**	**
	Flood	**			**	**	***	**
	Heat/cold		***			**		***
	Hail	*			*	**	**	
	Unseasonal rain	**			**	***	**	
Productivity	FAPAR/LAI					***	**	**
	Biomass/Yield			**		***	**	*

Applicability: 0 (blank), 1 (*), 2 (**), 3 (***)

FIGURE 2.16 Geospatial data and their strength in depicting crop information for insurance purpose (scale reflects authors' view on applicability ranging from '0': not applicable to '3': highly applicable). Map lines delineate study areas and do not necessarily depict accepted national boundaries.

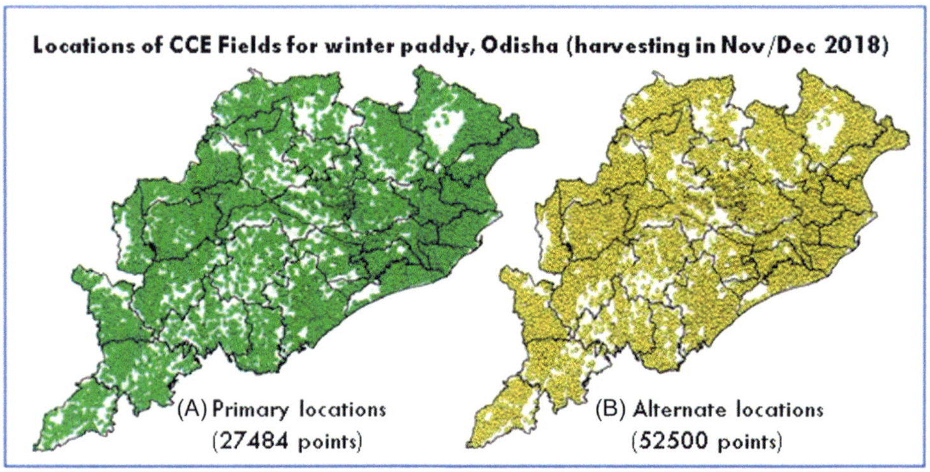

FIGURE 2.17 Smart sampling points generated from satellite-based yield proxy for paddy crop in Odisha state. Map lines delineate study areas and do not necessarily depict accepted national boundaries.

maintaining accuracy or accuracy can be increased without altering the number of CCEs. Fig. 2.17 shows the sampled location of CCE plots in Odisha state based on the smart sampling principle.

2.7.4 Challenges in applying remote sensing technology for crop insurance

The average land parcel sizes in India are quite small, that is, less than 1 acre, which requires high spatial resolution datasets for accurately assessing and monitoring crops at the local scale. Capturing the crop damages due to a natural disaster requires high temporal resolution data to capture timely and relevant information. Remote sensing data needs to be calibrated and validated using ground truth data to ensure reliable yield estimates. This involves collecting accurate and representative information from the actual crop fields, which can be time-consuming and resource-intensive. Remote sensing data can be affected by various sources of noise, including atmospheric interference, cloud cover, and sensor limitations. These factors can introduce inaccuracies in the analysis and interpretation of the data, potentially impacting the reliability of crop insurance assessments. Higher-resolution satellite data, which is often necessary for accurately delineating individual crop plots, can be expensive to acquire. Further, timely, and continuous data delivery is crucial for effective crop insurance monitoring and assessment.

2.8 Conclusions and future perspective

The rapid advancements within the domain of geospatial technologies, notably remote sensing and GIS, have assumed pivotal roles in facilitating the sustainable management of natural resources. These technologies are instrumental in procuring precise and pertinent data, thereby contributing to the conservation of invaluable natural resources for the benefit of future generations.

Remote sensing, operating across the optical, microwave, thermal, and hyperspectral spectra, has emerged as a potent tool capable of assessing crop and soil attributes across various spatial and temporal scales, all while maintaining cost-effectiveness. Remote sensing satellite imagery proves exceptionally valuable for diverse applications, including but not limited to crop growth monitoring, evaluation of crop health, estimation of crop extent and yield, precision agriculture, determination of crop biomass, identification of pest and disease outbreaks, soil surveying and mapping, evaluation of land degradation, estimation of soil moisture content, and assessment of soil quality.

GIS, recognized as a vital decision support tool within a geo-information context, play a pivotal role in solving complex problems related to spatial data. The integration of these technologies facilitates the development of optimal management strategies and action plans tailored to preserve agricultural sustainability within any given region. This innovative approach not only conserves energy resources directly and indirectly but also reduces ecological footprints, promoting the eco-intensification of natural resources. These efforts collectively contribute to ensuring food security, nutritional well-being, environmental preservation, and economic stability for our ever-expanding global population.

References

Aggarwal, P. K., Chand, R., Bhutani, A., Kumar, V., Goel, S. K., Rao, K. N., & Lochan, R. (2016). *Report of the task force on 'Enhancing Technology Use in Agriculture Insurance'*. NITI (National Institute for Transforming India) Aayog, Government of India. https://www.niti.gov.in/sites/default/files/2023-02/EnhancingTechnology-Use-in-AgricultureInsurance-30-07-21.pdf. (accessed 09.10.24).

Alexandratos, N., & Bruinsma, J. (2012). World agriculture towards 2030/2050: The 2012 revision, *ESA working paper no. 12–03*. FAO, Rome. https://www.fao.org/3/ap106e/ap106e.pdf (Accessed 09.10.24).

Asrar, G. Q., Fuchs, M., Kanemasu, E. T., & Hatfield, J. L. (1984). Estimating absorbed photosynthetic radiation and leaf area index from spectral reflectance in wheat 1. *Agronomy Journal*, *76*(2), 300–306. Available from https://doi.org/10.2134/agronj1984.00021962007600020029x.

Ban, H. Y., Ahn, J. B., & Lee, B. W. (2019). Assimilating MODIS data-derived minimum input data set and water stress factors into CERES-maize model improves regional corn yield predictions. *PLoS One*, *14*(2), e0211874. Available from https://doi.org/10.1371/journal.pone.0211874.

Banakara, K. B., Sharma, N., Sahoo, S., Dubey, S. K., & Chowdary, V. M. (2023). Evaluation of weather parameter-based pre-harvest yield forecast models for wheat crop: a case study in Saurashtra region of Gujarat. *Environmental Monitoring and Assessment*, *195*, 51. Available from https://doi.org/10.1007/s10661-022-10552-4.

Bang, S., Bishnoi, R., Chauhan, A. S., Dixit, A. K., & Chawla, I. (2019). Fuzzy logic based crop yield prediction using temperature and rainfall parameters predicted through ARMA, SARIMA, and ARMAX models. In *Twelfth international conference on contemporary computing (IC3)*, Noida, India (pp. 1–6). https://doi.org/10.1109/IC3.2019.8844901.

Baret, F., & Guyot, G. (1991). Potentials and limits of vegetation indices for LAI and APAR assessment. *Remote Sensing of Environment*, *35*(2-3), 161–173. Available from https://doi.org/10.1016/0034-4257(91)90009-U.

Becker, M., & Johnson, D. E. (2001). Improved water control and crop management effects on lowland rice productivity in West Africa. *Nutrient Cycling in Agroecosystems*, *59*, 119–127. Available from https://doi.org/10.1023/A:1017585328904.

Becker-Reshef, I., Justice, C., Barker, B., Humber, M., Rembold, F., Bonifacio, R., … Pound, J. (2020). Strengthening agricultural decisions in countries at risk of food insecurity: The GEOGLAM crop monitor for early warning. *Remote Sensing of Environment*, *237*111553. Available from https://doi.org/10.1016/j.rse.2019.111553.

Becker-Reshef, I., Justice, C., Sullivan, M., Vermote, E., Tucker, C., Anyamba, A., … Hansen, M. (2010). Monitoring global croplands with coarse resolution earth observations: The Global Agriculture Monitoring (GLAM) project. *Remote Sensing*, *2*(6), 1589–1609. Available from https://doi.org/10.3390/rs2061589.

Benami, E., Jin, Z., Carter, M. R., Ghosh, A., Hijmans, R. J., Hobbs, A., … Lobell, D. B. (2021). Uniting remote sensing, crop modelling and economics for agricultural risk management. *Nature Reviews Earth & Environment*, *2*(2), 140–159. Available from https://doi.org/10.1038/s43017-020-00122-y.

Berger, K., Atzberger, C., Danner, M., D'Urso, G., Mauser, W., Vuolo, F., & Hank, T. (2018). Evaluation of the PROSAIL model capabilities for future hyperspectral model environments: A review study. *Remote Sensing*, *10*(1), 85. Available from https://doi.org/10.3390/rs10010085.

Bernardes, T., Moreira, M. A., Adami, M., Giarolla, A., & Rudorff, B. F. T. (2012). Monitoring biennial bearing effect on coffee yield using MODIS remote sensing imagery. *Remote Sensing*, *4*(9), 2492–2509. Available from https://doi.org/10.3390/rs4092492.

Bhattacharya, B. K., Chowdary, V. M., Das, A., Mukesh, K., Srikanth, P., Kumari, M., ... Saroj, M. (2024). Inventory and mapping of Kharif crops using machine learning with EOS-04 time series SAR data. *Current Science (00113891)*, *126*(9), p1050. Available from https://doi.org/10.18520/cs/v126/i9/1050-1060.

Birthal, P. S., Khan, T., Negi, D. S., & Agarwal, S. (2014). Impact of climate change on yields of major food crops in India: Implications for food security. *Agricultural Economics Research Review*, *27*(2), 145–155. Available from https://doi.org/10.5958/0974-0279.2014.00019.6.

Brisson, N., Ruget, F., Gate, P., Lorgeou, J., Nicoullaud, B., Tayot, X., ... Mary, B. (2002). STICS: A generic model for simulating crops and their water and nitrogen balances. II. Model validation for wheat and maize. *Agronomie*, *22*(1), 69–92. Available from https://doi.org/10.1051/agro:2001005.

Brock Porth, C., Porth, L., Zhu, W., Boyd, M., Tan, K. S., & Liu, K. (2020). Remote sensing applications for insurance: A predictive model for pasture yield in the presence of systemic weather. *North American Actuarial Journal*, *24*(2), 333–354. Available from https://doi.org/10.2139/ssrn.3195389.

Campos-Taberner, M., García-Haro, F. J., Camps-Valls, G., Grau-Muedra, G., Nutini, F., Crema, A., & Boschetti, M. (2016). Multitemporal and multiresolution leaf area index retrieval for operational local rice crop monitoring. *Remote Sensing of Environment*, *187*, 102–118. Available from https://doi.org/10.1016/j.rse.2016.10.009.

Canisius, F., Shang, J., Liu, J., Huang, X., Ma, B., Jiao, X., ... Walters, D. (2018). Tracking crop phenological development using multi-temporal polarimetric Radarsat-2 data. *Remote Sensing of Environment*, *210*, 508–518. Available from https://doi.org/10.1016/j.rse.2017.07.031.

Ceccato, P., Flasse, S., & Gregoire, J. M. (2002). Designing a spectral index to estimate vegetation water content from remote sensing data: Part 2. Validation and applications. *Remote Sensing of Environment*, *82*(2-3), 198–207. Available from https://doi.org/10.1016/S0034-4257(02)00036-6.

Ceccato, P., Flasse, S., Tarantola, S., Jacquemoud, S., & Grégoire, J. M. (2001). Detecting vegetation leaf water content using reflectance in the optical domain. *Remote Sensing of Environment*, *77*(1), 22–33. Available from https://doi.org/10.1016/S0034-4257(01)00191-2.

Chakraborty, A., Seshasai, M. V. R., & Dadhwal, V. K. (2014). Geo-spatial analysis of the temporal trends of kharif crop phenology metrics over India and its relationships with rainfall parameters. *Environmental Monitoring and Assessment*, *186*, 4531–4542. Available from https://doi.org/10.1007/s10661-014-3717-6.

Chakraborty, A., Seshasai, M. V. R., Reddy, C. S., & Dadhwal, V. K. (2018). Persistent negative changes in seasonal greenness over different forest types of India using MODIS time series NDVI data (2001–2014). *Ecological Indicators*, *85*, 887–903. Available from https://doi.org/10.1016/j.ecolind.2017.11.032.

Chakraborty, A., Srikanth, P., Murthy, C. S., Rao, P. V. N., & Chowdhury, S. (2021). Assessing lodging damage of jute crop due to super cyclone Amphan using multi-temporal Sentinel-1 and Sentinel-2 data over parts of West Bengal, India. *Environmental Monitoring and Assessment*, *193*(8), 464. Available from https://doi.org/10.1007/s10661-021-09220-w.

Challinor, A. J., Parkes, B., & Ramirez-Villegas, J. (2015). Crop yield response to climate change varies with cropping intensity. *Global Change Biology*, *21*(4), 1679–1688. Available from https://doi.org/10.1111/gcb.12808.

Chattopadhyay, C., Agrawal, R., Kumar, A., Meena, R. L., Faujdar, K., Chakravarthy, N. V. K., ... Shekhar, C. (2011). Epidemiology and development of forecasting models for white rust of *Brassica juncea* in India. *Archives of Phytopathology and Plant Protection*, *44*(8), 751–763. Available from https://doi.org/10.1080/03235400903458571.

Chipanshi, A., Zhang, Y., Kouadio, L., Newlands, N., Davidson, A., Hill, H., ... Reichert, G. (2015). Evaluation of the Integrated Canadian Crop Yield Forecaster (ICCYF) model for in-season prediction of crop yield across the Canadian agricultural landscape. *Agricultural and Forest Meteorology*, *206*, 137–150. Available from https://doi.org/10.1016/j.agrformet.2015.03.007.

Choudhary, K. K., Chakraborty, A., Murthy, C. S., & Poddar, M. K. (2022). Satellite-based assessment of hailstorm-affected potato crop for insurance purpose. *Natural Hazards*, *114*(1), 787–810. Available from https://doi.org/10.1007/s11069-022-05412-w.

Clevers, J. G. P. W., Büker, C., Van Leeuwen, H. J. C., & Bouman, B. A. M. (1994). A framework for monitoring crop growth by combining directional and spectral remote sensing information. *Remote Sensing of Environment*, *50*(2), 161–170. Available from https://doi.org/10.1016/0034-4257(94)90042-6.

Collet, K. P., & Abednego, B. (1987). L'évaluationd'unpaysageurbainétudié à l'aide d'un indicedevégétation. *Télédétectionsatellitaireetespaceurbain, Société géographique de Liège, Belgique, notes de recherches*, *9*, 3–16.

Dalrymple, D. G. (1971). Survey of multiple cropping in less developed nations *(No. 91)*. Foreign Economic Development Service, US Department of Agriculture, https://ageconsearch.umn.edu/record/145628/files/faer91.pdf. (Accessed 09.10.24).

De Leeuw, J., Vrieling, A., Shee, A., Atzberger, C., Hadgu, K. M., Biradar, C. M., ... Turvey, C. (2014). The potential and uptake of remote sensing in insurance: A review. *Remote Sensing*, *6*(11), 10888–10912. Available from https://doi.org/10.3390/rs61110888.

Deblonde, G., & Cihlar, J. (1993). A multiyear analysis of the relationship between surface environmental variables and NDVI over the Canadian landmass. *Remote Sensing Reviews*, *7*(2), 151–177. Available from https://doi.org/10.1080/02757259309532172.

Delegido, J., Vergara, C., Verrelst, J., Gandía, S., & Moreno, J. (2011). Remote estimation of crop chlorophyll content by means of high-spectral-resolution reflectance techniques. *Agronomy Journal*, *103*(6), 1834–1842. Available from https://doi.org/10.2134/agronj2011.0101.

Dkhar, D. K., Feroze, S. M., Singh, R., & Ray, L. I. (2017). Effect of rainfall variability on rice yield in north eastern hills of India: A case study. *Agricultural Research*, *6*, 341–346. Available from https://doi.org/10.1007/s40003-017-0276-4.

Durbha, S. S., King, R. L., & Younan, N. H. (2007). Support vector machines regression for retrieval of leaf area index from multiangle imaging spectroradiometer. *Remote Sensing of Environment*, *107*(1-2), 348–361. Available from https://doi.org/10.1016/j.rse.2006.09.031.

Eklundh, L., & Jönsson, P. (2017). *TIMESAT 3.3 with seasonal trend decomposition and parallel processing –- Software Manual* (92). Lund University, https://web.nateko.lu.se/timesat/docs/TIMESAT33_SoftwareManual.pdf. (Accessed 09.10.24).

Fang, H., & Liang, S. (2003). Retrieving leaf area index with a neural network method: Simulation and validation. *IEEE Transactions on Geoscience and Remote Sensing, 41*(9), 2052–2062. Available from https://doi.org/10.1109/TGRS.2003.813493.

FAO. (2017). *The future of food and agriculture – Trends and challenges*. Rome. https://www.fao.org/3/i6583e/i6583e.pdf (Accessed 09.10.24).

Fensholt, R., Sandholt, I., & Rasmussen, M. S. (2004). Evaluation of MODIS LAI, fAPAR and the relation between fAPAR and NDVI in a semi-arid environment using in situ measurements. *Remote Sensing of Environment, 91*(3-4), 490–507. Available from https://doi.org/10.1016/j.rse.2004.04.009.

Féret, J. B., Gitelson, A. A., Noble, S. D., & Jacquemoud, S. (2017). PROSPECT-D: Towards modeling leaf optical properties through a complete lifecycle. *Remote Sensing of Environment, 193*, 204–215. Available from https://doi.org/10.1016/j.rse.2017.03.004.

Fisher, J. I., Richardson, A. D., & Mustard, J. F. (2007). Phenology model from surface meteorology does not capture satellite-based greenup estimations. *Global Change Biology, 13*(3), 707–721. Available from https://doi.org/10.1111/j.1365-2486.2006.01311.x.

Galford, G. L., Mustard, J. F., Melillo, J., Gendrin, A., Cerri, C. C., & Cerri, C. E. (2008). Wavelet analysis of MODIS time series to detect expansion and intensification of row-crop agriculture in Brazil. *Remote Sensing of Environment, 112*(2), 576–587. Available from https://doi.org/10.1016/j.rse.2007.05.017.

Gao, B. C. (1996). NDWI—A normalized difference water index for remote sensing of vegetation liquid water from space. *Remote Sensing of Environment, 58*(3), 257–266. Available from https://doi.org/10.1016/S0034-4257(96)00067-3.

Gerhards, M., Schlerf, M., Mallick, K., & Udelhoven, T. (2019). Challenges and future perspectives of multi-/Hyperspectral thermal infrared remote sensing for crop water-stress detection: A review. *Remote Sensing, 11*(10), 1240. Available from https://doi.org/10.3390/rs11101240.

Ghazaryan, G., Dubovyk, O., Löw, F., Lavreniuk, M., Kolotii, A., Schellberg, J., & Kussul, N. (2018). A rule-based approach for crop identification using multi-temporal and multi-sensor phenological metrics. *European Journal of Remote Sensing, 51*(1), 511–524. Available from https://doi.org/10.1080/22797254.2018.1455540.

Gitelson, A. A., Merzlyak, M. N., & Lichtenthaler, H. K. (1996). Detection of red edge position and chlorophyll content by reflectance measurements near 700 nm. *Journal of Plant Physiology, 148*(3-4), 501–508. Available from https://doi.org/10.1016/S0176-1617(96)80285-9.

Goel, N. S. (1988). Models of vegetation canopy reflectance and their use in estimation of biophysical parameters from reflectance data. *Remote Sensing Reviews, 4*(1), 1–212. Available from https://doi.org/10.1080/02757258809532105.

Goward, S. N., & Huemmrich, K. F. (1992). Vegetation canopy PAR absorptance and the normalized difference vegetation index: An assessment using the SAIL model. *Remote Sensing of Environment, 39*(2), 119–140. Available from https://doi.org/10.1016/0034-4257(92)90131-3.

Goyal, A., Mohit, K., Neeti, N., Maity, S., Mukesh, K., Chowdary, V. M., & Jha, C. S. (2022). Crop classification in the mixed cropping environment using SAR data and machine learning algorithms. In C. S. Jha, A. Pandey, V. Chowdary, & V. Singh (Eds.), *Geospatial technologies for resources planning and management. Water science and technology library* (p. 115). Cham: Springer, https://doi.org/10.1007/978-3-030-98981-1_10.

Gu, Y., Brown, J. F., Verdin, J. P., & Wardlow, B. (2007). A five-year analysis of MODIS NDVI and NDWI for grassland drought assessment over the central Great Plains of the United States. *Geophysical Research Letters, 34*(6), L06407. Available from https://doi.org/10.1029/2006GL029127.

Guo, Y., Xiang, H., Li, Z., Ma, F., & Du, C. (2021). Prediction of rice yield in East China based on climate and agronomic traits data using artificial neural networks and partial least squares regression. *Agronomy, 11*(2), 282. Available from https://doi.org/10.3390/agronomy11020282.

Haboudane, D., Miller, J. R., Pattey, E., Zarco-Tejada, P. J., & Strachan, I. B. (2004). Hyperspectral vegetation indices and novel algorithms for predicting green LAI of crop canopies: Modeling and validation in the context of precision agriculture. *Remote Sensing of Environment, 90*(3), 337–352. Available from https://doi.org/10.1016/j.rse.2003.12.013.

Hakkim, V. A., Joseph, E. A., Gokul, A. A., & Mufeedha, K. (2016). Precision farming: The future of Indian agriculture. *Journal of Applied Biology and Biotechnology, 4*(6), 068–072. Available from https://doi.org/10.7324/JABB.2016.40609.

Heumann, B. W., Seaquist, J. W., Eklundh, L., & Jönsson, P. (2007). AVHRR derived phenological change in the Sahel and Soudan, Africa, 1982–2005. *Remote Sensing of Environment, 108*(4), 385–392. Available from https://doi.org/10.1016/j.rse.2006.11.025.

Heupel, K., Spengler, D., & Itzerott, S. (2018). A progressive crop-type classification using multitemporal remote sensing data and phenological information. *PFG–Journal of Photogrammetry, Remote Sensing and Geoinformation Science, 86*, 53–69. Available from https://doi.org/10.1007/s41064-018-0050-7.

Holben, B., & Fraser, R. S. (1984). Red and near-infrared sensor response to off-nadiir viewing. *International Journal of Remote Sensing, 5*(1), 145–160. Available from https://doi.org/10.1080/01431168408948795.

Holben, B. N., Kaufman, Y. J., & Kendall, J. D. (1990). NOAA-11 AVHRR visible and near-IR inflight calibration. *International Journal of Remote Sensing, 11*(8), 1511–1519. Available from https://doi.org/10.1080/01431169008955109.

Hosseini, M., & McNairn, H. (2017). Using multi-polarization C-and L-band synthetic aperture radar to estimate biomass and soil moisture of wheat fields. *International Journal of Applied Earth Observation and Geoinformation, 58*, 50–64. Available from https://doi.org/10.1016/j.jag.2017.01.006.

Huang, J., Gómez-Dans, J. L., Huang, H., Ma, H., Wu, Q., Lewis, P. E., … Zhao, F. (2019). Assimilation of remote sensing into crop growth models: Current status and perspectives. *Agricultural and Forest Meteorology, 276*107609. Available from https://doi.org/10.1016/j.agrformet.2019.06.008.

Huete, A., Didan, K., Miura, T., Rodriguez, E. P., Gao, X., & Ferreira, L. G. (2002). Overview of the radiometric and biophysical performance of the MODIS vegetation indices. *Remote Sensing of Environment, 83*(1-2), 195–213. Available from https://doi.org/10.1016/S0034-4257(02)00096-2.

Huete, A. R. (1988). A Soil-Adjusted Vegetation Index (SAVI). *Remote Sensing of Environment*, *25*, 295–309. Available from https://doi.org/10.1016/0034-4257(88)90106-X.

Jacquemoud, S., & Baret, F. (1990). PROSPECT: A model of leaf optical properties spectra. *Remote Sensing of Environment*, *34*(2), 75–91. Available from https://doi.org/10.1016/0034-4257(90)90100-Z.

Jacquemoud, S., Baret, F., Andrieu, B., Danson, F. M., & Jaggard, K. (1995). Extraction of vegetation biophysical parameters by inversion of the PROSPECT + SAIL models on sugar beet canopy reflectance data. Application to TM and AVIRIS sensors. *Remote Sensing of Environment*, *52* (3), 163–172. Available from https://doi.org/10.1016/0034-4257(95)00018-V.

Jensen, J. R. (1996). *Introductory digital image processing: A remote sensing perspective* (2nd Edition). Upper Saddle River, NJ: Prentice Hall, Inc.

Jha, S. N., Vishwakarma, R. K., Ahmad, T., Rai, A., & Dixit, A. K. (2015). Report on assessment of quantitative harvest and post-harvest losses of major crops and commodities in India. *All India Coordinated Research Project on Post-Harvest Technology*, ICAR-CIPHET, ICAR-CIPHET, P. O.-PAU, Ludhiana-141 004. https://www.mofpi.gov.in/sites/default/files/phl_2010_ciphet.pdf (Accessed 09.10.24).

Jones, C. A., Kiniry, J. R., & Dyke, P. T. (1986). *CERES-Maize: A simulation model of maize growth and development*. Texas A&M University Press, https://www.ars.usda.gov/ARSUserFiles/30980500/CERES-Maize%20Book.pdf. (Accessed 09.10.24).

Jones, J. W., Hoogenboom, G., Porter, C. H., Boote, K. J., Batchelor, W. D., Hunt, L. A., ... Ritchie, J. T. (2003). The DSSAT cropping system model. *European Journal of Agronomy*, *18*(3-4), 235–265. Available from https://doi.org/10.1016/S1161-0301(02)00107-7.

Justice, C. O., Townshend, J. R. G., Vermote, E. F., Masuoka, E., Wolfe, R. E., Saleous, N., ... Morisette, J. T. (2002). An overview of MODIS Land data processing and product status. *Remote Sensing of Environment*, *83*(1-2), 3–15. Available from https://doi.org/10.1016/S0034-4257(02)00084-6.

Kaufman, Y. J., & Holben, B. N. (1993). Calibration of the AVHRR visible and near-IR bands by atmospheric scattering, ocean glint and desert reflection. *International Journal of Remote Sensing*, *14*(1), 21–52. Available from https://doi.org/10.1080/01431169308904320.

Kaufman, Y. J. (1984). Atmospheric effects on remote sensing of surface reflectance. *Remote Sensing: Critical Review of Technology*, *475*, 20–33. Available from https://doi.org/10.1117/12.966238.

Keating, B. A., Carberry, P. S., Hammer, G. L., Probert, M. E., Robertson, M. J., Holzworth, D., ... McLean, G. (2003). An overview of APSIM, a model designed for farming systems simulation. *European Journal of Agronomy*, *18*(3-4), 267–288. Available from https://doi.org/10.1016/S1161-0301(02)00108-9.

Ker, A. P., & Goodwin, B. K. (2000). Nonparametric estimation of crop insurance rates revisited. *American Journal of Agricultural Economics*, *82*(2), 463–478. Available from https://doi.org/10.1111/0002-9092.00039.

Kira, O., Nguy-Robertson, A. L., Arkebauer, T. J., Linker, R., & Gitelson, A. A. (2016). Informative spectral bands for remote green LAI estimation in C3 and C4 crops. *Agricultural and Forest Meteorology*, *218*, 243–249. Available from https://doi.org/10.1016/j.agrformet.2015.12.064.

Kumar, P., Prasad, R., Gupta, D. K., Mishra, V. N., Vishwakarma, A. K., Yadav, V. P., ... Avtar, R. (2018). Estimation of winter wheat crop growth parameters using time series Sentinel-1A SAR data. *Geocarto International*, *33*(9), 942–956. Available from https://doi.org/10.1080/10106049.2017.1316781.

Kumar, S. N., Aggarwal, P. K., Rani, S., Jain, S., Saxena, R., & Chauhan, N. (2011). Impact of climate change on crop productivity in Western Ghats, coastal and northeastern regions of India. *Current Science*, 332–341. Available from https://www.jstor.org/stable/24078512.

Kumar, V., Kumari, M., & Saha, S. K. (2013). Leaf area index estimation of lowland rice using semi-empirical backscattering model. *Journal of Applied Remote Sensing*, *7*(1), 073474. Available from https://doi.org/10.1117/1.JRS.7.073474, 073474.

Kumari, M., Chakraborty, A., Chakravarathi, V., & Roy, P. S. (2023). Spatio-temporal trend of crop phenology, SPEI, and their interactions over different agro-ecological regions of India. *Theoretical and Applied Climatology*, 1–19. Available from https://doi.org/10.1007/s00704-023-04557-3.

Landsberg, J. J., & Waring, R. H. (1997). A generalised model of forest productivity using simplified concepts of radiation-use efficiency, carbon balance and partitioning. *Forest Ecology and Management*, *95*(3), 209–228. Available from https://doi.org/10.1016/S0378-1127(97)00026-1.

Lillesand, T., Kiefer, R. W., & Chipman, J. (2015). *Remote sensing and image interpretation*. 7th ed. John Wiley & Sons. ISBN: 978-1-119-12846-5. https://www.wiley.com/en-fr/Remote + Sensing + and + Image + Interpretation%2C + 7th + Edition-p-9781119128465 (Accessed 09.10.24).

López-Lozano, R., & Baruth, B. (2019). An evaluation framework to build a cost-efficient crop monitoring system. Experiences from the extension of the European crop monitoring system. *Agricultural Systems*, *168*, 231–246. Available from https://doi.org/10.1016/j.agsy.2018.04.002.

Lopez-Sanchez, J. M., Cloude, S. R., & Ballester-Berman, J. D. (2011). Rice phenology monitoring by means of SAR polarimetry at X-band. *IEEE Transactions on Geoscience and Remote Sensing*, *50*(7), 2695–2709. Available from https://doi.org/10.1109/TGRS.2011.2176740.

Lyon, J. G., Yuan, D., Lunetta, R. S., & Elvidge, C. D. (1998). A change detection experiment using vegetation indices. *Photogrammetric Engineering and Remote Sensing*, *64*(2), 143–150, https://doi.org/0099-1112/98/6402-143$3.00/0.

MacDonald, R. B., & Hall, F. G. (1980). Global crop forecasting. *Science (New York, N.Y.)*, *208*(4445), 670–679. Available from https://doi.org/10.1126/science.208.4445.670.

Mandal, D., Bhattacharya, A., & Rao, Y. S. (2021). *Radar remote sensing for crop biophysical parameter estimation*. Singapore: Springer. Available from https://doi.org/10.1007/978-981-16-4424-5.

Mandal, D., Kumar, V., Lopez-Sanchez, J. M., Bhattacharya, A., McNairn, H., & Rao, Y. S. (2020). Crop biophysical parameter retrieval from Sentinel-1 SAR data with a multi-target inversion of Water Cloud Model. *International Journal of Remote Sensing*, *41*(14), 5503–5524. Available from https://doi.org/10.1080/01431161.2020.1734261.

Meena, R. S., Mitran, T., Kumar, S., Yadav, G., Bohra, J. S., & Datta, R. (2018). Application of remote sensing for sustainable agriculture and forest management. *Information Process in Agriculture*, *5*, 295–297. Available from https://doi.org/10.1016/J.INPA.2018.03.003.

Meroni, M., Marinho, E., Sghaier, N., Verstrate, M. M., & Leo, O. (2013). Remote sensing based yield estimation in a stochastic framework—Case study of durum wheat in Tunisia. *Remote Sensing*, *5*(2), 539–557. Available from https://doi.org/10.3390/rs5020539.

Mirik, M., Ansley, R. J., Steddom, K., Rush, C. M., Michels, G. J., Workneh, F., ... Elliott, N. C. (2014). High spectral and spatial resolution hyperspectral imagery for quantifying Russian wheat aphid infestation in wheat using the constrained energy minimization classifier. *Journal of Applied Remote Sensing*, *8*(1), 083661. Available from https://doi.org/10.1117/1.JRS.8.083661, 083661.

Mitran, T., Meena, R. S., & Chakraborty, A. (2020). Geospatial technologies for crops and soils: An Overview. In T. Mitran, R. S. Meena, & A. Chakraborty (Eds.), *Geospatial Technologies for Crops and Soils*. Singapore: Springer. Available from https://doi.org/10.1007/978-981-15-6864-0_1.

Mohammed, G. H., Colombo, R., Middleton, E. M., Rascher, U., van der Tol, C., Nedbal, L., ... Joiner, J. (2019). Remote sensing of solar-induced chlorophyll fluorescence (SIF) in vegetation: 50 years of progress. *Remote Sensing of Environment*, *231*111177. Available from https://doi.org/10.1016/j.rse.2019.04.030.

Mohit, K., Akash, G., Neeti, N., Maity, S., Mukesh, K., Chowdary, V. M., ... Kumar, R. (2022). Machine learning-based meta-classifier for Kharif Bajra (pearl millet) discrimination in the mixed croppingenvironment using multi-temporal SAR data. *Geocarto International*, *37*(27), 16671–16686. Available from https://doi.org/10.1080/10106049.2022.2113452.

Monteith, J. L. (1972). Solar radiation and productivity in tropical ecosystems. *Journal of Applied Ecology*, *9*(3), 747–766. Available from https://doi.org/10.2307/2401901.

Mulianga, B., Bégué, A., Simoes, M., & Todoroff, P. (2013). Forecasting regional sugarcane yield based on time integral and spatial aggregation of MODIS NDVI. *Remote Sensing*, *5*(5), 2184–2199. Available from https://doi.org/10.3390/rs5052184.

Mulla, D. J. (2013). Twenty five years of remote sensing in precision agriculture: Key advances and remaining knowledge gaps. *Biosystems Engineering*, *114*(4), 358–371. Available from https://doi.org/10.1016/j.biosystemseng.2012.08.009.

Pareeth, S., Karimi, P., Shafiei, M., & De Fraiture, C. (2019). Mapping agricultural landuse patterns from time series of Landsat 8 using random forest based hierarchial approach. *Remote Sensing*, *11*(5), 601. Available from https://doi.org/10.3390/rs11050601.

Parihar, J. S., & Oza, M. P.FASAL: An integrated approach for crop assessment and production forecasting. (2006). *Agriculture and Hydrology Applications of Remote Sensing* (6411, p. 641101)SPIE, https://doi.org/10.1117/12.713157.

Paulraj, A. P., & Easwaran, N. (2020). Evaluation of 'Revamped'Crop Insurance Pradhan Mantri Fasal Bima Yojana (PMFBY) among Paddy Farmers in Tamil Nadu, India. *Current Journal of Applied Science and Technology*, *39*(34), 66–77. Available from https://doi.org/10.9734/cjast/2020/v39i3431037.

Pearson, R. L., & Miller, L. D. (1972). Remote mapping of standing crop biomass for estimation of the productivity of the shortgrass prairie. *Remote Sensing of Environment*, *VIII*, 1355, [Bibcode: 1972rse.conf.1355P] (Accessed 09.10.24).

Peñuelas, J., Gamon, J. A., Fredeen, A. L., Merino, J., & Field, C. B. (1994). Reflectance indices associated with physiological changes in nitrogen- and water-limited sunflower leaves. *Remote Sensing of Environment*, *48*(2), 135–146. Available from https://doi.org/10.1016/0034-4257(94)90136-8.

Phung, H. P., Lam-Dao, N., Nguyen-Huy, T., Le-Toan, T., & Apan, A. A. (2020). Monitoring rice growth status in the Mekong Delta, Vietnam using multitemporal Sentinel-1 data. *Journal of Applied Remote Sensing*, *14*(1), 014518. Available from https://doi.org/10.1117/1.JRS.14.014518, 014518.

Piao, S., Fang, J., Zhou, L., Ciais, P., & Zhu, B. (2006). Variations in satellite-derived phenology in China's temperate vegetation. *Global Change Biology*, *12*(4), 672–685. Available from https://doi.org/10.1111/j.1365-2486.2006.01123.x.

PMFBY (2020) *Operational guidelines of Pradhan Mantri Fasal Bima Yojana (PMFBY)*, (File Number – 13015/03/2018 Credit II, Published on 17th August 2020, https://pmfby.gov.in/pdf/Revamped%20Operational%20Guidelines_17th%20August%202020.pdf (Accessed 09.10.24).

Potter, C. S., Randerson, J. T., Field, C. B., Matson, P. A., Vitousek, P. M., Mooney, H. A., & Klooster, S. A. (1993). Terrestrial ecosystem production: A process model based on global satellite and surface data. *Global Biogeochemical Cycles*, *7*(4), 811–841. Available from https://doi.org/10.1029/93GB02725.

Prince, S. D., & Goward, S. N. (1995). Global primary production: a remote sensing approach. *Journal of Biogeography*, 815–835. Available from https://doi.org/10.2307/2845983.

Prince, S. D., & Tucker, C. J. (1986). Satellite remote sensing of rangelands in Botswana II. NOAA AVHRR and herbaceous vegetation. *International Journal of Remote Sensing*, *7*(11), 1555–1570. Available from https://doi.org/10.1080/01431168608948953.

Pu, R., & Cheng, J. (2015). Mapping forest leaf area index using reflectance and textural information derived from WorldView-2 imagery in a mixed natural forest area in Florida, US. *International Journal of Applied Earth Observation and Geoinformation*, *42*, 11–23. Available from https://doi.org/10.1016/j.jag.2015.05.004.

Racine, J., & Ker, A. (2006). Rating crop insurance policies with efficient nonparametric estimators that admit mixed data types. *Journal of Agricultural and Resource Economics*, 27–39. Available from https://doi.org/10.22004/ag.econ.10146.

Ratha, D., Mandal, D., Kumar, V., McNairn, H., Bhattacharya, A., & Frery, A. C. (2019). A generalized volume scattering model-based vegetation index from polarimetric SAR data. *IEEE Geoscience and Remote Sensing Letters*, *16*(11), 1791–1795. Available from https://doi.org/10.1109/LGRS.2019.2907703.

Raut, N., Sitaula, B. K., Aune, J. B., & Bajracharya, R. M. (2011). Evolution and future direction of intensified agriculture in the central mid-hills of Nepal. *International Journal of Agricultural Sustainability*, *9*(4), 537–550. Available from https://doi.org/10.1080/14735903.2011.609648.

Rembold, F., Atzberger, C., Savin, I., & Rojas, O. (2013). Using low resolution satellite imagery for yield prediction and yield anomaly detection. *Remote Sensing*, *5*(4), 1704–1733. Available from https://doi.org/10.3390/rs5041704.

Rembold, F., Meroni, M., Urbano, F., Csak, G., Kerdiles, H., Perez-Hoyos, A., ... Negre, T. (2019). ASAP: A new global early warning system to detect anomaly hot spots of agricultural production for food security analysis. *Agricultural Systems*, *168*, 247–257. Available from https://doi.org/10.1016/j.agsy.2018.07.002.

Ripple, W. J. (1986). Spectral reflectance relationships to leaf water stress. *Photogrammetric Engineering and Remote Sensing*, *52*, https://doi.org/0099-1112/86/5210-1669$02.25/0.

Rotairo L., Durante A. C., Lapitan P., & Rao, L. N. (2019). *Use of remote sensing to estimate paddy area and production: A handbook*. Asian Development Bank. Available from https://doi.org/10.22617/TIM190098-3.

Rouse, J. W., Haas, R. W., Schell, J. A., Deering, D. W., & Harlan, J. C. (1974). *Monitoring the vernal advancement and retrogradation (Greenwave effect) of natural vegetation NASA/GSFCTType III final report*, Greenbelt, MD: NASA/ GSFC. https://ntrs.nasa.gov/citations/19730017588 (Accessed 09.10.24).

Roy, P. S. (1989). Spectral reflectance characteristics of vegetation and their use in estimating productive potential. *Proceedings: Plant Sciences*, *99*, 59–81. Available from https://doi.org/10.1007/BF03053419.

Running, S. W., Nemani, R., Glassy, J. M., & Thornton, P. E. (1999) *MODIS daily photosynthesis (PSN) and annual net primary production (NPP) product (MOD17), algorithm theoretical basis document, version 3.0*. https://lpdaac.usgs.gov/documents/95/MOD17_ATBD.pdf (Accessed 09.10.24).

Running, S. W., Nemani, R. R., Heinsch, F. A., Zhao, M., Reeves, M., & Hashimoto, H. (2004). A continuous satellite-derived measure of global terrestrial primary production. *Bioscience*, *54*(6), 547–560. Available from https://doi.org/10.1641/0006-3568(2004)054[0547:ACSMOG]2.0.CO;2.

Sakamoto, T., Yokozawa, M., Toritani, H., Shibayama, M., Ishitsuka, N., & Ohno, H. (2005). A crop phenology detection method using time-series MODIS data. *Remote sensing of environment*, *96*(3-4), 366–374. Available from https://doi.org/10.1016/j.rse.2005.03.008.

Sarvia, F., De Petris, S., & Borgogno Mondino, E. (2020). Multi-scale remote sensing to support insurance policies in agriculture: From mid-term to instantaneous deductions. *GIScience& Remote Sensing*, 1–15. Available from https://doi.org/10.1080/15481603.2020.1798600.

Sawant, S., Mohite, J., Sakkan, M., & Pappula, S. (2019). Near real time crop loss estimation using remote sensing observations. *8th International Conference on Agro-Geoinformatics (Agro-Geoinformatics)*, Istanbul, Turkey (pp. 1–5). https://doi.org/10.1109/Agro-Geoinformatics.2019.8820217.

Sawasawa, H. (2013). Rice yield estimation: Integrating RS, GIS, and management factors. A case study of Birkoor and Kortigiri Mandals-Nizamabad District, India. *Master's Thesis*. International Institute for Geo-Information Science and Earth Observation (ITC), Enschede, The Netherlands. https://webapps.itc.utwente.nl/librarywww/papers_2003/msc/nrm/sawasawa.pdf (Accessed 09.10.24).

Schaber, J., & Badeck, F. W. (2003). Physiology-based phenology models for forest tree species in Germany. *International Journal of Biometeorology*, *47*, 193–201. Available from https://doi.org/10.1007/s00484-003-0171-5.

Schlerf, M., & Atzberger, C. (2006). Inversion of a forest reflectance model to estimate structural canopy variables from hyperspectral remote sensing data. *Remote Sensing of Environment*, *100*(3), 281–294. Available from https://doi.org/10.1016/j.rse.2005.10.006.

Schlund, M., & Erasmi, S. (2020). Sentinel-1 time series data for monitoring the phenology of winter wheat. *Remote Sensing of Environment*, *246*111814. Available from https://doi.org/10.1016/j.rse.2020.111814.

Schmedtmann, J., & Campagnolo, M. L. (2015). Reliable crop identification with satellite imagery in the context of common agriculture policy subsidy control. *Remote Sensing*, *7*(7), 9325–9346. Available from https://doi.org/10.3390/rs70709325.

Sellers, P. J. (1985). Canopy reflectance, photosynthesis and transpiration. *International Journal of Remote Sensing*, *6*(8), 1335–1372. Available from https://doi.org/10.1080/01431168508948283.

Shastry, A., Sanjay, H. A., & Bhanusree, E. (2017). Prediction of crop yield using regression techniques. *International Journal of Soft Computing*, *12* (2), 96–102, https://medwelljournals.com/abstract/?doi = ijscomp.2017.96.102. (Accessed 09.10.24).

Shen, Y., Wu, L., Di, L., Yu, G., Tang, H., Yu, G., & Shao, Y. (2013). Hidden Markov models for real-time estimation of corn progress stages using MODIS and meteorological data. *Remote Sensing*, *5*(4), 1734–1753. Available from https://doi.org/10.3390/rs5041734.

Sherrod, L. A., Peterson, G. A., Westfall, D. G., & Ahuja, L. R. (2003). Cropping intensity enhances soil organic carbon and nitrogen in a no-till agroecosystem. *Soil Science Society of America Journal*, *67*(5), 1533–1543. Available from https://doi.org/10.2136/sssaj2003.1533.

Silleos, N., Perakis, K., & Petsanis, G. (2002). Assessment of crop damage using space remote sensing and GIS. *International Journal of Remote Sensing*, *23*(3), 417–427. Available from https://doi.org/10.1080/01431160110040026.

Srikanth, P., Biswal, A., Sahay, B., Chowdary, V. M., Sreenivas, K., & Chauhan, P. (2024). Mapping of Kharif sown area using temporal RISAT-1A SAR and optical data. *Journal of the Indian Society of Remote Sensing*. Available from https://doi.org/10.1007/s12524-024-01977-0.

Steduto, P., Hsiao, T. C., Raes, D., & Fereres, E. (2009). AquaCrop—The FAO crop model to simulate yield response to water: I. Concepts and underlying principles. *Agronomy Journal*, *101*(3), 426–437. Available from https://doi.org/10.2134/agronj2008.0139s.

Steele-Dunne, S. C., McNairn, H., Monsivais-Huertero, A., Judge, J., Liu, P. W., & Papathanassiou, K. (2017). Radar remote sensing of agricultural canopies: A review. *IEEE Journal of Selected Topics in Applied Earth Observations and Remote Sensing*, *10*(5), 2249–2273. Available from https://doi.org/10.1109/JSTARS.2016.2639043.

Stöckle, C. O., Donatelli, M., & Nelson, R. (2003). CropSyst, a cropping systems simulation model. *European Journal of Agronomy*, *18*(3-4), 289–307. Available from https://doi.org/10.1016/S1161-0301(02)00109-0.

Tapia-Silva, F. O., Itzerott, S., Foerster, S., Kuhlmann, B., & Kreibich, H. (2011). Estimation of flood losses to agricultural crops using remote sensing. *Physics and Chemistry of the Earth, Parts A/B/C*, *36*(7-8), 253–265. Available from https://doi.org/10.1016/j.pce.2011.03.005.

Tiwari, R., Chand, K., & Anjum, B. (2020). Crop insurance in India: A review of Pradhan Mantri Fasal Bima Yojana (PMFBY). *FIIB Business Review*, *9*(4), 249–255. Available from https://doi.org/10.1177/2319714520966.

Tolpin, V. A., Bartalev, S. A., Elkina, E. S., Kashnitskiy, A. V., Konstantinova, A. M., Loupian, E. A., ... Sunil, J. K. (2019). The VEGA-GEOGLAM information system: A tool for the development of methods and approaches to using satellite remote sensing data in problem-solving tasks of global agricultural monitoring. *SovrProbl DZZ Kosm*, *16*, 183–197. Available from https://doi.org/10.21046/2070-7401-2019-16-3-183-197.

Townshend, J. R., & Justice, C. O. (1986). Analysis of the dynamics of African vegetation using the normalized difference vegetation index. *International Journal of Remote Sensing*, *7*(11), 1435–1445. Available from https://doi.org/10.1080/01431168608948946.

Trendov, M., Varas, S., & Zeng, M. (2019). Digital technologies in agriculture and rural areas: Status report. *Digital technologies in agriculture and rural areas: Briefing paper*. FAO, Rome. https://www.fao.org/3/ca4887en/ca4887en.pdf (Accessed 09.10.24)

Tucker, C. J. (1980). Remote sensing of leaf water content in the near infrared. *Remote Sensing of Environment*, *10*(1), 23–32. Available from https://doi.org/10.1016/0034-4257(80)90096-6.

Van Diepen, C. V., Wolf, J. V., Van Keulen, H., & Rappoldt, C. (1989). WOFOST: A simulation model of crop production. *Soil Use and Management*, *5*(1), 16–24. Available from https://doi.org/10.1111/j.1475-2743.1989.tb00755.x.

Van Hoolst, R., Eerens, H., Haesen, D., Royer, A., Bydekerke, L., Rojas, O., ... Racionzer, P. (2016). FAO's AVHRR-based Agricultural Stress Index System (ASIS) for global drought monitoring. *International Journal of Remote Sensing*, *37*(2), 418–439. Available from https://doi.org/10.1080/01431161.2015.1126378.

Verhoef, W. (1985). Earth observation modeling based on layer scattering matrices. *Remote Sensing of Environment*, *17*(2), 165–178. Available from https://doi.org/10.1016/0034-4257(85)90072-0.

Veroustraete, F., Sabbe, H., & Eerens, H. (2002). Estimation of carbon mass fluxes over Europe using the C-Fix model and Euroflux data. *Remote Sensing of Environment*, *83*(3), 376–399. Available from https://doi.org/10.1016/S0034-4257(02)00043-3.

Weiss, M., Baret, F., Myneni, R., Pragnère, A., & Knyazikhin, Y. (2000). Investigation of a model inversion technique to estimate canopy biophysical variables from spectral and directional reflectance data. *Agronomie*, *20*(1), 3–22. Available from https://doi.org/10.1051/agro:2000105.

Wu, B., Gommes, R., Zhang, M., Zeng, H., Yan, N., Zou, W., ... Van Heijden, A. (2015). Global crop monitoring: A satellite-based hierarchical approach. *Remote Sensing*, *7*(4), 3907–3933. Available from https://doi.org/10.3390/rs70403907.

Wu, B., Zhang, M., Zeng, H., Tian, F., Potgieter, A. B., Qin, X., ... Boken, V. (2023). Challenges and opportunities in remote sensing-based crop monitoring: A review. *National Science Review*, *10*(4), nwac290, https://doi.org/10.1093/nsr/nwac290.

Xiao, X., Zhang, Q., Hollinger, D., Aber, J., & Moore, B., III (2005). Modeling gross primary production of an evergreen needleleaf forest using MODIS and climate data. *Ecological Applications*, *15*(3), 954–969. Available from https://doi.org/10.1890/04-0470.

Yao, F., Feng, L., & Zhang, J. (2014). Corn area extraction by the integration of MODIS-EVI time series data and China's environment satellite (HJ-1) data. *Journal of the Indian Society of Remote Sensing*, *42*, 859–867. Available from https://doi.org/10.1007/s12524-014-0377-5.

Yuan, W., Liu, S., Yu, G., Bonnefond, J. M., Chen, J., Davis, K., ... Suyker, A. E. (2010). Global estimates of evapotranspiration and gross primary production based on MODIS and global meteorology data. *Remote Sensing of Environment*, *114*(7), 1416–1431. Available from https://doi.org/10.1016/j.rse.2010.01.022.

Zhang, J., Huang, Y., Pu, R., Gonzalez-Moreno, P., Yuan, L., Wu, K., & Huang, W. (2019). Monitoring plant diseases and pests through remote sensing technology: A review. *Computers and Electronics in Agriculture*, *165*, 104943, https://doi.org/10.1016/j.compag.2019.104943.

Zhang, M., Wu, B., Yu, M., Zou, W., & Zheng, Y. (2014). Crop condition assessment with adjusted NDVI using the uncropped arable land ratio. *Remote Sensing*, *6*(6), 5774–5794. Available from https://doi.org/10.3390/rs6065774.

Zhao, M., Heinsch, F. A., Nemani, R. R., & Running, S. W. (2005). Improvements of the MODIS terrestrial gross and net primary production global data set. *Remote Sensing of Environment*, *95*(2), 164–176. Available from https://doi.org/10.1016/j.rse.2004.12.011.

Chapter 3

Agroforestry for sustainable livelihoods: role of geo-informatics, ICT, and citizen science

R.K. Singh[1,2], Mukunda Dev Behera[2], C.M. Biradar[1], S.K. Dhyani[1], A.J. Prakash[2], Pulakesh Das[3], A. Dogra[1], A.H. Rizvi[1], A. Singh[1], F. Ahmad[1], R. Choudhary[1] and J. Rizvi[1]

[1]*World Agroforestry (CIFOR-ICRAF), New Delhi, India,* [2]*Centre for Ocean, River, Atmosphere and Land Sciences (CORAL), Indian Institute of Technology Kharagpur, Kharagpur, West Bengal, India,* [3]*School of Forest Resources, University of Maine, Orono, ME, United States*

3.1 Introduction

Although agroforestry is an age-old practice, it is relevant to contemporary socioeconomics and climate change adaptive land use practices that integrate 13 out of 17 sustainable development goals (SDG), linking food production systems with health, environment, and livelihoods. The increasing need for a food supply requires much higher agricultural productivity and a climate-resilient system. Other cumulative factors that pressure agricultural production include urban expansion, land degradation, extreme weather events, and water scarcity (Behera, Gupta, Barik, Das, & Panda, 2018; Matin, Ghosh, & Behera, 2019). These pressures also affect natural ecosystems, water, and energy balance, causing landscape deterioration and highlighting the need for sustainable land use practices (Schwilch et al., 2011). The ecologically unsustainable use of natural resources degrades ecosystems and threatens food and environmental security vis-à-vis economy and sustainability. Agroforestry is a viable land use option that meets the food and fodder requirements of the human and livestock populations and is also important for climate change mitigation. Tree integration in farming system provisions enhances resilience and buffers farm households against climate-related risks. Again, agroforestry is a low-hanging fruit as far as ecosystem services are concerned. Historically, traditional agroforestry systems (AFS) have been treasure houses for biological diversity vis-à-vis traditional knowledge systems for ecological sustainability.

Agroforestry is an integrated land use system in which woody perennials are deliberately unified with crops and/or animals on the same land management unit. According to the World Agroforestry (ICRAF), "Agroforestry is the integration of woody components with agricultural and pastoral operations on the same piece of land, either in a geographical or temporal sequence, so that both ecological and economic interactions occur." It is a sustainable method of land use management in which agriculture and forestry are combined into an integrated production system to maximize advantages (Kidd & Pimental, 2012). Agroforestry has gained recognition for its potential to enhance sustainability and productivity as an integrated land management practice combining trees, crops, or livestock (Nair, 2012). In the present context of land degradation and natural resource crunch, agroforestry befits the operational ease of a farmer for reaping long−term environmental benefits. The United Nations (UN) declared the next decade, starting in 2021, as the International Decade of Restoration, which succeeded the International Decade of Biodiversity (2011−2020).

3.1.1 What is agroforestry?

Agroforestry is an agroecological approach that integrates diverse crops, trees, and livestock at multiple scales, including trees on farms, farming in forests, trees outside forests, forest margins, and intercropping with agriculture. It leverages the ability of trees to store carbon, improve biodiversity, maintain soil organic matter and carbon, regulate hydrological conditions, and record the climate history (Nair, 2018). Since the term agroforestry was coined in the late 1970s to describe the work of the World Agroforestry (ICRAF), the concept has evolved greatly. The CIFOR-ICRAF approach

Sustainable Development Perspectives in Earth Observation. DOI: https://doi.org/10.1016/B978-0-443-14072-3.00018-6

addresses the complexity of the interactions between people and ecological systems through a holistic system approach (Cifor-icraf.org, 2023). Agroforestry is recognized as the ultimate tool for enhancing resilience to climate change and reducing the carbon footprint of developmental activities. Global adoption of agroforestry practices also supports the United Nations' SDG, promoting environmental sustainability and economic well-being (Chaturvedi, Pandey, & Singh, 2017; Nair, 2013). To meet their intended nationally determined contributions, 23 nations prioritize agroforestry for mitigation, while 29 countries prioritize adaptation (https://ccafs.cgiar.org/agricultures-prominence-indcs-data-and-maps#.Wfa1uohx200).

Agroforestry has immense potential to fulfill the social, economic, and ecological goals of large populations and simultaneously improve land quality by enhancing soil fertility (Dollinger & Jose, 2018; Nair, 1985), enhancing ecosystem services (Kuyah et al., 2016), boosting food security, addressing environmental degradation (Waldron et al., 2017), and mitigating the impact of climate change (Ospina, 2017). Agroforestry directly or indirectly can contribute significantly to some of the most challenging SDGs, which can be achieved using existing land use systems with appropriate strategies by orienting higher levels of integration of financial and human resources, policy, or institutional space within existing frameworks (van Noordwijk et al., 2018).

3.1.2 Ecological benefits of agroforestry

Agroforestry is a sustainable agricultural practice that offers a range of ecological benefits with significant implications for climate change mitigation and adaptation. One key advantage is carbon sequestration, in which trees in AFS capture and store carbon dioxide through photosynthesis, thereby reducing atmospheric CO_2 levels. Lal (2005) has demonstrated the contribution of agroforestry to climate change mitigation through carbon sequestration. Dhyani, Ram, and Dev (2016) have reviewed the potential of AFS in carbon sequestration in India. AFS promote biodiversity by creating diverse habitats for various species, supporting more plant and animal diversity than monoculture agriculture (Jose, 2009). According to Nair et al. (2009a; 2009b), tree roots in these systems play an important role in minimizing soil erosion and improving soil structure, making them more robust to climate change impacts while improving soil fertility through nutrient cycling and moisture retention. According to Jose (2009), agroforestry trees provide shade, reduce temperature extremes, and protect crops from heat stress, windbreaks, and shelterbelt formation, safeguarding crops from extreme weather events. Additionally, agroforestry practices positively affect water management by improving water quality and reducing runoff (Singh, Garg, Wani, Tewari, & Dhyani, 2014, 2021), which is crucial for mitigating the effects of climate change on water resources, as Zhang et al. (2010) emphasized. Moreover, certain tree species in AFS attract pollinators and pest predators, contributing to natural pest control and increased crop yields, as Bommarco (2013) highlighted.

Additionally, AFS are more resilient to environmental stressors such as droughts and pests, owing to the protective role of trees as windbreaks and shelterbelts. The integration of trees and shrubs can also enhance land's esthetic and biodiversity value, potentially increasing its long-term value (Josiah, 2014). Agroforestry is growing globally and continues to play a crucial role in agriculture. Agroforestry is carried out on an estimated 43% of the total agricultural area (over 1 billion hectares) worldwide and employs over 900 million people (approximately 30%) in rural regions (Zomer, Trabucco, Coe, Place, van Noordwijk, & Xu, 2014, 2022). However, wood fuels still account for two-thirds of all energy sources other than human and animal energy used in Africa, nearly one-third in Asia, one-fifth in Latin America, and 6% in the Near East. It has been estimated that more than 10 million km^2 of agricultural land globally has a tree cover of over 10% (Zomer, Trabucco, Coe, & Place, 2009). Agroforestry can return 8–13 times more profit than conventional agriculture, with increased ecosystem services.

3.1.3 Socioeconomic benefits of agroforestry

As a versatile and dynamic approach, agroforestry has been adapted to various ecological and social contexts, making it a valuable solution to address the challenges of a changing world (Sileshi et al., 2019). Agroforestry contributes to sustainable livelihoods for farmers by diversifying income sources and reducing vulnerability to market fluctuations (Garrity et al., 2010; Garrity, Akinnifesi, Ajayi, & Weldesemayat, 2010). Agroforestry can empower communities in developing countries by offering sustainable wood and nonwood forest products (Nair, 2012). Furthermore, diverse vegetation in AFS yields multiple valuable products such as medicinal plants, nontimber forest products (NTFPs), fruits, flowers, honey, and mushrooms. Thus it allows farmers to earn from tree products alongside traditional crops and livestock, providing a financial safety net against market fluctuations. It further improves the productivity of associated crops; for example, integrating *Faidherbia albida* trees with maize in West Africa leads to a 280% increase in maize yield (Garrity, Akinnifesi, et al., 2010; Garrity et al., 2010). Furthermore, agroforestry aids in carbon sequestration and climate change mitigation, potentially offering opportunities for carbon credits and payments for ecosystem services (Nair et al., 2009a; 2009b). Trees in AFS can help mitigate the effects of climate change by sequestering carbon, regulating water flow, and providing protection against soil erosion (Mukhlis, Rizaludin, & Hidayah, 2022). By improving soil health, agroforestry reduces the need for synthetic fertilizers and pesticides, resulting in cost savings for farmers and supporting sustainable agricultural practices. AFS yield a wider variety of products compared to monoculture farming. Farmers can harvest fruits, nuts, timber, fodder, and other NTFPs alongside their crops, generating additional income streams (Mukhlis et al., 2022). The sale of various products from AFS, including fruits, timber, and NTFPs, can significantly boost household income, especially for small-scale farmers (Kiptot, Sinclair, & Place, 2022). Agroforestry practices can create new job opportunities in areas like processing, marketing, and value addition of harvested products (Kiptot et al., 2022). By promoting both environmental and socioeconomic well-being, agroforestry offers a promising solution for sustainable development.

3.2 Agroforestry in India

India is often called the cradle of agroforestry, practiced for generations. Although it has lost priority in the last few decades, it has gained importance owing to several factors and the enabling environment. India's agroforestry practices align with the country's diverse agroclimatic zones, addressing challenges such as soil degradation, water scarcity, and climate change resilience. The area under agroforestry in 15 agroclimatic zones is estimated to be 28.247 million hectares (Arunachalam, Rizvi, Handa, & Ramanan, 2022). The country fulfills approximately 70% of its timber needs through agroforestry, valued at approximately INR 14,000 crores annually (Shrivastava & Saxena, 2017). The primary raw materials in approximately 26,500 wood-based industries were sourced from agroforestry. Agroforestry meets almost half the need for fuelwood, around two-thirds of the small timber, 70%–80% wood for plywood, 60% raw material for paper pulp, and 9%–11% raw material for paper pulp cents of the green fodder requirements of the country. Tree-based systems produce lac, gum, resins, and medicinal products in various industries. In recent decades, reliance on trees (agroforestry) for traditional tree-based needs, such as firewood, has declined owing to changes in society's social and economic structure and trees attaining the status of cash crops for sale (Fig. 3.1).

Agroforestry is particularly important to small and marginal farmers. Agroforestry employs a relatively lower investment, as well as for unskilled farmers. Regarding household income, central Indian upland rice fields provide an illuminating economy (Viswanath, Nair, Kaushik, & Prakasam, 2000). Approximately, 30 million popular trees produce 1125 million m^3 of industrial wood annually in Uttar Pradesh, Uttarakhand, Haryana, Punjab, Himachal Pradesh, Jammu, and Kashmir in combination with crops (Dhiman, 2012). This is equivalent to 60,000 ha of pure poplar plantations (500 trees/ha). It is estimated that approximately 25,000 ha of equivalent poplar plantations are established annually in

FIGURE 3.1 Field photographs and examples of the tree being integrated into a rice-based farming system in Odisha state, India. *From ICRAF.*

association with several agricultural and horticultural crops under a rotation of 6–8 years. It can only create much-needed employment for 40 million people in plantation activities (Chandra, 2001). Many studies have indicated that AFS can provide employment requirements of 450 person-days/ha per year (Dhyani, Samra, Ajit, Handa, & Uma, 2007). In addition to existing agroforestry practices, there is also a broad scope of employment generation with improved AFS, resulting in 943-million-person days annually from 25.4 million ha of agroforestry area (NRCAF, 2013). The potential of agroforestry for rural development and employment generation is 5.763 million human days per year in the Indian Himalayas alone (Dhyani, Sharda, & Samra, 2005). Tewari (1992) estimated that 1 ha of bamboo plantation with 500 clumps generates 3.9 person-days of employment for unskilled labor and 47.3 person-days for supervisors annually over 30 years. In addition, agroforestry promotes multiple allied industries, such as the plywood and plyboard manufacturing units established in the twin cities of Yamunanagar and Jagadhari, Haryana.

The scope of agroforestry in India is extensive, with several ongoing activities and government initiatives, such as the Mahatma Gandhi National Rural Employment Guarantee Act (MGNREGA), which includes provisions for agroforestry (MGNREGA, 2005). The Sub-Mission on Agroforestry also promoted agroforestry practices in India. Furthermore, the National Bamboo Mission plays a significant role in bamboo-based agroforestry (NBM, 2018). Research institutions such as the Indian Council of Agricultural Research (ICAR) and the World Agroforestry Center (ICRAF) actively engage in agroforestry research and extension, contributing to India's expansion and development of agroforestry practices. Agroforestry research in India started in 1983–1984 with the All India Coordinated Research Project launched by the ICAR. Agroforestry research spanning 37 centers include agricultural universities and institutes under the Indian Council of Forestry Research and Education.

The policy-level promotion of agroforestry in India began in the year 2000, when the National Agriculture Policy (2020) stated quite clearly, "Agriculture has become a relatively unrewarding profession due to generally unfavourable price regime and low-value addition, causing the abandonment of farming and increasing migration from rural areas." As a result, the policy emphasizes that "farmers will be encouraged to take up farm/agroforestry for more significant revenue generation by improving technology, extension, and financial assistance packages and reducing impediments to agroforestry growth. On February 10, 2014, the Government of India adopted a new agroforestry policy with the primary goal of identifying measures to extend the use of agroforestry to alleviate pressure on existing forests and enhance the amount of tree cover outside the bounds of natural forests (NAP, 2014). Agroforestry in India is governed by various policies, with the 2014 National Agroforestry Policy as a pivotal reference (NAP, 2014). This policy promotes agroforestry practices nationwide and emphasizes the need to integrate trees with crops. At the state level, the distribution and implementation of agroforestry practices varies widely. Different states have adopted various approaches and programs to encourage agroforestry based on their needs. For instance, states such as Madhya Pradesh and Chhattisgarh have made significant strides in the promotion of agroforestry (Mishra, Kumar, & Singh, 2018). A sizable number of forestlands in northeast Indian states are periodically deforested for shifting cultivation practices every year (Das et al., 2021). Various governmental and nongovernmental efforts have been put forward to reduce the dependency on shifting cultivation to more sustainable land use practices, including agroforestry. For example, the Indo-German Developmental Corporation project implemented in Tripura has significantly replaced the past shifting cultivation lands with multistory AFS (Behera et al., 2018). Monoculture plantations and rubber-based intercropping agroforestry have expanded in several states in this landscape in the last few decades, significantly reducing shifting cultivation practices (Das, Panda, et al., 2022; Das, Behera, et al., 2022). NITI Aayog, Pant, Tiwari, and Choudhury (2018) recommended promoting different agroforestry interventions such as home gardens as a viable alternative to reduce shifting cultivation. NITI Aayog et al. (2018) also suggested considering AFS as a land use class in national land use land cover (LULC) mapping and allocating funds to promote agroforestry in shifting cultivation areas.

3.3 Type of agroforestry practices

Various agroforestry practices that address specific ecological and economic needs exist worldwide. Some notable practices include the following.

1. Alley Cropping: Planting rows of trees within agricultural fields to provide shade and windbreak protection for crops such as maize and pigeon peas.
2. Silvopasture: Combining trees, forage, and livestock to offer shade, fodder, and timber for cattle farming and production.
3. Windbreaks and shelterbelts: In windbreaks, trees and shrubs are planted as barriers to shield crops from strong winds and are mainly used in windy areas such as arid, semiarid, and coastal regions. Shelterbelts are strips of

vegetation composed of trees and shrubs that grow along the coast to protect coastal areas from high-velocity winds. Shelterbelts also serve as binders to prevent sand erosion.

4. Taungya fystem: Simultaneous cultivation of crops and trees, mainly teak, on forestlands in India, prevalent in states such as Karnataka and Tamil Nadu.
5. Fruit plant-based agroforestry: Growing fruit trees alongside annual or perennial crops, grasses, and legumes in plant-based agroforestry, for instance, mango and guava orchards combined with crops and vegetables.
6. Home gardens: Integrating trees with vegetables, spices, and medicinal plants, common in rural areas as well as many places in urban areas.
7. Agro-silvopastoral systems: Complexly interrelate crops, trees, and livestock, contributing to both food and timber production.

Prominent AFS practiced in India:

1. Agrisilviculture: This system involves growing crops alongside protected tree crops and is widespread across seven agroclimatic regions in India.
2. Agri-horticulture: In this system, fruit trees are combined with crops and is practiced in six agroclimatic regions.
3. Agri-silvi-horticulture: This system combines trees, fruit trees, and crops and is practiced in two agroclimatic regions.
4. Agri-silvi-pasture: A woody component (trees or shrubs) is deliberately associated with cattle on the same land unit in two agroclimatic regions.
5. Boundary plantation: Trees are planted at a considerable distance from agricultural fields.
6. Block plantation: This involves planting trees with closed spacing in compact blocks of more than 0.1 ha, mainly in the eastern plateau, hills, and the central plateau and hills.
7. Energy plantation: Trees are initially grown with crops, primarily in the lower Gangetic Plain.
8. Alley cropping: Shrubs or trees are planted with close spacing within rows, leaving room for herbaceous cropping in the alleys between rows.
9. Silvi-olericulture: Combines trees and vegetables.
10. Horti-pasture: Fruit trees are combined with pasture or animals.
11. Horti-olericulture: This system combines fruits and vegetables.
12. Silvi-pasture: Livestock, forage production, and trees are integrated into the same land management unit.
13. Forage forestry: Combination of forage trees and pastures.
14. Shelter belts: Trees are combined with crops, primarily in the East Coast plateau, hills, and western dry regions.
15. Windbreaks: Trees are combined with crops to break the wind force.
16. Live fence: shrubs and trees form boundaries.
17. Silvi or horti-sericulture: Trees or fruit trees are combined with sericulture.
18. Hortiapiculture: Fruit trees are combined with honeybee cultivation.
19. Aqua-forestry: Trees are combined with fisheries.
20. Homestead/home gardens: Various combinations of trees, fruit trees, and vegetables are integrated. Additionally, there are specialized systems such as Jhum (shifting cultivation) in the Eastern Himalayas and home gardens in specific regions like the East coast plateau and hills, West coast plains, ghats, and Island.

3.4 Use of geo-informatics and data analytics

Geo-informatics encompasses a range of technologies and methods for collecting, analyzing, and managing geographical data. This field is crucial for applications in various domains, including sustainable land use planning and agroforestry research. The components of geo-informatics are as follows:

1. Remote sensing: Remote sensing involves gathering information about the Earth's surface without direct physical contact. Remote sensing involves acquiring information about Earth's surface without direct physical contact. It utilizes sensors on satellites, aircraft, or ground-based instruments to capture data such as imagery and spectral information. Remote sensing enables monitoring of land cover, climate change, and natural disasters (Lillesand & Kiefer, 2015).
2. Geographic Information System (GIS): GIS is a system for capturing, storing, analyzing, and visualizing spatial data. It integrates various data types, allowing users to create maps, perform spatial analyses, and make informed decisions regarding urban planning and resource management (Longley, Goodchild, Maguire, & Rhind, 2015).

3. Geolocation: Global Positioning System and Navigation with Indian Constellation are satellite-based navigation systems that provide accurate positioning and timing information. It has widespread applications, ranging from smartphone navigation to georeferencing data for field surveys and mapping (Kaplan & Hegarty, 2006). The Differential Global Positioning System (DGPS) is a satellite-based navigation system that enhances geolocation accuracy by providing real-time correction data to receivers. These correction data help mitigate errors caused by factors such as atmospheric interference, satellite clock errors, and signal multipath. The DGPS compares the position that a reference station calculates with a known, precise location and then transmits the correction information to nearby geolocation receivers.
4. Data-processing algorithms/statistics data-processing algorithms and statistical methods are essential for extracting meaningful information from geospatial data. These techniques enable the analysis of trends, patterns, and correlations within geographic datasets, aiding decision-making in various applications (Chang, 2008).

3.4.1 Use of geo-informatics in agroforestry

Geo-informatics is pivotal in enhancing agroforestry practices by providing essential tools and data-driven insights into various aspects of this sustainable land management approach. Using geo-informatics in agroforestry presents numerous challenges and promising pathways for the future (Rizvi et al., 2020). Remote sensing technologies offer promising tools for mapping and monitoring AFS because they capture spatial and spectral information over large areas (Thenkabail, Schull, Turral, & Gassman, 2019). Various remote sensing platforms, including satellites, aerial platforms, and unmanned aerial vehicles (UAVs), are employed for agroforestry mapping (Puliti et al., 2020). These platforms provide diverse data types such as optical, radar, and light detection and ranging (LiDAR), each offering unique advantages and limitations for mapping AFS (Joshi, Pandit, & Joshi, 2018). For instance, optical sensors are valuable for identifying tree species and assessing vegetation structure, whereas radar sensors penetrate cloud cover and provide information on biomass and canopy structure (Lhermitte, Verbesselt, Verstraeten, & Coppin, 2019).

Geo-informatics has a few broad applications in agroforestry:

1. Agroforestry system mapping: Geo-informatics facilitates precise mapping and delineation of AFS within a landscape. Using remote sensing and GIS technologies enables the identification of specific land areas where agroforestry practices are in use, thus aiding in strategic planning and management of these systems (Smith, van Noordwijk, Hoang, & Kumar, 2015).
2. Species mapping: Geo-informatics aids in identifying and mapping tree and crop species used in AFS. Satellite imagery and field data collection allow for the creation of species distribution maps, which are crucial for selecting suitable species and optimizing species combinations for agroforestry practices (Poudel, Yamanaka, & Shrestha, 2019).
3. Site-suitability analysis: Geo-informatics helps conduct site-suitability analyses by considering various environmental factors, such as soil type, climate, topography, and land use. GIS-based suitability assessments enable the identification of the most suitable locations for AFS, ensuring optimal utilization of available resources (Schwendenmann, Veldkamp, Moser, Koehler, & Clough, 2017).
4. Biomass/carbon stock estimation: Geo-informatics can effectively estimate biomass and carbon stocks in AFS (Singh, Rizvi, Behera, & Biradar, 2021). Remote sensing technology can be employed to monitor changes in vegetation over time, and these data can then be used to calculate carbon sequestration rates, assisting in assessing agroforestry's climate change mitigation potential (Makalew, 2018).
5. Green/Tree cover change assessment: Geo-informatics is instrumental in monitoring changes in green and tree cover changes in green and tree cover over time. Through satellite imagery analysis and GIS, the expansion or contraction of trees and green cover within agroforestry areas can be tracked, thereby providing insights into the effectiveness of agroforestry practices in promoting environmental sustainability (Sannigrahi, Mishra, & Behera, 2020).

Remote sensing technology has been instrumental in capturing diverse surface characteristics with distinct reflectance values, enabling the discrimination of different land use classes (AFS, forestry, agriculture) based on their spectral signatures (Ballanti, Blesius, Hines, & Kruse, 2016). However, interpreting these data requires knowledge about shapes, textures, patterns, and site contexts to derive meaningful information about land use activities (Blaschke et al., 2014). Mapping AFS extent necessitates a meticulous and systematic approach, along with expertise and is heavily reliant on factors such as the definition of AFS, scale, spatial dataset, type of analysis (classifier), timing of dataset, and availability of ground control points (Den Herder et al., 2017; Rizvi, Newaj, Handa, Sridhar, & Kumar, 2019). Pixel-based classification methods like maximum likelihood and support vector machines (SVM) are commonly employed for mapping

AFS based on spectral signatures (Simic-Milas, Balla, Veverec, Lukic, & Savic, 2017). Additionally, object-based image analysis techniques are gaining traction because they integrate spatial information and contextual relationships into the classification process (Lu, Batzel, Moran, & Dutra, 2020). Recent advancements in machine learning (ML) and deep learning (DL) algorithms have shown promise in enhancing the accuracy of agroforestry mapping by leveraging large datasets and complex spatial patterns (Kussul, Lavreniuk, Skakun, & Shelestov, 2021). For instance, Lu et al. (2022) successfully mapped the abundance of *F. albida* trees in Senegal using a multilayer perception feedforward artificial neural network (ANN) from Sentinel-2 time-series satellite images.

In India, the Center for Agroforestry Research and Innovation (CAFRI), Jhansi, took the lead in 2007 by utilizing geospatial technology to outline the extent of AFS using medium-resolution satellite data (LISS III, 23.5 m). This pioneering effort was followed by subsequent studies (Rizvi et al., 2009). Later, nationwide estimates of AFS extent were provided by the Forest Survey of India (FSI) in 2013, as well as by Dhyani and Handa (2013) and Rizvi, Dhyani, Newaj, Karmakar, and Saxena (2014). Among various methodologies, Rizvi, Newaj, Karmakar, Saxena, and Dhyani (2016) developed a widely accepted approach for delineating AFS areas. According to CAFRI, Jhansi, the country's agroforestry area is 13.75 million hectares (mha). However, the FSI estimated 11.54 mha, or 3.39% of the country's geographical area (ISFR, 2013). Most agroforestry expansion has been reported in Maharashtra, Gujarat, and Rajasthan. The total area under agroforestry in Odisha state was mapped as 0.51 mha by FSI and 0.57 mha by CAFRI. The FSI estimate excludes several agroforestry interventions, such as block plantations, which underestimates the CAFRI estimate. According to Dhyani and Handa (2013), the approximate area under agroforestry is projected to be 25.32 mha, or 8.2% of India's geographical area. Recently, the estimated area under agroforestry in 15 agroclimatic zones was 28.247 mha (Arunachalam et al., 2022).

3.4.2 Artificial intelligence-based, web-based platform, mobile, and chatbot application in agroforestry systems

The application of artificial intelligence (AI) through ML and DL has enabled rapid and accurate data analysis and prediction. Multisource satellite images, geo/agro-tagging of in situ data, and various environmental and socioeconomic conditions allow the monitoring of AFS, identification of suitable areas, and interventions for agroforestry interventions (Singh et al., 2022). Recently, Singh et al. (2021) studied the aboveground biomass (AGB) in AFS in the Belpada block of Balangir District, Odisha. They used three ML regression models (i.e., Random Forest [RF], SVM, and ANN) for AGB prediction using satellite data and derived indicators. They reported a high species diversity (35 species with dominant *Mangifera indica*, *Tectona grandis*, and *Eucalyptus globulus*) and an AGB stock of 0.02–119 megagrams per hectare (Mg ha^{-1}). Singh et al. (2022) used the existing agroforestry interventions and environmental variables such as soil, topography, climate, and socioeconomic factors to generate the agroforestry suitability map in this landscape through ML models and weighted overlay GIS operation.

Information and Communications Technology (ICT)-enabled citizens have become an essential tool for providing real-time analytics to accelerate the use of sophisticated systems in agroforestry. Citizen-science-driven approaches and crowdsourced platforms provide interactive data to farmers and decision-makers and help in agroforestry scaling. Such platforms provide site-specific information to users as a reward for providing data to fill gaps in the platform and processing timely information for precision decisions that support the scaling of tailored interventions. Digital tools are used to inform multistakeholder decision-making toward effective transformation to more sustainable and resource-efficient agroecosystems.

The "Agroforestry" app developed by CIFOR-ICRAF as a digital solution for building awareness and scaling agroforestry has the following key features:

- The Earth Observation System and ML-based real-time quantification of farming system dynamics focus on the utilization of agricultural fallows by growing pulses and vegetables as additional crops.
- The fusion of agronomy and agroecological frameworks provides site- and context-specific recommendations for effective agricultural practices.
- To overcome the critical data gaps in in-season digital advisories, implementing geotagging and agro-tagging for on-farm data collection, curation, and reporting for in-synch activities.
- The app helps extension agents, development partners, and progressive farmers to select the right agroforestry system and package-associated practices (including tree species, accompanying crops, and the nearest nursery to procure quality planting materials).

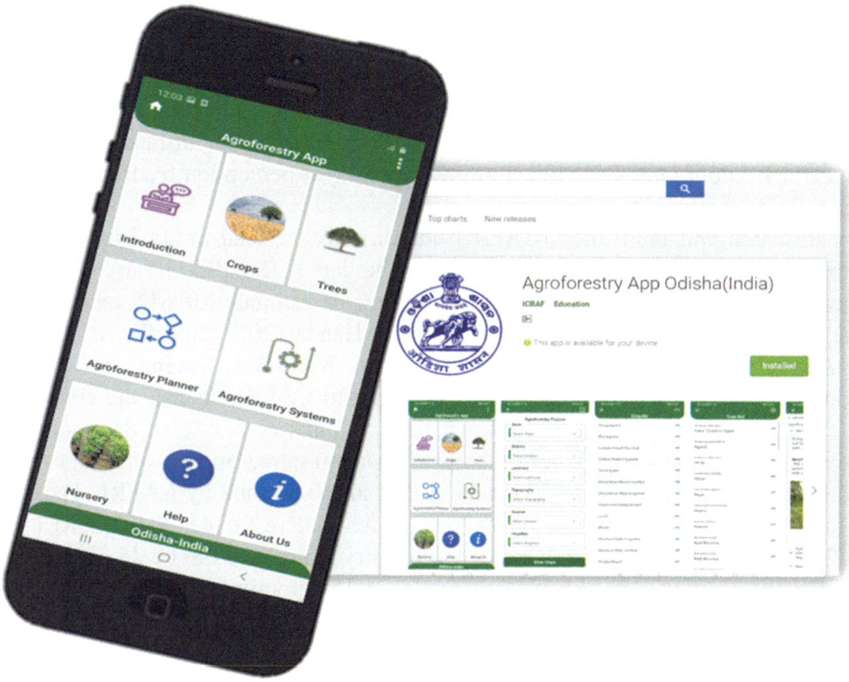

FIGURE 3.2 Agroforestry Assistant App can be downloaded from the Google Play store—https://play.google.com/store/apps/details?id = org.icraf.agroforestryappindia &hl = hi&gl = US.

- Mobile phone-based digital applications will accelerate sustainable agroforestry intensification for site-specific intervention. A similar application has been developed and named as Agroforestry Assitant (Fig. 3.2).

3.4.3 Challenges in agroforestry monitoring

Mapping the dominance and extent of species in AFS is complex because of the heterogeneity of mixed tree, crop, and livestock systems, and their close association. This complexity is further exacerbated in windbreaks and boundary planting practices, where trees are spaced at significant distances. While freely available remote sensing data facilitate mapping agroforestry at medium resolution (Rizvi et al., 2016), challenges persist owing to the lack of high-resolution data, appropriate procedures, and robust techniques for the accurate mapping of various tree species or dominance in AFS. Furthermore, the scarcity of systematically collected ground reference information has hampered the development of predictive models for estimating agroforestry areas. Although effective in detecting individual trees, high-resolution satellite or aerial imagery is costly and lacks the spatial coverage required for regional-to-national-scale agroforestry assessments. Alternatively, the integration of 10 m Sentinel-2 multispectral data and Sentinel-1 microwave data with field observations enhances the mapping of dominant tree species over 30 m resolution Landsat optical data. Incorporating local knowledge of soil, rainfall, climate, and topography with satellite-derived information aids in developing predictive models using fuzzy modeling, ML, and multicriteria decision analysis (Chen et al., 2015).

The major challenges are as follows.

- Data accessibility: Access to quality geospatial data, including high-resolution satellite imagery and accurate geolocation, can be costly and challenging for small-scale agroforestry practitioners (FAO, 2020).
- Integration of multiple data sources: Agroforestry involves diverse tree and crop species. Integrating data from various sources and formats can be complex and time-consuming (van Noordwijk et al., 2019).
- Technical skills: Adequate training and skills are required to effectively utilize geo-informatics tools that may not be readily available to all practitioners (2016).
- Data accuracy: Ensuring the accuracy and reliability of geospatial data, particularly in remote or resource-constrained areas, can be challenging (Sola et al., 2018).
- Data security: Safeguarding sensitive data related to agroforestry practices and land use from unauthorized access or misuse is a growing concern (UN Nations, 2018).
- Tree detection is insufficient; however, approaches for identifying trees and land use systems must be integrated (WAC, 2015).

In a recent review, besides investment and funding support for the promotion of agroforestry, the following points need to be considered for the use of geo-informatics and ICTs in agroforestry:

1. Digital campus and capacity building: Taking into account user access to smartphones and the wide spread of digital technology-enabled citizens science, there is an urgent need for investment in the development of digital campuses and capacity building programs, self-help training programs, field schools, and workshops to effectively equip agroforestry practitioners with the necessary skills to use geo-informatics tools (2016).
2. Open data initiatives: Promote open-access geospatial data repositories, making valuable resources accessible to all (FAO, 2020). SAR imagery must be integrated with optical, LiDAR data, or models to enhance the assessment of tree parameters or AGB. NASA's NISAR (NASA-ISRO SAR) mission, utilizing the L and S bands, along with the P band of the BIOMASS mission scheduled for launch in 2022 and 2023, respectively, presents an opportunity for integration with existing SAR data sources like the L band of ALOS-2 PALSAR-2 and the C-band of Sentinel-1. This integration could significantly enhance the accurate quantification of biomass in AFS (Ningthoujam, Singh, & Kumar, 2016, 2017).
3. Public-private partnerships: Collaboration between governments, private sector entities, and research institutions can facilitate data sharing and investment in technology infrastructure (UN, 2018).
4. Blockchain technology: Implement blockchain for secure and transparent data management, ensuring data integrity and preventing unauthorized access (van Noordwijk et al., 2019).
5. ML and DL: Leverage ML and DL algorithms to automate data processing and analysis, reducing the technical expertise required (Sola et al., 2018). These technologies offer the potential to improve the accuracy of structural and functional analysis of AFS, including the estimation of carbon sequestration potential. Establishing a comprehensive framework is essential for determining various aspects of AFS, such as area delineation, canopy structure assessment, biomass estimation, evaluation of biogeochemical cycles, and climate modeling, which includes mapping the potential for carbon sequestration at a national level.
6. Unmanned aerial vehicles (UAVs) or drones offer innovative and efficient avenues for monitoring and assessing agroforestry systems.
7. Local knowledge integration: Combine indigenous and local knowledge with geo-informatics for a more holistic approach to agroforestry planning and management (2016).
8. Creating a tree spectral library entails identifying a range of species suitable for AFS, which aligns with the National AFS Policy of 2014 in India. Initially, this involves selecting major tree species deemed appropriate for AFS.
9. There are significant opportunities to utilize multicriteria decision-making approaches, such as the analytic hierarchy process (AHP) and fuzzy AHP, to enhance the accuracy and reliability of agroforestry system suitability assessments. It is crucial to analyze and monitor the potential for carbon sequestration and assess the environmental benefits.
10. At the same time, monitoring the health of AFS will emerge as a critical concern in the future, which could potentially be addressed through the utilization of hyperspectral and thermal datasets.
11. There is a need to explore the integration of optical and soil moisture observations within AFS to understand their role in the hydrological cycle better and develop strategies for sustainable water management (Kumar, Singh, & Patel, 2021).

3.5 Case studies

The chapter includes two case studies on (1) assessing ecosystem services (carbon stock assessment) in AFS and (2) agroforestry intervention planning using geo-informatics in the Belpada block of Odisha.

3.5.1 Aboveground biomass in agroforestry systems using remote sensing and machine learning

In an agroforestry system, the AGB, a proxy for carbon stock, was assessed in the Belpada block of the Bolangir District in Odisha, India. The project "Enabling Smallholders in Odisha to Produce and consume more nutritious food through Agroforestry Systems," implemented by the ICRAF with financial support from the Government of Odisha, developed many agroforestry sites in the study area (Singh et al., 2022). The case study discusses the AGB estimation conducted by Singh et al. (2023) in agroforestry interventions in the Belpada block, Balangir, Odisha, using geo-informatics. This study used Sentinel-1 (synthetic aperture radar; SAR) and Sentinel-2 (optical) data to predict and extrapolate AGB at the landscape level. High-resolution satellite data such as Sentinel-1 (microwave backscatter) and Sentinel-2 (surface reflectance) at 10 and 20 m resolution were used as predictor variables. Using an allometric equation, the plot-level AGB was estimated using field data on diameter and wood-specific density. ML methods, including RF, SVM, and ANN, were employed for AGB prediction using satellite-derived surface reflectance, vegetation indices, backscatter values, and other proxies as predictor variables. Ground data on various agroforestry interventions were captured during the field survey (Fig. 3.3). A total of 196 sampling plots were studied, of which 134 were mixed plantations and 62 were monoculture plots. A stratified sampling method was adopted for 30×30 m plots, and all planted trees were measured for girth at breast height.

ML models were executed on the R programming platform, wherein the caret package was used. The implemented RF model demonstrated superior performance compared to the SVM and ANN models. Random sampling was applied for model fitting and evaluation, wherein 70% of the data was used for model building and 30% for validation. Various error statistics, such as the coefficient of determination (R^2), mean absolute error (MAE), and root mean square error (RMSE), were used for validation. In addition, we calculated a plant value index (PVI), which indicates the contribution of various tree species to the total biomass and carbon stock. The field-based AGB varied significantly, and a correlation was found between the AGB and diversity parameters. The importance of predictor variables was assessed, and the combination of Sentinel-1 and Sentinel-2 data significantly improved model accuracy.

The most abundant species were *M. indica* (39.51%), *T. grandis* (33.17%), and *E. globulus* (19.51%). The data indicated a reverse trend between species richness and age class, highlighting the recent introduction of more diverse species. The highest PVI was observed for *T. grandis* (88.35), followed by *E. globulus* (56.87), *M. indica* (53.75), and *Azadirachta indica* (15.45). The AGB in the plot-level field inventories ranged between 0.02 and 119 Mg ha^{-1}. About 68% of the plots exhibited low biomass (<1 Mg ha^{-1}), while AGB above 10 Mg ha^{-1} AGB was observed in eight plots.

The RF model exhibited the highest prediction accuracy for the AGB estimation ($R^2 = 0.69$; RMSE = 17.07 Mg ha^{-1}). The RF model predicted the AGB map, variable importance, and validation plot, as shown in Fig. 3.4. Various vegetation indices followed by microwave backscatter performed well in estimating the AGB. The uncertainty in AGB estimations was found to range between 0.14% and 15%. This study demonstrated the effectiveness of combining optical and SAR data with ML techniques for accurate AGB estimation in agroforestry systems, emphasizing the importance of variable selection and model validation.

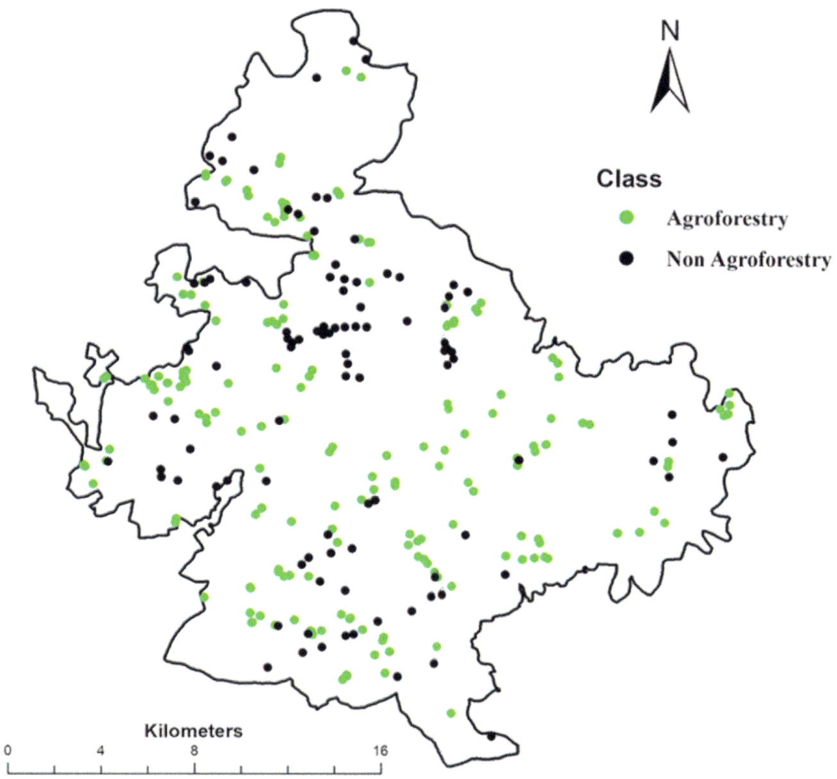

FIGURE 3.3 Ground data points of existing agroforestry patches. Map lines delineate study areas and do not necessarily depict accepted national boundaries.

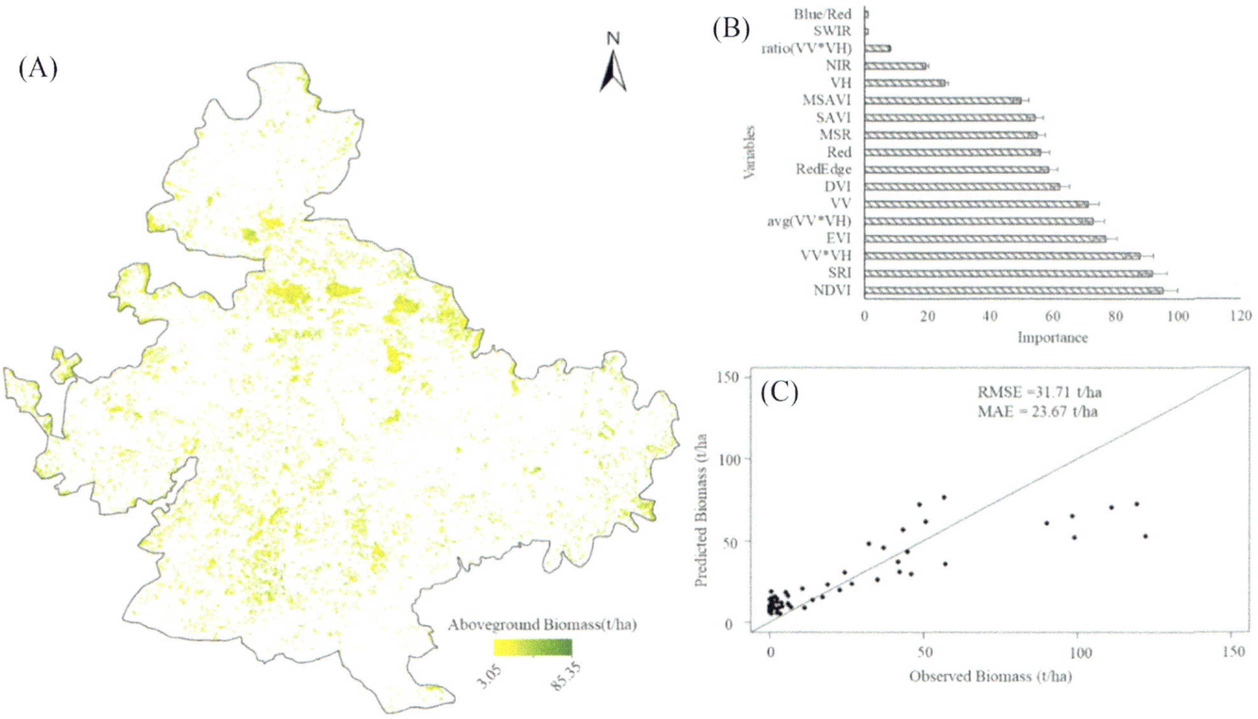

FIGURE 3.4 RF model (A) Random Forest AGB prediction map, (B) variable importance, and (C) predicted versus observed of Sentinel-1 and Sentinel-2. *AGB*, Aboveground biomass; *RF*, random forest. Map lines delineate study areas and do not necessarily depict accepted national boundaries.

3.5.2 Agroforestry implementation planning using geo-informatics and machine learning

The second case study describes a methodology for identifying agroforestry suitability in the Belapada region, using environmental variables such as soil, topography, climate, and socioeconomic factors (Singh et al., 2022). This study employed data from the National Bureau of Soil Survey and Land Use Planning for soil information, the WorldClim database for climate data, and the ALOS-PALSAR Digital Elevation Model for topographic variables. Ground data on various agroforestry interventions, including single and mixed species, were collected during field surveys. The majority of the agroforestry interventions included wadi (agro-horti-forestry), bund and boundary plantations, and home gardens. The field data from 200 samples included the species information and geolocation of single (142) species (Eucalyptus, Mango, Mahua, Teak) and mixed (58) species (Teak, Gamahari, Khajur, Mahua, Chahar). The selection of variables for agroforestry land suitability analysis involves soil type, depth, texture, drainage, pH, slope, annual mean temperature, rainfall, and socioeconomic factors. Considering their importance in agroforestry, these factors were standardized and ranked using fuzzy set modeling. The methodology includes calculating weights for each variable based on their importance using RF regression and ground observations.

The agroforestry suitability model was generated through a multicriteria analysis approach-weighted linear combination method, integrating soil, topography, climate, and socioeconomic factors. The final suitability index map was validated using ground observations (Fig. 3.5). The suitability map was then used to identify the intervention-wise suitability maps of the study area employing the LULC map (Table 3.1).

The RF model-generated suitability map indicated 0.87 suitability based on the validation data points, which exhibited a high modeling accuracy. The importance of the variables showed the dominance of socioeconomic variables over soil and climate variables. More than 90% of the agricultural land in the Belpada block was identified as suitable for agroforestry interventions. The maximum potential was estimated for bund and boundary plantations in rice fallow lands, followed by cotton fallow lands. The potential areas for bund and boundary plantations on double-cropping land were relatively lower. However, vacant land around settlements and wastelands was prescribed for home gardens and block plantations.

Agroforestry suitability and intervention maps can help farmers and decision-makers implement agroforestry in the Belpada block. The interventions can be categorised as (A) bund and boundary plantation in double-crop (paddy-pulses), (B) bund plantation, boundary and single intercropping crop (paddy-fallow), (C) bund and boundary plantation

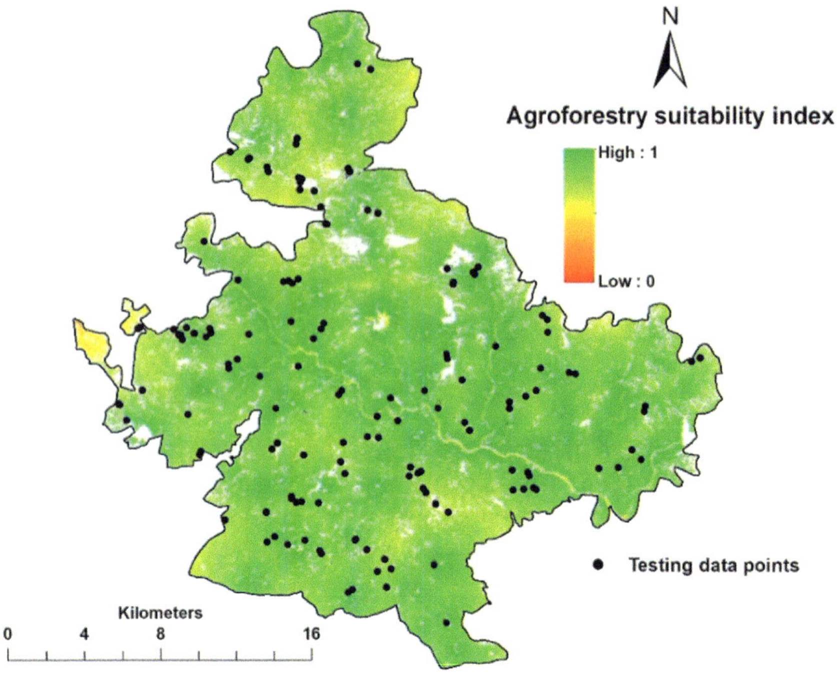

FIGURE 3.5 Agroforestry suitability map with the validation ground reference data points overlaid. Map lines delineate study areas and do not necessarily depict accepted national boundaries.

TABLE 3.1 Criteria adopted for various agroforestry interventions.

LULC class	Criteria table			
	LULC	Interventions	Species	Examples
Agricultural land	(a) Double crop: paddy-pulses	Bund plantation	Fruit and wood (teak) plantation	Apple ber teak and bamboo
	(b) Single crop: paddy-fallow	Intercropping and bund plantation	Bund plantation-wood plantation	Apple ber teak and bamboo
	(c) Single crop: cotton-fallow	Intercropping and bund plantation	Intercropping-orchard plantation	Intercropping with mango with cotton
Built-up	Settlement areas	Home garden	Fruit and multipurpose plantation	Orchard, amla, lemon
Wasteland	Wasteland	Block/bulk plantation	Bamboo plantation	Bamboo

LULC, Land use land cover.

and intercropping in single crop (cotton-fallow), (D) for home gardens, and (E) block/bulk plantations (Fig. 3.6). Moreover, such analyses can help in policy framing and state-level analysis, including the assessment of ecosystem services provided by agroforestry systems .

3.6 Challenges and way forward

Agroforestry represents a promising strategy that combines agriculture and forestry to create sustainable livelihoods while promoting environmental conservation and resilience in the face of global challenges. Farmers traditionally practice various agroforestry systems in different agroclimatic zones. The land under agroforestry is expanding annually in India due to various government and individual initiatives, policies, and schemes. Successful agroforestry models exist for different agroecological regions to achieve crop diversification objectives and improve farmers' returns. The selection of a suitable agroforestry model is highly influenced by the landholding pattern, soil productivity, and marketability

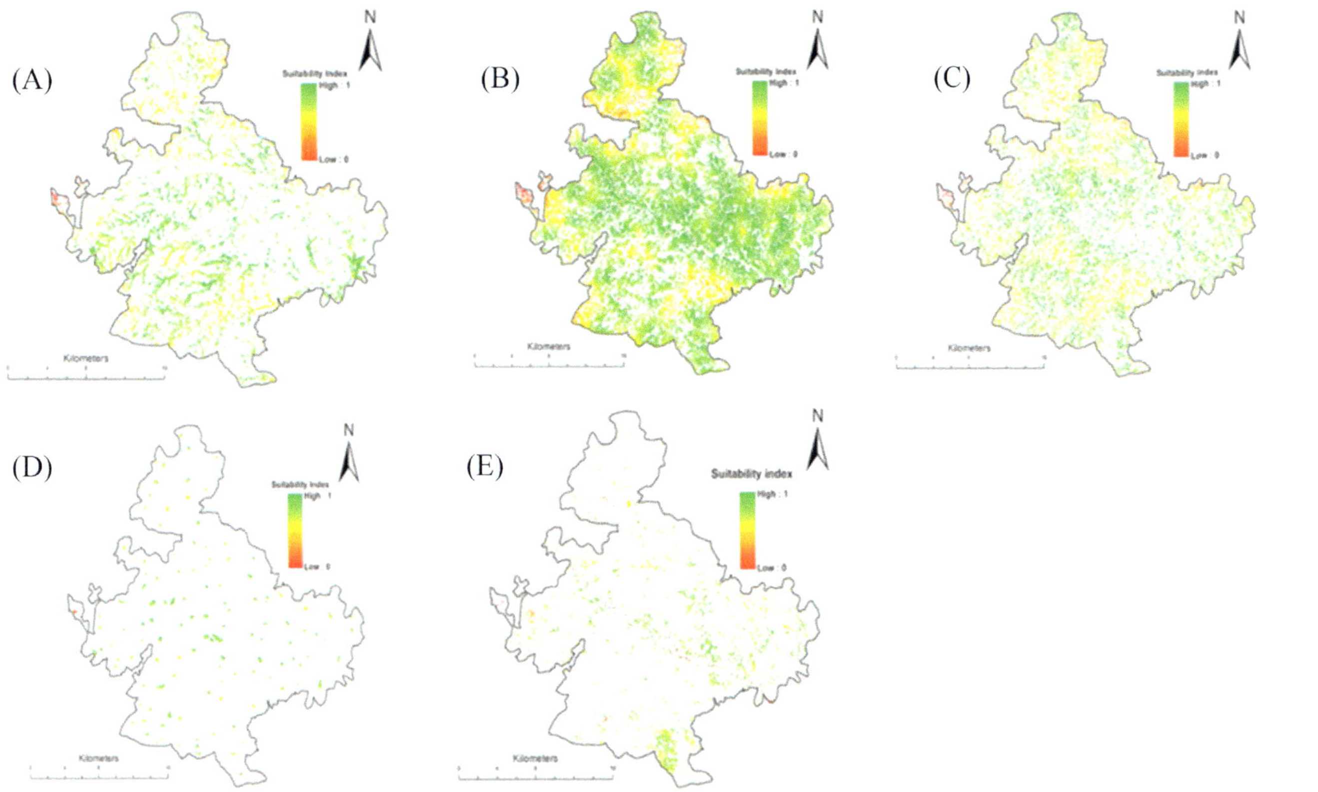

FIGURE 3.6 Agroforestry suitability map for (A) bund and boundary plantation in double-crop (paddy-pulses), (B) bund plantation, boundary and single intercropping crop (paddy-fallow), (C) bund and boundary plantation and intercropping in single crop (cotton-fallow), (D) for home gardens, and (E) block/bulk plantations. Map lines delineate study areas and do not necessarily depict accepted national boundaries.

of the final product (Handa et al., 2019). Despite the countless benefits of agroforestry, farmers are less concerned about adopting agroforestry models, as they cannot obtain a reasonable price for their final product because of the absence of a suitable local market outlet. To provide a new push for agroforestry in the country, agroforestry research needs to focus on (1) developing farmer-friendly agroforestry technologies for fragile ecosystems such as the Himalayan region and coastal, arid, semiarid, and water-logged areas to ensure sustainability and high productivity through economically viable natural resource management options, (2) quantification of differential ecosystem services for different practices, (3) introducing tree genomics for the identification and characterization of quality germplasm, and (4) developing decision support mechanisms and extension methodologies for large-scale adoption of agroforestry. Blockchain technology may also be introduced to enhance the treatability of small timber from farms for interstate and cross-border trade of agroforestry products.

References

Arunachalam, A., Rizvi, R. H., Handa, A. K., & Ramanan, S. S. (2022). Agroforestry in India: Area estimates and methods. *Current Science*, *123*(6), 743–744.

Ballanti, L., Blesius, L., Hines, E., & Kruse, B. (2016). Tree species classification using hyperspectral imagery: A comparison of two classifiers. *Remote Sensing*, *8*(6), 445.

Behera, M. D., Gupta, A. K., Barik, S. K., Das, P., & Panda, R. M. (2018). Use of satellite remote sensing as a monitoring tool for land and water resources development activities in an Indian tropical site. *Environmental Monitoring and Assessment*, *190*, 1–21.

Blaschke, T., Hay, G. J., Kelly, M., Lang, S., Hofmann, P., Addink, E., … … Tiede, D. (2014). Geographic object-based image analysis – Towards a new paradigm. *ISPRS Journal of Photogrammetry and Remote Sensing*, *87*, 180–191.

Bommarco, R. (2013). Ecological enhancement of farmland by agri-environmental schemes does not necessarily increase the diversity of farmland arthropods. *Proceedings of the Royal Society B*, *280*(1754), 20122197.

Chandra, J. (2001). Scope of Poplar Cultivation. *The Indian Forester*, *127*, 51–60.

Chen, J., Chen, J., Liao, A., Cao, X., Chen, L., Chen, X., He, C., Han, G., Peng, S., Lu, M., Zhang, W., Tong, X., & Mills, J. (2015). Global land cover mapping at 30 m resolution: A POK-based operational approach. *ISPRS Journal of Photogrammetry and Remote Sensing*. Available from https://doi.org/10.1016/j.isprsjprs.2014.09.002.

Chang, K. (2008). *Introduction to geographic information systems*. McGraw-Hill Education.

Chaturvedi, O. P., Pandey, A., & Singh, B. (2017). Agroforestry in India: A review. *Agroforestry Systems, 91*(4), 901–929.

Das, P., Behera, M. D., Barik, S. K., Mudi, S., Jagadish, B., Sarkar, S., ... Chauhan, P. S. (2022). Shifting cultivation induced burn area dynamics using ensemble approach in Northeast India. *Trees, Forests and People, 7*, 100183.

Das, P., Mudi, S., Behera, M. D., Barik, S. K., Mishra, D. R., & Roy, P. S. (2021). Automated mapping for long-term analysis of shifting cultivation in Northeast India. *Remote Sensing, 13*(6), 1066.

Das, P., Panda, R. M., Dash, P., Jana, A., Jana, A., Ray, D., ... Kolluru, V. (2022). Multi-decadal mapping and climate modelling indicates eastward rubber plantation expansion in India. *Sustainability, 14*(13), 7923.

Den Herder, M., Moreno, G., Mosquera-Losada, R. M., Palma, J. H., Sidiropoulou, A., Freijanes, J. J. S., ... Burgess, P. J. (2017). Current extent and stratification of agroforestry in the European Union. *Agriculture, Ecosystems & Environment, 241*, 121–132.

Dhyani, S. K., & Handa, A. K. (2013). Area under agroforestry in India: An assessment for present status and future perspective. *Indian Journal of Agroforestry, 15*(1), 1–11.

Dhyani, S. K., Ram, A., & Dev, I. (2016). Potential of agroforestry systems for carbon sequestration in India. *Indian Journal of Agricultural Sciences, 86*(9), 1103–1112.

Dhiman, R. C. (2012). Status of poplar culture in India. *Forestry Bulletin, 12*(1), 15–32.

Dhyani, S. K., Samra, J. S., Ajit., Handa, A. K., & Uma. (2007). Forestry to support increased agricultural production: focus on employment generation and rural development. *Agricultural Economics Researh Review, 20*(2), 179–202.

Dhyani, S. K., Sharda, V. N., & Samra, J. S. (2005). Agroforestry for sustainable management of soil, water and environmental quality: Looking back to think ahead. *Range Management \& Agroforestry, 26*(1), 71–83.

Dollinger, J., & Jose, S. (2018). Agroforestry for soil health. *In Agroforestry Systems*. Available from https://doi.org/10.1007/s10457-018-0223-9.

FAO. (2020). Geo-Informatics and Earth Observations in Support of Agroforestry.

Garrity, D. P., et al. (2010). Evergreen agriculture: A robust approach to sustainable food security in Africa. *Food Security, 2*(3), 197–214.

Garrity, D., Akinnifesi, F., Ajayi, O., & Weldesemayat, S. (2010). Evergreen agriculture: A robust approach to sustainable food security in Africa. *Food Security, 2*(3), 197–214.

Handa, A.K., Dev, I., Rizvi, R.H., Kumar, N., Ram, A., Kumar, D., Kumar, A., Bhaskar, S., Dhyani, S.K., & Rizvi, J. (2019). Successful Agroforestry Models for Different Agro-Ecological Regions in India.

ISFR. (2013). *India state forest report 2013*. https://fsi.nic.in/forest-report-2013.

Jose, S. (2009). Agroforestry for ecosystem services and environmental benefits: An overview. *Agroforestry Systems, 76*(1), 1–10.

Joshi, N. P., Pandit, M., & Joshi, P. K. (2018). Remote sensing-based assessment of agroforestry systems: A review. *International Journal of Remote Sensing, 39*(16), 5279–5302.

Josiah, S. J. (2014). The Influence of agroforestry practice on the economic development of rural farmers. *International Journal of Agriculture and Biology, 16*(4), 726–732.

Kaplan, E. D., & Hegarty, C. J. (2006). *Understanding GPS: Principles and applications*. Artech House.

Kidd, C. V., & Pimental, D. (Eds.), (2012). *Integrated resource management: Agroforestry for Development*. Elsevier.

Kiptot, E., Sinclair, F., & Place, F. (2022). Trees for life: Socio-economic benefits of agroforestry. *World Agroforestry*.

Kussul, N., Lavreniuk, M., Skakun, S., & Shelestov, A. (2021). Remote sensing and machine learning for tree species classification in agroforestry landscapes. *Remote Sensing, 13*(7), 1331.

Kuyah, S., Öborn, I., Jonsson, M., Dahlin, A. S., Barrios, E., Muthuri, C., ... Sinclair, F. L. (2016). Trees in agricultural landscapes enhance provision of ecosystem services in Sub-Saharan Africa. *In International Journal of Biodiversity Science, Ecosystem Services and Management*. Available from https://doi.org/10.1080/21513732.2016.1214178.

Kumar, A., Singh, R., & Patel, S. (2021). Integrating optical and soil moisture data for improved water management in agroforestry systems. *Agricultural Water Management, 240*106297. Available from https://doi.org/10.1016/j.agwat.2020.106297.

Lal, R. (2005). Forest soils and carbon sequestration. *Forest Ecology and Management, 220*(1-3), 242–258.

Lhermitte, S., Verbesselt, J., Verstraeten, W. W., & Coppin, P. (2019). A review of remote sensing approaches for monitoring species composition and diversity in tropical and subtropical forests. *Remote Sensing of Environment, 233*, 111355.

Lillesand, T. M., & Kiefer, R. W. (2015). *Remote sensing and image interpretation*. John Wiley & Sons.

Longley, P. A., Goodchild, M. F., Maguire, D. J., & Rhind, D. W. (2015). *Geographic information science and systems*. John Wiley & Sons.

Lu, D., Batzel, J., Moran, E., & Dutra, L. (2020). Object-based image analysis for remote sensing of forests. *Remote Sensing of Environment, 251*, 112088.

Lu, T., Brandt, M., Tong, X., Hiernaux, P., Leroux, L., Ndao, B., & Fensholt, R. (2022). Mapping the abundance of multipurpose agroforestry *Faidherbia albida* trees in Senegal. *Remote Sensing, 14*(3), 662.

Makalew, J. A. (2018). Analysis of carbon stock and sequestration potential of agroforestry systems in the Central Rift Valley of Ethiopia. *Land Use Policy, 77*, 647–657.

Matin, S., Ghosh, S., & Behera, M. D. (2019). Assessing land transformation and associated degradation of the west part of Ganga River Basin using forest cover land use mapping and residual trend analysis. *Journal of Arid Land*. Available from https://doi.org/10.1007/s40333-018-0106-y.

MGNREGA 2005. Mahatma Gandhi National Rural Employment Guarantee Act, 2005: An Act to provide for the enhancement of livelihood security of the rural population. Ministry of Rural Development, Government of India. Retrieved from http://www.rural.nic.in.

Mishra, P., Kumar, A., & Singh, R. (2018). State-level initiatives in promoting agroforestry: A case study of Madhya Pradesh and Chhattisgarh. *Agroforestry Systems*, *92*(6), 1573–1585. Available from https://doi.org/10.1007/s10457-018-0219-2.

Mukhlis, I., Rizaludin, M. S., & Hidayah, I. (2022). Understanding socio-economic and environmental impacts of agroforestry on rural communities. *Forests*, *13*(4), 556.

Nair, P. K. R. (1985). Classification of agroforestry systems. *Agroforestry Systems*. Available from https://doi.org/10.1007/BF00122638.

Nair, P. K. R. (2012). Agroforestry systems and environmental quality: Introduction. *Agroforestry Systems*, *86*(1), 1–6.

Nair, P. K. R. (2013). *An introduction to agroforestry*. Springer Science & Business Media.

Nair, P. K. R. (2018). Agroforestry as a strategy for carbon sequestration. *Journal of Plant Nutrition and Soil Science*, *181*(4), 237–248.

Nair, P. K. R., et al. (2009a). Agroforestry as a strategy for carbon sequestration. *Journal of Plant Nutrition and Soil Science*, *172*(1), 10–23.

Nair, P. K. R., et al. (2009b). Carbon sequestration in agroforestry systems. *Advances in Agronomy*, *108*, 237–307.

NAP. (2014). *National Agroforestry Policy of India*. Ministry of Agriculture, Government of India Retrieved from. Available from https://agricoop.nic.in.

National Bamboo Mission (NBM). (2018). *National Bamboo Mission operational guidelines*. Ministry of Agriculture and Farmers Welfare, Government of India Retrieved from. Available from https://nbm.nic.in.

Ningthoujam, S., Singh, R., & Kumar, A. (2016). Integrating SAR imagery with optical and LiDAR data for enhanced biomass estimation in agroforestry systems. *Remote Sensing of Environment*, *187*, 186–199. Available from https://doi.org/10.1016/j.rse.2016.10.024.

Ningthoujam, S., Singh, R., & Kumar, A. (2017). The potential of NASA's NISAR mission for agroforestry biomass quantification: A multi-sensor approach. *Journal of Geophysical Research: Biogeosciences*, *122*(12), 3124–3136. Available from https://doi.org/10.1002/2017JG003747.

NRCAF. (2013). *The Vision-2050*.

ICRAF. (2016). Harnessing geo-informatics tools for agroforestry applications: A guide to technical skills and training. Retrieved from https://www.icraf.org

NITI Aayog, Pant, R. M., Tiwari, B. K., & Choudhury, D. (2018). Report of Working Group III, shifting cultivation: Towards a transformational approach. New Delhi.

Ospina, C. (2017). Climate and economic benefits of agroforestry systems. https://climate.org/%0Awp-content/uploads/2017/03/Agroforestry-Article-3.6.17.

Poudel, K. P., Yamanaka, T., & Shrestha, B. B. (2019). Assessment of tree and shrub diversity and their potential for agroforestry system in Mid-Hill area of Nepal. *Applied Geography*, *111*, 102088.

Puliti, S., Orka, H. O., Nelson, R., Gobakken, T., Næsset, E., & Solberg, S. (2020). Opportunities and challenges for operational agroforestry mapping at national scales using multispectral satellite sensor data. *International Journal of Applied Earth Observation and Geoinformation*, *90*, 102113.

Rizvi, R. H., Agarwal, A., Chaturvedi, O. P., & Singh, R. (2020). Using geo-informatics in agroforestry: Challenges and promising pathways. *Current Science, 118*(2), 206–212.

Rizvi, R. H., Dhyani, S. K., Newaj, R., Karmakar, P. S., & Saxena, A. (2014). Mapping agroforestry area in India through remote sensing and preliminary estimates. *Indian Farming*, *63*(11).

Rizvi, R. H., Newaj, R., Handa, A. K., Sridhar, K. B., & Kumar, A. (2019). Agroforestry mapping in India through geospatial technologies: Present status and way forward. *Technical Bulletin*, *1*(2019), 1–35.

Rizvi, R. H., Newaj, R., Karmakar, P. S., Saxena, A., & Dhyani, S. K. (2016). Remote sensing analysis of agroforestry in Bathinda and Patiala districts of Punjab using sub-pixel method and medium resolution data. *Journal of the Indian Society of Remote Sensing*, *44*(4), 657–664.

Rizvi, R. H., Yadav, R. S., Singh, R., Datt, K., Khan, I. A., & Dhyani, S. K. (2009). Spectral analysis of remote sensing image for assessment of agroforestry areas in Yamunanagar district of Haryana. In National Symposium on "Advances in Geo-spatial Technologies with Special Emphasis on Sustainable Rainfed Agriculture," RRSSC, Nagpur (Vol. 7).

Sannigrahi, S., Mishra, B., & Behera, S. (2020). Remote sensing and GIS-based evaluation of green cover dynamics and its relation with climate variability in a part of Eastern India. *Environment, Development and Sustainability*, *22*(3), 2669–2690.

Schwendenmann, L., Veldkamp, E., Moser, G., Koehler, L., & Clough, Y. (2017). Carbon sequestration and emissions from a home garden in the tropics: A case study from Sulawesi, Indonesia. *Agroforestry Systems*, *91*(2), 269–283.

Schwilch, G., Bestelmeyer, B., Bunning, S., Critchley, W., Herrick, J., Kellner, K., . . . Winslow, M. (2011). Experiences in monitoring and assessment of sustainable land management. *Land Degradation and Development*. Available from https://doi.org/10.1002/ldr.1040.

Shrivastava, S., & Saxena, A. k (2017). Wood is Good: But, is India doing enough to meet its present and future needs? http://cdn.cseindia.org/attachments/0.39269300_1505297987_wood-is-good.pdf.

Sileshi, G. W., Akinnifesi, F. K., Debusho, L. K., Beedy, T. L., Ajayi, O. C., & Mong'omba, S. (2019). Agroforestry: A versatile and dynamic approach for sustainable development in a changing world. *Sustainability*, *11*(3), 915. Available from https://doi.org/10.3390/su11030915.

Simic-Milas, A., Balla, D., Veverec, I., Lukic, T., & Savic, S. (2017). Pixel-based and object-based image analysis for classification of agroforestry systems using Landsat 8 OLI satellite imagery. *Computers and Electronics in Agriculture*, *139*, 30–39.

Singh, R., Garg, K. K., Anantha, K. K., Venkataradha, A., Dev, I., Dixit, S., & Dhyani, S. K. (2021). Building resilient agricultural system through groundwater management interventions in degraded landscapes of Bundelkhand region, Central India. *Journal of Hydrology: Regional Studies*, *37*. Available from https://doi.org/10.1016/j.ejrh.2021.100929.

Singh, R. K., Behera, M. D., Das, P., Rizvi, J., Dhyani, S. K., & Biradar, Ç. M. (2022). Agroforestry suitability for planning site-specific interventions using machine learning approaches. *Sustainability*, *14*(9), 5189.

Singh, R. K., Biradar, C. M., Behera, M. D., Prakash, A. J., Das, P., Mohanta, M. R., ... Rizvi, J. (2023). Optimizing carbon fixation through agroforestry: Estimation of aboveground biomass using multi-sensor data synergy and machine learning. *Ecological Informatics*102408. Available from https://doi.org/10.1016/j.ecoinf.2023.102408.

Singh, R. K., Rizvi, J., Behera, M. D., & Biradar, C. (2021). Automated crop type mapping using time-weighted dynamic time warping – A basis to derive inputs for enhanced food and nutritional security. *Current Research in Environmental Sustainability*, *3*, 100032.

Singh, R., Garg, K. K., Wani, S. P., Tewari, R. K., & Dhyani, S. K. (2014). Impact of water management interventions on hydrology and ecosystem services in Garhkundar-Dabar watershed of Bundelkhand region, Central India. *Journal of Hydrology*, *509*, 132–149.

Smith, J., van Noordwijk, M., Hoang, M. H., & Kumar, B. M. (2015). Learning from agroforestry and farming systems. *Current Opinion in Environmental Sustainability*, *14*, 127–133.

Sola, G., et al. (2018). *Geospatial technologies for sustainable land management.*

Tewari, D. N. (1992). *A monograph on bamboo* (p. 498) Dehra Dun: International Book Distributors. Valley Offset Printers and Publishers.

Thenkabail, P. S., Schull, M. A., Turral, H., & Gassman, P. W. (2019). The role of remote sensing in precision agriculture. *American Journal of Agricultural Economics*, *101*(3), 894–922.

UN Nations. (2018). *Safeguarding land and agricultural data: Challenges and strategies.* United Nations Department of Economic and Social Affairs Retrieved from. Available from https://www.un.org.

UN. (2018). *Geospatial technologies for sustainable development.*

van Noordwijk, M., Duguma, L. A., Dewi, S., Leimona, B., Catacutan, D. C., Lusiana, B., ... Minang, P. A. (2018). SDG synergy between agriculture and forestry in the food, energy, water and income nexus: Reinventing agroforestry. *In Current Opinion in Environmental Sustainability.* Available from https://doi.org/10.1016/j.cosust.2018.09.003.

van Noordwijk, M., Mulia, R., Dewi, S., Lusiana, B., Minang, P. A., & Leimona, B. (2019). How tree cover and precipitation influence water use in agroforestry in Northern China. *Hydrology and Earth System Sciences*, *23*(3), 1235–1251. Available from https://doi.org/10.5194/hess-23-1235-2019.

Viswanath, S., Nair, P. K. R., Kaushik, P. K., & Prakasam, U. (2000). Acacia nilotica trees in rice fields: A traditional agroforestry system in central India. *Agroforestry Systems*, *50*(2), 157–177.

WAC. (2015). *Integrating tree detection and land use systems: Advancing agroforestry research.* Nairobi, Kenya: World Agroforestry Centre Retrieved from. Available from https://www.worldagroforestry.org.

Waldron, A., Garrity, D., Malhi, Y., Girardin, C., Miller, D. C., & Seddon, N. (2017). Agroforestry can enhance food security while meeting other sustainable development goals. *Tropical Conservation Science.* Available from https://doi.org/10.1177/1940082917720667.

Zhang, X., et al. (2010). Effects of alley cropping and no-tillage management on soil properties in Southwest China. *Journal of Environmental Management*, *91*(4), 837–844.

Zomer, R. J., Bossio, D. A., Trabucco, A., van Noordwijk, M., & Xu, J. (2022). Global carbon sequestration potential of agroforestry and increased tree cover on agricultural land. *Circular Agricultural Systems*, *2*(1), 1–10.

Zomer, R.J., Trabucco A., Coe, R., Place, F., van Noordwijk, M., Xu, J. (2014). Trees on farms: an update and reanalysis of agroforestry's global extent and socio-ecological characteristics. Working Paper 179, February, 54. http://www.worldagroforestry.org/downloads/publications/PDFs/WP14064.PDF.

Zomer, R. J., Trabucco, A., Coe, R., & Place, F. (2009). Trees on farm: Analysis of global extent and geographical patterns of agroforestry. *Agriculture, Ecosystems & Environment*, *56*(3), 157–167. Available from http://www.worldagroforestry.org/downloads/publications/pdfs/b15012.pdf%5Cnhttp://www.hindawi.com/journals/isrn.forestry/2012/867249/%5Cn, http://linkinghub.elsevier.com/retrieve/pii/S0167880915002339%5Cnhttp://books.google.co.jp/books?id = sZYfAQAAIAAJ.

Chapter 4

Sustaining floriculture and floral fragrance in a changing climate

Saroj Kanta Barik[1], Mukunda Dev Behera[2] and Dibyendu Adhikari[3]

[1]*Centre for Advanced Studies in Botany, North-Eastern Hill University, Shillong, India,* [2]*Centre for Ocean, River, Atmosphere and Land Sciences (CORAL), Indian Institute of Technology Kharagpur, Kharagpur, West Bengal, India,* [3]*Plant Ecology and Climate Change Science, CSIR-National Botanical Research Institute, Lucknow, Uttar Pradesh, India*

4.1 Introduction

Floral fragrance plays a crucial role in plant ecology and evolution, serving as a key component of plant-pollinator interactions and reproductive success. The scent emitted by flowers acts as a vital channel of communication between flowering plants and their pollinators, influencing pollinator behavior and the formation of plant-pollinator networks (Hetherington-Rauth & Ramírez, 2016; Raguso, 2008). Studies have shown that floral fragrance is a putatively important character in the evolution of flowering plants, with natural selection on scent being a significant but understudied aspect (Parachnowitsch, Raguso, & Kessler, 2012). Furthermore, the evolution of floral scents is correlated with chemical communication in insects, indicating the interconnectedness of floral traits and their impact on plant evolution (Schiestl, 2010). Floral fragrance not only attracts pollinators but also serves as a defense mechanism against herbivores, highlighting its multifaceted role in plant ecology and evolution (Ramos & Schiestl, 2020; Theis, Kesler, & Adler, 2009). Additionally, the variation in floral fragrance compounds may act as a pollination strategy, disrupting associative learning processes and inhibiting pollinators' ability to recognize non-rewarding flowers (Moya & Ackerman, 1993). Overall, the importance of floral fragrance in plant ecology and evolution is underscored by its role in pollinator attraction, plant defense, and the intricate interplay between plants and their surrounding environment.

4.1.1 Biochemical pathways of fragrance production in flowers

The biochemical pathways of fragrance production in flowers have been a subject of extensive research, shedding light on the intricate processes underlying the synthesis of floral scents. Studies have employed various approaches, including metabolic flux analysis, transcriptomic data integration, and bioinformatics analysis, to unravel the pathways responsible for fragrance production in diverse plant species. For instance, investigations into petunia petal tissue have utilized metabolic flux analysis with stable isotope labeling to elucidate the biochemical pathways leading to the formation of benzenoid compounds, providing insights into the interconnections between intermediates within the benzenoid network (Boatright et al., 2004). Similarly, transcriptomic data have been leveraged to propose pathways underlying ornamental traits such as petal coloration, double flowers, and fragrance production in carnation flowers, highlighting the integration of genomic, biochemical, and morphological patterns (Zhang et al., 2022). Studies on *Phalaenopsis* orchids have delved into the chemical components of floral scent, identified floral expressed sequence tags (ESTs), and deduced the pathways of floral scent biosynthesis, offering valuable insights into the genetic and biochemical mechanisms governing fragrance production in orchid flowers (Hsiao et al., 2006). These endeavors have contributed to a comprehensive understanding of the biosynthetic pathways and regulatory mechanisms underlying the production of floral scents, paving the way for the development of new cultivars with desirable traits through molecular breeding approaches (Ramya et al., 2020).

Sustainable Development Perspectives in Earth Observation. DOI: https://doi.org/10.1016/B978-0-443-14072-3.00007-1

4.1.2 Correlation between environmental factors and floral fragrance variation

The production and composition of floral fragrances are profoundly influenced by environmental and climatic conditions. Extensive research has demonstrated that variables such as temperature, soil quality, and plant stress significantly impact the generation and emission of floral volatiles. For example, elevated temperature has been observed to adversely affect scent production in certain species (Cna'ani et al., 2015). Moreover, environmental stressors, including physical damage to plants, can lead to changes in the emitted volatile compounds, which may have implications for attracting pollinators and the overall reproductive success of the plant (Cna'ani et al., 2015; Effmert et al., 2005). The stability and release of floral fragrances are also influenced by factors like light, oxygen, and humidity, affecting their photoactivation and the sensitivity of certain chemical groups (Karlberg et al., 2013; Wu et al., 2023). These findings underscore the complexity of the relationship between environmental factors and the regulation of floral scent emissions.

4.1.3 Diversity in floral scent and its climate linkages

The diversity and variability of floral scents have been a subject of considerable research, with studies establishing a clear connection between climatic conditions and the spectrum of floral fragrances. Environmental elements such as temperature variation, drought stress, and nutrient levels play a pivotal role in shaping the composition and emission patterns of floral scents (Cna'ani et al., 2021; Fu et al., 2017; Luizzi, Friberg, & Petrén, 2021). Research indicates that temperature changes can markedly affect floral scent emissions, showcasing the adaptive nature of scent production in response to climatic shifts (Fu et al., 2017). Furthermore, drought-induced stress has been shown to result in phenotypic changes in floral scent, suggesting that environmental factors can induce variations in fragrance beyond genetic predispositions (Luizzi et al., 2021). The diurnal emission patterns of floral scents are also determined by a combination of phylogenetic lineage and abiotic conditions, highlighting the multifactorial influence of the environment on floral scent dynamics (Cna'ani et al., 2021).

4.1.4 Insights from prior research on floral volatiles

Previous studies have provided in-depth insights into the mechanisms and influences shaping the volatile composition and emission patterns in flowers. Effmert et al. (2005) explored the volatile composition and emission patterns in *Mirabilis jalapa*, offering perspectives on the spatial-temporal dynamics of scent production. Additionally, Verdonk, Haring, Tunen, and Schuurink (2005) elucidated the regulatory function of the *ODO1* gene in fragrance biosynthesis in Petunia, shedding light on the genetic underpinnings of scent production and its implications for ecological niche modeling (ENM) studies. In a similar vein, Xu, Wang, Hong, and Zhou (2019) conducted a transcriptomic analysis of the scent biosynthesis pathway in *Cymbidium faberi*, functionally characterizing its jasmonic acid carboxyl methyltransferase gene, thereby enhancing our understanding of the genetic and biochemical pathways involved in floral fragrance production. These studies collectively emphasize the profound impact of climatic factors on floral scent variation, offering key insights into the ecological and evolutionary significance of fragrance diversity in the plant kingdom.

4.1.5 Impact of climate change on flowers

Floriculture faces significant challenges due to climate change marked by shifts in temperature and weather patterns. These changes may critically impact the fragrance of flowers, as the process of creating volatile organic compounds (VOCs), vital for floral scent, is directly influenced by rising temperature and fluctuating precipitation. Elevated temperature may hasten the release of these compounds, potentially resulting in a more potent but ephemeral scent. On the other hand, some VOCs could be diminished under thermal stress, diminishing the overall intensity of the fragrance. Moreover, varying weather patterns, including more frequent occurrences of extreme conditions like drought and heavy rainfall, impose additional stress on plants. This can affect VOC synthesis, thereby modifying the scent profiles of flowers. Understanding how climate change affects floral fragrance is therefore crucial, not only for the sustainability of the floriculture sector but also for maintaining the delicate ecological equilibrium involving these plants.

Research indicates that climate factors can shape floral scents, adapting their physicochemical properties to better suit the environmental conditions (Farré-Armengol, Fernández-Martínez, Filella, Junker, & Peñuelas, 2020). Extended droughts and increased nitrogen usage in agriculture, both outcomes of climate change, can alter floral scent by modifying soil water and nitrogen levels (Höfer, Ayasse, & Kuppler, 2022). Moreover, there is a strong link between temperature and flowering patterns, suggesting that climate-induced shifts in flowering time could affect reproductive success and ecological dynamics (Rawal, Kasel, Keatley, & Nitschke, 2014). Furthermore, elevated temperatures have been shown to influence the quality and quantity of floral scents in various species (Cordeiro & Dötterl, 2023). These findings underscore the complex interplay between climate change and floral scent, highlighting the necessity to anticipate and comprehend the effects of climate change on the fragrance of flowers (Park & Mazer, 2018).

4.1.6 Impact of changing climate on fragrance quality and quantity

The impact of changing climate on the fragrance quality and quantity of flowers is a complex and multifaceted issue. Climate change can significantly change the flowering phenology, with some species being more sensitive to these changes than others (Willis, Ruhfel, Primack, Miller-Rushing, & Davis, 2008). Higher temperature due to climate change may have significant negative impacts on flower development, affecting both the quantity and quality of fragrance (Praveen & Ramach, 2017). Studies have shown that temperature has a strong effect on the quantity of fragrance emitted by flowers (Jakobsen & Olsen, 1994). Furthermore, global climate change could affect the cues used by plants to begin flowering and the number of flowers produced (Inouye, Saavedra, & Lee-Yang, 2003). Contrary to expectation, damage from herbivory increases fragrance production in male flowers (Theis et al., 2009). Additionally, factors such as light availability, hormone production, climatic conditions, developmental stage of the flower, and activity of pollinators can affect the rhythmic emission of fragrance from flowers (Khan et al., 2022).

The response of flowering times to climate change is complex, with evidence of strong directional selection for early flowering in response to climate change (Iler, Høye, Inouye, & Schmidt, 2013). Observations on how individual plants alter the timing of flowering in response to local temperature or rainfall provide a way to evaluate the impacts of climate variation on the world's flora (Park & Mazer, 2018). Climate warming, which advances the flowering onset of many plant species, could negatively affect the reproductive success of certain plants if co-flowering plants advance their flowering onset more strongly (Kehrberger & Holzschuh, 2019). The timing of the first flower has been explained on the basis of a number of factors such as snowmelt, elevation, and growing degree days (Lessard-Therrien, Bolmgren, & Davies, 2014). These findings highlight the intricate relationship between climate change and flowering phenology. In addition to the timing of flowering, the fragrance quality and quantity of flowers are also influenced by climate change. Studies have shown that the floral fragrance of certain plant species follows a clear circadian rhythm, with volatile emission increasing in the evening and reaching its maximum at night, influenced by environmental factors such as temperature and light availability (Stashenko & Martínez, 2018). Varietal performance in terms of fragrance has been observed to vary significantly in different climatic conditions (Shahrin, Roni, Taufique, Mehraj, & Uddin, 2015). Thus, understanding these complex interactions is crucial for predicting and managing the potential effects of climate change on floral fragrance.

4.1.7 Linking fragrance levels to ecological niches

The existing research on fragrance levels and their connection to ecological niches reveals several gaps that require further investigation. Gross, Sun, and Schiestl (2016) demonstrated a region-specific selection on floral scent in a terrestrial orchid, indicating the influence of geographical factors on scent variation. However, the specific climatic and environmental drivers shaping these region-specific differences remain to be elucidated. Parachnowitsch et al. (2012) highlighted the need for biologically relevant hypotheses and advanced statistical tools, such as path analyses, to determine the selection of VOCs and their relationship to ecological niches. Furthermore, Anderson (2016) emphasized the importance of biotic interactions in niche and distribution models, suggesting that the exclusion of biotic interactors may limit the comprehensive understanding of niche dynamics. These studies collectively underscore the need for a more integrated approach that considers the influence of climatic, geographic, and biotic factors on floral scent variation within ecological niches, providing a foundation for future research to bridge these gaps and enhance our understanding of the complex interplay between fragrance levels and ecological niches.

4.1.8 Chemical composition vis-à-vis sensory detection level

The sensory detection of fragrance in flowers is a complex interplay of chemical composition and human perception. The levels of fragrance in flowers are determined by the concentration and combination of VOCs that they emit. These VOCs are responsible for the characteristic scents of each flower. The sensory detection of these fragrances involves olfactory receptors in humans, which can detect and distinguish a wide range of aromatic molecules. For example, jasmine's scent is primarily due to compounds like jasmonates and benzyl acetate, which give it a rich, sweet, and highly potent fragrance. Tuberose, on the other hand, emits a complex blend of compounds,including methyl benzoate and tuberose polyanthines, contributing to its intense and heady aroma. The detection of these fragrances is not just a binary presence/absence scenario. It involves a gradient of intensity, where the strength of the perceived scent can vary depending on the concentration of VOCs, the environmental conditions (such as humidity and temperature), and the sensitivity of the individual's olfactory system. This gradient of fragrance intensity can be mapped and modeled using ENM.

4.1.9 ENM based on fragrance level

ENMs is used to predict the distribution of species based on environmental variables. However, in the context of fragrance studies, ENM can be adapted to understand the distribution of fragrance intensity. By incorporating the data on environmental conditions such as temperature, light and precipitation that affect VOC production , ENM can help predicting areas where the flowers' fragrances would be at their most potent. This modeling approach can be particularly useful for horticultural planning and for understanding how changes in climate could impact the sensory qualities of these flowers. Thus, ENM provides a novel approach to map and understand these interactions, offering valuable insights into the future of floriculture in changing environmental conditions.

4.1.10 Species distribution and ecological niche modeling

ENM has emerged as a valuable tool offering insights into species distribution, habitat suitability, and respons of species to environmental changes (Adhikari et al., 2019). By integrating species occurrence data with environmental GIS layers, ENM provides a quantitative approach to understand niche evolution and ecological diversity within plant clades Warren, Glor, and Turelli (2008). This modeling technique has been applied to predict the potential distribution of invasive plant species, shedding light on their ecological niche shifts and rapid expansion (Manzoor, Griffiths, Obiakara, Esparza-Estrada, & Lukáč, 2020). Furthermore, ENM has been widely used in plant ecology to assess suitable environmental conditions for different plant species at a continental scale, highlighting its broad applicability in understanding the niche and distribution of species (Velazco, Galvão, Villalobos, & Marco, 2017). We hypothesize that ENM can be employed to identify areas with a high concentration of fragrant compounds in flowers, based on the premise that environmental conditions play a significant role in shaping the production and distribution of floral scents. By integrating climatic variables along with known occurrences of fragrant flower species, the ecological niche model can predict suitable habitats with optimal conditions for the production of fragrant compounds.

Considering the above, the present study was undertaken with the following objectives: (1) to characterize the climatic niche of fragrant floral species through ENM by integrating climatic variables along with known occurrences of fragrant flower species, and (2) to predict potential geographic areas where flowers with high fragrance levels are likely to thrive under different climate scenarios and drought-prone areas.

4.2 Materials and methods

4.2.1 Selection of flower species

For the present study, we selected two open field floriculture species with exquisite fragrance, viz., jasmine (*Jasminum officinale* L.) and tuberose (*Polianthes tuberosa* L.). Jasmines are cultivated for their sweet floral fragrance, and the emitted scent volatiles vary among different species and cultivars, contributing to their unique aroma (Barman & Mitra, 2018; Bera, Kotamreddy, Samanta, Maiti, & Mitra, 2015; Yu et al., 2017). The fragrance of jasmine is highly valued in various industries, including perfumery, medicine, cosmetics, and in the food and beverage industry, such as infusion into green tea (Joulain, 2021; Qi et al., 2022; Xu et al., 2021). The jasmine fragrance remains one of the most sought-after botanical scents for centuries. The oil is evocative and powerful and a popular ingredient of modern perfumery. About 83% of women's cosmetics and 33% of men's perfumes contain jasmine fragrance, a unisex odor. The flower yields its full odor at night after the day's temperature cools down. The *Jasminum* genus (family: Oleaceae) distributed in tropical and subtropical regions of Asia and Europe, has more than 200 species with varied shapes and sizes belonging to both deciduous and evergreen functional groups, and a few of them are poisonous as well. Jasminaldehyde is the aroma compound responsible for jasmine's characteristic scent. The other chemicals present in Jasmine flower include benzaldehyde, benzoic acid, alpha-terpineol, benzyl acetate, farnesol, benzyl alcohol, eugenol, geraniol, nerolidol, jasmone, linalyl acetate, salicylic acid, and vanillin. While tuberose is cherished for the aromatic oils extracted from its fragrant white flowers, its characteristic compounds, such as methyl isoeugenol and benzyl benzoate, are highly sought after in the perfume and fragrance industry (El-Shanhorey & Ahmed, 2020; Fragoso-Jiménez et al., 2019). Thus, both jasmine and tuberose have significant commercial value due to their distinct and captivating fragrances, making them highly sought after floral crop in the fragrance market (Li, Luo, Ma, & Zeng, 2018; Zhang et al., 2020; Zhou, 2023).

4.2.2 Occurrence data

Occurrence data for the two species were collected through field surveys, compilation from literature, and from the database of the Global Biodiversity Information Facility (GBIF, 2021a,b). Overall, 453 occurrence records were compiled for jasmine, while 17 for tuberose.

4.2.3 Climatic variables

We used 19 bioclimatic variables representing the current climatic conditions (baseline) to delineate the niches of jasmine and tuberose, and predict the potential areas of cultivation of these two flower species in India (Table 4.1). These variables are derivatives of the average monthly climate data for minimum, mean, and maximum temperature and precipitation of the last 30 years, i.e., 1970–2000 (Fick & Hijmans, 2017). They describe the annual temperature and precipitation trends, including seasonality and extremities-factors which may impose physiological constraints on

TABLE 4.1 Analysis of variable contributions.

Variable	Jasmine		Tuberose	
	Percent contribution	Permutation importance	Percent contribution	Permutation importance
bio_1: Annual mean temperature	7.7	0.4	10.9	8.1
bio_2: Mean diurnal range (mean of monthly (max temp–min temp))	0.4	0.3	2.5	0.2
bio_3: Isothermality (bio_2/bio_7) (*100)	0.9	4.5	0	0
bio_4: Temperature seasonality (standard deviation*100)	3.4	3.1	3.7	19.1
bio_5: Max Temperature of warmest month	3.7	5.8	2.5	2.6
bio_6: Min temperature of coldest month	34.1	12.8	2.5	0
bio_7: Temperature annual range (bio_5-bio_6)	0.7	5.5	7.7	0
bio_8: Mean temperature of wettest quarter	12.5	10.5	1.1	5
bio_9: Mean temperature of driest quarter	1.4	0.8	0.1	0
bio_10: Mean temperature of warmest quarter	1.4	5.6	0.2	7.7
bio_11: Mean temperature of coldest quarter	0.9	2.1	31.4	27.5
bio_12: Annual precipitation	1.4	7.7	0	0
bio_13: Precipitation of wettest month	7.8	1.8	2.8	2.9
bio_14: Precipitation of driest month	0.6	1.4	3.8	2.4
bio_15: Precipitation seasonality (coefficient of variation)	3.2	6.9	6.3	0.4
bio_16: Precipitation of wettest quarter	13	24.3	0	0
bio_17: Precipitation of driest quarter	0.4	0.4	0.6	1.1
bio_18: Precipitation of warmest quarter	4.1	3	10.4	19.9
bio_19: Precipitation of coldest quarter	2.4	3.4	13.4	11

species and affect their geographical distributions (O'Donnel & Ignizio, 2012). The dataset was downloaded as geoTiff files of 2.5 min spatial resolution from the WorldClim website (https://www.worldclim.org/).

4.2.4 Ecological niche model parameterization

The study employed MaxEnt software version 3.4.3 to model the ecological niche of jasmine and tuberose. MaxEnt distinguishes the environmental conditions preferred by a species by combining occurrence records with the characterization of available environments within the accessible area. The model's parameters included a limit of 10,000 background points, 500 iterations, and a convergence threshold of 0.00001. Feature types such as hinge, product, linear, and quadratic were employed to manage model complexity. Default regularization with a value of one was used to control overfitting. Consistency of the model was evaluated through 20 replicated runs using cross-validation in MaxEnt. The model's effectiveness was measured by area under the curve (AUC) of MaxEnt, while the significance of bioclimatic predictors was determined through variable contribution analyses, jackknife testing, and response curve examination. These analyses involved ranking variables based on their contribution, assessing model gains with the inclusion or exclusion of individual variables, and evaluating the influence of variables across climatic gradients. The fitness of the niche model was assessed using AUC, which is a statistical metric that quantifies the model's ability to differentiate between presence and absence scenarios. This involves plotting the true positive rate against the false positive rate at various thresholds, forming the receiver operating characteristic (ROC) curve. The AUC value, ranging between 0 and 1, represents the area under this curve, where 0.5 indicates no predictive power and 1.0 signifies perfect discrimination.

4.2.5 Climatic niche characterization

Climatic niche characterization is essential in deciphering the intricate interactions between a species and the climatic factors shaping its habitat. For the selected fragrant flower species, this process involved a comprehensive assessment through three key methods: variable contribution analysis, the jackknife test of variable importance, and the examination of response curves. The variable contribution analysis step entailed an evaluation of how each climatic variable individually affects the predictive accuracy of the niche model. Within the Maxent framework, the contribution of a variable is gauged by the improvement in model efficacy—specifically, the increase in the AUC-value, when that variable is incorporated. By comparing these increments across various variables, it becomes possible to pinpoint the climatic elements most critical to the flower species distribution.

The jackknife test is a resampling method that ascertains the significance and reliability of each climatic variable within the model. By sequentially excluding each variable and reassessing the model's performance, this test discerns which variables are pivotal predictors of the species presence. A variable whose absence leads to a notable decline in model accuracy is considered highly significant. In contrast, a variable whose removal does not detrimentally affect or even enhance the model's performance may be less critical or redundant, possibly due to overlaps with other variables.

The response curves provide a visual representation of how the likelihood of species presence varies with changes in each climatic variable. These curves are instrumental in understanding how modification in a specific environmental factor influences the species' distribution probability. For example, a response curve might illustrate an increase in presence probability with a rise in temperature to a certain threshold, beyond which the probability diminishes. This helps in identifying the species' optimal environmental ranges and its tolerances to varying climatic conditions. Together, these methods form a robust approach to understand the species' climatic niche, offering vital insights into its environmental preferences and thresholds.

4.2.6 Distribution of potential areas

The prediction of the potential distribution areas for jasmine and tuberose was based on 'cloglog' outputs. This output was then grouped into five classes, viz., marginal (<0.2), low (0.2–0.4), moderate (0.4–0.6), high (0.6–0.8), and very high (>0.8) suitability classes.

4.2.7 Distribution under different climate scenarios

For forecasting the future prevalence of jasmine and tuberose in India's varying climates, the present study utilized the HadGEM3-GC31-LL and IPSL-CM6A-LR general circulation models (GCMs), incorporating two Shared Socioeconomic Pathways (SSPs)—ssp126 and ssp585, that anticipate different greenhouse gas emission scenarios. The study focused on the projection period from 2041 to 2060, using datasets with a spatial resolution of 2.5 arc minutes, corresponding to an approximate 4.6x4.6 km grid cell at the equator. This granularity is critical for precise local prediction, accommodating the diverse and intricate agricultural landscapes where jasmine and tuberose are cultivated in India.

4.2.8 Identifying drought prone areas of India

Drought-prone regions in India were delineated using the Global Aridity Index and Potential Evapotranspiration (ET0) Database Version 3 Database (Global-AI_PET_v3) (Trabucco & Zomer, 2018; https://figshare.com/articles/dataset/Global_Aridity_Index_and_Potential_Evapotranspiration_ET0_Climate_Database_v2/7504448/4). This database presents the Global Aridity Index (Global-AI) and Global Reference Evapotranspiration (Global-ET0) (Global-AI_PET_v3). It offers detailed global raster data on evapotranspiration and potential vegetative growth's rainfall deficit from 1970 to 2000, based on the FAO-56 Penman-Monteith Reference Evapotranspiration (ET0) equation. Monthly aridity index data from January to December for India were extracted and compiled into a composite image, representing the entire year. Subsequently, a principal component analysis was conducted to distill the primary information. The top three principal components were visualized using RGB color bands, and an ISO cluster unsupervised classification with 20 arbitrary classes was applied to map the humidity-aridity gradient. We believe that using 20 classes in the classification captures the intricate variations in India's diverse climatic zones, reflecting the complexity of the data and ensuring proper differentiation of aridity and humidity levels. Moreover, the high resolution of the global aridity index data warrants the use of more categories to preserve subtle distinctions, enhancing the statistical representation of principal component analysis when visualized through RGB bands. Thus, this level of categorization may strike a balance between detailed mapping of ecological zones for effective planning and the practicality of interpreting and managing the classification. All these processes were executed in the Q-GIS software environment.

4.3 Results

4.3.1 Model performance

The average test AUCs of the replicate runs for jasmine were 0.865 ± 0.013 SD while for tuberose it was 0.894 ± 0.123 SD. The high AUC values show satisfactory model performance.

4.3.2 Characterization of climate niche

Based on the variable contributions and permutation importance for jasmine, it seems that specific climatic factors are particularly influential for the plant's growth and survival. The minimum temperature of the coldest month (bio_6) is the most significant factor, indicating that the climatic niche of jasmine is heavily influenced by the minimum temperatures it can endure. This suggests that it may have a lower threshold for cold temperatures and might require protection or specific cultivation practices in colder climate. Precipitation of the wettest quarter (bio_16) also has a high contribution and permutation importance, showing that the water availability during the wettest period of the year is crucial for jasmine. Both mean temperature of wettest quarter (bio_8) and precipitation of wettest month (bio_13) have moderate contributions and permutation importance, indicating that both temperature and precipitation during the wettest period are important. While annual mean temperature (bio_1) although has a lower contribution and permutation importance, the overall annual temperature plays a role in defining the climatic niche. The permutation importance values suggest how changes in these variables might impact the presence or abundance of jasmine in a given area. Variables with higher permutation importance, like bio_6 and bio_16, are likely to be more critical in determining the plant's distribution (Table 4.1).

The climatic niche of tuberose is largely defined by the temperature and precipitation patterns during the coldest and warmest quarters of the year, with annual averages also playing a significant role. The plant is likely to be sensitive to extremes in temperature and moisture, especially in the cold period of the year. The mean temperature of the coldest quarter (bio_11) has the highest percent contribution and a high permutation importance, indicating that the temperature during the coldest part of the year is very significant for the tuberose's climatic niche. Precipitation of the coldest quarter (bio_19) also has a significant percent contribution and permutation importance, suggesting that the amount of precipitation during the coldest quarter is crucial for the species. Annual mean temperature (bio_ 1) and precipitation of the warmest quarter (bio_18) also have notable contributions and permutation importance, highlighting the relevance of average annual conditions and precipitation during the warmest part of the year. The other variables have lower contributions and permutation importance, but still, they provide additional context on how tuberose interacts with its environment (Table 4.1).

The outcome from the jackknife test for variable significance in defining the ecological niche of jasmine revealed that the mean temperature of the wettest quarter (bio_8) was the most informative when evaluated singularly, implying that it carries the most useful information independently. Additionally, the removal of bio_8 results in the most substantial decline in model accuracy, indicating that it possesses vital information for understanding the climatic preferences of jasmine that the other variables do not provide (Fig. 4.1).

The results of the jackknife test for variable importance revealed that the climatic variable that contributes most significantly to the model when assessed independently is the mean temperature of the coldest quarter (bio_11), indicating that it encompasses the most valuable information on its own for the climatic niche of tuberose. Conversely, the precipitation of the coldest quarter (bio_19) shows the greatest reduction in model performance when excluded, suggesting that it contains unique information critical to understanding the climatic preferences of tuberose that is not captured by the other variables (Fig. 4.2).

The response curves for jasmine illustrate how the predicted suitability for jasmine varies with changes in a particular environmental factor (Fig. 4.3). The curve pertaining to bio_1 shows that the suitability for jasmine increases with temperature up to a threshold, beyond which it plateaus, indicating a preference for warmer climates but with a limit to the benefit of increased temperature. The curves for bio_10 and bio_6 show that the responses for these variables exhibit peaks at certain ranges, suggesting jasmine has optimal temperature ranges during the warmest and coldest times of the year. The curve for bio_16 has a relatively flat response with a slight increase in suitability at higher precipitation levels, indicating that jasmine may tolerate a wide range of precipitation during the wettest quarter but potentially prefers more abundant rainfall. The curve for bio_8 shows a peak in suitability within a specific temperature range during the wettest quarter, which might be critical for jasmine's growth and flowering. While for bio_19, the curve suggests a decrease in suitability with higher precipitation in the coldest quarter, indicating that jasmine may be susceptible to cold and wet conditions. Overall, the climatic niche of jasmine is characterized by a preference for specific temperature range throughout the year and certain precipitation patterns, particularly during the wettest and coldest periods. These factors should be carefully considered in jasmine cultivation to ensure the plant is grown under appropriate climatic conditions.

The response curves for tuberose based on a single environmental variable indicate how the predicted suitability changes with variations in each specific bioclimatic factor (Fig. 4.4). The response curve for bio_1 suggests that tuberose has a high

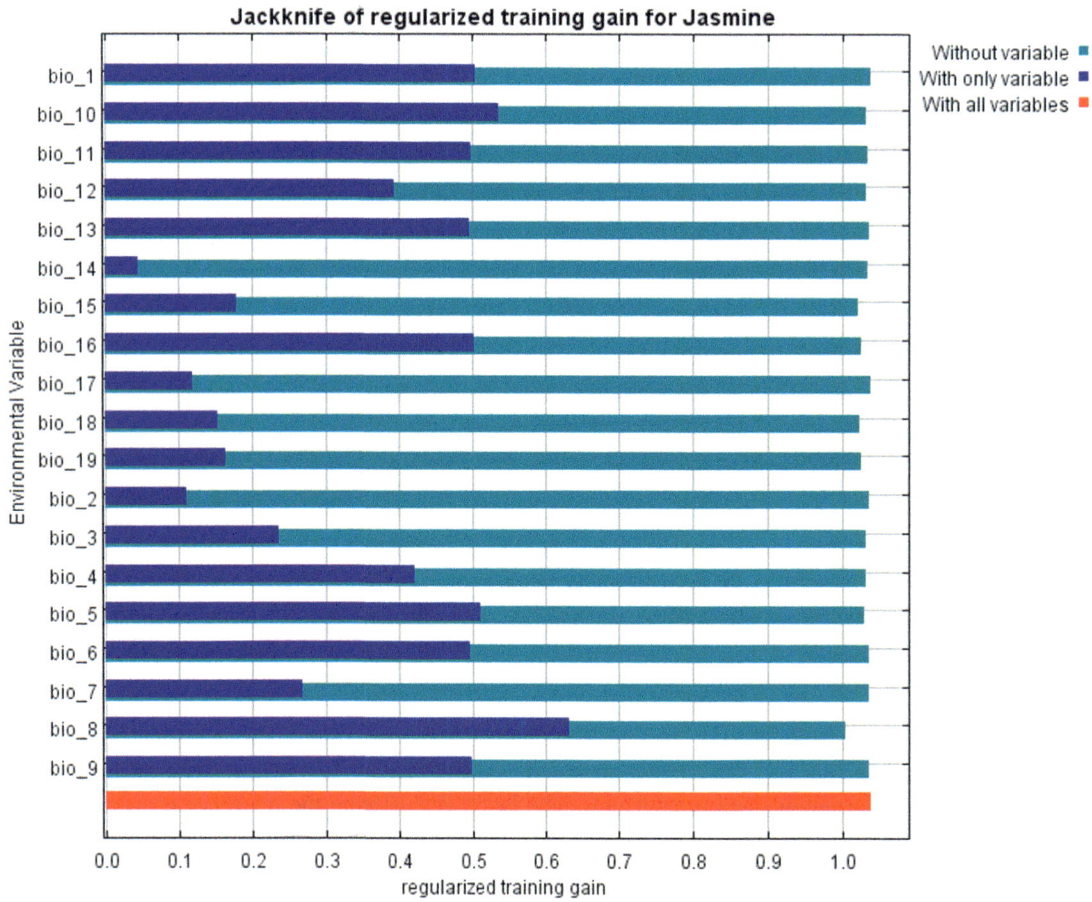

FIGURE 4.1 Jackknife of regularized training gain for jasmine.

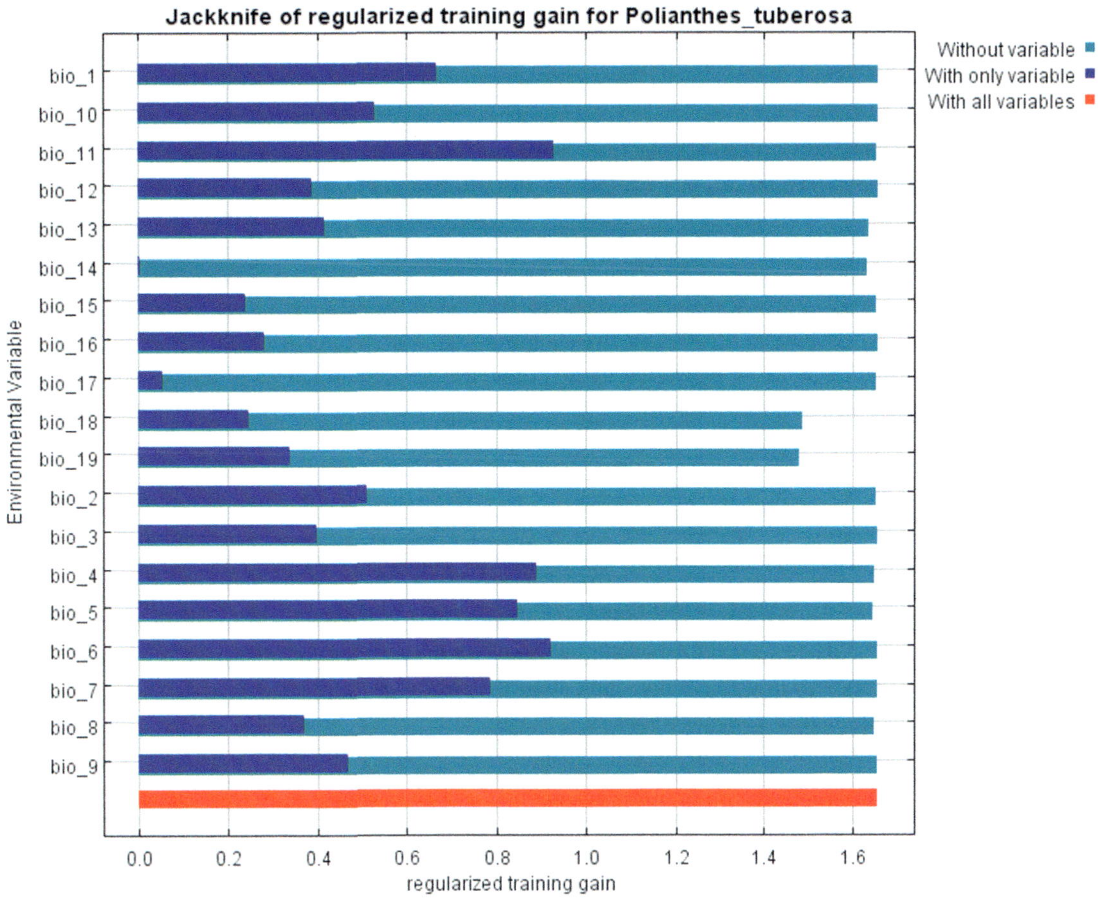

FIGURE 4.2 Jackknife of regularized training gain for tuberose.

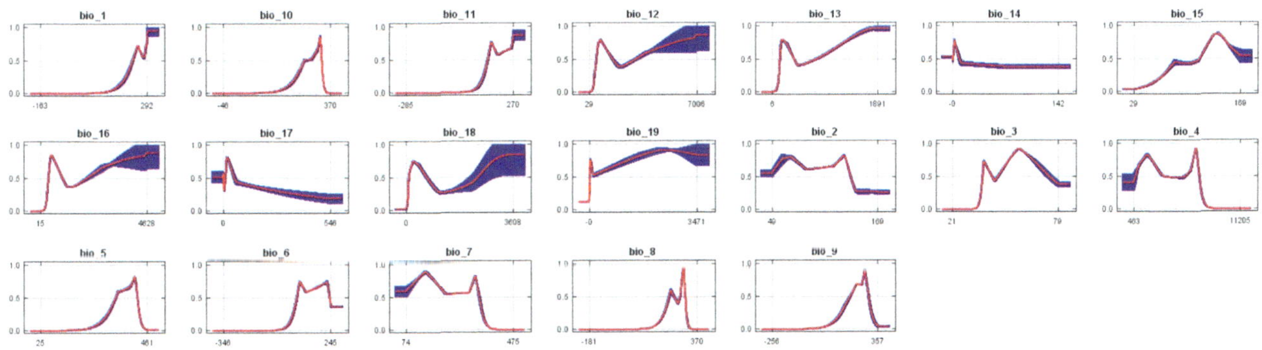

FIGURE 4.3 Response curves for jasmine showing the dependence of predicted suitability both on the selected variable and on dependencies induced by correlations between the selected variable and other variables. The blue bands around the curves represent confidence intervals, reflecting the variability or uncertainty in the model predictions.

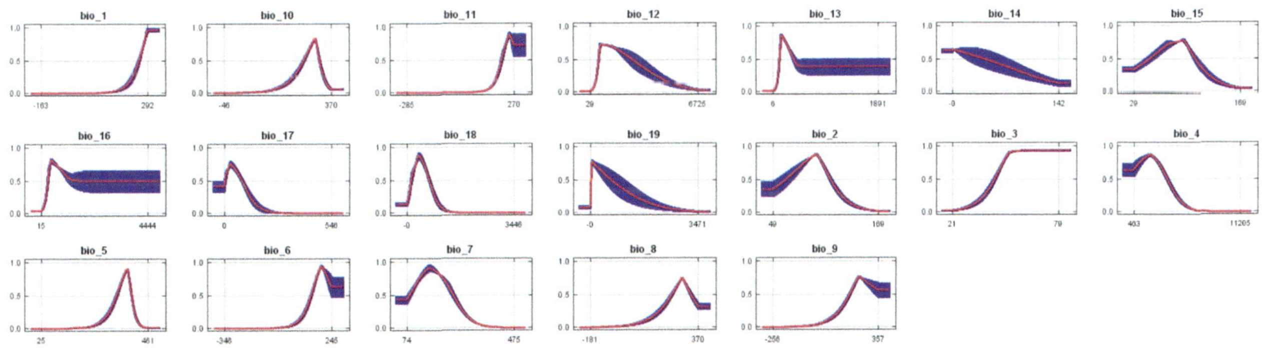

FIGURE 4.4 Response curves for tuberose showing the dependence of predicted suitability both on the selected variable and on dependencies induced by correlations between the selected variable and other variables. The blue bands around the curves represent confidence intervals, reflecting the variability or uncertainty in the model predictions.

suitability in areas with a higher annual mean temperature, as suitability sharply increases and then plateaus at higher temperature values. The curves for bio_10, bio_12, bio_13, bio_14, bio_15, bio_2, bio_3, bio_4, bio_5, bio_6, bio_7, bio_8, and bio_9 show varying degrees of suitability across their respective ranges, with peaks and troughs indicating the optimal conditions for tuberose. For instance, the suitability often peaks at certain intermediate values, suggesting its preference for moderate of temperature and precipitation conditions. For bio_16, bio_17, bio_18, and bio_19, these appear to reflect specific seasonal precipitation patterns that influence tuberose suitability. The wide bands on some curves may indicate a range of values where suitability is less sensitive to changes in the environmental variable, or it could reflect uncertainty in model prediction. In general, the niche for tuberose is defined by its preference for certain ranges of temperature and precipitation patterns throughout the year. The species seems to favor moderate climatic conditions, avoiding extremes that might reduce its viability.

4.3.3 Distribution of potential areas of cultivation in India

The climatic suitability map for jasmine and tuberose across India was categorized into five levels based on the probability levels, viz., marginal, low, moderate, high, and very high (Figs. 4.5 and 4.6). The baseline scenario shows a broad moderate to high suitability for jasmine cultivation throughout India. The highest suitability is observed in the northern and coastal regions, with particularly favorable conditions along the west coast, northeast, and far north. The central region presents a mix of moderate to low suitability, indicating that while jasmine can be cultivated here, it may not be as prolific or may require more intensive management.

Under the IPSL SSP 126 scenario, which assumes a mitigation pathway toward lower greenhouse gas emissions, the suitability for jasmine cultivation shifts noticeably. The northern regions and the west coast maintain high suitability, while central India shows increased suitability compared to the baseline. This suggests a potential expansion of optimal cultivation zones for jasmine in this scenario. Conversely, the IPSL SSP 585 scenario, which represents a high-emission

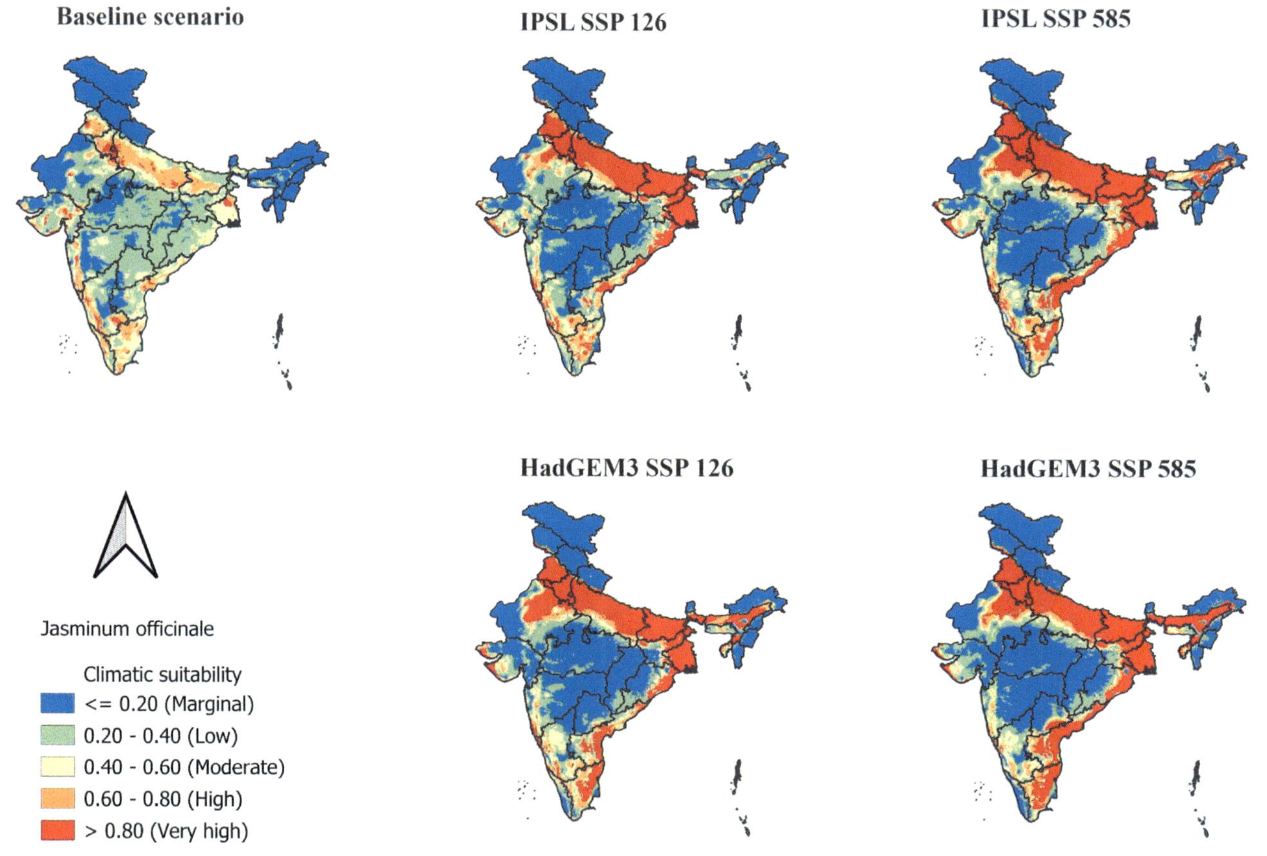

FIGURE 4.5 Distribution of climatically suitable potential areas of distribution for jasmine in India. Map lines delineate study areas and do not necessarily depict accepted national boundaries.

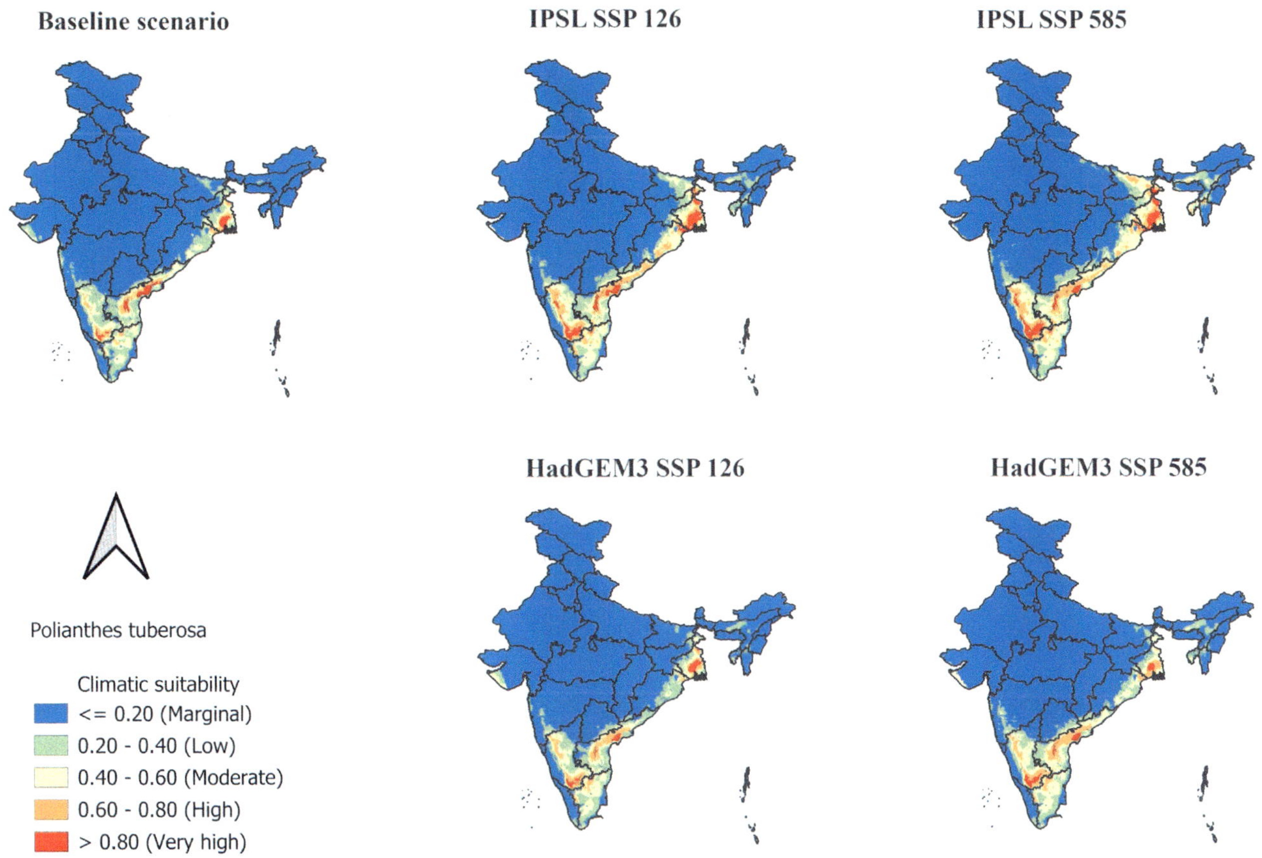

FIGURE 4.6 Distribution of climatically suitable potential areas of cultivation for tuberose in India. Map lines delineate study areas and do not necessarily depict accepted national boundaries.

pathway, projects a dramatic shift with a significant reduction in areas of high suitability, particularly in the interior regions. The northern and coastal regions still retain moderate to high suitability, but with contraction of optimal areas, especially in the central and southern regions.

The HadGEM3 model scenarios present similar trends. Under SSP 126, there is a preservation or minor expansion of suitable areas for jasmine cultivation compared to the baseline, especially in the northern parts. However, under SSP 585, the map shows a stark decrease in suitable areas, with high suitability confined to the northern and coastal peripheries, while central India faces a substantial reduction. Thus, the results indicate that jasmine cultivation could potentially benefit from climate change mitigation efforts (SSP 126), with even some expansion of suitable areas. However, if high-emission scenarios prevail (SSP 585), the cultivation of jasmine could become more challenging, with the need for adaptive strategies to overcome reduced suitability in many regions. It highlights the importance of sustainable practices and potential agricultural adjustments to maintain jasmine production in the face of climate change (Fig. 4.5).

The current and projected future distribution of tuberose in India under various climate change scenarios are contrasting (Fig. 4.6). The baseline scenario indicates that the climatic suitability for tuberose cultivation in India is predominantly moderate to high across much of the country. The highest suitability (very high) is concentrated along the coastal regions of Western and Eastern India, reflecting the ideal climatic conditions for tuberose in these areas. There are also significant areas with moderate climatic suitability extending into the interior, suggesting that current conditions across many regions are favorable for tuberose cultivation. Under the IPSL SSP 126 scenario, which represents a pathway of lower greenhouse gas emissions, the climatic suitability for tuberose appears to decrease slightly, particularly in central India. However, coastal areas remain highly suitable, with some regions showing an increase in suitability. In contrast, under the high-emission IPSL SSP 585 scenario, marked reduction in areas with high climatic suitability for tuberose was observed, particularly in the inland areas. This suggests that increased greenhouse gas emissions could significantly affect tuberose cultivation in these regions. The HadGEM3 model projects similar trends. Under the SSP 126 scenario, tuberose retains much of its current climatic suitability, although with some reduction in central regions. Under the high-emission SSP 585 scenario, the suitability drastically decreases, especially in the interior, indicating a much more constrained environment

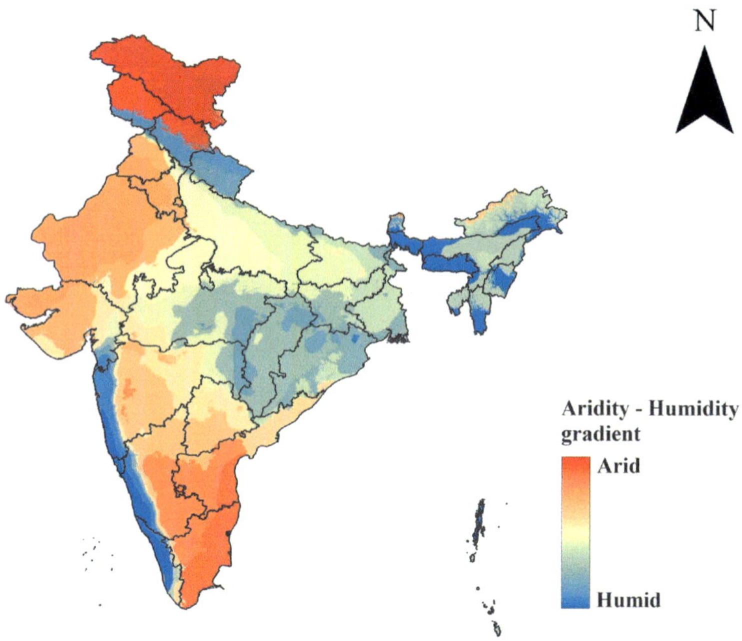

FIGURE 4.7 Aridity–humidity gradient map indicating potential drought-prone areas in India. Map lines delineate study areas and do not necessarily depict accepted national boundaries.

for tuberose cultivation due to the predicted changes in climate. Overall, the results suggest that if greenhouse gas emissions are curtailed (as in SSP 126 scenarios), the impact on tuberose cultivation may be manageable, with coastal regions continuing to provide a conducive environment. However, under a high-emission trajectory (SSP 585 scenarios), the future of tuberose cultivation could face significant challenges, with suitable areas shrinking considerably. This underscores the need for adaptive strategies in the floriculture sector to address the potential impacts of climate change on crop distribution.

4.3.4 Drought prone areas in India vis-à-vis cultivation of fragrant flowers

Potential drought-prone areas in India were identified based on an aridity-humidity gradient (Fig. 4.7). The potential drought-prone areas are predominantly located in the western and central parts of India, as indicated by the shades of orange to red in the map (Fig. 4.6). These regions would include parts of Gujarat, Rajasthan, western Madhya Pradesh, Maharashtra, and some areas of Karnataka and Telangana. These areas typically experience lower rainfall and higher evaporation rates, which make them more susceptible to drought conditions. The cultivation of jasmine and tuberose, which are water-intensive crops, is likely to be challenging in these arid and semi-arid regions. These flowers require adequate water for optimal growth and fragrance production, and insufficient water availability can lead to stress, affecting both yield and quality.

4.4 Discussion

4.4.1 Climatic niche and geographic distribution

The climatic niche and geographic distribution of jasmine and tuberose, as indicated by the maps, suggest a correlation with the regions in India that provide optimal conditions for the production of their fragrant flowers. Both plants are known for their aromatic properties, which are highly valued in perfumery and other fragrance-related industries. Jasmine, with its broader climatic suitability across India, particularly in very high suitability areas along the coastal regions, benefits from the warm, humid climate, which is conducive to the production of its essential oils. Jasmine flowers are typically most fragrant at night, and the consistent temperatures of these regions may help in maintaining the production of the volatile compounds responsible for the fragrance. The moderate suitability areas, while still able to support jasmine cultivation, might experience a variation in fragrance intensity due to differing climatic conditions such as less humidity or cooler temperature. Tuberose, with a more selective distribution, shows a preference for very high suitability in the warm and humid coastal belts. Tuberose flowers release their fragrance at night as well, and the climate in these areas is likely to support the synthesis and accumulation of the scent compounds. The high fragrance levels associated with tuberose are potentially maximized in these optimal climatic conditions. In contrast, areas with low to marginal suitability might not only have lower yields but could also potentially produce flowers with less intense fragrance due to less-than-ideal growing conditions. For both plants, the intensity and quality of the fragrance are affected by the climatic conditions, particularly temperature and humidity, which influence the metabolic processes within the plant that produce the aromatic compounds. Warmer temperatures can increase the volatility of these compounds, enhancing the scent, while adequate moisture levels are necessary for the plant's overall health and for the production of oils. The climatic niche and geographic distribution of jasmine and tuberose in India are closely aligned with the regions known for producing high-quality, fragrant flowers. The optimal growing areas along the coasts provide the ideal conditions for maximizing the aromatic potential of these flowers, which is essential for their commercial value and cultural significance.

4.4.2 Adaptation and survival strategies in the context of climatic suitability

The mapping of climatically suitable regions for jasmine and tuberose in India, based on their climatic niches, offers deep insight into their adaptation and survival strategies within diverse biological and ecological frameworks. Both species have undergone physiological adaptation to specific climatic conditions that not only bolster their growth but also fine-tune the synthesis of fragrant compounds, a critical element of their reproductive strategy to attract pollinators. The varying distribution of jasmine points to its phenotypic plasticity, enabling it to endure and reproduce across a spectrum of climatic scenarios. This is a trait that may not be as prominent in tuberose due to its more limited climatic suitability range. This adaptability is a key factor in the face of climate change (Singh et al., 2023). Furthermore, the intensity of their fragrance, an adaptive reproductive trait, serves to lure pollinators, providing a competitive edge in environments where such agents are sparse or selective. Ecological interactions with other species also shape the survival strategies of jasmine and tuberose, implying that their geographical spread is attuned to areas where these interactions are most favorable. Lastly, a thorough understanding of their current climatic niches is vital to anticipate their reactions to climate change, which may necessitate shifts in their ranges or alterations in the phenology of flowering and scent production, impacting the survival and geographical distribution of these fragrant species.

The cultivation of jasmine and tuberose in India, both of which are known for their high water demands, presents a significant challenge in arid and semi-arid regions. These flowers thrive and produce their characteristic fragrances most effectively when they receive sufficient water. However, in areas where water is scarce, plant stress can lead to diminished yields and compromised quality. Addressing this challenge requires a multifaceted approach to ensure the successful cultivation of these crops. Efficient irrigation strategies are crucial in drought-prone areas to provide jasmine and tuberose with the water they need. Techniques such as drip irrigation or other water-conservation practices can maximize the utility of available water resources. When considering the expansion of cultivation, selecting sites in more humid regions, as demarcated by the blue hues on climate maps, can leverage natural precipitation to meet the crops' water needs. Furthermore, breeding programs aimed at developing drought-resistant jasmine and tuberose varieties can be key to maintain production levels in regions susceptible to water stress. Such programs can yield varieties that are better adapted to withstand periods of low water availability without a significant loss of yield or fragrance intensity.

Anticipating the impacts of climate change is also critical. By interpreting aridity indicators on maps and preparing for possible changes in climate patterns, growers can implement mitigation strategies to safeguard their crops. This proactive approach can encompass everything from adjusting the timing of planting to adapting irrigation schedules and selecting drought-tolerant plant varieties. Regular monitoring of soil moisture levels and weather patterns is another essential component of this adaptive strategy. It enables farmers to make informed decisions that align with the immediate conditions of their environment, thereby optimizing the chances of a successful harvest. Hence, strategic planning and adaptive management are key to ensuring the sustainable cultivation of jasmine and tuberose in India's drought-prone western and central regions. By employing tailored agricultural practices and leveraging technology, growers can overcome the challenges posed by these drier climates.

4.4.3 Potential applications in floral variety conservation and horticulture

The detailed understanding of climatic niches and suitable geographic distributions for jasmine and tuberose is invaluable for varietal conservation planning and improving horticultural practices. It provides a scientific basis for making informed decisions that will benefit the conservation of plant species/varieties, optimization of their cultivation practices, and sustainability of their use in fragrant plant production. In conservation, the delineation of areas with high climatic suitability is pivotal for the protection and restoration of natural habitats where jasmine and tuberose are indigenous and flourish. This will enable targeted efforts to shield these environments and aid in the rehabilitation of regions where these species have diminished. Such climatic suitability maps are invaluable for biodiversity management, as they can pinpoint locales for the establishment of reserves or protected areas, thus safeguarding the species and their ecosystems. They also inform on climate change resilience by revealing the plants' climatic preferences, which is essential for evaluating their susceptibility to global climatic shifts and for devising strategies to bolster their resilience—be it through the selection of genotypes with a wider tolerance for climatic variability or by facilitating the relocation of species to regions expected to remain habitable in the face of impending climate change. Furthermore, this data is crucial for ex-situ conservation, guiding the formation of botanical gardens or seed banks and indicating the optimal conditions for sustaining living specimens or seed preservation, ensuring the continued existence of these species beyond their natural settings.

In the field of horticulture, the climatic niche maps provide a valuable tool for horticulturists and farmers to identify the most appropriate cultivation sites for jasmine and tuberose, potentially enhancing yields and flower quality. Such data are crucial in cultivar development, aiding to the creation of new varieties that are resistant/resilient to specific climatic conditions or extreme weather, thus broadening the potential cultivation areas. These insights also play a crucial role in shaping cultural practices, enabling the adjustment of planting dates, irrigation, and nutrient management to meet the climatic needs of these plants, ensuring optimal growth and fragrance production. Furthermore, understanding the geographic distribution of climatically suitable areas equips growers with predictive insights for pest and disease management, fostering proactive strategies and integrated management approaches. This knowledge also opens up possibilities for market expansion, as growers can explore new regions for cultivation, introducing economic opportunities. Additionally, these findings serve an educational and public engagement function, illustrating the impact of climate on plant distribution and emphasizing the importance of plant placement, which is a key concept in horticulture and landscape design.

4.4.4 Limitations of the study and future prospects

While the study of climatic niches and geographic distributions provides essential insights, the present study has certain limitations, including reliance on static climate models that do not account for the changing dynamics of climate change, potentially altering the suitability zones for the species in question. It also omitted key non-climatic factors, such as soil types, land use changes, and biotic interactions like pests, diseases, and pollinators, which could lead to an oversimplified understanding of the plants' habitat requirements. The role of intraspecific genetic variation, including varieties and landraces, in species' adaptability to diverse climates was not considered, potentially overlooking important aspects of plant resilience. Additionally, the scale and resolution of the geographic data used may not have been sufficient to capture microclimates and local adaptations, which are vital for detailed conservation and cultivation strategies. The study also neglected the complex web of ecological interactions that influence plant survival, including competition, mutualism, and predation. Finally, the possible impact of human activities—such as urbanization, agriculture, and pollution—on climatic suitability might have been underrepresented, overlooking significant anthropogenic effects on the environment.

Future research on the climatic niches of plants like jasmine and tuberose should encompass dynamic climate modeling to anticipate changes in the distribution of suitable areas over time, in light of evolving climate scenarios. It should also aim to integrate multifactorial data, broadening the scope to include various environmental variables, particularly edaphic factors, to refine the accuracy of predictive models. Delving into genetic studies to understand the basis of climatic adaptation could aid in developing resilient and adaptable breeding programs. Additionally, fine-scale analysis is crucial to account for local variations, leading to more targeted recommendations for both conservation and cultivation practices. Examining species interactions is essential to comprehend fully how biotic relationships impact plant growth and survival. Assessing anthropogenic impacts is equally important, as human activities continuously reshape environmental patterns and, consequently, the climatic suitability for these species. Consideration of economic and sociocultural dynamics will provide a holistic view of the factors influencing plant cultivation and conservation. Lastly, a thorough evaluation of existing conservation strategies in the context of a changing climate will be necessary to ensure their ongoing effectiveness and to make informed adjustments as required. By addressing these limitations and exploring these areas, future research can provide a more comprehensive understanding of the factors that influence the distribution, fragrance level, and cultivation of jasmine and tuberose, thereby significantly improving the varietal conservation and associated horticultural practices.

References

Adhikari, D., Tiwary, R., Singh, P. P., Upadhaya, K., Singh, B., Haridasan, K. E., ... Barik, S. K. (2019). Ecological niche modeling as a cumulative environmental impact assessment tool for biodiversity assessment and conservation planning: A case study of critically endangered plant Lagerstroemia minuticarpa in the Indian Eastern Himalaya. *Journal of environmental management*, *243*, 299–307.

Anderson, R. (2016). When and how should biotic interactions be considered in models of species niches and distributions? *Journal of Biogeography*, *44*(1), 8–17. Available from https://doi.org/10.1111/jbi.12825.

Barman, M., & Mitra, A. (2018). Temporal relationship between emitted and endogenous floral scent volatiles in summer- and winter-blooming *Jasminum* species. *Physiologia Plantarum*, *166*(4), 946–959. Available from https://doi.org/10.1111/ppl.12849.

Bera, P., Kotamreddy, J., Samanta, T., Maiti, S., & Mitra, A. (2015). Inter-specific variation in headspace scent volatiles composition of four commercially cultivated jasmine flowers. *Natural Product Research*, *29*(14), 1328–1335. Available from https://doi.org/10.1080/14786419.2014.1000319.

Boatright, J., Negre, F., Chen, X., Kish, C., Wood, B., Peel, G., . . . Dudareva, N. (2004). Understanding in vivo benzenoid metabolism in petunia petal tissue. *Plant Physiology*, *135*(4), 1993–2011. Available from https://doi.org/10.1104/pp.104.045468.

Cna'ani, A., Dener, E., Ben-Zeev, E., Günther, J., Köllner, T., Tzin, V., . . . Seifan, M. (2021). Phylogeny and abiotic conditions shape the diel floral emission patterns of desert *Brassicaceae* species. *Plant, Cell & Environment*, *44*(8), 2656–2671. Available from https://doi.org/10.1111/pce.14045.

Cna'ani, A., Mühlemann, J., Ravid, J., Masci, T., Klempien, A., Nguyen, T., . . . Vainstein, A. (2015). Petunia × hybrida floral scent production is negatively affected by high-temperature growth conditions. *Plant, Cell & Environment*, *38*(7), 1333–1346. Available from https://doi.org/10.1111/pce.12486.

Cordeiro, G., & Dötterl, S. (2023). Floral scents in bee-pollinated buckwheat and oilseed rape under a global warming scenario. *Insects*, *14*(3), 242. Available from https://doi.org/10.3390/insects14030242.

Effmert, U., Große, J., Röse, U., Ehrig, F., Kägi, R., & Piechulla, B. (2005). Volatile composition, emission pattern, and localization of floral scent emission in *Mirabilis jalapa* (*Nyctaginaceae*). *American Journal of Botany*, *92*(1), 2–12. Available from https://doi.org/10.3732/ajb.92.1.2.

El-Shanhorey, N., & Ahmed, S. (2020). Effect of foliar applied tryptophan on tuberose plants for decreasing the harmful effect of some heavy metals pollution in the irrigation water (a) effect of cadmium treatments. *Alexandria Journal of Agricultural Sciences*, *65*(3), 189–200. Available from https://doi.org/10.21608/alexja.2020.109675.

Farré-Armengol, G., Fernández-Martínez, M., Filella, I., Junker, R., & Peñuelas, J. (2020). Deciphering the biotic and climatic factors that influence floral scents: A systematic review of floral volatile emissions. *Frontiers in Plant Science*, *11*. Available from https://doi.org/10.3389/fpls.2020.01154.

Fick, S. E., & Hijmans, R. J. (2017). WorldClim 2: new 1-km spatial resolution climate surfaces for global land areas. *International journal of climatology*, *37*(12), 4302–4315.

Fragoso-Jiménez, J., Tapia-Campos, E., Estarrón-Espinosa, M., Barba-Gonzalez, R., Castañeda-Saucedo, M., & Castillo-Herrera, G. (2019). Effect of supercritical fluid extraction process on chemical composition of *Polianthes tuberosa* flower extracts. *Processes*, *7*(2), 60. Available from https://doi.org/10.3390/pr7020060.

Fu, J., Hou, D., Zhang, C., Bao, Z., Zhao, H., & Hu, S. (2017). The emission of the floral scent of four *Osmanthus fragrans* cultivars in response to different temperatures. *Molecules (Basel, Switzerland)*, *22*(3), 430. Available from https://doi.org/10.3390/molecules22030430.

GBIF.org. (2021a). GBIF occurrence download. Available from https://doi.org/10.15468/dl.rengd6.

GBIF.org. (2021b). GBIF occurrence download. Available from https://doi.org/10.15468/dl.jc4neg.

Gross, K., Sun, M., & Schiestl, F. (2016). Why do floral perfumes become different? Region-specific selection on floral scent in a terrestrial orchid. *PLoS One*, *11*(2), e0147975. Available from https://doi.org/10.1371/journal.pone.0147975.

Hetherington-Rauth, M., & Ramírez, S. (2016). Evolution and diversity of floral scent chemistry in the euglossine bee-pollinated orchid genus *Gongora*. *Annals of Botany*, *118*(1), 135–148. Available from https://doi.org/10.1093/aob/mcw072.

Hsiao, Y., Tsai, W., Kuoh, C., Huang, T., Wang, H., Wu, T., . . . Chen, H. (2006). Comparison of transcripts in *Phalaenopsis bellina* and *Phalaenopsis equestris* (*Orchidaceae*) flowers to deduce monoterpene biosynthesis pathway. *BMC Plant Biology*, *6*(1). Available from https://doi.org/10.1186/1471-2229-6-14.

Höfer, R., Ayasse, M., & Kuppler, J. (2022). Water deficit, nitrogen availability, and their combination differently affect floral scent emission in three *Brassicaceae* species. *Journal of Chemical Ecology*, *48*(11-12), 882–899. Available from https://doi.org/10.1007/s10886-022-01393-z.

Iler, A., Høye, T., Inouye, D., & Schmidt, N. (2013). Long-term trends mask variation in the direction and magnitude of short-term phenological shifts. *American Journal of Botany*, *100*(7), 1398–1406. Available from https://doi.org/10.3732/ajb.1200490.

Inouye, D., Saavedra, F., & Lee-Yang, W. (2003). Environmental influences on the phenology and abundance of flowering by *Androsace septentrionalis* (primulaceae). *American Journal of Botany*, *90*(6), 905–910. Available from https://doi.org/10.3732/ajb.90.6.905.

Jakobsen, H., & Olsen, C. (1994). Influence of climatic factors on emission of flower volatiles in situ. *Planta*, *192*(3). Available from https://doi.org/10.1007/bf00198572.

Joulain, D. (2021). *Jasminum grandiflorum* flowers—Phytochemical complexity and its capture in extracts: A review. *Flavour and Fragrance Journal*, *36*(5), 526–553. Available from https://doi.org/10.1002/ffj.3672.

Karlberg, A., Börje, A., Johansen, J., Lidén, C., Rastogi, S., Roberts, D., . . . White, I. (2013). Activation of non-sensitizing or low-sensitizing fragrance substances into potent sensitizers – Prehaptens and prohaptens. *Contact Dermatitis*, *69*(6), 323–334. Available from https://doi.org/10.1111/cod.12127.

Kehrberger, S., & Holzschuh, A. (2019). How does timing of flowering affect competition for pollinators, flower visitation and seed set in an early spring grassland plant? *Scientific Reports*, *9*(1). Available from https://doi.org/10.1038/s41598-019-51916-0.

Khan, Z., Razzaq, A., Sattar, T., Ahmed, A., Khan, S., & Ghouri, M. (2022). Understanding floral biology for crispr-based modification of color and fragrance in horticultural plants. *F1000Research*, *11*, 854. Available from https://doi.org/10.12688/f1000research.122453.1.

Lessard-Therrien, M., Bolmgren, K., & Davies, T. (2014). Predicting flowering phenology in a subarctic plant community. *Botany*, *92*(10), 749–756. Available from https://doi.org/10.1139/cjb-2014-0026.

Li, H., Luo, L., Ma, M., & Zeng, L. (2018). Characterization of volatile compounds and sensory analysis of jasmine scented black tea produced by different scenting processes. *Journal of Food Science*, *83*(11), 2718–2732. Available from https://doi.org/10.1111/1750-3841.14340.

Luizzi, V., Friberg, M., & Petrén, H. (2021). Phenotypic plasticity in floral scent in response to nutrient, but not water, availability in the perennial plant arabis alpina. *Functional Ecology*, *35*(8), 1655–1665. Available from https://doi.org/10.1111/1365-2435.13866.

Manzoor, S., Griffiths, G., Obiakara, M., Esparza-Estrada, C., & Lukáč, M. (2020). Evidence of ecological niche shift in *Rhododendron ponticum* (L.) in britain: Hybridization as a possible cause of rapid niche expansion. *Ecology and Evolution*, *10*(4), 2040–2050. Available from https://doi.org/10.1002/ece3.6036.

Moya, S., & Ackerman, J. (1993). Variation in the floral fragrance of *Epidendrum ciliare* (orchidaceae). *Nordic Journal of Botany*, *13*(1), 41–47. Available from https://doi.org/10.1111/j.1756-1051.1993.tb00009.x.

O'Donnel, M. S., & Ignizio, D. A. (2012). Bioclimatic predictors for supporting ecological applications in the conterminous United States (No. 691). *US Geological Survey*.

Parachnowitsch, A., Raguso, R., & Kessler, A. (2012). Phenotypic selection to increase floral scent emission, but not flower size or colour in bee-pollinated *Penstemon digitalis*. *New Phytologist*, *195*(3), 667–675. Available from https://doi.org/10.1111/j.1469-8137.2012.04188.x.

Park, I., & Mazer, S. (2018). Overlooked climate parameters best predict flowering onset: Assessing phenological models using the elastic net. *Global Change Biology*, *24*(12), 5972–5984. Available from https://doi.org/10.1111/gcb.14447.

Praveen, D., & Ramach, A. (2017). Development of regional crop simulation modeling for jasmine under the purview of climate change: A perspective from tamil nadu, india. *Horticulture International Journal*, *1*(2). Available from https://doi.org/10.15406/hij.2017.01.00009.

Qi, X., Wang, H., Chen, S., Feng, J., Chen, H., Qin, Z., … Deng, Y. (2022). The genome of single-petal jasmine (*Jasminum sambac*) provides insights into heat stress tolerance and aroma compound biosynthesis. *Frontiers in Plant Science*, *13*. Available from https://doi.org/10.3389/fpls.2022.1045194.

Raguso, R. (2008). Wake up and smell the roses: The ecology and evolution of floral scent. *Annual Review of Ecology Evolution and Systematics*, *39* (1), 549–569. Available from https://doi.org/10.1146/annurev.ecolsys.38.091206.095601.

Ramos, S., & Schiestl, F. (2020). Evolution of floral fragrance is compromised by herbivory. *Frontiers in Ecology and Evolution*, *8*. Available from https://doi.org/10.3389/fevo.2020.00030.

Ramya, M., Jang, S., An, H., Lee, S., Park, P., & Park, P. (2020). Volatile organic compounds from orchids: From synthesis and function to gene regulation. *International Journal of Molecular Sciences*, *21*(3), 1160. Available from https://doi.org/10.3390/ijms21031160.

Rawal, D., Kasel, S., Keatley, M., & Nitschke, C. (2014). Herbarium records identify sensitivity of flowering phenology of eucalypts to climate: Implications for species response to climate change. *Austral Ecology*, *40*(2), 117–125. Available from https://doi.org/10.1111/aec.12183.

Schiestl, F. (2010). The evolution of floral scent and insect chemical communication. *Ecology Letters*, *13*(5), 643–656. Available from https://doi.org/10.1111/j.1461-0248.2010.01451.x.

Shahrin, S., Roni, M., Taufique, T., Mehraj, H., & Uddin, A. (2015). Study on flowering characteristics and categorization of rose cultivars for color, fragrance and usage. *Journal of Bioscience and Agriculture Research*, *4*(1), 20–30. Available from https://doi.org/10.18801/jbar.040115.39.

Singh, P. P., Behera, M. D., Rai, R., Shankar, U., Upadhaya, K., Nonghuloo, I. M., … Tiwary, R. (2023). Morpho-physiological and demographic responses of three threatened Ilex species to changing climate aligned with species distribution models in future climate scenarios. *Environmental Monitoring and Assessment*, *195*(1), 139.

Stashenko, E., & Martínez, J. (2018). The expression of biodiversity in the secondary metabolites of aromatic plants and flowers growing in Colombia. *Potential of Essential Oils*, 196, https://doi.org/10.5772/intechopen.78001.

Theis, N., Kesler, K., & Adler, L. (2009). Leaf herbivory increases floral fragrance in male but not female *Cucurbita pepo* subsp. texana (cucurbitaceae) flowers. *American Journal of Botany*, *96*(5), 897–903. Available from https://doi.org/10.3732/ajb.0800300.

Trabucco, A., & Zomer, R.J. (2018). Global aridity index and potential evapo-transpiration (ET0) climate database v2. CGIAR consortium for spatial information (CGIAR-CSI). Published online, available from the CGIAR-CSI GeoPortal at https://cgiarcsi.community.

Velazco, S., Galvão, F., Villalobos, F., & Marco, P. (2017). Using worldwide edaphic data to model plant species niches: An assessment at a continental extent. *PLoS One*, *12*(10), e0186025. Available from https://doi.org/10.1371/journal.pone.0186025.

Verdonk, J., Haring, M., Tunen, A., & Schuurink, R. (2005). ODORANT1 regulates fragrance biosynthesis in petunia flowers. *The Plant Cell*, *17*(5), 1612–1624. Available from https://doi.org/10.1105/tpc.104.028837.

Warren, D., Glor, R., & Turelli, M. (2008). Environmental niche equivalency versus conservatism: Quantitative approaches to niche evolution. *Evolution; International Journal of Organic Evolution*, *62*(11), 2868–2883. Available from https://doi.org/10.1111/j.1558-5646.2008.00482.x.

Willis, C., Ruhfel, B., Primack, R., Miller-Rushing, A., & Davis, C. (2008). Phylogenetic patterns of species loss in thoreau's woods are driven by climate change. *Proceedings of the National Academy of Sciences*, *105*(44), 17029–17033. Available from https://doi.org/10.1073/pnas.0806446105.

Wu, J., Guo, W., Wang, Y., Liu, J., Wang, H., Zheng, Z., … Kaplan, D. (2023). Stabilization and sustained release of fragrances using silk-peg microspheres. *Acs Biomaterials Science & Engineering*. Available from https://doi.org/10.1021/acsbiomaterials.3c00064.

Xu, Q., Wang, S., Hong, H., & Zhou, Y. (2019). Transcriptomic profiling of the flower scent biosynthesis pathway of *Cymbidium faberi* rolfe and functional characterization of its jasmonic acid carboxyl methyltransferase gene. *BMC Genomics*, *20*(1). Available from https://doi.org/10.1186/s12864-019-5501-z.

Xu, S., Ding, Y., Jin, S., Zhang, Z., Wu, Z., Tiezhao, Y., … Xue, G. (2021). A high-quality genome assembly of *Jasminum sambac* provides insight into floral trait formation and oleaceae genome evolution. *Molecular Ecology Resources*, *22*(2), 724–739. Available from https://doi.org/10.1111/1755-0998.13497.

Yu, Y., Lyu, S., Chen, D., Lin, Y., Chen, J., Chen, G., … Ye, N. (2017). Volatiles emitted at different flowering stages of *Jasminum sambac* and expression of genes related to α-farnesene biosynthesis. *Molecules (Basel, Switzerland)*, *22*(4), 546. Available from https://doi.org/10.3390/molecules22040546.

Zhang, J., Li, J., Wang, J., Sun, B., Liu, Y., & Huang, M. (2020). Characterization of aroma-active compounds in jasminum sambac concrete by aroma extract dilution analysis and odour activity value. *Flavour and Fragrance Journal*, *36*(2), 197–206. Available from https://doi.org/10.1002/ffj.3631.

Zhang, X., Lin, S., Peng, D., Wu, Q., Liao, X., Xiang, K., … Fu, X. (2022). Integrated multi-omic data and analyses reveal the pathways underlying key ornamental traits in carnation flowers. *Plant Biotechnology Journal*, *20*(6), 1182–1196. Available from https://doi.org/10.1111/pbi.13801.

Zhou, C. (2023). Multiomics analysis reveals the involvement of jslhy in controlling aroma production in jasmine flowers. *Journal of Agricultural and Food Chemistry*. Available from https://doi.org/10.1021/acs.jafc.3c05768.

Chapter 5

Crop insurance using geo-information: a strategy for climate change mitigation

Sourav Kumar and Bikash Ranjan Parida
Department of Geoinformatics, School of Natural Resources Management, Central University of Jharkhand, Ranchi, India

5.1 Introduction

Agriculture sector plays a crucial role in India's economy, and its contribution toward gross domestic product (GDP) is 20.2% in the year 2020–2021 (NSO, 2021). The agriculture sector accounts for 45.5% of the country's employed workforce (2021–2022), which was 54.6% in the year 2011. Farming is not just an occupation for Indian farmers, it is also a necessity for sustaining their livelihood (Bharati et al., 2014). As compared to the 1950s, the contribution of agriculture sector to GDP reduced by 34.8% (2020–2021). The sharp reduction of GDP is associated with several factors, such as migration of rural people toward urban areas, mechanization in agriculture led to unemployment, climate change, and competition from other sectors like manufacturing and services, and government policies. Furthermore, the erratic climate and various hydro-metrological hazards have caused a loss of crop production and thereby, it becomes unprofitable (Guresci, 2022). Farming communities have not granted crop insurance which leads to loss of income and distress to farmers. Crop insurance constitutes a yield-based procedure that compensates farmers for production-related losses. It covers losses caused by extreme weather events (e.g., cyclonic rains, floods, drought, fire, landslide, pests, diseases, etc.) both before and after harvest. These losses reduce crop yield, which has an impact on farmer income. Crop insurance, in other words, secures farmers from monetary losses caused by crop failure caused by natural disasters such as agriculture fires, pests, diseases, droughts, floods, and so on. The total amount spent multiplied by crop earnings or some percentage of estimated crop earnings is the sum insured (Ghosh, Gupta, Singh, & Ward, 2019). The difference between the guaranteed yield and average yield (or called threshold yield) is used to calculate indemnification. When the yield loss is verified, the claims are approved. In the recent years, Government of India (GoI) took various steps against it and crop insurances were one of its steps.

In India, crop insurance schemes operated are the National Agricultural Insurance Scheme (NAIS), the pilot Modified National Agricultural Insurance Scheme (MNAIS), the pilot Weather-Based Crop Insurance Scheme (WBCIS), and the pilot Coconut Palm Insurance Scheme (CPIS). A more comprehensive scheme, NAIS, was founded in 1999 to support all farmers whether they were loanee or nonloanee. The NAIS replaced the erstwhile Comprehensive Crop Insurance Scheme formulated in 1985. In 2010 the NAIS was replaced by MNAIS to make the scheme easier (LARRDIS, 2014). The "National Crop Insurance Programme" (NCIP) was introduced from Rabi 2013–2014 to render the crop protection scheme more farmer friendly. Agriculture in India is rich in diversity and vulnerable to a variety of threats. A large proportion of farmers are smaller and marginal. Most agriculture is rainfed, which increases yield fluctuation and risk. Crop insurance, which is required for the vast majority of farmers, attempts to manage yield risk but is hampered by structural, layout, and financial issues. As a result, crop insurance plans confront several challenges. In response to similar issues, area-based schemes were created in the 1980s. Some insurance schemes are weather-based and employ an area-based strategy. A pilot WBCIS was founded in 2007 and operated in over 20 states (LARRDIS, 2014). Aside from the Crop Insurance Company of India, other private businesses have been granted permission to operate the Scheme. The WBCIS and RWBCIS (Restructured WBCIS) are meant to offer farmers with insurance protection against unfavorable weather events, such as shortfall and excess rainfall, hot or low temperature, humidity, and so on, which are thought to have a negative influence on crop productivity. Multiple agencies and organizations are taking part in crop insurance plans due to the diverse nature of the nation where a large number of small and marginal

Sustainable Development Perspectives in Earth Observation. DOI: https://doi.org/10.1016/B978-0-443-14072-3.00012-5

farmers use area-based techniques. As a result, coordinated efforts are essential for the successful implementation of crop insurance schemes.

Crop insurance covered around 10% of Indian farmers in 2011 (Biswal & Bahinipati, 2022) and increased to 26% by 2018–2019, although the GoI aimed to cover up to 50% under insurance. Some ongoing challenges with the crop insurance system include (1) lack of knowledge about insurance schemes, (2) insufficient insurance coverage, (3) determining the level of crop losses, and (4) prompt claim processing. The Standing Committee on Finance has suggested that agricultural damage be assessed and compensation be transferred directly into farmers' accounts as soon as possible. Furthermore, in order to decrease the use of unproductive loans, the government should raise understanding about which crops should be cultivated in different locations based on soil quality, rainfall frequency, and so on. The national government introduced the Pradhan Mantri Fasal Bima Yojana (PMFBY) in 2016. The scheme attempts to provide farmers with crop failure insurance and stabilize farmers' income while motivating farmers to adopt modern agricultural techniques. The scheme applies to all farmers, including subsistence farmers and farm laborers, who cultivate notified crops in notified regions. The crops include cereals, pulses, oilseeds, vegetables, and spices. In the 2019–2020 financial year, the scheme covered 22.36 million farmers as compared to 15.65 million farmers during 2016–2017 (PMFBY, 2020).

5.2 Need of insurance

It is one option for farmers to manage the risk of yield loss and a strategy for mitigating the impact of income loss on farmers. In other words, it protects farmers from yield fluctuations caused by hazards from all natural factors. According to the NAIS (2023), the level of risk (indemnity) of yield losses is calculated based on the shortfall of yields and threshold yields that can be written as:

$$\text{Indemnity} = \frac{\text{Shortfall in Yield}}{\text{Threshold Yield}} \times \text{Sum Insured for the Farmer}$$

where shortfall in yield is the gap between threshold yield and actual yield for the defined area. This is a traditional method of determining different levels of indemnity, such as low risk, medium risk, and high risk areas, which heavily depend on the Crop Cutting Experiment that is expensive in nature. As such there is no suitable method to identify the farmers for giving access to crop insurance (Biswal & Bahinipati 2022). In the last 5 years, researchers proposed deploying the remote sensing-based method to locate the plots that are damaged by hydro-metrological hazards like floods, droughts, landslides, cyclones, and hailstorms. Droughts caused by climate variability have a significant impact on agriculture. Thus the ability to meet the needs of growing populations will be challenged by both local and global impacts on food production. The world's most vulnerable and food-insecure regions require financial tools to reduce the risk of food production shortages while opening up opportunities for productivity growth. Farmers should be able to access insurance products across diverse regions and agroecosystems by using index-based insurance (Clement, Wouter Botzen, Brouwer, & Aerts, 2018).

Hydro-metrological hazards were typically common in India but the occurrences were often in the state of Bihar. It is a flood-prone state due to heavy rains in the catchment area of the major North Indian River, and sudden rise in water discharge causes floods over Bihar almost every year. Due to this, Bihar has to suffer huge agricultural and economic losses. Around 80% of the total population of Bihar is employed in agricultural and allied activities which exceed the national average. The agricultural sector contributes significantly to the overall development of the state economy. As a result, agriculture is the backbone of the Bihar economy. Furthermore, nearly 80% of the total geographical area of Bihar is used for agricultural purposes (Hoda, Gulati, Jose, & Rajkhowa, 2021). Bihar's soil and abundant water resources, particularly groundwater, form the fundamental basis of agriculture. Bihar has a fertile alluvial plain of Gangetic Valley. Variations in relief, precipitation, vegetation, and parent rocks have resulted in various soil types due to the region's relief features.

5.3 Types of crop insurance

Crop insurance refers to a type of protection policy that insures farmers and crop producers against anticipated losses due to natural disasters, such as floods, droughts, insects, and pests. Crop insurance is classified into two types, namely crop-yield insurance and crop-revenue insurance (Coble, Heifner, & Zuniga, 2000). Crop-yield insurance secures the anticipated income or revenue and pays an indemnity if yields and the volume of a crop's harvest are insufficient. Multiple Peril Crop Insurance is one of the examples of crop-yield insurance available for most of the states in India.

Crop-revenue insurance is an amalgamation of crop-yield insurance and price insurance, which protects the producers or farmers in years when crop yield is inadequate and/or crops have low prices.

During 1947–1985, there were isolated schemes of crop insurance implemented in India. For example, two types of approaches were practiced on crop Insurance, such as the Individual approach and the Homogenous Area approach (NIAS, 2023). During the period from 1972–1979, the Individual approach was implemented for crops such as cotton, groundnut, and potato, while during the period of 1979–1985, the Homogenous Area approach was implemented mainly for food crops and oilseeds. In the Individual approach, farmers and crop producers get compensation based on their previous products. The amount of premium and indemnity of each farmer has to be estimated individually. In the Homogenous Area approach, the compensation and premium of products are not ascertained by individual farmers. The homogeneous area would be made up of villages that are similar in terms of crop production and have similar annual variability in crop yields. Various agro-climatically homogeneous areas are treated as units, with individual farmers in those area units paying the same premium and receiving the same benefits, regardless of individual yield loss. Therefore all farmers and crop producers have to pay an indemnity of a uniform rate. The insurance was given based on what was per the conventional system. Subsequently, these approaches were replaced by NAIS in 1999. The conventional systems have lots of disadvantages as it does not identify the real farmers whose crops are damaged. This was improved further, and the Modified NAIS was introduced as a pilot program in Rabi 2010–2011. It continued to be a pilot program until Kharif 2014. It later evolved into a full-fledged program under the NCIP that was introduced from Rabi 2013 to 2014. It later became a full-fledged scheme under the PMFBY since Kharif 2016–2017.

5.4 Remote sensing and crop insurance

Crop damage valuations have traditionally been difficult to make because they are done manually after an insurable event such as floods, drought, or pest infestation. In this context, remote sensing and geographic information system (GIS) have enormous potential in the field of crop insurance market. All the parameters of crop insurance can be estimated through remote sensing, such as flood inundation assessment, crop damage due to drought, and other extreme weather events (Banerjee & Pandey, 2019; Bokusheva, Kogan, Vitkovskaya, Conradt, & Batyrbayeva, 2016). It can be useful for vulnerability and risk mapping due to crop losses at the farm scale. Hence, satellite-derived index-based insurance is becoming more popular because of its low-cost ability, and it is easy to implement on croplands. The majority of index-based insurance products (e.g., rainfall and/or heat insurance) rely on weather information collected via meteorological stations (Barrett et al., 2007). This makes insurance products reliant on robust meteorological facilities which is frequently lacking in many of the developing countries that are the target clients for this agricultural insurance innovation. Satellite data, particularly normalized difference vegetation index (NDVI), is increasingly being promoted as a means of overcoming these weather information gaps (Chantarat, Mude, Barrett, & Turvey, 2009). As a result, there are currently programs in Northern America (Canada, United States), Europe, Africa, and India that use NDVI-based crop insurance (Leblois & Quirion, 2013). Several task forces are also recommended for improving technology use in crop insurance (Aggarwal, Chand, Bhutani, Kumar, & Goel, 2016). Remote sensing technologies (e.g., satellites and unmanned aerial vehicles [UAVs]), smartphones, digital cameras, analytical and modeling approaches have all been proposed. Optical-based satellite data (e.g., Landsat series, Sentinel-2) were widely used in crop damage assessment globally and thus to make decisions on crop insurance (Murthy et al., 2022; Vroege, Dalhaus, & Finger, 2019). The concept of parametric or index-based insurance has been developed in recent years (Afshar et al., 2021; Surminski, Bouwer, & Linnerooth-Bayer, 2016). Index insurance relies heavily on the selection and validation of a payout index for the hazard of interest. Index-based insurance is designed to offer monetary assistance in the event of a major damage. Index insurance uses data that includes weather information and satellites to determine payout contracts when predefined thresholds are crossed (Benami et al., 2021; Tellman, Lall, Islam, & Bhuyan, 2022). To make sure there is a high correlation between the index and yield losses, payout thresholds must be determined using precise probability distributions (Clement, Wouter Botzen, et al., 2018). Index-based flood insurance (IBFI) is a novel approach for effective payout schemes for low-income communities in flood-prone areas which was developed by the International Water Management Institute (Malik, Amarnath, Malik, & Amarnath, 2021). It combined high-tech modeling and satellite imagery with other data to predict flood thresholds, which could result in quick compensation payouts. End-to-end solutions have been developed in collaboration with a diverse group of organizations. As a result, it contributed to long-term approaches to IBFI that can assist smallholders in managing production risks.

Remotely sensing-based indices such as NDVI, vegetation health (VH), vegetation condition index (VCI), enhanced vegetation index (EVI), temperature condition index (TCI), and crop health factor (CHF) have been deployed effectively for index-based insurance (Bokusheva et al., 2016; McLaurin & Turvey, 2011; Murthy et al., 2022). In the aforementioned studies, most of the researchers have used optical data for developing crop insurance frameworks. However, usages of synthetic aperture radar (SAR) have more potential (Boryan, Yang, Sandborn, Willis, & Haack, 2018; Vijayakumar, Kumar, Sundaram, & Subramanian, 2022), especially during the monsoon season for developing crop insurance models as optical data have limitations. Because of microwave scattering's unique sensitivity to crop structure, many studies have used SAR backscatter for crop monitoring (Ranjan & Parida, 2019; Schumann et al., 2007). In this context, the present study aimed to investigate crop insurance due to damage by flood by capturing the inundation pattern using SAR data during the monsoon season in 2019 in Bihar.

5.5 Study area

The state of Bihar has been selected for developing a crop insurance framework. It is located in the northeastern part of India with a geographical location from 24° 20' 10″ N to 27° 31' 15″ N latitude and 82° 19' 50″ E to 88° 17' 40″ E longitude with an average elevation of 53 m (above mean sea level (MSL)). The total geographical area of Bihar is 94,163.00 km^2 with 38 districts and 101 subdivisions. Bihar lies in the river plains of Ganga that pass through the middle of Bihar from west to east which divides Bihar into two unequal halves such as North Bihar and South Bihar. The entire North Bihar is drained by the Ghaghra, the Gandak, the Bagmati, the Kamla-Balan, the Kosi, and the Mahananda, which merged with the Ganga river. The rivers that drained through South Bihar are the Sone, the Punpun, the Mohane, and the Gumani. North Bihar is characterized by recurring flood events every year. The districts that are more prone to floods are Supaul, Madhubani, Darbhanga, and Sitamarhi (Tripathi, Parida, & Pandey, 2019). Floods typically occur in the monsoon season. In contrast, South Bihar is characterized by droughts due to erratic rainfall and the years 2009 and 2010 were reported as droughts and, consequently, damaged standing crops (Yaduvanshi, Srivastava, & Pandey, 2015). Bihar has three distinct seasons, namely the winter season (December to February), the summer season (March to June), and the monsoon season (July to September). Most of the rainfall occurred during the Southwest monsoon (June–September) and therefore crops grown during the monsoon season are called Kharif crops.

5.6 Materials and methods

Dual-polarization C-band SAR data of the Sentinel-1 were used to demarcate the flood inundation of the year 2017 and 2019 (Table 5.1). In 2019 North Bihar faced the maximum area affected by floods in July, while South Bihar in September (Tripathi, Pandey, Parida, & Kumar, 2020). Due to this, flood inundation for the year was assessed separately for these two regions. The calculation of crop damage was based on the respective flood durations in each area. Sentinel-2 images were utilized for crop damage assessment for 2019. In this study, cropland identification was performed on Sentinel-2 data. Standard false color composite was prepared using Lantsat-8 data to show the geographical context of the area (Fig. 5.1). ArcGIS was used to analyze the geospatial dataset, and the boundary of the study area was demarcated by environmental systems research institute (ESRI) shapefile.

TABLE 5.1 Data used in this study.

Data used	Acquisition before flood		Acquisition during/after flood		Characteristics
Sentinel-1A	Mar. 13, 2017	Mar. 03, 2019	Aug. 27, 2017	Jul. 20, 2019	20 × 22 m (az. × rg.)
	Mar. 12, 2017	Mar. 10, 2019	Aug. 23, 2017	Jul. 24, 2019	12 days
	Mar. 20, 2017	Mar. 14, 2019	Aug. 28, 2017	Jul. 25, 2019	VH (polarization)
				Sept. 30, 2019	
				Oct. 07, 2019	
Sentinel-2	Median of data available between May and June 2019		Median of data available between Aug. and Sept. 2019		10 × 10 m
					05 days
	Median of data available between Aug. and Sept. 2019		Median of data available between Oct. and Nov. 2019		

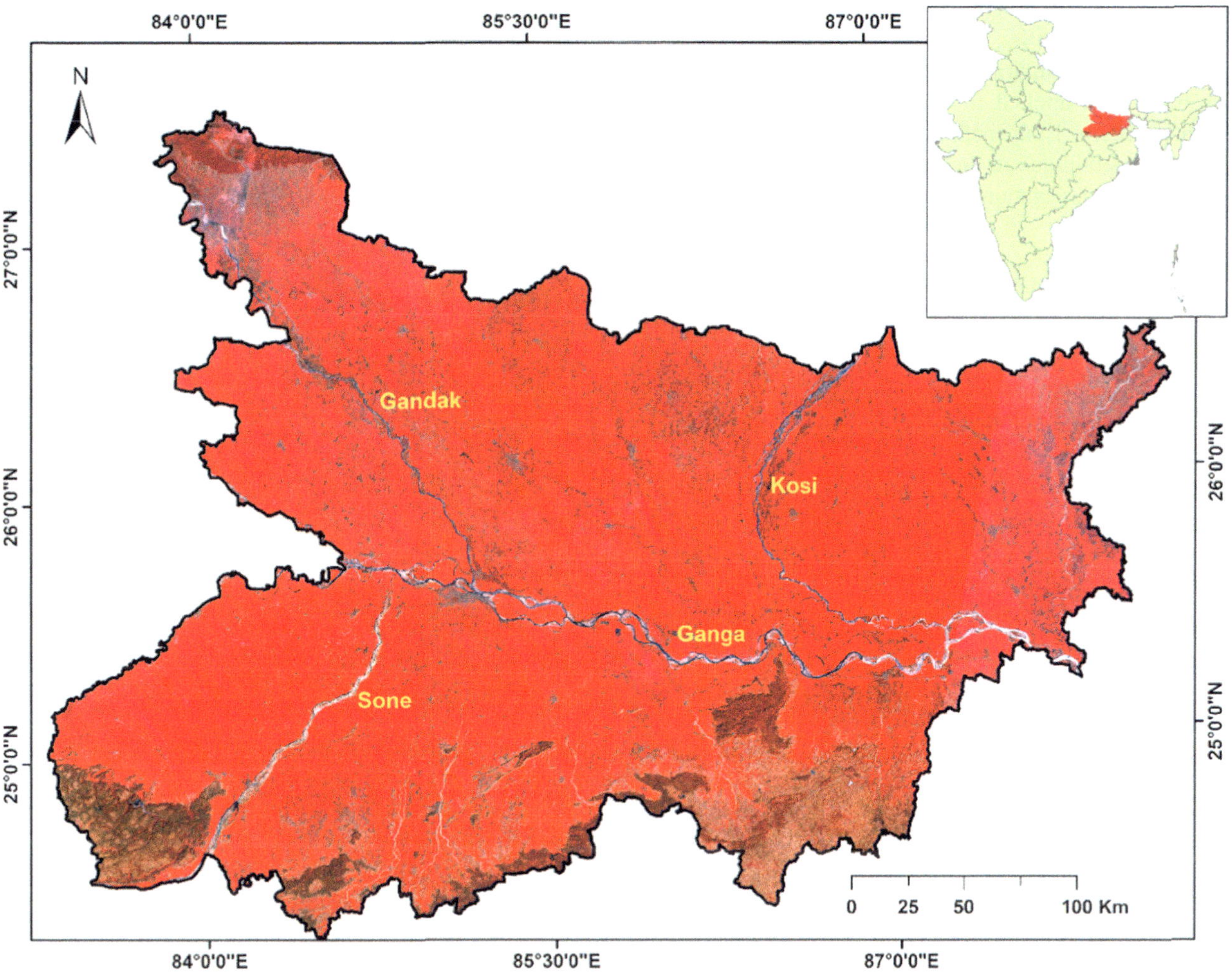

FIGURE 5.1 Study area with standard False Color Composite (FCC) image. Map lines delineate study areas and do not necessarily depict accepted national boundaries. *(Source: Landsat-8).*

Cropland identification was conducted by computing the mean images of one year to enhance classification accuracy within Google Earth Engine (GEE). This study employed pixel-based supervised classification algorithms for the identification of cropland. We utilize the discrepancy between preflood and postflood to assess crop damage caused by floods. To mitigate the impact of limited data availability due to cloud cover, the median values of the data are utilized to calculate the index.

5.7 Preprocessing of synthetic aperture radar data

Several literature studies suggested that VH polarizations are adequate for extracting the water pixels. Polarization like VV has the potential to identify partially submerged features (Clement, Kilsby, & Moore, 2018; Manjusree, Prasanna Kumar, Bhatt, Rao, & Bhanumurthy, 2012). Water pixels were extracted through SAR data in GEE by using ground range detected (GRD) SARA data of Sentinel-1. The preprocessing of Sentinel-1 data involves several steps (orbit file correction, radiometric calibration, multilooking, speckle filtering, and terrain correction) to enhance the quality and usability of SAR imagery. The "COPERNICUS/S1_GRD" Sentinel-1 data in GEE comprises Level-1 GRD scenes that have been processed to represent backscatter coefficients (σ°) in decibels (dB). It is converted to dB using $10*\log10\sigma^\circ$. These coefficients quantify the radar cross-section of targets per unit ground area. All the other preprocessing, like applying orbit file to update the orbit state of satellites (providing accurate satellite position and velocity information), calibration is the process that transforms digital pixel values into radiometrically calibrated SAR backscatter and speckle filtering that enhances the image quality by reducing speckle which appears in the SAR image due to interference of the waves reflected from the features. Orthorectification addresses topographic variations using Shuttle Radar

Topography Mission Digital Elevation Models. All the preprocessing steps are implemented by GEE using the Sentinel-1 toolbox.

Thresholding algorithms extracted surface water pixels on the backscatter. Binarization and change detection methods have been applied to compare their respective results in delineating surface water (Parida, Pandey, Kumar, & Tripathi, 2022). Both methods have almost similar results. However, the binarization technique has advantages over change detection due to the single backscatter, whereas change detection combined backscatter threshold. The result of the binarization technique was utilized for further analysis.

5.8 Extraction of water pixels

Binarization Method: To distinguish between water and nonwater surfaces, the histogram of the filtered backscatter coefficient was analyzed. Lower backscatter values were associated with the water class, while higher values were indicative of the nonwater class (Martinis & Rieke, 2015). Multiple thresholds were applied to VH polarization to differentiate water pixels from nonwater pixels. These thresholds, values −22.5, −23.2, −22.7, and −23.5 dB were necessary because the area of interest was covered by multiple dates and swaths of data.

Change Detection Method: Change detection is introduced as a hybrid methodology based on a combined backscatter threshold. A reference image, representing the presituation, was selected for all datasets. To identify water pixels, the difference between postevent and preevent images was calculated by dividing the postevent values by the preevent values (Ajadi, Meyer, & Webley, 2016; Parida, Pandey, et al., 2022). In this case, a backscatter threshold of 1.35, 1.32, and 1.29 was applied to the VH polarization image.

5.9 Crop damage assessment

Crop damage assessment is to identify the level of damage and areas at risk where mitigation actions are essential. This study describes a vegetation index-based method for crop damage assessment. The levels of crop damage were calculated using Sentinel-2 images of pre- and postflood incidents. NDVI is a preferable index for vegetation health monitoring (Rouse, Haas, Schell, & Deering, 1974).

$$\text{dNDVI} = \text{Post NDVI} - \text{Pre NDVI}$$

where dNDVI is the difference of NDVI. To determine the extent of flood damage, different damage classes were assigned to negative NDVI values (Table 5.2) (Banerjee & Pandey, 2019).

5.10 GIS model generation for crop risk analysis

The crop insurance model was developed in QGIS using insurance parameters and the detailed flowchart is given in Fig. 5.2. The parameter and weight factor used in this study played a crucial role in assessing crop insurance risk. We used flood inundation maps of 2 years along with the crop vulnerability information about the historical flood extent and its potential impact on crop areas. Additionally, the crop damage assessment, classified into five distinct classes, allowed for a detailed understanding of the varying degrees of damage caused by floods in the study area. To calculate the risk, a weight overlay approach was applied, where the weight factor was assigned to each class of crop damage

TABLE 5.2 Crop damage classes with respect to normalized difference vegetation index (NDVI) values.

Damage classes	dNDVI values	Damage percentage
Very severe damage	−0.75 to −0.98	75%–100%
Severe damage	−0.50 to −0.75	50%–75%
Moderate damage	−0.25 to −0.50	25%–50%
Low damage	0.00 to −0.25	1%–25%
No damage	≥0.00	0%

and flood-inundated area based on its significance in the overall risk assessment. This strategic combination of parameters and weight factors allowed for a comprehensive evaluation of flood-induced crop insurance risk.

The model generated the crop risk map. The resulting raster was then divided into three categories: low risk, moderate risk, and high risk. Crop insurance risk assessment is a vital tool that benefits both farmers and insurance companies. The area having high crop damage and regular flood inundation was associated with high risk, and this region have high crop premium and insurance payout (Fig. 5.2).

5.11 Results and discussion

The crop insurance risk map generated through the integration of crop damage and flood inundation data provides a comprehensive depiction of vulnerability to flood-induced losses in agricultural areas. By overlaying information on flood-prone regions with the severity of crop damage assessment, the map highlights areas with higher risk.

5.11.1 Cropland identification

Cropland identification was carried out by using supervised image classification. All cropland pixels are classified into agriculture and other land use land covers such as water bodies, wasteland, and settlements were classified in one class as nonagriculture (Fig. 5.3). The area and percentage of each class are shown in Table 5.3.

5.11.2 Crop vulnerability assessment

Earth observation satellites have the unique capability of providing real-time and periodic synoptic and multitemporal coverage over large areas. As a result, it is useful for continuous monitoring of surface parameters associated with flood inundation and its dynamics. It is only space technology, which has the potential to provide all the reliable information about the flooded region, damage assessment, watershed areas, and river behavior with weather forecast at regular intervals. This information is very essential for flood management and crop insurance.

Crop vulnerability resulting from natural disasters such as floods, droughts, and hailstorms plays an important role in crop insurance. Every year, there are huge losses in the field of agriculture due to these calamities. Flood is the most common disaster in Bihar. So, it is very important to monitor the agricultural flood for food security and economic

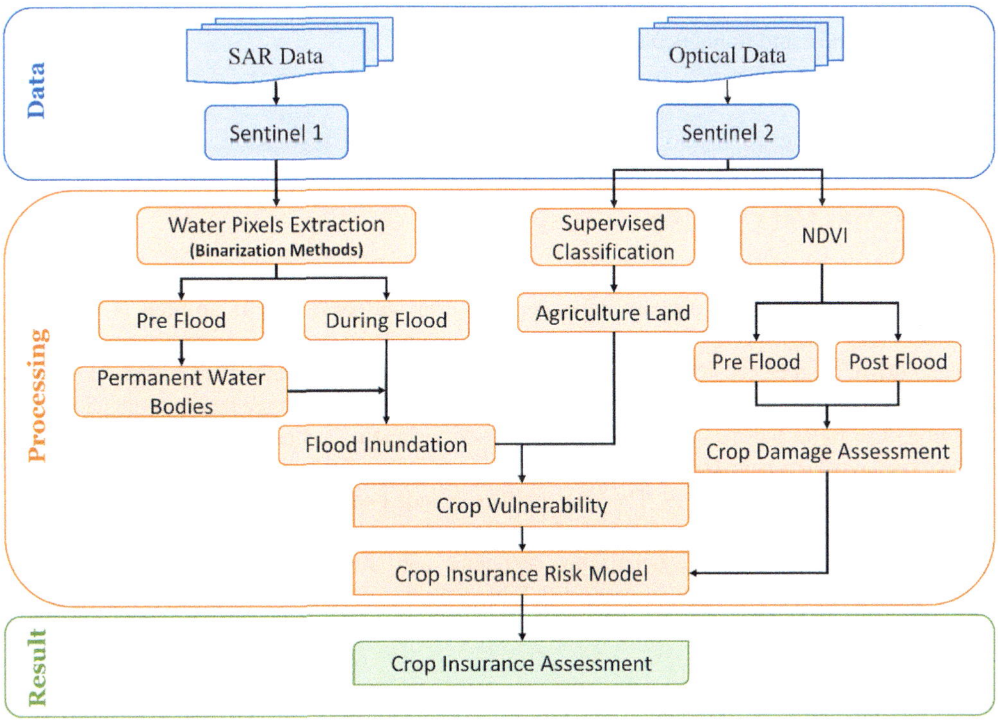

FIGURE 5.2 Flowchart of the methodology used in this study.

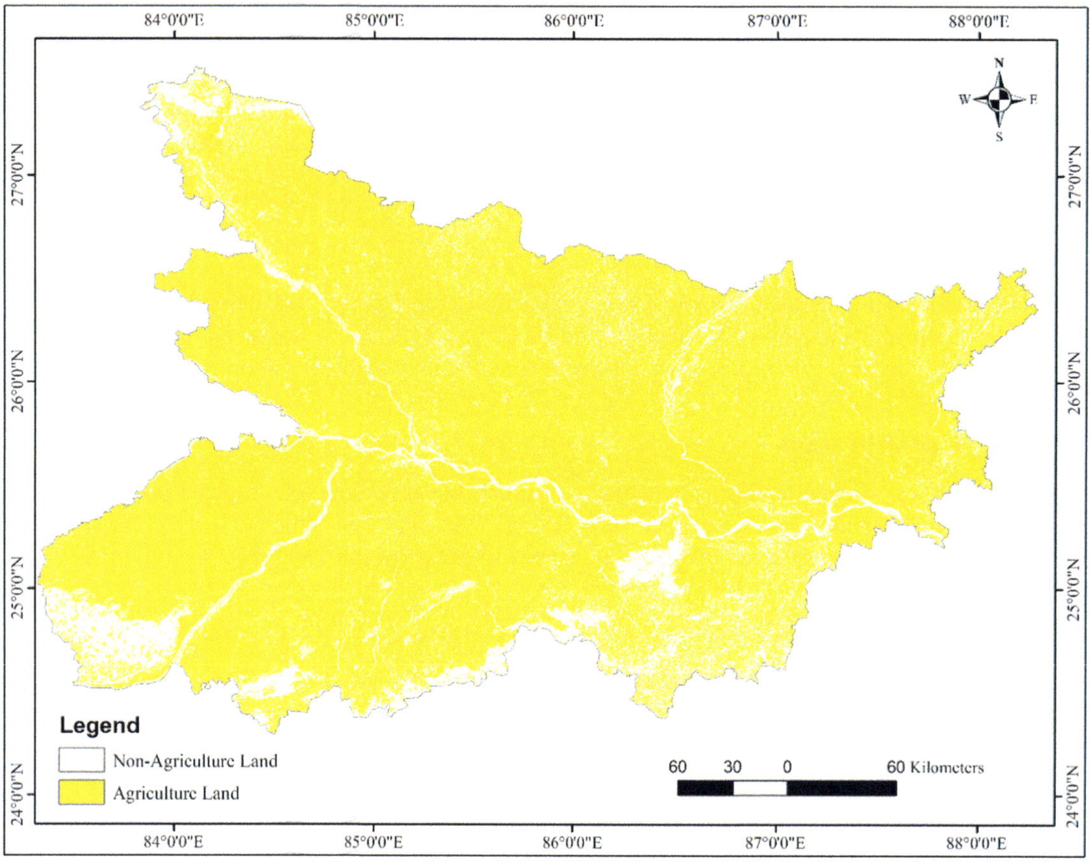

FIGURE 5.3 Map showing the distribution of agriculture in Bihar. Map lines delineate study areas and do not necessarily depict accepted national boundaries.

TABLE 5.3 Area of agriculture and nonagriculture land.

Classes	Area (km^2)	Percentage (%)
Agriculture land	77,213.66	82%
Nonagriculture land	16,950.69	18%

stability of the state. Crop premium is directly dependent on crop vulnerability. If an area is facing regular natural calamities, then it becomes a high vulnerability zone and premiums will be higher.

The crop vulnerability of the area is calculated through flood inundation by two methods (binarization technique and change detection) using the SAR data. SAR data are advantageous over optical data because they penetrate the cloud and provide clear data even in monsoon. The following methods, namely binarization technique and change detection, are utilized for mapping inundation, and their spatial pattern is given in Figs. 5.4–5.7. Area statistics of flood inundation of both methods are shown in Table 5.4 for the flood events in 2017 and 2019. The area statistics include total flooded areas and the flooded agriculture fields. There is a slight difference in the area estimated because change detection is a hybrid method. Total flood inundated in cropland were 6347.81 and 5869.84 km^2 in 2017 by binarization technique and change detection method, respectively, and which increased by 2355.4 and 2334.34 km^2 in 2019, respectively (Table 5.4).

5.11.3 Crop damage assessment

The health of crops is indicated by the ratio of absorption and reflection in the red and near-infrared (NIR) regions, respectively (Gamon et al., 1995; Rouse et al., 1974). Healthier crops have high reflectance in NIR and low reflectance

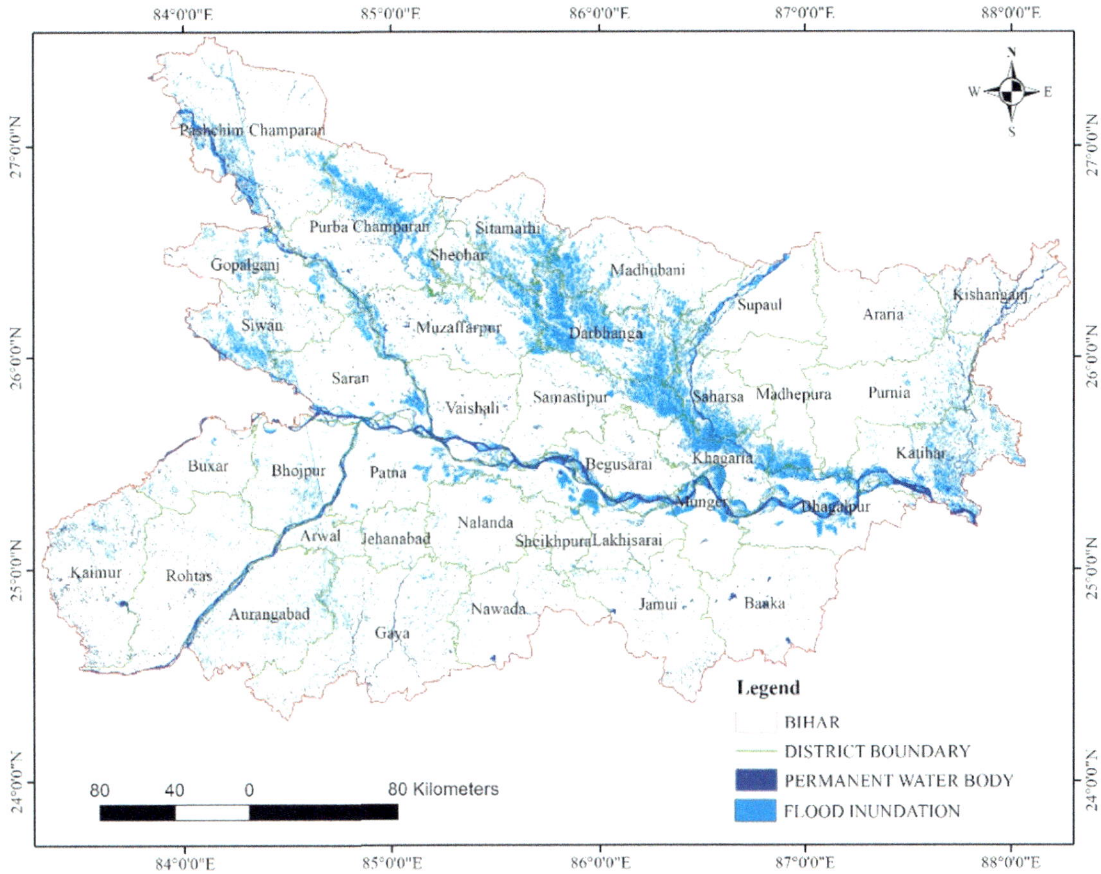

FIGURE 5.4 SAR-based flood inundation map of Bihar (2017) by Binarization technique. Map lines delineate study areas and do not necessarily depict accepted national boundaries.

in red bands in comparison to damaged crops due to less chlorophyll content (Funk & Budde, 2009). In this study, crop damage assessment was performed only for the year 2019 by using NDVI. Crop damage was estimated by differential NDVI (dNDVI), generating a flood damage map of the crop fields. This map represents the different degrees of damage by flood water. The crop fields damaged by flooding were classified into five classes (i.e., very severe damage, severe damage, moderate damage, low damage, and no damage) based on the dNDVI value (Table 5.2). The negative value of the dNDVI is the indicator of crop damage. Area statistics, damage percentage, and NDVI values of all the classes are shown in Table 5.5, which showes very few areas under the category of severe to very severe damage. Nearly 2176 km^2 was found under moderate damage.

Crop damages are directly proportional to the duration of the flood, growth stages of plants, and floodwater depth (Parida, Tripathi, Pandey, & Kumar, 2022). Floods that occurred in July and August over North Bihar affected crops, which was the start of Kharif crops, whereas, in South Bihar, it affected in the first week of October, which coincided with the maturity growth stage of plants. The majority of cropland fell into the moderate and low damage classes, with only a small portion affected by the very severe damage category. But this small area may affect food security and stability. Insurance premium is based on crop risk but payout by insurance is based on crop damages (Figs. 5.8 and 5.9). We also performed a GIS analysis by creating a buffer of 5 and 15 km along the river over the flooded region. The flood water was found mostly in the low-lying area for a long duration along the river and caused severe crop damage.

5.11.4 Crop insurance model

The crop insurance model was created using crop risk and insurance parameters. Crop risk maps are extremely effective and useful for insurance payout and premium determination. (Fig. 5.10). The area having high crop damage and regular flood inundation was associated with high risk but if the crop damage is reduced then the area comes under the moderate otherwise low risk zone. Area statistics of crop risk are shown in Table 5.6.

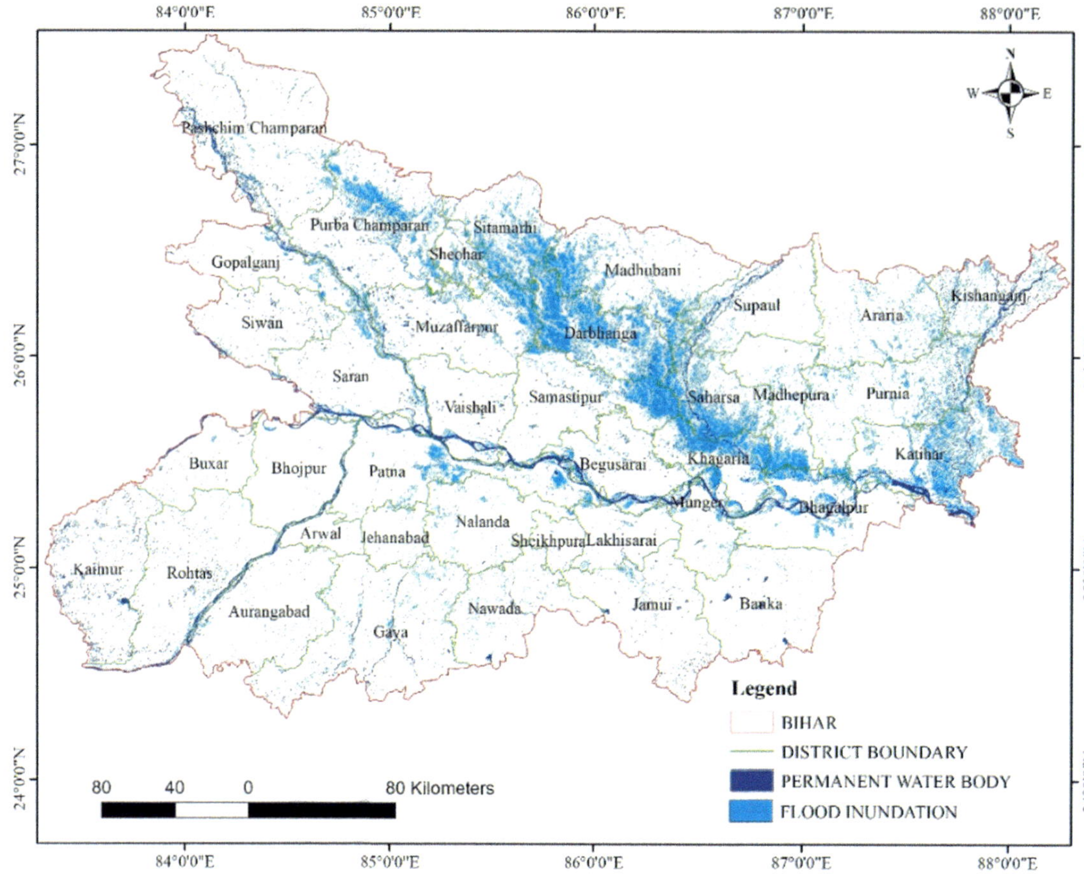

FIGURE 5.5 SAR-based flood inundation map of Bihar (2017) by change detection. Map lines delineate study areas and do not necessarily depict accepted national boundaries.

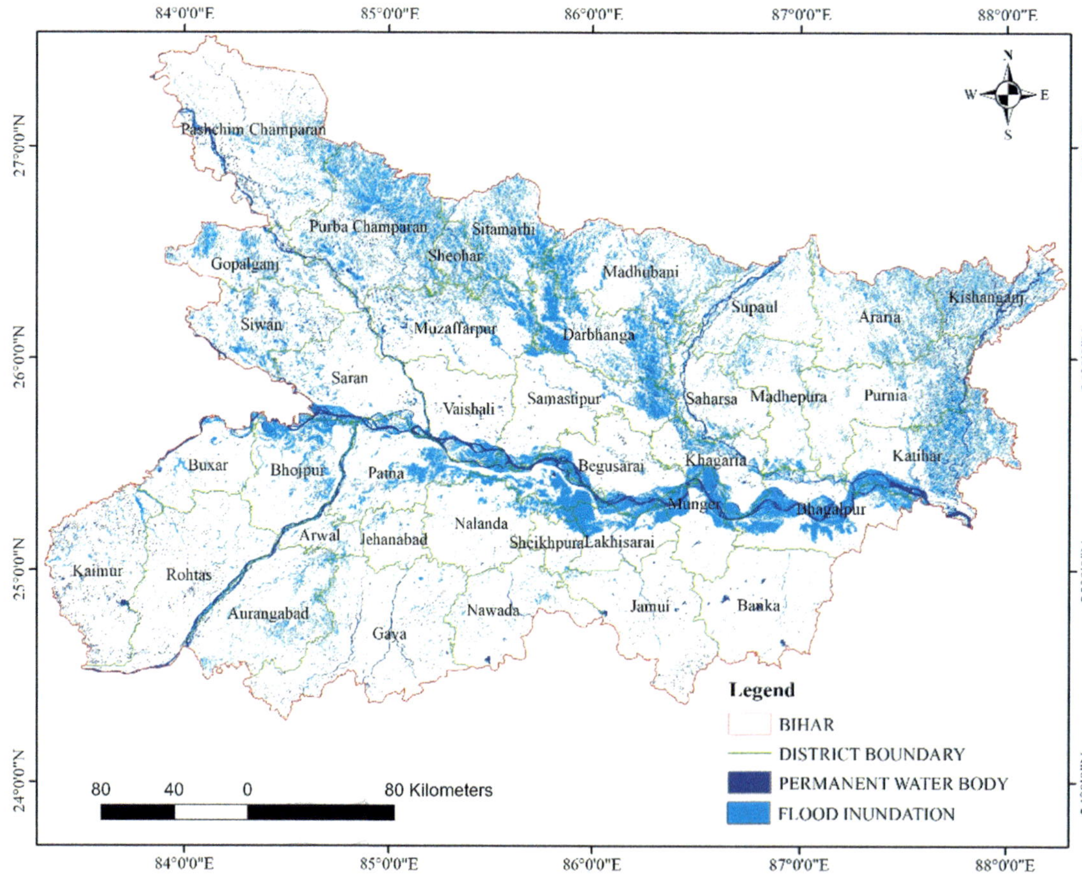

FIGURE 5.6 SAR-based flood inundation map of Bihar (2019) by Binarization technique. Map lines delineate study areas and do not necessarily depict accepted national boundaries.

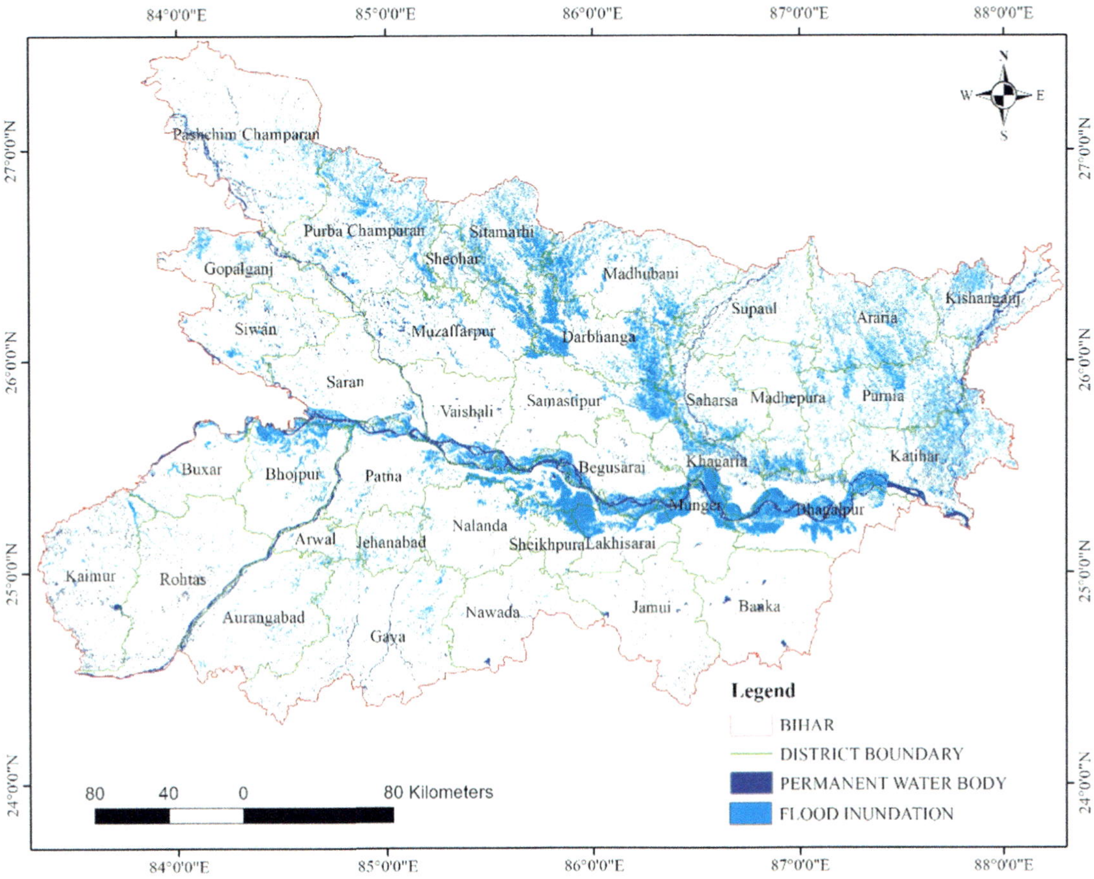

FIGURE 5.7 SAR-based flood inundation map of Bihar (2019) by change detection. Map lines delineate study areas and do not necessarily depict accepted national boundaries.

TABLE 5.4 Area of flood inundation in Bihar 2017 and 2019 (km^2) by binarization technique and change detection methods.

	Flood inundation area		Flood inundation in cropland		Differences (km^2)	
	Binarization technique	Change detection	Binarization technique	Change detection	Area	Cropland
2017	7551.04	6801.27	6347.81	5869.84	749.77	477.97
2019	9463.65	9292.42	8703.21	8204.18	171.23	499.03

TABLE 5.5 Area of crop damaged under each damage class.

Damage classes	dNDVI values	Damage percentage	Area (km^2)
Very severe damage	−0.75 to −0.98	75%–100%	3.23
Severe damage	−0.50 to −0.75	50%–75%	165.58
Moderate damage	−0.25 to −0.50	25%–50%	2175.87
Low damage	0.00 to −0.25	1%–25%	22,693.20
No damage	≥0	0%	52,173.22

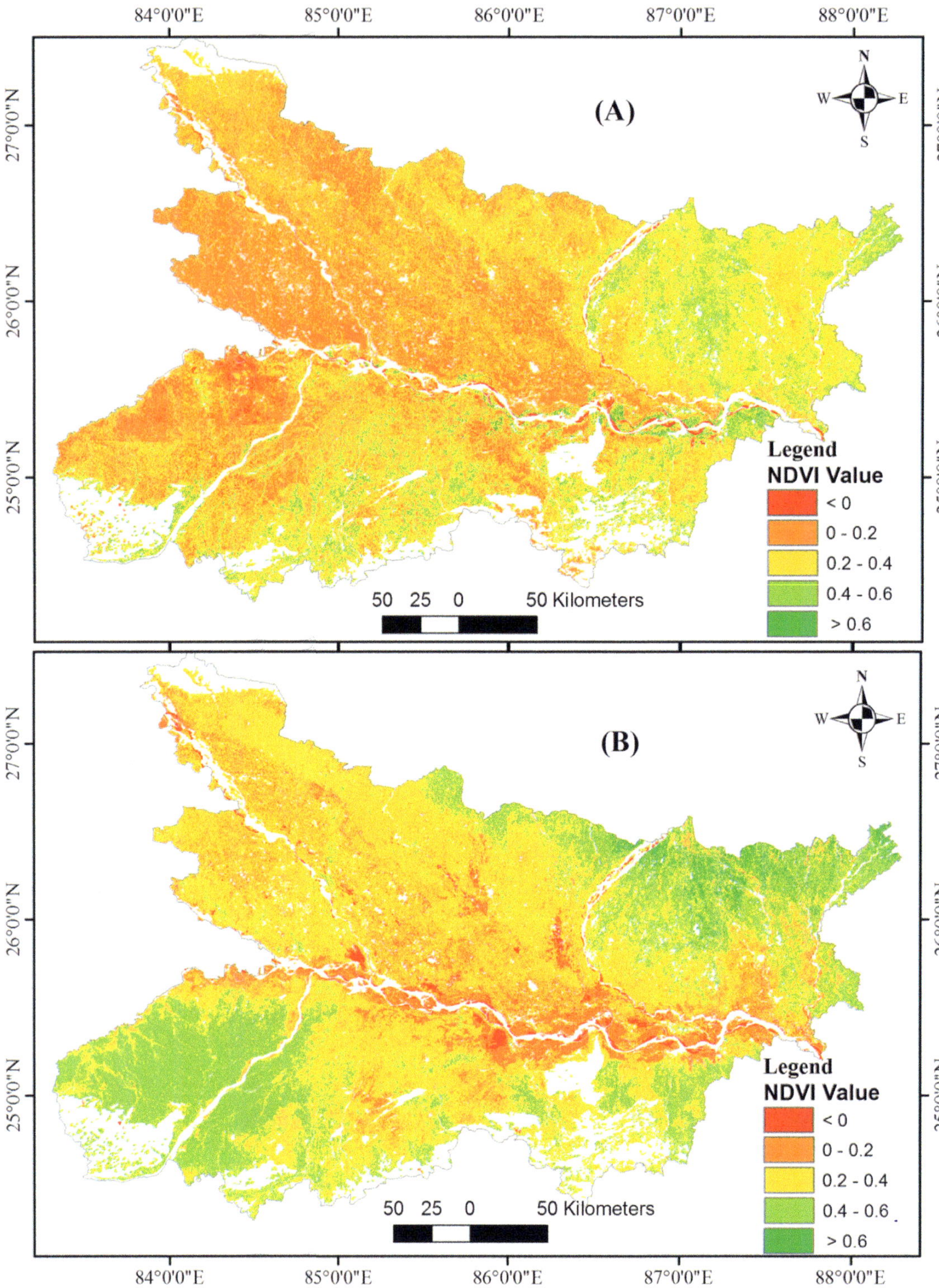

FIGURE 5.8 (A) Pre-flood Situation of Bihar 2019; (B) Post-flood Situation of Bihar 2019. Map lines delineate study areas and do not necessarily depict accepted national boundaries.

This map serves as a valuable resource for insurers, policymakers, and farmers alike, as it enables targeted risk mitigation measures, tailored insurance coverage, and informed land-use planning in regions where the agricultural sector is most susceptible to flood-related losses. The farmer should have to pay more premiums in the case of high-risk zones as well as company have to pay more insurance money in the case of losses or damage due to natural calamity. Crop insurance not only provides compensation for crop damage but it acts as a resilient for poor or small-scale farmers.

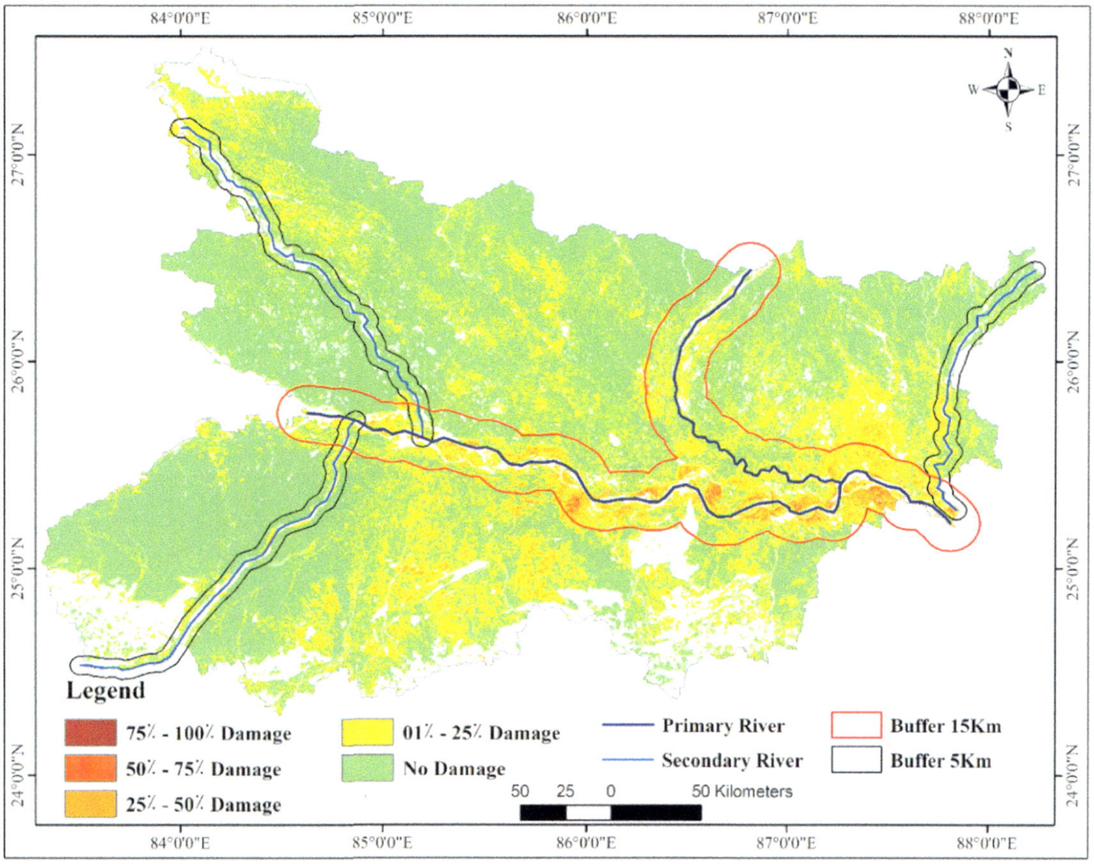

FIGURE 5.9 Crop damage percentage in the vicinity of river worth buffer of 15 km and 5 km. Map lines delineate study areas and do not necessarily depict accepted national boundaries.

In this study, our primary objective was to develop a precise crop insurance risk map that might be useful for farmers and insurance companies. The map, which was developed by integrating data on crop damage and flood inundation, provides a tactical way to evaluate and manage the risks related to flooding events in agricultural areas. We intend to present farmers with crucial information for making knowledgeable decisions about insurance coverage and risk mitigation techniques by assessing flood-induced crop loss and evaluating the relevant insurance risk levels. Moreover, this risk map can enable more effective and fairer crop insurance policies by assisting insurance providers in streamlining their risk assessment procedures. In this study, we encountered data limitations, particularly in cloudy situations. SAR has the capability to penetrate the cloud cover and provide a valuable asset in mapping flood extent. Optical data are sensitive to cloud cover and atmospheric conditions, which inherent challenges to evaluate crop damage. Future research in this field could explore the integration of SAR and optical data to develop a more comprehensive and robust approach to crop damage assessment, further enhancing the reliability of our crop insurance risk mapping efforts. There have been attempts to prepare a crop insurance model based on UAV and drone data, along with village-level farmer databases to identify the plot scale damage at ultrahigh spatial resolutions (i.e., centimeter-level) (Iwahashi et al., 2022; Panday, Pratihast, Aryal, & Kayastha, 2020). Drone-based images are a significant contribution to crop yield estimation and validation models. Therefore remote sensing data collected by drones will assist farmers in receiving precise details on crop conditions and losses, as well as shortening crop insurance claim timelines. Despite these advantages, drone-based data solutions face some obstacles such as administrative hurdles for obtaining drone flight permissions including the bureaucratic process across the world. Furthermore, its application is limited due to the scarcity of data processing hardware and software. The majority of drone operations are still restricted to academics. The ability to transfer learned skills to the level of farmers in making informed decisions requires additional scaffolding and a long way to go.

Crop insurance is an affirming safeguard to climate variability for smallholder farmers. Climate change has resulted in more episodes of riverine floods in Bihar, which may be harmful, particularly to smallholder farmers, by causing crop losses or failures. With climate change, agricultural risks will increase and hence the development of new mechanisms and measures are necessary to address such risks. To deal with the production risk associated with climatic aberrations, ex ante measures

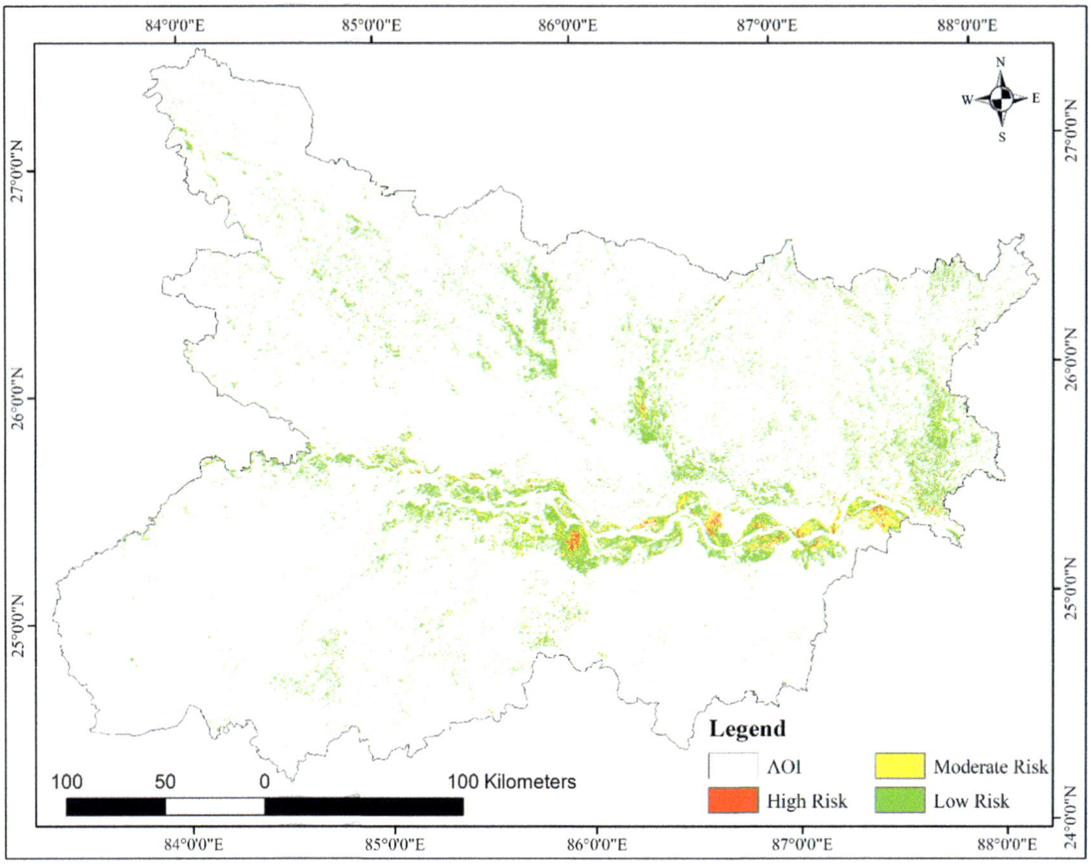

FIGURE 5.10 Crop risk map of Bihar. Map lines delineate study areas and do not necessarily depict accepted national boundaries.

TABLE 5.6 Area statistics of crop risk.

Risk class	Area (km^2)
Low risk	5080.90
Moderate risk	992.49
High risk	138.55

are vital before crop loss occurs and further ex post measures such as crop insurance are beneficial after crop loss occurs (Swain, 2014). By inducing proper proactive (e.g., use of drought-resistant variety and seed variety, integrated pest management, efficient irrigation systems, etc.) and reactive (e.g., payout of indemnity) responses in insurance users, the insurance industry can help both in mitigation as well as adaptation to climate change. Mitigation strategies include incentivizing the use of green technologies, promoting climate-friendly cropping patterns, sustainable farming, and less energy-intensive agriculture. Crop insurance is a risk-management tool. The best way to deal with agricultural losses caused by climate variability may be through insurance schemes and prices rather than crop yields (Bozzola & Swanson, 2014; Falco, Adinolfi, Bozzola, & Capitanio, 2014). Crop insurance can play a significant role in climate change adaptation strategy, and an optimal policy intervention might be formulated to deal with insurance mechanisms.

5.12 Conclusions

There are vast and diverse opportunities for remote sensing in the field of crop insurance. Remote sensing has the potential to do any type of insurance for any natural calamity. It is very helpful for mapping and analyzing a region with a large spatial extent. Many satellites have multitemporal data with high resolution which have the potential to monitor any natural resources. In this study, Sentinel-1 and 2 satellite data were analyzed to determine the crop insurance. We found that the frequency of flood inundation is very high in the northern part of Bihar in comparison to South Bihar due to the recurring riverine flooding. Along the river Ganga, agricultural areas are facing high crop damage due to floods in low-lying regions. The crop insurance model was developed to determine the crop risk zone by using flood inundation and crop damage. This is very helpful for crop premiums and payouts by insurance companies. The purpose of this paper is to discuss the potential benefit of insurance as a climate change adapting mechanism in agriculture. The paper adds to the limited understanding of opinions on climate-related risks for paying out crop insurance in Bihar, India.

Acknowledgments

The authors acknowledge the Copernicus and European Space Agency (ESA) for making the SAR data freely available that was used in the present study. The corresponding author would like to express gratitude to the University Grants Commission (UGC) of India under the start-up grant (F.4–5(209-FRP)/2015/BSR) for necessary support.

References

Afshar, M. H., Foster, T., Higginbottom, T. P., Parkes, B., Hufkens, K., Mansabdar, S., ... Kramer, B. (2021). Improving the performance of index insurance using crop models and phenological monitoring. *Remote Sensing*, *13*, 924. Available from https://doi.org/10.3390/rs13050924.

Aggarwal, P., Chand, R., Bhutani, A., Kumar, V., Goel, S., 2016. Report of the Task Force on 'Enhancing Technology Use in Agriculture Insurance'. NITI (National Institute for Transforming India) Aayog, Government of India.

Ajadi, O., Meyer, F., & Webley, P. (2016). Change detection in synthetic aperture radar images using a multiscale-driven approach. *Remote Sensing*, *8*, 482. Available from https://doi.org/10.3390/rs8060482.

Banerjee, S., & Pandey, A. C. (2019). Crop insurance model to consolidate academia-industry cooperation: A case study over Assam, India. *Spatial Information Research*, *27*, 719–731. Available from https://doi.org/10.1007/s41324-019-00291-z.

Barrett, C. B., Barnett, B. J., Carter, M. R., Chantarat, S., Hansen, J. W., Mude, A. G., ... Ward, M. N. (2007). Poverty traps and climate risk: Limitations and opportunities of index-based risk financing. *SSRN Journal*. Available from https://doi.org/10.2139/ssrn.1141933.

Benami, E., Jin, Z., Carter, M. R., Ghosh, A., Hijmans, R. J., Hobbs, A., ... Lobell, D. B. (2021). Uniting remote sensing, crop modelling and economics for agricultural risk management. *Nature Reviews Earth & Environment*, *2*, 140–159. Available from https://doi.org/10.1038/s43017-020-00122-y.

Bharati, R., Azad, N., Singh, K., Chakraborti, S., Chandra, N., & Singh, S. (2014). Factors affecting adoption of crop insurance in Bihar. *Journal of AgriSearch*, *1*, 102–107.

Biswal, D., & Bahinipati, C. S. (2022). Why are farmers not insuring crops against risks in India? A review. *Progress in Disaster Science*, *15*100241. Available from https://doi.org/10.1016/j.pdisas.2022.100241.

Bokusheva, R., Kogan, F., Vitkovskaya, I., Conradt, S., & Batyrbayeva, M. (2016). Satellite-based vegetation health indices as a criteria for insuring against drought-related yield losses. *Agricultural and Forest Meteorology*, *220*, 200–206. Available from https://doi.org/10.1016/j.agrformet.2015.12.066.

Boryan, C. G., Yang, Z., Sandborn, A., Willis, P., & Haack, B. (2018). Operational agricultural flood monitoring with Sentinel-1 synthetic aperture radar, . *Presented at the IGARSS 2018-2018 IEEE international geoscience and remote sensing symposium*, IEEE, Valencia (pp. 5831–5834). https://doi.org/10.1109/IGARSS.2018.8519458.

Bozzola, M., & Swanson, T. (2014). Policy implications of climate variability on agriculture: Water management in the Po river basin, Italy. *Environmental Science & Policy*, *43*, 26–38. Available from https://doi.org/10.1016/j.envsci.2013.12.002.

Chantarat, S., Mude, A. G., Barrett, C. B., & Turvey, C. G. (2009). The performance of index based livestock insurance: Ex ante assessment in the presence of a poverty trap. *SSRN Journal*. Available from https://doi.org/10.2139/ssrn.1844751.

Clement, K. Y., Wouter Botzen, W. J., Brouwer, R., & Aerts, J. C. J. H. (2018). A global review of the impact of basis risk on the functioning of and demand for index insurance. *International Journal of Disaster Risk Reduction*, *28*, 845–853. Available from https://doi.org/10.1016/j.ijdrr.2018.01.001.

Clement, M. A., Kilsby, C. G., & Moore, P. (2018). Multi-temporal synthetic aperture radar flood mapping using change detection: Multi-temporal SAR flood mapping using change detection. *Journal of Flood Risk Management*, *11*, 152–168. Available from https://doi.org/10.1111/jfr3.12303.

Coble, K., Heifner, R., & Zuniga, M. (2000). Implications of crop yield and revenue insurance for producer hedging. *Journal of Agricultural and Resource Economics*, *25*, 432–452.

Falco, S. D., Adinolfi, F., Bozzola, M., & Capitanio, F. (2014). Crop insurance as a strategy for adapting to climate change. *Journal of Agricultural Economics*, *65*, 485–504. Available from https://doi.org/10.1111/1477-9552.12053.

Funk, C., & Budde, M. E. (2009). Phenologically-tuned MODIS NDVI-based production anomaly estimates for Zimbabwe. *Remote Sensing of Environment*, *113*, 115–125. Available from https://doi.org/10.1016/j.rse.2008.08.015.

Gamon, J. A., Field, C. B., Goulden, M. L., Griffin, K. L., Hartley, A. E., Joel, G., ... Valentini, R. (1995). Relationships between NDVI, canopy structure, and photosynthesis in three Californian vegetation types. *Ecological Applications*, *5*, 28–41. Available from https://doi.org/10.2307/1942049.

Ghosh, R., Gupta, S., Singh, V., & Ward, P. (2019). Is there a market for multi-peril crop insurance in developing countries moving beyond subsidies? Evidence from India. *International Food Policy Research Institute IFPRI Discussion Paper* 1820 (pp. 1–43).

Guresci, E. (2022). Agricultural factors as the root cause of rural migration from a global perspective. *Cuadernos de Desarrollo Rural*, *19*. Available from https://doi.org/10.11144/Javeriana.cdr19.afrc.

Hoda, A., Gulati, A., Jose, S., & Rajkhowa, P. (2021). Sources and drivers of agricultural growth in Bihar. In A. Gulati, R. Roy, & S. Saini (Eds.), *Revitalizing Indian agriculture and boosting farmer incomes, India studies in business and economics* (pp. 211–246). Singapore: Springer Nature Singapore. Available from https://doi.org/10.1007/978-981-15-9335-2_8.

Iwahashi, Y., Sigit, G., Utoyo, B., Lubis, I., Junaedi, A., Trisasongko, B. H., ... Homma, K. (2022). Drought damage assessment for crop insurance based on vegetation index by unmanned aerial vehicle (UAV) multispectral images of paddy fields in Indonesia. *Agriculture*, *13*, 113. Available from https://doi.org/10.3390/agriculture13010113.

LARRDIS (2014). Parliament Library and Reference, Research, Documentation and Information Service (LARRDIS). Crop Insurance in India. No. 17 /RN/Ref./ November /2014.

Leblois, A., & Quirion, P. (2013). Agricultural insurances based on meteorological indices: Realizations, methods and research challenges. *Meteorological Applications*, *20*, 1–9. Available from https://doi.org/10.1002/met.303.

Malik, R. P. S., Amarnath, G., Malik, R. P. S., & Amarnath, G. (2021). Economics of Index-based flood insurance (IBFI). https://doi.org/10.22004/AG.ECON.316618.

Manjusree, P., Prasanna Kumar, L., Bhatt, C. M., Rao, G. S., & Bhanumurthy, V. (2012). Optimization of threshold ranges for rapid flood inundation mapping by evaluating backscatter profiles of high incidence angle SAR images. *International Journal of Disaster Risk Science*, *3*, 113–122. Available from https://doi.org/10.1007/s13753-012-0011-5.

Martinis, S., & Rieke, C. (2015). Backscatter analysis using multi-temporal and multi-frequency SAR data in the context of flood mapping at River Saale, Germany. *Remote Sensing*, *7*, 7732–7752. Available from https://doi.org/10.3390/rs70607732.

McLaurin, M. K., & Turvey, C. G. (2011). Applicability of the normalized difference vegetation index in index-based crop insurance design. *SSRN Journal*. Available from https://doi.org/10.2139/ssrn.1776467.

Murthy, C. S., Poddar, M. K., Choudhary, K. K., Pandey, V., Srikanth, P., Ramasubramanian, S., & Senthil Kumar, G. (2022). Paddy crop insurance using satellite-based composite index of crop performance. *Geomatics, Natural Hazards and Risk*, *13*, 310–336. Available from https://doi.org/10.1080/19475705.2021.2025155.

NIAS (2023). Revised National Agricultural Insurance Scheme (NAIS) (Rashtriya Krishi Bima Yojana-RKBY). Available online: https://pmfby.gov.in/pdf/NAIS_SCHEME.pdf.

NSO (2021). National Statistical Office (NSO), M/o Statistics & PI. Available online: https://www.pib.gov.in/PressReleasePage.aspx?PRID = 1741942.

Panday, U. S., Pratihast, A. K., Aryal, J., & Kayastha, R. B. (2020). A review on drone-based data solutions for cereal crops. *Drones*, *4*, 41. Available from https://doi.org/10.3390/drones4030041.

Parida, B. R., Pandey, A. C., Kumar, S., & Tripathi, G. (2022). *Comparative flood area analysis based on change detection and binarization methods using Sentinel-1 synthetic aperture radar data. Radar remote sensing* (pp. 93–108). Elsevier. Available from https://doi.org/10.1016/B978-0-12-823457-0.00001-X.

Parida, B. R., Tripathi, G., Pandey, A. C., & Kumar, A. (2022). Estimating floodwater depth using SAR-derived flood inundation maps and geomorphic model in Kosi river basin (India). *Geocarto International*, *37*, 4336–4360. Available from https://doi.org/10.1080/10106049.2021.1899298.

PMFBY (2020). Pradhan Mantri Fasal Bima Yojana, Ministry Of Agriculture & Farmers Welfare. Available online: https://pmfby.gov.in/stateWiseDataPage.

Ranjan, A. K., & Parida, B. R. (2019). Paddy acreage mapping and yield prediction using sentinel-based optical and SAR data in Sahibganj district, Jharkhand (India). *Spatial Information Research*, *27*, 399–410. Available from https://doi.org/10.1007/s41324-019-00246-4.

Rouse, J. W., Haas, R. H., Schell, J. A., & Deering, D. W. (1974). Monitoring vegetation systems in the Great Plains with ERTS. *NASA Special Publication*, *351*, 309.

Schumann, G., Matgen, P., Hoffmann, L., Hostache, R., Pappenberger, F., & Pfister, L. (2007). Deriving distributed roughness values from satellite radar data for flood inundation modelling. *Journal of Hydrology*, *344*, 96–111. Available from https://doi.org/10.1016/j.jhydrol.2007.06.024.

Surminski, S., Bouwer, L. M., & Linnerooth-Bayer, J. (2016). How insurance can support climate resilience. *Nature Climate Change*, *6*, 333–334. Available from https://doi.org/10.1038/nclimate2979.

Swain, M. (2014). *Crop insurance for adaptation to climate change in India*. Asia Research Center working paper 61, London, United Kingdom.

Tellman, B., Lall, U., Islam, A. K. M. S., & Bhuyan, M. A. (2022). Regional index insurance using satellite-based fractional flooded area. *Earth's Future*, *10*e2021EF002418. Available from https://doi.org/10.1029/2021EF002418.

Tripathi, G., Pandey, A. C., Parida, B. R., & Kumar, A. (2020). Flood inundation mapping and impact assessment using multi-temporal optical and SAR satellite data: A case study of 2017 flood in Darbhanga District, Bihar, India. *Water Resources Management*, *34*, 1871–1892. Available from https://doi.org/10.1007/s11269-020-02534-3.

Tripathi, G., Parida, B. R., & Pandey, A. C. (2019). Spatio-temporal rainfall variability and flood prognosis analysis using satellite data over North Bihar during the August 2017 flood event. *Hydrology*, *6*, 38. Available from https://doi.org/10.3390/hydrology6020038.

Vijayakumar, S., Kumar, R., Sundaram, R., & Subramanian, P. (2022). Remote sensing-based transformative crop insurance for rice. *Current Science*, *123*, 254–255.

Vroege, W., Dalhaus, T., & Finger, R. (2019). Index insurances for grasslands – A review for Europe and North-America. *Agricultural Systems*, *168*, 101–111. Available from https://doi.org/10.1016/j.agsy.2018.10.009.

Yaduvanshi, A., Srivastava, P. K., & Pandey, A. C. (2015). Integrating TRMM and MODIS satellite with socio-economic vulnerability for monitoring drought risk over a tropical region of India. *Physics and Chemistry of the Earth, Parts A/B/C*, *83–84*, 14–27. Available from https://doi.org/10.1016/j.pce.2015.01.006.

Chapter 6

Terrestrial water cycle in future climate over India

Pinaki Das[1], Pulakesh Das[2], Swapnil Vyas[1] and Pramit Kumar Deb Burman[3,4]

[1]*Department of Geography, Savitribai Phule Pune University, Pune, Maharashtra, India,* [2]*School of Forest Resources, University of Maine, Orono, ME, United States,* [3]*Centre for Climate Change Research, Indian Institute of Tropical Meteorology, Ministry of Earth Sciences, Pune, Maharashtra, India,* [4]*Department of Atmospheric and Space Sciences, Savitribai Phule Pune University, Pune, Maharashtra, India*

6.1 Introduction

Evapotranspiration (ET) is a crucial component of the global hydrological cycle (Katul, Oren, & Manzoni, 2012; Shukla & Mintz, 1982). It represents the process of water transfer from the Earth's surface to the atmosphere comprising two primary components: evaporation and transpiration (Kool, Agam, & Lazarovitch, 2014), and thus transports more than 60% of the precipitated water on land to the atmosphere. Evaporation converts liquid water from different sources like lakes, rivers, soil, or canopy-intercepted rainfall into water vapor when exposed to sunlight and heat (Wang & Dickinson, 2012). Transpiration, on the other hand, refers to the release of water vapor by plants, primarily through the small openings called stomata on their leaves (Nelson, Pérez-Priego, & Zhou, 2020). As water vapor contains the latent heat of vaporization, the energy transported via this avenue forms a significant component of the global energy budget (Trenberth, Fasullo, & Mackaro, 2011). Thus ET plays a vital role in climate science and ecology by regulating water movement through ecosystems, affecting factors like photosynthesis, soil moisture, atmospheric stability and mixing, boundary-layer evolution, regional climate pattern, and overall water availability. Therefore systematic and accurate ET estimation is essential in various fields, including agriculture, hydrology, and climate science (Frank & Inouye, 1994; Pereira, Perrier, Allen, & Alves, 1999).

Any terrestrial ecosystem's carbon and water cycles are intricately coupled through the stomatal gas exchange mechanism (Gentine, Green, & Guérin, 2019). The water cost paid by the ecosystems for carbon uptake or photosynthesis is defined as water use efficiency (WUE), which is essential to managing the water requirement in agriculture and forest ecosystems to maintain ecosystem productivity (Lawson & Vialet-Chabrand, 2019). Plants may sequester more carbon in a CO_2-rich warmer climate and yield higher biomass (Ainsworth & Rogers, 2007; Davis, Sohngen, & Lewis, 2022). However, the larger uptake may come with an increasing water cost (Condon, Richards, Rebetzke, & Farquhar, 2004; Hamdy, Ragab, & Scarascia-Mugnozza, 2003), thereby further burdening the already strained global water cycle (Hu, Teng, Zhang, & Chen, 2019). Also, a diminishing trend in soil moisture may further downregulate the carbon uptake (Lenka & Lal, 2012). Hence, assessing the future changes in ET in response to climate change and warming remains imperative for diverse applications and studies, including water resource management, agriculture, and climate change research (Oki & Kanae, 2006).

ET is regulated by several environmental and biological factors, including climate type (Tabari & Talaee, 2014), vegetation physiology (Denmead, 1984), soil features (Wetzel & Chang, 1987), and land use (Li, Zhang, & Jing, 2017). At field level, it can be measured using a variety of methods, including pan evaporation (Roderick, Rotstayn, Farquhar, & Hobbins, 2007), lysimeter (Bala, Rawat, Misra, & Srivastava, 2016), micrometeorological gradient measurement (Liu & Xu, 2018), surface renewal technique (Snyder, Spano, & Pawu, 1996), surface energy balance (Wagle, Bhattarai, Gowda, & Kakani, 2017), Bowen ratio method (Irmak, Skaggs, & Chatterjee, 2014), scintillometer (McGloin, McGowan, & McJannet, 2014), and eddy covariance (EC) technique (Deb Burman, Sarma, & Morrison, 2019). In recent years, satellite-based methods have gained prominence for estimating ET over large areas by utilizing remotely sensed surface temperature, vegetation type and conditions, and meteorological variables. These techniques

Sustainable Development Perspectives in Earth Observation. DOI: https://doi.org/10.1016/B978-0-443-14072-3.00014-9

offer a valuable alternative for assessing periodic water fluxes at large spatial scales (Fisher, Malhi, & Bonal, 2009; Mallick, Toivonen, & Trebs, 2018).

Potential evapotranspiration (PET) is often used as an alternate measure of canopy water loss without recourse to the direct measurement of ET (Xiang, Li, Horton, & Feng, 2020). PET is a theoretical concept representing the maximum amount of water that could be lost through evaporation and transpiration if environmental factors were ideal without any stress to the ecosystem (Kristensen & Jensen, 1975). It is often estimated using empirical formulations like the Thornthwaite method (Thornthwaite, 1948), Shuttleworth-Wallace equation (Shuttleworth & Wallace, 1985), Hargreaves–Samani equation (Hargreaves and Samani, 1985), Priestley–Taylor equation (Priestley & Taylor, 1972), and Penman–Monteith equation (Monteith et al., 1965; Penman, 1948) using the meteorological variables as inputs. There are several intercomparison studies conducted to assess the relative performance of these methods in providing the most accurate estimate of atmospheric water loss by any vegetation canopy (Landeras, Ortiz-Barredo, & López, 2008; Stannard, 1993; Sumner & Jacobs, 2005). However, the Penman–Monteith equation is widely endorsed by the Food and Agricultural Organization (FAO) of the United States for this purpose (Allen, Pereira, Raes, & Smith, 1998). However, these models have limitations, especially in the data-scarce regions (Hao, Zhang, & Li, 2018; McAneney & Itier, 1996).

Carbon and water exchanges of terrestrial ecosystems can be simulated by the process-based dynamic global vegetation models (DGVMs) that rely on our understanding of the underlying processes and drivers controlling such exchanges in the soil-plant-atmosphere continuum (Best, Pryor, & Clark, 2011; Deb Burman, A.g., & Chakraborty, 2024; Sitch, Smith, & Prentice, 2003). Such models are parameterized based on the in situ observations for specific biome or plant functional types (Stöckli, Lawrence, & Niu, 2008; Williams, Richardson, & Reichstein, 2009). Whereas simplified modeling schemes lack the desired accuracy, process-based models demand extensive data for calibration and validation, posing challenges in data-scarce regions (Lawrence, Fisher, & Koven, 2019).

Widely biodiverse forest types exist over the large geographical span of India, experiencing divergent local climates (Forest Survey of India, 2021). Similar studies have reported the water exchange in different ecosystems, such as mixed teak deciduous forests in central India (Jha, Thumaty, & Rodda, 2013), a semiurban location in southern peninsular India (Patil, Waghmare, Dharmaraj, & Chinthalu, 2016), the mangrove in Indian Sundarbans (Rodda, Thumaty, Jha, & Dadhwal, 2016), a semievergreen deciduous forest in vegetation-rich Northeast India (Deb Burman et al., 2019), a grassland in the Indo-Gangetic Plain (Chakraborty, Sarangi, & Krishnan, 2019), a coniferous evergreen forest in the Western Himalayan region (Deb Burman, Launiainen, & Mukherjee, 2021), a mangrove along the Bay of Bengal coast in southern India (Deb Burman, Chakraborty, & El-Madany, 2022), tropical moist deciduous forest in the Himalayan foothills of Northwest India (Srinet, Nandy, & Watham, 2022), among others. However, India is mostly agrarian, with vast tracts of agricultural fields relying on monsoonal rainfall and irrigation activities for year-round production (Gadgil & Gadgil, 2006; McDermid, Nocco, & Lawston-Parker, 2023). The EC measurements that report the ET of these ecosystems with the maximum possible accuracy are sparse over the vast agricultural lands in India. The prominent studies on agriculture include rainfed crops like chickpeas and cotton in central India (Chakraborty, Choudhary, & Srikanth, 2021; Chakraborty, Venugopalan, & Mani, 2022). Bhat, Morrison, and Taylor (2020) described the EC measurement of ET at several stations in India to improve the forecast of monsoon processes by numerical weather prediction and land-surface models. Among the other studies, Singh, Patel, and Bhattacharya (2014) measured ET, its variability and ecosystem WUE in the western Himalayan forest in India using micrometeorological and remote sensing measurements. Tyagi, Satyanarayana, Rajvanshi, and Mandal (2014) measured the ET using micrometeorological measurements over southern West Bengal, and Raja, Singh, and Srinivas (2018) measured the ET over the Thar desert, an arid region, using the Bowen ratio method. Several researchers modeled the ET from remote sensing measurements (Bhattacharya, Mallick, Patel, & Parihar, 2010; Eswar, Sekhar, & Bhattacharya, 2017; Mallick, Bhattacharya, & Rao, 2009). These observations are subsequently used by Bhattacharya, Mallick, and Desai (2022) for remote sensing-based model development for estimating ET. Anand, Satheesh, and Moorthy (2022) improved the representation of surface energy fluxes in a modeled dataset using multiyear EC measurement of ET at a semiarid location in southern peninsular India. Satyanarayana, Lykossov, and Mohanty (2000) modeled the ET over a semiarid region in western India using a one-dimensional model forced by profile measurements of meteorological variables on the surface. Patil, Waghmare, Halder, and Dharmaraj (2011) tuned a land-surface model using in situ tower-based meteorological observations to simulate the ET over a semiarid region in Gujarat. Deb Burman et al. (2024) simulated the ET at the Kaziranga National Park in Northeast India using a DGVM forced by EC observations. However, the changes in ET over this region, in response to the changes in atmospheric and soil moisture demand and availability remain underexplored, and hence its impact on the terrestrial ecosystems native to this region remains uncertain (Krishna Kumar, Rupa Kumar, & Ashrit, 2004; Mathur & AchutaRao, 2020).

The coupled model intercomparison project (CMIP) by the World Climate Research Programme (WCRP) is an intermodel assessment project aimed at producing multiple realizations of the climate variables by different coupled Earth system models pertaining to different climate change scenarios (Meehl, Boer, & Covey, 2000). Thus it aims to solve the grand challenges in climate science for mitigation strategy and adaptation plan development. Diverse studies have been conducted using this dataset, such as changes in temperature and precipitation (Almazroui, Nazrul Islam, & Saeed, 2017; Grose, Narsey, & Delage, 2020), the occurrence of extreme events (Akinsanola, Kooperman, & Pendergrass, 2020; Chen, Wang, & Miao, 2021; Klutse, Quagraine, & Nkrumah, 2021), etc. Over the Indian region, CMIP6 data have been used to assess the occurrences of future heat waves in the country (Das & Umamahesh, 2022), to check the spatiotemporal variations in future rainfall at specific regions (Naskar, Singh, Pattanaik, & Katyar, 2023), to study the extreme rainfall (Chaubey & Mall, 2023), flood (Nanditha & Mishra, 2022), and drought events (Aadhar & Mishra, 2020; Zhai, Mondal, & Fischer, 2020), to estimate the terrestrial carbon uptake in future (Uchale, Deb Burman, & Tiwari, 2023), and to study Northeast monsoon variability (Tiwari, Mishra, Rai, & Pandey, 2023). However, there is a lack of a comprehensive study assessing the spatiotemporal changes in both precipitation and ET under changing climates. In this research gap, we conducted our present study with the following threefold objectives: first, to check the trends in ET and precipitation over India under different climate warming scenarios; second, to compare the intraannual variabilities of these two variables, and finally, to analyze the projected changes in water balance over the Indian landmass.

6.2 Materials and methods

6.2.1 Study region

Our study area is the Indian mainland (Fig. 6.1), a vast and densely populated country located in South Asia, spanning approximately 3.3 million km^2, extending from 8.4°N to 37.6°N, and from 68.7°E to 97.4°E.

6.2.2 Data used

6.2.2.1 Satellite record of actual evapotranspiration

The ET records used in this study were obtained from the moderate resolution imaging spectroradiometer (MODIS) satellite, spanning the years from 2000 to 2015. This dataset was acquired at 500 m spatial resolution. The data source

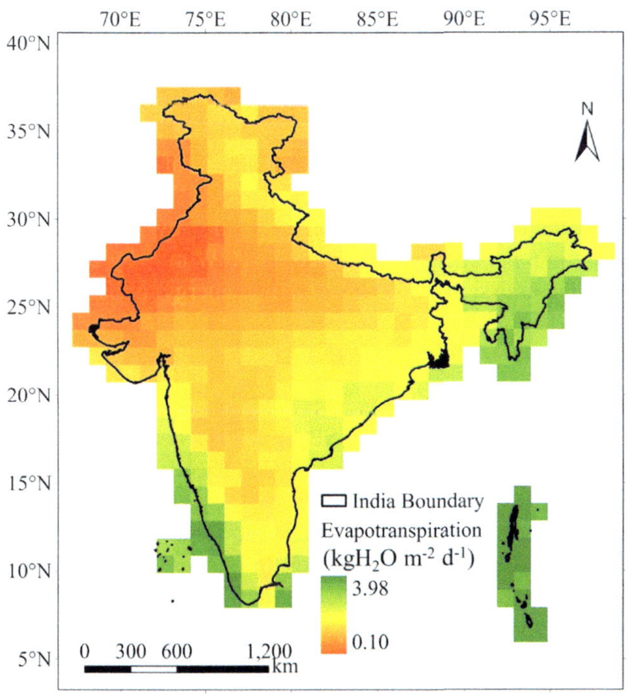

FIGURE 6.1 Annual mean daily evapotranspiration (ET) from 2000 to 2015 from moderate resolution imaging spectroradiometer (MODIS) terra ET product (MOD16A2GF Version 6.1). Map lines delineate study areas and do not necessarily depict accepted national boundaries.

specifically employed the MOD16A3 Version 6.1 Evapotranspiration/Latent Heat Flux product available at https://lpdaac.usgs.gov/products/mod16a3v006/, which is a yearly composite dataset. The Penman–Monteith equation is employed for this product development. It integrates daily meteorological reanalysis data and leverages a suite of MODIS remotely sensed data products, including variables on vegetation property, like albedo and land cover.

6.2.2.2 Ground observations

We gathered rainfall data from 2000 to 2022 for this study from a network of 1384 surface weather stations in India, which is maintained by the India Meteorological Department (IMD), available at https://www.imdpune.gov.in/cmpg/Griddata/Rainfall_1_NetCDF.html. This dataset, with a spatial resolution of $1^\circ \times 1^\circ$, includes historical daily rainfall records spanning more than a century from various weather stations in India (Rajeevan, Bhate, & Jaswal, 2008).

6.2.2.3 Shared socioeconomic pathways

The shared socioeconomic pathways (SSPs) are a set of possible future climate scenarios considering both emission and mitigation in the backdrop of probable future socioeconomic developments and their impact on climate change. The SSPs include assumptions about future energy use, land use, and greenhouse gas emissions, which are used to project future climate change impacts (O'Neill, Kriegler, & Riahi, 2014). These assumptions are translated into quantitative data using integrated assessment models (Meehl, Boer, & Covery, 2000; Taylor, Stouffer, & Meehl, 2012). The SSPs range from a world with high inequality and fossil fuel use, to a world with strong cooperation and low emissions (Riahi, van Vuuren, & Kriegler, 2017).

The following specific variables related to the energy and climate aspects are considered to design the SSPs. First, different levels of greenhouse gas emissions from human activities, such as CO_2, CH_4, and N_2O, are typically expressed in terms of concentration pathways or emission scenarios. Second, the energy usage patterns, including the type and energy sources utilized, include fossil fuel and renewable energy sources. The third component is land use practices and changes, such as deforestation, afforestation, urbanization, and agricultural practices. These changes can affect surface albedo, which, in turn, impacts the Earth's energy budget. Finally, the implementation of adaptation and mitigation strategies to combat climate change are also considered in designing the SSPs. These strategies can influence the trajectory of future climate impacts and net energy imbalances.

As an outcome of this, the seven different SSPs selected for climate modeling are (Nazarenko et al., 2022):

- SSP1–1.9: A low-end scenario resulting in 1.9 W m^{-2} of net energy imbalance by 2100.
- SSP1–2.6: A scenario with low radiative forcing resulting in a net energy imbalance of 2.6 W m^{-2} by 2100.
- SSP2–4.5: A scenario with medium radiative forcing resulting in a net energy imbalance of 4.5 W m^{-2} by 2100.
- SSP3–7.0: A scenario with a medium to high radiative forcing resulting in a net energy imbalance of 7.0 W m^{-2} by 2100.
- SSP4–3.4: A scenario with low radiative forcing resulting in a net energy imbalance of 3.4 W m^{-2} by 2100.
- SSP4–6.0: A scenario with medium radiative forcing resulting in a net energy imbalance of 6.0 W m^{-2} by 2100.
- SSP5–8.5: A scenario with high radiative forcing from high emissions producing a net energy imbalance of 8.5 W m^{-2} by 2100.

6.2.2.4 Climate model simulations

We have used the actual ET, potential ET, and precipitation records simulated for the SSPs described above by several climate models in CMIP6, available open-access at the Earth System Grid Federation (ESGF) archive (https://esgf-node.llnl.gov/search/cmip6/). These models are listed in Table 6.1. The r, l, p, and f flags in the variant-id label of any CMIP6 data refer to the particular realization, initialization method, perturbed physics number and forcing number used in generating that data, respectively (Eyring, Bony, & Meehl, 2016). To ensure consistency in our analysis involving multiple climate models, scenarios, and experiments, we have used only the datasets belonging to "r1i1p1f1," that is, all the flags being set to one. All the modeled data products are downloaded at a monthly resolution. The future analysis period is defined from 2015 to 2100, whereas the performance of the models is tested against the recorded datasets of ET from MODIS and precipitation ground measurement of IMD from 2000 to 2022.

TABLE 6.1 CMIP6 models used in the present study.

Sr. no.	Model	Variant label	Simulation period	Spatial resolution (° × °)	Temporal resolution	Institution (country)	References
1	ACCESS-CM2	r1i1p1f1	2015–2100	1.9° × 1.3°	Monthly	CSIRO-ARCCSS (Australia)	Bi, Dix, and Marsland (2020)
2	ACCESS-ESM1–5	r1i1p1f1	2015–2100	1.9° × 1.2°	Monthly	CSIRO (Australia)	Ziehn, Chamberlain, and Law (2020)
3	AWI-CM-1–1-MR	r1i1p1f1	2015–2100	0.9° × 0.9°	Monthly	AWI (Germany)	Semmler, Danilov, and Gierz (2020)
4	CAMS-CSM1–0	r1i1p1f1	2015–2100	1.1° × 1.1°	Monthly	CAMS (China)	Rong, Li, and Chen (2019)
5	CanESM5	r1i1p1f1	2015–2100	2.8° × 2.8°	Monthly	CCCMA (Canada)	Swart, Cole, and Kharin (2019)
6	CanESM5–1	r1i1p1f1	2015–2100	2.8° × 2.8°	Monthly	CCCMA (Canada)	Swart et al. (2019)
7	CAS-ESM2–0	r1i1p1f1	2015–2100	2.25° × 2.25°	Monthly	CAS (China)	Zhang, Zhang, and Jin (2020)
8	CESM2-WACCM	r1i1p1f1	2015–2100	1.3° × 0.9°	Monthly	NCAR (USA)	Danabasoglu, Lamarque, and Bacmeister (2020)
9	CMCC-CM2-SR5	r1i1p1f1	2015–2100	1.3° × 0.9°	Monthly	CMCC (Italy)	Cherchi, Fogli, and Lovato (2019)
10	CMCC-ESM2	r1i1p1f1	2015–2100	1.3° × 0.9°	Monthly	CMCC (Italy)	Lovato, Peano, and Butenschön (2022)
11	E3SM-1–0	r1i1p1f1	2015–2100	0.9° × 0.9°	Monthly	E3SM (USA)	Golaz, Caldwell, and Van Roekel (2019)
12	E3SM-1–1	r1i1p1f1	2015–2100	0.9° × 0.9°	Monthly	E3SM (USA)	Golaz et al. (2019)
13	E3SM-1–1-ECA	r1i1p1f1	2015–2100	0.9° × 0.9°	Monthly	E3SM (USA)	Golaz et al. (2019)
14	EC-Earth3-Veg	r1i1p1f1	2015–2100	0.7° × 0.7°	Monthly	EC-Earth Consortium (Europe)	Döscher, Acosta, and Alessandri (2022)
15	EC-Earth3-Veg-LR	r1i1p1f1	2015–2100	1.1° × 1.1°	Monthly	EC-Earth-Consortium (Europe)	Döscher et al. (2022)
16	FGOALS-f3-L	r1i1p1f1	2015–2100	1.3° × 1°	Monthly	CAS (China)	He et al. (2020)
17	FGOALS-g3	r1i1p1f1	2015–2100	2° × 2.3°	Monthly	CAS (China)	Lin, Zhao, and Wei (2022)
18	GFDL-ESM4	r1i1p1f1	2015–2100	1.3° × 1°	Monthly	NOAA GFDL (USA)	Dunne, Horowitz, and Adcroft (2020)
19	GFDL-CM4	r1i1p1f1	2015–2100	1.3° × 1°	Monthly	NOAA GFDL (USA)	Held, Guo, and Adcroft (2019)
20	IITM-ESM	r1i1p1f1	2015–2100	1.9° × 1.9°	Monthly	CCCR-IITM (India)	Swapna, Krishnan, and Sandeep (2018)
21	INM-CM4–8	r1i1p1f1	2015–2100	2° × 1.5°	Monthly	INM (Russia)	Volodin, Mortikov, and Kostrykin (2018)

(Continued)

TABLE 6.1 (Continued)

Sr. no.	Model	Variant label	Simulation period	Spatial resolution (° × °)	Temporal resolution	Institution (country)	References
22	INM-CM5–0	r1i1p1f1	2015–2100	2° × 1.5°	Monthly	INM (Russia)	Volodin et al. (2018)
23	IPSL-CM5A2-INCA	r1i1p1f1	2015–2100	2.25° × 2.25°	Monthly	IPSL (France)	Sepulchre, Caubel, and Ladant (2020)
24	IPSL-CM6A-LR	r1i1p1f1	2015–2100	2.5° × 1.3°	Monthly	IPSL (France)	Boucher, Servonnat, and Albright (2018)
25	MIROC6	r1i1p1f1	2015–2100	1.4° × 1.4°	Monthly	MIROC (Japan)	Kataoka, Tatebe, Koyama, and Mochizuki (2020)
26	MPI-ESM1–2-HR	r1i1p1f1	2015–2100	0.9° × 0.9°	Monthly	MPI-M (Germany)	Mauritsen, Bader, and Becker (2019)
27	MPI-ESM1–2-LR	r1i1p1f1	2015–2100	1.9° × 1.9°	Monthly	MPI-M (Germany)	Mauritsen et al. (2019)
28	MPI-ESM1–2-HR	r1i1p1f1	2015–2100	0.9° × 0.9°	Monthly	DKRZ (German)	Mauritsen et al. (2019)
29	MRI-ESM2–0	r1i1p1f1	2015–2100	1.1° × 1.1°	Monthly	MRI (Japan)	Yukimoto, Kawai, and Koshiro (2019)
30	NESM3	r1i1p1f1	2015–2100	1.9° × 1.9°	Monthly	NUIST (China)	Cao, Wang, and Yang (2018)
31	NorESM2-LM	r1i1p1f1	2015–2100	2.5° × 1.9°	Monthly	NCC (Norway)	Seland, Bentsen, and Olivié (2020)
32	NorESM2-MM	r1i1p1f1	2015–2100	0.9° × 1.3°	Monthly	NCC (Norway)	Seland et al. (2020)

6.2.2.4.1 Potential evapotranspiration

The Penman–Monteith equation is described below, which is used to compute PET (Allen et al., 1998).

$$\mathrm{ET} = \frac{0.408\Delta(R_n - G) + \gamma\frac{900}{T_{\mathrm{mean}} + 273}\quad u_2(e_s - e_a)}{\Delta + \gamma(1 + 0.34\quad u_2)} \tag{6.1}$$

where ET is the evapotranspiration (mm), R_n is the net surface radiation (MJ m^{-1} d^{-1}), G is the soil heat flux (MJ m^{-2} d^{-1}), T_{mean} is the air temperature at 2-m height (°C), u_2 is the wind speed at 2 m height (m s^{-1}), e_s is the saturation vapor pressure (kPa), e_a is actual vapor pressure (kPa), Δ is the slope of the vapor pressure curve (kPa °C^{-1}), and γ is the psychometric constant (kPa °C^{-1}).

6.2.3 Methodology

6.2.3.1 Climate data operators

The climate data operators (CDO) is an open-access command-line tool to analyze climate and weather data, which is available at https://code.mpimet.mpg.de/projects/cdo/. It is an open-source software package developed by the Max Planck Institute for Meteorology (MPI-M), Germany. CDO allows users to manipulate and analyze large, gridded climate datasets by offering a comprehensive suite of operators for performing various types of operations on climate data, such as statistical analysis, filtering, interpolation, and conversion between different file formats. It is supported on various computing platforms and can be used with a wide range of file formats, including NetCDF, GRIB, HDF, and binary formats. Different datasets used in our work have different spatial resolutions. For an efficient analysis, we have regridded all of those parameters to a common $1^\circ \times 1^\circ$ resolution following Anav, Friedlingstein, and Beer (2015). This is achieved using the "remapcon" command in CDO (Jones, 1999).

6.2.3.2 Extraction of data

We used the India administrative boundary layer in Python3 to extract the data from the CMIP6 netCDF files. The xarray (https://docs.xarray.dev/en/stable/), geopandas (https://geopandas.org/en/stable/), glob (https://docs.python.org/3/library/glob.html), and shapely (https://pypi.org/project/shapely/) packages were used in to process the netCDF files and extract the climate variable data for the region of interest. The tool—Application for Extracting and Exploring Analysis Ready Samples (AppEEARS) (https://appeears.earthdatacloud.nasa.gov/) is used for processing and retrieval of geospatial data. The MODIS data were preprocessed using the quality control layer, wherein the best quality data were used for further analysis. The scale factor (0.1) is applied to retrieve the actual ET value from the MODIS datasets.

6.2.3.3 Model ensemble mean time series

The monthly data of actual evapotranspiration (AET), PET, and precipitation (P_n) were extracted for all the available models within the CMIP6 dataset. Additionally, mean values for each month within the time span of 2015 to 2100 were computed, and monthly averages were aggregated to obtain yearly values from 2015 to 2100. These calculations were performed for each file derived from the various models under distinct SSPs. The standard deviation (SD) was computed for each of these files to assess the variability. Thus spatiotemporal multimodel ensemble mean time series of ET and P_n over India are obtained for each SSP. The error bars in these trends were set as one SD about the mean value. The units for both these variables were kept as kg H_2O m^{-2} d^{-1} for an efficient comparison among multiple datasets and their easy interpretation. Linear regression was applied to analyze temporal trends in ET and P_n across different SSPs. The technique revealed the direction and significance of changes, while error bars representing uncertainty. This approach provided data-driven insights into how ET and P_n respond to socioeconomic and environmental variations. The rates of increase in ET and P_n for each SSP computed this way are listed in Tables 6.2 and 6.3, respectively. The coefficient of determination (R^2) and *P*-values are also shown for the fits in these models. A simplified version of the

TABLE 6.2 Annual trend of evapotranspiration (ET) over India from 2015 to 2100.

SSP	Mean trend (10^{-3} kg H_2O m^{-2} y^{-1})	Maximum trend (10^{-3} kg H_2O m^{-2} y^{-2})				Minimum trend (10^{-3} kg H_2O m^{-2} y^{-2})			
		Value	Model	R^2	*P*-value	Value	Model	R^2	*P* value
SSP1−1.9	1.148 ± 0.828	2.366	GFDL-ESM4	0.356	.000	−0.23	FGOALS-g3	0.007	.456
SSP1−2.6	1.707 ± 1.06	4.21	CMCC-CM2-SR5	0.27	.000	−0.008	IPSL-CM5A2-INCA	0.000	.986
SSP2−4.5	2.359 ± 1.256	4.961	CanESM5	0.32	.000	−0.13	MRI-ESM2−0	0.001	.803
SSP3−7.0	2.228 ± 1.429	5.783	CanESM5	0.42	.000	0.341	GFDL-ESM4	0.009	.379
SSP4−3.4	1.648 ± 1.78	4.698	CanESM5	0.33	.000	0.182	MRI-ESM2−0	0.000	.77
SSP4−6.0	2.969 ± 5.448	7.069	CanESM5	0.000	.605	0.771	FGOALS-g3	0.089	.005
SSP5−8.5	3.883 ± 2.045	8.865	NESM3	0.75	.000	1.48	NorESM-MM	0.066	.017

TABLE 6.3 Annual trend of precipitation (P_n) over India from 2015 to 2100.

SSP	Mean trend (10^{-3} kg H_2O m^{-2} y^{-1})	Maximum trend (10^{-3} kg H_2O m^{-2} y^{-2})				Minimum trend (10^{-3} kg H_2O m^{-2} y^{-2})			
		Value	Model	R^2	*P*-value	Value	Model	R^2	*P*-value
SSP1–1.9	1.219 ± 1.442	3.415	GFDL-ESM4	0.116	.001	−0.34	CAMS-CSM1–0	0.002	.678
SSP1–2.6	2.319 ± 2.02	8.677	CMCC-CM2-SR5	0.172	.000	−0.42	IPSL-CM5A2-INCA	0.002	.685
SSP2–4.5	3.753 ± 2.525	8.991	CMCC-CM2-SR5	0.228	.000	−0.45	FGOALS-f3-L	0.001	.796
SSP3–7.0	5.406 ± 2.626	8.333	INM-CM5–0	0.385	.000	0.822	CAS-ESM2–0	0.01	.350
SSP4–3.4	3.204 ± 3.028	8.142	CanESM5	0.214	.000	0.591	MRIESM2–0	0.003	.644
SSP4–6.0	6.173 ± 4.995	14.019	CanESM5	0.479	.000	1.184	FGOALS-g3	0.018	.224
SSP5–8.5	9.574 ± 3.651	17.229	CanESM5	0.554	.000	4.393	MRI-ESM2–0	0.102	.003

water balance was computed by subtracting ET from P_n at an annual scale over each grid point (Allan, 2023; Zhang & Sun, 2012).

6.2.3.4 *Taylor diagram*

Taylor diagram is a graphical tool used in the field of statistics and meteorology to assess the similarity between observed data and data generated by models or simulations (Taylor, 2001). It consists of a polar plot where the distance from the origin to data points represents the SD of model data relative to the observed data, while the angle from the reference axis signifies the correlation coefficient.

6.2.3.5 *Biogeographic zones*

Biogeographic zones or eco-regions represent the regions wherein the climate, species, and ecosystems are uniform. India is classified into 10 broad biogeographic zones (Rodgers & Panwar, 1988). However, in the current study, the Indian Himalayan region is divided into two parts, namely the Eastern and Western Himalayan regions (Das & Behera, 2019; Mathur, Sivakumar, & Onial, 2014).

6.3 Results and discussion

The daily mean MODIS ET data retrieved from 2000 to 2015 is shown in Fig. 6.1. The MODIS ET data were resampled at $1^\circ \times 1^\circ$. The mean daily ET map indicated a higher value in forest-dominated wet regions in Andaman and Nicobar Islands, Northeast India, and the Western Ghats, followed by a moderate value in the Eastern Ghats. In comparison, a low value is observed in the Western Himalayan region and the Deccan Peninsula region. The lowest ET value was recorded in the arid region.

6.3.1 Performance of CMIP6 model simulations

This study assessed the performance of CMIP6 model simulations in replicating critical hydroclimatic variables, such as ET and P_n. Fig. 6.2 shows the mean annual time series of ET and P_n generated by various CMIP6 models across distinct SSPs. These simulations are compared against observational data, MODIS and IMD. All CMIP6 models exhibit increasing trends for ET in all scenarios from 2015 to 2100, with the highest trend being observed for the SSP5–8.5 scenario. The MODIS data demonstrate an increasing trend in the ET values from 2000 to 2022 (Fig. 6.2A). A similar

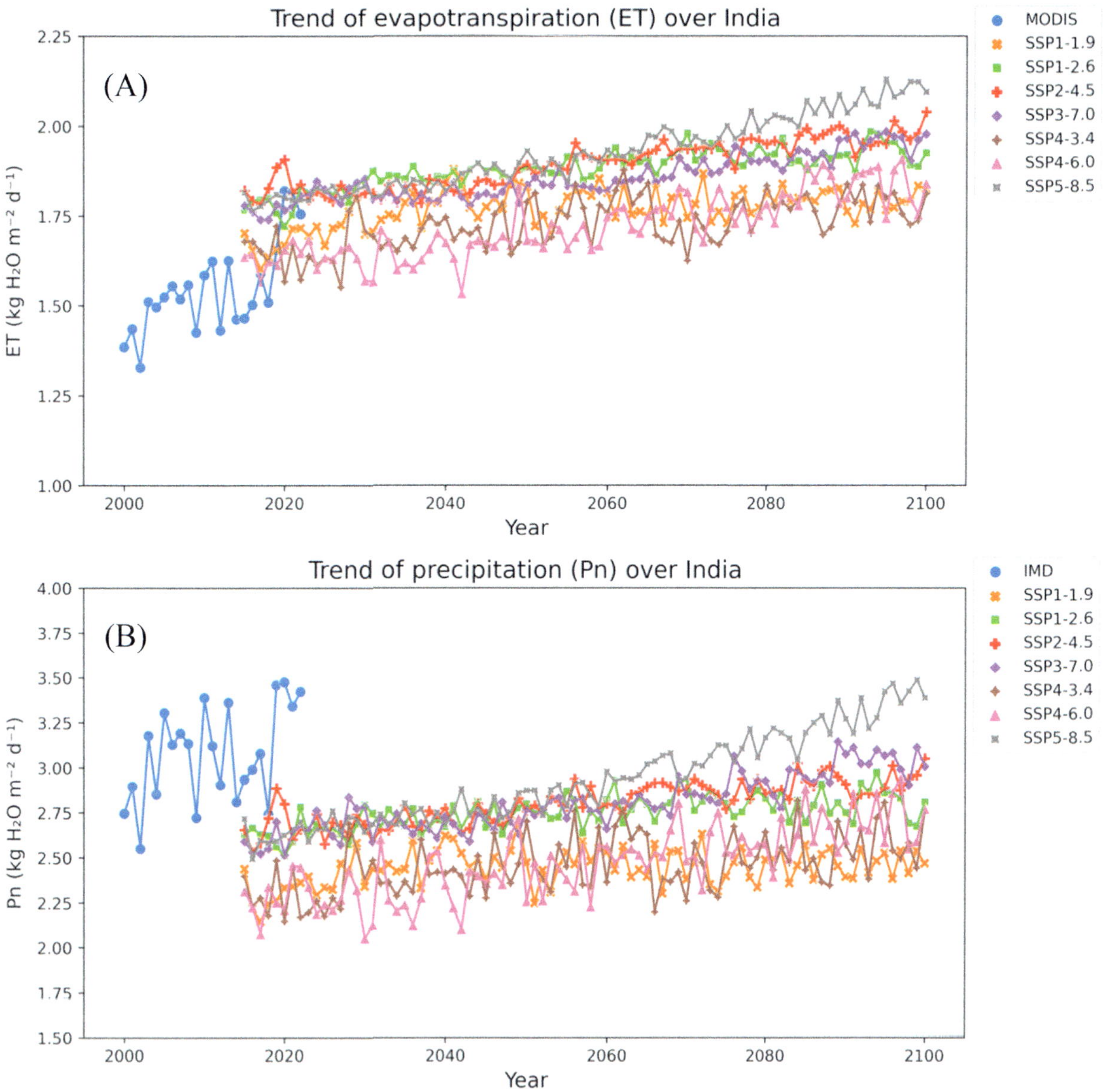

FIGURE 6.2 Annual values of (A) evapotranspiration and (B) precipitation projected by multimodel ensemble means pertaining to different shared socioeconomic pathways (SSPs), estimated from moderate resolution imaging spectroradiometer (MODIS) and measured by the India Meteorological Department (IMD). The y-axes scales are kept different between the two panels for better interpretation.

observation is found for P_n. CMIP6 models show increasing trends for P_n in all scenarios, with the maximum trend being recorded for SSP5−8.5. Observational P_n record by IMD also registered an increasing trend (Fig. 6.2B). However, the modeled P_n in CMIP6 underestimates the actual records, whereas minor overestimation is observed in the ET values (Fig. 6.2).

6.3.1.1 Potential evapotranspiration

The performance of the CMIP6 model was assessed by comparing MODIS ET data to showcase the robustness and reliability of CMIP6 data in capturing the ET dynamics across various scenarios (Fig. 6.3A). Correlation coefficients were computed for each scenario of SSP1−1.9 to SSP5−8.5, ranging from 0.9678 to 0.9941, indicating strong linear associations between the model outputs of CMIP6 and MODIS data. Notably, SSP2−4.5 demonstrated the highest correlation coefficient (0.9941), underlining its exceptional agreement with observed ET values and positioning it prominently on the Taylor diagram. Furthermore, SSP5−8.5, with a correlation coefficient of 0.9899, exhibited remarkable alignment

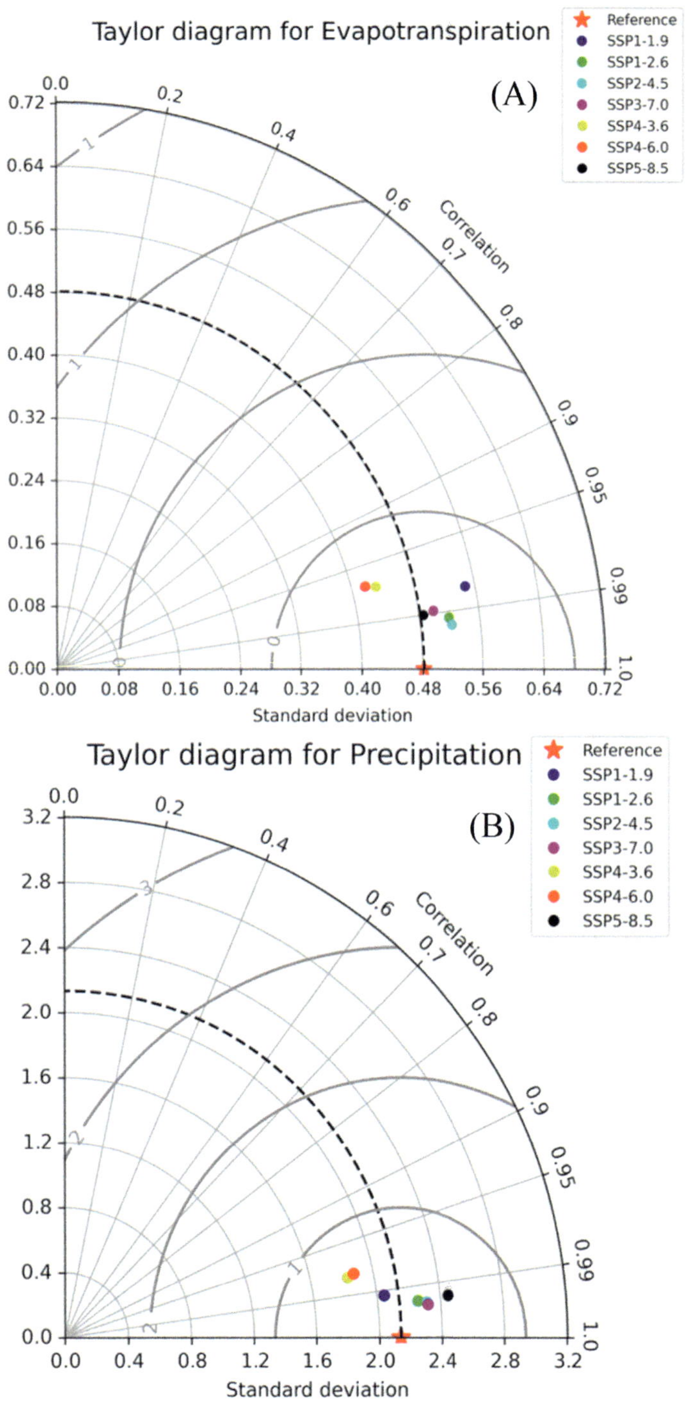

FIGURE 6.3 Taylor diagrams summarizing the standard deviation and correlation coefficient of the models from 2015 to 2100 for (A) evapotranspiration and (B) precipitation. The reference data are taken as the MODIS record in (A) and the India Meteorological Department measurements in (B).

with observed data, placing it prominently on the Taylor diagram. Even scenarios like SSP4−3.4, with a correlation coefficient of 0.9703, found their place on the diagram, further attesting to the model's versatility across diverse environmental scenarios. The inclusion of the Taylor diagram as in Fig. 6.3A adds an insightful dimension to our model evaluation. It visually represents the model's performance in terms of correlation, SD, and amplitude ratio, offering a comprehensive view of its capabilities. The consistent clustering of data points near the reference point on the diagram underscores the model's reliability in reproducing observed patterns, irrespective of scenario variations. These findings affirm the credibility of the CMIP6 models in estimating ET across different socioeconomic and climatic pathways. The Taylor diagram reinforces the model's ability to provide valuable insights into future water resource dynamics.

Overall, the various model's exceptional performance, as depicted on the Taylor diagram, strengthens its utility as a valuable tool for assessing the impacts of climate change and socioeconomic developments on ET, aiding in informed decision-making and sustainable water resource management.

6.3.1.2 Precipitation

The performance evaluation of the CMIP6 model for precipitation w.r.t IMD data, incorporating insights from the Taylor diagram as in Fig. 6.3B, reaffirms its precision and versatility in replicating precipitation patterns across a spectrum of socioeconomic and climatic scenarios. The computed correlation coefficients for each of SSP1−1.9 to SSP5−8.5 scenario ranged impressively from 0.9778 to 0.9962, illustrating substantial linear agreements between the CMIP6 model's precipitation outputs and IMD data. Particularly noteworthy is SSP3−7.0, exhibiting the highest correlation coefficient at 0.9962, firmly establishing its position on the Taylor diagram. This outstanding correlation underscores the model's proficiency in simulating precipitation dynamics, particularly in scenarios where radiative forcing reaches 7.0 W m^{-2} by 2100. Moreover, other scenarios like SSP2−4.5, SSP126, and SSP5−8.5, each with correlation coefficients exceeding 0.995, are also prominently featured on the Taylor diagram, demonstrating their notable alignment with observed precipitation data. Even scenarios such as SSP4−3.4, with a correlation coefficient of 0.9797, maintain a notable presence on the diagram, further substantiating the model's adaptability across diverse environmental contexts. The inclusion of the Taylor diagram, as in Fig. 6.3A, provides a comprehensive visual representation of our model's performance, encompassing not only correlation but also SD and amplitude ratio. The tight clustering of data points near the reference point on the diagram signifies the model's consistency in reproducing observed precipitation patterns, regardless of the specific scenario considered. The model's exceptional performance, as revealed through the Taylor diagram, solidifies its reputation as a powerful tool for assessing the impacts of climate change and socioeconomic developments on precipitation. Its ability to consistently capture precipitation dynamics across various scenarios underscores its value in informing decisions related to water resource management and climate change adaptation.

6.3.2 Future projections

6.3.2.1 Trends in evapotranspiration and precipitation

The time series of ET and P_n from 2015 to 2100 are shown in Fig. 6.2. Additionally, the spatiotemporal mean time series of ET computed from MODIS and P_n computed from IMD records are also included in this figure. In the absence of any suitable ground measurement of ET, the MODIS product is used to assess the performance of CMIP6 models in simulating ET, whereas the IMD record is used for the ground validation purpose for the CMIP models in predicting P_n. Both ET and P_n are seen to increase in this period for all the SSPs. However, the increase in P_n is larger than ET for all SSPs (Fig. 6.2). By 2100, the annual ET varies from 1.50 to 2.00 kg H_2O m^{-2} d^{-1} for different SSPs (Fig. 6.2A) whereas the variation in P_n for the same period is from 2.25 to 3.50 H_2O m^{-2} d^{-1} (Fig. 6.2B). The minimum increase in ET is recorded for SSPs 1−1.9, 4−3.4, and 4−6.0 (Fig. 6.2A) whereas the minimum increase for P_n is registered for SSP1−1.9 (Fig. 6.2B). The maximum increase in both ET and P_n is observed for SSP5−8.5 (Fig. 6.2).

6.3.2.1.1 Trend in evapotranspiration

For SSP1−1.9, the mean trend of ET is $(1.148 \pm 0.828) \times 10^{-3}$ kg H_2O m^{-2} y^{-2} (Table 6.2). Among the different models participating in this scenario, the maximum trend in ET is predicted by GFDL-ESM4 to be 2.366×10^{-3} kg H_2O m^{-2} y^{-2}, whereas the minimum trend is predicted to be -0.23×10^{-3} kg H_2O m^{-2} y^{-2} by FGOALS-g3 (Table 6.2). The mean trend of ET under SSP1−2.6 is $(1.707 \pm 1.06) \times 10^{-3}$ kg H_2O m^{-2} y^{-2} (Table 6.2), that varies from a minimum of -0.008×10^{-3} kg H_2O m^{-2} y^{-2} by IPSL-CM5A2-INCA to a maximum of 1.707 ± 1.06 by CMCC-CM2-SR5 (Table 6.2).

For SSP2−4.5, the average trend in ET is $(2.359 \pm 1.256) \times 10^{-3}$ kg H_2O m^{-2} y^{-2} (Table 6.2). However, for this scenario, the maximum and minimum trends are 4.961×0^{-3} kg H_2O m^{-2} y^{-2} and -0.33×10^{-3} kg H_2O m^{-2} y^{-2}, respectively, predicted by CanESM5 and MRI-ESM2, correspondingly (Table 6.2). It is interesting to note here that the modeled trends for these three scenarios vary from positive to negative, signifying a net increase or decrease in ET over India, correspondingly by the end of this century.

For SSP3−7.0, ET increases at a mean rate of $(2.228 \pm 1.429) \times 10^{-3}$ kg H_2O m^{-2} y^{-2} among all the models; maximum and minimum trends in ET for this scenario are 5.783×10^{-3} kg H_2O m^{-2} y^{-2} and 0.341×10^{-3} kg H_2O m^{-2} y^{-2}, respectively, projected by CanESM5 and GFDL-ESM4, correspondingly (Table 6.2).

For SSP4−3.4, maximum and minimum trends in ET are 4.698×10^{-3} kg H_2O m^{-2} y^{-2} by CanESM5 and 0.182×10^{-3} kg H_2O m^{-2} y^{-2} by MRI-ESM2, with the model ensemble mean being $(1.648 \pm 1.78) \times 10^{-3}$ kg H_2O m^{-2} y^{-2} (Table 6.2). For SSP4−6.0, mean increasing trend in ET is $(2.969 \pm 5.448) \times 10^{-3}$ kg H_2O m^{-2} y^{-2}, with the maximum and minimum trends being projected by CanESM5 and FGOALS-g3, respectively, as 7.069×10^{-3} kg H_2O m^{-2} y^{-2} and 0.771×10^{-3} kg H_2O m^{-2} y^{-2} (Table 6.2). Although a net mean increasing trend in ET is projected for this scenario, it has to be treated with the caveat that a high error bar can swing it to be negative (Table 6.2).

6.3.2.1.2 Trend in precipitation

Similar to ET, increasing trends are also recorded in P_n for all the SSPs (Fig. 6.2B). The maximum increasing trend in P_n is $(9.574 \pm 3.651) \times 10^{-3}$ kg H_2O m^{-2} y^{-2}, observed for SSP5−8.5, whereas the minimum trend in P_n is $(1.219 \pm 1.442) \times 10^{-3}$ kg H_2O m^{-2} y^{-2} for SSP1−1.9 (Table 6.3). However, for SSP1−1.9, the maximum trend in P_n is 3.415×10^{-3} kg H_2O m^{-2} y^{-2} simulated by GFDL-ESM4, but the minimum trend is decreasing with a rate of -0.34×10^{-3} kg H_2O m^{-2} y^{-2} by CAMS-CSM1−0 (Table 6.3). For SSP1−2.6, the ensemble mean annual rate of increase in P_n is $(2.319 \pm 2.02) \times 10^{-3}$ kg H_2O m^{-2} y^{-2} which varies between 8.677×10^{-3} kg H_2O m^{-2} y^{-2} by CMCC-CM2-SR5 and $- -0.42 \times 10^{-3}$ kg H_2O m^{-2} y^{-2} by IPSL-CM5A2-INCA (Table 6.3). For SSP2−4.5, the mean annual rate of P_n increase is $(3.753 \pm 2.525) \times 10^{-3}$ kg H_2O m^{-2} y^{-2} (Table 6.3). For this scenario also, the maximum rate of increase is positive (8.991×10^{-3} kg H_2O m^{-2} y^{-2} by CMCC-CM2-SR5) but the minimum rate is negative (-0.45×10^{-3} kg H_2O m^{-2} y^{-2} by FGOALS-f3) projected by the individual models (Table 6.3).

However, for the remaining scenarios, all the models project an unabashed increase in P_n in future (Table 6.3). For SSPs 3−7.0, 4−3.4, 4−6.0, and 5−8.5, the mean rates of increase in P_n are $(5.406 \pm 2.626) \times 10^{-3}$ kg H_2O m^{-2} y^{-2}, $(3.204 \pm 3.028) \times 10^{-3}$ kg H_2O m^{-2} y^{-2}, $(6.173 \pm 4.995) \times 10^{-3}$ kg H_2O m^{-2} y^{-2}, and $(9.574 \pm 3.651) \times 10^{-3}$ kg H_2O m^{-2} y^{-2}, respectively (Table 6.3). The maximum projected rates of increase in P_n for these scenarios are 8.142×10^{-3} kg H_2O m^{-2} y^{-2}, 14.019×10^{-3} kg H_2O m^{-2} y^{-2}, and 17.229×10^{-3} kg H_2O m^{-2} y^{-2}, all projected by CanESM5, except SSP3−7.0 for which the maximum projected rate of P_n increase is 8.333×10^{-3} kg H_2O m^{-2} y^{-2} by INM-CM5−0 (Table 6.3). For these scenarios, the minimum increase in P_n is projected to be 0.822×10^{-3} kg H_2O m^{-2} y^{-2} by CAS-ESM2−0, 0.591×10^{-3} kg H_2O m^{-2} y^{-2} by MRI-ESM2−0, 1.184×10^{-3} kg H_2O m^{-2} y^{-2} by FGOALS-g3, and 4.393×10^{-3} kg H_2O m^{-2} y^{-2} by MRI-ESM2−0 (Table 6.3).

Hence, we find ET and P_n both to register overall increasing trends commensurate with warming. However, several individual models predict a decrease in both of these for some SSPs with lower degrees of warming. Our finding is supported by Almazroui, Saeed, and Saeed (2020), who reported the average P_n over India to increase by 27.3% by 2100 under the SSP5−8.5 scenario. According to Abbas, Ullah, and Ullah (2022), the precipitation to increase over Pakistan, another South Asian country, in future climates for all SSPs. Using CMIP5 data, Almazroui et al. (2017) found the precipitation to increase over the southern Arabian peninsula but decline in its northern part in future. Using remote sensing data and machine learning models, Liu, Jiang, and Wang (2022) showed that ET will increase in SSP5−8.5 over the southeastern part of Asia and the southeastern and eastern parts of Africa.

6.3.2.2 *Intraannual variabilities*

Our extensive analysis of CMIP6 models, encompassing the evaluation of ET and P_n trends across a spectrum of SSP scenarios, demonstrated a consistent increase in both these variables. The trends became larger as we transitioned from SSP1−1.9 to SSP5−8.5. Notably, on average, P_n surpassed ET on an annual scale, suggesting an overall increase in surface water availability in these scenarios.

However, prominent seasonality in both these variables is found in different locations in India (Deb Burman et al., 2019; Srinet et al., 2022). Hence, it is not sufficient to study the variabilities of these variables at an annual scale to infer about the local water budget. Rather, it is imperative to study the variabilities of ET and P_n within a year in conjunction with each other. If the period of higher water loss is supplemented with higher rainfall amounts, it will be helpful for addressing the local water demand for agricultural and allied purposes. On the other hand, if a period of higher water demand is not supported by increased rainfall, which might peak at other times during the year, a strategic water management plan may be required to store the excess rainfall and distribute it efficiently.

To effectively study this, we have checked the average monthly variations in ET and Pn from 2015 to 2100 for two extreme climate scenarios, namely SSP1−1.9 and SSP5−8.5. Monthly boxplots of ET and Pn simulated by selected models for these two scenarios are shown in Figs. 6.4 and 6.5, respectively. In a boxplot, each box shows the variability in data for each month over multiple years. Inside each box, the central, lower, and upper lines indicate the median, first

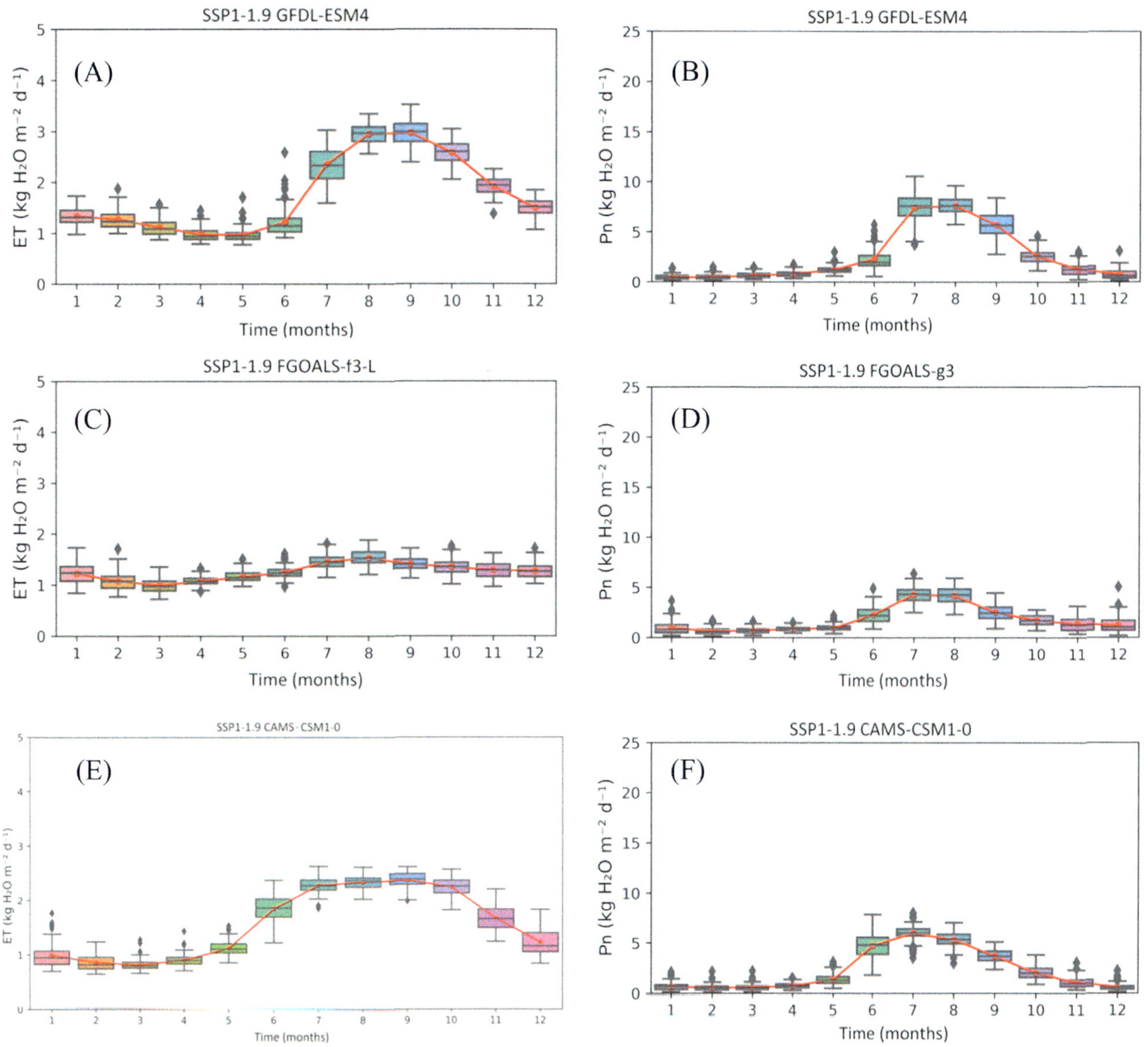

FIGURE 6.4 Box plots showing the monthly variability in daily evapotranspiration (ET_d in kg H_2O m^{-2} d^{-1}) and precipitation (P_n in kg H_2O m^{-2} d^{-1}) over India from 2015 to 2100 in SSP1−1.9 scenario as projected for (A) ET_d by GFDL-ESM4, (B) P_n by GFDL-ESM4, (C) ET_d by FGOALS-f3-L, (D) P_n by FGOALS-f3-L, (E) ET_d by CAMS-CSM1−0, and (F) P_n by CAMS-CSM1−0. Each box represents the variability in data for each month spanning multiple years; inside each box, the central, lower, and upper lines indicate the median, first quartile, and third quartile in the data. The lowest and highest whiskers of the box indicate the minimum and maximum of the data, respectively. Points located outside the whiskers are outliers.

quartile, and third quartile in the data. The lowest and highest whiskers of the box indicate the minimum and maximum values in the data record, respectively. Points located outside the whiskers are outliers.

For SSP1−1.9, GFDL-ESM4 predicts two peaks in ET in a year; annually, the maximum ET is predicted in September, followed by December; whereas the minimum ET is predicted in May (Fig. 6.4A). However, the same model predicts the P_n to peak in July and plummet to the lowest in January (Fig. 6.4B); the average P_n during the peak is $\sim 7.5 \times 10^{-3}$ kg H_2O m^{-2} y^{-1}, whereas the average ET during its peak is $\sim 3 \times 10^{-3}$ kg H_2O m^{-2} y^{-1}. For the same SSP, FGOALS-f3-L predicts maximum and minimum ET in August and March, with the corresponding values being 1.5×10^{-3} kg H_2O m^{-2} y^{-1} and 1×10^{-3} kg H_2O m^{-2} y^{-1}, respectively (Fig. 6.4C). Hence, the predicted intraannual variability in ET by FGOALS-f3-L is much subdued compared to GFDL-ESM4. However, FGOALS-f3-L also predicts a second smaller peak in ET in January with the overall distribution being bimodal in nature (Fig. 6.4C), similar to GFDL-ESM4. FGOALS-f3-L predicts the P_n to peak at $\sim 4 \times 10^{-3}$ kg H_2O m^{-2} y^{-1} in July and drop to the

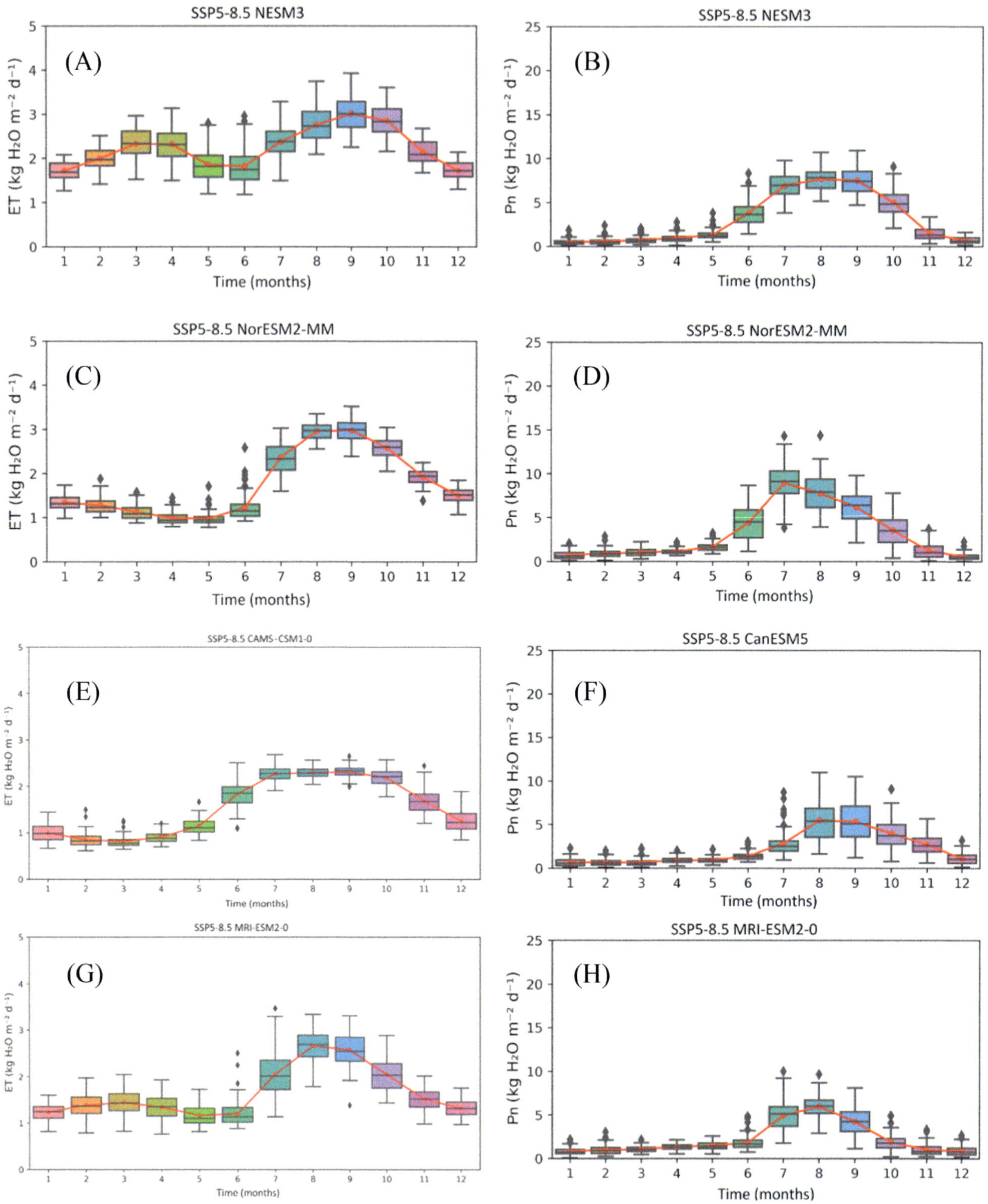

FIGURE 6.5 Box plots showing the monthly variability in daily evapotranspiration (ET_d in kg H_2O m^{-2} d^{-1}) and precipitation (P_n in kg H_2O m^{-2} d^{-1}) over India from 2015 to 2100 in SSP5–5.85 scenario as projected for (A) ET_d by NESM3, (B) P_n by NESM3, (C) ET_d by NorESM2-MM, (D) P_n by NorESM2-MM, (E) ET_d by CAMS-CSM1–0, (F) P_n by CanESM5, (G) ET_d by MRI-ESM2–0, and (H) Pn by MRI-ESM2–0. Each box represents the variability in data for each month spanning multiple years; inside each box, the central, lower, and upper lines indicate the median, first quartile, and third quartile in the data. The lowest and highest whiskers of the box indicate the minimum and maximum of the data, respectively. Points located outside the whiskers are outliers.

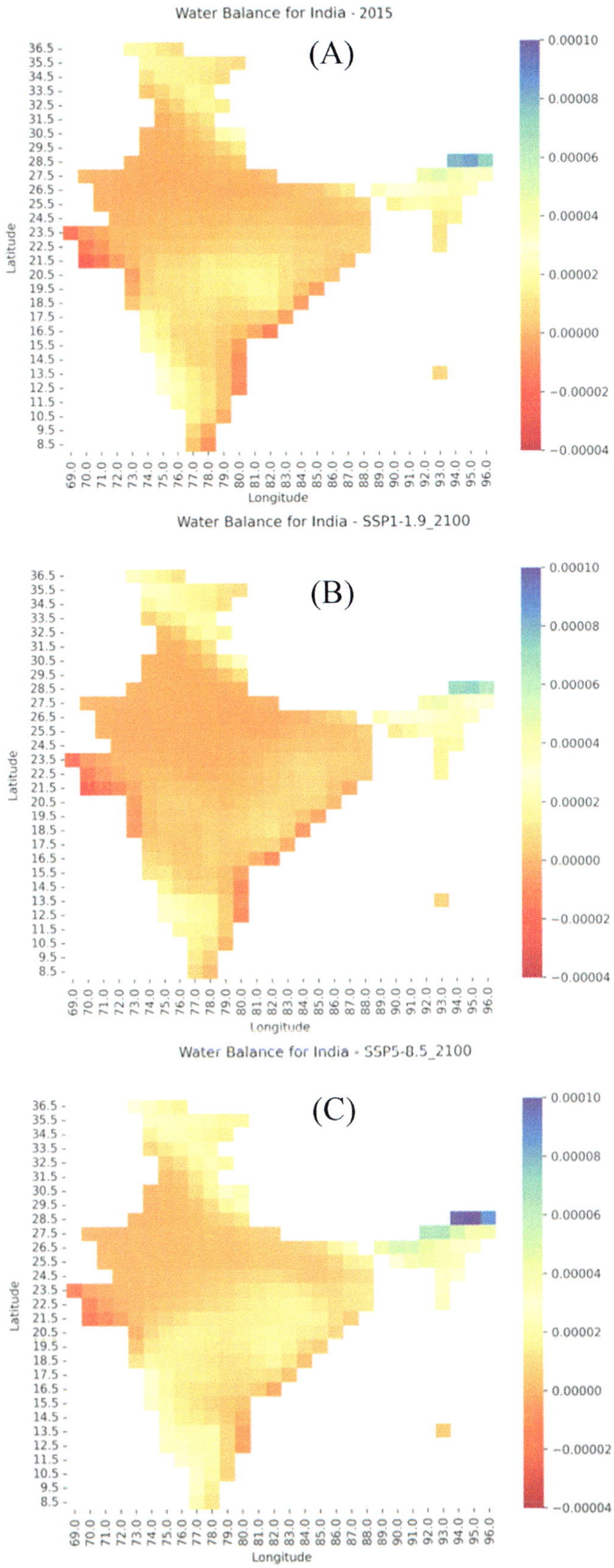

FIGURE 6.6 Annual average water balance (P_n - ET in kg H_2O m^{-2} y^{-1}) map over India in (A) 2015, (B) 2100 for SSP1−1.9 scenario, (C) 2100 for SSP5−8.5 scenario. Map lines delineate study areas and do not necessarily depict accepted national boundaries.

minimum in January (Fig. 6.4D). On the contrary, CAMS-CSM1−0 predicts a unimodal distribution for ET with peak and drop in September and March at ~2.5×10^{-3} kg H_2O m^{-2} y^{-1} and ~0.8×10^{-3} kg H_2O m^{-2} y^{-1}, respectively (Fig. 6.4E). The same model predicts the peak and lowest values in P_n in July at 7.5×10^{-3} kg H_2O m^{-2} y^{-1} (Fig. 6.4C) and March at 1×10^{-3} kg H_2O m^{-2} y^{-1} (Fig. 6.4C).

Hence, summarily, it is seen that different models predict diverse distributions of ET and P_n in a year, even if they correspond to the same climate scenario. The same model can predict a bimodal distribution for ET but a unimodal distribution for P_n. The number of peaks and their timings differ between water gain and loss by the land, which has important implications for regional water budget and management. For SSP5−8.5, a prominent bimodal distribution in ET is predicted by NESM3, with the primary and secondary peaks being registered in September and March, respectively, with the corresponding values being 3×10^{-3} kg H_2O m^{-2} y^{-1} and 2.5×10^{-3} kg H_2O m^{-2} y^{-1} (Fig. 6.5A); the year-round minimum ET is recorded in June at 1.0×10^{-3} kg H_2O m^{-2} y^{-1} (Fig. 6.5A). The same model, however, shows a unimodal distribution in P_n with the maximum being recorded in August at 7.5×10^{-3} kg H_2O m^{-2} y^{-1} and the minimum being recorded in January at 1×10^{-3} kg H_2O m^{-2} y^{-1} (Fig. 6.5A).

6.3.2.3 Water budget

A water budget, often referred to as a hydrological budget, is a systematic approach used to analyze the inflow, outflow, and storage of water within a specific region, watershed, or ecosystem. This assessment is crucial for several reasons. Firstly, it helps quantify the total water resources of an area, revealing whether there is a surplus or deficit. Secondly, it aids in identifying the sources of water and how it is allocated for various uses, including agriculture, industry, and domestic consumption. Furthermore, water budgets are indispensable for predicting and managing water resources in response to changing climate conditions. They can pinpoint areas susceptible to water scarcity or excessive flooding, guiding the development of strategies like constructing reservoirs or implementing water conservation measures. In essence, a water budget serves as a balance sheet for water in a given region.

The water budget over India is computed as (P_n − ET) at an annually averaged scale (Fig. 6.6). In 2015 the most prominent surplus water budget was observed in Eastern Himalaya (EH) (Fig. 6.6A). In addition to this, the North-East (NE), Western Himalaya (WH), Western Ghats (WG), and part of the Deccan Peninsula (DP) regions of India register marginally positive water budget. However, the water budget remains in deficit in arid and semiarid (AS), Gangetic Plain (GP), and the Eastern Coastal regions of India (Fig. 6.6A). Under the SSP1−1.9 scenario, the surplus water budget in EH, NE, and WG weakens by 2100, whereas no change is projected in WH (Fig. 6.6B). Moreover, in the rest of India, a deficit water budget can be seen, excluding some parts of WG (Fig. 6.6B). On the contrary, under the SSP5−8.5 scenario, the surplus water budget will increase. Most parts in EH, WH, WG, and parts of DP regions are projected to have surplus water budgets. The remaining regions in AS and the Eastern Coast of India are projected to have deficit water budgets (Fig. 6.6C).

Confounding opinion exists regarding the change in aridity over the Indian region, whereas, a slow but steady increase in aridity is observed over India in multiple studies (Choudhary, Mahato, & Roy, 2023; Ramarao, Sanjay, & Krishnan, 2018), several researchers have reported otherwise (Ambika & Mishra, 2020). However, our findings have further implications for the natural vegetation native to these eco-regions. Das, Behera, Bhaskaran, and Roy (2022) predicted the forest covers in wet climatic zones to be more resilient under climate change than the semiarid ones. Moreover, the dry tropical forests of the Eastern Ghats and Deccan peninsula are projected to have less resilience, which may trigger a forest to scrubland or grassland transition in future.

6.4 Conclusions

This study explored the spatiotemporal variabilities of ET and precipitation over India in the context of climate change. Overall, the climate models exhibited a steady increase in ET and precipitation, consistent with the findings from satellite-derived and ground-measured datasets in the present era. Both of these increasing trends are commensurate with the warming and continue till the end of the present century under all climate change scenarios; maximum trends are recorded for the SSP5−8.5 scenario. The multimodel ensemble mean trend of ET for SSP-1.19 and SSP5−5.85 are $(1.148 \pm 0.828) \times 10^{-3}$ kg H_2O m^{-2} y^{-1} and $(3.883 \pm 2.045) \times 10^{-3}$ kg H_2O m^{-2} y^{-1}, respectively. For the same scenarios, the multimodel ensemble mean trends of precipitation are $(1.219 \pm 1.442) \times 10^{-3}$ kg H_2O m^{-2} y^{-1} and $(9.574 \pm 6.671) \times 10^{-3}$ kg H_2O m^{-2} y^{-1}. Significant discord is noted among the modeled results. Individual ensemble members even predict decreasing trends in ET and precipitation, in contradiction to the predictions of many other ensemble members. The annual distribution, peak and lowest values of ET and precipitation vary widely. Peaks in ET and precipitation occur in different months. While all the models unanimously show an unimodal trend for rainfall, with the maximum values being concentrated in monsoon months, ET registers bimodal patterns in several cases. This has implications for local water storage and management. Overall, precipitation increases more than ET at the country scale.

The overall water budgets in most parts of India may experience a reduction under the SSP1−1.9 scenario, whereas they may be positive or surplus in the SSP5−8.5 scenario. Furthermore, the Indian Himalayan region and Western Ghats are projected to experience positive water budgets in the future. Our findings underscore the substantial impact that climate change is likely to exert on water resources in India. Moreover, the intricate interplay between variables across different climate scenarios and model outcomes highlights the need for a nuanced and adaptable approach to address these challenges. It is crucial to acknowledge the limitations of data availability as we seek to comprehend and mitigate the impending consequences of climate change on water resources and design more in-depth and region-specific studies to address these questions.

Acknowledgments

We gratefully acknowledge the Director of IITM Pune for his constant motivation and support. IITM Pune is an autonomous research institute fully funded by the Ministry of Earth Sciences (MoES), Government of India (GoI). We express our appreciation to the WCRP for making the CMIP6 dataset freely available for research and acknowledge the ESGF for archiving this data and facilitating access. We also thank the IMD for providing the gridded rainfall record over India. Pinaki Das acknowledges the Savitribai Phule Pune University for facilitating his master's project work at IITM Pune.

Author contribution

P.K.D.B. conceptualized the study. P.K.D.B. and S.V. designed the methodology. PI.D., PU.D., and P.K.D.B. analyzed the data. P.K.D.B. and PI.D. wrote the original draft of the manuscript. PU.D. and S.V. reviewed and edited the manuscript. P.K.D.B. supervised the overall work and communicated the manuscript.

Funding

Funding information is not applicable to this work.

Data availability

All the data used in this work are available open-access and their sources have been described in Section 2 of this manuscript.

References

Aadhar, S., & Mishra, V. (2020). On the projected decline in droughts over South Asia in CMIP6 multimodel ensemble. *J Geophys Res Atmos, 125*. Available from https://doi.org/10.1029/2020JD033587, e2020JD033587.

Abbas, A., Ullah, S., Ullah, W., Waseem, M., Dou, X., Zhao, C., … Ali, G. (2022). Evaluation and projection of precipitation in Pakistan using the Coupled Model Intercomparison Project Phase 6 model simulations. *Int J Climatol, 42*, 6665−6684. Available from https://doi.org/10.1002/joc.7602.

Ainsworth, E. A., & Rogers, A. (2007). The response of photosynthesis and stomatal conductance to rising [CO2]: mechanisms and environmental interactions. *Plant, cell \& Environ, 30*, 258−270. Available from https://doi.org/10.1111/j.1365-3040.2007.01641.x.

Akinsanola, A. A., Kooperman, G. J., Reed, K. A., Pendergrass, A. G., & Hannah, W. M. (2020). Seasonal representation of extreme precipitation indices over the United States in CMIP6 present-day simulations. *Environ Res Lett, 15*, 94003. Available from https://doi.org/10.1088/1748-9326/abb397.

Allan, R. P. (2023). Amplified seasonal range in precipitation minus evaporation. *Environ Res Lett, 18*, 94004. Available from https://doi.org/10.1088/1748-9326/acea36.

Allen, R. G., Pereira, L. S., Raes, D., & Smith, M. (1998). *Crop evapotranspiration: Guidelines for computing crop requirements*. Food and Agriculture Organization of the United Nations.

Almazroui, M., Islam, M. N., Saeed, S., Alkhalaf, A. K., & Dambul, R. (2017). Assessment of uncertainties in projected temperature and precipitation over the Arabian Peninsula using three categories of Cmip5 multimodel ensembles. *Earth Syst Environ, 1*, 1−20. Available from https://doi.org/10.1007/s41748-017-0027-5.

Almazroui, M., Saeed, S., Saeed, F., Islam, M. N., & Ismail, M. (2020). Projections of Precipitation and Temperature over the South Asian Countries in CMIP6. *Earth Syst Environ, 4*, 297−320. Available from https://doi.org/10.1007/s41748-020-00157-7.

Ambika, A. K., & Mishra, V. (2020). Substantial decline in atmospheric aridity due to irrigation in India. *Environ Res Lett*, *15*, 124060. Available from https://doi.org/10.1088/1748-9326/abc8bc.

Anav, A., Friedlingstein, P., Beer, C., Ciais, P., Harper, A., Jones, C., ... Zhao, M. (2015). Spatiotemporal patterns of terrestrial gross primary production: A review. *Reviews of Geophysics (Washington, D.C. : 1985)*, *53*, 1–34. Available from https://doi.org/10.1002/2015RG000483.

Bala, A., Rawat, K. S., Misra, A. K., & Srivastava, A. (2016). Assessment and validation of evapotranspiration using SEBAL algorithm and Lysimeter data of IARI agricultural farm, India. *Geocarto International*, *31*, 739–764. Available from https://doi.org/10.1080/10106049.2015.1076062.

Best, M. J., Pryor, M., Clark, D. B., Rooney, G. G., Essery, R. L. H., Ménard, C. B., ... Harding, R. J. (2011). The Joint UK Land Environment Simulator (JULES), model description – Part 1: Energy and water fluxes. *Geosci Model Dev*, *4*, 677–699. Available from https://doi.org/10.5194/gmd-4-677-2011.

Bhat, G. S., Morrison, R., Taylor, C. M., Bhattacharya, B. K., Paleri, S., Desai, D., ... Sisodiya, A. (2020). Spatial and temporal variability in energy and water vapour fluxes observed at seven sites on the Indian subcontinent during 2017. *Q J R Meteorol Soc*, *146*, 2853–2866. Available from https://doi.org/10.1002/qj.3688.

Bhattacharya, B. K., Mallick, K., Desai, D., Bhat, G. S., Morrison, R., Clevery, J. R., ... Baldocchi, D. (2022). A coupled ground heat flux–surface energy balance model of evaporation using thermal remote sensing observations. *Biogeosciences*, *19*, 5521–5551. Available from https://doi.org/10.5194/bg-19-5521-2022.

Bhattacharya, B. K., Mallick, K., Patel, N. K., & Parihar, J. S. (2010). Regional clear sky evapotranspiration over agricultural land using remote sensing data from Indian geostationary meteorological satellite. *J Hydrol*, *387*, 65–80. Available from https://doi.org/10.1016/j.jhydrol.2010.03.030.

Bi, D., Dix, M., Marsland, S., O'Farrell, S., Sullivan, A., Bodman, R., ... Heerdegen, A. (2020). Configuration and spin-up of ACCESS-CM2, the new generation Australian Community Climate and Earth System Simulator Coupled Model. *J South Hemisph Earth Syst Sci*, *70*, 225–251. Available from https://doi.org/10.1071/ES19040.

Boucher, O., Servonnat, J., Albright, A. L. A., Aumont, O., Balkanski, Y., Bastrikov, V., ... Vuichard, N. (2018). Presentation and Evaluation of the IPSL - CM6A - LR Climate Model. *Journal of Advances in Modeling Earth Systems*, *12*, 1–52. Available from https://doi.org/10.1029/2019MS002010.

Cao, J., Wang, B., Yang, Y. M., Ma, L., Li, J., Sun, B., ... Wu, L. (2018). The NUIST Earth System Model (NESM) version 3: description and preliminary evaluation. *Geosci Model Dev*, *11*, 2975–2993. Available from https://doi.org/10.5194/gmd-11-2975-2018.

Chakraborty, A., Choudhary, K. K., Srikanth, P., Ramana, K. V., Seshasai, M. V. R., Narayanarao, K., ... Annapurna, G. (2021). CO2, H2O and energy fluxes from chickpea crop grown under residual soil moisture condition in rainfed peninsular India using eddy covariance techniques. *F Crop Res*. Available from https://doi.org/10.1016/j.fcr.2021.108307.

Chakraborty, A., Venugopalan, M. V., Mani, J. K., Bagadkar, A. J., & Manikandan, A. (2022). Rainfed cotton crop in central India is a strong net CO2 sink: An eddy covariance-based analysis of ecosystem fluxes. *F Crop Res*, *286*, 108595. Available from https://doi.org/10.1016/j.fcr.2022.108595.

Chakraborty, T., Sarangi, C., Krishnan, M., Tripathi, S. N., Morrison, R., & Evans, J. (2019). Biases in model-simulated surface energy fluxes during the Indian monsoon onset period. *Boundary-Layer Meteorol*, *170*, 323–348. Available from https://doi.org/10.1007/s10546-018-0395-x.

Chattopadhyay, N., & Hulme, M. (1997). Evaporation and potential evapotranspiration in India under conditions of recent and future climate change. *Agric For Meteorol*, *87*, 55–73. Available from https://doi.org/10.1016/S0168-1923(97)00006-3.

Chaubey, P. K., & Mall, R. K. (2023). Intensification of extreme rainfall in Indian river basin: Using bias corrected CMIP6 climate data. *Earth's Futur*, *11*. Available from https://doi.org/10.1029/2023EF003556, e2023EF003556.

Chen, L., Wang, G., Miao, L., Gnyawali, K. R., Li, S., Amankwah, S. O. Y., ... Zhan, M. (2021). Future drought in CMIP6 projections and the socioeconomic impacts in China. *Int J Climatol*, *41*, 4151–4170. Available from https://doi.org/10.1002/joc.7064.

Cherchi, A., Fogli, P. G., Lovato, T., Peano, D., Iovino, D., Gualdi, S., ... Navarra, A. (2019). Global Mean Climate and Main Patterns of Variability in the CMCC-CM2 Coupled Model. *J Adv Model Earth Syst*, *11*, 185–209. Available from https://doi.org/10.1029/2018MS001369.

Choudhary, A., Mahato, S., Roy, P. S., Pandey, D. N., & Joshi, P. K. (2023). Analyzing the long-term variability and trend of aridity in India using non-parametric approach. *Stoch Environ Res Risk Assess 1–18*. Available from https://doi.org/10.1007/s00477-023-02483-4.

Condon, A. G., Richards, R. A., Rebetzke, G. J., & Farquhar, G. D. (2004). Breeding for high water-use efficiency. *Journal of Experimental Botany*, *55*, 2447–2460. Available from https://doi.org/10.1093/jxb/erh277.

Danabasoglu, G., Lamarque, J. F., Bacmeister, J., Bailey, D. A., DuVivier, A. K., Edwards, J., ... Strand, W. G. (2020). The Community Earth System Model Version 2 (CESM2). *J Adv Model Earth Syst*, *12*, 1–35. Available from https://doi.org/10.1029/2019MS001916.

Das, J., & Umamahesh, N. V. (2022). Heat wave magnitude over India under changing climate: Projections from CMIP5 and CMIP6 experiments. *Int J Climatol*, *42*, 331–351. Available from https://doi.org/10.1002/joc.7246.

Das, P., & Behera, M. D. (2019). Can the forest cover in India withstand large climate alterations? *Biodivers Conserv*, *28*, 2017–2033. Available from https://doi.org/10.1007/s10531-019-01759-y.

Das, P., Behera, M. D., Bhaskaran, P. K., & Roy, P. S. (2022). Forest cover resilience to climate change over India using the MC2 dynamic vegetation model. *Environmental Monitoring and Assessment*, *194*, 903. Available from https://doi.org/10.1007/s10661-022-10545-3.

Davis, E. C., Sohngen, B., & Lewis, D. J. (2022). The effect of carbon fertilization on naturally regenerated and planted US forests. *Nat Commun*, *13*, 5490. Available from https://doi.org/10.1038/s41467-022-33196-x.

Deb Burman, P. K., Prajeesh, A. G., Chakraborty, S., Tiwari, Y. K., Sarma, D., & Gogoi, N. (2024). Simulating the ecosystem-atmosphere carbon, water and energy fluxes at a subtropical Indian forest using an ecosystem model. *Ecological Modelling*, *490*, 110637. Available from https://doi.org/10.1016/j.ecolmodel.2024.110637.

Deb Burman, P. K., Chakraborty, S., El-Madany, T. S., Ramasubramanian, R., Gogoi, N., Gnanamoorthy, P., ... Karipot, A. (2022). A comparative study of ecohydrologies of a tropical mangrove and a broadleaf deciduous forest using eddy covariance measurement. *Meteorol Atmos Phys, 134*, 4. Available from https://doi.org/10.1007/s00703-021-00840-y.

Deb Burman, P. K., Launiainen, S., Mukherjee, S., Chakraborty, S., Gogoi, N., Murkute, C., ... Kumar, K. (2021). Ecosystem-atmosphere carbon and water exchanges of subtropical evergreen and deciduous forests in India. *For Ecol Manage, 495*, 119371. Available from https://doi.org/10.1016/j.foreco.2021.119371.

Deb Burman, P. K., Sarma, D., Morrison, R., Karipot, A., & Chakraborty, S. (2019). Seasonal variation of evapotranspiration and its effect on the surface energy budget closure at a tropical forest over north-east India. *J Earth Syst Sci, 128*, 127. Available from https://doi.org/10.1007/s12040-019-1158-x.

Denmead, O. T. (1984). *Plant physiological methods for studying evapotranspiration: problems of telling the forest from the trees. Developments in Agricultural and Managed Forest Ecology* (pp. 167–189). Elsevier. Available from https://doi.org/10.1016/0378-3774(84)90052-0.

Dunne, J. P., Horowitz, L. W., Adcroft, A. J., Ginoux, P., Held, I. M., John, J. G., ... Zhao, M. (2020). The GFDL Earth System Model Version 4.1 (GFDL-ESM 4.1): Overall Coupled Model Description and Simulation Characteristics. *J Adv Model Earth Syst, 12*. Available from https://doi.org/10.1029/2019MS002015.

Döscher, R., Acosta, M., Alessandri, A., Anthoni, P., Arsouze, T., Bergman, T., ... Zhang, Q. (2022). The EC-Earth3 Earth system model for the Coupled Model Intercomparison Project 6. *Geosci Model Dev, 15*, 2973–3020. Available from https://doi.org/10.5194/gmd-15-2973-2022.

Eswar, R., Sekhar, M., & Bhattacharya, B. K. (2017). Comparison of three remote sensing based models for the estimation of latent heat flux over India. *Hydrol Sci J, 62*, 2705–2719. Available from https://doi.org/10.1080/02626667.2017.1404067.

Eyring, V., Bony, S., Meehl, G. A., Senior, C. A., Stevens, B., Stouffer, R. J., ... Taylor, K. E. (2016). Overview of the Coupled Model Intercomparison Project Phase 6 (CMIP6) experimental design and organization. *Geosci Model Dev, 9*, 1937–1958. Available from https://doi.org/10.5194/gmd-9-1937-2016.

Fisher, J. B., Malhi, Y., Bonal, D., Da Rocha, H. R., De Araújo, A. C., Gamo, M., ... Von Randow, C. (2009). The land–atmosphere water flux in the tropics. *Glob Chang Biol, 15*, 2694–2714. Available from https://doi.org/10.1111/j.1365-2486.2008.01813.x.

Forest Survey of India (2021) India State of Forest Report 2021. Ministry of Environment Forest and Climate Change, Dehradun, India, https://fsi.nic.in/forest-report-2021-details.

Frank, D. A., & Inouye, R. S. (1994). Temporal Variation in Actual Evapotranspiration of Terrestrial Ecosystems: Patterns and Ecological Implications. *J Biogeogr, 21*, 401–411. Available from https://doi.org/10.2307/2845758.

Gadgil, S., & Gadgil, S. (2006). The Indian monsoon, GDP and agriculture. *Econ Polit Wkly 4887–4895*. Available from https://doi.org/10.2307/4418949.

Gentine, P., Green, J. K., Guérin, M., Humphrey, V., Seneviratne, S. I., Zhang, Y., & Zhou, S. (2019). Coupling between the terrestrial carbon and water cycles—a review. *Environ Res Lett, 14*, 83003. Available from https://doi.org/10.1088/1748-9326/ab22d6.

Golaz, J. C., Caldwell, P. M., Van Roekel, L. P., Petersen, M. R., Tang, Q., Wolfe, J. D., ... Zhu, Q. (2019). The DOE E3SM Coupled Model Version 1: Overview and Evaluation at Standard Resolution. *J Adv Model Earth Syst, 11*, 2089–2129. Available from https://doi.org/10.1029/2018MS001603.

Grose, M. R., Narsey, S., Delage, F. P., Dowdy, A. J., Bador, M., Boschat, G., ... Power, S. (2020). Insights from CMIP6 for Australia's future climate. *Earth's Futur, 8*e2019EF001469. Available from https://doi.org/10.1029/2019EF001469.

Hargreaves, H., & Samani, Z. A. (1985). Reference Crop Evapotranspiration from Temperature. *Applied Engineering in Agriculture, 1*(2), 96–99. Available from https://doi.org/10.13031/2013.26773.

Hamdy, A., Ragab, R., & Scarascia-Mugnozza, E. (2003). Coping with water scarcity: water saving and increasing water productivity. Irrig Drain J Int Comm. *Irrig Drain, 52*, 3–20. Available from https://doi.org/10.1002/ird.73.

Hao, X., Zhang, S., Li, W., Duan, W., Fang, G., Zhang, Y., & Guo, B. (2018). The uncertainty of Penman-Monteith method and the energy balance closure problem. *J Geophys Res Atmos, 123*, 7433–7443. Available from https://doi.org/10.1029/2018JD028371.

HE, B., YU, Y., BAO, Q., LIN, P., LIU, H., LI, J., ... XIE, J. (2020). CAS FGOALS-f3-L model dataset descriptions for CMIP6 DECK experiments. *Atmos Ocean Sci Lett, 13*, 582–588. Available from https://doi.org/10.1080/16742834.2020.1778419.

Held, I. M., Guo, H., Adcroft, A., Dunne, J. P., Horowitz, L. W., Krasting, J., ... Zadeh, N. (2019). Structure and Performance of GFDL's CM4.0 Climate Model. *J Adv Model Earth Syst, 11*, 3691–3727. Available from https://doi.org/10.1029/2019MS001829.

Hu, B., Teng, Y., Zhang, Y., & Chen, Z. (2019). The projected hydrologic cycle under the scenario of 936 ppm CO_2 in 2100. *Hydrogeol J, 27*, 31–53. Available from https://doi.org/10.1007/s10040-018-1844-9.

Irmak, S., Skaggs, K. E., & Chatterjee, S. (2014). A review of the Bowen ratio surface energy balance method for quantifying evapotranspiration and other energy fluxes. *Trans ASABE, 57*, 1657–1674. Available from https://doi.org/10.13031/trans.57.10686.

Jha, C. S., Thumaty, K. C., Rodda, S. R., Sonakia, A., & Dadhwal, V. K. (2013). Analysis of carbon dioxide, water vapour and energy fluxes over an Indian teak mixed deciduous forest for winter and summer months using eddy covariance technique. *J Earth Syst Sci, 122*, 1259–1268. Available from https://doi.org/10.1007/s12040-013-0350-7.

Jones, P. W. (1999). First-and second-order conservative remapping schemes for grids in spherical coordinates. *Mon Weather Rev, 127*, 2204–2210. Available from https://doi.org/10.1175/1520-0493(1999)127%3C2204:FASOCR%3E2.0.CO;2.

Kataoka, T., Tatebe, H., Koyama, H., & Mochizuki, T. (2020). Seasonal to Decadal Predictions With MIROC6: Description and Basic Evaluation. *Journal of Advances in Modeling Earth Systems*, 1–25. Available from https://doi.org/10.1029/2019MS002035.

Katul, G. G., Oren, R., Manzoni, S., Higgins, C., & Parlange, M. B. (2012). Evapotranspiration: a process driving mass transport and energy exchange in the soil-plant-atmosphere-climate system. *Reviews of Geophysics (Washington, D.C. : 1985)*, *50*, RG3002. Available from https://doi.org/10.1029/2011RG000366.

Klutse, N. A. B., Quagraine, K. A., Nkrumah, F., Quagraine, K. T., Berkoh-Oforiwaa, R., Dzrobi, J. F., & Sylla, M. B. (2021). The climatic analysis of summer monsoon extreme precipitation events over West Africa in CMIP6 simulations. *Earth Syst Environ*, *5*, 25–41. Available from https://doi.org/10.1007/s41748-021-00203-y.

Kool, D., Agam, N., Lazarovitch, N., Heitman, J. L., Sauer, T. J., & Ben-Gal, A. (2014). A review of approaches for evapotranspiration partitioning. *Agric For Meteorol*, *184*, 56–70. Available from https://doi.org/10.1016/j.agrformet.2013.09.003.

Krishna Kumar, K., Rupa Kumar, K., Ashrit, R. G., Deshpande, N. R., & Hansen, J. W. (2004). Climate impacts on Indian agriculture. *Int J Climatol*, *24*, 1375–1393. Available from https://doi.org/10.1002/joc.1081.

Kristensen, K. J., & Jensen, S. E. (1975). A model for estimating actual evapotranspiration from potential evapotranspiration. *Hydrol Res*, *6*, 170–188. Available from https://doi.org/10.2166/nh.1975.0012.

Landeras, G., Ortiz-Barredo, A., & López, J. J. (2008). Comparison of artificial neural network models and empirical and semi-empirical equations for daily reference evapotranspiration estimation in the Basque Country (Northern Spain). *Agric water Manag*, *95*, 553–565. Available from https://doi.org/10.1016/j.agwat.2007.12.011.

Lawrence, D. M., Fisher, R. A., Koven, C. D., Oleson, K. W., Swenson, S. C., Bonan, G., ... Zeng, X. (2019). The Community Land Model Version 5: Description of New Features, Benchmarking, and Impact of Forcing Uncertainty. *J Adv Model Earth Syst*, *11*, 4245–4287. Available from https://doi.org/10.1029/2018MS001583.

Lawson, T., & Vialet-Chabrand, S. (2019). Speedy stomata, photosynthesis and plant water use efficiency. *The New Phytologist*, *221*, 93–98. Available from https://doi.org/10.1111/nph.15330.

Lenka, N. K., & Lal, R. (2012). Soil-related constraints to the carbon dioxide fertilization effect. CRC. *Crit Rev Plant Sci*, *31*, 342–357. Available from https://doi.org/10.1080/07352689.2012.674461.

Li, G., Zhang, F., Jing, Y., Liu, Y., & Sun, G. (2017). Response of evapotranspiration to changes in land use and land cover and climate in China during 2001–2013. *The Science of the Total Environment*, *596*, 256–265. Available from https://doi.org/10.1016/j.scitotenv.2017.04.080.

Lin, P., Zhao, B., Wei, J., Liu, H., Zhang, W., Chen, X., ... Zhou, T. (2022). The Super-large Ensemble Experiments of CAS FGOALS-g3. *Advances in Atmospheric Sciences*, *39*, 1746–1765.

Liu, S., & Xu, Z. (2018). Micrometeorological methods to determine evapotranspiration. *Obs Meas Ecohydrol Process*. Available from https://doi.org/10.1007/978-3-662-47871-4_7-2.

Liu, Y., Jiang, Q., Wang, Q., Jin, Y., Yue, Q., Yu, J., ... Yao, X. (2022). The divergence between potential and actual evapotranspiration: An insight from climate, water, and vegetation change. *The Science of the Total Environment*, *807*, 150648. Available from https://doi.org/10.1016/j.scitotenv.2021.150648.

Lovato, T., Peano, D., Butenschön, M., Materia, S., Iovino, D., Scoccimarro, E., ... Navarra, A. (2022). CMIP6 Simulations With the CMCC Earth System Model (CMCC-ESM2). *J Adv Model Earth Syst*, *14*. Available from https://doi.org/10.1029/2021MS002814.

Mallick, K., Bhattacharya, B. K., Rao, V. U. M., Reddy, D. R., Banerjee, S., Venkatesh, H., ... Patel, N. K. (2009). Latent heat flux estimation in clear sky days over Indian agroecosystems using noontime satellite remote sensing data. *Agric For Meteorol*, *149*, 1646–1665. Available from https://doi.org/10.1016/j.agrformet.2009.05.006.

Mallick, K., Toivonen, E., Trebs, I., Boegh, E., Cleverly, J., Eamus, D., ... Garcia, M. (2018). Bridging Thermal Infrared Sensing and Physically-Based Evapotranspiration Modeling: From Theoretical Implementation to Validation Across an Aridity Gradient in Australian Ecosystems. *Water Resources Research*, *54*, 3409–3435. Available from https://doi.org/10.1029/2017WR021357.

Mathur, R., & AchutaRao, K. (2020). A modelling exploration of the sensitivity of the India's climate to irrigation. *Clim Dyn*, *54*, 1851–1872. Available from https://doi.org/10.1007/s00382-019-05090-8.

Mathur, V., Kuppusamy, S., Onial, M., Pande, A., Singh, Y., Jasmine, B., ... Bhattacharya, I. (2014). India's Fifth National Report to Convention on Biological Diversity. *Minist Environ For Clim Chang Gov India*.

Mauritsen, T., Bader, J., Becker, T., Behrens, J., Bittner, M., Brokopf, R., ... Roeckner, E. (2019). Developments in the MPI-M Earth System Model version 1.2 (MPI-ESM1.2) and Its Response to Increasing CO2. *J Adv Model Earth Syst*, *11*, 998–1038. Available from https://doi.org/10.1029/2018MS001400.

McAneney, K. J., & Itier, B. (1996). Operational limits to the Priestley-Taylor formula. *Irrig Sci*, *17*, 37–43. Available from https://doi.org/10.1007/s002710050020.

McDermid, S., Nocco, M., Lawston-Parker, P., Keune, J., Pokhrel, Y., Jain, M., ... Yokohata, T. (2023). Irrigation in the Earth system. *Nat Rev Earth Environ*. Available from https://doi.org/10.1038/s43017-023-00438-5.

McGloin, R., McGowan, H., McJannet, D., Cook, F., Sogachev, A., & Burn, S. (2014). Quantification of surface energy fluxes from a small water body using scintillometry and eddy covariance. *Water Resources Research*, *50*, 494–513. Available from https://doi.org/10.1002/2013WR013899.

Meehl, G. A., Boer, G. J., Covey, C., Latif, M., & Stouffer, R. J. (2000). The Coupled Model Intercomparison Project (CMIP). *Bull Am Meteorol Soc*, *81*, 313–318. Available from https://doi.org/10.1175/1520-0477(2000)081%3C0313:TCMIPC%3E2.3.CO;2.

Monteith, J. L., & others. (1965). Evaporation and environment. *In: Symp. Soc. Exp. Biol.*, 4.

Nanditha, J. S., & Mishra, V. (2022). Multiday precipitation is a prominent driver of floods in Indian River Basins. *Water Resources Research*, *58*. Available from https://doi.org/10.1029/2022WR032723, e2022WR032723.

Naskar, P. R., Singh, G. P., Pattanaik, D. R., & Katyar, S. (2023). CMIP6 projections of spatiotemporal changes in rainfall and droughts over India. *J Earth Syst Sci*, *132*, 123. Available from https://doi.org/10.1007/s12040-023-02143-9.

Nelson, J. A., Pérez-Priego, O., Zhou, S., Poyatos, R., Zhang, Y., Blanken, P. D., ... Jung, M. (2020). Ecosystem transpiration and evaporation: Insights from three water flux partitioning methods across FLUXNET sites. *Glob Chang Biol*, *26*, 6916–6930. Available from https://doi.org/10.1111/gcb.15314.

Oki, T., & Kanae, S. (2006). Global hydrological cycles and world water resources. *Science (New York, N.Y.)*, *313*, 1068–1072. Available from https://doi.org/10.1126/science.1128845.

O'Neill, B. C., Kriegler, E., Riahi, K., Ebi, K. L., Hallegatte, S., Carter, T. R., ... Van Vuuren, D. P. (2014). A new scenario framework for climate change research: the concept of shared socioeconomic pathways. *Climatic Change*, *122*, 387–400. Available from https://doi.org/10.1007/s10584-013-0905-2.

Patil, M. N., Waghmare, R. T., Dharmaraj, T., & Chinthalu, G. R. (2016). The influence of wind speed on surface layer stability and turbulent fluxes over southern Indian peninsula station. *J Earth Syst Sci*, *125*, 1399–1411. Available from https://doi.org/10.1007/s12040-016-0735-5.

Patil, M. N., Waghmare, R. T., Halder, S., & Dharmaraj, T. (2011). Performance of Noah land surface model over the tropical semi-arid conditions in western India. *Atmos Res*, *99*, 85–96. Available from https://doi.org/10.1016/j.atmosres.2010.09.006.

Penman, H. L. (1948). Natural evaporation from open water, bare soil and grass. *Proceedings of the Royal Society of London. Series A: Mathematical and Physical Sciences*, *193*, 120–145. Available from https://doi.org/10.1098/rspa.1948.0037.

Pereira, L. S., Perrier, A., Allen, R. G., & Alves, I. (1999). Evapotranspiration: concepts and future trends. *J Irrig Drain Eng*, *125*, 45–51. Available from https://doi.org/10.1061/(ASCE)0733-9437(1999)125:2(45).

Priestley, C. H. B., & Taylor, R. J. (1972). On the Assessment of Surface Heat Flux and Evaporation Using Large-Scale Parameters. *Mon Weather Rev*, *100*, 81–92. Available from https://doi.org/10.1175/1520-0493(1972)100%3C0081:otaosh%3E2.3.co;2.

Raja, P., Singh, N., Srinivas, C. V., Singhal, M., Chauhan, P., Singh, M., ... Sinha, N. K. (2018). Analyzing energy–water exchange dynamics in the Thar desert. *Clim Dyn*, *50*, 3281–3300. Available from https://doi.org/10.1007/s00382-017-3804-9.

Rajeevan, M., Bhate, J., & Jaswal, A. K. (2008). Analysis of variability and trends of extreme rainfall events over India using 104 years of gridded daily rainfall data. *Geophysical Research Letters*, *35*, 1–6. Available from https://doi.org/10.1029/2008GL035143.

Ramarao, M. V. S., Sanjay, J., Krishnan, R., Mujumdar, M., Bazaz, A., & Revi, A. (2018). On observed aridity changes over the semiarid regions of India in a warming climate. *Theor Appl Climatol 1–10*. Available from https://doi.org/10.1007/s00704-018-2513-6.

Riahi, K., van Vuuren, D. P., Kriegler, E., Edmonds, J., O'Neill, B. C., Fujimori, S., ... Tavoni, M. (2017). The Shared Socioeconomic Pathways and their energy, land use, and greenhouse gas emissions implications: An overview. *Glob Environ Chang*, *42*, 153–168. Available from https://doi.org/10.1016/j.gloenvcha.2016.05.009.

Rodda, S., Thumaty, K., Jha, C., & Dadhwal, V. (2016). Seasonal Variations of Carbon Dioxide, Water Vapor and Energy Fluxes in Tropical Indian Mangroves. *Forests*, *7*, 35. Available from https://doi.org/10.3390/f7020035.

Roderick, M. L., Rotstayn, L. D., Farquhar, G. D., & Hobbins, M. T. (2007). On the attribution of changing pan evaporation. *Geophysical Research Letters*, *34*, L17403. Available from https://doi.org/10.1029/2007GL031166.

Rodgers, W. A., & Panwar, H. S. (1988). *Planning a wildlife protected network in India*. Wildlife Institute of India.

Rong, X., Li, J., Chen, H., Su, J., Hua, L., Zhang, Z., & Xin, Y. (2019). Introduction of CAMS-CSM model and its participation in CMIP6. *Adv Clim Chang Res*, *15*, 540. Available from https://doi.org/10.12006/J.ISSN.1673-1719.2019.186.

Satyanarayana, A. N. V., Lykossov, V. N., & Mohanty, U. C. (2000). A study on atmospheric boundary-layer characteristics at Anand, India using LSP experimental data sets. *Boundary-layer Meteorol*, *96*, 393–419. Available from https://doi.org/10.1023/A:1002699522932.

Seland Ø., Bentsen M., Olivié D., Toniazzo T., Gjermundsen A., Seland Graff L., ... Schulz M. (2020) Overview of the Norwegian Earth System Model (NorESM2) and key climate response of CMIP6 DECK, historical, and scenario simulations, https://doi.org/10.5194/gmd-13-6165-2020

Semmler, T., Danilov, S., Gierz, P., Goessling, H. F., Hegewald, J., Hinrichs, C., ... Jung, T. (2020). Simulations for CMIP6 With the AWI Climate Model AWI-CM-1-1. *J Adv Model Earth Syst*, *12*. Available from https://doi.org/10.1029/2019MS002009, e2019MS002009.

Sepulchre, P., Caubel, A., Ladant, J.-B., Bopp, L., Boucher, O., Braconnot, P., ... Tardif, D. (2020). IPSL-CM5A2 - An Earth system model designed for multi-millennial climate simulations. *Geosci Model Dev*, *13*, 3011–3053. Available from https://doi.org/10.5194/gmd-13-3011-2020.

Shukla, J., & Mintz, Y. (1982). Influence of land-surface evapotranspiration on the earth's climate. *Science (New York, N.Y.)*, *215*, 1498–1501. Available from https://doi.org/10.1126/science.215.4539.1498.

Shuttleworth, W. J., & Wallace, J. S. (1985). Evaporation from sparse crops-an energy combination theory. *Q J R Meteorol Soc*, *111*, 839–855. Available from https://doi.org/10.1002/qj.49711146910.

Singh, N., Patel, N. R., Bhattacharya, B. K., Soni, P., Parida, B. R., & Parihar, J. S. (2014). Analyzing the dynamics and inter-linkages of carbon and water fluxes in subtropical pine (Pinus roxburghii) ecosystem. *Agric For Meteorol*, *197*, 206–218. Available from https://doi.org/10.1016/j.agrformet.2014.07.004.

Sitch, S., Smith, B., Prentice, I. C., Arneth, A., Bondeau, A., Cramer, W., ... Venevsky, S. (2003). Evaluation of ecosystem dynamics, plant geography and terrestrial carbon cycling in the LPJ dynamic global vegetation model. *Glob Chang Biol*, *9*, 161–185. Available from https://doi.org/10.1046/j.1365-2486.2003.00569.x.

Snyder, R. L., Spano, D., & Pawu, K. T. (1996). Surface renewal analysis for sensible and latent heat flux density. *Boundary-Layer Meteorol*, *77*, 249–266. Available from https://doi.org/10.1007/BF00123527.

Srinet, R., Nandy, S., Watham, T., Padalia, H., Patel, N. R., & Chauhan, P. (2022). Measuring evapotranspiration by eddy covariance method and understanding its biophysical controls in moist deciduous forest of northwest Himalayan foothills of India. *Trop Ecol*, *63*, 387–397. Available from https://doi.org/10.1007/s42965-021-00216-8.

Stannard, D. I. (1993). Comparison of Penman-Monteith, Shuttleworth-Wallace, and modified Priestley-Taylor evapotranspiration models for wildland vegetation in semiarid rangeland. *Water Resources Research*, *29*, 1379–1392. Available from https://doi.org/10.1029/93WR00333.

Stöckli, R., Lawrence, D. M., Niu, G.-Y., Oleson, K. W., Thornton, P. E., Yang, Z.-L., ... Running, S. W. (2008). Use of FLUXNET in the Community Land Model development. *Journal of Geophysical Research*, *113*, G01025. Available from https://doi.org/10.1029/2007JG000562.

Sumner, D. M., & Jacobs, J. M. (2005). Utility of Penman–Monteith, Priestley–Taylor, reference evapotranspiration, and pan evaporation methods to estimate pasture evapotranspiration. *J Hydrol*, *308*, 81–104. Available from https://doi.org/10.1016/j.jhydrol.2004.10.023.

Swapna, P., Krishnan, R., Sandeep, N., Prajeesh, A. G., Ayantika, D. C., Manmeet, S., & Vellore, R. (2018). Long-Term Climate Simulations Using the IITM Earth System Model (IITM-ESMv2) With Focus on the South Asian Monsoon. *J Adv Model Earth Syst*, *10*, 1127–1149. Available from https://doi.org/10.1029/2017MS001262.

Swart, N. C., Cole, J. N. S., Kharin, V. V., Lazare, M., Scinocca, J. F., Gillett, N. P., ... Winter, B. (2019). The Canadian Earth System Model version 5 (CanESM5.0.3). *Geosci Model Dev*, *12*, 4823–4873. Available from https://doi.org/10.5194/gmd-12-4823-2019.

Tabari, H., & Talaee, P. H. (2014). Sensitivity of evapotranspiration to climatic change in different climates. *Global and Planetary Change*, *115*, 16–23. Available from https://doi.org/10.1016/j.gloplacha.2014.01.006.

Taylor, K. E. (2001). Summarizing multiple aspects of model performance in a single diagram. *Journal of Geophysical Research*, *106*, 7183–7192. Available from https://doi.org/10.1029/2000JD900719.

Thornthwaite, C. W. (1948). An approach toward a rational classification of climate. *Geography Review*, *38*, 55–94. Available from https://doi.org/10.2307/210739.

Tiwari, R., Mishra, A. K., Rai, S., & Pandey, L. K. (2023). Evaluation and projection of northeast monsoon precipitation over India under higher warming scenario: a multimodel assessment of CMIP6. *Theor Appl Climatol*, *151*, 859–870. Available from https://doi.org/10.1007/s00704-022-04299-8.

Trenberth, K. E., Fasullo, J. T., & Mackaro, J. (2011). Atmospheric moisture transports from ocean to land and global energy flows in reanalyses. *J Clim Trenberth*, 4907–4924. Available from https://doi.org/10.1175/2011JCLI4171.1.

Tyagi, B., Satyanarayana, A. N. V., Rajvanshi, R. K., & Mandal, M. (2014). Surface energy exchanges during pre-monsoon thunderstorm activity over a tropical station Kharagpur. *Pure Appl Geophys*, *171*, 1445–1459. Available from https://doi.org/10.1007/s00024-013-0682-x.

Uchale, G., Deb Burman, P. K., Tiwari, Y. K., Datye, A., & Sarkar, A. (2023). Investigating Terrestrial Carbon Uptake Over India Using Multimodel Simulations of Gross Primary Productivity and Satellite-Based Biophysical Product. *J Geophys Res Biogeosciences*, *128*. Available from https://doi.org/10.1029/2023JG007468, e2023JG007468.

Volodin, E. M., Mortikov, E. V., Kostrykin, S. V., Galin, V. Y., Lykossov, V. N., Gritsun, A. S., ... Emelina, S. V. (2018). Simulation of the modern climate using the INM-CM48 climate model. *Russ J Numer Anal Math Model*, *33*, 367–374. Available from https://doi.org/10.1515/rnam-2018-0032.

Wagle, P., Bhattarai, N., Gowda, P. H., & Kakani, V. G. (2017). Performance of five surface energy balance models for estimating daily evapotranspiration in high biomass sorghum. *ISPRS J Photogramm Remote Sens*, *128*, 192–203. Available from https://doi.org/10.1016/j.isprsjprs.2017.03.022.

Wang, K., & Dickinson, R. E. (2012). A review of global terrestrial evapotranspiration: observation, modelling, climatology, and climatic variability. *Reviews of Geophysics (Washington, D.C. : 1985)*, *50*, 1–54. Available from https://doi.org/10.1029/2011RG000373.

Wetzel, P. J., & Chang, J.-T. (1987). Concerning the relationship between evapotranspiration and soil moisture. *J Appl Meteorol Climatol*, *26*, 18–27. Available from https://doi.org/10.1175/1520-0450(1987)026%3C0018:CTRBEA%3E2.0.CO;2.

Williams, M., Richardson, A. D., Reichstein, M., Stoy, P. C., Peylin, P., Verbeeck, H., ... Wang, Y.-P. (2009). Improving land surface models with FLUXNET data. *Biogeosciences*, *6*, 1341–1359. Available from https://doi.org/10.5194/bg-6-1341-2009.

Xiang, K., Li, Y., Horton, R., & Feng, H. (2020). Similarity and difference of potential evapotranspiration and reference crop evapotranspiration–a review. *Agric Water Manag*, *232*, 106043. Available from https://doi.org/10.1016/j.agwat.2020.106043.

Yukimoto, S., Kawai, H., Koshiro, T., Oshima, N., Yoshida, K., Urakawa, S., ... Ishii, M. (2019). The meteorological research institute Earth system model version 2.0, MRI-ESM2.0: Description and basic evaluation of the physical component. *J Meteorol Soc Japan*, *97*, 931–965. Available from https://doi.org/10.2151/jmsj.2019-051.

Zhai, J., Mondal, S. K., Fischer, T., Wang, Y., Su, B., Huang, J., ... Uddin, M. J. (2020). Future drought characteristics through a multi-model ensemble from CMIP6 over South Asia. *Atmos Res*, *246*, 105111. Available from https://doi.org/10.1016/j.atmosres.2020.105111.

Zhang, H., Zhang, M., Jin, J., Fei, K., Ji, D., Wu, C., ... Zhu, J. (2020). Description and Climate Simulation Performance of CAS-ESM Version 2. *J Adv Model Earth Syst*, *12*. Available from https://doi.org/10.1029/2020MS002210, e2020MS002210.

Zhang, Y., & Sun, J. (2012). Model projections of precipitation minus evaporation in China. *Acta Meteorol Sin*, *26*, 376–388. Available from https://doi.org/10.1007/s13351-012-0309-7.

Ziehn, T., Chamberlain, M. A., Law, R. M., Lenton, A., Bodman, R. W., Dix, M., ... Srbinovsky, J. (2020). The Australian Earth System Model: ACCESS-ESM1.5. *J South Hemisph Earth Syst Sci*, *70*, 193–214. Available from https://doi.org/10.1071/ES19035.

Chapter 7

Impacts of ocean atmospheric phenomena on hydroclimate extremes

Sarmistha Singh[1,2], Chinju Saju[1] and K. Athira[1]

[1]Department of Civil Engineering, Indian Institute of Technology Palakkad, Palakkad, Kerala, India, [2]ESSENCE, Indian Institute of Technology Palakkad, Palakkad, Kerala, India

7.1 Introduction

Short-term fluctuations in hydroclimatic conditions arise from natural climate variability. The primary source of these climate variations is the differential heating of the sun on Earth, which leads to atmospheric and oceanic circulations. The atmosphere and ocean circulate and interact with each other three-dimensionally. The atmosphere moves faster than the ocean, but the ocean stores a large amount of heat and releases it slowly over a long period. Therefore, in this cycle, the ocean serves as a memory. (Ramamasy & Baas, 2007). These large-scale climatic oscillations or ocean atmospheric (OA) circulations result from the complex interplay between the Earth's oceans and the atmosphere that operate across many geographical regions and time intervals. The alterations in atmospheric pressure, oceanic temperature, and wind orientation throughout the oceans exert a discernible influence on worldwide weather patterns. The periodic pattern (seasonal, interannual, and decadal) of OA circulations refers to the natural variations in the Earth's climate. These large-scale OA phenomena have strong influences on the meteorological and hydrological variables such as rainfall, temperature, and streamflow (Chowdary, Parekh, & Gnanaseelan, 2021; Curry & Webster, 1998; Mahendra et al., 2021; Maleski & Martinez, 2018; Martinez, Baigorria, & Jones, 2009; Mohammadi, Vaheddoost, & Mehr, 2020). These OA phenomena also play a pivotal role in influencing hydroclimatic extremes, which include occurrences such as floods (Athira, Singh, & Abebe, 2023; Jha, Das, & Goyal, 2021; Meena, Athira, & Singh, 2023), droughts (Mitra, Srivastava, Singh, & Yates, 2014; Singh, Srivastava, Abebe, & Mitra, 2015; Veetil and Singh, 2022), heat waves (Sanderson, Economou, Salmon, & Jones, 2017; Singh, Yadav, & Goyal, 2023; Trenberth & Fasullo, 2012; Wright, De Beurs, & Henebry, 2014) and other severe weather conditions. Considering the significant impact of hydroclimatic extremes on both human society and ecosystems, it is important to examine their linkage with local and global factors (Sharma, Patel, & Jothiprakash, 2020).

Hydroclimatic events and global circulation phenomena are linked through "hydroclimatic teleconnection," despite their vast distance. Teleconnection of hydrometeorological variables refers to the correlation or linkage between local meteorological variables in a specific place and global atmospheric circulations that are geographically distant (Adarsh & Janga Reddy, 2019). El Niño Southern Oscillation (ENSO), Pacific decadal oscillation (PDO), Atlantic multidecadal oscillation (AMO), North Atlantic oscillation (NAO), Indian Ocean Dipole (IOD), Equatorial Indian Ocean Oscillations (EQUINOO), and Madden Julian Oscillation (MJO) are some of the major climatic oscillations that modulate the global climatic conditions (Dai & Wigley, 2000; Gadgil & Gadgil, 2006; Hrudya, Varikoden, & Vishnu, 2021; Rasmusson and Carpenter, 1983; Schlesinger and Ramankutty, 1994). Interestingly, the researchers have observed that the regional precipitation patterns do not correspond to global temperature change due to the potential masking effect of local climatic variability. Large-scale climatic oscillations are instead the main cause of these fluctuations. De Beurs, Henebry, Owsley, and Sokolik (2018) and Wang, Huang, He, and Guan (2014) observed that the phase shifts in climatic oscillations influence the teleconnection between precipitation and other meteorological variables, thereby enhancing or suppressing the occurrence of rainfall. Recognizing and studying these variations is crucial for predicting and preparing for the associated impacts on water resources and agricultural production, thereby affecting the food—water—energy nexus of the region. Determining the hydroclimatic teleconnection between large-scale atmospheric—oceanic circulations and

Sustainable Development Perspectives in Earth Observation. DOI: https://doi.org/10.1016/B978-0-443-14072-3.00003-4

basin-scale hydrologic variables is also crucial for effective water resource management at a local scale (Maity & Nagesh Kumar, 2009). Moreover, regional flood and drought patterns can be better understood by studying large-scale climate oscillations and how they interact with decadal to multidecadal oscillations.

7.1.1 El Niño Southern Oscillation and its indices

The Pacific Walker Circulation (PWC) is a zonal circulation cell that occurs in the equatorial region due to the difference in sea surface temperature (SST) between the cold pool area in the eastern equatorial Pacific and the warm pool area in the western equatorial Pacific (Philander, 1989). During typical years, the SST in the western region of the equatorial Pacific Ocean is consistently higher than that in the eastern region. El Niño refers to the phenomenon where the SST in the eastern region of the equatorial Pacific Ocean becomes higher than usual. However, in certain atypical years, this condition undergoes a significant and sudden transformation. The term used to describe a scenario that is antithetical to the normal condition is La Niña. The ENSO phenomenon naturally takes place in the tropical Pacific, oscillating between abnormally warm (El Niño) and cool (La Niña) phases. The two phases are linked to the Walker Circulation and they are the result of the pressure gradient force developed by the coexistence of low-pressure systems in Indonesia and high-pressure systems in the eastern Pacific Ocean.

Quantitative ENSO indices are calculated by averaging SSTs and sea level pressures (SLPs) within specific regions that cover the equatorial Pacific (Stenseth et al., 2003). ENSO indices are used to monitor and quantify the different phases of the ENSO cycle such as El Niño and La Niña events. These indices help researchers and meteorologists track the variations in SSTs and atmospheric conditions in the equatorial Pacific Ocean. Some of the ENSO indices are mentioned below.

7.1.1.1 *Southern Oscillation Index*

The Southern Oscillation Index (SOI) quantifies the magnitude or strength of the Walker Circulation. It is a crucial atmospheric indicator used to measure the intensity of El Niño and La Niña phenomena and their possible effects on the Australian region. The SOI quantifies the disparity in atmospheric pressure at the Earth's surface between Tahiti (located in the Pacific Ocean) and Darwin (situated in Australia). The most accurate representation of the index is achieved by using monthly (or longer) averages. This is because daily or weekly SOI readings can vary significantly due to short-lived, day-to-day weather patterns, especially when a tropical storm is present. SOI values consistently above around $+8$ indicate the occurrence of a La Niña event, whereas values consistently below approximately -8 indicate the occurrence of an El Niño event (Anderson & Lucas, 2018; Stone, Hammer, & Marcussen, 1996; Trenberth, 1984; Yan et al., 2011).

7.1.1.2 *Oceanic Niño Index*

The Oceanic Niño Index (ONI) is derived from anomalies in SSTs in the central equatorial Pacific. El Niño is officially recognized when the ONI surpasses a specific threshold value, generally $+0.5°C$, for a continuous period of time, typically five consecutive overlapping three-month seasons. On the other hand, La Niña is officially recognized when the ONI drops below a particular level, typically $-0.5°C$, for a comparable duration (Glantz & Ramirez, 2020; Hafez, 2016; Prasetyo & Nabilah, 2017; Van Oldenborgh et al., 2021).

7.1.1.3 *Equatorial Southern Oscillation Index*

The Equatorial Southern Oscillation Index (EQSOI) integrates data on SSTs and atmospheric variables to generate an exhaustive indicator for ENSO. El Niño conditions are correlated with positive values, while La Niña conditions are correlated with negative values (Bouri, Gupta, Pierdzioch, & Salisu, 2021; Hassani, 2021; L'Heureux, 2016).

7.1.1.4 *Niño 1+2, Niño 3, Niño 3.4, and Niño 4 indices*

The dynamics of ENSO are defined by the anomalies in SST in various equatorial parts of the Pacific Ocean, specifically Niño 1 + 2, Niño 3, Niño 3.4, and Niño 4 (Mokhov & Timazhev, 2016). The Niño 1 + 2 zone is the smallest and easternmost area among the Niño SST areas. It is located along the coast of South America, where the phenomenon of El Niño was initially identified by the local communities. This index exhibits the highest variability among the Niño SST indices. The Niño 3 region comprises the central Pacific, while the Niño 3.4 region, which is frequently applied to ONI estimates, encompasses the central-western Pacific. The Niño 4 region, on the other hand, comprises the western

Pacific. Positive anomalies signify elevated SSTs, commonly linked to El Niño, while negative anomalies suggest reduced temperatures, connected with La Niña (Mokhov & Timazhev, 2013; Trenberth, 1997).

7.1.1.5 Trans-Niño Index

The Trans-Niño Index quantifies the normalized SST differential between the Niño 1 + 2 and Niño 4 regions. The Niño 1 + 2 region is situated adjacent to the Galapagos Islands, off the western shores of Peru and Ecuador. The Niño 4 zone extends over the International Date Line in the equatorial Pacific. The index is established to describe the progression of an El Niño phenomenon; it may also have practical use for making predictions. Positive values are associated with El Niño conditions, while negative values are associated with La Niña conditions (Kennedy, Garen, & Koch, 2009; Trenberth & Stepaniak, 2001).

7.1.1.6 Multivariate ENSO Index

The Multivariate ENSO Index (MEI) is a commonly employed climatic index that offers a complete assessment of the ENSO phenomenon. The MEI stands out from other ENSO indices by considering several variables to accurately represent the intricate characteristics of ENSO. It is the first principal component that represents the dominant pattern of variability among six key variables seen in the tropical Pacific: SLP, zonal and meridional components of surface wind, SST, surface air temperature (SAT), and cloudiness of the sky. The MEI monthly data are normalized using a reference period from 1950 to 1993. They are then given as fractions of the standard deviation, with a mean of 0 and a standard deviation of 1. This makes it an invaluable tool for evaluating the intensity and phases of ENSO events (Mazzarella, Giuliacci, & Scafetta, 2013; Wolter & Timlin, 2011; Zhang et al., 2019).

7.1.2 Pacific decadal oscillation

The PDO combines two North Pacific SST variability categories. Evidence of PDO implications in the southern hemisphere includes considerable surface climate anomalies over the mid-latitude South Pacific Ocean, Australia, and South America (Mantua & Hare, 2002). PDO is closely linked to the interactions between teleconnection dynamics and tropical regions, specifically the interactions between meridional modes in the extratropics and ENSO, as well as the linkages between ENSO teleconnections and the prevailing modes of atmospheric variability in the mid-latitude (Liu & Di Lorenzo, 2018). The North Pacific index is partially derived from tree-ring data from North America that was previously utilized to reconstruct the PDO. Currently, the PDO index is the major component of monthly SST anomalies poleward of 20° N in the North Pacific Ocean (Mantua, Hare, Zhang, Wallace, & Francis, 1997). Based on the positive and negative values of SST anomalies, the phases of PDO index were defined as warm/positive and cold/negative phase, respectively (Goddard, 2014).

7.1.3 North Atlantic oscillation

The NAO is yet another major meteorological and climatic phenomenon in the northern hemisphere that influences hydroclimate extremes. The NAO is part of the atmospheric response to ocean oscillations, which affect tropical rainfall and global warming (Hoerling, Hurrell, & Xu, 2001). The NAO is a redistribution of atmospheric mass between the Arctic and the subtropical Atlantic that causes major changes in SAT, winds, storminess, and precipitation throughout the Atlantic and nearby continents. The positive phase of the NAO is characterized by stronger westerly winds, milder and wetter conditions in northern Europe, and drier conditions in southern Europe, while the negative phase is associated with weaker westerly winds, colder and drier conditions in northern Europe, and warmer, wetter conditions in southern Europe (Moura, Mendes, Aníbal, & Gomes, 2017). The NAO significantly influences winter variations in climate of Europe, the Mediterranean, parts of the Middle East, and eastern North America across various time scales, ranging from short-term to long-term (Deser, Hurrell, & Phillips, 2017).

The NAO index is a commonly used metric to assess the strength of the westerlies over the eastern North Atlantic and western Europe. It is particularly useful for predicting winter climatic patterns in Europe. The NAO index exhibits a strong correlation with numerous environmental variables connected to the atmosphere, particularly in the winter season (Wanner et al., 2001). The NAO indices based on instrumental records utilize three types of variables: SLP, SAT, and SST (Jianping & Wang, 2003).

7.1.4 Atlantic multidecadal oscillations

The AMO is a consistent pattern of oscillating variations in North Atlantic SST that occurs approximately every 60–90 years. The indices of the AMO have historically been derived from the average annual SST anomalies in the North Atlantic region (Dijkstra, Te Raa, Schmeits, & Gerrits, 2006; Knudsen, Seidenkrantz, Jacobsen, & Kuijpers, 2011). The 1900s–1920s and 1960s–1980s had cool AMO phases, whereas a warm phase occurred in the 1930s–1950s (Knight, Folland, & Scaife, 2006). The AMO index is the primary empirical orthogonal function (EOF) of SST anomalies in the North Atlantic region, spanning from 20° N to 65° N and from 100° W to 0° E. The temperature anomalies in the North Atlantic exhibit an inverse sign compared to those in the South Atlantic (Norel, Kałczyński, Pińskwar, Krawiec, & Kundzewicz, 2021). The comparison plot of interannual (ENSO), decadal (PDO), and multidecadal (AMO) SST anomalies of different climatic oscillations is provided in Fig. 7.1.

7.1.5 Madden Julian Oscillation

Large-scale signals in atmospheric circulation, deep convection, and other processes that move eastward from the Indian to Pacific Oceans dominate tropical intraseasonal variability. This is due to the phenomenon known as MJO (Zhang, 2013). The MJO is a tropical phenomenon characterized by intense precipitation and strong winds spanning approximately 1500 km in latitude and 4500 km in longitude. The phenomenon commences above the Indian Ocean and progresses toward the Pacific Ocean in approximately one month. As it progresses toward the east, it exerts an impact on meteorological and climatic phenomena over various regions of the globe (Zhang, Adames, Khouider, Wang, & Yang, 2020). MJO events in the equatorial Indian and western Pacific oceans involve a large-scale, eastward-moving center of strong deep convection and precipitation ("active" or "enhanced phase"), surrounded by weak zones to the east and west (''inactive'' or ''suppressed phases''). The MJO's two phases are linked by vertical zonal circulations that span the troposphere. Abnormally strong westerly winds to the west of the large-scale convective center and easterly winds to the east are present in the lower troposphere (below 10 km, typically at 1.5 km or 850 hPa level) near the surface. Zonal winds reverse directions in the upper troposphere (above 10 km, usually at 13 km or 200 hPa). The link between the large-scale circulation and convective center is crucial for MJO development (Zhang, 2005).

A seasonally invariant indicator for monitoring the MJO is developed. The method relies on two EOFs derived from the combined data of near-equatorially averaged 850-hPa zonal wind, 200-hPa zonal wind, and satellite-observed outgoing longwave radiation (OLR). The index consists of two principal component time series known as the Real-time Multivariate MJO series 1 (RMM1) and 2 (RMM2). Although RMM1 and RMM2 models depict the MJO's evolution

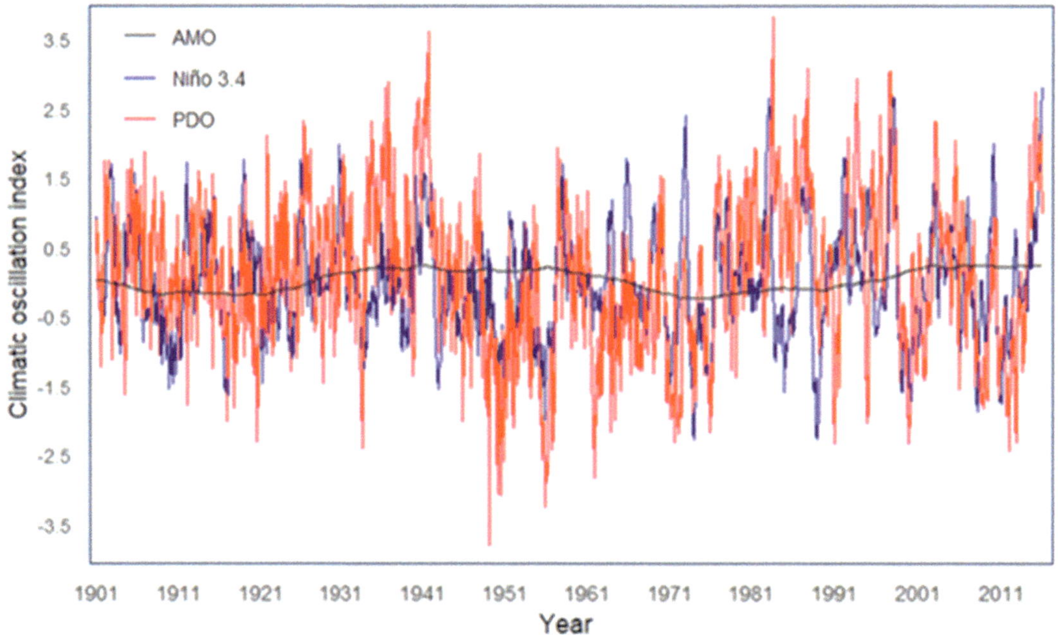

FIGURE 7.1 The time series of ENSO, AMO, and PDO indices.

around the equator as being unaffected by the season, the consistent behavior away from the equator shows a significant dependence on the season. Regardless of the season, the eastward propagation in the equatorial belt is monitored to reflect both the southerly extreme of convection into the Australian monsoon and the northward propagating behavior in the Indian monsoon (Wheeler & Hendon, 2004). Several enhancements have been made to the RMM MJO index. An example of a development is the integration of OLR into the index. The new RMM index shows improvements in spatial organization, power spectra, and standard deviation when compared to the original RMM (Liu et al., 2016).

7.1.6 Indian Ocean Dipole

The IOD is a climatic phenomenon distinguished by the disparity in SSTs between two specific areas in the Indian Ocean: the western Indian Ocean (in proximity to the Arabian Sea) and the eastern Indian Ocean (near the Bay of Bengal) (Li, Wang, Chang, & Zhang, 2003). Positive IOD episodes are triggered by the occurrence of intensified ocean upwelling between the Java and Sumatra coasts in the eastern Indian Ocean. This phenomenon causes abnormally low SSTs in the eastern Indian Ocean, leading to intensified easterly winds along the Equator, intensified ocean warming in the western Indian Ocean, and a corresponding rise (fall) in atmospheric convection over the western (eastern) tropical Indian Ocean. The IOD index is calculated by subtracting the average SST of the eastern Indian Ocean (90° E to 110° E, 10° S to 0° latitude) from the western Indian Ocean (50° E to 70° E, 10° S to10° N). IOD exhibits pronounced seasonal phase-locking properties, typically initiating during summer, peaking in autumn, and rapidly diminishing in winter (Li, Feng, Sun, & Zhang, 2022; Saji, Goswami, Vinayachandran, & Yamagata, 1999). This seasonal phase-locking pattern is intrinsic to the Indian Ocean, and unrelated to the ENSO. Positive IOD events in the Indian Ocean are characterized by warm SST anomalies in the western and cold anomalies in the southeastern equatorial Indian Ocean. Conversely, negative IOD events exhibit the reverse conditions (Rao, Luo, Behera, & Yamagata, 2009; Saji et al., 1999; Vinayachandran, Murty, & Ramesh Babu, 2002). Adverse IOD occurrences can result in contrasting outcomes, characterized by heightened precipitation in the western Indian Ocean area and reduced precipitation in the eastern Indian Ocean area (Abram et al., 2020). The IOD exerts a substantial influence on local climatic patterns, such as rainfall and temperature, and can affect weather conditions in neighboring regions (Saji & Yamagata, 2003).

7.1.7 Equatorial Indian Ocean Oscillation

EQUINOO is the atmospheric constituent of the IOD mode. In the summer monsoon season (June–September), Eastern Equatorial Indian Ocean (EEIO) convection is inversely associated with the Western Equatorial Indian Ocean (WEIO). The SLP and zonal surface wind anomalies near the equator match the convection anomalies. Increased convection over the WEIO leads to a westward gradient in anomalous surface pressure, resulting in easterly surface winds over the equator and the EQUINOO refers to the oscillation between these states (Maity & Nagesh Kumar, 2009; Maity, Kumar, & Nanjundiah, 2007).

7.1.7.1 EQWIN index

EQWIN utilizes surface wind data from NCEP (National Centres for Environmental Prediction) reanalysis. The computation of EQWIN involves determining the long-term mean and standard deviation using wind data from NCEP (spanning from 1958 to 2011). The primary driving force behind surface wind over the tropical oceans is assumed to be the heating caused by the condensation of water vapor in deep convection. There is a strong correlation between the anomalies in OLR over the EEIO and the WEIO with EQWIN (Francis & Gadgil, 2013). EQUINOLR and EQUIRAIN are two further EQUINOO indices that quantify the disparity in the abnormality of OLR and precipitation, respectively, between the WEIO and EEIO (Francis & Gadgil, 2013).

7.2 Impacts of OA phenomena on hydroclimatic extremes

7.2.1 Precipitation

ENSO is the dominant driver affecting the fluctuations in global precipitation (Dai & Wigley, 2000), including regional patterns like Indian Summer Monsoon Rainfall (ISMR). It is observed that ENSO phases are responsible for about 29% of the interannual variability of ISMR. The phase change in ENSO from El Niño to La Niña is usually linked to less (or more) rainfall during the ISMR periods (Fig. 7.2) (Chakraborty & Singhai, 2021; Meena et al., 2023). Along with ENSO, a decadal signal like PDO has also modified the variability of the tropical Pacific Ocean and affected regional-scale precipitation (Barnett, Pierce,

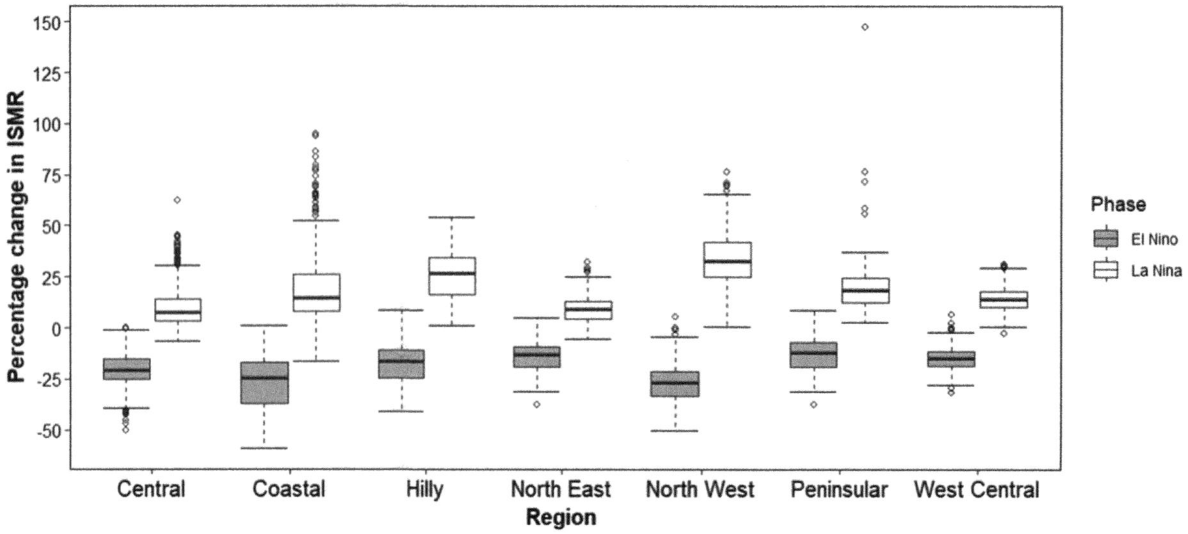

FIGURE 7.2 Box and whisker plot showing the percentage change (from the long-term median) during ISMR periods across different climatic regions in India for the El Niño and La Niña phases of ENSO.

Latif, Dommenget, & Saravanan, 1999; Krishnamurthy & Krishnamurthy, 2017; Power et al., 1999). For instance, a warm SST anomaly in the central equatorial Pacific contributed to low-level divergence and decreased rainfall over India in June. The combined effects of PDO and ENSO altered Thailand's overall and extreme precipitation patterns, usually resulting in devastating flash floods (Limsakul & Singhruck, 2016). In addition to climatic oscillations in the Pacific Ocean, studies have identified that EQUINOO positive (negative) phases enhance (diminish) the ISMR (Gadgil, Rajeevan, Zubair, & Yadav, 2011; Meena et al., 2023; Vishnu, Francis, Ramakrishna, & Shenoi, 2019). Besides that, some of the studies indicated that the positive (negative) phase of AMO has enhanced (weakened) ISMR (Luo et al., 2018; Wang, Yang, Ljungqvist, & Zhao, 2013). The physical mechanisms responsible for these relationships are the atmospheric teleconnection or the ocean-atmospheric interaction (Goswami, Madhusoodanan, Neema, & Sengupta, 2006; Zhou, Li, Luo, Gao, & Furevik, 2015).

SST variations have a significant role in shaping the intensity and frequency of tropical cyclones and the occurrence of extreme precipitation events across the globe (Wang et al., 2014). Ratna, Cherchi, Osborn, Joshi, and Uppara (2021) illustrated that an abnormally strong IOD linked to anomalous low-level divergence over the EEIO and convergence over India. This contributes to unusual rainfall during the Indian Summer Monsoon, often occuring towards the end of the season. Research suggested that the main factors influencing the rainfall on the Maritime Continent are low-level circulation in Indonesia's northeast and southwest; however, Kurniadi, Weller, Min, and Seong (2021) discovered that they are more comparable to the independent occurrence of ENSO and IOD. ENSO independent of IOD impacts rainfall extremes more in the northern and eastern regions of Indonesia, while the IOD independent ENSO explains the rainfall extremes over the southern and western regions. Huang et al. (2019) found that the phase shift of AMO from a cold to a warm phase around 1995 appears to play a significant role in boosting northeast extreme precipitation through enhanced tropical cyclone activity. However, Asian monsoons exhibit a weak correlation with AMO (Khadgarai, Kumar, & Pradhan, 2021).

7.2.2 Temperature

In addition to the precipitation, climatic oscillations display a strong teleconnection with temperature as well (De Beurs et al., 2018). Studies illustrated those significant effects from a diverse set of climate oscillation patterns on the temperature in Central Asia (Syed, Iqbal, Syed, & Rasul, 2014). An increase in the temperature causes forest fires, and Chen, Wang, Hu, Zhou, and Hu (2018) identified a significant correlation between El Niño and burned areas in the grasslands of Central Asia. Many studies implied that temperature variations are significantly influenced by climatic oscillations in Middle Eastern countries. For example, in the Saudi Arabia region, Almazroui, Islam, Jones, Athar, and Rahman (2012) showed a significant correlation between surface temperature and ENSO and NAO. Similarly, in Oman, Charabi and Abdul-Wahab (2009) found the complementary effect of ENSO and IOD on the Southern Oman Summer Monsoon. Additionally, Chandran, Basha, and Ouarda (2016) indicated that the strong variability of temperature in the United Arab Emirates is closely associated with the North Atlantic Oscillation (NAO) index and the East Atlantic Oscillation (EAO) index during April–October and July–December.

7.2.3 Heat waves

The potential impact of climate variability modes on extreme weather events poses intricate challenges for our environment. These climatic modes can govern the frequency, severity, and duration of heat waves on a worldwide scale. Murari, Sahana, Daly, and Ghosh (2016) investigated the correlation between ENSO and characteristics of heat waves in India, such as duration, annual number of heat-wave days, and maximum temperature during heat-wave events. Their findings indicated that the shift in circulation patterns caused by El Niño weakened south-westerlies in the Arabian Sea, leading to an increased number of clear sky days over India. This phenomenon elucidates the occurrence of warmer and prolonged heat waves during El Niño years. Utilizing a large multimember ensemble of a physical climate mode, Reddy, Perkins-Kirkpatrick, and Sharples (2021) investigated the relationship between contiguous heat waves in Australia and the interactions of the ENSO and IOD. They found that areal magnitude, total area, median duration, and maximum area of large and extremely large contiguous heat waves in Australia are significantly higher (lower) during strong El Niño cooccurring with strong IOD positive (strong La Niña cooccurring with the strong IOD negative) years. In the United Kingdom, heat waves display multidecadal variations linked to AMO while the NAO has less impact on heat-wave modulation (Sanderson et al., 2017). The combined effects of the ENSO, AMO, and IOD account for 62.35%−70.01% of the observed intensification of heat waves in China (Wei, Han, Wang, Zhang, & Rajagopalan, 2023).

7.2.4 Cold extremes

On the other hand, studies suggested that the rise in cold extremes in the northern hemisphere is also linked to the naturally occurring, large-scale climate patterns in the atmosphere and oceans, and that these patterns are distinct from those that are believed to be a primary driver of the global warming slowdown (Johnson, Xie, Kosaka, & Li, 2018). Research on the relationship between wintertime cold extremes in East Asia and large-scale OA teleconnections revealed that the Arctic Oscillations (AOs) play a significant role in the occurrence of cold extremes, while the influence of ENSO is relatively weak (Yang, Xie, & Gao, 2019). Studies imply that the Siberian high acts as a conduit for interaction between the AO and the frequency of cold waves in East Asia. It is noted that the occurrence of prolonged winter cold waves in northern China and contemporaneous AO conditions from 1960 to 2001 are inversely related. Cold surges during a negative AO are influenced by the southward expansion of the Siberian high originating in the Arctic regions. Conversely, cold surges during a positive AO occur when the Siberian high extends south-eastward across the Eurasian continent. In northern and western Europe, the NAO has been responsible for several severe cold spells. An illustrative instance is the period of December 2009 to February 2010 characterized by a sequence of cold weather events and unusual snow accumulation in multiple northern hemisphere countries.

7.2.5 Drought

Several studies have evaluated the teleconnection between droughts and climatic oscillations globally by taking the drought index as well as drought characteristics such as onset, duration, severity, and intensity (Abiy, Melesse, Abtew, & Whitman, 2019; Gupta, Kumar Jain, & Singh, 2020; Kumar, Rajeevan, Pai, Srivastava, & Preethi, 2013; Raziei, Saghafian, Paulo, Pereira, & Bordi, 2009; Singh et al., 2015). The distribution of drought features varies significantly across different climate states of ENSO as shown by Wong, Lambert, Leonard, and Metcalfe (2010). In Australian districts, more severe droughts occur during El Niño states as compared to La Niña. A parallel observation was made on Borneo Island, where El Niño was found to nearly double the joint probability of dry conditions compared to normal conditions (Najib, Nurdiati, & Sopaheluwakan, 2022). Similarly, the PDO and AMO exert notable influence on the decadal and multidecadal variability of drought across various regions globally (Qian, Yu, & Chen, 2014; Schoennagel, Veblen, Romme, Sibold, & Cook, 2005). For example, McCabe, Betancourt, Gray, Palecki, and Hidalgo (2008) noted that more than half of the spatial and temporal multidecadal variability of drought in the conterminous United States could be successfully explained by PDO and AMO. Reddy and Ganguli (2013) studied the influence of ENSO on drought characteristics and reported that droughts during El Niño states are more severe than that during La Niña and neutral phases. These studies strongly advocate for the incorporation of climatic oscillation information in drought frequency analysis to enhance the reliability of drought risk assessments and drought prediction accuracy.

7.2.6 Compound extremes

The concurrent extreme events of compound drought and heat waves (CDHs) have captured attention due to their potential to intensify the harmful impacts caused by univariate extremes. An illustrative case is the summer of 2010

TABLE 7.1 Region-wise observations associated with climatic oscillations and hydroclimatic extremes.

Region	Inference	Citation
North America (NA)	A major influence from AMO on temperature and ENSO on annual precipitation across most of NA is observed. However, temperature and precipitation are not much affected by PDO. In the southeast part, the influence of La Niña on temperature and precipitation (higher temperature and lower precipitation) is enhanced during a positive AMO and PDO, while El Niño combined with a positive PDO or AMO would result in below-normal precipitation. The influence of AMO and ENSO in duration, frequency, and magnitude of heat waves are also reported. The individual effects of phases of ENSO, AMO, and NAO as well as coupled effects of ENSO–PDO and ENSO–AMO seemed to have a significant impact on droughts.	Keellings & Waylen, 2015; Maleski & Martinez, 2018; Ropelewski & Halpert, 1987; Singh et al., 2015; Singh, Abebe, & Srivastava, 2018; Singh, Abebe, Srivastava, & Chaubey, 2021; Zhao & Brissette, 2022
South America (SA)	In South America, there is a significant and varied correlation between precipitation, temperature, and the ENSO. For instance, El Niño is commonly associated with drought in the Amazon and northeastern South America causing floods in the tropical west coast and southeastern South America, resulting in significant socioeconomic consequences. Eastern South American climate and weather are also affected by the tropical Atlantic SSTs meridional gradient. Warm tropical North Atlantic surface waters have caused droughts in Amazonia and northeastern Brazil. South American climatic variability may also be affected by high-latitude forcing like the Antarctic Oscillation and North Atlantic Oscillation.	Cai et al., 2020; Czaja & Frankignoul, 2002; Garreaud, Vuille, Compagnucci, & Marengo, 2009; Giannini, Saravanan, & Chang, 2004; Kucharski, Syed, Burhan, Farah, & Gohar, 2015; Kug & Kang, 2006; Satyamurty, Nobre, & Silva Dias, 1998
Asia	The monthly variation in the Arctic Oscillation along with the warm or cold phases of the ENSO are the reason for winter climate anomalies in East Asia. In the central part of Asia, aridity is reported due to ENSO. The occurrence of heat waves with the influence of NAO in the central part of Asia is also observed. A positive Indian Ocean Dipole (IOD) event is also examined in the continent which induces aridity in Indonesia, heightened precipitation in India, Bangladesh, and Vietnam, and a dry and scorching summer in Japan, Korea, and the eastern region of China. Analyzes of Indian summer monsoon rainfall indicate that extreme precipitation occurrences are influenced by ENSO, EQUINOO, PDO, and AMO on spatiotemporal scales.	Athira et al., 2023; Chen, Lan, Wang, & Ma, 2013; De Beurs et al., 2018; Gadgil et al., 2011; Meena et al., 2023; Vishnu et al., 2019; Wright et al., 2014
Europe	A strong correlation exists between European climate and ENSO. The North Pacific region, the tropical Atlantic, and the stratosphere are the main players in the several chains of mechanisms along with ENSO that may influence the North Atlantic–European sector. The winter anomalies are also influenced by the phases of NAO. Another finding is the abnormal cyclonic (anticyclonic) circulation and a southward (northward) displacement of the North Atlantic jet stream are intrinsic to the positive (negative) East Atlantic Oscillation (EAO) phase, respectively.	Brönnimann, 2007; Deser et al., 2017; Mikhailova & Yurovsky, 2016; Rodó, Baert, & Comin, 1997; Rodwell, Rowell, & Folland, 1999; Rogers, 1997
Australia	ENSO associations seem to be robust throughout the year, especially during spring in the southern parts and during summer in the eastern and northern parts of Australia. South Annular Mode (SAM) affects the climatic pattern of the northern and eastern parts of the region, especially	Bauer-Marschallinger, Dorigo, Wagner, & Van Dijk, 2013; Marshall, Wheeler, & Cowan, 2023; Meehl & Arblaster, 1998; Murphy & Timbal, 2008

(Continued)

TABLE 7.1 (Continued)

Region	Inference	Citation
	during summer. ENSO is also responsible for the reduction of precipitation in South Eastern Australia. The IOD impacts spring and winter climatic conditions in the north, east, southeast, and west. The combined effects of ENSO and MJO phase result in warmth over southern Australia in spring and overnight cooling over southern Queensland in winter.	
Africa	The main cause of rainfall appears to be tropical Atlantic and Indian Ocean sea surface temperature changes, vaguely connected to ENSO. Hence, rather than ENSO, IOD is the predominant climate mode. The AMO is another influential factor in causing drought occurrences.	Ward, 1998; Nakamura, et al., 2011; Yang, Seager, Cane, & Lyon, 2015; Rowell, Booth, Nicholson, & Good, 2015; Nicholson, 2019; Ndehedehe, Agutu, Ferreira, and Getirana (2020)
Antarctica	The connections between ENSO and the changes in Mean Sea Level Pressure anomalies over the Southern Ocean and the coastline region of Antarctica are observed. A pattern of pressure anomalies was discovered that propagated between the southern and northern polar regions over a period of 10 years. Antarctic Oscillations are also prevalent in this region and it is responsible for Antarctic winter season temperature abnormalities.	Cullather, Bromwich, & Van Woert, 1996; Turner, 2004; Van Den Broeke & van Lipzig 2004; Van Den Broeke & Van Lipzig, 2003

when Europe and western Russia faced unprecedented heat waves. This was accompanied by droughts triggered by reduced precipitation and elevated temperatures (Hauser et al., 2017), resulting in human casualties and a significant decline in Russian food production (Grumm, 2011). Similarly, the Yangtze River basin in China witnessed a record-breaking heatwave in the summer of 2022, with the prolonged extreme heat leading to a severe drought that had cascading negative effects on agriculture, hydropower, and various industries (Ma & Yuan, 2023). Atmospheric blocking due to the influence of large-scale climatic oscillations mainly caused the occurrence of such extreme compound events during the summer period (Pan et al., 2024). Mukherjee, Ashfaq, and Mishra (2020) found that ENSO is the most influential natural mode of variability associated with the seasonal occurrences of these compound events over tropical regions in the southern hemisphere, while PDO influences their occurrence over western North America in the northern hemisphere during boreal summer. However, NAO association with CDH events is relatively weak (Table 7.1).

7.3 Conclusions

The influence of climatic oscillations on climatic extremes is a complex and crucial component of comprehending the dynamics of Earth's climate system. These oscillations have a major influence on the frequency and severity of hydroclimatic extremes, such as floods, droughts, wet extremes, cold waves, heat waves, compound droughts, etc. Adequate knowledge of hydroclimatic extremes is essential for the structural design and management of water resources. Major hydroclimatic extremes have substantially increased in recent decades. These extremes are influenced by climate change and climate variability associated with large-scale oscillations. ENSO has been identified as the most influential forcing factor that controls the occurrence of extremes on an interannual scale. Both droughts and floods are significantly influenced by climatic oscillations like ENSO, PDO, AMO, IOD, etc., globally, while cold waves in the southern hemisphere and southern Asia are modulated by AOs. Given the concurrent occurrence of drought and heat waves, CDHs are considered the most hazardous event among all extremes. Studies indicate that CDH affects various sectors, and major climatic variables controlling its characteristics include ENSO, PDO, and NAO. Understanding the teleconnection between climatic oscillations and extremes is crucial for early warning systems and the operation of hydraulic structures.

Although there have been notable progressions in our comprehension of the function of climatic oscillations, there are still several areas of exploration that remain unfilled. There is a requirement for more accurate and geographically targeted research to understand the specific effects of oscillations in local areas. This is because extreme weather events typically vary from region to region, and these variations may not be fully accounted for in global-scale models. Additionally, further research could explore the intricate interplay of various oscillations, as their collective impacts may result in nonlinear reactions that remain incompletely understood. Moreover, the study of climatic oscillations in relation to climate change is still a developing field of investigation. Evaluating the impact of human activities on the occurrence and strength of oscillations, as well as how this may affect extreme weather events, is essential for creating effective measures to address climate change.

The investigation of the effects of ocean-atmospheric phenomena on climate extremes delves into the complex interplay between these enormous and ever-changing parts of our planet's ecosystem. The development of measures to adapt to and minimize the impacts of a changing climate can be advanced if scientists can improve their capacity to predict and respond to extreme weather events by understanding the mechanisms underlying them. As the climate continues to evolve, proactive measures informed by scientific insights become increasingly important to address the challenges posed by hydroclimatic extremes. In summary, the examination of the significance of climatic oscillations in relation to climatic extremes is an ever-developing and dynamic discipline that safeguards our knowledge of climate variability and transformation. Its findings have far-reaching consequences for the promotion of sustainable environmental conservation and strengthening of societies.

References

Abiy, A. Z., Melesse, A. M., Abtew, W., & Whitman, D. (2019). Rainfall trend and variability in Southeast Florida: Implications for freshwater availability in the Everglades. *PLoS One*, *14*(2), e0212008.

Abram, N. J., Hargreaves, J. A., Wright, N. M., Thirumalai, K., Ummenhofer, C. C., & England, M. H. (2020). Palaeoclimate perspectives on the Indian Ocean dipole. *Quaternary Science Reviews*, *237*, 106302.

Adarsh, S., & Janga Reddy, M. (2019). *Links between global climate teleconnections and Indian monsoon rainfall. Climate Change Signals and Response: A Strategic Knowledge Compendium for India* (pp. 61–72). Singapore: Springer.

Almazroui, M., Islam, M. N., Jones, P. D., Athar, H., & Rahman, M. A. (2012). Recent climate change in the Arabian Peninsula: Seasonal rainfall and temperature climatology of Saudi Arabia for 1979–2009. *Atmospheric Research*, *111*, 29–45.

Anderson, T. R., & Lucas, M. I. (2018). Upwelling ecosystems. In K. N. G. Elsevier (Ed.), *Encyclopedia of ecology* (pp. 700–710). New York, NY: Elsevier Science.

Athira, K., Singh, S., & Abebe, A. (2023). Impact of individual and combined influence of large-scale climatic oscillations on Indian summer monsoon rainfall extremes. *Climate Dynamics*, *60*(9-10), 2957–2981.

Barnett, T. P., Pierce, D. W., Latif, M., Dommenget, D., & Saravanan, R. (1999). Interdecadal interactions between the tropics and midlatitudes in the Pacific basin. *Geophysical Research Letters*, *26*(5), 615–618.

Bauer-Marschallinger, B., Dorigo, W. A., Wagner, W., & Van Dijk, A. I. (2013). How oceanic oscillation drives soil moisture variations over mainland Australia: An analysis of 32 years of satellite observations. *Journal of Climate*, *26*(24), 10159–10173.

Bouri, E., Gupta, R., Pierdzioch, C., & Salisu, A. A. (2021). El Niño and forecastability of oil-price realized volatility. *Theoretical and Applied Climatology*, *144*, 1173–1180.

Brönnimann, S. (2007). Impact of El Niño–southern oscillation on European climate. *Reviews of Geophysics*, *45*(3).

Cai, W., McPhaden, M. J., Grimm, A. M., Rodrigues, R. R., Taschetto, A. S., Garreaud, R. D., ... Vera, C. (2020). Climate impacts of the El Niño–southern oscillation on South America. *Nature Reviews Earth & Environment*, *1*(4), 215–231.

Chakraborty, A., & Singhai, P. (2021). Asymmetric response of the Indian summer monsoon to positive and negative phases of major tropical climate patterns. *Scientific reports*, *11*(1), 22561.

Chandran, A., Basha, G., & Ouarda, T. B. M. J. (2016). Influence of climate oscillations on temperature and precipitation over the United Arab Emirates. *International Journal of Climatology*, *36*(1), 225–235.

Charabi, Y., & Abdul-Wahab, S. A. (2009). Synoptic aspects of the summer monsoon of southern Oman and its global teleconnections. *Journal of Geophysical Research: Atmospheres*, *114*(D7).

Chen, W., Lan, X., Wang, L., & Ma, Y. (2013). The combined effects of the ENSO and the Arctic Oscillation on the winter climate anomalies in East Asia. *Chinese Science Bulletin*, *58*, 1355–1362.

Chen, X., Wang, S., Hu, Z., Zhou, Q., & Hu, Q. (2018). Spatiotemporal characteristics of seasonal precipitation and their relationships with ENSO in Central Asia during 1901–2013. *Journal of Geographical Sciences*, *28*, 1341–1368.

Chowdary, J. S., Parekh, A., & Gnanaseelan, C. (2021). *Indian summer monsoon variability: El Niño-teleconnections and beyond*. Elsevier.

Cullather, R. I., Bromwich, D. H., & Van Woert, M. L. (1996). Interannual variations in Antarctic precipitation related to El Niño-southern oscillation. *Journal of Geophysical Research: Atmospheres*, *101*(D14), 19109–19118.

Curry, J. A., & Webster, P. J. (1998). *Thermodynamics of atmospheres and oceans*. Elsevier.

Czaja, A., & Frankignoul, C. (2002). Observed impact of Atlantic SST anomalies on the North Atlantic Oscillation. *Journal of Climate*, *15*(6), 606–623.

Dai, A., & Wigley, T. M. L. (2000). Global patterns of ENSO-induced precipitation. *Geophysical Research Letters*, *27*(9), 1283–1286.

De Beurs, K. M., Henebry, G. M., Owsley, B. C., & Sokolik, I. N. (2018). Large scale climate oscillation impacts on temperature, precipitation and land surface phenology in Central Asia. *Environmental Research Letters, 13*(6), 065018.

Deser, C., Hurrell, J. W., & Phillips, A. S. (2017). The role of the North Atlantic Oscillation in European climate projections. *Climate Dynamics, 49*, 3141–3157.

Dijkstra, H. A., Te Raa, L., Schmeits, M., & Gerrits, J. (2006). On the physics of the Atlantic multidecadal oscillation. *Ocean Dynamics, 56*, 36–50.

Francis, P. A., & Gadgil, S. (2013). A note on new indices for the equatorial Indian Ocean oscillation. *Journal of Earth System Science, 122*, 1005–1011.

Gadgil, S., & Gadgil, S. (2006). The Indian monsoon, GDP and agriculture. *Economic and Political Weekly*, 4887–4895.

Gadgil, S., Rajeevan, M.N., Zubair, L., & Yadav, P. (2011). Interannual variation of the south Asian monsoon: Links with ENSO and EQUINOO. In *The Global Monsoon System: Research and Forecast* (pp. 25–42).

Garreaud, R. D., Vuille, M., Compagnucci, R., & Marengo, J. (2009). Present-day South American climate. *Palaeogeography, Palaeoclimatology, Palaeoecology, 281*(3-4), 180–195.

Giannini, A., Saravanan, R., & Chang, P. (2004). The preconditioning role of tropical Atlantic variability in the development of the ENSO teleconnection: Implications for the prediction of Nordeste rainfall. *Climate Dynamics, 22*, 839–855.

Glantz, M. H., & Ramirez, I. J. (2020). Reviewing the Oceanic Niño Index (ONI) to enhance societal readiness for El Niño's impacts. *International Journal of Disaster Risk Science, 11*, 394–403.

Goddard, L. (2014). Heat hide and seek. *Nature Climate Change, 4*(3), 158–161.

Goswami, B. N., Madhusoodanan, M. S., Neema, C. P., & Sengupta, D. (2006). A physical mechanism for North Atlantic SST influence on the Indian summer monsoon. *Geophysical Research Letters, 33*(2).

Grumm, R. H. (2011). The central European and Russian heat event of July–August 2010. *Bulletin of the American Meteorological Society, 92*(10), 1285–1296.

Gupta, V., Kumar Jain, M., & Singh, V. P. (2020). Multivariate modeling of projected drought frequency and hazard over India. *Journal of hydrologic engineering, 25*(4), 04020003.

Hafez, Y. (2016). Study on the relationship between the oceanic nino index and surface air temperature and precipitation rate over the Kingdom of Saudi Arabia. *Journal of Geoscience and Environment Protection, 4*(05), 146.

Hassani, H. (2021). *The ENSO cycle and forecastability of global inflation and output growth: Evidence from standard and mixed-frequency multivariate singular spectrum analyses. The ENSO cycle and forecastability of global inflation and output growth: Evidence from standard and mixed-frequency multivariate singular spectrum analyses. Hassani, Hossein.* Pretoria, South Africa: Department of Economics, University of Pretoria.

Hauser, M., Gudmundsson, L., Orth, R., Jézéquel, A., Haustein, K., Vautard, R. ..., ... Seneviratne, S. I. (2017). Methods and model dependency of extreme event attribution: The 2015 European drought. *Earth's Future, 5*(10), 1034–1043.

Hoerling, M. P., Hurrell, J. W., & Xu, T. (2001). Tropical origins for recent North Atlantic climate change. *Science, 292*(5514), 90–92.

Hrudya, P. H., Varikoden, H., & Vishnu, R. (2021). A review on the Indian summer monsoon rainfall, variability and its association with ENSO and IOD. *Meteorology and Atmospheric Physics, 133*, 1–14.

Huang, D., Dai, A., Yang, B., Yan, P., Zhu, J., & Zhang, Y. (2019). Contributions of different combinations of the IPO and AMO to recent changes in winter East Asian jets. *Journal of Climate, 32*(5), 1607–1626.

Jha, S., Das, J., & Goyal, M. K. (2021). Low frequency global-scale modes and its influence on rainfall extremes over India: Nonstationary and uncertainty analysis. *International Journal of Climatology, 41*(3), 1873–1888.

Jianping, L., & Wang, J. X. (2003). A new North Atlantic Oscillation index and its variability. *Advances in Atmospheric Sciences, 20*, 661–676.

Johnson, N. C., Xie, S. P., Kosaka, Y., & Li, X. (2018). Increasing occurrence of cold and warm extremes during the recent global warming slowdown. *Nature Communications, 9*(1), 1724.

Keellings, D., & Waylen, P. (2015). Investigating teleconnection drivers of bivariate heat waves in Florida using extreme value analysis. *Climate Dynamics, 44*, 3383–3391.

Kennedy, A. M., Garen, D. C., & Koch, R. W. (2009). The association between climate teleconnection indices and Upper Klamath seasonal streamflow: Trans-Niño Index. *Hydrological Processes: An International Journal, 23*(7), 973–984.

Khadgarai, S., Kumar, V., & Pradhan, P. K. (2021). The connection between extreme precipitation variability over Monsoon Asia and large-scale circulation patterns. *Atmosphere, 12*(11), 1492.

Knight, J. R., Folland, C. K., & Scaife, A. A. (2006). Climate impacts of the Atlantic multidecadal oscillation. *Geophysical Research Letters, 33*(17).

Knudsen, M. F., Seidenkrantz, M. S., Jacobsen, B. H., & Kuijpers, A. (2011). Tracking the Atlantic Multidecadal Oscillation through the last 8,000 years. *Nature Communications, 2*(1), 178.

Krishnamurthy, L., & Krishnamurthy, V. (2017). Indian monsoon's relation with the decadal part of PDO in observations and NCAR CCSM4. *International Journal of Climatology, 37*(4), 1824–1833.

Kucharski, F., Syed, F. S., Burhan, A., Farah, I., & Gohar, A. (2015). Tropical Atlantic influence on Pacific variability and mean state in the twentieth century in observations and CMIP5. *Climate Dynamics, 44*, 881–896.

Kug, J. S., & Kang, I. S. (2006). Interactive feedback between ENSO and the Indian Ocean. *Journal of Climate, 19*(9), 1784–1801.

Kumar, K. N., Rajeevan, M., Pai, D. S., Srivastava, A. K., & Preethi, B. (2013). On the observed variability of monsoon droughts over India. *Weather and Climate Extremes, 1*, 42–50.

Kurniadi, A., Weller, E., Min, S. K., & Seong, M. G. (2021). Independent ENSO and IOD impacts on rainfall extremes over Indonesia. *International Journal of Climatology, 41*(6), 3640–3656.

L'Heureux, M. (2016). An overview of the El Niño-Southern Oscillation (ENSO) since 2014. *Climate Prediction S&T Digest, 4.*

Li, C., Feng, Y., Sun, T., & Zhang, X. (2022). Long term Indian Ocean Dipole (IOD) index prediction used deep learning by convLSTM. *Remote Sensing, 14*(3), 523.

Li, T., Wang, B., Chang, C. P., & Zhang, Y. (2003). A theory for the Indian Ocean dipole–zonal mode. *Journal of the Atmospheric Sciences, 60*(17), 2119–2135.

Limsakul, A., & Singhruck, P. (2016). Long-term trends and variability of total and extreme precipitation in Thailand. *Atmospheric Research, 169*, 301–317.

Liu, P., Zhang, Q., Zhang, C., Zhu, Y., Khairoutdinov, M., Kim, H. M. ..., ... Zhang, M. (2016). A revised real-time multivariate MJO index. *Monthly Weather Review, 144*(2), 627–642.

Liu, Z., & Di Lorenzo, E. (2018). Mechanisms and predictability of Pacific decadal variability. *Current Climate Change Reports, 4*, 128–144.

Luo, F., Li, S., Gao, Y., Keenlyside, N., Svendsen, L., & Furevik, T. (2018). The connection between the Atlantic multidecadal oscillation and the Indian summer monsoon in CMIP5 models. *Climate Dynamics, 51*, 3023–3039.

Ma, F., & Yuan, X. (2023). When will the unprecedented 2022 summer heat waves in Yangtze River basin become normal in a warming climate? *Geophysical Research Letters, 50*(4), e2022GL101946.

Mahendra, N., Chowdary, J. S., Darshana, P., Sunitha, P., Parekh, A., & Gnanaseelan, C. (2021). Interdecadal modulation of interannual ENSO-Indian summer monsoon rainfall teleconnections in observations and CMIP6 models: Regional patterns. *International Journal of Climatology, 41*(4), 2528–2552.

Maity, R., & Nagesh Kumar, D. (2009). Hydroclimatic influence of large-scale circulation on the variability of reservoir inflow. *Hydrological Processes: An International Journal, 23*(6), 934–942.

Maity, R., Kumar, D. N., & Nanjundiah, R. S. (2007). Review of hydroclimatic teleconnection between hydrologic variables and large-scale atmospheric circulation patterns with Indian perspective. *ISH Journal of Hydraulic Engineering, 13*(1), 77–92.

Maleski, J. J., & Martinez, C. J. (2018). Coupled impacts of ENSO AMO and PDO on temperature and precipitation in the Alabama–Coosa–Tallapoosa and Apalachicola–Chattahoochee–Flint river basins. *International Journal of Climatology, 38*, e717–e728.

Mantua, N. J., & Hare, S. R. (2002). The Pacific decadal oscillation. *Journal of Oceanography, 58*, 35–44.

Mantua, N. J., Hare, S. R., Zhang, Y., Wallace, J. M., & Francis, R. C. (1997). A Pacific interdecadal climate oscillation with impacts on salmon production. *Bulletin of the American Meteorological Society, 78*(6), 1069–1080.

Marshall, A. G., Wheeler, M. C., & Cowan, T. (2023). Madden–Julian oscillation impacts on Australian temperatures and extremes. *Journal of Climate, 36*(2), 335–357.

Martinez, C. J., Baigorria, G. A., & Jones, J. W. (2009). Use of climate indices to predict corn yields in southeast USA. *International Journal of Climatology: A Journal of the Royal Meteorological Society, 29*(11), 1680–1691.

Mazzarella, A., Giuliacci, A., & Scafetta, N. (2013). Quantifying the Multivariate ENSO Index (MEI) coupling to CO2 concentration and to the length of day variations. *Theoretical and Applied Climatology, 111*, 601–607.

McCabe, G. J., Betancourt, J. L., Gray, S. T., Palecki, M. A., & Hidalgo, H. G. (2008). Associations of multi-decadal sea-surface temperature variability with US drought. *Quaternary International, 188*(1), 31–40.

Meehl, G. A., & Arblaster, J. M. (1998). The Asian–Australian monsoon and El Nino–Southern Oscillation in the NCAR climate system model. *Journal of Climate, 11*(6), 1356–1385.

Meena, D., Athira, K., & Singh, S. (2023). Effects of Large-Scale Climatic Oscillations on the Variability of the Indian Summer Monsoon Rainfall. *Journal of Meteorological Research, 37*(3), 341–352.

Mikhailova, N. V., & Yurovsky, A. V. (2016). The East Atlantic oscillation: Mechanism and impact on the European climate in winter. *Physical Oceanography* (4), 25–33.

Mitra, S., Srivastava, P., Singh, S., & Yates, D. (2014). Effect of ENSO-induced climate variability on groundwater levels in the lower Apalachicola-Chattahoochee-Flint River Basin. *Transactions of the ASABE, 57*(5), 1393–1403.

Mohammadi, B., Vaheddoost, B., & Mehr, A. D. (2020). A spatiotemporal teleconnection study between Peruvian precipitation and oceanic oscillations. *Quaternary International, 565*, 1–11.

Mokhov, I. I., & Timazhev, A. V. (2013). *Climatic anomalies in Eurasia from El Niño/La Niña effects. Doklady Earth Sciences* (pp. 1141–1145). Springer. (453, 1).

Moura, D., Mendes, I., Aníbal, J., & Gomes, A. (2017). Guadiana River Estuary. Investigating the past, present and future.

Mokhov, I. I., & Timazhev, A. V. (2016). Weather-climate anomalies in Russian regions: El Niño-associated predictability. *Research Activities in Atmospheric and Oceanic Modelling, 15*, 6–09.

Mukherjee, S., Ashfaq, M., & Mishra, A. K. (2020). Compound drought and heatwaves at a global scale: The role of natural climate variability-associated synoptic patterns and land-surface energy budget anomalies. *Journal of Geophysical Research: Atmospheres, 125*(11), e2019JD031943.

Murari, K. K., Sahana, A. S., Daly, E., & Ghosh, S. (2016). The influence of the El N iño Southern Oscillation on heat waves in India. *Meteorological Applications, 23*(4), 705–713.

Murphy, B. F., & Timbal, B. (2008). A review of recent climate variability and climate change in southeastern Australia. *International Journal of Climatology: A Journal of the Royal Meteorological Society, 28*(7), 859–879.

Najib, M. K., Nurdiati, S., & Sopaheluwakan, A. (2022). Copula-based joint distribution analysis of the ENSO effect on the drought indicators over Borneo fire-prone areas. *Modeling Earth Systems and Environment*, 1–10.

Nakamura, N., Kayanne, H., Iijima, H., McClanahan, T. R., Behera, S. K., & Yamagata, T. (2011). Footprints of IOD and ENSO in the Kenyan coral record. *Geophysical Research Letters*, *38*(24).

Ndehedehe, C. E., Agutu, N. O., Ferreira, V. G., & Getirana, A. (2020). Evolutionary drought patterns over the Sahel and their teleconnections with low frequency climate oscillations. *Atmospheric Research*, *233*, 104700.

Nicholson, S. E. (2019). A review of climate dynamics and climate variability in Eastern Africa. *Limnology, climatology and paleoclimatology of the East African lakes* (pp. 25–56).

Norel, M., Kałczyński, M., Pińskwar, I., Krawiec, K., & Kundzewicz, Z. W. (2021). Climate variability indices—A guided tour. *Geosciences*, *11*(3), 128.

Pan, X., Wang, W., Shao, Q., Wei, J., Li, H., Zhang, F., ... Yang, L. (2024). Compound drought and heat waves variation and association with SST modes across China. *Science of the Total Environment*, *907*, 167934.

Philander, S. G. (1989). El Niño, La Niña, and the southern oscillation. *International Geophysics Series*, *46*, X-289.

Power, S., Tseitkin, F., Mehta, V., Lavery, B., Torok, S., & Holbrook, N. (1999). Decadal climate variability in Australia during the twentieth century. *International Journal of Climatology: A Journal of the Royal Meteorological Society*, *19*(2), 169–184.

Prasetyo, Y., & Nabilah, F. (2017). Pattern analysis of El Nino and La Nina phenomenon based on sea surface temperature (SST) and rainfall intensity using oceanic Nino Index (ONI) in West Java Area. In *IOP Conference Series: Earth and Environmental Science* (Vol. 98, No. 1, p. 012041). IOP Publishing.

Qian, C., Yu, J. Y., & Chen, G. (2014). Decadal summer drought frequency in China: The increasing influence of the Atlantic Multi-decadal Oscillation. *Environmental Research Letters*, *9*(12), 124004.

Ramamasy, S., & Baas, S. (2007). Climate variability and change: adaptation to drought in Bangladesh A resource book and training guide. In *Case Study*, . (9). Dhaka: Asian Development and Preparedness Centre and Food Agriculture Organization.

Rao, S. A., Luo, J. J., Behera, S. K., & Yamagata, T. (2009). Generation and termination of Indian Ocean dipole events in 2003, 2006 and 2007. *Climate Dynamics*, *33*, 751–767.

Rasmusson, E. M., & Carpenter, T. H. (1983). The relationship between eastern equatorial Pacific sea surface temperatures and rainfall over India and Sri Lanka. *Monthly Weather Review*, *111*(3), 517–528.

Ratna, S. B., Cherchi, A., Osborn, T. J., Joshi, M., & Uppara, U. (2021). The extreme positive Indian Ocean dipole of 2019 and associated Indian summer monsoon rainfall response. *Geophysical Research Letters*, *48*(2), e2020GL091497.

Raziei, T., Saghafian, B., Paulo, A. A., Pereira, L. S., & Bordi, I. (2009). Spatial patterns and temporal variability of drought in western Iran. *Water Resources Management*, *23*, 439–455.

Reddy, M. J., & Ganguli, P. (2013). Spatio-temporal analysis and derivation of copula-based intensity–area–frequency curves for droughts in western Rajasthan (India). *Stochastic Environmental Research and Risk Assessment*, *27*, 1975–1989.

Reddy, P. J., Perkins-Kirkpatrick, S. E., & Sharples, J. J. (2021). Interactive influence of ENSO and IOD on contiguous heatwaves in Australia. *Environmental Research Letters*, *17*(1), 014004.

Rodó, X., Baert, E., & Comin, F. A. (1997). Variations in seasonal rainfall in Southern Europe during the present century: Relationships with the North Atlantic Oscillation and the El Niño-Southern Oscillation. *Climate Dynamics*, *13*, 275–284.

Rodwell, M. J., Rowell, D. P., & Folland, C. K. (1999). Oceanic forcing of the wintertime North Atlantic Oscillation and European climate. *Nature*, *398*(6725), 320–323.

Rogers, J. C. (1997). North Atlantic storm track variability and its association to the North Atlantic Oscillation and climate variability of northern Europe. *Journal of Climate*, *10*(7), 1635–1647.

Ropelewski, C. F., & Halpert, M. S. (1987). Global and regional scale precipitation patterns associated with the El Niño/Southern Oscillation. *Monthly Weather Review*, *115*(8), 1606–1626.

Rowell, D. P., Booth, B. B., Nicholson, S. E., & Good, P. (2015). Reconciling past and future rainfall trends over East Africa. *Journal of Climate*, *28*(24), 9768–9788.

Saji, N. H., Goswami, B. N., Vinayachandran, P. N., & Yamagata, T. (1999). A dipole mode in the tropical Indian Ocean. *Nature*, *401*(6751), 360–363.

Saji, N. H., & Yamagata, T. J. C. R. (2003). Possible impacts of Indian Ocean dipole mode events on global climate. *Climate Research*, *25*(2), 151–169.

Sanderson, M. G., Economou, T., Salmon, K. H., & Jones, S. E. (2017). Historical trends and variability in heat waves in the United Kingdom. *Atmosphere*, *8*(10), 191.

Satyamurty, P., Nobre, C. A., & Silva Dias, P. L. (1998). South America. *Meteorology of the Southern Hemisphere*, 119–139.

Schlesinger, M. E., & Ramankutty, N. (1994). An oscillation in the global climate system of period 65–70 years. *Nature*, *367*(6465), 723–726.

Schoennagel, T., Veblen, T. T., Romme, W. H., Sibold, J. S., & Cook, E. R. (2005). ENSO and PDO variability affect drought-induced fire occurrence in Rocky Mountain subalpine forests. *Ecological Applications*, *15*(6), 2000–2014.

Sharma, P. J., Patel, P. L., & Jothiprakash, V. (2020). Hydroclimatic teleconnections of large-scale oceanic-atmospheric circulations on hydrometeorological extremes of Tapi Basin, India. *Atmospheric Research*, *235*, 104791.

Singh, S., Abebe, A., & Srivastava, P. (2018). Evaluation of nonparametric and parametric statistical procedures for modeling and prediction of cluster-correlated hydroclimatic data. *Water Resources Research*, *54*(9), 6948–6964.

Singh, S., Abebe, A., Srivastava, P., & Chaubey, I. (2021). Effect of ENSO modulation by decadal and multi-decadal climatic oscillations on contiguous United States streamflows. *Journal of Hydrology: Regional Studies*, *36*, 100876.

Singh, S., Srivastava, P., Abebe, A., & Mitra, S. (2015). Baseflow response to climate variability induced droughts in the Apalachicola–Chattahoochee–Flint River Basin, USA. *Journal of Hydrology*, *528*, 550–561.

Singh, S., Yadav, A., & Goyal, M. K. (2023). Univariate and bivariate spatiotemporal characteristics of heat waves and relative influence of large-scale climate oscillations over India. *Journal of Hydrology*, 130596.

Stenseth, N. C., Ottersen, G., Hurrell, J. W., Mysterud, A., Lima, M., Chan, K. S., ... Ådlandsvik, B. (2003). Studying climate effects on ecology through the use of climate indices: The North Atlantic Oscillation, El Nino Southern Oscillation and beyond. *Proceedings of the Royal Society of London. Series B: Biological Sciences*, *270*(1529), 2087–2096.

Stone, R. C., Hammer, G. L., & Marcussen, T. (1996). Prediction of global rainfall probabilities using phases of the Southern Oscillation Index. *Nature*, *384*(6606), 252–255.

Syed, F. S., Iqbal, W., Syed, A. A. B., & Rasul, G. (2014). Uncertainties in the regional climate models simulations of South-Asian summer monsoon and climate change. *Climate Dynamics*, *42*, 2079–2097.

Trenberth, K. E. (1984). Signal versus noise in the Southern Oscillation. *Monthly Weather Review*, *112*(2), 326–332.

Trenberth, K. E. (1997). The definition of el nino. *Bulletin of the American Meteorological Society*, *78*(12), 2771–2778.

Trenberth, K. E., & Fasullo, J. T. (2012). Climate extremes and climate change: The Russian heat wave and other climate extremes of 2010. *Journal of Geophysical Research: Atmospheres*, *117*(D17).

Trenberth, K. E., & Stepaniak, D. P. (2001). Indices of el Niño evolution. *Journal of Climate*, *14*(8), 1697–1701.

Turner, J. (2004). The el nino–southern oscillation and antarctica. *International Journal of Climatology: A Journal of the Royal Meteorological Society*, *24*(1), 1–31.

Van Den Broeke, M. R., & Van Lipzig, N. P. (2003). Response of wintertime Antarctic temperatures to the Antarctic Oscillation: Results of a regional climate model. *Antarctic Peninsula Climate Variability: Historical and Paleoenvironmental Perspectives*, *79*, 43–58.

Van Den Broeke, M. R., & Van Lipzig, N. P. (2004). Changes in Antarctic temperature, wind and precipitation in response to the Antarctic Oscillation. *Annals of Glaciology*, *39*, 119–126.

Van Oldenborgh, G. J., Hendon, H., Stockdale, T., L'Heureux, M., De Perez, E. C., Singh, R., & Van Aalst, M. (2021). Defining El Niño indices in a warming climate. *Environmental Research Letters*, *16*(4), 044003.

Veetil, S. V., & Singh, S. (2022). Multi-scalar association between large-scale climatic pattern and droughts in India. EGU General Assembly 2022, Vienna, Austria, 23–27 May 2022, EGU22-551. https://doi.org/10.5194/egusphere-egu22-551.

Vinayachandran, P. N., Murty, V. S. N., & Ramesh Babu, V. (2002). Observations of barrier layer formation in the Bay of Bengal during summer monsoon. *Journal of Geophysical Research: Oceans*, *107*(C12), SRF-19.

Vishnu, S., Francis, P. A., Ramakrishna, S. S. V. S., & Shenoi, S. S. C. (2019). On the relationship between the Indian summer monsoon rainfall and the EQUINOO in the CFSv2. *Climate Dynamics*, *52*, 1263–1281.

Wang, J., Yang, B. A. O., Ljungqvist, F. C., & Zhao, Y. A. N. (2013). The relationship between the Atlantic Multidecadal Oscillation and temperature variability in China during the last millennium. *Journal of Quaternary Science*, *28*(7), 653–658.

Wang, S., Huang, J., He, Y., & Guan, Y. (2014). Combined effects of the Pacific decadal oscillation and El Nino-southern oscillation on global land dry–wet changes. *Scientific reports*, *4*(1), 6651.

Wanner, H., Brönnimann, S., Casty, C., Gyalistras, D., Luterbacher, J., Schmutz, C., ... Xoplaki, E. (2001). North Atlantic Oscillation—concepts and studies. *Surveys in Geophysics*, *22*, 321–381.

Ward, M. N. (1998). Diagnosis and short-lead time prediction of summer rainfall in tropical North Africa at interannual and multidecadal timescales. *Journal of Climate*, *11*(12), 3167–3191.

Wei, J., Han, W., Wang, W., Zhang, L., & Rajagopalan, B. (2023). Intensification of heatwaves in China in recent decades: Roles of climate modes. *npj Climate and Atmospheric Science*, *6*(1), 98.

Wheeler, M. C., & Hendon, H. H. (2004). An all-season real-time multivariate MJO index: Development of an index for monitoring and prediction. *Monthly Weather Review*, *132*(8), 1917–1932.

Wolter, K., & Timlin, M. S. (2011). El Niño/Southern Oscillation behaviour since 1871 as diagnosed in an extended multivariate ENSO index (MEI. ext). *International Journal of Climatology*, *31*(7), 1074–1087.

Wong, G., Lambert, M. F., Leonard, M., & Metcalfe, A. V. (2010). Drought analysis using trivariate copulas conditional on climatic states. *Journal of Hydrologic Engineering*, *15*(2), 129–141.

Wright, C. K., De Beurs, K. M., & Henebry, G. M. (2014). Land surface anomalies preceding the 2010 Russian heat wave and a link to the North Atlantic oscillation. *Environmental Research Letters*, *9*(12), 124015.

Yan, H., Sun, L., Wang, Y., Huang, W., Qiu, S., & Yang, C. (2011). A record of the Southern Oscillation Index for the past 2,000 years from precipitation proxies. *Nature Geoscience*, *4*(9), 611–614.

Yang, W., Seager, R., Cane, M. A., & Lyon, B. (2015). The annual cycle of East African precipitation. *Journal of Climate*, *28*(6), 2385–2404.

Yang, Y., Xie, N., & Gao, M. (2019). The relationship between the wintertime cold extremes over East Asia with large-scale atmospheric and oceanic teleconnections. *Atmosphere*, *10*(12), 813.

Zhang, C. (2005). Madden-julian oscillation. *Reviews of Geophysics*, *43*(2).

Zhang, C. (2013). Madden–Julian oscillation: Bridging weather and climate. *Bulletin of the American Meteorological Society*, *94*(12), 1849–1870.

Zhang, C., Adames, Á. F., Khouider, B., Wang, B., & Yang, D. (2020). Four theories of the Madden-Julian oscillation. *Reviews of Geophysics*, *58*(3), e2019RG000685.

Zhang, T., Hoell, A., Perlwitz, J., Eischeid, J., Murray, D., Hoerling, M., & Hamill, T. M. (2019). Towards probabilistic multivariate ENSO monitoring. *Geophysical Research Letters*, *46*(17-18), 10532 10540.

Zhao, C., & Brissette, F. (2022). Impacts of large-scale oscillations on climate variability over North America. *Climatic Change*, *173*(1-2), 4.

Zhou, X., Li, S., Luo, F., Gao, Y., & Furevik, T. (2015). Air–sea coupling enhances the East Asian winter climate response to the Atlantic Multidecadal Oscillation. *Advances in Atmospheric Sciences*, *32*, 1647–1659.

Chapter 8

Geospatial technology for coastal water resources management

Sabyasachi Swain[1], Y.R. Satyaji Rao[2], V.S. Jeyakanthan[2] and R. Venkata Ramana[2]

[1]*Department of Civil Engineering, Malaviya National Institute of Technology, Jaipur, Rajasthan, India,* [2]*Deltaic Regional Centre, National Institute of Hydrology, Kakinada, Andhra Pradesh, India*

8.1 Introduction

8.1.1 Importance of coastal water resources management

Coastal water resources management is critical for biodiversity and ecosystem services. As the coastal water resources encompass a variety of ecosystems, including estuaries, coral reefs, and mangroves, which are hotspots for biodiversity, supporting a wide range of marine species. They are essential for the life cycles of many aquatic organisms, offering breeding and feeding grounds crucial for their survival. The health of these ecosystems is directly linked to the overall health of our oceans and the abundance of marine life (Dugan et al., 2011).

Coastal regions play a pivotal role in the global economy. They support key industries such as fishing, tourism, and transportation. The fisheries sector, in particular, is a major source of livelihood for millions of people worldwide. Coastal tourism, a rapidly growing sector, contributes significantly to local and national economies, especially in developing countries. The economic value of goods and services provided by coastal ecosystems is estimated to be in the billions of dollars annually (Costanza et al., 1997). Moreover, coastal ecosystems like mangroves and coral reefs provide natural defense mechanisms against environmental threats. Mangroves, for instance, act as buffers against storm surges and coastal erosion, protecting inland communities from the devastating impacts of hurricanes and tsunamis (Mondal & Das, 2023). Coral reefs help in dissipating wave energy, thereby safeguarding shorelines. These natural barriers are critical in a world where climate change is increasing the frequency and intensity of extreme weather events, as evident from several prior studies (Alongi, 2008; Sharma, Swain, Mishra, & Pandey, 2024; Swain, 2017; Swain, Mishra, & Pandey, 2021, 2022b; Swain, Mishra, Pandey, & Dayal, 2021, 2022a, 2022b; Swain, Mishra, Pandey, Dayal, & Srivastava, 2022). Besides, coastal ecosystems are significant carbon sinks. Mangroves, salt marshes, and seagrasses capture and store large amounts of carbon, playing a crucial role in mitigating climate change. The degradation of these ecosystems not only leads to loss of biodiversity and protective services but also results in the release of stored carbon, exacerbating global warming (Mcleod et al., 2011).

Despite their importance, coastal ecosystems are among the most threatened environments on the planet. They face numerous challenges including pollution, overfishing, habitat destruction, and climate change-induced sea-level rise. Protecting these ecosystems requires a multifaceted approach that balances conservation efforts with the sustainable use of resources (Barbier et al., 2011). The sustainable management of coastal water resources is essential for the preservation of biodiversity, economic prosperity, and the protection of communities from environmental hazards. Integrating modern geospatial technology with traditional knowledge can offer innovative solutions for the challenges faced by coastal ecosystems.

8.1.2 Challenges in coastal water resources management

One of the primary challenges in coastal water resource management is the overexploitation of marine resources, particularly fisheries. Overfishing has led to the depletion of many fish stocks, with significant consequences for marine ecosystems and the people who depend on them for livelihoods. According to the Food and Agriculture Organization

Sustainable Development Perspectives in Earth Observation. DOI: https://doi.org/10.1016/B978-0-443-14072-3.00001-0

(FAO), over 33% of the world's fish stocks are fished at biologically unsustainable levels (FAO, 2020). Moreover, coastal areas are often heavily polluted due to industrial, agricultural, and urban runoff. This pollution includes a range of contaminants like plastics, heavy metals, and nutrients, which can lead to eutrophication and harmful algal blooms, severely impacting marine life and human health. Additionally, habitat destruction due to coastal development, aquaculture, and other human activities has led to the loss of critical ecosystems such as mangroves and coral reefs (Halpern et al., 2008). Further, climate change poses a significant threat to coastal ecosystems. Rising sea levels can lead to increased coastal erosion, salinization of freshwater resources, and loss of habitat. Furthermore, ocean acidification and warming are causing widespread coral bleaching and altering species distributions and ecosystem dynamics (IPCC, 2019). Coastal regions are often densely populated, and the communities living there can be particularly vulnerable to changes in coastal ecosystems. The loss of livelihoods due to depleted fish stocks or damaged ecosystems, coupled with inadequate policy frameworks and management strategies, exacerbates the socio-economic challenges in these areas (Adger et al., 2005). More importantly, effective governance and management are crucial for the sustainable use of coastal resources; however, many coastal areas suffer from inadequate governance structures, lack of integrated management approaches, and insufficient enforcement of regulations. This leads to resource conflicts and hampers effective conservation and sustainable use of coastal resources (Österblom & Folke, 2015). In general, achieving a balance between economic development and environmental conservation in coastal areas is a complex task. The need for economic growth and infrastructure development often conflicts with the need to preserve sensitive ecosystems and maintain the ecological integrity of coastal waters (Beatley, 2009). Addressing all the aforementioned challenges requires integrated management approaches that combine scientific understanding with traditional knowledge, stakeholder engagement, and strong governance frameworks. Recognizing and addressing these challenges is essential for ensuring the sustainable management of coastal water resources.

8.1.3 Overview of traditional ecological knowledge and modern geospatial technology

Regarding traditional knowledge required for sustainable coastal water management, the traditional ecological knowledge (TEK) is a key factor. TEK refers to the knowledge systems developed by indigenous and local communities over generations through direct contact with their environment (Berkes, 1993; Huntington, 2000; Loch & Riechers, 2021). TEK encompasses an understanding of ecological processes, species behavior, seasonal changes, and sustainable resource management practices. It is rooted in the cultural and spiritual dimensions of indigenous societies and is often passed down orally through generations (Nugroho et al., 2023). TEK is not only about the ecological world but also involves a deep understanding of the interconnections between humans and nature (Berkes, Colding, & Folke, 2000).

Modern geospatial technology encompasses tools like geographic information systems (GIS), remote sensing, and satellite imageries that enable the collection, analysis, and visualization of spatial data. These technologies have revolutionized environmental management by providing detailed and accurate information about the Earth's surface, facilitating the monitoring and assessment of environmental changes, resource management, and planning (Sharma, Mishra, Pandey, Kumre, & Swain, 2020; Swain, Mishra, Pandey, Pandey, et al., 2022; Swain, Mishra, & Pandey, 2022a; Swain, Mishra, Pandey, & Kalura, 2022). GIS and remote sensing technologies are instrumental in analyzing spatial patterns and temporal changes, offering invaluable insights into environmental processes and human impacts (Wright & Wang, 2011).

Integrating TEK with modern geospatial technology represents a promising approach to enhance coastal water resources management. This integration allows for the combination of detailed local knowledge with large-scale, quantitative data. The fusion of TEK with modern scientific methods can lead to more holistic and culturally sensitive management strategies, enhancing the sustainability of resource use and conservation efforts (Johnson et al., 2016). Such an integration can improve the accuracy of environmental assessments, enrich scientific understanding with local insights, and facilitate community engagement in resource management (Boafo et al., 2016; de Freitas, Bastos, & Dias, 2018; Jiao et al., 2012; Parlee, 2005). It also aids in identifying and protecting culturally significant areas, managing resources according to traditional practices, and ensuring that local communities benefit from and participate in conservation efforts (Fernández-Llamazares & Cabeza, 2018). However, integrating these diverse knowledge systems may face challenges. The issues such as intellectual property rights, data sensitivity, and ensuring respect for indigenous knowledge and practices must be carefully navigated. Further, differences in methodologies and epistemologies between traditional knowledge and modern science can pose barriers to effective integration (Turner & Clifton, 2009). Overall, the approach of integrating traditional knowledge with modern geospatial technology offers a novel and promising pathway to enhance the management and conservation of coastal water resources as it recognizes the value of indigenous and local knowledge systems while harnessing the power of modern technology, leading to more effective and inclusive environmental management practices.

8.2 Current coastal restoration management

8.2.1 Existing strategies and practices

The current state of coastal restoration management involves employing a variety of strategies and practices to address the complex challenges faced by coastal ecosystems. These include coastal habitat restoration, integrated coastal zone management (ICZM), marine protected areas (MPAs), ecosystem-based management (EBM), coastal erosion control, pollution control, and fisheries management, which are discussed as follows. Coastal habitat restoration involves rehabilitating degraded coastal ecosystems, such as mangroves, coral reefs, and salt marshes (Mondal & Das, 2023; Proulx, Ross, Macdonald, Fitzsimmons, & Smit, 2021). These efforts often include replanting vegetation, removing invasive species, and rebuilding natural structures. According to the National Oceanic and Atmospheric Administration (NOAA), habitat restoration is critical for maintaining ecosystem services and protecting biodiversity (NOAA, 2019). ICZM is a process that uses an integrated approach to manage the coastal zone, addressing environmental, social, cultural, and economic dimensions. It involves various stakeholders, including local communities, in decision-making processes. ICZM aims to achieve sustainable development of coastal areas while minimizing conflicts between different uses (Cicin-Sain & Knecht, 1998). Similarly, MPAs are regions where human activities are restricted to protect marine ecosystems. The establishment of MPAs is a widely used strategy for conserving marine biodiversity and restoring fish stocks. MPAs can be effective in preserving marine habitats and species, provided they are well managed and enforced (Agardy, Di Sciara, & Christie, 2011). EBM is another approach that considers the entire ecosystem, including humans, when making management decisions. It emphasizes the interconnectedness of ecosystem components and aims to maintain ecological integrity. EBM seeks to balance ecological, social, and economic goals (Long, Charles, & Stephenson, 2017).

Coastal erosion control strategies include both hard (e.g., seawalls, groins) and soft (e.g., beach nourishment, dune restoration) engineering solutions. These practices aim to protect coastal areas from erosion and sea-level rise. However, hard engineering solutions can sometimes exacerbate erosion problems elsewhere (Pilkey & Cooper, 2014). Similarly, efforts to reduce coastal pollution involve regulating industrial discharges, improving sewage treatment facilities, and controlling agricultural runoff. Enhancing water quality is crucial for the health of coastal ecosystems and public health. The Water Framework Directive (WFD) by the European Union is an example of a legislative framework aimed at achieving good water quality status in all waters (EU WFD, 2000). Further, sustainable fisheries management practices include setting catch limits, implementing seasonal closures, and establishing gear restrictions. These measures are designed to prevent overfishing and ensure the long-term sustainability of fish stocks. The FAO's Code of Conduct for Responsible Fisheries provides guidelines for sustainable fisheries management (FAO, 1995).

8.2.2 Limitations of current approaches

It is important to understand the limitations of current approaches. One of the key limitations of current coastal management practices is the lack of integration and coordination among different management strategies and across governance levels. As reported by Olsen, Page, and Ochoa (2009), fragmented governance and sectoral approaches often lead to conflicts and inefficiencies. Despite the emphasis on stakeholder involvement, local communities are often inadequately involved in decision-making processes. This can lead to resistance and noncompliance with management measures. Ensuring meaningful community participation is essential for the success of coastal management initiatives (Jentoft, 2000). Effective coastal management requires significant financial and human resources; however, many coastal restoration and management programs suffer from inadequate funding and resource constraints, limiting their effectiveness and scope (Bennett, Blythe, Tyler, & Ban, 2016). The impacts of climate change, such as sea-level rise and increased storm intensity, present significant challenges to coastal management. Traditional management approaches may not be sufficient to address these dynamic and complex issues (Nicholls et al., 2007). Enforcement of regulations and compliance with management measures is often weak, particularly in areas with limited governance capacity. This undermines the effectiveness of conservation efforts and sustainable resource use (Edgar et al., 2014). Besides, coastal ecosystems are complex and dynamic, and our understanding of these systems is still incomplete. Additionally, the social dimensions of coastal management, including conflicting interests and cultural values, add to the complexity of effective management (Charles, 2012).

It is clear that there have been substantial efforts toward coastal restoration management through methods including coastal habitat restoration, ICZM, MPAs, and EBM. However, despite these efforts, the limitations of current approaches are becoming increasingly apparent. One of the primary challenges is the gap between the theoretical frameworks of these strategies and their practical implementation. Issues such as insufficient funding, political hurdles, and a lack of coordinated efforts often hinder the effectiveness of these management practices. Furthermore, the reliance on traditional management approaches sometimes overlooks the dynamic nature of coastal ecosystems and the need for adaptive management strategies that can respond to changing environmental conditions and evolving scientific understanding.

Overall, the integration of TEK with modern scientific approaches, particularly geospatial technology, presents an opportunity to enhance the effectiveness of coastal restoration efforts (Bethel et al., 2011, 2014; Drew, 2005; Hutton & Allen, 2020). TEK offers valuable insights into long-term environmental changes and sustainable resource management practices, which, when combined with the precision and analytical capabilities of modern technology, can lead to more nuanced and effective conservation strategies. So, while significant strides have been made in coastal restoration management (Barbier, 2020; Foale, 2006; Noble, Harasti, Fulton, & Doran, 2020; Stori, Peres, Turra, & Pressey, 2019; Teixeira et al., 2013), the field is constantly evolving. Addressing the limitations of current approaches and embracing the integration of diverse knowledge systems will be crucial for the effective and sustainable management of coastal ecosystems in the face of ongoing environmental challenges.

8.3 The role of local knowledge in coastal restoration

8.3.1 Definition and essence of traditional ecological knowledge

TEK is a cumulative body of knowledge, practices, and beliefs, evolving through adaptive processes and handed down through generations, about the relationships of living beings (including humans) with one another and with their environment. As defined by Berkes et al. (2000), TEK is not merely a set of empirical observations; it encompasses the cultural, spiritual, and historical dimensions of indigenous and local peoples' interactions with their environment. TEK is dynamic, continually adapting to changes in the environment and incorporating new experiences and understandings.

The essence of TEK lies in its holistic approach to understanding ecological processes. It is grounded in a deep sense of place and the recognition of the interconnectedness of all life forms. This understanding is often embedded in cultural practices, rituals, stories, and laws, reflecting a relationship with the environment that is based on respect, reciprocity, and responsibility (Turner & Berkes, 2006).

8.3.2 Values of integrating traditional ecological knowledge in coastal management

Integrating TEK into coastal management brings several benefits. Firstly, it enriches scientific understanding with nuanced local knowledge that is often absent in conventional management approaches. For instance, TEK can provide detailed information on species behavior, local ecological changes, and sustainable harvesting techniques, which can enhance the effectiveness of conservation strategies (Armitage et al., 2011). Moreover, the integration of TEK fosters community engagement and stewardship. When local communities see their knowledge and practices being recognized and valued in coastal management, it enhances their sense of ownership and responsibility toward conservation efforts (Nadasdy, 2003). This participatory approach leads to more sustainable and culturally appropriate management outcomes. Further, TEK-based approaches often embody adaptive management practices. They are inherently flexible, allowing for adjustments based on continual observations and learning. This adaptability is particularly crucial in the face of climate change and other environmental uncertainties (Folke et al., 2005). In addition, integrating TEK can address issues of environmental justice and rights of indigenous and local communities. Recognizing and valuing TEK in coastal management not only contributes to conserving biodiversity but also respects and upholds the cultural and intellectual rights of these communities (Tsing et al., 1999).

In coastal management, TEK manifests in various forms and has been instrumental in sustaining ecosystems for millennia. There are several examples of TEK in coastal management. For instance, indigenous communities in the Pacific Northwest have managed salmon fisheries using complex systems of taboos, ceremonial practices, and regulated fishing rights, which have maintained the sustainability of these resources over time (Lertzman & Vredenburg, 2005). In the Caribbean, traditional practices such as the rotation of fishing grounds and the use of selective fishing gear have been effective in conserving fish populations and habitats (Mahon et al., 2008). Another example is the use of traditional weather and ocean condition forecasting by indigenous navigators in the Pacific Islands. This knowledge, passed down through generations, involves reading wind patterns, cloud formations, and ocean swells, and is crucial for safe navigation and sustainable use of marine resources (Hau'ofa, 1993).

Overall, it can be inferred that the role of local knowledge in coastal restoration is vital. TEK offers a rich, place-based understanding of ecological processes, complementing scientific knowledge with depth and context. Through various examples, it is evident that TEK can significantly enhance coastal management practices. Its integration not only improves ecological outcomes but also supports the rights and participation of indigenous and local communities, fostering a more inclusive and sustainable approach to coastal ecosystem conservation.

8.4 Modern geospatial technology in coastal management

8.4.1 Overview of geospatial technologies

Remote sensing technology refers to the practice of obtaining information about objects or areas from a distance, typically from aircraft or satellites. In the context of coastal management, remote sensing is utilized for its ability to provide large-scale, continuous, and detailed observations of coastal areas. This technology captures data in various spectral bands, which allows for the detailed analysis of coastal vegetation, land use patterns, and changes in coastal morphology over time (Jensen, 2007). The evolution of remote sensing, from early aerial photography to advanced satellite systems, has dramatically enhanced our ability to monitor and analyze coastal environments.

GIS is a powerful tool for managing, analyzing, and visualizing spatial data. It allows for the integration of various datasets, including those obtained from remote sensing, to create detailed maps and models of coastal areas. GIS applications in coastal management include mapping habitat areas, assessing erosion and sedimentation patterns, and planning for coastal development and conservation. The strength of GIS lies in its capacity to handle complex spatial data and provide insights into the spatial relationships of coastal features (Wright et al., 2011).

Additional technologies such as global positioning system (GPS), light detection and ranging (LiDAR), and unmanned aerial vehicles (UAVs) also play significant roles in coastal management. GPS provides precise location data, essential for field surveys and mapping. LiDAR offers high-resolution topographic data, which is crucial for understanding coastal topography and changes in sea level. UAVs are increasingly used for their flexibility and ability to capture high-resolution imagery of coastal areas, aiding in detailed and up-to-date assessments (Anderson & Gaston, 2013).

8.4.2 Applications in coastal water resources management

Geospatial technologies are extensively used in the monitoring and assessment of coastal ecosystems. Remote sensing data, for example, is instrumental in tracking changes in coastal habitats, such as wetland loss or coral reef degradation. GIS models are used to predict the impacts of sea-level rise on coastal areas, aiding in the planning and implementation of adaptation strategies. Additionally, these technologies facilitate the monitoring of coastal water quality and the assessment of pollution sources (Phinn et al., 2008).

In the event of natural disasters such as hurricanes or tsunamis, geospatial technologies are crucial for emergency response and recovery efforts. They provide rapid assessments of the impacted areas, helping in directing resources and aid effectively. Postdisaster, these technologies aid in damage assessment and are vital in the planning and rebuilding of more resilient coastal infrastructure (Cutter et al., 2012). Further, remote sensing datasets have the benefit of near-real-time availability, which makes them instrumental to mitigate the impacts of floods and droughts. In case of droughts, the propagation of meteorological to hydrological or agricultural droughts can be monitored using geospatial datasets. For instance, the remote sensing-based indices like normalized difference vegetation index (NDVI), vegetation condition index (VCI), and vegetation health index (VHI) are widely used to characterize agricultural drought characteristics for their inherent nature of representing vegetation conditions/health. Similarly, global-scale streamflow datasets viz., global flood awareness system (GloFAS), have been developed in the recent years, which facilitates hydrological drought characterization and assessments over the coastal regions of developing countries, which are typically data-scarce (Swain, Mishra, Pandey, Srivastava, & Nandi, 2023). Amani, Mehravar et al. (2022) and Amani, Moghimi et al. (2022) have described oceanic remote sensing techniques and their applications in detail. Fig. 8.1 represents various applications of remote sensing in coastal water management, as referred from Amani, Mehravar et al. (2022).

GIS and remote sensing play key roles in identifying critical habitats, assessing their conditions, and planning conservation and restoration efforts. For instance, they help in mapping spawning grounds and nursery habitats for fisheries management. LiDAR and UAV data are used to create detailed models of habitats, which are crucial for restoration

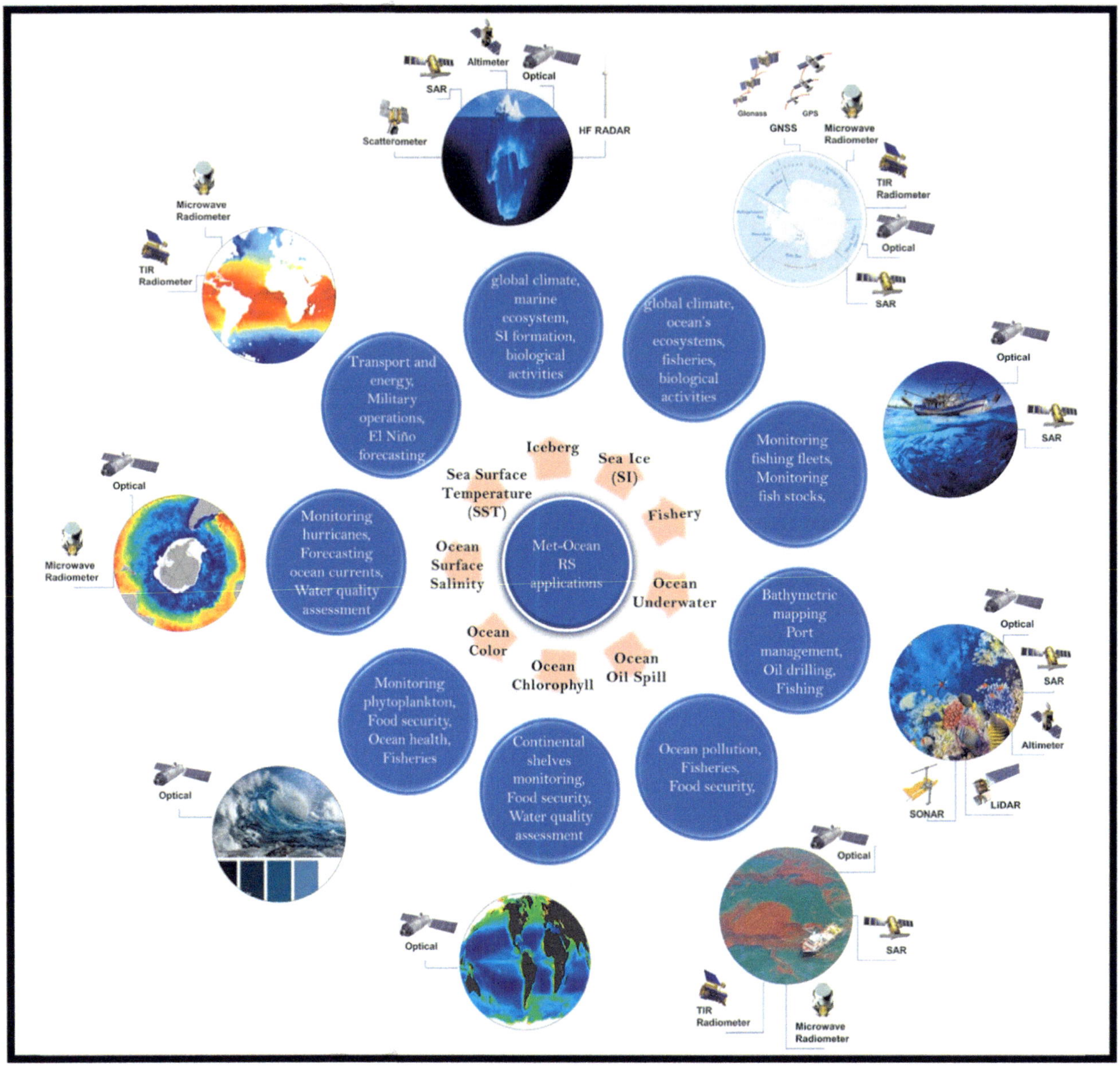

FIGURE 8.1 Various applications of remote sensing in coastal water management. *From Amani, M., Mehravar, S., Asiyabi, R. M., Moghimi, A., Ghorbanian, A., Ahmadi, S. A., … Jin, S. (2022). Ocean remote sensing techniques and applications: A review (part ii).* Water, 14*(21), 3401.*

planning, such as in mangrove reforestation projects (Mumby & Edwards, 2014). Geospatial technologies are indispensable in understanding and addressing the impacts of climate change on coastal regions. They provide valuable data for modeling sea-level rise scenarios and assessing the vulnerability of coastal communities. This information is critical for developing adaptation measures such as coastal defense structures, managed retreat, or the establishment of buffer zones (Nicholls & Cazenave, 2010). Effective coastal planning and development require a thorough understanding of the coastal environment, which geospatial technologies provide. They are used in delineating coastal development zones, assessing the suitability of areas for various activities, and ensuring that development activities are sustainable and do not adversely impact the coastal ecosystem (Bartlett & Smith, 2013).

The gravity recovery and climate experiment (GRACE) satellites, a joint mission by National Aeronautics and Space Administration (NASA) and the German Aerospace Center, have been instrumental in providing valuable data for various aspects of Earth's water resources, including those relevant to coastal regions. There are various ways in which GRACE data can aid in coastal water management, such as monitoring groundwater changes, understanding sea level rise, ecosystem health, integrated coastal zone management, providing insights on climate change and hydrological cycle (Swain et al., 2022). GRACE satellites measure changes in Earth's gravity field, which are influenced by mass changes such as those resulting from groundwater depletion or recharge. In coastal areas, GRACE data can help identify trends in groundwater storage, crucial for managing coastal aquifers and preventing issues like saltwater intrusion. Further, overextraction of groundwater in coastal regions can lead to saltwater intrusion, where seawater encroaches into freshwater aquifers (Dhal & Swain, 2022). GRACE's ability to monitor groundwater depletion helps in assessing and managing these risks. GRACE data contribute to understanding global sea level rise by providing insights into the amounts of water entering the oceans from melting ice sheets and glaciers. This information is vital for coastal planning and adaptation strategies in response to sea level rise. GRACE helps in understanding regional variations in sea level rise, which can be due to factors like ocean currents, temperature changes, and gravitational effects. Such information is critical for region-specific coastal management and mitigation plans. Coastal wetlands and estuaries are critical ecosystems that can be monitored using GRACE data. Changes in water storage in these areas can be indicative of ecological health and inform conservation and restoration efforts. GRACE data provide insights into changes in the global hydrological cycle under the influence of climate change, including alterations in precipitation patterns and river discharge. Such data help in predicting and managing flood risks in coastal areas. For coastal regions prone to drought, GRACE can help in monitoring and managing water scarcity by providing data on changes in total water storage, including surface waters and aquifers. Besides, GRACE data, integrated with other satellite and ground-based observations, provide comprehensive information useful for ICZM. It helps in making informed decisions regarding water resource allocation, land use planning, and environmental protection. Therefore GRACE's unique ability to monitor changes in Earth's gravity field related to water mass provides invaluable data for managing water resources in coastal regions. These data are crucial for addressing various challenges, including groundwater depletion, sea level rise, climate change impacts, and ecosystem conservation. It aids policymakers, scientists, and environmental managers in developing strategies for sustainable and resilient coastal water resource management.

Overall, modern geospatial technologies, including remote sensing, GIS, GPS, LiDAR, GRACE, and UAVs, have become integral tools in coastal water resources management. Their applications range from monitoring and assessment, disaster management, habitat conservation, climate change adaptation, to coastal planning and development. These technologies provide critical data and insights, enabling more informed decision-making and effective management strategies to preserve and sustainably use coastal ecosystems.

8.5 Rationale for data integration to support decision making

8.5.1 Benefits of integrating traditional ecological knowledge and scientific knowledge

Integrating TEK with scientific data provides a comprehensive understanding of coastal ecosystems. While scientific data offer quantitative insights, TEK contributes toward qualitative and experiential knowledge. This combination enriches ecological assessments and enhances the understanding of long-term environmental changes, thus leading to more informed decision making (Reid et al., 2006). Hence, TEK provides complementary insights.

Secondly, integrating TEK with scientific methodologies can lead to more effective (or enhanced) resource management strategies. TEK's holistic view, which often includes sustainable harvesting practices and an understanding of ecological interdependencies, complements scientific approaches that focus on specific ecological parameters. This integrated approach can result in more sustainable and effective resource management plans (Berkes et al., 2000).

The integration of these knowledge systems also aids in the development of more effective environmental monitoring programs. TEK can provide historical baselines and contextual understanding of ecological changes, which, when combined with scientific monitoring techniques, offer a more nuanced view of ecosystem health and resilience (Folke et al., 2005).

Further, TEK contributes toward an inclusive policymaking. Incorporating TEK into coastal management policies ensures the inclusion of indigenous and local communities in decision-making processes. This inclusivity not only recognizes the rights of these communities but also ensures that management strategies are culturally appropriate and more likely to be supported and adhered to by local stakeholders (Nadasdy, 2003).

Integrating TEK, which is inherently adaptive, with scientific knowledge supports the development of adaptive management practices. Such practices are vital in the face of climate change and other environmental uncertainties, as they allow for flexible and responsive management strategies (Armitage et al., 2011).

8.5.2 Challenges and considerations in data integration

The integration of TEK with scientific knowledge faces challenges due to differences in methodologies and epistemologies. Scientific approaches are often quantitative and reductionist, while TEK is qualitative, holistic, and contextual. Bridging these methodological differences requires innovative approaches and mutual respect for the validity of both knowledge systems (Agrawal, 2002). Secondly, maintaining the authenticity and respect for TEK in the integration process is crucial. There is a risk of misinterpretation or decontextualization of TEK when it is filtered through scientific frameworks. Ensuring that knowledge holders are directly involved in the integration process is vital for maintaining the integrity of TEK (Turner & Clifton, 2009). Issues of intellectual property rights and data sensitivity are also significant when integrating TEK with scientific data. It is essential to establish clear agreements on the ownership, use, and sharing of TEK to protect the rights of indigenous and local communities (Tsing et al., 1999).

Integrating these knowledge systems often requires capacity building, particularly within local communities, to engage effectively with scientific data and tools. Additionally, sufficient resources and funding are necessary to support the integration process, including research, data collection, and the development of integrated management tools (Johnson et al., 2016). Further, developing collaborative frameworks that support the integration of TEK and scientific knowledge is essential. Such frameworks should facilitate mutual learning, respect diverse knowledge systems, and promote comanagement approaches. This requires the establishment of institutional structures that support collaboration and knowledge exchange (Fernández-Llamazares & Cabeza, 2018).

Overall, the rationale for integrating TEK with scientific data in coastal management is compelling. This integration offers numerous benefits, including comprehensive ecological insights, enhanced resource management, inclusive policymaking, and adaptive management capabilities. However, it also presents challenges, such as bridging methodological differences, ensuring respect for TEK, addressing intellectual property issues, and building collaborative frameworks. Addressing these challenges is crucial for realizing the full potential of data integration in supporting effective and sustainable coastal management decision making.

8.6 Methodological framework

8.6.1 Data collection and analysis techniques

Data collection in coastal environments employs a variety of techniques, reflecting the diverse nature of coastal ecosystems. Remote sensing and GIS are commonly used for large-scale data collection, providing spatial and temporal insights into coastal dynamics, habitat changes, and anthropogenic impacts (Jensen, 2007). Field surveys, including biodiversity assessments and water quality measurements, are crucial for ground-truthing remote sensing data and understanding localized phenomena (Phinn et al., 2008).

Advanced analytical techniques, such as spatial analysis in GIS, statistical modeling, and machine learning algorithms, are applied to interpret the collected data. These techniques help in identifying patterns, predicting future trends, and assessing the impacts of various coastal management strategies. In the realm of TEK, participatory mapping and ethnographic methods are employed to document local knowledge and insights (Johnson et al., 2016).

The integration of diverse datasets involves combining remotely sensed data, GIS layers, field observations, and TEK. This process requires careful consideration of the scale, resolution, and compatibility of different data types. Data integration allows for a more holistic understanding of coastal systems, supporting comprehensive management strategies (Wright et al., 2011).

8.6.2 Integrating traditional ecological knowledge and geospatial technology

Integrating TEK with geospatial technology involves developing frameworks that respect both knowledge systems while leveraging their strengths. Participatory GIS (PGIS) is one such approach, where community members contribute their local knowledge to GIS databases, enriching spatial data with cultural and historical context (Rambaldi et al., 2006).

Technological tools such as mobile GIS, web-based platforms, and interactive mapping interfaces facilitate the integration process. These tools enable the documentation, visualization, and analysis of TEK alongside scientific data, making the information accessible to both local communities and scientists.

Successful examples of this integration include the use of TEK in mapping critical habitats and cultural sites, which can inform protected area planning and resource management. Another example is the integration of indigenous seasonal calendars with climate data in GIS to understand the impacts of climate change on traditional livelihoods and ecosystems (Turner & Clifton, 2009).

8.6.3 Analyzing cases of integration specific to costal problems

The integration of TEK with modern geospatial technologies demands approaches specific to problems in the coastal regions. Studying and analyzing the prior cases of such integration may be helpful. The case studies will certainly enhance the understanding of the advantages and challenges of such integration in the context of coastal regions, which will help in framing better approaches. For example, a case study in coastal erosion management might involve integrating TEK regarding historical shoreline changes with satellite imagery analysis of recent coastal erosion patterns. This approach can identify vulnerable areas and inform the development of erosion mitigation strategies, such as managed retreat or the construction of natural barriers. On the other hand, in a fisheries management case study, local fishermen's knowledge about fish migration patterns and spawning grounds can be combined with remote sensing data on ocean conditions and GIS-based habitat maps. This integrated dataset can inform sustainable fishing practices and the establishment of no-take zones in MPAs. Similarly, a climate change adaptation case study could integrate community observations of changing weather patterns and sea-level rise impacts with scientific climate models and sea-level rise projections in GIS. This integration can aid in identifying vulnerable communities and developing locally appropriate adaptation strategies, such as the construction of climate-resilient infrastructure.

Overall, the methodological framework for integrating TEK with modern geospatial technology in coastal water resources management involves a multifaceted approach encompassing data collection, analysis, and integration. By combining diverse techniques and tools, and analyzing case studies, this framework supports the development of more effective, inclusive, and adaptive coastal management strategies.

8.7 Case studies and examples

8.7.1 Specific instances of integration in various regions

Studying and analyzing the prior cases of integrating TEK and modern geospatial technologies may be useful, as discussed in the previous section. A few significant case studies from different parts of the world are briefly discussed below.

Coastal management in the Pacific Northwest: In the Pacific Northwest, indigenous communities have integrated TEK with scientific data to manage salmon fisheries. This integration has involved combining traditional knowledge of salmon life cycles and migration patterns with scientific assessments of fish populations. The result has been the development of sustainable fishing practices that align with both ecological needs and cultural traditions (Lertzman & Vredenburg, 2005).

Coral reef conservation in the Great Barrier Reef: The Great Barrier Reef in Australia presents an instance where TEK of the indigenous communities, particularly their understanding of marine species and ecological indicators, has been integrated with scientific research in coral monitoring. This collaborative approach has led to more effective strategies for reef conservation and management, considering both ecological and cultural values (Olsson et al., 2008).

Controlling coastal erosion in the Maldives: In the Maldives, local knowledge about historical shoreline changes, coupled with scientific data on sea-level rise, has informed coastal erosion control measures. This integration has guided the implementation of adaptive strategies, such as building sea walls and promoting natural vegetation growth along coastlines (Brown et al., 2011).

8.7.2 Lessons learned and best practices

Integrating TEK with modern geospatial technology in coastal water resources management involves recognizing the value of TEK for its depth and context. Building trust and collaboration between indigenous communities and scientists is essential, as is balancing the methodologies and epistemologies of both knowledge systems. Adaptive management approaches and ensuring cultural sensitivity in the integration process are critical for sustainable outcomes. Documentation and sharing of best practices, along with capacity building and empowerment of local communities, enhance the efficacy and inclusiveness of this approach. Finally, establishing long-term monitoring and evaluation mechanisms ensures the sustainability and adaptability of management practices, taking into account both ecological and socio-cultural impacts. Fig. 8.2 represents a summary of lessons learned and best practices for integrating TEK with modern geospatial technologies.

A key lesson is the recognition of TEK as a valuable source of environmental knowledge. Integrating TEK with scientific data can provide a more comprehensive understanding of ecosystems, as local knowledge often includes information gained through generations of observation and interaction with the environment.

Effective integration requires building trust and collaboration between scientists and local communities. This involves respecting indigenous knowledge systems, ensuring open communication, and actively involving community members in research and decision-making processes (Armitage et al., 2011).

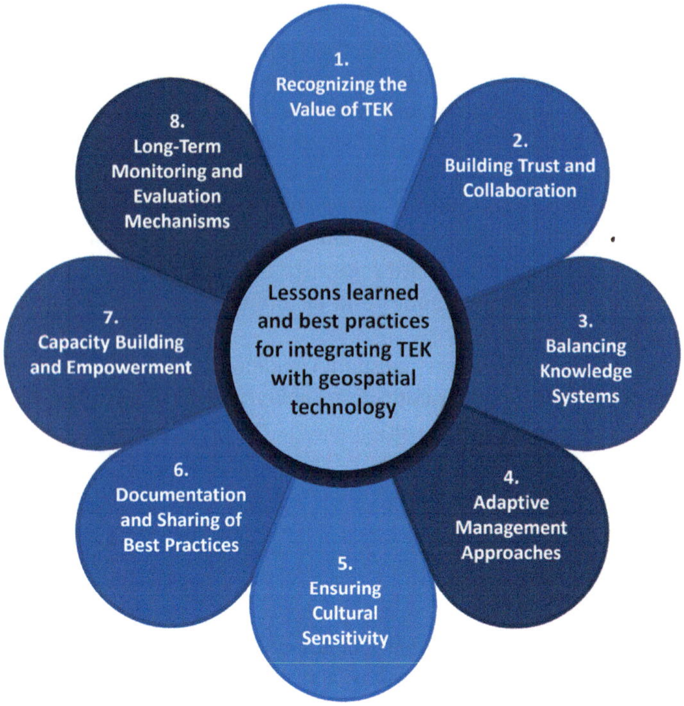

FIGURE 8.2 Flower chart summarizing the lessons learned and best practices for integrating TEK with modern geospatial technologies.

It is crucial to balance TEK and scientific knowledge without prioritizing one over the other. This balance can be achieved by establishing frameworks that allow for the coproduction of knowledge, where both TEK and scientific methods inform and complement each other (Johnson et al., 2016).

Adaptive management is essential in dealing with the uncertainties of environmental changes. Integrating TEK, with its inherent adaptability and long-term perspective, with scientific data enables the development of flexible management strategies that can be adjusted in response to new information or changing conditions (Berkes et al., 2000).

Cultural sensitivity in the integration process is paramount. This includes understanding and respecting the cultural context of TEK, ensuring that its use does not infringe on intellectual property rights or exploit indigenous communities (Tsing et al., 1999).

Documenting and sharing best practices from successful integrations can guide future efforts. Case studies that detail the processes, challenges, and outcomes of integration projects can serve as valuable resources for other regions and communities undertaking similar initiatives (Reid et al., 2006).

Investing in capacity building within local communities is crucial. Empowering community members with skills in data collection, analysis, and interpretation can enhance their participation in the management of their natural resources (Fernández-Llamazares & Cabeza, 2018).

Establishing long-term monitoring and evaluation mechanisms facilitates continuous assessment, createing feedback loops to inform stakeholders, integrating diverse data sources, and evaluating ecological and cultural impacts. This allows for the tracking of changes over time, providing insights that might not be evident in short-term studies.

Overall, the case studies and examples from various regions demonstrate the potential and effectiveness of integrating TEK with modern scientific approaches in coastal management. The lessons learned and best practices highlight the importance of respecting and valuing TEK, building collaborative relationships, maintaining cultural sensitivity, establishing monitoring mechanisms, and adopting adaptive management approaches. These insights can guide future efforts in integrating diverse knowledge systems for effective and sustainable coastal resource management.

8.8 Challenges and limitations

The challenges and limitations of such integration can be broadly categorized into two categories, that is, (1) ethical considerations and data sensitivity, (2) balancing indigenous and scientific perspectives. One of the primary ethical considerations in integrating traditional and scientific knowledge systems is ensuring respect for the intellectual property and cultural rights of indigenous communities. There is a risk of misappropriation or misuse of traditional knowledge without proper consent or acknowledgment (Tsing et al., 1999). Ethical integration requires transparent communication, explicit consent, and clear agreements on how TEK will be used and shared. Further, data collected from indigenous and local communities often contain sensitive information, including cultural practices, sacred sites, and traditional resource use patterns. Managing these data with care to protect community privacy and prevent potential exploitation is crucial. This involves adhering to protocols and guidelines that ensure data confidentiality and control over data access (Nadasdy, 2003). Ensuring informed consent and adopting participatory approaches in data collection and integration processes are vital. This means involving community members not just as data sources but as active participants in decision-making processes, respecting their autonomy and decision-making rights (Agrawal, 2002).

Regarding a balance between indigenous and scientific perspectives, overcoming epistemological differences is a major challenge, although it is crucial. Bridging the gap between indigenous knowledge systems, which are often holistic and qualitative, and scientific methods, which are typically quantitative and reductionist, poses significant challenges. This requires a mutual understanding and respect for the value and limitations of each knowledge system (Berkes et al., 2000). Moreover, there is a tendency in environmental management to prioritize scientific knowledge over TEK, which can lead to the marginalization of indigenous perspectives. Avoiding this hierarchization requires acknowledging the legitimacy and value of TEK as a complementary source of ecological understanding (Johnson et al., 2016). Besides, power imbalances between scientific researchers and indigenous communities can hinder effective knowledge integration. Addressing these imbalances involves empowering communities through capacity building, ensuring equitable participation in research, and recognizing the self-determination rights of indigenous peoples (Turner & Clifton, 2009). Further, incorporating TEK into environmental policy and management practices is challenging due to differing methodologies, scales, and applications. It requires innovative approaches that can translate qualitative TEK into formats compatible with scientific and policy frameworks (Reid et al., 2006).

Overall, integrating TEK with modern scientific approaches in coastal management faces several challenges and limitations, particularly concerning ethical considerations, data sensitivity, and the balance between indigenous and scientific perspectives. Addressing these challenges requires ethical integrity, mutual respect, and innovative methodologies to ensure effective and equitable integration. By overcoming these challenges, a more holistic and inclusive approach to coastal resource management can be achieved.

8.9 Future directions and recommendations

The future directions toward enhancing coastal management through such integration of knowledge includes embracing interdisciplinary approaches, advancing technologies for better integration, and fostering adaptive comanagement. Future coastal management strategies should increasingly embrace interdisciplinary approaches that integrate TEK with scientific data. This integration not only enriches ecological understanding but also fosters more comprehensive and sustainable management practices. Developing methodologies that can effectively combine qualitative TEK with quantitative scientific data is crucial for this integration (Berkes et al., 2000). Similarly, investment in advancing technologies, such as GIS and remote sensing, should be geared toward facilitating the incorporation of TEK. This involves creating user-friendly platforms that allow for the input and analysis of TEK alongside scientific data, making integration more accessible and effective (Wright et al., 2011). Besides, adaptive comanagement, which combines community-based management with adaptive strategies that respond to ecological feedback, is a promising direction. This approach benefits from the localized, detailed knowledge of TEK and the broad, system-wide understanding from scientific data, enhancing the adaptability and resilience of coastal management strategies (Armitage et al., 2011).

On the other hand, policy implications and community involvement are equally vital. The policy frameworks supporting integration, involving communities in policy making, capacity building for local communities, and promoting collaborative research and learning can be recommended for such integrated systems. Policy frameworks need to evolve to support the integration of TEK and scientific knowledge (specifically geospatial technologies) in coastal management. This requires policies that recognize the value of TEK, protect intellectual property rights, and encourage participatory approaches in environmental decision making (Agrawal, 2002). Also, community involvement in policy development is essential. Policies that are codeveloped with local communities, respecting their knowledge and needs, are more likely to be effective and sustainable. This participatory approach also ensures that management strategies are culturally appropriate and more readily accepted by local stakeholders (Johnson et al., 2016). In addition, strengthening the capacity of local communities to engage with scientific data and participate in management processes is vital. Providing education and training opportunities can empower communities, ensuring they have the skills and knowledge necessary to actively contribute to coastal management (Turner & Clifton, 2009). Moreover, encouraging collaborative research and learning between scientists and indigenous communities can bridge gaps in understanding and foster mutual respect. Such collaborations should be designed to provide benefits to all parties involved, contributing to the coproduction of knowledge (Reid et al., 2006).

Overall, it can be said that the future of coastal management lies in the successful integration of TEK with scientific approaches, necessitating interdisciplinary methods, advanced technologies, and adaptive comanagement strategies. Policy frameworks must evolve to support this integration, ensuring community involvement and capacity building. By embracing these directions and recommendations, coastal management can become more effective, sustainable, and inclusive, ultimately leading to healthier and more resilient coastal ecosystems.

8.10 Summary and conclusions

The chapter has comprehensively explored the integration of traditional ecological knowledge with modern geospatial technology for effective coastal water resources management. It started with highlighting the importance of coastal resources, emphasizing their ecological, economic, and social significance. The current strategies in coastal management were examined, revealing the need for more inclusive and adaptive approaches. The chapter then delved into the role of local knowledge in coastal restoration, underscoring the value of TEK in enriching scientific understanding and management practices. Data integration was discussed as a crucial element in supporting decision making, with a focus on the benefits and challenges of combining TEK with scientific knowledge. Methodological frameworks for this integration were outlined, including data collection techniques, tools for integration, and relevant case studies. These examples provided practical insights into the application of integrated approaches in various coastal contexts.

The key findings of this chapter include the recognition that TEK, when integrated with scientific data, provides a more holistic understanding of coastal ecosystems. This integration enhances coastal management by incorporating long-term, place-based knowledge with advanced scientific methods. The importance of ethical considerations, data sensitivity, and the balance between indigenous and scientific perspectives emerged as crucial factors in this process. The chapter also identified best practices and lessons learned from specific instances of integration, guiding future efforts in this field. Integrating TEK with scientific approaches offers significant implications for sustainable coastal management. This integrated approach can lead to more effective and adaptable management strategies, capable of addressing the complex challenges faced by coastal ecosystems. It supports the development of management practices that are not only ecologically sound but also culturally sensitive and inclusive of local communities.

Several studies in recent years have focussed on fulfilling sustainable development goals (SDGs), especially in developing countries (Guptha, Swain, Al-Ansari, Taloor, & Dayal, 2021, 2022; Nandi & Swain, 2023, 2024; Nandi, Patel, & Swain, 2024; Patel et al., 2022; Sahoo, Majumder, Swain, Pateriya, & Al-Ansari, 2022; Sahoo, Swain, Goswami, Sharma, & Pateriya, 2021; Swain et al., 2024; Swain, Sahoo, & Taloor, 2022; Swain, Sahoo, Taloor, Mishra, & Pandey, 2022). The integrated approach for coastal water resources management is particularly relevant to the fulfillment of several SDGs. Specifically, it aligns with SDG 14 (Life Below Water), SDG 13 (Climate Action), SDG 15 (Life on Land), SDG 11 (Sustainable Cities and Communities), and SDG 17 (Partnerships for the Goals). By enhancing the sustainable use and conservation of marine resources through informed and inclusive management practices, SDG 14 can be fulfilled. The development of adaptive management strategies that respond to climate change impacts on coastal ecosystems contributes directly to fulfillment of SDG 13. As coastal management often involves the conservation of terrestrial habitats and biodiversity, such integration is absolutely relevant to SDG 15. Such an integrated system also contributes to SDG 11 by promoting resilient and sustainable urban development, particularly in coastal cities vulnerable to climate change. Further, as the integration of TEK and scientific knowledge requires collaboration among various stakeholders, including indigenous communities, scientists, policymakers, and conservation organizations, it is a direct and significant contributor to fulfilling SDG 17. The integration of TEK with modern scientific knowledge in general, and modern geospatial technologies in particular, can also indirectly contribute to fulfilling other SDGs, especially in the context of developing countries.

It can be concluded that the integration of TEK with modern geospatial technology represents a forward-thinking approach in coastal water resources management. This chapter underscores the necessity of this integration for the development of sustainable, adaptable, and inclusive management practices. Such approaches not only enhance the health and resilience of coastal ecosystems but also contribute significantly to the achievement of key SDGs, paving the way for a more sustainable and equitablefuture.

References

Adger, W. N., et al. (2005). Social-ecological resilience to coastal disasters. *Science (New York, N.Y.)*, *309*(5737), 1036–1039.

Agardy, T., Di Sciara, G. N., & Christie, P. (2011). Mind the gap: addressing the shortcomings of marine protected areas through large scale marine spatial planning. *Marine Policy*, *35*(2), 226–232.

Agrawal, A. (2002). Indigenous knowledge and the politics of classification. *International Social Science Journal*, *54*(173), 287–297.

Alongi, D. M. (2008). Mangrove forests: Resilience, protection from tsunamis, and responses to global climate change. *Estuarine, Coastal and Shelf Science*, *76*(1), 1–13.

Amani, M., Mehravar, S., Asiyabi, R. M., Moghimi, A., Ghorbanian, A., Ahmadi, S. A., … Jin, S. (2022a). Ocean remote sensing techniques and applications: A review (part ii). *Water*, *14*(21), 3401.

Amani, M., Moghimi, A., Mirmazloumi, S. M., Ranjgar, B., Ghorbanian, A., Ojaghi, S., … Jin, S. (2022b). Ocean remote sensing techniques and applications: A review (part i). *Water*, *14*(21), 3400.

Anderson, K., & Gaston, K. J. (2013). Lightweight unmanned aerial vehicles will revolutionize spatial ecology. *Frontiers in Ecology and the Environment*, *11*(3), 138–146.

Armitage, D., et al. (2011). Co-management and the co-production of knowledge: Learning to adapt in Canada's Arctic. *Global Environmental Change*, *21*(3), 995–1004.

Barbier, E. B., et al. (2011). The value of estuarine and coastal ecosystem services. *Ecological Monographs*, *81*(2), 169–193.

Bartlett, D., & Smith, J. (2013). *GIS for coastal zone management*. CRC Press.

Beatley, T. (2009). *Planning for coastal resilience: Best Practices for calamitous times*. Island Press.

Berkes, F., Colding, J., & Folke, C. (2000). Rediscovery of traditional ecological knowledge as adaptive management. *Ecological Applications*, *10*(5), 1251–1262.

Bennett, N. J., Blythe, J., Tyler, S., & Ban, N. C. (2016). Communities and change in the anthropocene: understanding social-ecological vulnerability and planning adaptations to multiple interacting exposures. *Regional Environmental Change*, *16*, 907–926.

Brown, S., et al. (2011). Sea-level rise impacts on Africa and the effects of mitigation and adaptation: An application of DIVA. *Regional Environmental Change*, *11*(S1), 207–224.

Barbier, E. B. (2020). Estuarine and coastal ecosystems as defense against flood damages: An economic perspective. *Frontiers in Climate*, *2*, 594254.

Berkes, F. (1993). Traditional ecological knowledge in perspective. *Traditional Ecological Knowledge: Concepts and Cases*, *1*, 1–9.

Bethel, M. B., Brien, L. F., Danielson, E. J., Laska, S. B., Troutman, J. P., Boshart, W. M., … Phillips, M. A. (2011). Blending geospatial technology and traditional ecological knowledge to enhance restoration decision-support processes in coastal Louisiana. *Journal of Coastal Research*, *27*(3), 555–571.

Bethel, M. B., Brien, L. F., Esposito, M. M., Miller, C. T., Buras, H. S., Laska, S. B., … Parsons Richards, C. (2014). Sci-TEK: A GIS-based multidisciplinary method for incorporating traditional ecological knowledge into Louisiana's coastal restoration decision-making processes. *Journal of Coastal Research*, *30*(5), 1081–1099.

Boafo, Y. A., Saito, O., Kato, S., Kamiyama, C., Takeuchi, K., & Nakahara, M. (2016). The role of traditional ecological knowledge in ecosystem services management: The case of four rural communities in Northern Ghana. *International Journal of Biodiversity Science, Ecosystem Services and Management*, *12*(1-2), 24–38.

Charles, A. (2012). People, oceans and scale: governance, livelihoods and climate change adaptation in marine social–ecological systems. *Current Opinion in Environmental Sustainability*, *4*(3), 351–357.

Costanza, R., et al. (1997). The value of the world's ecosystem services and natural capital. *Nature*, *387*(6630), 253–260.

Çicin-Sain, B., & Knecht, R. W. (1998). *Integrated Coastal and Ocean Management: Concepts and Practices*. Island Press.

Cutter, S. L., et al. (2012). A geospatial approach to the conceptual framework of disaster resilience. *Annals of the Association of American Geographers*, *103*(6), 1036–1054.

de Freitas, J. G., Bastos, M. R., & Dias, J. A. (2018). *Traditional ecological knowledge as a contribution to climate change mitigation and adaptation: The case of the Portuguese coastal populations.. Handbook of climate change communication: Vol. 3: Case studies in climate change communication* (pp. 257–269). .

Dhal, L., & Swain, S. (2022). Understanding and modeling the process of seawater intrusion: A review. *Advances in Remediation Techniques for Polluted Soils and Groundwater*, 269–290. Available from https://doi.org/10.1016/B978-0-12-823830-1.00009-2.

Drew, J. A. (2005). Use of traditional ecological knowledge in marine conservation. *Conservation biology*, *19*(4), 1286–1293.

Dugan, J. E., et al. (2011). *Estuarine and coastal ecosystems and their services. Treatise on estuarine and coastal science*. .. .

Edgar, G. J., Stuart-Smith, R. D., Willis, T. J., Kininmonth, S., Baker, S. C., Banks, S., … Thomson, R. J. (2014). Global conservation outcomes depend on marine protected areas with five key features. *Nature*, *506*(7487), 216–220.

EU WFD. (2000). Directive 2000/60/EC of the European Parliament and of the Council Establishing a Framework for the Community Action in the Field of Water Policy. European Union Water Framework Directive. *Official Journal of the European Communities, L 327*, 1–72.

FAO. (1995). *Code of Conduct for Responsible Fisheries*. Rome: Fisheries and Aquaculture Department, Food and Agriculture Organization.

FAO. (2020). *The state of world fisheries and aquaculture 2020*. Food and Agriculture Organization of the United Nations.

Fernández-Llamazares, Á., & Cabeza, M. (2018). Rediscovering the potential of indigenous storytelling for conservation practice. *Conservation Letters*, *11*(3), e12398.

Foale, S. (2006). The intersection of scientific and indigenous ecological knowledge in coastal Melanesia: Implications for contemporary marine resource management. *International Social Science Journal*, *58*(187), 129–137.

Folke, C., et al. (2005). Adaptive governance of social-ecological systems. *Annual Review of Environment and Resources*, *30*, 441–473.

Guptha, G. C., Swain, S., Al-Ansari, N., Taloor, A. K., & Dayal, D. (2022). Assessing the role of SuDS in resilience enhancement of urban drainage system: A case study of Gurugram City, India. *Urban Climate*, *41*, 101075. Available from https://doi.org/10.1016/j.uclim.2021.101075.

Guptha, G. C., Swain, S., Al-Ansari, N., Taloor, A. K., & Dayal, D. (2021). Evaluation of an urban drainage system and its resilience using remote sensing and GIS. *Remote Sensing Applications: Society and Environment*, *23*, 100601. Available from https://doi.org/10.1016/j.rsase.2021.100601.

Halpern, B. S., et al. (2008). A global map of human impact on marine ecosystems. *Science (New York, N.Y.)*, *319*(5865), 948–952.

Hau'ofa, E. (1993). Our sea of islands. *The Contemporary Pacific*, *6*(1), 148–161.

Huntington, H. P. (2000). Using traditional ecological knowledge in science: Methods and applications. *Ecological Applications*, *10*(5), 1270–1274.

Hutton, N. S., & Allen, T. R. (2020). The role of traditional knowledge in coastal adaptation priorities: The Pamunkey Indian Reservation. *Water*, *12* (12), 3548.

IPCC. (2019). Special report on the ocean and cryosphere in a changing climate. Intergovernmental Panel on Climate Change.

Jiao, Y., Li, X., Liang, L., Takeuchi, K., Okuro, T., Zhang, D., & Sun, L. (2012). Indigenous ecological knowledge and natural resource management in the cultural landscape of China's Hani Terraces. *Ecological Research*, *27*, 247–263.

Jentoft, S. (2000). The community: a missing link of fisheries management. *Marine policy*, *24*(1), 53–60.

Jensen, J. R. (2007). *Remote sensing of the environment: An earth resource perspective*. Prentice Hall.

Johnson, J. T., et al. (2016). Weaving indigenous and sustainability sciences to diversify our methods. *Sustainability Science*, *11*(1), 1–11.

Lertzman, K., & Vredenburg, H. (2005). Indigenous peoples, resource extraction and sustainable development: An ethical approach. *Journal of Business Ethics*, *56*(3), 239–254.

Long, R. D., Charles, A., & Stephenson, R. L. (2017). Key principles of ecosystem-based management: the fishermen's perspective. *Fish. and Fisheries*, *18*(2), 244–253.

Loch, T. K., & Riechers, M. (2021). Integrating indigenous and local knowledge in management and research on coastal ecosystems in the Global South: A literature review. *Ocean and Coastal Management*, *212*, 105821.

Mahon, R., et al. (2008). Governing fisheries as complex adaptive systems. *Marine Policy*, *32*(1), 104–112.

Mcleod, E., et al. (2011). A blueprint for blue carbon: toward an improved understanding of the role of vegetated coastal habitats in sequestering CO2. *Frontiers in Ecology and the Environment*, *9*(10), 552–560.

Mumby, P.J., EdwardsA.J., 2014. Mapping marine habitats with remote sensing and GIS. In Marine ecology.

Mondal, B. K., & Das, R. (2023). *Appliance of indigenous knowlege in mangrove conservation and sustaining livelihood in Indian Sundarban Delta: A geospatial analysis. Traditional ecological knowledge of resource management in Asia* (pp. 77–101). Cham: Springer International Publishing.

Nandi, S., Patel, P., & Swain, S. (2024). IMDLIB: An open-source library for retrieval, processing and spatiotemporal exploratory assessments of gridded meteorological observation datasets over India. *Environmental Modelling and Software*, *171*, 105869. Available from https://doi.org/10.1016/j.envsoft.2023.105869.

Nandi, S., & Swain, S. (2024). Role of groundwater systems in fulfilling sustainable development goals: A focus on SDG6 and SDG13. *Current Opinion in Environmental Science & Health*, *42*100576. Available from https://doi.org/10.1016/j.coesh.2024.100576.

Noble, M. M., Harasti, D., Fulton, C. J., & Doran, B. (2020). Identifying spatial conservation priorities using traditional and local ecological knowledge of iconic marine species and ecosystem threats. *Biological Conservation*, *249*, 108709.

Nugroho, H. Y. S. H., Sallata, M. K., Allo, M. K., Wahyuningrum, N., Supangat, A. B., Setiawan, O., ... Najib, N. N. (2023). Incorporating traditional knowledge into science-based sociotechnical measures in upper watershed management: Theoretical framework, existing practices and the way forward. *Sustainability*, *15*(4), 3502.

Nandi, S., & Swain, S. (2023). Analysis of heatwave characteristics under climate change over three highly populated cities of South India: A CMIP6-based assessment. *Environmental Science and Pollution Research*, *30*, 99013–99025. Available from https://doi.org/10.1007/s11356-022-22398-x.

Nadasdy, P. (2003). Reevaluating the co-management success story. *Arctic*, *56*(4), 367–380.

Nicholls, R. J., Wong, P. P., Burkett, V. R., Codignotto, J. O., Hay, J. E., McLean, R. F., Ragoonaden, S., & Woodroffe, C. D. (2007). *Coastal systems and low-lying areas. Climate Change 2007: Impacts, Adaptation and Vulnerability. Contribution of Working Group II to the Fourth Assessment Report of the Intergovernmental Panel on Climate Change* (pp. 315–356). Cambridge University Press.

Nicholls, R. J., & Cazenave, A. (2010). Sea-level rise and its impact on coastal zones. *Science (New York, N.Y.)*, *328*(5985), 1517–1520.

NOAA. (2019). *Coastal Zone Management Act 309 Guidance 2021-2025 Enhancement Cycle*. National Oceanic and Atmospheric Administration.

Olsen, S.B., Page, G.G., & Ochoa, E. (2009). *The Analysis of Governance Responses to Ecosystem Change: A Handbook for Assembling a Baseline*. LOICZ Reports & Studies No. 34, GKSS Research Center, Geesthacht, Germany.

Olsson, P., et al. (2008). Enhancing the fit through adaptive co-management: Creating and maintaining bridging functions for matching scales in the Kristianstads Vattenrike Biosphere Reserve, Sweden. *Ecology and Society*, *13*(1).

Österblom, H., & Folke, C. (2015). Globalization, marine regime shifts and the Soviet Union. *Philosophical Transactions of the Royal Society B: Biological Sciences* (1659), 370.

Parlee, B. (2005). A place for traditional ecological knoinledge in resource management. *Breaking Ice: Renewable Resource and Ocean Management in the Canadian North*, *7*, 141.

Patel, P., Thakur, P. K., Aggarwal, S. P., Garg, V., Dhote, P. R., Nikam, B. R., ... Al-Ansari, N. (2022). Revisiting 2013 Uttarakhand flash floods through hydrological evaluation of precipitation data sources and morphometric prioritization. *Geomatics, Natural Hazards and Risk*, *13*(1), 646–666. Available from https://doi.org/10.1080/19475705.2022.2038696.

Proulx, M., Ross, L., Macdonald, C., Fitzsimmons, S., & Smit, M. (2021). Indigenous traditional ecological knowledge and ocean observing: A review of successful partnerships. *Frontiers in Marine Science*, *8*, 703938.

Pilkey, O. H., & Cooper, J. A. G. (2014). *The Last Beach*. Duke University Press. Available from https://doi.org/10.1515/9780822375944.

Phinn, S., et al. (2008). Remote sensing of mangroves and estuarine communities in Central Queensland, Australia. *Wetlands Ecology and Management, 16*(5), 379–392.

Rambaldi, G., et al. (2006). Participatory spatial information management and communication in developing countries. *The Electronic Journal on Information Systems in Developing Countries, 25*(1), 1–9.

Reid, W. V., et al. (2006). *Bridging scales and knowledge systems: Concepts and applications in ecosystem assessment.* Island Press.

Sahoo, S., Majumder, A., Swain, S., Pateriya, B., & Al-Ansari, N. (2022). Analysis of decadal land use changes and its impacts on urban heat island (UHI) using remote sensing-based approach: A smart city perspective. *Sustainability, 14*(19), 11892. Available from https://doi.org/10.3390/su141911892.

Sahoo, S., Swain, S., Goswami, A., Sharma, R., & Pateriya, B. (2021). Assessment of trends and multi-decadal changes in groundwater level in parts of the Malwa region, Punjab, India. *Groundwater for Sustainable Development, 14*, 100644. Available from https://doi.org/10.1016/j.gsd.2021.100644.

Sharma, I., Mishra, S. K., Pandey, A., Kumre, S. K., & Swain, S. (2020). Determination and verification of antecedent soil moisture using soil conservation service curve number method under various land uses by employing the data of small Indian experimental farms. *Watershed Management 2020*, 141–150. Available from https://doi.org/10.1061/9780784483060.013, Reston, VA: American Society of Civil Engineers.

Sharma, I., Swain, S., Mishra, S. K., & Pandey, A. (2024). Investigating climate and land use change impacts on design flood estimation over Indian tropical catchments. *Tropical Ecology, 65*, 483–507. Available from https://doi.org/10.1007/s42965-024-00323-2.

Stori, F. T., Peres, C. M., Turra, A., & Pressey, R. L. (2019). Traditional ecological knowledge supports ecosystem-based management in disturbed coastal marine social-ecological systems. *Frontiers in Marine Science, 6*, 571.

Swain, S. (2017). Hydrological modeling through soil and water assessment tool in a climate change perspective a brief review. In *2017 2nd International conference for convergence in technology (I2CT)*, pp. 358–361. IEEE. https://doi.org/10.1109/I2CT.2017.8226151.

Swain, S., Mishra, P. K., Nandi, S., Pradhan, B., Sahoo, S., & Al-Ansari, N. (2024). A simplistic approach for monitoring meteorological drought over arid regions: a case study of Rajasthan, India. *Applied Water Science, 14*(2)36. Available from https://doi.org/10.1007/s13201-023-02085-z.

Swain, S., Mishra, S. K., & Pandey, A. (2022a). Appraisal of land use/land cover change over tehri catchment using remote sensing and GIS. *Geospatial Technologies for Land and Water Resources Management*, 37–51. Available from https://doi.org/10.1007/978-3-030-90479-1_3, Springer, Cham.

Swain, S., Mishra, S. K., & Pandey, A. (2022b). Assessing spatiotemporal variation in drought characteristics and their dependence on timescales over Vidarbha Region, India. *Geocarto International, 37*(27), 17971–17993. Available from https://doi.org/10.1080/10106049.2022.2136260.

Swain, S., Mishra, S. K., Pandey, A., & Dayal, D. (2022a). Assessment of drought trends and variabilities over the agriculture-dominated Marathwada Region, India. *Environmental Monitoring and Assessment, 194*(12), 883. Available from https://doi.org/10.1007/s10661-022-10532-8.

Swain, S., Mishra, S.K., Pandey, A., & Dayal, D. (2022b). Spatiotemporal assessment of precipitation variability, seasonality, and extreme characteristics over a Himalayan catchment. *Theoretical and Applied Climatology, 147*, 817–833. Available from https://doi.org/10.1007/s00704-021-03861-0.

Swain, S., Mishra, S. K., Pandey, A., & Dayal, D. (2021). Identification of meteorological extreme years over central division of Odisha using an index-based approach. *Hydrological Extremes*, 161–174. Available from https://doi.org/10.1007/978-3-030-59148-9_12, Springer, Cham.

Swain, S., Mishra, S. K., Pandey, A., & Kalura, P. (2022). Inclusion of groundwater and socio-economic factors for assessing comprehensive drought vulnerability over Narmada River Basin, India: A geospatial approach. *Applied Water Science, 12*(2), 14. Available from https://doi.org/10.1007/s13201-021-01529-8.

Swain, S., Mishra, S. K., Pandey, A., Dayal, D., & Srivastava, P. K. (2022). Appraisal of historical trends in maximum and minimum temperature using multiple non parametric techniques over the agriculture-dominated Narmada Basin, India. *Environmental Monitoring and Assessment, 194* (12), 893. Available from https://doi.org/10.1007/s10661-022-10534-6.

Swain, S., Mishra, S. K., Pandey, A., Pandey, A. C., Jain, A., Chauhan, S. K., & Badoni, A. K. (2022). Hydrological modelling through SWAT over a Himalayan catchment using high-resolution geospatial inputs. *Environmental Challenges, 8*, 100579. Available from https://doi.org/10.1016/j.envc.2022.100579.

Swain, S., Sahoo, S., & Taloor, A. K. (2022). Groundwater quality assessment using geospatial and statistical approaches over Faridabad and Gurgaon districts of National Capital Region, India. *Applied Water Science, 12*(4), 75. Available from https://doi.org/10.1007/s13201-022-01604-8.

Swain, S., Sahoo, S., Taloor, A. K., Mishra, S. K., & Pandey, A. (2022). Exploring recent groundwater level changes using Innovative Trend Analysis (ITA) technique over three districts of Jharkhand, India. *Groundwater for Sustainable Development, 18*, 100783. Available from https://doi.org/10.1016/j.gsd.2022.100783.

Swain, S., Taloor, A. K., Dhal, L., Sahoo, S., & Al-Ansari, N. (2022). Impact of climate change on groundwater hydrology: A comprehensive review and current status of the Indian hydrogeology. *Applied Water Science, 12*(6), 120. Available from https://doi.org/10.1007/s13201-022-01652-0.

Swain, S., Mishra, S. K., Pandey, A., Srivastava, P. K., & Nandi, S. (2023). Characterization and assessment of hydrological droughts using GloFAS streamflow data for the Narmada River Basin, India. *Environmental Science and Pollution Research*, 1–14. Available from https://doi.org/10.1007/s11356-023-27036-8.

Tsing, A. L., et al. (1999). Words in motion: The cultural politics of human rights and environmentalism in Indonesian Borneo. *American Ethnologist, 26*(4), 905–924.

Turner, N. J., & Berkes, F. (2006). Coming to understanding: Developing conservation through incremental learning in the Pacific Northwest. *Human Ecology, 34*(4), 495–513.

Turner, N. J., & Clifton, H. (2009). It's so different today": Climate change and indigenous lifeways in British Columbia, Canada.". *Global Environmental Change*, *19*(2), 180–190.

Teixeira, J. B., Martins, A. S., Pinheiro, H. T., Secchin, N. A., de Moura, R. L., & Bastos, A. C. (2013). Traditional ecological knowledge and the mapping of benthic marine habitats. *Journal of Environmental Management*, *115*, 241–250.

Wright, D. J., et al. (2011). *Marine and coastal geographical information systems*. CRC Press.

Wright, D. J., & Wang, S. (2011). The emergence of spatial cyberinfrastructure. *Proceedings of the National Academy of Sciences*, *108*(14), 5488–5491.

Chapter 9

Comprehensive insights into dam break analysis: a mini review

Roniki Anjaneyulu and Ratnakar Swain

Department of Civil Engineering, National Institute of Technology (NIT), Rourkela, Odisha, India

9.1 Introduction

9.1.1 General

The study of dam break analysis is considered significant in predicting the peak discharge during dam failure. Dams play substantial roles in flood control, water supply, irrigation, hydropower generation, navigation, and recreation benefits. However, because of water storage and inundation, immense potential energy is generated, dramatically threatening the downstream (Peng & Zhang, 2013). Eight leading indices, that is, capacity of reservoir (CR), dam height (HD), economy at risk (ER), population at risk (PR), understanding of dam breach (UB), industry type (TI), warning time (TW), and building vulnerability (VB), are selected to establish an evaluation index system (Ge, Jaio, Sun, & Li, 2019). Therefore dams are an essential element of infrastructure for any country. Over 45,000 large dams are being used worldwide (Hjorth & Bengtsson, 2012), and 800,000 dams have been constructed. The US Army Corps of Engineering Hydrologic Engineering Center (HEC) research document lists 13 dam failure causes as (1) earthquake, (2) landslide, (3) piping, (4) extreme storm, (5) equipment malfunctioning, (6) structure damage, (7) foundation failure and sabotage. Different case studies show that dam failure is due to reasons ranging from seepage or piping, overtopping foundation defects, sliding, natural disasters, and equipment failure (Kyaw, 2020).

Overtopping failure: Generally, during an overtopping failure of an earthen dam, a head-cut erosion process will first start on the dam embankment's downstream side. While water goes over the dam crest, the dam crest acts like a broad-crested weir. The head cut begins to cut into the dam crest, the weir crest length will become shorter, and the appropriate weir coefficient will trend toward a sharp-crested weir value (Temple, Hanson, & Neilsen, 2006). When the head cut reaches the upstream side of the dam crest, a mass failure of the upstream crest may occur, and the hydraulic control section will act very much like a sharp-crested weir. The head cut will continue to erode upstream through the dam embankment and the dam and widen simultaneously (Brunner, 2014).

Piping failure: Water seeps through the dam at a significant enough rate, so it initially erodes material and transports it out of the dam. A large hole is formed as the material is eroded, thus carrying more water and eroding more material. During this process, the water movement through the dam is modeled as a pressurized orifice-type flow. During the piping nine flow processes, erosion and head cutting will occur on the dam's downstream side due to the flow exiting the pipe. As the piping hole grows, materials above the hole will slough off and fall into moving water (Brunner, 2014). The head-cutting and sloughing processes will continue pushing back toward the dam's upstream side while the piping hole grows simultaneously. If the piping hole is large enough, the weight of the material above the hole may be too great to be maintained, and a mass caving of material will occur. This will result in a large rise in outflow through the breach and will accelerate the breaching process (Wang, Kahawita, Mokhtari, Phat, & Quach, 2006). Also, at this point, the hydraulics of the flow transitions from orifice-type flow to open-type flow. The head-cutting and erosion process then continues back through the dam and downward.

Additionally, the breach will be widening. Depending on the volume of water behind the dam, the breach may continue to cut down and widen until the natural channel bed is reached. Then, the breach will go into a widening phase (Brunner, 2014).

Sustainable Development Perspectives in Earth Observation. DOI: https://doi.org/10.1016/B978-0-443-14072-3.00019-8

9.1.2 Impacts of dam breach failures

After identifying the dam failure consequences for each asset inside the inundation area, consider the impacts they are:

Economic impacts include repair costs, debris removal, replacement of contents, and the loss of jobs. Financial results are not limited to assets in the inundation area but may extend to infrastructure and resources that serve a much broader scope. In addition to direct damage from dam failure, economic impacts include the time required to repair, replace, and reopen businesses, governmental and nonprofit agencies, and industrial facilities (Security, 2011).

The dam failure scenario's social impacts can range from out-migration of people who have lost jobs to in-migration of workers to help repair and rebuild. Social impacts may include a shortage of safe, affordable housing. They may consist of loss of confidence in public officials and eventually the election of new officials, loss of social cohesiveness, loss of recreational opportunities, decreased level of government-supported community services, and diminished quality of life, as well as identification with new communities and discovery of new social opportunities (Dunning & Durden, 2009). To identify potential social impacts, convene one or more groups of community residents and discuss the possible effects on housing, jobs, social organizations, loss of government service, recreation, culture, and mental and physical health.

The environmental impact of a dam failure scenario depends heavily on the information gathered regarding assets in the inundation area and the identification of assets with potentially large quantities of pollutants. For example, an industry may store large amounts of hazardous, radioactive, or other toxic materials on site. No adverse environmental impacts will result if these materials are stored in containers that would not rupture under the modeled dam scenario. However, if the water releases and transports such materials, possible impacts might include disruption to wildlife and their habitats. Positive environmental impacts of a dam failure scenario might, but would not necessarily, include new silt and nutrient deposits on farmland or increased water levels downstream that would improve opportunities for fishing.

Other potential environmental impacts of a dam failure scenario are (1) sedimentation, resulting from flood waters rushing across land and moving silt, sand, soil, and rocks downstream. It causes sediments to infiltrate and fill water resources, such as ponds and wetlands, and threaten the health of aquatic plants and fish (Dunning & Durden, 2009). (2) Erosion contributes to sedimentation as river banks erode from rushing water. It leads to the collapse or undermining of structures or infrastructure. (3) Thermal effects: Changes in the temperature of the water downstream could negatively affect fish and fisheries, and deep impoundments could release very cold water. In contrast, shallow impoundments could release warmer water.

9.2 Historical studies on dam break analysis

The characteristics and contents of dam break research have changed with the research background since studies of dam break problems were started in France in the 1850s. Based on different factors and contents of dam break research at different times, three periods are divided: the first period (from 1850 to 1950), the second period (between 1950 and 1990), and the third period (from 1990 until now) (You, Li, Min, & Xiaolei, 2012).

Because there was no computer before 1948, dam break studies in the first period were mainly focused on theoretical solutions for dam break wave problems and related physical model tests. For example, Ritter neglected the bed shear friction and riverbed topography and obtained a simplified solution for rectangular instantaneous total collapse with dry riverbed downstream for the first time by using the characteristic form of Saint-Venant equations in 1892 (Ritter, 1982). Ignoring the riverbed topography and friction, Stocker divided the flow patterns of the dam site into three types: continuous wave flow, critical flow, and discontinuous wave flow, and extended Ritter solution to the wet-bed situation (Stoker, 1957). Research on analytical solutions for dam break wave problems tends to be mature nowadays but is only limited in some special cases (You et al., 2012).

Since large dams were constructed in the 1950s, causal factors of dam break disasters and threats of dam break floods in downstream cities or towns were paid much attention to during the second research period. For example, the USA and Austria conducted laboratory experimental tests and field tests of earth-rock dams to study the causal factors of dam break disasters and the timing scale for the dam break process (Xie, 1993). For the design of different projects, several research institutions modeled the dam break process to investigate dam break peak discharge, outflow hydrograph, and flood inundation (Lu, 1995). As the computer had been invented and used in dam break research in this period, many modeling technologies had been developed, and lots of dam break mathematical models had been built, such as the Cristofano model, HW (Harris-Wagner), DAMBRK, BREACH, BEED, SMPDBK, BRDAM, MIKE, DHI, and so on (You et al., 2012).

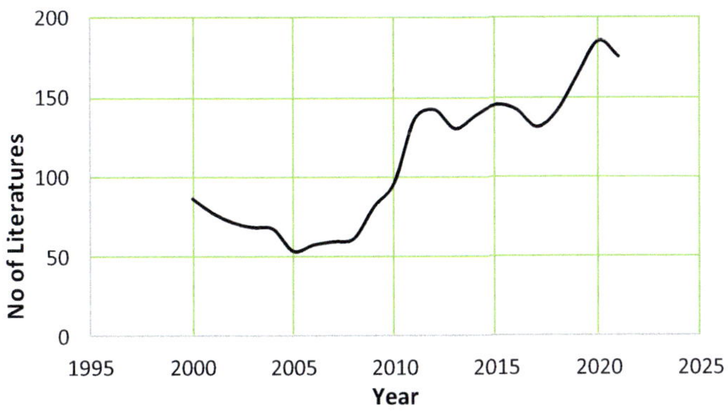

FIGURE 9.1 Year-wise number of literature published.

In the third period, dam safety problems were highlighted by significant dam break incidents in nearly several decades and the development of science, technology, and society. Characteristics of dam break research in this period were: (1) emphasizing cooperation between different countries and regions, such as NDSP (National Dam Safety Program held by the United States), CADAM (Concert Action on Dam Break Modeling, held by Europe Union, 1998.2−2000.1), and IMPACT (Investigation of Extreme Flood Processes and Uncertainty, held by Europe Union) (Bradbook, Waller, & Morris, 2005); (2) putting more attention to quantitative description, multiple processes coupling and system integration in dam break research (Hall, Sayers, & Dawson, 2005). For example, inundation hazard assessment models were built, comprising hydrodynamic models for river channels, dam breach models, and inundation models for the downstream areas of the failed dam (You et al., 2012). From the last 22 years, the number of papers published in different journals year-wise can be shown in the graphical representation in Fig. 9.1.

9.3 Literature review

9.3.1 General

In this study, a new comprehensive evaluation model was proposed on hydrodynamic factors, and the evacuation potential of the population at risk (PAR) 20 has significant impacts on the loss of life (LOL) caused by dam breaches (Ge et al., 2022). This paper provides a high-level overview of simplified to comprehensive methodologies for tailings dam breach study (TDBS) in practice (Liu & Henderson, 2020). The main objectives of this study are the determination of the outflow hydrograph and the peak discharge, estimation of dam breach parameters using appropriate empirical formulae, and routing of peak discharge and prediction of hydrograph at different sections downstream up to the point of consideration (Sawai, Shyamal, & Kumar, 2019). This study proposes that the lateral enlargement process can be modeled using a hyperbolic relationship with sufficient accuracy. Consequently, they can also perform the dam breach analysis and the sensitive study for a target case within 1 hour in an Excel spreadsheet (Chen, Ping, Wang, Yu, & Chen, 2019). This study recommended an artificial neural network (ANN) as an alternate method for estimating dam breach parameters. The ANN method is proposed due to its accurate prediction when it was applied to similar other cases in water resources (Sammen, Mohamed, Ghazali, Sidek, & El-Shafie, 2017). This paper is devoted to improving the estimation of embankment dam breach parameters. A new set of empirical equations to estimate average breach width, depth to final breach elevation, and breach formation time is generated using the data from 166 failed embankment dams (Soliman, Bekhit, Hamed, & El-Zwahry, 2014). This paper mainly focuses on risk, analysis, management, mitigation, and emergency planning (Viseu & de Almedia, 2009).

The research paper on Hidkal dam's dam break analysis in Karnataka, India, highlights overtopping failure as the critical scenario, emphasizing the need for proactive measures to safeguard downstream areas (Bharath, Shivapur, Hiremath, & Maddamsetty, 2021). The generated inundation maps aid in emergency response planning and flood mitigation (Xiong Yi, 2011; Lariyah et al., 2013) while sensitivity analysis assesses the influence of breach parameters (Froehlich, 2008; Sidek & Ros, 2011) on flood dynamics and protection measures.

The rapid dam break flood reproduction system, which leverages remote sensing data and digital elevation models (DEMs), effectively acquires key parameters for simulating dam break floods, such as the landslide body and affected buildings (Wang, et al., 2018). Utilizing a 2D cellular automata model, this system enables the dynamic reproduction of dam break spatiotemporal processes, offering advantages in terms of parallelism and computational simplicity (Ernst et al., 2010). Moreover, it constrains the flood's evolution with postdisaster information, ensuring a more accurate representation of the flood's impact (Huang, Huang, Ju, Xu, & He, 2015). To maximize the efficiency of flood disaster information transmission and perception, the system integrates dynamic visualization within a virtual geographical environment (VGE) (Chen & Lin, 2018; Zhu et al., 2014). This approach shows great promise in flood disaster management and response efforts.

9.3.2 HEC-RAS

In this study, the analysis was carried out using the HEC-RAS and HEC-Geo RAS models integrated with geographic information system (GIS). The unsteady flow simulation was done in the HEC-RAS model. HEC-Geo RAS model prepares geometric data of the Legadadi River in GIS. Input data for the HEC-RAS model and dam breach parameters were calculated using the empirical method of Froehlich (2008) and Gudeta (2022). The purpose of this study is to investigate the impact of a possible failure of the Bramianos dam, comprising of dam break simulations in both overtopping and piping states, the behavior of the flood wave that will result from a potential collapse of the dam, applying HEC-RAS 2D on two different elevation profiles: a digital elevation model (DEM) and an unmanned aircraft system derived digital surface model (UAS-derived DSM) (Psomiadis et al., 2021). The work in this paper focuses on dam breach parameters and their catastrophic consequences for human lives and the economy. This paper proposes a catastrophic evaluation method for fast evaluation of the potential implications of dam breach for eight leading indices (Ge et al., 2019).

This review paper explores dam failure scenarios in Iran's Vahdat dam basin using HEC-RAS modeling (Amini, Ali, Ghazali, & Huat, 2009). It covers a range of scenarios, emphasizing the increased risk of downstream dam failure when the upstream dam fails (Xiong Yi, 2011). The study highlights the importance of appropriate dam location, safety measures, and the critical volume for downstream dam reservoirs to prevent overtopping failure (Amini et al., 2009). Hydraulic parameters, such as flow velocities and water levels, are detailed along the flow pathway (Khosravi, Rostaminejad, Cooper, Mao, & Melesse, 2019), with flood zoning maps indicating areas at risk. This review provides valuable insights for risk assessment and disaster management in the region (Froehlich, 2008).

The review paper synthesizes key findings from diverse studies on flood risk assessment and management (Kundzewicz & Schellnhuber, 2004; Youssef, Sefry, Pradhan, & Alfadail, 2016). These studies emphasize the importance of accurate rainfall data and GIS techniques for understanding flood patterns (Elfadaly, Shams eldein, & Lasaponara, 2019). Dam break analysis using tools like HEC-RAS provides critical insights into the potential consequences of dam failures (Bharath et al., 2021). Integration of GIS with remote sensing and hydrologic modeling proves effective for assessing and mitigating flood risks in various regions (Abdelkarim, Awawdeh, Alogayell, & Al-Alola, 2020; Bhusal, Parajuli, Regmi, & Kalra, 2022).

The study comprehensively assesses flood risks in the Hirakud Reservoir area, incorporating climate change scenarios (Anjaneyulu, Swain, & Behera, 2023). Calibration and validation of the HEC-HMS model demonstrated its accuracy, and predictions showed increasing future flow. Dam breach analysis highlighted potential inundation areas and flood depths (Parhi, Mishra, Singh, & Tripathi, 2012). The research emphasizes the need for proactive flood mitigation strategies to safeguard the region (Anjaneyulu, Swain, & Behera, 2023).

9.3.3 DAMBRK

The study employed BOSS DAMBRK software to model dam break scenarios and assess downstream hazards (Fread, 1984). Results revealed a peak discharge of 60 $m^3\ s^{-1}$ and highlighted areas in the "danger zone" (FEMA, 1989). The study acknowledged uncertainties, including volume losses and two-dimensional flow effects (Gundlach & Thomas, 1977). Enhancing modeling accuracy may require addressing these factors (Kho et al., 2009).

In this study conducted at Cengklik Dam in Central Java, Indonesia, the importance of reservoirs for various functions, including water supply, hydroelectrical power, and flood control, was emphasized (Limantara, 2010). The potential consequences of dam break events were investigated, which could lead to fast flood wave propagation downstream, endangering lives and property. The research utilized the Boss Dambrk Program to analyze key flood parameters, such as flood wave time travel (Tb), flood peak time travel (Tp), and flood peak depth (Hp) (Limantara & Juowono, 2012). Empirical models were derived to understand these parameters in the context of dam break events, offering valuable insights for developing early warning systems (Wahl, 2004).

9.3.4 FLDWAV

This review paper presents a comprehensive dam break analysis utilizing the FLDWAV model and the dynamic wave method based on the Saint-Venant equations, focusing on the Idukki Dam in Kerala, India. The FLDWAV model offers valuable insights into the potential consequences of a dam failure, emphasizing the critical need for flood forecasting and warnings to safeguard lives and property (Afsal, Peter, Jobe, Vinaya, & Binoy Alias, 2016; Fread, 1988; Modi & Seth, 2019).

The methodology presented in this study offers a comprehensive approach for assessing dam break flood hazards using GIS, HEC-RAS, and FLO-2D models. Combining one-dimensional and two-dimensional models accounts for complex terrain and urban areas. Sensitivity analysis reveals the significant impact of different breaching modes on flood characteristics. This research provides valuable insights for flood hazard assessment and emergency planning, enhancing our understanding of dam break scenarios and their potential consequences (Bellos & Tsakiris, 2015; Chen, Evans, Djordjević, & Savić, 2012; Dottori & Todini, 2013; Schubert & Sanders, 2012).

9.3.5 MIKE 11

The reviewed literature exhibits a collective emphasis on utilizing the MIKE 11 program for sophisticated numerical modeling and analysis in the domain of dam safety assessment and hydraulic modeling. Each selected paper contributes significantly to understanding dam break scenarios, focusing on meticulous methodologies and advanced modeling techniques, particularly within the MIKE 11 framework.

Lariyah et al. (2013) meticulously investigated dam failure dynamics, with a specific focus on Kahang Dam. Employing a systematic approach involving field data collection and computer simulations, the study intricately examines breach parameters and their interplay in predicting outflow hydrographs. The findings underscore the imperative need for precise emergency response plans, emphasizing the nuanced intricacies inherent in dam failure scenarios.

Shah et al. (2019) make a substantial contribution by rigorously analyzing the hydraulic and dam break dynamics of Kenyir Dam using advanced MIKE models, including MIKE 11 and MIKE 21. The study derives breach outflow hydrographs and conducts floodplain modeling, particularly emphasizing the probable maximum flood (PMF) scenario to offer a robust assessment of worst-case scenarios. The identified peak outflow is a crucial benchmark for informed decision-making and effective risk mitigation. Yu, Feng, and Glotko (2016) provide a notable contribution to a dam safety assessment by employing the MIKE 11 program for numerical modeling of dam break scenarios at the Nizhny Novgorod control structure. Integrating GISs (ARCGIS10) and a DEM based on SRTM data enhances model accuracy. The study's technical depth, meticulous methodology, and application to aging water-engineering systems make it a valuable resource for dam safety assessment and emergency planning.

Sidek and Ros (2011) focus on modeling the dam break of Saddle Dam A in the Kenyir reservoir using MIKE 11. The study involves dam break modeling for breach under two scenarios, namely the PMF scenario and the clear day scenario. Dam breach parameters are predicted using Froehlich and Macdonald–Langridge-Monopolis (MDLM) predictor equations, showcasing the versatility of the MIKE 11-1D hydrodynamic model developed by the Danish Hydraulics Institute. Finally, Benjankar, Tonina, and McKean (2015) contribute to the literature by studying three central Idaho river systems using both 1D and 2D hydrodynamic models, specifically MIKE 11 and MIKE 21. The study underscores the significance of finer cross-sections in obtaining superior results for hydraulic variables in 1D modeling, highlighting the adaptability and applicability of MIKE 11 across different modeling scenarios.

In synthesis, these studies collectively underscore the pivotal role of the MIKE 11 program in advancing our comprehension of dam safety, risk mitigation, and emergency planning through rigorous numerical modeling and comprehensive analyzes. The adaptability of MIKE 11 across diverse dam break scenarios and river systems solidifies its status as an indispensable tool in hydraulic modeling and safety assessment.

9.4 Some important models of dam break analysis

9.4.1 Dam break forecasting model (DAMBRK)

The DAMBRK model, developed by the National Weather Service and outlined in studies by Fread (1984), stands at the forefront of dam break analysis, representing a sophisticated tool for predicting the dynamics of dam break wave formation and downstream progression. Its significance lies in its three essential components: a detailed description of the dam failure mode encompassing temporal and geometric aspects, computation of the outflow hydrograph through the dam break, and the hydraulic routing of this hydrograph through downstream channels or valleys (Browne, 2011).

At the core of the DAMBRK model are the complete one-dimensional Saint-Venant equations governing unsteady flow. These equations, coupled with internal boundary equations for structures like dams and bridges/embankments and external boundary equations at the upstream and downstream ends of the routing reach, form the basis for the model's predictive capabilities. The system of equations is solved through a nonlinear weighted four-point implicit finite difference method, allowing for the simulation of both subcritical and supercritical flows. Notably, DAMBRK is versatile enough to handle Newtonian (clear water) and non-Newtonian fluids, with parameters such as fluid rheology being user-specified for accurate representation (Gildeh1, 2021).

The governing equation of DAMBRK:

$$\frac{\partial h}{\partial t} + \frac{\partial (hu)}{\partial x} = 0$$

$$\frac{\partial (hu)}{\partial t} + \frac{\partial}{\partial x}\left(hu^2 + \frac{1}{2}gh^2\right) + gh\frac{\partial Z}{\partial x} = 0$$

where h is the flow depth, u is the flow velocity, g is the acceleration due to gravity, t is time, x is the spatial coordinate along the channel, and Z is the bed elevation.

The applicability of DAMBRK is expansive, catering to a range of input data, from rough estimates to detailed specifications. It excels in accessibility, requiring minimal computational effort on microcomputers and mainframe facilities. Its utility extends beyond dam break scenarios to include routing mud and debris flows or rainfall/snowmelt floods using specified upstream hydrographs. However, as with any model, DAMBRK has its limitations. While it can simulate a variety of scenarios, the accuracy of its predictions is contingent on the quality of input data, and it may face challenges in highly complex or site-specific situations.

In practice, DAMBRK proves invaluable in dam break analysis. It begins by allowing users to input parameters defining the geometric and temporal characteristics of the reservoir and dam break. The model then computes the peak outflow and outflow hydrograph using these inputs. Subsequently, a dynamic wave method is employed to simulate the downstream progression of the dam break wave through the channel or valley. The model accommodates factors such as downstream dams, flow constrictions, tributary inflows, river sinuosity, levees, and tidal effects during the flood's propagation. Outputs from DAMBRK include critical information like high water profiles along the valley, flood arrival times, and hydrographs at user-selected locations, making it an indispensable tool for assessing dam break scenarios (Fread, 2014).

9.4.2 Hydrologic engineering center-river analysis system

Hydrologic engineering center-river analysis system (HEC-RAS) 2D, developed by the US Army Corps of Engineers, is a powerful tool for hydraulic modeling, particularly in dam break analysis. This software facilitates unsteady two-dimensional flow routing using the Saint-Venant equations, allowing the simulation of downstream effects resulting from hypothetical dam failures. HEC-RAS 2D requires input parameters such as ultimate breach geometry, time to breach, failure mode, and other critical factors to generate a dam breach hydrograph and simulate the resulting flood wave. It is extensively employed in dam safety studies and flood damage analysis (Gundlach & Thomas, 1977).

The core of HEC-RAS 2D lies in its solution of the Saint-Venant equations, which form the basis for hydrodynamic modeling. These partial differential equations govern the conservation of mass and momentum, capturing the unsteady flow behavior in river channels. The finite volume method and flexible mesh models, such as DAMBRK and HECRAS, are employed to solve these equations, ensuring an accurate representation of dam break scenarios. While HEC-RAS 2D does not currently model non-Newtonian fluids, it incorporates the fundamental principles necessary for robust dam break analysis.

The Saint-Venant equations can be written as (Liggett, 1975) as follows:

Conservation of mass (continuity) equation:

$$B\frac{\partial y}{\partial t} + UB\frac{\partial y}{\partial x} + A\frac{\partial U}{\partial x} + U\frac{\partial A}{\partial x} = q$$

Conservation of momentum equation:

$$\frac{\partial U}{\partial t} + U\frac{\partial U}{\partial x} + g\frac{\partial U}{\partial y} = \frac{q(v_{\mathrm{L}} - U)}{A} + g(S_O - S_{\mathrm{f}})$$

where B = cross-section top width (m); y = flow depth (m); U = longitudinal velocity of the flow (m s^{-1}); A = cross-section area (m_2); q − lateral contribution per unit of channel length (m^3 s^{-1} m^{-1}); g = acceleration of gravity (m s^{-2}); v_L = input velocity of the lateral contribution in the longitudinal direction (m s^{-1}); S_o = channel bottom slope (m m^{-1}), and S_f = friction slope (m m^{-1}).

In the HEC-RAS model, the Saint-Venant equations are solved in terms of discharge Q (USACE, 2010):

$$\frac{\partial A}{\partial t} + \frac{\partial Q}{\partial x} = q$$

$$\frac{\partial Q}{\partial t} + \frac{\partial QU}{\partial x} + gA\left(\frac{\partial Z}{\partial x} + S_f\right) = 0$$

The term $\partial Z/\partial x$ refers tohe water surface slope (Fernandes & Gomes, 2017).

HEC-RAS 2D is widely applicable in dam break analysis, offering a comprehensive suite of features. It allows flow initiation from a hydrograph, sediment transport and water quality modeling in one dimension (1D), sediment transport in two dimensions (2D), and the development of finite volume methods for dam break simulations. The model's capability to generate dam breach hydrographs and simulate downstream consequences positions it as a crucial tool for assessing the potential impacts of dam failures on surrounding areas (Purwanto, Juwono, & Asmaranto, 2017).

While HEC-RAS 2D is a valuable tool, it does have some limitations. One notable constraint is the current lack of support for modeling non-Newtonian fluids. This restricts its application in scenarios where the fluid properties deviate from traditional Newtonian behavior. Additionally, the accuracy of simulations is contingent on the precision of input parameters, and uncertainties in these parameters may affect the reliability of the results (Mani, Kumar, & Patra, 2020).

Conducting a dam break analysis using HEC-RAS involves providing key input parameters such as the dam's location, failure mode, shape and progression, formation time, trigger condition, and weir and pipe flow coefficients. By leveraging the software's capabilities, users can generate dam breach hydrographs and simulate the resulting flood wave, allowing for a comprehensive understanding of downstream consequences. The time growth template for breaks provided by HEC-RAS assists in forecasting the evolution of the breach over time, aiding in the assessment of potential risks and impacts associated with dam failures. Overall, HEC-RAS is a robust tool for dam break analysis, contributing significantly to the field of hydraulic modeling and dam safety (Raman & Liu, 2019).

9.4.3 MIKE 11

MIKE 11, developed by the Danish Hydrologic Institute (DHI), is a powerful software package designed for simulating flows, water quality, and sediment transport in various water bodies, including estuaries, rivers, and irrigation channels. Recognized for its versatility and reliability, MIKE 11 serves as a comprehensive tool in hydrological studies and engineering applications worldwide. MIKE 11 employs a finite difference scheme to solve the Saint-Venant equations, which form the basis for simulating flow dynamics. These equations, encompassing principles of mass and momentum conservation, enable MIKE 11 to represent water flow in different scenarios accurately. Additionally, the model incorporates a mechanism for modeling breaches, allowing the simulation of dam breaks. The breach growth is described through time series for breach width, crest level, and side slope. An erosion model based on the Engelund–Hansen sediment transport equation further enhances its capabilities (Dat, Tri, Truong, & Hoa, 2019).

The governing equation of MIKE 11:

$$\frac{\partial Q}{\partial x} + \frac{\partial (A + A_O)}{\partial t} - q = 0$$

and

$$\frac{\partial Q}{\partial t} + \frac{\partial\left(\frac{Q^2}{A}\right)}{\partial x} + gA\frac{\partial h}{\partial x} + S_f + S_c = 0$$

where Q = discharge or flow; h = water surface elevation; A = active flow area; A_o = inactive flow area; q = lateral outflow; g = gravitational acceleration; t = time; x = distance along waterway; S_c = expansion contraction slope, and S_f = friction slope.

MIKE 11 model applicability extends to many water bodies, making it a valuable tool for dam break analysis. The model is adept at simulating complex scenarios involving breaches and can describe breach growth over time. It is particularly suited for applications where a detailed understanding of flow dynamics, water quality, and sediment transport is crucial. The lateral link feature facilitates the connection to other MIKE models, such as MIKE 21 (2D), enabling a comprehensive analysis of coupled 1D-2D systems (Vanderkimpen, Melger, & Peeters, 2009).

While MIKE 11 offers advanced capabilities, it is essential to consider its limitations. The model's accuracy is contingent on the quality of input data and the appropriateness of assumptions made during simulations. Users should know the inherent uncertainties associated with modeling breach growth and erosion. Like any modeling tool, MIKE 11 may have limitations in capturing all nuances of real-world scenarios, and its results should be interpreted with an understanding of these constraints (Vanderkimpen & Peeters, 2008).

In the context of dam break analysis, MIKE 11 proves to be a valuable asset. The model allows for the explicit simulation of breaches, considering parameters such as breach width, crest level, and side slope over time. Using the Saint-Venant equations accurately represents flow dynamics during dam break scenarios. The model's versatility is further enhanced by its ability to dynamically link with MIKE 21 (2D), enabling the creation of coupled 1D-2D models. This coupling is crucial for a comprehensive analysis of inundation patterns, considering both the river channel and floodplain, and provides a detailed understanding of the potential impact of dam breaks on downstream areas (Dat et al., 2019).

9.4.4 FLDWAV

The FLDWAV model, developed by the National Weather Service (NWS), is a robust computational tool explicitly designed for analyzing flows during dam breaks and other significant flood events. An evolution from the DAMBRK model, FLDWAV encompasses advanced capabilities to simulate flows in single and multiple channels. It exhibits versatility in handling various channel configurations, from straight to meandering channels, making it a comprehensive solution for flood modeling (Gildeh1, 2021).

At the heart of the FLDWAV model are the complete one-dimensional Saint-Venant unsteady flow equations. These equations, an expanded version of those developed by Barré de Saint-Venant in 1871, include equations of mass and momentum. The choice of the dynamic wave method for hydraulic routing is based on the model's ability to simulate unsteady flood waves accurately. This method considers acceleration effects associated with dam break waves. It considers downstream unsteady backwater effects caused by various factors such as channel constrictions, dams, bridge-road embankments, and tributary inflows.

The governing equation of FLDWAV:

$$\frac{\partial}{\partial x} + \frac{\partial S_{co}(A + A_O)}{\partial t} - q = 0$$

and

$$\frac{\partial (S_m Q)}{\partial t} + \frac{\partial \left(\frac{\beta Q^2}{A}\right)}{\partial x} + gA\left(\frac{\partial h}{\partial x} + S_f + S_e + S_i\right) + L + W_f B = 0$$

where Q = discharge or flow; h = depth of water; A = active cross-sectional area; A_o = inactive area; q = lateral inflow at any cross-section; g = gravitational acceleration; t = time; x = longitudinal distance along the river reach; S_{co} = sinuosity factor; S_m = sinuosity factors; β = momentum coefficient for velocity distribution; S_f = channel and floodplain friction slope, S_e = expansion/contraction slope, S_i = additional friction slope due to internal viscous dissipation of non-Newtonian fluids such as mud or debris flows; L = momentum effect of lateral flow; B = active channel top width at water surface depth h and W_f = effect of wind resistance on the surface of the flow (Zhou, Judge, & Donnelly, 2004).

FLDWAV exhibits a wide range of applicability in dam break analysis. It can model free surface flows in subcritical, supercritical, mixed flow regimes and conduit flow under pressure. The model is adept at simulating clear water and mud/debris fluid flows, making it suitable for diverse flow compositions. Additionally, FLDWAV accommodates the modeling of off-channel flow areas that may act as flow storage during high flows, offering a comprehensive approach to floodplain analysis (Kunkel, 2011).

While FLDWAV is a powerful tool, it is essential to recognize its limitations. The model's accuracy is contingent on various factors, including the quality of input data and the assumptions inherent in its underlying equations. Like any modeling approach, FLDWAV has inherent simplifications and may not capture all the complexities of real-world

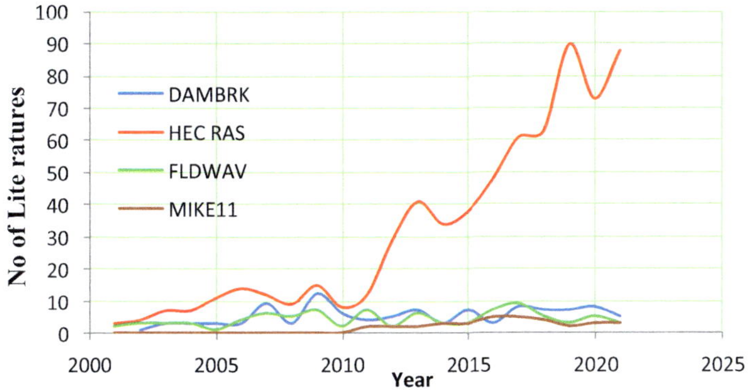

FIGURE 9.2 Year-wise number of literatures on DAMBRK, HEC-RAS, FLDWAV, and MIKE1 models.

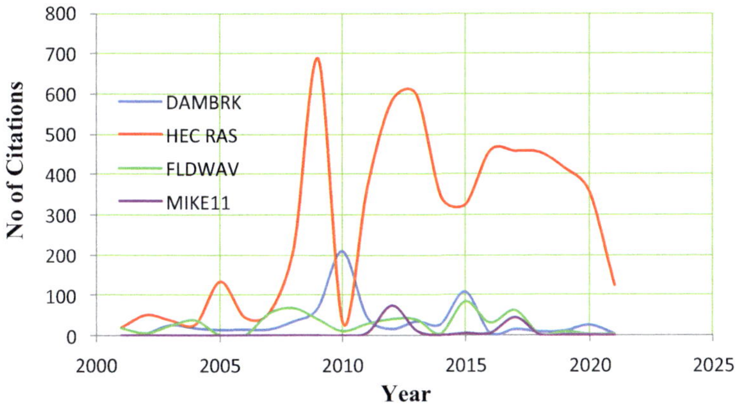

FIGURE 9.3 Year-wise number of citations on literature published on DAMBRK, HEC-RAS, FLDWAV, and MIKE1 models.

scenarios. Users should exercise caution and consider potential uncertainties in the results, especially when input data are limited or uncertain (Gee & Brunner, 2005).

In the context of dam break analysis, FLDWAV proves invaluable. Leveraging the dynamic wave method, it accounts for acceleration effects and downstream unsteady backwater effects produced by dam breaks. The model allows for the simulation of time-dependent dam breaks and control structures, providing a realistic representation of the dynamic behavior of flood waves. By utilizing the complete one-dimensional Saint-Venant equations, FLDWAV enables a thorough and accurate assessment of the consequences of dam breaks, aiding in developing effective flood management strategies and infrastructure planning (Zhou, Judge, & Donnelly, 2004).

The number of literatures and their citations of three models from the last 20 years are shown in Figs. 9.2 and 9.3.

9.4.5 Other dam break models

Other dam break models are provided in Table 9.1.

9.5 Integration of remote sensing techniques in dam break analysis

Integrating remote sensing techniques has emerged as a critical advancement in dam break analysis, specifically in optimizing floodplain delineation processes. One notable contribution in this domain is exemplified by Yang, Townsend, and Daneshfar (2006), who pioneered a direct processing approach by seamlessly combining the GIS with the HEC-RAS. Their study, conducted in Ottawa, Ontario, Canada, focused on the South Nation River system and involved numerical model simulations for six distinct design storm events. Through the integration of water surface profiles with DEM data, a triangular irregular network (TIN) terrain model was established. This model facilitated the accurate reproduction of floodplain zones for various storm events, showcasing the capability of HEC-RAS to enhance computational routines and support GIS data manipulation for improved visualization (Yang et al., 2006).

TABLE 9.1 Comparison table of other dam break models (Derdous, Djemili, Bouchehed, & Bouguerra, 2015; O'Brien & Garcia, 2009; Shahraki, Zadbar, Motevalli, & Aghajani, 2012).

Feature	FLO-2D	SMPDBK
Software developer	Not specified	NWS, USA
Primary purpose	Flood hazard, mudflow, debris flow, dam break analysis	Dam classification, dam break risk
Fluid type modeling capability	Sediment-laden and without sediment flow	Unsteady flow, sediment transport, and various add-on modules
Geometric and temporal input	Grid system, elevation, roughness, discharge	Quick and easy to use, simplified cases
Solution method	Finite volume method	Not specified
Computational capabilities	Flood hazard, mudflow, debris flow	Dam classification, downstream inundation
Additional modeling capabilities	Predicting flow path and area, discharge estimation based on depth of flow	Downstream inundation areas, efficient simulation, no high-spec computer requirement
Advantages	Good in predicting flow path and area, sediment-laden and without sediment flow modeling	Quick and easy to use, efficient for simpler cases, provides similar results to DAMBRK
Limitations	Accuracy is dependent on grid density and available data	Quick and easy to use but may lack detailed parameters
Model complexity	Intermediate	Quick and easy to use

Building upon this foundation, researchers such as Salajegheh, Bakhshaei, Chavoshi, Keshtkar, and Najafi Hajivar (2009) and Asnaashari, Meredith, and Scruton (2014) have further advanced the integration of GIS and HEC-RAS for floodplain analysis. Salajegheh et al. (2009) focused on the Polasjan river basin in Iran's central plateau, employing a direct approach to process HEC-RAS output in the ArcGIS platform. Their findings underscored the effectiveness of GIS in floodplain mapping and analysis, particularly when integrated with HEC-RAS. The resulting model proved user-friendly and offered digital outputs, contributing to resource savings and ease of comparison with other digital data (Salajegheh et al., 2009).

Additionally, Qi and Altinakar (2012) and Asnaashari et al. (2014) have extended the integration of remote sensing techniques to enhance dam break flood management. Qi and Altinakar (2012) developed a GIS-based decision support system for Sinclair Dam in Milledgeville, Georgia, USA, incorporating two-dimensional flood simulations based on classified remote sensing images. The system, validated using HEC-RAS and HEC-FDA software, demonstrated its versatility and reliability in estimating flood damage, emphasizing the potential for improved decision-making in flood prevention design (Qi & Altinakar, 2012). Asnaashari et al. (2014) conducted two hypothetical dam breach studies in Canada, employing HEC-RAS and GIS tools. Their approach, focusing on estimating dam break outflow hydrographs and conducting incremental consequence analyzes, showcased the utility of the integrated model for improving dam breach analyzes (Asnaashari et al., 2014). These technical contributions collectively underscore the evolving landscape of dam break analysis through the integration of remote sensing techniques, offering enhanced precision and efficiency in floodplain delineation and management strategies.

Syifa, Park, Achmad, Lee, and Eom (2019) introduced a comprehensive study in Iraq's climatically sensitive northern region, employing advanced remote sensing, GIS, and multicriteria decision-making techniques for dam site selection. Leveraging high-resolution data, including Landsat imageries and ASTER DEM, the study rigorously derived 12 critical conditioning factors for precise dam site suitability assessment. Validation against the analytic hierarchy process (AHP) underscored the superior accuracy of fuzzy logic, particularly in spatially clustered scenarios, emphasizing the instrumental role of remote sensing and GIS in navigating complex water management challenges.

Samsudin, Zuhaily, Setu, Muda, and Amin (2023) enriched the landscape of dam break analysis by integrating hydrodynamic modeling via HEC-RAS with advanced geospatial technologies, including GIS, quantum GIS (QGIS), and OpenStreetMap (OSM). Despite acknowledging challenges related to OSM data, the research provided a commendable and cost-effective solution for rapid decision making during dam-related crises, contributing significantly to advancing the integration of remote sensing and GIS in dam break analysis and disaster management. Cacciuttolo and Cano (2023) systematically explored mine tailings management through real case analysis, employing advanced remote sensing techniques and GIS. The study investigated the spatial and temporal dynamics of the supernatant process water pond within tailings storage facilities, emphasizing the critical importance of monitoring these ponds to mitigate the risks associated with dam breaks. Sun et al. (2023) focused on check dam detection on the Loess Plateau, China, employing a sophisticated framework integrating high-resolution remote sensing, deep learning, and geospatial analysis. Their study achieved a commendable 87.0% recall rate for 521 check dams, providing crucial insights for efficient watershed management and proactive dam break risk assessment. Tebbs, Avery, and Chadwick (2020) employed advanced satellite products to develop predictive models for lake chlorophyll based on environmental variables. The study, focusing on Lake Turkana impacted by altered ecological flows from the Gibe III dam, highlighted the adaptability of remote sensing and GIS for evaluating phytoplankton dynamics in lakes and emphasized its utility in dam break risk assessments and water management scenarios.

In conclusion, these diverse technical contributions collectively underscore the evolving landscape of dam break analysis by integrating remote sensing techniques. Each study makes significant strides in advancing the technical robustness and practical applicability of remote sensing and GIS in the nuanced domain of dam break analysis and disaster management.

9.5.1 Monitoring surface water change due to dam break using satellite remote sensing

Remote sensing data and derived products are powerful tools for monitoring land surface change and allied processes. Among various remote sensing platforms, satellite remote sensing captures the land surface reflectance or backscatter signals from space at periodic intervals. Several space agencies provide remote sensing images publicly for research and developmental activities, such as linear imaging and self scanning (LISS) data by Indian Space and Research Organization (ISRO), Landsat data by the National Aeronautics and Space Administration (NASA), Sentinel data by European Space Agency (ESA).

The Sentinel-2 multispectral remote sensing image contains 13 spectral bands in the optical wavelength. The blue, green, red, and near-infrared (NIR) spectral bands in the Sentinel-2 image have 10 m spatial resolution. The spectral enhancement uses two or more bands to generate an index map, highlighting a few features in an image and suppressing other features. Green and NIR are used in normalized difference water indexes (NDWI = [green − NIR]/[green + NIR]), which enhances the surface water bodies in an image (McFeeters, 1996). The NDWI value ranges from −1 to +1, wherein the higher values indicate surface water bodies and the lower values indicate other land surface features. The NDWI maps surface water areas by applying a threshold value (e.g., NDWI ≥ 0.3), which can be optimized by comparing the map with reference information. The multitemporal NDWI maps are suitable for rapid surface water change monitoring using remote sensing images (Das et al., 2021).

The change in surface water area in the Chungthang Dam was studied before and after the dam break using satellite remote sensing. The satellite images were processed in the Google Earth Engine (GEE) webGIS platform, providing access to geospatial data, including Sentinel-2 images. The GEE platform can be accessed through a web browser, allowing data processing without any dedicated geospatial data processing software on the computer. The generated maps can be downloaded, reducing the raw data storage space, or they can be distributed with others by creating an application on GEE (Das et al., 2021).

The cloud-free top-of-atmosphere reflectance images of January 2022 and January 2024 were used. The standard false color composite (SFCC: NIR, red, and green bands were passed through red, green, and blue channels) images of January 2022 and January 2024 used in the study are shown in Fig. 9.4A and B, respectively. The blue features in the image indicate waterbodies, while red and white indicate vegetation and bare soil (or sand) in the image. The NDWI maps of January 2022 and January 2024 are shown in Fig. 9.4C and D, respectively. The surface water change was identified in a 200-ha area around the Chungthang Dam. The change detection was performed to map the surface water area using the following criteria:

$$\text{Surface Water Change} = \text{NDWI\_Jan\_2022} > 0.1 \text{ and } \text{NDWI\_Jan\_2024} < 0.05 \text{ and } \text{NDWI\_Change} > 0.2$$

The generated surface water change is shown in Fig. 9.4E, which indicates a net change in surface water area of 9.93 ha.

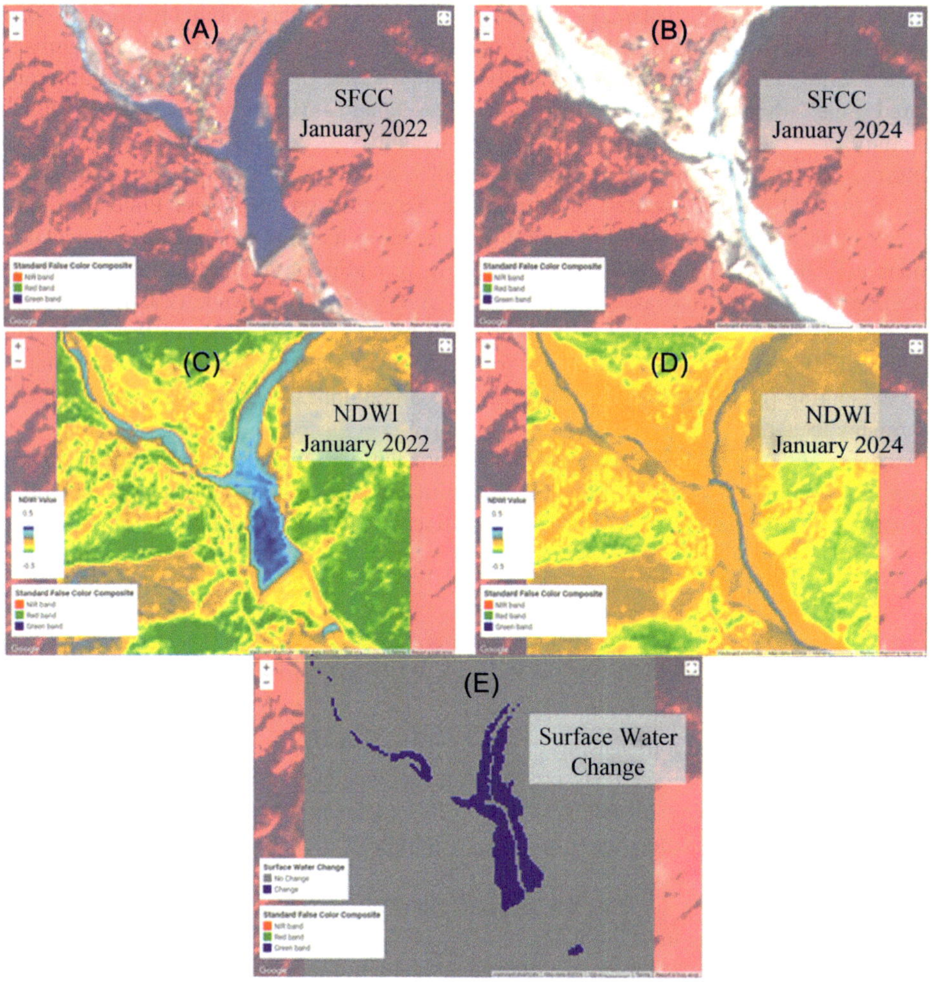

FIGURE 9.4 Standard false color composite (SFCC) images of (A) January 2022 and (B) January 2024, normalized difference water index (NDWI) maps of (C) January 2022 and (D) January 2024, and (E) surface water change from January 2022 to January 2024. Map lines delineate study areas and do not necessarily depict accepted national boundaries.

9.6 Dam breach parameters

Historical dam failures and laboratory studies are the data sources used in this paper. Failure data of 166 dams are assembled from several publications. The available data is categorized into two categories (Soliman et al., 2014).

i. The first category is dam characteristics, which include:
- ✓ Dam height (H): the difference between the reservoir's lowest and dam crest levels.
- ✓ Crest width (W_{crest}): the dam width at the crest.
- ✓ Dam length (L): the dam length at the dam crest.
- ✓ Average width (W_{avg}): the average width between the dam crest and the dam width at the bottom.
- ✓ Downstream slope (Z_{ds}): the slope of the dam downstream face (H:V).
- ✓ Storage volume (V): the reservoir volume corresponding to the dam crest level.

ii. The second category is dam breach parameters, which include:
- ✓ Average breach width (B_{avg}): two definitions are available for this parameter depending on the case studies. The first is the average breach width (B_{avg}) measured at the midpoint of breach height. At the same time, the second is the top width of the breach opening measured at the dam crest.
- ✓ Depth to final breach elevation (H_f): the difference in elevation between the dam crest and final breach elevation.
- ✓ Formation time (T_f): the duration required to form the final breach opening starting from the first erosion of the upstream face in case of piping failure or the first erosion of the downstream face in case of overtopping failure (Soliman et al., 2014).

9.6.1 Dam breach parameters calculation

9.6.1.1 *Froehlich (1995)*

Froehlich utilized 63 earthen, zoned earthen, earthen with a core wall (i.e., clay), and rock fill data sets to develop equations to predict average breach width, side slopes, and failure time. The data that Froehlich used for his regression analysis had the following ranges:

- ✓ Height of the dams: 3.66−92.96 m (12−305 feet)
  (With 90% <30 m, and 76% <15 m)
- ✓ The volume of water at breach time: 0.0130−660.0 $m^3 \times 10^6$ (11−535,000 acre-feet)
  (with 87% <25.0 $m^3 \times 10^6$, and 76% <15.0 $m^3 \times 10^6$)

Froehlich's regression equations for average breach width and failure time are:

$$B_{avg} = 0.1803 K_O V_w^{0.32} h_b^{0.19}$$

$$t_f = 0.00254 V_w^{0.53} h_b^{-0.90} \text{ (Froehlich, 1995)}$$

where:

- ✓ B_{avg} = average breach width (m)
- ✓ K_o = constant (1.4 for overtopping failures, 1.0 for piping)
- ✓ V_w = reservoir volume at time of failure (cubic meters)
- ✓ h_b = height of the final breach (m)
- ✓ t_f = breach formation time (hours)

Froehlich states that the average side slopes should be:

- ✓ 1.4 H:1 V overtopping failures
- ✓ 0.9 H:1 V otherwise (i.e., piping/seepage)

While not clearly stated in Froehlich's paper, the height of the breach is normally calculated by assuming the breach goes from the top of the dam all the way down to the natural ground elevation at the breach location (Froehlich, 1995).

9.6.1.2 *Froehlich (2008)*

In 2008 Dr. Froehlich updated his breach equations based on adding new data. Dr. Froehlich utilized 74 earthen, zoned earthen, earthen with a core wall (i.e., clay), and rock fill data sets to develop equations to predict average breach width, side slopes, and failure time. The data that Froehlich used for his regression analysis had the following ranges:

- ✓ Height of the dams: 3.05−92.96 m (10−305 feet)
  (With 93% <30 m, and 81% <15 m)
- ✓ Volume of water at breach time: 0.0139−660.0 $m^3 \times 10^6$ (11.3−535,000 acre-feet)
  (with 86% <25.0 $m^3 \times 10^6$, and 82% <15.0 $m^3 \times 10^6$)

Froehlich's regression equations for average breach width and failure time are:

$$B_{avg} = 0.1803 K_O V_w^{0.32} h_b^{0.04}$$

$$t_f = 63.2\sqrt{\frac{V_w}{g h_b^2}} \text{ (Froehlich, 2008)}$$

where:

- ✓ B_{ave} = average breach width (m)
- ✓ K_o = constant (1.3 for overtopping failures, 1.0 for piping)
- ✓ V_w = reservoir volume at time of failure (cubic meters)
- ✓ h_b = height of the final breach (m)
- ✓ g = gravitational acceleration (9.80665 m per second squared)
- ✓ t_f = breach formation time (seconds)

Froehlich's 2008 paper states that the average side slopes should be:

- ✓ 1.0 H:1 V overtopping failures
- ✓ 0.7 H:1 V otherwise (i.e., piping/seepage)

While not clearly stated in Froehlich's paper, the height of the breach is normally calculated by assuming the breach goes from the top of the dam down to the natural ground elevation at the breach location (Froehlich, 2008).

9.6.1.3 MacDonald and Langridge-Monopolis (1984)

MacDonald and Langridge-Monopolis utilized 42 data sets (predominantly earth fill dams, earth fill dams with a clay core, and rock fill dams) to develop a relationship for what they call the "Breach Formation Factor." The Breach Formation Factor is a product of the volume of water coming out of the dam and the height of water above the dam. MacDonald and Langridge-Monopolis then related the breach formation factor to the volume of material eroded from the dam's embankment. The data that MacDonald and Langridge-Monopolis used for their regression analysis had the following ranges:

- ✓ Height of the dams: 4.27–92.96 m (14–305 feet)
  (With 76% <30 m, and 57% <15 m)
- ✓ Breach outflow volume: 0.0037–660.0 $m^3 \times 10^6$ (3–535,000 acre-feet)
  (with 79% <25.0 $m^3 \times 10^6$, and 69% <15.0 $m^3 \times 10^{6)}$

The following is the MacDonald and Langridge-Monopolis equation for volume of material eroded and breach formation time, as reported by Wahl (2004):

For earth fill dams:

$V_{\text{eroded}} = 0.026\,(V_{\text{out}} \times h_w)^{0.769}$

$t_f = 0.0179\,(V_{\text{eroded}})^{0.364}$ (MacDonald & Monopolis; 1984)

For earth fill with clay core or rock fill dams:

$V_{\text{eroded}} = 0.00348\,(V_{\text{out}} \times h_w)^{0.852}$ (MacDonald & Monopolis; 1984)

where:

- ✓ V_{eroded} = volume of material eroded from the dam embankment (cubic meters).
- ✓ V_{out}= volume of water that passes through the breach (cubic meters); for example, storage volume at the time of breach plus volume of inflow after breach begins, minus any spillway and gate flow after the breach begins.
- ✓ h_w = water depth above the bottom of the breach (m).
- ✓ t_f = breach formation time (hours).

MacDonald and Langridge-Monopolis stated that the breach should be trapezoidal with side slopes of 0.5 H:1 V. The breach size is computed by assuming the breach erodes vertically to the bottom of the dam and erodes horizontally until the maximum amount of material has been eroded or the dam's abutments have been reached. The base width of the breach can be computed from the dam geometry with the following equation:

$$W_b = \frac{V_{\text{eroded}} - h_b^2\left(CZ_b + h_b Z_b Z_3/3\right)}{Z_b\left(C + h_b Z_3/2\right)} \quad \text{(MacDonald \& Monopolis; 1984)}$$

where:

- ✓ W_b = bottom width of the breach (m).
- ✓ h_b = height from the top of the dam to bottom of breach (m).
- ✓ C = crest width of the top of dam (m).
- ✓ $Z_3 = Z_1 + Z_2$
- ✓ Z_1 = average slope (Z_1:1) of the upstream face of dam.
- ✓ Z_2 = average slope (Z_2:1) of the downstream face of dam.
- ✓ Z_b = side slopes of the breach (Z_b:1), 0.5 for the MacDonald method (MacDonald & Monopolis, 1984).

9.7 Challenges in dam break analysis

In the context of dam break analysis, numerous challenges contribute to the complexity of the undertaking. One prominent challenge involves the limitations in geometric data, where the unavailability of a high-resolution DEM necessitates using a 30×30 m grid resolution DEM, introducing potential errors in the analysis (Abimeal, 2015). Hydrological uncertainties pose another significant hurdle, as predicting extreme rainfall events and adapting to changing precipitation patterns, particularly in light of climate change, remains a challenging task (Chang, Luo, Wu, & Xiang, 2022). Hydraulic modeling complexities, including accurately simulating breach initiation and flow propagation, add further layers of intricacy. Structural and geotechnical considerations, encompassing the assessment of dam integrity and failure mechanisms, present additional challenges (McGlue, et al., 2023). The absence of comprehensive emergency response plans covering evacuation procedures and interagency coordination exacerbates the challenges associated with potential dam failures. Addressing these multifaceted challenges requires a holistic, interdisciplinary approach that considers technical, social, economic, and regulatory dimensions (Abimeal, 2015). These challenges underscore the need for ongoing research and advancements to enhance the precision and reliability of dam break analyzes.

9.8 Dam breach risk management analysis and mitigation plan

9.8.1 Dam breach risk management analysis

The primary purpose of dam break risk management and hazard mitigation is to reduce the expected human losses and the downstream damages related to dam accidents through structural and nonstructural measures.

There are two important phases in dam break risk management: (1) predicting the losses or damages and their likelihood through risk assessment and (2) finding the appropriate mitigation measures when residual risks are unacceptable (Viseu & de Almedia, 2009).

Risk analysis provides an understanding of the nature and the extent of the uncertainty concerning the conditions under which the dam will be required to perform, as well as the uncertainty in the response of the dam to those conditions. Risk analysis for dam safety is a structured process to identify the extent and the likelihood of consequences associated with dam or dam component failures. A typical dam risk analysis incorporates different uncertainties (Viseu & de Almedia, 2009).

Risk evaluation is the process of understanding and judging the risk's significance, including comparing the estimated risk level with acceptance risk criteria. The primary role of the risk evaluation phase is to generate decision guidance against which the risk analysis results can be compared (Slunga, 2001). The process of supporting risk-based decisions requires a statement of the owner's safety management principles, values, and preferences and those of the public, including consideration of the prevailing financial, legal, and regulatory conditions (Viseu & de Almedia, 2009).

9.8.2 Mitigation plan

The notification system is the communication system to be implemented between the dam owner and the agencies responsible for the dam safety (e.g., the dam safety authority), including the downstream civil protection services. There are two key elements for effective notification: identifying the individuals to alert in the agencies and designing the notification devices (Viseu & de Almedia, 2009).

The notification systems warning the population at risk, including public warning using audible systems as well as adopting visible systems (strobe lights and billboards) (Duston & Myers, 2002); personal direct notification via telephone or cell phones, including door-to-door warning and, nowadays, the use of short message service (SMS) in mobile telephones via cellular diffusion and television or radio station news broadcasts (Santos, Goncalves, Fernandes, & Serrano, 2005; Viseu & de Almedia, 2009).

9.9 Summary

Based on the present literature review, it has been found that for almost all types of dams, HEC-RAS Software is best for breach analysis and is preferable as it is readily available. Software like NWS DAMBRK, HECRAS, and MIKE 11 models are available for dam break modeling. All these software essentially carry out hydrodynamic modeling involving the solution of the basic St. Venant's partial differential equations of conservation of mass and momentum. Different case studies show that most dam failures are due to piping and overtopping foundation defects. Few papers have also done a risk analysis following mitigation plans, which are very important after dam failure; notification systems are widely used in mitigation plans.

References

Abimeal, L. (2015). Dam breach analysis using HEC-RAS and HEC-GeoRAS: The case of Kesem Kebena Dam. MS Thesis.

Abdelkarim, A., Awawdeh, M. M., Alogayell, H. M., & Al-Alola, S. S. (2020). Integration of remote sensing and hydrologic, hydraulic modelling on assessment flood risk and mitigation: Al-Lith city. *KSA. GEOMATE Journal*, *18*(70), 252–280.

Afsal, K., Peter, A., Jobe, P., Vinaya, S., & Binoy Alias, M. (2016). Dam break analysis of Idukki Dam using FLDWAV river mechanics. *International Research Journal of Engineering and Technology (IRJET)*.

Amini, A., Ali, T., Ghazali, A., & Huat, B. (2009). Adjustment of peak streamflows of a tropical river for urbanization. *American Journal of Environmental Sciences*, *5*(3), 285.

Asnaashari, A., Meredith, D., & Scruton, M. (2014). Dam breach inundation analysis using HEC-RAS and GIS - two case studies in British Columbia. Canada: Canadian Dam Association Annual Conference, Banff, Alberta.

Anjaneyulu, R., Swain, R., & Behera, M. D. (2023). Future projections of worst floods and dam break analysis in Mahanadi River Basin under CMIP6 climate change scenarios. *Environmental Monitoring and Assessment*, *195*(10), 1173.

Bellos, V., & Tsakiris, G. (2015). Comparing various methods of building representation for 2D flood modelling in built-up areas. *Water Resources Management*, *29*, 379–397.

Benjankar, R., Tonina, D., & McKean, J. (2015). One-dimensional and two-dimensional hydrodynamic modeling derived flow properties: Impacts on aquatic habitat quality predictions. *Earth Surface Processes and Landforms*, *40*(3), 340–356.

Bharath, A., Shivapur, A., Hiremath, C., & Maddamsetty, R. (2021). Dam break analysis using HEC-RAS and HEC-GeoRAS: A case study of Hidkal dam, Karnataka state, India. *Environmental Challenges*, *5*, 100401.

Bhusal, A., Parajuli, U., Regmi, S., & Kalra, A. (2022). Application of machine learning and process-based models for rainfall-runoff simulation in Dupage River basin, Illinois. *Hydrology*, *9*(7), 117.

Bradbook, K., Waller, S., & Morris, D. (2005). National floodplain mapping and methods. *Natural Hazards*.

Browne, R. (2011). *Tailings dam break models in Nevada – A practitioner's perspective. Denver: USSD Workshop*.

Brunner, G. (2014). Using HEC-RAS for dam break studies. U.S. Army corps of engineers Hydrologic Engineering Center.

Cacciuttolo, C., & Cano, D. (2023). Spatial and temporal study of supernatant process water pond in tailings storage facilities: Use of remote sensing techniques for preventing mine tailings dam failures. *Sustainability*, *15*(6), 4984.

Chang, M., Luo, C., Wu, B., & Xiang, L. (2022). Catastrophe process of outburst debris flow triggered by the landslide dam failure. *Journal of Hydrology*, *609*, 127729.

Chen, A. S., Evans, B., Djordjević, S., & Savić, D. A. (2012). A coarse-grid approach to representing building blockage effects in 2D urban flood modelling. *Journal of Hydrology*, *426*, 1–16.

Chen, M., & Lin, H. (2018). Virtual geographic environments (VGEs): Originating from or beyond virtual reality (VR). *International Journal of Digital Earth*, *11*(4), 329–333.

Chen, Z., Ping, Z., Wang, N., Yu, S., & Chen, S. (2019). An approach to quick and easy evaluation of the dam breach flood. *Scince China Technological Sciences*.

Das, P., Pandey, V., & Dutta, D. (2021). Land surface water resource monitoring and climate change. In *Mapping, monitoring, and modeling land and water resources* (pp. 311–326). CRC Press.

Dat, T., Tri, D., Truong, D., & Hoa, N. (2019). Application of mike flood model in inundation simulation with the dam-break scenarios: A case study of Dak-Drinh Reservoir in Vietnam. *International Journal of Earth Sciences and Engineering*, *12*(01), 60–70.

Derdous, O., Djemili, L., Bouchehed, H., & Bouguerra, H. (2015). Dam break flood forecasting via MIKE 11: Case study of Zardezas reservoir in Skikda, Algeria. *Dam Engineering*, *25*(3), 91–101.

Dottori, F., & Todini, E. (2013). Testing a simple 2D hydraulic model in an urban flood experiment. *Hydrological Processes*, *27*(9), 1301–1320.

Dunning, M., & Durden, S. (2009). *Handbook on applying "other social effects" factors in corps of engineers water resources planning. Institute for Water Resources*.

Duston, G., & Myers, B. (2002). Silver Creek dam early warning system. Preliminary Design Report, City of Silverton.

Elfadaly, A., Shams eldein, A., & Lasaponara, R. (2019). Cultural heritage management using remote sensing data and GIS techniques around the archaeological area of ancient Jeddah in Jeddah City, Saudi Arabia. *Sustainability*, *12*(1), 240.

Ernst, J., Dewals, B., Detrembleur, S., Archambeau, P., Erpicum, S., & Pirotton, M. (2010). Micro-scale flood risk analysis based on detailed 2D hydraulic modelling and high resolution geographic data. *Natural Hazards*, *55*, 181–209.

FEMA. (1989). Dam safety: Know the potential hazards. Federal Emergency Management Agency, Booklet L152, USA.

Fernandes, C., & Gomes, J. (2017). Verification of Saint-Venant equations solution based on the lax diffusive method for flow routing in natural channels. *RBRH*.

Fread, D. (1984). *The NWS dam-break food forecasting model. Ofce of hydrology. National Weather Service*.

Fread, D. (2014). National Weather Service models to forecast dam-breach floods. *Hydrology of Disasters*, 192–211.

Fread, Da (1988). *FLDWAV: "A generalised flood routing model". ASCE, Proceedings of national conference on hydraulic engineering* (p. 6) Colorado: Colorado Springs.

Froehlich, D. (1995). Embankment dam breach parameters revisited. *Conference on Water Resources Engineering*.

Froehlich, D. C. (2008). Embankment dam breach parameters and their uncertainties. *Journal of Hydraulic Engineering*, *134*(12), 1708–1721.

Ge, W., Jaio, Y., Sun, H., & Li, Z. (2019). A method for fast evalution of potential consequences of dam breach. *Water*.

Ge, W., Jiao, Y., Wu, M., Li, Z., Wang, T., Li, W., & Van Gelder, P. (2022). Estimating loss of life caused by dam breaches based on the simulation of floods routing and evacuation potential of population at risk. *Journal of Hydrology*.

Gee, D., & Brunner, G. (2005). Dam break flood routing using HEC-RAS and NWS-FLDWAV. *Impacts of Global Climate Change*, 1–9.

Gildeh1, H. K. (2021). Tailings dam breach analysis: A review of methods, practices. *Mine Water and the Environment, 23*.

Gudeta, S. (2022). Dam break modelling by using HEC-RAS and HEC-GEO RAS. *International Journal of Engineering Technology and Scientific Innovation*.

Gundlach, D., & Thomas, W. (1977). Guidelines for calculating and routing a dam-break flood (No. 5). Department of Defense, Department of the Army, Corps of Engineers, Hydrologic Engineering Center.

Hall, J., Sayers, P., & Dawson, R. (2005). National-scale assessment of current and future flood risk in England and Wales. *Natural Hazards*.

Hjorth, P., & Bengtsson, L. (2012). Large dams, statistics and critical review. In L. Bengtsson, R. W. Herschy, & R. W. Fairbridge (Eds.), *Encyclopedia of Lakes and Reservoirs. Encyclopedia of Earth Sciences Series*. Dordrecht: Springer. Available from https://doi.org/10.1007/978-1-4020-4410-6_28.

Huang, J., Huang, R., Ju, N., Xu, Q., & He, C. (2015). 3D WebGIS-based platform for debris flow early warning: A case study. *Engineering Geology, 197*, 57–66.

Kho, F., Law, P., Lai, S., Oon, Y., Ngu, L., & Ting, H. (2009). Quantitative dam break analysis on a reservoir earth dam. *International Journal of Environmental Science & Technology, 6*, 203–210.

Khosravi, K., Rostaminejad, M., Cooper, J., Mao, L., & Melesse, A. (2019). *Dam break analysis and flood inundation mapping: The case study of Sefid-Roud Dam, Iran. Extreme hydrology and climate variability* (pp. 395–405). *Elsevier*.

Kundzewicz, Z. W., & Schellnhuber, H. J. (2004). Floods in the IPCC TAR perspective. *Natural Hazards, 31*, 111–128.

Kunkel, J. (2011). *Downstream modeling of tailings fow from failure of a 380-ft high tailing dam*. Denver, CO: USSD Workshop.

Kyaw, P. (2020). Modeling approach for Earthen dam breach analysis in North Yamar dam, Myanmar. *American Scientific Research Journal for Engineering, Technology, and Sciences*.

Lariyah, M., Vikneswaran, M., Hidayah, B., Thiruchelvam, S., Abd Isham, A., & Rohani, H. (2013). Numerical modelling dam break analysis for water supply project. *IOP Conference Series: Earth and Environmental Science, 16*(1), 012044.

Liggett, J. (1975). *Basic equations of unsteady flow - unsteady flow in open channels. Water Resources Publications*.

Limantara, L. (2010). Calibration of synthetic unit hydrograph: Case study in Brantas Hulu subwatershed, Province East Java-Indonesia. *International Journal of Academic Research, 2*(5), 146–148.

Limantara, L., & Juowono, T. (2012). Flood behavior analysis due to dam break (case study in Cengklik Dam, Center Java of Indonesia). *International Journal of Emerging Technology and Advanced Engineering, 2*(11), 3–4.

Liu, S., & Henderson, M. (2020). An overview on methodologies for tailing dam breach Study. Tailing and Mine Waste 2020 International Conference.

Lu, X. (1995). Review on protective study of Three Gorges Project. *China Three Gorges Construction*.

MacDonald, T., & Monopolis, J. (1984). Breaching characteristics of dam failures. *Jpurnel of Hydraulic Engineering*.

Mani, P., Kumar, R., & Patra, J. (2020). *Dam break flood hazard assessment: A case study for a small dam at source stream of River Ganga in Uttarakhand. Roorkee Water Conclave*.

McFeeters, S. K. (1996). The use of the Normalized Difference Water Index (NDWI) in the delineation of open water features. *International Journal of Remote Sensing*, 17(7), 1425–1432.

McGlue, M., Dilworth, J., Johnson, H., Whitehead, S., Thigpen, J., Yeager, K., & Swanger, W. (2023). Effect of dam emplacement and water level changes on sublacustrine geomorphology and recent sedimentation in Jackson Lake, Grand Teton National Park, Wyoming, United States. *Earth Science, Systems and Society, 3*, 10066.

Modi, P. N., & Seth, S. M. (2019). *Hydraulics and fluid mechanics including hydraulics machines*. Rajsons Publications Pvt. Ltd.

O'Brien, J., & Garcia, R. (2009). FLO-2D reference manual. Available from: www.flo-2d.com, 595.

Parhi, P. K., Mishra, S. K., Singh, R., & Tripathi, V. K. (2012). Floods in Mahanadi river basin, Orissa (India): a critical review. *India Water Week-Water Energy and Food Security*.

Peng, M., & Zhang, L. (2013). Dynamic decision making for dam-break emergency management. *Natural Hazards and Earth System Sciences*.

Psomiadis, E., Tomains, L., Kavvadias, A., Soulis, K., Charizopoulos, N., & Michas, S. (2021). Potential dam breach analysis and flood wave risk assessment using HEC-RAS and remote sensing data: A multicriteria approach. *Water*.

Purwanto, P., Juwono, P., & Asmaranto, R. (2017). Analisa Keruntuhan Bendungan Tugu Kabupaten Trenggalek. *Journal of Water Resources Engineering, 8*(2), 222–230.

Qi, H., & Altinakar, M. (2012). GIS-based decision support system for dam break flood management under uncertainty with two-dimensional numerical simulations. *Journal of Water Resources Planning and Management, 138*(4), 334–341.

Raman, A., & Liu, F. (2019). An investigation of the Brumadinho Dam Break with HEC RAS simulation. arXiv preprint arXiv:1911.05219.

Ritter, A. (1982). The propagation of water waves. *Zeitschrift des Vereines Deutsher Ingenieure*.

Salajegheh, A., Bakhshaei, M., Chavoshi, S., Keshtkar, A., & Najafi Hajivar, M. (2009). Floodplain mapping using HEC-RAS and GIS in semi-arid regions of Iran. *Desert, 14*(1), 83–93.

Sammen, S., Mohamed, T., Ghazali, A., Sidek, L., & El-Shafie, A. (2017). An evaluation of existent methods for estimation of embankment dam breach parameters. *Natural Hazards*.

Samsudin, S., Zuhaily, N., Setu, A., Muda, R., & Amin, M. (2023). Dam Break flood mapping and analysis using open source GIS tool and data. *IOP Conference Series: Earth and Environmental Science, 1167*(1), 012022.

Santos, M. A., Goncalves, A., Fernandes, J. P., & Serrano, H. (2005). Instrumentos para a minimizacao do risco de cheias induzidas. *Proceedings of the seminar "Barragens: Tecnologia, Seguranca e Interaccao com a Sociedade"*.

Sawai, A., Shyamal, D., & Kumar, L. (2019). Dam break analysis-review of literature. *International Journal for Research in Engineering Application & Management.*

Security, U. D. (2011). Estimating economic consequences for dam failure scenarios.

Schubert, J. E., & Sanders, B. F. (2012). Building treatments for urban flood inundation models and implications for predictive skill and modeling efficiency. *Advances in Water Resources*, *41*, 49–64.

Shah, M., Sidek, L., Yalit, M., Marufuzzaman, M., Basri, H., & Yaacob, M. (2019). *2D Hydraulic modelling of dam break analysis using MIKE FLOOD for Kenyir Dam. ICDSME 2019: Proceedings of the first international conference on dam safety management and engineering* (p. 214) *Springer Nature.*

Shahraki, A., Zadbar, A., Motevalli, M., & Aghajani, F. (2012). Modeling of earth dam break with SMPDBK case study: Bidekan earth dam. *World Applied Sciences Journal*, *19*(3), 376–386.

Sidek, L., & Ros, F. (2011). Hydrodynamic Dam breach modelling of earthfill saddle dam. *Journal of Energy and Environment*, *3*(1).

Slunga, E. (2001). *Concept and bases of risk analysis for dam with an example application on Kyrkosjarvi dam. Development of rescue actions based on dam-break flood analysis. Finland: Publication of the RESCDAM Project.*

Soliman, A., Bekhit, H., Hamed, K., & El-Zwahry, A. (2014). Prediction of dam breach parameters and their uncertainties. *Journel of Engineering Applied Scince.*

Stoker, J. (1957). *Water wave. New York: Interscience Publishers.*

Sun, L., Guo, H., Chen, Z., Yin, Z., Feng, H., Wu, S., & Siddique, K. (2023). Check dam extraction from remote sensing images using deep learning and geospatial analysis: A case study in the Yanhe River Basin of the Loess Plateau, China. *Journal of Arid Land*, *15*(1), 34–51.

Syifa, M., Park, S., Achmad, A., Lee, C., & Eom, J. (2019). Flood mapping using remote sensing imagery and artificial intelligence techniques: A case study in Brumadinho, Brazil. *Journal of Coastal Research*, *90*(SI), 197–204.

Tebbs, E., Avery, S., & Chadwick, M. (2020). Satellite remote sensing reveals impacts from dam-associated hydrological changes on chlorophyll-a in the world's largest desert lake. *River Research and Applications*, *36*(2), 211–222.

Temple, D., Hanson, G., & Neilsen, M. (2006). Analysis of overtopping earth embankment dams. *ASABE Annual International Meeting Proceedings.*

USACE. (2010). *HEC-RAS – river analysis system. Hydrologic engineering center.*

Vanderkimpen, P., & Peeters, P. (2008). Flood modeling for risk evaluation: A MIKE FLOOD sensitivity analysis. *River Flow*, 2295–2304.

Vanderkimpen, P., Melger, E., & Peeters, P. (2009). Flood modeling for risk evaluation—a MIKE FLOOD vs. SOBEK 1D2D benchmark study. *Flood Risk Management: Research and Practice*, 77–84.

Viseu, T., & de Almedia, A. (2009). Dam-break risk management and hazard mitigation. *WIT Transactions on State of the Art in Science and Engineering.*

Wahl, T. (2004). Uncertainty of predictions of embankment dam breach parameters. *Journal of Hydraulic Engineering*, *130*(5), 389–397.

Wang, P., Kahawita, R., Mokhtari, A., Phat, T., & Quach, T. (2006). Modelling breach formation in embankments due to overtopping. *ICOLD Conference.*

Wang, Y., Chen, A., Fu, G., Djordjević, S., Zhang, C., & Savić, D. (2018). An integrated framework for high-resolution urban flood modelling considering multiple information sources and urban features. *Environmental Modelling and Software*, *107*, 85–95.

Xie, R. (1993). *Hydraulics of dam break. Jinan: Shandong Science and Technology Press.*

Xiong Yi, F. (2011). A dam break analysis using HECs-RAS. *Journal of Water Resource and Protection.*

Yang, J., Townsend, R., & Daneshfar, B. (2006). Applying the HEC-RAS model and GIS techniques in river network floodplain delineation. *Canadian Journal of Civil Engineering*, *33*(1), 19–28.

You, L., Li, C., Min, X., & Xiaolei, T. (2012). Review of dam-break research of earth-rock dam combining with dam safety management. *Procedia Engineering.*

Youssef, A. M., Sefry, S. A., Pradhan, B., & Alfadail, E. A. (2016). Analysis on causes of flash flood in Jeddah city (Kingdom of Saudi Arabia) of 2009 and 2011 using multi-sensor remote sensing data and GIS. *Geomatics, Natural Hazards and Risk*, *7*(3), 1018–1042.

Yu, Z., Feng, N., & Glotko, A. (2016). The forecast of flooding resulting from dam break near Nizhny Novgorod. *Procedia Engineering*, *154*, 1306–1313.

Zhu, J., Yin, L., Wang, J., Zhang, H., Hu, Y., & Liu, Z. (2014). Dam-break flood routing simulation and scale effect analysis based on virtual geographic environment. *IEEE Journal of Selected Topics in Applied Earth Observations and Remote Sensing*, *8*(1), 105–113.

Zhou, R. D., Judge, D. G., & Donnelly, C. R. (2004). Comparison of HEC-RAS with FLDWAV and DAMBRK models for dam break analysis. Canada.

Section B

Forest and land cover systems and climate change

Chapter 10

Remote sensing of biodiversity and climate change: adaptation and mitigation prospective

Mukunda Dev Behera[1], S. Mahanand[1], P. Tripathi[2], M.P. Menachery[1], Soumit K. Behera[3], Saroj Kanta Barik[4] and Mrutyunjaya Mohapatra[5]

[1]*Centre for Ocean, River, Atmosphere and Land Sciences (CORAL), Indian Institute of Technology Kharagpur, Kharagpur, West Bengal, India,* [2]*International Centre for Integrated Mountain Development, Kathmandu, Nepal,* [3]*CSIR-National Botanical Research Institute, Lucknow, Uttar Pradesh, India,* [4]*Centre for Advanced Studies in Botany, North-Eastern Hill University, Shillong, India,* [5]*Indian Meteorological Department, Government of India, New Delhi, India*

10.1 Introduction

The structural dissimilarity of the environment (environmental heterogeneity) influences taxa diversity by many direct or indirect mechanisms in time and space (Willig Michael, Dawn, & Richard, 2003). Geographical variety in species richness is caused by resource partitioning, decreased competition and specialization, increased space availability, and multiple habitats (Currie et al., 2004). The species diversity improves with the availability of suitable space; and unique feature of the habitats at local scale (Shmida & Wilson, 1985). Remote sensing has a substantial influence on recognizing those patterns with various direct and indirect approaches. The term direct approach describes the use of aerial or satellite sensors, such as high spatial resolution and hyperspectral sensors, to directly observe specific organisms, species assemblages, or ecological communities. Indirect methods use environmental variables as proxies that are obtained from remotely sensed data (Wang, Franklin, Guo, & Cattet, 2010). Earth observation-based analysis is useful for measurement of community diversity, like species richness and abundance, at landscape to global scales, but for detailing with species level and identifying specific species traits or processes often requires airborne hyperspectral sensors.

While considering the trend in species composition, in general a decreasing nature is evident in species richness with increase in latitude is diagnosed among taxa, space, scale, and habitat. However, the temporal variation is still lacking consistency. The declining species richness from tropics to temperate along latitude is interesting and debatable (Giehl & Jarenkow, 2012); and for this trend can be explained by environmental stability, habitat heterogeneity, and seasonality in climate (Rohde, 1992) and area explaining the richness in mid-latitudes with higher landmasses, not in tropics has been reported (Mannion, Upchurch, Benson, & Goswami, 2014). Simultaneously, elevation adds value to niche concepts (Grinnell, 1917) and community diversity (Whittaker, 1967). The elevation pattern is mostly hump shaped with mid-peaking (Karger et al., 2011; McCain, 2009) and considered as an indirect variable correlated with other constraining predictors influencing species diversity (Austin, 2002). The decrease in area with increase in elevation negatively affects the species richness and overlapping niche at mid-elevation favors the species accumulation (Colwell, Rahbek, & Gotelli, 2004). The climate change catalyzes the variability in biodiversity which directly contributes toward the imbalance of ecosystem. In the face of the current biodiversity crisis, conservation strategies that address climate change are essential to protect species and genetic diversity, as shifting species distributions and shrinking habitats threaten their survival, also in protected areas (Thuiller et al., 2008; Velásquez-Tibatá, Salaman, & Graham, 2013). Satellite remote sensing (RS) has significantly advanced our understanding of climate change by monitoring the atmosphere, land, and oceans, and many researchers have emphasized its potential to enhance conservation biology and ecology through synergies with remote sensing science (Pettorelli, Safi, & Turner, 2014; Yang et al., 2013). Incorporation of in situ observations along with RS data enhance the calibration and accuracy assessment by taking advantage of their complementary nature (Geller et al., 2017).

Sustainable Development Perspectives in Earth Observation. DOI: https://doi.org/10.1016/B978-0-443-14072-3.00025-3

10.1.1 Species–environment relationships

Species diversity influenced by environmental heterogeneity and climate and their interacting mechanisms within the ecosystems is not very clear. The climate (temperature and precipitation) is one of the strongest drivers; and warm and wet climate prevail over cold or dry conditions in shaping species diversity. The species diversity varies with temperature and nonlinear pattern of precipitation along elevation (Odland & Birks, 1999). The precipitation pattern affects moisture gradient, a key factor in species distribution in the tropics (Brown, 1988). The influence of topography, soil, and topography-driven soil moisture are important determinants of species diversity (Moeslund, Arge, Bøcher, Dalgaard, & Svenning, 2013). The elevation which indirectly shapes climate and soil is the significant topographic determinant (Bushing, White, & MacKende, 1992). The aspect or orientation and slope which distribute solar radiation is responsible for structure, composition, and diversity of vegetation (Titshall, O'Connor, & Morris, 2000). The close relation between rich species diversity along moisture gradient and wide range of temperature and precipitation anomalies with vegetation up to 5500 m fascinated researchers to explore Himalayan species–environment relationships along a small geographical extent (Behera & Kushwaha, 2007). Despite the fact that the Himalayan region is rich in biodiversity, the changes in many species' habitats brought about by climate change highlight the wider effects of climate change on biodiversity in the Himalayas (Tiwary, Singh, Adhikari, Behera, & Barik, 2024).

10.1.2 Understanding diversity distribution pattern

Numerous theories, including those pertaining to energy availability, water availability, seasonality, habitat heterogeneity, and disturbance hypotheses, have been put up to explain the diversity-distribution pattern. While considering different species communities in a landscape level, phenomenon like landscape fragmentation and destructions, occurring through the majority human interventions, change the natural pattern of existence (Chitale, Behera, & Roy, 2014). This consistent change in ecosystem affects the species richness in a great manner. Many studies focus on the remote sensing-enabled fragmentation indices. However, studies have frequently observed varying primary drivers in determining species richness for the same taxa at different spatial scales at same or different regions (Behera, Kushwaha, & Roy, 2005). Understanding these patterns with the most reliable empirical models and species distribution models have opened up new opportunities to the policy makers for the optimization of conservation efforts, for example, by identifying hotspots of species richness (Dubuis, 2013; Parviainen, Marmion, Luoto, Thuiller, & Heikkinen, 2009). Species richness can be modeled in two distinct approaches: (1) the "assemble first, predict later" method, which treats species as a final community where the number of species in inventoried plots is statistically linked to a set of environmental predictors; and (2) the "predict first, assemble later" method, which involves modeling individual species initially and subsequently assembling them to reconstruct species richness, with the presence and absence of each species statistically associated with set of environmental predictors. The envelope of all appropriate habitats effectively represents the species' realized environmental niche. The model's fitted relationship is used to estimate the prospective distribution of suitable habitats for the species. Moreover, challenges arise from the reliance on a single method for forecasting, as projections from alternative models can vary significantly, potentially undermining even basic evaluations of whether species distributions are likely to contract or expand under specific climate scenarios (Pearson et al., 2006). Using many models, referred to as "ensembles," and applying the proper methods to examine the range of projections that result is one way to address this intermodal variation. As a collection of tools, deep learning is ideally positioned to take use of the hitherto unseen accessibility of satellite data. It offers chances to combine data from various scales and identify patterns that could stimulate novel theories and approaches to complicated ecological systems (Pettorelli, Williams, Schulte to Bühne, & Crowson, 2024).

10.1.3 Scale disparity

One set of parameters at one resolution and another set at a different resolution may provide the best explanation for the spatial variance in species richness (Whittaker, Willis, & Field, 2001). Predictors' effects, however, likely contribute to the pattern seen at any one scale since they operate at several scales (resolutions or extents). Assessing species diversity at multiple scales is challenging which needs multiple hierarchical partitioning of diversity. The study of ecosystems has always been highly influenced by the spatial scale showing the factors respond differently at different scale. Zuo, Wang, Zhao, and Lian (2014) observed the scale dependency of species richness along the gradient of dune stabilization. Rahbek and Graves (2001) observed precipitation to be most influential at finer scale whereas cloud cover and area were prominent at coarser scale. Keil, Biesmeijer, Barendregt, Reemer, and Kunin (2011) tested the scale dependency of species richness at spatial and temporal changes, and observed that rates of species richness changes at different spatial scales. The difference of species richness in The Netherland and United Kingdom showed shift with scale and the direction of change differed with an increase in hoverfly species richness in The Netherlands at fine scales that disappears at coarser scale indicating a decrease in beta diversity. Conversely, in the United Kingdom there were fine scale losses in species richness shifting to coarse scale gains, indicating an increase in beta diversity. Lennon, Koleff, Greenwood, and Gaston (2001) observed a northward shift in species-rich areas with coarsening scale. All above studies put a focus on the serious implications of scale sensitivity for conservation policy.

10.1.4 Sampling sufficiency

The observed multiscale patterns of species richness also depend on the quality of data, where incomplete sampling and sampling bias can have profound effect on conservation strategies, potentially influencing the choice of priority conservation areas. Common examples of sampling bias include proximity to roads or collector's residence, this might mislead the analysis after data compilation because (1) within inventories, there is no indication of completeness of species richness list and (2) between inventories, the comparison and combining of data become complicated because of incompatible sampling methods, terminology or data handling systems. Various ways to overcome such difficulties and assessing inventory completeness has been used in the past. The three main approaches to standardize the sampling data include: (1) lognormal distribution fitting to species abundance data and estimating the part of the distribution to the left of the veil line; (2) asymptotic curve fitting to randomized accumulation curve; and (3) use of non parametric estimators based on the abundance or incidence of rare species. Walther and Moore (2005) demonstrated the superior performance of asymptotic curve fitting. The species accumulation curves have been applied to field inventories using standardized sampling. The rate at which new species are added to inventories can also be described using accumulation curves. The species accumulation curve slope in these asymptote models characterizes this rate and can be used as a gauge for survey completeness (Hortal & Lobo, 2005). A larger slope value indicates a greater sampling effort needed for a complete inventory, whereas a smaller slope value indicates a more complete survey. The use of asymptote model has been widely used, while few studies have reported the slope value and are much less used criteria to assess sampling completeness.

10.1.5 Species accumulation curve

Species accumulation curve analysis consists of the recording of new species of a given community or site as a function of sampling effort (Chacoff et al., 2011). The curves of accumulation models reach asymptoteness when the probability of adding a new species to the list approaches zero, but are nonasymptotic if this probability does not reach zero. The following two asymptotic models are often used for data fitting.

1. **Clench model:** According to a modified version of the Michaelis–Menten equation, the likelihood of adding a species to the list rises with time but falls with the number of species that have previously been identified. According to Colwell (2009), this model, which is frequently employed in studies on biodiversity, is suitable for a comparatively small intermediate number of samples. It can be summed up as follows:

$$S(x) = \mathrm{ax}/(1 + \mathrm{bx})$$

2. **Linear dependency model (LDM):** Also referred to as the negative exponential model, it postulates that the likelihood of adding new species decreases exponentially with increasing species list size.

$$S(x) = \mathrm{a/b}[1 - \exp(-\mathrm{bx})]$$

Here, is a measure of sampling effort, () is the predicted number of species at effort, **a** represents the rate of increase at the beginning of the sampling, **b** is a parameter related to the mode of accumulation of new species during sampling, **a/b** is the asymptote.

These models were implemented in statistical software.

10.1.6 Biodiversity and international targets

The assessment and monitoring of biodiversity from various platforms have been of research of high priority. Biodiversity loss has become a social and political issue over the last few decades, and its protection is a major subject within national nature conservation policies as well as international conventions, conservation targets, and political programs, for example, the Convention on Biological Diversity (CBD). Biodiversity monitoring programs suffer from three main constraints: (1) incomplete taxonomic and spatial coverage, (2) lack of compatibility between data sets owing to different collection methodologies; and (3) insufficient integration at different scales. The establishment of a monitoring scheme based on an appropriate set of indicators is vital for precise assessment of the effectiveness of measures applied within biodiversity protection (e.g., action plans for endangered species, agro-environmental and landscape protection programs). Many indicators of biodiversity change have been proposed, but their representativeness and applicability frequently suffer from poor available data or local circumstances. In CBD Decision XI/3, the Group on Earth Observation Biodiversity Observation Network (GEO-BON) continues to work on the identification of essential biodiversity variables (EBVs). EBVs aim to help observation communities harmonize monitoring, by identifying how variables should be sampled and measured. Nearly 100 indicators have been proposed for the 2020 CBD targets (Pereira, 2013). These variables are considered essential to capturing a minimum set of critical variables needed to monitor, report, and manage biodiversity change (Pereira et al., 2017).

The CBD Subsidiary Body on Scientific, Technical, and Technological Advice (SBSTTA) suggests employing satellite remote sensing data to track changes in the area of particular biomes, ecosystems, and habitats, as well as the connectivity and fragmentation of ecosystems. Further requirements are outlined in the CBD Focal Area B, Indicators for Immediate Testing, which includes the following: (1) status and trends of the components of biological diversity, (2) trends in extent of selected biomes, ecosystems, and habitats, (3) trends in abundance and distribution of selected species, (4) sustainable use of forest area, agricultural and aquaculture ecosystems under sustainable management numbers, and cost of alien invasions, (5) ecosystem integrity and ecosystem goods and services, (6) connectivity/fragmentation of ecosystems, (7) incidence of human-induced ecosystem failure, and (8) health and well-being of people living in biodiversity.

10.1.7 Biodiversity and Earth observation

Ecosystem and biodiversity monitoring and characterization using satellite remote sensing has a long history. Even before Earth observation began, there was a need for remotely sensed data to monitor biodiversity, and for decades, this need has been largely met. Data from remote sensing can be used to track changes in ecosystem status and distribution at a reasonable cost. Monitoring biodiversity indicators can be carried out using ground-based sample data or remotely sensed data, or both (Sommerville, Milner-Gulland, & Jones, 2011). Remote sensing data are readily used to provide spatial information, but needed expertise for handling and processing. Contemporarily, lots of remote sensing data products are freely available from coarse (MODIS data) to fine resolution (Landsat PAN) imageries for analysis and observation. Additionally, a few commercial satellites, such as Planet Lab, facilitate ecological research by providing limited access to data for teaching and research (https://www.planet.com/; Njomaba et al., 2024). Hence for large-scale biodiversity monitoring, remote sensing offers another alternative data source to provide near real-time information. It has been discovered that integrating EBVs with Remote Sensing of Earth Observations (RS-EO) is a useful method for measuring and tracking changes in biodiversity over time and space (Reddy et al., 2024). The subset of EBVs whose surveillance is primarily or exclusively dependent on satellite-based data is referred to as SRS-EBVs. Therefore variables that could be SRS-EBVs do exist for which monitoring depends on combining measurements from sensors on satellites with in situ and aerial observations, models and simulations, and classification procedures (Pettorelli et al., 2016). To get various indicators of natural habitats, it is necessary to integrate the usage of remote sensing data sources with varying temporal, geographical, and spectral resolutions (Alleaume et al., 2018) (Table 10.1).

Predictive spatial modeling and monitoring has been used increasingly to relate sparse diversity survey or collection data to remotely mapped environmental attributes, thereby allowing distributions of biological entities to be extrapolated across an entire region of interest. Commonly used predictors include terrain indices, long-term average climate surfaces, edaphic variables, land-cover variables and spectral bands or indices from remote sensing. Modeling and monitoring at community-level (1) combines data from multiple species at some stage in the analytical process and (2) produces information on spatial patterns in the distribution of biodiversity at a collective (or emergent) level. Two types of input data relating to this grid are: (1) diversity survey or collection data, typically for scattered locations across the grid, and (2) remotely derived environmental predictors, covering the entire grid. These data are normally stored within, or linked directly to a geographical information system (GIS) and web-enabled geospatial database (Fig. 10.1).

10.1.7.1 Biodiversity surveillance and satellite imaging

Global biodiversity protocols deliver adequate and defensible biodiversity information for conservation policy. They use robust indicators to demonstrate the state of biodiversity pressure on it, and responses to those pressures attached

TABLE 10.1 List of environmental heterogeneity drivers or indices derived for biodiversity study.

EH subject area	Measure category	Variables
Satellite indices/product	Vegetation indices (applicable for EVI, NDVI)	Normalized dispersion of EVI Evenness of EVI Contrast Dissimilarity Entropy Homogeneity
	LAI	LAI
	Surface reflectance	Surface reflectance
	Forest fire	Forest fire
Land cover map	Land cover proportion	% Cover of forest % Cover of grassland
	Land cover density	Land cover types Shannon index of land cover types
	Edge density	Mean patch size
Vegetation types	Plant diversity	Plant species no. or percentage Shannon index of tree species
	Vegetation complexity	Foliage height diversity PCA of vegetation variables
	Vegetation dimension	Trunk perimeter Density of plant
	Nearest forest	Distance to nearest forest
Climate	Climate	Precipitation Temperature range Aridity index
Soil	Soil diversity	Soil types Shannon index of soil types
	Soil variables	Soil moisture Soil carbon Soil nitrogen Soil pH
Topography	Elevation diversity	Elevation range
	Profile	Curvature profile Slope Aspect
Socio-economic	Population	Population density
	Road network	Road buffer
	Built-up area	Distance to built-up
Mixed effect	Composite heterogeneity index	Combination of all drivers

with it (Chiarucci, Bacaro, & Scheiner, 2011). CBD mostly regulate all the agenda and protocol for biodiversity monitoring. The first target established by CBD in 2002 to reduce biodiversity loss and increase conservation measures by 2010 was a failure (Butchart et al., 2010). The next target set was "Aichi Target-2020" by parties and stakeholders that includes 20 targets under five major strategic goals. Geijzendorffer et al. (2016) elaborate on the EBV as a tool for bridging the gap between biodiversity data and policy reporting. The seventeenth meetings of CBD at Montreal emphasize more on the use of remotely sensed derived indicator use for biodiversity monitoring purpose.

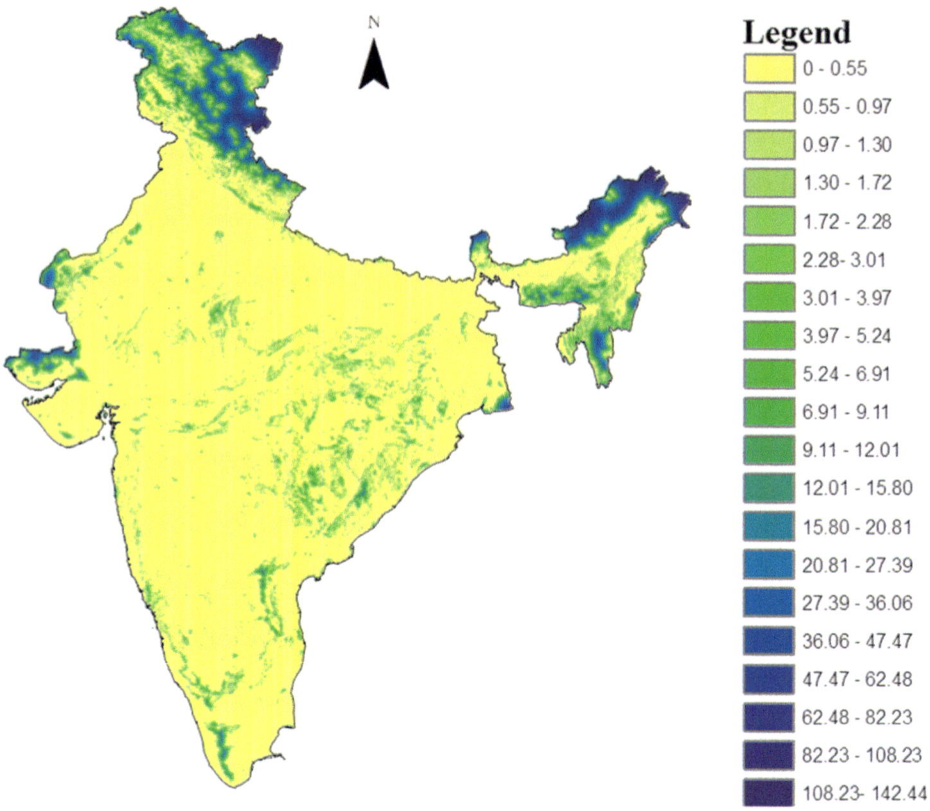

FIGURE 10.1 Map showing distance to nearest cropland (in km) indicates higher urban disturbance to forest and its plant diversity content with increasing proximity. It is observed that high elevation range along trans-Himalayan and Eastern Himalaya are away from crop land serving as an indicator to more plant diversity content. Accordingly, it indicates the less human interference over there as well as less disturbed natural forest. Map lines delineate study areas and do not necessarily depict accepted national boundaries.

Most of the essential indicators for biodiversity monitoring can be derived from remotely sensing imageries. As now-a-days number of satellite products are freely available. Petrou, Manakos, and Stathaki (2015) reviewed and enlisted the free available satellite information and extraction of biodiversity indicators. Turner et al. (2015) summarized the advantages of free and open access RS data for biodiversity monitoring; also emphasize three factors: continuity, affordability, and access. So, rapid application of RS in monitoring, data management and analysis; that decrease long-term monitoring costs, increase the frequency of assessments, and extend coverage to remote places (Nagendra et al., 2013). Researchers are also trying to use high resolution RS data to derive environmental heterogeneity parameters (Szulkin, Zelazowski, Marrot, & Charmantier, 2015), tipping point range prediction studies (Das, Behera, Roy, & Barik, 2024). CBD report of Montreal reviewed the use of RS for monitoring biodiversity and tracking toward achieving Aichi target (Hagerman, Campbell, Gray, & Pelai, 2021).

10.1.8 Statistical models for biodiversity and climate change studies

10.1.8.1 Species distribution model

No species can exist in absolute isolation. In ecology, the community environment interface and interspecific relationship are fundamental to species distribution. The recent trends in ecological research concerns effect of biotic interactions (Zimmermann, Edwards, Graham, Pearman, & Svenning, 2010), and large-scale species patterns depend on both biotic interaction and abiotic predictors of the environment (Meier et al., 2010). Incorporation of biotic factors in addition to abiotic components constraining community structure and species distribution using SDMs is a new field of research. SDMs include occurrences, counts, or frequencies (Leathwick & Austin, 2001) or as competition coefficients with abiotic predictors. The influences of biotic interaction vary with scale. Meier et al. (2010) report the variable importance (both biotic and abiotic) on tree species and variability in species patterns. They see little overlap between two predictor sets also supported by the findings of Pellissier, Brathen, et al. (2010) on exploring the effect of dominant shrub species as predictor on other artic-alpine species. And addition of biotic interaction to abiotic predictors improves the variable importance by enhancing the SDMs predictive power (Le Roux, Lenoir, Pellissier, Wisz, & Luoto, 2013).

It is also evidenced that horizontal and vertical biotic interactions closely associated with the species distribution patterns (Sexton, McIntyre, Angert, & Rice, 2009; Zarnetske, Skelly, & Urban, 2012) and hence posit importance in investigating species distribution at both local and regional scales (Austin & Van Neil, 2011; Kissling et al., 2012). However, the importance of biotic interactions in species range determination (Arujo & Lauto, 2007; Boulangeat, Gravel, & Thuiller, 2012; Heikkinen, Luoto, Virkkala, Pearson, & Körber, 2007) using SDMs is slow (Austin & Van Neil, 2011) and studies have explicitly examine the influence of single species and/or similar interaction (Van der Putten, Macel, & Visser, 2010). The species multiple interactions are less emphasized. Studying underlying causes of invasion and assessment of ecosystem damage is crucial (Richardson, 2011). Exploring their interaction with the community providing inputs for planning and protection of nature is unique and probably scarce in literature.

10.1.8.2 Machine learning and multivariate statistics

SDMs use species occurrences as response variables in addition to vegetation types (Baselga & Araújo, 2009), species richness (Pineda & Lobo, 2009), or functional characteristics of species assemblages (Küster, Bierman, Klotz, & Kühn, 2011; Pellissier, Pottier, Vittoz, Dubuis, & Guisan, 2010). As a general rule, environmental variables are being fitted as constraints to quantify species—environment relationship. SDMs based on ordination, regression, mechanistic, and machine learning methods are being used, and based on data suitability and the nature of the dataset, the type of methods are being chosen for analysis. Predicting single species fitting "multiresponse" models, for example, vector generalized additive models (Yee & Mackenzie, 2002) have also been used. RDA, a linear-based multivariate ordination technique, is being widely used for similar analysis of two matrices of species and their predictors. CCA, a nonlinear multivariate technique, is the most popular multiple direct gradient analysis that fits appropriately to community composition and environmental variables (ter Braak, 1986). Both the methods are asymmetric canonical methods based on constrained ordination quantifying species responses with the predictors of distribution. For community level analysis, mixed model approach is being applied and generalized linear mixed model (GLMM) and generalized additive mixed model (GAMM) are currently in use. Maxent (maximum entropy) is a machine learning species distribution method most popularly used (Phillips, Anderson, & Schapire, 2006) to predict future niche expansion/reduction on basis of future climate change scenarios. Stacking individual species by assembling many models to predict the distribution is also being used. Using total species pool data for predicting distribution is being used mostly to correlate the community structure with the environment. Generalized linear model (GLM), a parametric-based linear model fitting poison distribution, is a popular regression technique while generalized additive model (GAM), a nonparametric regression technique is being found suitable for the purpose. In addition, using classification techniques (boosted regression tree (BRT) and random forest (RF)) are adding value to predict species distribution; and comparison of many methods and their assembling yields better predictability.

Statistical models GLM, GAM, and geographically weighted regression (GWR) are implemented to assess the relationship between species richness and predictor variables. Modeling species richness statistically involves selecting a preferred technique from a range of alternatives, where species distribution models are increasingly employed to assess the potential impacts of climate change on species distributions. Regression-based two species distribution models, that is, GLM and GAM have been used widely in various studies (Dubuis, 2013; Virkkala, Marmion, Heikkinen, Thuiller, & Luoto, 2010) to model species richness with a good predictive accuracy and significance. Virkkala et al. (2010) observed statistically significant prediction of bird range shift in Finland and Finnmark for GLM and GAM. Parviainen et al. (2009) also observed the best predictive performance of these models for species richness and hotspot maps when compared to eight modeling techniques, when applied to threatened plant species richness in north-eastern Finland.

10.1.8.2.1 Generalized linear model

Although they allow nonlinear fits between response and predictor variables, GLM models are an extension of linear models. The dependent variable values in GLMs are predicted using a combination of predictor variables that are connected to the response variable by a link function. The response variable's distribution might be nonnormal and need not be continuous (Nogués-Bravo, 2009).

10.1.8.2.2 Generalized additive model

GAMs are similar to GLMs, but they make the assumption that a nonlinear, nonparametric smoothing function links the response variable to the environmental predictors. The most used distribution for count data is the Poisson distribution (Guisan & Zimmermann, 2000).

10.1.8.2.3 Geographically weighted regression

For every focused cell, this method supports all regression statistics, including partial slopes and coefficient of determination (R2). A technique based on spatial proximity is used for the fitting. According to Brunsdon, Fotheringham, and Charlton (2002), proximity is defined as the adjacent places within a specific bandwidth (b) of a preset radius and shape. Each observation's contribution to the GWR analysis at a given place depends on how far away it is from that site geographically; observations that are farther away have a less impact than those that are closer. Spatial kernel function to perform this geographical weighting, and the analysis scale is determined by the function's bandwidth. The regression equation is

$$\text{Yi} = \beta_0(u_i v_i) + \sum_{j=1}^{k} \left(x_{ij}\beta_j(u_i, v_i) + \varepsilon_i\right)$$

where $j = 1$, k explanatory variables, ε_i is a random error term and the location for each observation is defined by the coordinates (u_i, v_i). $\beta_0 - \beta_k$ are the parameters of the model with β_j (u_i, v_i) a realization of the continuous function $\beta_j(u_i, v_i)$ at location i.

10.1.8.2.4 Ensemble modeling

A potential critical issue in applying a model is the impact of modeling approach and interpretation between model differences for the same species and different climate scenarios (Pearson et al., 2006). The models can vary in their prediction and thus it is unclear which model performs better at predicting future distributions under a changing climate (Pearson et al., 2006). Furthermore, studies (e.g., Pearson et al., 2006) have shown that projections by alternative models are so variable as to compromise even the simplest assessment of whether species distributions should be expected to contract or expand for any given climate scenario. A solution to intermodel variations that has been penetrated in ecology field is very recent termed as "ensemble forecasting" and use suitable techniques to understand the resulting range of projections (Thuiller, Lavorel, Araujo, Sykes, & Prentice, 2005; Thuiller, Richardson, et al., 2005). The results of ensemble modeling, according to Tripathi, Behera, and Roy (2024), are more reliable than those of individual models, highlighting the potential of this approach for conservation efforts in the face of a rapidly altering climate.

10.1.8.2.5 Direct or macro ecological versus cumulative approach

In direct approach, individual species counts in a grid cells will directly be related to statistical models. However, in cumulative approach presence and absence of each species will individually be related to a set of predictor variables to predict species' suitable habitats and then potential maps will be piled up (cumulated) to produce a species richness map. Thus both the predicted output may be compared along with the comparison of statistical methodology.

10.1.9 Adaptation and mitigation strategies: biodiversity in climate change studies using EO

During last three decades, global temperature rises approximately by 0.68°C (Hansen et al., 2006) and is expected to continue up to 68°C by 2100 (IPCC, 2007). However, the information is lacking on the effect of species interactions on species diversity and distribution in climate change scenarios even though we foresee anthropogenic effect on climate on species, communities, and ecosystems (Parmesan & Yohe, 2003) in form of habitat loss, gain and shift; and change in phenology and physiological behavior (Hughes, 2000; Walther et al., 2002), and possible impacts on biotic interactions (Hughes, 2000). Predictions reveal greater range shift in future may be devastating and affect both global fauna and flora diversity (Peterson et al., 2002; Thomas et al., 2004; Thuiller, Lavorel, et al., 2005; Thuiller, Richardson, et al., 2005). A natural community restoration plan grounded in scientific insights is essential for sustainable development. An ecosystem level approach provides local and regional protection of plants, animals, and habitats seems appropriate for conservation. Plan should be need-based and should not check economic growth and development. Involving people for long-term benefit of nature for the betterment of the community is very much necessary.

The report given by IPCC in 2022, that is AR6, shows that a rise in temperature is inevitable. And the climate risk under no adaptation is high. Losses and damages will be significantly larger if intervention is not handled properly and in an unplanned manner. Climate change threats to biodiversity (Kreft & Jetz, 2007; Sala et al., 2000) under future climate-change scenarios which needs assessment. SDMs provide solution to forecast species richness at species and community level studies to estimate future with the current species distribution project consequences under climate change (La Sorte, Lee, Wilman, & Jetz, 2009; Lemoine, Schaefer, & Böhning-Gaese, 2007). There is confirmation that change in the climate has an array of effects on life on this planet. Temporal shifting in flowering of plants (Høye, Post, Meltofe, Schmidt, & Forchhammer, 2007; Parmesan, 2006) also influences species composition, distributions and their assemblages (Root et al., 2003; Thuiller et al., 2008) result in range expansion, reduction or shifting of range (Parmesan, 2006; Walther et al., 2002) and stressful environment from unsuitable sites lead to extinction and reduces range size (Thomas et al., 2004). As a result of rising global temperatures, certain varied tropical lowland areas experience a lowland biotic attrition (Colwell, Brehm, Cardelús, Gilman, & Longino, 2008) which generates changes species turnover and richness (Peterson et al., 2002). SDMs used to evaluate the difference between recent and potential future distribution provides valuable information on possible range shifts (Guisan & Thuiller, 2005) and status of the potential risk. SDMs map potential area of species diversity and distribution pattern (Barthlott, Mutke, Rafiqpoor, Kier, & Kreft, 2005; Mutke & Barthlott, 2005), establishes species–environment relationships (Ricklefs, Qian, & White, 2004) or predict future climate change scenarios (Thuiller, Lavorel, et al., 2005; Thuiller, Richardson, et al., 2005). Recent studies reveal the key role of climate and environment in species richness with significant differences with respect to biogeographical regions (Qian, 2009).

Since climate change is a global issue rather than a local one, every effort taken to improve adaptation must prioritize international cooperation. Effective adaptation depends on the integration of local organizations and people, therefore localize solutions are also required. Nature-based solutions, like maintaining intact ecosystems, managing working lands, and restoring native cover, are some of the most important strategies for reducing climate change and protecting biodiversity (Wudu, Abegaz, Ayele, & Ybabe, 2023). By providing information and insights that guide conservation efforts, sustainable land management techniques, and climate change mitigation tactics, approaches using EBVs help to accomplish the sustainable development goals (Life on Land and Climate Action).

References

Alleaume, S., Dusseux, P., Thierion, V., Commagnac, L., Laventure, S., Lang, M., … Luque, S. (2018). A generic remote sensing approach to derive operational essential biodiversity variables (EBVs) for conservation planning. *Methods in Ecology and Evolution*, *9*(8), 1822–1836.

Arujo, M. B., & Lauto, M. (2007). The importance of biotic interactions for modelling species distributions under climate change. *Global Ecology and Biogeography*, *1616*, 743–753.

Austin, M. P. (2002). Spatial prediction of species distribution: An interface between ecological theory and statistical modelling. *Ecological Modelling*, *157*, 101–118.

Austin, M. P., & Van Neil, K. P. (2011). Improving species distribution models for climate change studies: Variable selection and scale. *Journal of Biogeography*, *38*, 1–8.

Barthlott, W., Mutke, J., Rafiqpoor, M. D., Kier, G., & Kreft, H. (2005). Global centres of vascular plant diversity. *Nova Acta Leopoldina*, *92*, 61–83.

Baselga, A., & Araújo, M. B. (2009). Individualistic vs community modelling of species distributions under climate change. *Ecography*, *32*, 55–65.

Behera, M. D., & Kushwaha, S. P. S. (2007). An analysis of altitudinal behavior of tree species in Subansiri district, Eastern Himalaya. *Biodiversity and Conservation*, *16*, 1851–1865.

Behera, M. D., Kushwaha, S. P. S., & Roy, P. S. (2005). Geo-spatial modeling for rapid biodiversity assessment in Eastern Himalayan region. *Forest Ecology and Management*, *207*, 363–384.

Boulangeat, I., Gravel, D., & Thuiller, W. (2012). Accounting for dispersal and biotic interactions to disentangle the drivers of species distributions and their abundances. *Ecology Letters*, *15*, 584–593.

ter Braak, C. J. F. (1986). Canonical correspondence analysis: A new eigenvector technique for multivariate direct gradient analysis. *Ecology*, *67*, 1167–1179.

Brown, J. H. (1988). In A. A. Myers, & P. S. Giller (Eds.), *Species diversity: Analytical biogeography*. London: Chapman & Hall.

Brunsdon, C., Fotheringham, A. S., & Charlton, M. (2002). Geographically weighted summary statistics—a framework for localised exploratory data analysis. *Computers, Environment and Urban Systems*, *26*(6), 501–524.

Bushing, R. T., White, P. S., & MacKende, M. D. (1992). Gradient analysis of old spruce-fir forest of the Great Smokey Mountains circa 1935. *Canadian Journal of Botany*, *71*, 951–958.

Butchart., et al. (2010). Global biodiversity: Indicators of recent declines. *Science (New York, N.Y.)*, *328*, 1164–1168.

Chacoff, N. P., Va´zquez, D. P., Loma´scolo, S. B., Stevani, E. L., Dorado, J., & Padro´, B. (2011). Evaluating sampling completeness in a desert plant–pollinator network. *Journal of Animal Ecology*, *81*, 190–200.

Chiarucci, A., Bacaro, G., & Scheiner, S. M. (2011). Old and new challenges in using species diversity for assessing biodiversity. *Philosophical Transactions of the Royal Society of London B: Biological Sciences*, *366*, 2426–2437.

Chitale, V. S., Behera, M. D., & Roy, P. S. (2014). Future of endemic flora of biodiversity hotspots in India. *PLoS ONE*, *9*(12), 1–15.

Colwell, R. K. (2009). Biodiversity: Concepts, patterns, and measurement. *The Princeton Guide to Ecology*, *663*, 257–263.

Colwell, R. K., Brehm, G., Cardelús, C. L., Gilman, A. C., & Longino, J. T. (2008). Global warming, elevational range shifts, and lowland biotic attrition in the wet tropics. *Science (New York, N.Y.)*, *322*(5899), 258–261.

Colwell, R. K., Rahbek, C., & Gotelli, N. J. (2004). The mid-domain effect and species richness patterns: What have we learned so far? *The American Naturalist*, *163*, E1–E23.

Currie, D. J., Mittelbach, G. G., Cornell, H. V., Field, R., Guégan, J. F., Hawkins, B. A., ... Turner, J. R. G. (2004). Predictions and tests of climate-based hypotheses of broad-scale variation in taxonomic richness. *Ecology letters*, *7*(12), 1121–1134.

Das, P., Behera, M. D., Roy, P. S., & Barik, S. K. (2024). Predicting tipping points of vegetation resilience as a response to precipitation: Implications for understanding impacts of climate change in India. *Biodiversity and Conservation*, 1–18.

Dubuis, A. (2013). Predicting spatial patterns of plant biodiversity: From species to communities. PhD thesis.

Geijzendorffer, I. R., Regan, E. C., Pereira, H. M., Brotons, L., Brummitt, N., Gavish, Y., ... Walters, M. (2016). Bridging the gap between biodiversity data and policy reporting needs: An essential biodiversity variables perspective. *Journal of Applied Ecology*, *53*(5), 1341–1350.

Geller, G. N., Halpin, P. N., Helmuth, B., Hestir, E. L., Skidmore, A., Abrams, M. J., ... Williams, K. (2017). Remote sensing for biodiversity. *The GEO handbook on biodiversity observation networks*, 187–210.

Giehl, E. L. H., & Jarenkow, J. A. (2012). Niche conservatism and the differences in species richness at the transition of tropical and subtropical climates in South America. *Ecography*, *35*, 933–943.

Grinnell, J. (1917). The niche-relationships of the California Thrasher. *The Auk*, *34*, 427–433.

Guisan, A., & Thuiller, W. (2005). Predicting species distribution: Offering more than simple habitat models. *Ecology Letters*, *8*, 993–1009.

Guisan, A., & Zimmermann, N. E. (2000). Predictive habitat distribution models in ecology. *Ecological Modelling*, *135*, 147–186.

Hagerman, S. M., Campbell, L. M., Gray, N. J., & Pelai, R. (2021). Knowledge production for target-based biodiversity governance. *Biological Conservation*, *255*, 108980.

Hansen, J., Sato, M., Ruedy, R., Lo, K., Lea, D. W., & Medina-Elizade, M. (2006). Global temperature change. *Proceedings of the National Academy of Sciences*, *103*(39), 14288–14293.

Heikkinen, R. K., Luoto, M., Virkkala, R., Pearson, R. G., & Körber, J. H. (2007). Biotic interactions improve prediction of boreal bird distributions at macro-scales. *Global Ecology and Biogeography*, *16*(6), 754–763.

Hortal, J., & Lobo, J. M. (2005). An ED-based protocol for optimal sampling of biodiversity. *Biodiversity and Conservation*, *14*, 2913–2947.

Hughes, L. (2000). Biological consequences of global warming: Is the signal already apparent? *Trends in Ecology and Evolution (Personal Edition)*, *15*, 56–61.

Høye, T. T., Post, E., Meltofe, H., Schmidt, N. M., & Forchhammer, M. C. (2007). Rapid advancement of spring in the High Arctic. *Current Biology*, *17*, R449–R451.

IPCC. (2007). Climate change 2007: The physical science basis. Contribution of working group I to the fourth assessment report of the intergovernmental panel on climate change. In S. Solomon, D. Qin, M. Manning, Z. Chen, M. Marquis, K. B. Averyt, M. Tignor, & H. L. Miller (Eds.), Cambridge University Press, Cambridge, UK and New York, NY, USA, p. 996.

Karger, D. N., Kluge, J., Krömer, T., Hemp, A., Lehnert, M., & Kessler, M. (2011). The effect of area on local and regional elevational patterns of species richness. *Journal of Biogeography*, *38*, 1177–1185.

Keil, P., Biesmeijer, J. C., Barendregt, A., Reemer, M., & Kunin, W. E. (2011). Biodiversity change is scale-dependent: An example from Dutch and UK hoverflies (Diptera, Syrphidae). *Ecography*, *34*(3), 392–401.

Kissling, W. D., Dormann, C. F., Groeneveld, J., Hickler, T., Kühn, I., McInerny, G. J., ... O'Hara, R. B. (2012). Towards novel approaches to modelling biotic interactions in multispecies assemblages at large spatial extents. *Journal of Biogeography*, *39*(12), 2163–2178.

Kreft, H., & Jetz, W. (2007). Global patterns and determinants of vascular plant diversity. *Proceedings of the National Academy of Sciences of the United States of America*, *104*, 5925–5930.

Küster, E. C., Bierman, S. M., Klotz, S., & Kühn, I. (2011). Modelling the impact of climate and land use change on the geographical distribution of leaf anatomy in a temperate flora. *Ecography*, *34*(3), 507–518.

Leathwick, J. R., & Austin, M. P. (2001). Competitive interactions between tree species in New Zealand's oldgrowth indigenous forests. *Ecology*, *82*, 2560–2573.

Lemoine, N., Schaefer, H. C., & Böhning-Gaese, K. (2007). Species richness of migratory birds is influenced by global climate change. *Global Ecology and Biogeography*, *16*, 55–64.

Lennon, J. J., Koleff, P., Greenwood, J. J. D., & Gaston, K. J. (2001). The geographical structure of British bird distributions: Diversity, spatial turnover and scale. *Journal of Animal Ecology*, *70*(6), 966–979.

Mannion, P. D., Upchurch, P., Benson, R. B., & Goswami, A. (2014). The latitudinal biodiversity gradient through deep time. *Trends in Ecology and Evolution*, *29*(1), 42–50.

McCain, C. M. (2009). Global analysis of bird elevational diversity. *Global Ecology and Biogeography*, *18*, 346–360.

Meier, E. S., Kienast, F., Pearman, P. B., Svenning, J. C., Thuiller, W., Araújo, M. B., ... Zimmermann, N. E. (2010). Biotic and abiotic variables show little redundancy in explaining tree species distributions. *Ecography*, *33*(6), 1038–1048.

Moeslund, J. E., Arge, L., Bøcher, P. K., Dalgaard, T., & Svenning, J. C. (2013). Topography as a driver of local terrestrial vascular plant diversity patterns. *Nordic Journal of Botany*, *31*(2), 129–144.

Mutke, J., & Barthlott, W. (2005). Patterns of vascular plant diversity at continental to global scales. *BiolSkrif*, *55*, 521–537.

Nagendra, H., Lucas, R., Honrado, J. P., Jongman, R. H., Tarantino, C., Adamo, M., & Mairota, P. (2013). Remote sensing for conservation monitoring: Assessing protected areas, habitat extent, habitat condition, species diversity, and threats. *Ecological Indicators*, *33*, 45–59.

Njomaba, E., Ofori, J. N., Guuroh, R. T., Aikins, B. E., Nagbija, R. K., & Surový, P. (2024). Assessing forest species diversity in Ghana's tropical forest using PlanetScope data. *Remote Sensing*, *16*(3), 463.

Nogués-Bravo, D. (2009). Comparing regression methods to predict species richness patterns. *Web Ecology*, *9*(1), 58–67.

Odland, A., & Birks, H. J. B. (1999). The altitudinal gradient of vascular plant richness in Aurland, western Norway. *Ecography*, *22*, 548–566.

Parmesan, C. (2006). Ecological and evolutionary responses to recent climate change. *Annual Review of Ecology, Evolution, and Systematics*, *37*, 637–669.

Parmesan, C., & Yohe, G. (2003). A globally coherent fingerprint of climate change impacts across natural systems. *Nature*, *421*, 37–42.

Parviainen, M., Marmion, M., Luoto, M., Thuiller, W., & Heikkinen, R. K. (2009). Using summed individual species models and state-of-the-art modelling techniques to identify threatened plant species hotspots. *Biological Conservation*, *142*, 2501–2509.

Pearson, R. G., Thuiller, W., Araújo, M. B., Martinez-Meyer, E., Brotons, L., McClean, C., … Lees, D. C. (2006). Model-based uncertainty in species range prediction. *Journal of Biogeography*, *33*(10), 1704–1711.

Pellissier, L., Brathen, K. A., Pottier, J., Randin, C. F., Vittoz, P., Dubuis, A., … Guisan, A. (2010). Species distribution models reveal apparent competitive and facilitative effects of a dominant species on the distribution of tundra plants. *Ecography*, *33*, 1004–1014.

Pellissier, L., Pottier, J., Vittoz, P., Dubuis, A., & Guisan, A. (2010). Spatial pattern of floral morphology: Possible insight into the effects of pollinators on plant distributions. *Oikos*, *119*, 1805–1813.

Pereira, H. M. (2013). Essential biodiversity variables. *Science (New York, N.Y.)*, *339*, 277–278.

Pereira, H. M., Belnap, J., Böhm, M., Brummitt, N., Garcia-Moreno, J., Gregory, R., … van Swaay, C. (2017). Monitoring Essential Biodiversity Variables at the Species Level. In M. Walters, & R. Scholes (Eds.), *The GEO Handbook on Biodiversity Observation Networks*. Cham: Springer.

Peterson, A. T., Ortega-Huerta, M. A., Bartley, J., Sánchez-Cordero, V., Soberón, J., Buddemeier, R. H., & Stockwell, D. R. (2002). Future projections for Mexican faunas under global climate change scenarios. *Nature*, *416*(6881), 626–629.

Petrou, Z. I., Manakos, I., & Stathaki, T. (2015). Remote sensing for biodiversity monitoring: A review of methods for biodiversity indicator extraction and assessment of progress towards international targets. *Biodiversity Conservation*, *24*, 2333–2363.

Pettorelli, N., Safi, K., & Turner, W. (2014). Satellite remote sensing, biodiversity research and conservation of the future. *Philosophical Transactions of the Royal Society B: Biological Sciences*, *369*(1643), 20130190.

Pettorelli, N., Wegmann, M., Skidmore, A., Mücher, S., Dawson, T. P., Fernandez, M., … Geller, G. N. (2016). Framing the concept of satellite remote sensing essential biodiversity variables: Challenges and future directions. *Remote Sensing in Ecology and Conservation*, *2*(3), 122–131.

Pettorelli, N., Williams, J., Schulte to Bühne, H., & Crowson, M. (2024). Deep learning and satellite remote sensing for biodiversity monitoring and conservation. *Remote Sensing in Ecology and Conservation*.

Phillips, S. J., Anderson, R. P., & Schapire, R. E. (2006). Maximum Entropy Modeling of species geographic distributions. *Ecological Modeling*, *190*, 231–259.

Pineda, E., & Lobo, J. M. (2009). Assessing the accuracy of species distribution models to predict amphibian species richness patterns. *Journal of Animal Ecology*, *78*, 182–190.

Van der Putten, W. H., Macel, M., & Visser, M. E. (2010). Predicting species distributions and abundance responses to climate change: Why it is essential to include biotic interactions across tropic levels. *Philosophical Transactions of the Royal Society, B*, *365*, 2025–2034.

Qian, H. (2009). Global tests of regional effect on species richness of vascular plants and terrestrial vertebrates. *Ecography*, *32*, 553–560.

Rahbek, C., & Graves, G. R. (2001). Multiscale assessment of patterns of avian species richness. *PNAS*, *98*(8), 4534–4539.

Reddy, C. S., Satish, K. V., Saranya, K. R. L., Sri Surya, N. N., Neha, P. A., & Rajashekar, G. (2024). Harnessing essential biodiversity variables and remote sensing of earth observations-synthesizing biodiversity insights. *Spatial Information Research*, *32*(3), 265–276.

Richardson, D. M. (Ed.), (2011). *Fifty years of invasion ecology: The legacy of Charles Elton* (458). Wiley-Blackwell Publications Ltd.

Ricklefs, R. E., Qian, H., & White, P. S. (2004). The region effect on mesoscale plant species richness between eastern Asia and eastern North America. *Ecography*, *27*, 129–136.

Rohde, K. (1992). Latitudinal gradients in species diversity: The search for the primary cause. *Oikos*, *65*, 514–527.

Root, T. L., Price, J. T., Hall, K. R., Schneider, S. H., Rosenzweig, C., & Pounds, A. (2003). Fingerprints of global warming on wild animals and plants. *Nature*, *421*, 57–60.

Le Roux, P. C., Lenoir, J., Pellissier, L., Wisz, M. S., & Luoto, M. (2013). Horizontal, but not vertical, biotic interactions affect fine-scale plant distribution patterns in a low-energy system. *Ecology*, *94*(3), 671–682.

Sala, O. E., et al. (2000). Global biodiversity scenarios for the year 2100. *Science (New York, N.Y.)*, *287*, 1770–1774.

Sexton, J. P., McIntyre, P. J., Angert, A. L., & Rice, K. J. (2009). Evolution and ecology of species range limits. *Annual Review of Ecology, Evolution, and Systematics*, *40*, 415–436.

Shmida, A., & Wilson, M. W. (1985). Biological determinants of species diversity. *Journal of Biogeography*, *12*, 1–20.

Sommerville, M. M., Milner-Gulland, E. J., & Jones, J. P. G. (2011). The challenge of monitoring biodiversity in payment for environmental service interventions. *Biological Conservation*, *144*, 2832–2841.

La Sorte, F. A., Lee, T. M., Wilman, H., & Jetz, W. (2009). Disparities between observed and predicted impacts of climate change on winter bird assemblages. *Proceedings of the Royal Society B*, *276*, 3167–3174.

Szulkin, M., Zelazowski, P., Marrot, P., & Charmantier, A. (2015). Application of high resolution satellite imagery to characterize individual-based environmental heterogeneity in a wild blue tit population. *Remote Sensing*, *7*(10), 13319–13336.

Thomas, C. D., Cameron, A., Green, R. E., Bakkenes, M., Beaumont, L. J., Collingham, Y. C., …… Williams, S. E. (2004). Extinction risk from climate change. *Nature*, *427*(6970), 145–148.

Thuiller, W., Albert, C., Araújo, M. B., Berry, P. M., Cabeza, M., Guisan, A., … Zimmermann, N. E. (2008). Predicting global change impacts on plant species' distributions: Future challenges. *Perspectives in Plant Ecology, Evolution and Systematics*, *9*(3-4), 137–152.

Thuiller, W., Lavorel, S., Araujo, M. B., Sykes, M. T., & Prentice, I. C. (2005). Climate change threats to plant diversity in Europe. *Proceedings of the National Academy of Sciences of the United States of America*, *102*, 8245–8250.

Thuiller, W., Richardson, D. M., Pysek, P., Midgley, G. F., Hughes, G. O., & Rouget, M. (2005). Niche-based modelling as a tool for predicting the risk of alien plant invasions at a global scale. *Global Change Biology*, *11*, 2234–2250.

Titshall, L. W., O'Connor, T. G., & Morris, C. D. (2000). Effect of long-term exclusion of fire and herbivory on the soils and vegetation of sour grasslands. *African Journal of Range and Forage Science*, *17*, 70–80.

Tiwary, R., Singh, P. P., Adhikari, D., Behera, M. D., & Barik, S. K. (2024). Vulnerability assessment of Taxus wallichiana in the Indian Himalayan Region to future climate change using species niche models and global climate models under future climate scenarios. *Biodiversity and Conservation*, 1–20.

Tripathi, P., Behera, M. D., & Roy, P. S. (2024). Predicting the patterns of plant species distribution under changing climate in major biogeographic zones of mainland India. *Biodiversity and Conservation*, 1–21.

Turner, W., Rondinini, C., Pettorelli, N., Mora, B., Leidner, A. K., Szantoi, Z., ... Woodcock, C. (2015). Free and open-access satellite data are key to biodiversity conservation. *Biological Conservation*, *182*, 173–176.

Velásquez-Tibatá, J., Salaman, P., & Graham, C. H. (2013). Effects of climate change on species distribution, community structure, and conservation of birds in protected areas in Colombia. *Regional Environmental Change*, *13*, 235–248.

Virkkala, R., Marmion, M., Heikkinen, R. K., Thuiller, W., & Luoto, M. (2010). Predicting range shifts of northern bird species: Influence of modelling technique and topography. *Acta Oecologica*, *36*(3), 269–281.

Walther, B. A., & Moore, J. L. (2005). The concepts of bias, precision and accuracy, and their use in testing the performance of species richness estimators, with a literature review of estimator performance. *Ecography*, *28*(6), 815–829.

Walther, G. R., Post, E., Convey, P., Menzel, A., Parmesan, C., Beebee, T. J. C., ... Bairlein, F. (2002). Ecological responses to recent climate change. *Nature*, *416*(6879), 389–395.

Wang, K., Franklin, S. E., Guo, X., & Cattet, M. (2010). Remote sensing of ecology, biodiversity and conservation: A review from the perspective of remote sensing specialists. *Sensors*, *10*(11), 9647–9667.

Whittaker, R. H. (1967). Gradient analysis of vegetation. *Biological Reviews*, *42*, 207–264.

Whittaker, R. J., Willis, K. J., & Field, R. (2001). Scale and species richness: Towards a general, hierarchical theory of species diversity. *Journal of Biogeography*, *28*, 453–470.

Willig Michael, R., Dawn, M. K., & Richard, D. S. (2003). Latitudinal gradients of biodiversity: Pattern, process, scale, and synthesis. *Annual Review of Ecology, Evolution, and Systematics*, 273–309.

Wudu, K., Abegaz, A., Ayele, L., & Ybabe, M. (2023). The impacts of climate change on biodiversity loss and its remedial measures using nature based conservation approach: A global perspective. *Biodiversity and Conservation*, *32*(12), 3681–3701.

Yang, J., Gong, P., Fu, R., Zhang, M., Chen, J., Liang, S., ... Dickinson, R. (2013). The role of satellite remote sensing in climate change studies. *Nature Climate Change*, *3*(10), 875–883.

Yee, T. W., & Mackenzie, M. (2002). Vector generalized additive models in plant ecology. *Ecological Modelling*, *157*, 141–156.

Zarnetske, P. L., Skelly, D. K., & Urban, M. C. (2012). Biotic multipliers of climate change. *Science (New York, N.Y.)*, *336*, 1516–1518.

Zimmermann, N. E., Edwards, T. C., Jr, Graham, C. H., Pearman, P. B., & Svenning, J. C. (2010). New trends in species distribution modelling. *Ecography*, *33*(6), 985–989.

Zuo, X., Wang, S., Zhao, X., & Lian, J. (2014). Scale dependence of plant species richness and vegetation-environment relationship along a gradient of dune stabilization in Horqin Sandy Land, Northern China. *Journal of Arid Land*, *6*, 334–342.

Chapter 11

Linking meteorological and phenological observations in forests

A.J. Prakash[1], Mukunda Dev Behera[1], M. Mukhopadhyay[1], C.P. Singh[2], B.K. Bhattacharya[2], C. Jeganathan[3], R.K. Chaturvedi[4] and Mrutyunjaya Mohapatra[5]

[1]*Centre for Ocean, River, Atmosphere and Land Sciences (CORAL), Indian Institute of Technology Kharagpur, Kharagpur, West Bengal, India,* [2]*Space Application Center (ISRO-SAC), Ahmedabad, Gujarat, India,* [3]*Department of Remote Sensing, Birla Institute of Technology Mesra, Ranchi, Jharkhand, India,* [4]*Department of Humanities and Social Sciences, BITS Pilani, K K Birla Goa Campus, Vasco da Gama, Goa, India,* [5]*Indian Meteorological Department, Government of India, New Delhi, India*

11.1 Introduction

11.1.1 Background and significance

Forests play a critical role in mitigating global warming by sequestering and storing substantial amounts of carbon more effectively than most terrestrial ecosystems (Mo et al., 2023). In India, forests cover nearly a quarter of the nation's land area and store an estimated 7 billion tons of carbon. To meet its NetZero target by 2070, India is committed to enhancing the carbon sequestration capability of its forests, aiming to create an additional carbon sink of 2.5−3 billion tons by 2030 through improved forest quality and expanded green cover. However, climate change threatens the performance of these carbon sinks because rising temperatures can variably affect forest productivity and carbon dynamics (Das & Behera, 2019; Malhi et al., 2020). However, these ecosystems are becoming increasingly vulnerable to climate change effects, including altered temperature and precipitation patterns, increased extreme weather events, and ecological disturbances (Li, Wu, Liu, Zhang, & Li, 2018; Seidl, Rammer, & Spies, 2014). These changes threaten forest ecosystems and affect species distribution, ecosystem functioning, and the provision of services. Understanding the relationship between meteorological factors and forest phenology, such as budburst, flowering, and fruiting, is essential because these biological processes are influenced by environmental cues, such as temperature, precipitation, and net solar radiation (NSR) (Forrest & Miller-Rushing, 2010).

The impact of global warming, particularly the observed increase of 1.5°C, has significantly affected biodiversity and ecosystems, leading to shifts in species distribution, phenology, and population decline (Hoegh-Guldberg et al., 2019). The documentation of species extinctions directly attributed to human-caused climate change highlights the urgency to address these impacts. Earth observation (EO) data from satellites and airborne sensors provide invaluable insights into forest ecosystem dynamics at extensive temporal and spatial scales. This information is critical for developing conservation and management strategies to enhance ecosystem resilience in the face of climate change.

A multidisciplinary approach combining the expertize of climatologists, ecologists, and remote sensing specialists is essential for a comprehensive understanding of the complex interactions between climate variables and biological responses. Such collaborative efforts are vital for developing strategies for improving ecosystem resilience and sustainability. By monitoring phenological events and their climate sensitivities, informed decisions can be made to mitigate adverse effects, ensure the sustainability of forest ecosystems, and contribute to the global ecological balance. This integrated approach underscores the importance of forest conservation for combating climate change and preserving biodiversity.

Several noticeable changes have occurred at the species level because of climate change. These changes include shifts in species distribution, seasonal events (phenology), and population declines (Walther et al., 2002). It is alarming that the extinction of the first species directly linked to human-caused climate change has been observed (Weiskopf et al., 2020). EO data from satellites and airborne sensors are valuable for conducting extensive analyzes of these relationships over long periods and large geographical areas. A thorough understanding of forest ecosystem dynamics is essential to policymakers, land

Sustainable Development Perspectives in Earth Observation. DOI: https://doi.org/10.1016/B978-0-443-14072-3.00008-3

TABLE 11.1 Timeline history of phenological monitoring using various observational sources.

Year	Milestones in phenology monitoring	Meteorological and phenological observations in forests
1736	Carl Linnaeus records the flowering times of Swedish plants in his "Calendar of Flora."	–
1852	Henry David Thoreau begins detailed observations of plant and animal phenology in Concord, Massachusetts.	–
1875	The first publication of the "Nature's Calendar" in the United Kingdom, encouraging citizens to report seasonal observations.	–
1930	The International Phenological Gardens Network was established to connect gardens worldwide and monitor plant phenology.	–
1950	Advancements in satellite technology allow for remote sensing of vegetation greenness and phenological changes from space.	Satellite-derived climate data, such as temperature and precipitation, were correlated with remote sensing phenological data, enabling large-scale climate-phenology studies.
1960	The USA National Phenology Network was initiated to collect data on plant and animal phenology across the United States.	Meteorological stations within the network collect climate data to contextualize phenological observations and identify climate drivers.
1971	The International Biological Program (IBP) includes phenology as a key component, leading to increased research efforts.	IBP projects often integrated meteorological monitoring with phenological research to elucidate climate-phenology relationships.
1990	Digital cameras and computerized data collection methods revolutionize phenological monitoring.	Automated weather stations linked with digital cameras enable real-time data collection, facilitating detailed climate-phenology investigations.
2000	Citizen science projects like "Project BudBurst" and smartphone apps enable widespread participation in phenology monitoring.	Participants often recorded local weather conditions and phenological observations, fostering community-based climate-phenology studies.
2010	Integration of satellite and ground-based data provides comprehensive phenological information for global climate studies.	Climate data from satellites and ground-based phenological observations offer a holistic understanding of climate-phenology relationships at various scales.
2020	Emergence of AI and machine learning in analyzing complex phenological datasets. Integration of high-resolution drone imagery, IoT devices in phenological monitoring, and global initiatives for biodiversity monitoring underscore the technological evolution.	Advanced sensors, AI, and machine learning models enhance meteorological-phenological integration, contributing to precise predictions and understanding of phenological shifts due to climate change. Collaborative international research projects utilize big data analytics to study global phenological trends and their ecological impacts.

managers, and conservationists. This knowledge would allow them to make informed decisions that would improve the resilience and sustainability of such ecosystems, particularly in the face of climate change (Table 11.1).

11.1.2 Earth observation and climate change

EO plays a crucial role in understanding and monitoring climate change. Scientists have used remote sensing and related techniques to collect important information about the Earth's climate system (Guo, Zhang, & Zhu, 2015). Satellites equipped with specialized sensors can provide uninterrupted and consistent data. This allows for the long-term monitoring of climatic factors and the detection of any changes that occur over time. Remote sensing enables data collection from large and distant geographical areas, providing valuable insights into climate patterns and shifts at the regional, continental, and global scales. Recognizing spatial variations, trends, and anomalies is essential for evaluating the impact of climate change. EO platforms provide the advantage of regular data acquisition, which allows for the study of seasonal changes, long-term trends, and episodic events. The analysis of time helps us gain a better understanding of climate change and its impact on ecosystems. The use of EO data also has the potential to improve climate modeling by enhancing predictions, assessing the reliability of models, and providing valuable insights into future scenarios (Hollmann et al., 2013). Validating and refining climate models is crucial for a better understanding of climate processes. In addition, EO platforms are equipped with sensors that can collect data from different parts of the electromagnetic spectrum (Berger, Moreno, Johannessen, Levelt, & Hanssen, 2012). This allows the observation of various terrestrial phenomena. The process of combining data from various sources enabled us to conduct thorough analyzes of climate variables and how they interact with each other. The availability of EO data enhances the collaboration between scientists and policymakers, enabling them to conduct research and make informed decisions regarding climate change.

11.1.3 Indian contributions to phenological research

India has made substantial progress in phenological research, particularly by integrating remote sensing and ground-based observation networks. Research conducted across various tropical forest types including the tropical dry deciduous forests in Palpur Kuno, Madhya Pradesh (320 m), and Gir, Gujarat (300 m); the tropical moist deciduous forests in Similipal Biosphere Reserve, Odisha (800 m); the tropical semi-evergreen forests in Anshi, Goa (500 m); and the tropical dry deciduous forests in Dalma Wildlife Sanctuary, Jharkhand (700 m) has provided significant insights into how vegetation responds to seasonal and climatic changes (Jeganathan et al., 2024; Jose et al., 2023; Prakash et al., 2023). These studies are essential for understanding unique phenological patterns in tropical regions, where factors such as monsoon rainfall and prolonged dry periods strongly influence plant behaviour.

Findings from these diverse regions underscore the critical roles of temperature, precipitation, and soil moisture in regulating key phenophases, including budburst, leaf emergence, and senescence. Such Indian phenological studies provide crucial regional data, supporting global climate initiatives and guiding policymakers in developing targeted conservation strategies to bolster ecosystem resilience against climate change. By highlighting the specific challenges and responses of Indian forests, this research aligns with India's commitment to enhancing forest resilience and achieving its climate action goals (Prakash et al., 2023).

Expanding the PhenoCam network across India's varied forest types-spanning moist deciduous, dry deciduous, and semi-evergreen ecosystems—would significantly enhance phenological monitoring and offer crucial, localized insights into the effects of climate variability. A nationwide PhenoCam network could complement satellite data by capturing fine-scale observations of seasonal transitions across different ecological zones, such as leaf initiation and senescence. This initiative would provide critical data to understand better how Indian forests respond to monsoon-driven rainfall and seasonal dry spells, contributing valuable input to both local conservation efforts and global phenological models. By integrating PhenoCam data with broader Earth Observation datasets, India could substantially contribute to global forest management and climate resilience efforts.

11.2 Phenology and forest ecosystems

11.2.1 Definition and importance of phenology

Phenology is the study of recurring plant biological events, such as flowering, leafing, fruiting, and migration, which are observed in relation to seasonal and climatic changes. It is vital to understand forest ecosystems and their responses to climate change. Phenology reliably indicates climate variability and changes and is closely linked to temperature, precipitation, and photoperiod. Changes in event timing offer insight into climatic shifts and long-term trends. Additionally, phenology regulates ecosystem processes, including nutrient cycling, pollination, seed dispersal, and food availability, affecting species distribution and interactions within forests. Timing mismatches can disrupt these processes (Richardson et al., 2013). Understanding phenology helps to identify climate-vulnerable species, inform conservation, guide habitat restoration, and adapt management strategies to support species and overall ecosystem health (Gray & Ewers, 2021).

11.2.2 Key phenological events in forests

The timing of important events in forests, known as phenological events, can vary based on factors such as the specific region, climate conditions, and dominant tree species (Hoffmann & Sgrò, 2011). Several phenological events are commonly observed (Dronova & Taddeo, 2022). A budburst is an event that occurs during the growing season. Budburst is the initial stage of the season when leaf buds begin to swell and eventually open. Increasing temperatures and extended daylight hours commonly initiate this process. The subsequent phase after the budburst stage is known as leaf emergence. During this stage, the leaves begin to unfold and expand, significantly affecting the productivity and photosynthetic activity of the tree.

Various factors such as temperature, photoperiod, and specific traits of each species determine leaf emergence. Flowering is an essential process for numerous forest tree species because it allows the production of flowers for reproduction. Flowers are essential for several reasons, including their role in pollination, seed production, and preserving genetic diversity within ecosystems. Fruiting is a process that occurs once pollination and fertilization are accomplished. Fruits are important for seed dispersal because they attract various dispersers, including birds and mammals. Leaf senescence is a natural process in which the leaves age and eventually fall off. A decrease in daylight and lower temperatures initiated this process. This process enables plants to reabsorb nutrients efficiently and prepare for winter dormancy. Deciduous forests are renowned for the iconic transformation of leaf color during autumn. This change occurs as chlorophyll production decreases, signaling the onset of leaf senescence and shedding. Trees enter a dormancy period during the latter part of the growing season, forming buds on leaf axils or branch tips, which are crucial for survival under harsh winter conditions and resource scarcity (Xie & Wilson, 2020).

11.2.3 Linking phenological observations with climate variables

Understanding seasonal phenological processes is essential for maintaining a balanced hydrological cycle and an efficient energy exchange between terrestrial ecosystems and the atmosphere. These processes also offer valuable insights into how plants react to climate change, thus improving our understanding of large-scale climate patterns. Nevertheless, the advent of industrialization and the resulting disturbances caused by human activities have brought about imbalances that may affect plant phenology. A clear understanding of the complex connections between climatic factors and phenological phenomena is crucial to accurately assess the impact of these effects on forest phenology.

Recent studies by Negi, Joshi, Singh, and Joshi (2022) and Dugesar et al. (2022) provided valuable insights into the intricate relationships between meteorological factors and forest phenological events. These studies have enhanced our understanding of the impacts of climate variables on natural processes. Research has primarily focused on studying the influence of climatic factors, including temperature, precipitation, and photoperiod, on the timing and duration of phenological events in various forest types and species (Parmesan, 2007).

To further explore these findings, this chapter will examine the specific climatic factors that play a crucial role in influencing phenological patterns. These factors included temperature, precipitation, soil moisture, and NSR. Recent developments in remote sensing technologies and data analysis techniques have enabled more accurate and extensive observations of these phenomena. For example, satellite imagery and unmanned aerial vehicles (UAVs) with advanced sensors can offer detailed information about the timing of phenological events by detecting even slight changes in plant health and moisture levels. In addition, combining machine learning models with traditional ecological data can enhance the accuracy and effectiveness of predicting and modeling phenological changes under different climate scenarios. These models assist in synthesizing extensive datasets to reveal patterns that may not be immediately evident using traditional analytical methods.

Applying these advanced methodologies improves our understanding of ecological dynamics and helps in forest management and conservation efforts. Managers and policymakers can develop effective strategies to combat the negative impacts of climate change by understanding how climate variables affect forest phenology. Implementing various strategies, such as managing forest density, optimizing the timing of conservation actions, and choosing species for reforestation that are well-suited to changing conditions, can greatly enhance the effectiveness of conservation efforts.

11.3 Key meteorological variables influence forests

11.3.1 Role of temperature

Extensive research has been conducted on the relationship between temperature and key phenological events in forests, particularly budburst and leaf development (De Réaumur, 1735; Spafford, MacDougall, & Steenberg, 2023). It has been found that temperature plays a crucial role in regulating these events. Budburst is the term used to describe when buds open and new shoots emerge. It is influenced by temperature, particularly in spring. Warmer temperatures trigger physiological changes in the buds, eventually causing them to open. However, the specific temperature thresholds for budburst differ depending on the tree species. Temperature had a significant effect on leaf development. Warmer conditions tend to accelerate the growth and expansion of leaves, forming larger and fully developed leaves. Research has identified specific temperature thresholds for the initiation of budburst in different species. It has been observed that certain species require chilling periods during the winter for this process to occur. Temperature fluctuations can affect phenological timing, both within and between years. For example, hot periods in winter or early spring can cause buds to burst earlier. However, cold snaps or late frosts can result in delays or damage, which can cause shifts in phenological timing. These shifts could potentially affect tree growth and productivity.

Many studies have shown that climate change, marked by increasing temperatures, has caused earlier budburst and leaf development (Bajpai, Dutta, Singh, Chaudhary, & Pandey, 2020; Liu et al., 2016; Piao et al., 2019). The increase in global temperatures due to global warming is expected to facilitate the northward movement of species distribution, potentially leading to increased green vegetation in the Arctic and Antarctic regions. In tropical moist deciduous forests, both atmospheric and soil temperatures have been identified as key factors influencing the composition of forest communities, with variation depending on the species composition of the forest (Bajpai et al., 2020). Environmental temperature plays a crucial role in determining the timing of new leaf emergence and the end of the leafing season. Liu et al. (2016) found a positive correlation between preseason warming temperatures and the rate of leaf fall, although this relationship may not hold in arid or semiarid regions. Experimental studies have consistently demonstrated that higher environmental temperatures lead to earlier spring and later autumn (Menzel et al., 2006), whereas lower temperatures delay spring and accelerate the onset of autumn (Piao et al., 2019). However, the magnitude and consistency of these changes differ depending on the species, location, and ecosystem being studied. The relationship between temperature and phenology is crucial for predicting and evaluating the effects of climate change in forests. It helps us identify species that are at risk, anticipate potential mismatches, and develop strategies to adapt to changing climatic conditions.

11.3.2 Role of precipitation

Precipitation, which includes both rain and drought, plays a critical role in shaping the life cycles of forests (Cui, Xu, He, & Luo, 2022). Water and soil moisture availability determine how forests grow and thrive. When there is sufficient rainfall, it kick-starts vital processes, such as the sprouting of buds and the growth of leaves. The soil moisture level helps determine when these events occur and how intensely they occur. Having sufficient soil moisture encourages strong root development, which in turn helps trees absorb more water and supports leaf growth. However, a lack of moisture during drought can delay or disrupt these processes. Regular rainfall is essential for trees to develop reproductive structures and release pollen, whereas drought changes their growth patterns to save water and minimize water loss through leaves. Severe drought can cause trees to age their leaves prematurely and shed them as a strategy to conserve water. Different tree species respond to water stress in various ways, with some able to maintain their growth cycles even under dry conditions. The interaction between precipitation and plant growth patterns also depends heavily on the local climate and the specific characteristics of each ecosystem. For example, forests in Mediterranean climates have evolved to adapt their growth cycles to seasonal rain. Understanding how precipitation affects forest growth is key to predicting and managing the impact of climate change. Changes in rainfall patterns and increases in drought frequency can profoundly affect the growth, reproduction, and overall health of forests. By incorporating water and soil moisture assessments into forest growth studies, we can better understand how forests adapt to changing conditions. This knowledge is crucial for developing strategies to help forests cope with the challenges posed by our changing climate (Richardson & O'Keefe, 2009).

11.3.3 Role of soil moisture

Soil moisture profoundly influences forest phenology, shaping the timing of key events in the life cycle of trees and the overall health of forest ecosystems. Adequate soil moisture promotes early leaf emergence and bud break in trees (Richardson et al., 2013), whereas insufficiencies can delay these processes (Cleland, Chuine, Menzel, Mooney, & Schwartz, 2007). Soil moisture also affects flowering patterns and fruit production, affecting wildlife and pollinator communities. Moreover, soil moisture influences the timing of leaf senescence and fall, with dry conditions triggering premature leaf shedding (Vitasse, Lenz, & Körner, 2014). These shifts in phenology can have cascading effects on forest ecosystems, altering species composition and distribution and affecting ecosystem functioning (Walther et al., 2002). Because climate change alters precipitation patterns, understanding the intricate relationship between soil moisture and forest phenology has become increasingly critical for predicting and mitigating ecological consequences.

11.4 Key vegetation indices and data smoothing methods

11.4.1 Using vegetation indices as a proxy for phenology

Remote sensing technologies have greatly enhanced phenological studies by utilizing vegetation indices (VIs) that extract data from spectral bands sensitive to different plant characteristics, including pigments, water content, and structure. These indices facilitate the understanding of complicated spectral data by providing simple metrics that show the health and status of vegetation. In the past, indices such as the normalized difference vegetation index (NDVI), developed by Rouse, Haas, Schell, and Deering (1974), were preferred because they provide a comprehensive perspective on vegetation conditions, unlike direct spectral band data. NDVI utilizes red and near-infrared spectral bands to track chlorophyll content and internal leaf structure, which are crucial for phenological characterization. Owing to advancements in sensor technologies, such as MODIS and MERIS, scientists have developed new VIs that go beyond the capabilities of NDVI. These new indices were designed to capture specific plant traits such as chlorophyll levels, leaf area, and water content. Recent advancements have been in VI applications with new indices such as the wide dynamic range vegetation index (WDRVI) and enhanced vegetation index (EVI). The WDRVI enhances the sensitivity scale for estimating biomass in areas with dense vegetation. In contrast, the EVI includes a blue band sensitive to the atmosphere to correct for aerosol influences in the red band data (Table 11.2).

TABLE 11.2 Commonly used satellite-derived remote sensing vegetation indices used in phenological studies.

Class	Index	Calculation	Characteristic	References
Ratio and combined Index	NDVI	(ρNIR−ρRed)/(ρNIR + ρRed)	Sensitive to chlorophyll, but saturated in dense vegetation.	Rouse et al. (1974)
	WDRVI	(α*ρNIR−ρRed)/(α*ρNIR + ρRed)	Similar to NDVI, but still sensitive to dense vegetation.	Gitelson (2004)
	NDWI	(ρNIR−ρSWIR)/(ρNIR + ρSWIR)	Sensitive to water.	Gao (1996)
	GRVI	(ρGreen−ρRed)/(ρGreen + ρRed)	Sensitive to land cover types.	Tucker (1979)
	MTCI	(ρ735.75−ρ708.75)/(ρ708.75−ρ681.25)	Sensitive to high chlorophyll content and limited sensitivity to atmospheric effects or spatial resolution.	Dash and Curran (2004)
	PI	PI = {0 if NDVI < 0 or NDWI < 0, 2 otherwise}	Designed to remove the wetness (e.g., snow) and brightness effects (e.g., soil) on greenness.	Gonsamo and Pellikka (2012)
	PPI	PPI = −K* n(DVI)−M*DVI + M*DVI^2	Derived from radiative transfer theory.	Jin and Eklundh (2014)
	NDPI	α*(ρNIR−ρRed) + α*(ρSWIR)−ρ NIR− (ρRed)	Contrast vegetation from the background and minimize the difference among the backgrounds.	Cao et al. (2018)
Physically based index	LAI	By model fitting (e.g., linear, logistic)	Important structural parameter, but unable to be derived directly from satellite imagery.	Watson (1947)
	fAPAR	Various algorithms (e.g., JRC, MODIS)	Based on definite physiological significance but cannot be derived directly from satellite imagery.	Myneni, Knyazikhin, Glassy, Votava, and Shabanov (2003)
Adjusted VI	EVI	2.5*(ρNIR−ρRed)/((ρNIR + 6*ρRed−7.5*ρBlue) + 1)	Introducing an atmosphere-sensitive blue band to correct the red band for aerosol influences.	Huete, Justice, and van Leeuwen (1999)
	EVI2	G*(ρNIR−ρRed)/(ρNIR + c*ρRed + 1)	Similar to EVI, retaining the soil-noise adjustment function and maintaining improved sensitivity and linearity in high biomass regions, with the absence of a blue band.	Jiang, Huete, Didan, and Miura (2008)
	PVI	(ρNIR−a*ρRed−b)/(√(a^2 + 1))	Designed to minimize its sensitivity to soil reflectance	Richardson and Wiegand (1977)
	SAVI	(1 + L)*(ρNIR−ρRed)/(ρNIR + ρRed + L)	Designed to minimize soil brightness influences	Huete (1988)

TABLE 11.3 Summary of commonly used data smoothing methods.

Method	Description	Classification	Processing window	References
Maximum value compositing (MVC)	Selects the highest value to represent the condition for a certain period	Empirical method	Local	Holben (1986)
Locally weighted regression	Estimates a regression surface by fitting a function of the variables locally	Filtering method	Local	Cleveland and Devlin (1988)
Fast Fourier transform	Fits the first three harmonics using a least squares method	Data transformation	Whole	Sellers et al. (1994)
Empirical mode decomposition	Decomposes data into intrinsic mode functions (IMFs) by "sifting"	Data transformation	Whole	Huang et al. (1998)
Harmonic analysis of time series (HANT)	Decomposes data into trigonometric functions	Data transformation	Whole	Roerink, Menenti, and Verhoef (2000)
Asymmetrical Gaussian function-fitting	Fits asymmetric Gaussian functions	Curve fitting	Local	Jönsson, Cai, Melaas, Friedl, and Eklundh (2018)
Savitzky–Golay filter	Applies Savitzky–Golay filter iteratively to the upper envelope or flexible window	Curve fitting	Local	Chen et al. (2004)
High-order annual splines	Applies high-order polynomial splines with roughness damping	Curve fitting	Local	Hermance (2007)
Spatial-temporal Savitzky–Golay (STSG)	Uses neighboring pixels and multiyear data for noise reduction	Curve fitting	Local	Cao et al. (2018)

11.4.2 Data smoothing methods

The first step in temporal compositing involves creating a time-series VI dataset. However, this dataset may still contain noise from different sources such as cloud cover and atmospheric disturbances (Atkinson, Jeganathan, Dash, & Atzberger, 2012). Different smoothing methods have been used to improve the VI data and address this problem. The goal was to minimize the remaining noise and more precisely depict the vegetation conditions as they change over time. The choice of a specific smoothing method can substantially impact the accuracy of the phenological analysis derived from the smoothed data series (Atkinson et al., 2012; Zeng, Wardlow, Xiang, Hu, & Li, 2020). There are typically three main types of smoothing techniques identified by Atkinson et al. (2012): (1) empirical methods, (2) curve fitting methods, and (3) data transformations. Each category has its unique benefits and is selected based on the specific needs and objectives of the research (Table 11.3).

11.5 Challenges and opportunities

11.5.1 Data integration and analysis challenges

The integration of meteorological and phenological datasets poses significant challenges that require careful consideration when conducting meaningful analyzes. Ensuring the quality and reliability of data is of utmost importance. This is because differences in the methods used to collect data, the instruments utilized, and the protocols followed during observations can lead to uncertainties and biases. To address data quality issues effectively, it is essential to establish robust quality control measures, standardized protocols, and rigorous data validation procedures (Fan, Han, & Liu, 2014). Moreover, the disparity in spatial and temporal resolutions between meteorological and phenological data presents a substantial challenge. Meteorological data, such as from weather station networks, are typically collected over large areas. However, phenological data are often collected at smaller scales and sometimes even at the level of individual trees or plots. To overcome this challenge, it is necessary to coordinate and combine the different levels of detail carefully. This requires appropriate statistical methods (Cerlini, Saraceni, Orlandi, Silvestri, & Fornaciari, 2022).

Additionally, it is important to address compatibility issues arising from variations in data formats, units, and measurement scales across different datasets. This necessitates dedicated efforts to harmonize the data. To ensure seamless integration and comparability, it is necessary to standardize units, reconcile measurement scales, and convert data into a standard format. The challenge of missing data is a common issue that complicates integration efforts. Handling missing data carefully using imputation or statistical modeling techniques is important to address this challenge. These techniques help in estimating missing values. It is crucial to tackle these complex challenges to gain valuable insights from the meteorological and phenological data.

11.5.2 Methods for addressing these challenges

Several key strategies are crucial for successfully integrating meteorological and phenological datasets. Data harmonization is an essential aspect that requires standardized data formats, units, and measurement scales (Ruiz-Benito et al., 2020). This foundational pillar ensures consistency and coherence in the data analysis and interpretation. Developing shared data protocols and metadata standards can significantly enhance the efficiency of harmonization efforts. Furthermore, statistical modeling approaches provide an effective means of bridging the gaps between these datasets (Parmesan, 2007). Strong connections between meteorological variables and phenological events were established using regression models. This allows the estimation of missing or unavailable data points (Diez et al., 2014).

The use of remote sensing and EO data, specifically satellite imagery, is valuable. Remote sensing techniques enhance ground-based observations by overcoming the limitations of spatial coverage (Behera & Kushwaha, 2012). They provide a wider perspective on the ecosystem dynamics. In addition, the use of data assimilation techniques, such as Kalman filtering and Bayesian inference, allows the integration of meteorological and phenological data with ecosystem models. This led to improved estimates and predictions. These techniques carefully combine various data sources, consider uncertainties, and improve the accuracy of the results. We can fully utilize meteorological and phenological datasets by addressing these challenges and implementing appropriate data integration and analysis methods. This will lead to a better understanding of the complex connections between climatic variables and forest phenology. Enhanced integration plays a crucial role in improving the accuracy of predictions, conducting thorough assessments of climate change impacts, and making informed decisions for sustainable forest management and conservation (Das & Behera, 2019).

11.5.3 Limitations and uncertainties

Considering these limitations and uncertainties when connecting meteorological and phenological observations in forests is essential, primarily because of the difficulties caused by data gaps (Zeng et al., 2020). These gaps can arise because of the uneven distribution of meteorological stations and the difficulties in gathering comprehensive phenological data from various tree species, especially in remote or hard-to-reach forested regions. It is crucial to conduct a thorough analysis because various species have distinct reactions to environmental factors. This emphasizes the importance of collecting and analyzing extensive data.

Understanding how meteorological factors affect phenological events is challenging because of the complex interactions among different elements of the forest, including species diversity, soil conditions, and microclimates (De Frenne et al., 2021). In addition, the challenges posed by climate change, such as changes in temperature, rainfall patterns, and severe weather events, make it difficult to predict phenological responses. In addition, the effects of phenology can differ based on the observation scale, which makes it difficult to apply the findings to various spatial and temporal scales. It is crucial to conduct research and collect data to address the limitations and uncertainties associated with this variability. Highlighting the importance of consistent and extended monitoring is crucial for capturing and analyzing patterns over several years.

Recent technological advancements in remote sensing have opened new possibilities for gathering data in hard-to-reach locations. These advancements include the integration of drone technology and advanced satellite imaging, which have proven to be valuable tools for overcoming data-collection challenges (Prakash, Behera, Ghosh, Das, & Mishra, 2022). In addition, machine learning and big data analytics can improve the analysis and interpretation of intricate datasets, thereby enhancing our understanding of phenological patterns at various levels. To gain a more comprehensive understanding of phenological dynamics, it is essential to have a specific focus on studies tailored to individual species. These studies should encompass a wide variety of tree species, including both common and rare ones. By incorporating a range of data sources, such as remote sensing, citizen science initiatives, and ecological models, we can fill in the missing pieces and strengthen the trustworthiness of meteorological-phenological connections. Experimental studies that manipulate meteorological variables are extremely valuable for establishing causal relationships and validating the observed correlations. These studies help us better understand the impact of specific environmental factors on phenological processes.

Collaborative efforts involving researchers, organizations, and citizen science networks are crucial for promoting data sharing, standardization, and quality control. Researchers can gather larger datasets by working together, enabling comprehensive examination and verification of discoveries concerning the relationship between forest meteorology and phenology. By utilizing these advancements and adopting collaborative approaches, the field can progress toward a deeper understanding of how forests react to shifting climatic conditions. This will ultimately aid the development of more efficient conservation strategies and ecological forecasting.

11.5.4 Future directions and research opportunities

Understanding the intricate relationship between forest phenology and weather patterns has advanced recently and is expected to spur important future studies. UAVs, advanced ground-based sensors, and high-resolution satellite photography are only a few advanced remote sensing technologies that continue to offer previously unheard-of-data depths. These technologies improve our knowledge of forest ecosystem dynamics by providing comprehensive insights into phenological events and meteorological conditions at various scales.

Finely detailed data regarding forest ecosystems can now be collected thanks to technological advancements, improving our knowledge of local processes and how various species adapt to their surroundings. For example, ecological monitoring and modeling have been transformed by combining Light Detection and Ranging technology with conventional remote sensing techniques, yielding three-dimensional forest canopy structure data essential for evaluating biomass and biodiversity (Behera & Roy, 2002).

Merging EO data with on-ground phenological observations and meteorological measurements can significantly improve the accuracy and breadth of data integration. Together with enabling a deeper comprehension of the connections between meteorological factors and phenological events, this synergistic approach also helps verify the precision of phenological indicators obtained from satellite data. Big data analytics and improved computational models (machine learning and data assimilation techniques) will significantly improve the capacity to manage complicated datasets (Ghosh & Behera, 2018). These techniques effectively expose complex interconnections, link biological events and climatic factors, and enhance predictive models for predicting future ecological trends. Furthermore, species-specific phenological models provide a deep understanding of how particular species react to environmental changes by combining features specific to each species with local climatic data and remote sensing observations (Singh et al., 2024).

With these efforts, a new era in our knowledge of the timing of biological events in forest ecosystems at the species level is expected. Investigating the connections between plant and animal life cycle phenological stages and their effects on ecosystem processes are three important areas of research in this field that should concentrate on going forward. Long-term data analysis also helps reveal trends and vulnerabilities, making it easier to fully understand the ecological dynamics. In conclusion, data collection and analysis will be further improved over time, allowing for more accurate and real-time ecological forecasting and decision-making by integrating new technologies such as artificial intelligence in image processing and the growth of Internet of Things (IoT) applications in environmental monitoring.

11.6 Case study

11.6.1 Study area

The Similipal Biosphere Reserve (SBR) is a portion of the World Network of Biosphere Reserves recognized by UNESCO in 2009. It is situated in the northern area of the Eastern Ghats, extending between 86°03′E, 86°37′E, 21°28′N, and 22°15’ N. It lies under the administrative governance of the Mayurbhanj District of Odisha and covers an area of 2750 km^2. The SBR is home to a diverse range of flora and fauna owing to its significant variations in topography and ecology. The elevation of forest reserves ranges from 200 to 1168 m above mean sea level, with undulating rolling hills, slopes, and plains. Tropical forest ecosystems in the SBR include moist deciduous, dry deciduous, semi-evergreen, Sal forests, and many isolated grasslands (Reddy, Pattanaik, Dhal, & Biswal, 2006). Moist deciduous forests are the most prevalent forest type, covering an area of approximately 2045 km^2, followed by dry deciduous forests (approximately 565 km^2), Sal forests (approximately 300 km^2), and semi-evergreen forests (approximately 224 km^2) (Reddy et al., 2017; Saranya & Reddy, 2016). The region experiences a warm and humid tropical climate, with three distinct seasons (summer, winter, and monsoon). Rainfall occurs almost every month, with annual rainfall ranging from 120 to 210 cm. The temperature ranged from a minimum of 9°C in winter to a maximum of 34°C in summer. The reported flora, articles, and books in the region indicate the presence of 1076 vascular species and 240 tree species in the forest that exhibit an affinity for many northern and southern forest regimes in India. The dominant tree species in the SBR include *Shorea robusta*,

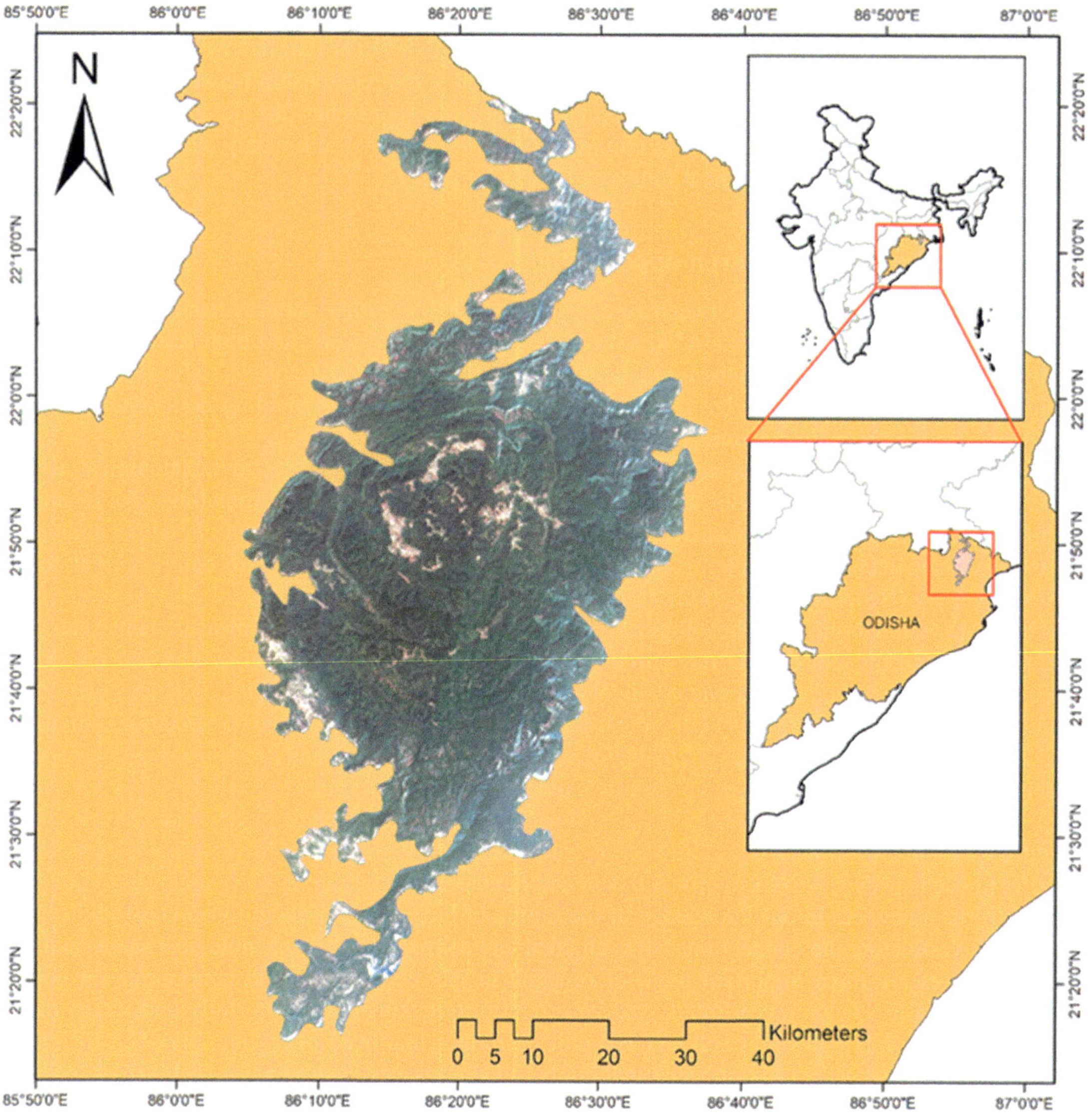

FIGURE 11.1 Study area of Similipal Biosphere Reserve in the context of the map of Odisha and India. Map lines delineate study areas and do not necessarily depict accepted national boundaries.

Protium serratum, *Xylia xylocarpa*, *Terminalia tomentos*a, *Haldina cordifolia*, and *Macaranga peltata* (Mohanta, Sahoo, & Sahu, 2021). The reserve includes a Critical Tiger Habitat of 1194.75 sq. km (core area) and a 1555.25 km^2 buffer zone, showcasing diverse forest types and rich biodiversity. Notably, this area contains a network of roads (597 km in the core and 139 km in the buffer zone) and houses a large human population. Within the broader SBR are 1265 villages with a total population of approximately 462,000, most belonging to minority groups, including a substantial proportion of Scheduled Tribes (91.77% in the SBR area). This demographic composition and the 61 villages within the sanctuary area (including one in the core area) represent a dynamic socio-ecological landscape (Fig. 11.1).

11.6.2 Datasets and methodology

This study involved time-series visualization and satellite-derived remote sensing data analysis to understand the relationship between local meteorological conditions and the forest phenological cycle. Standard remote sensing products, which define regional climate over larger temporal and spatial scales, were used to model the dynamics between forest phenology and local meteorological conditions (Table 11.4).

TABLE 11.4 Remote sensing and metrological data used for this study.

Instruments/Sensors	Data
ERA5-Land daily aggregated—ECMWF climate reanalysis	Daily 0.1 degree cumulative downward shortwave radiation (J m^{-2})
MOD11A1.061 Terra land surface temperature and emissivity daily global 1 km	Daily 1 km land surface temperature (°C)
ERA5-Land hourly—ECMWF climate reanalysis	Hourly 0.1 degree precipitation (m)
SPL3SMP_E.005 SMAP L3 radiometer global daily 9 km soil moisture L4, global 3 hourly 9 km gridded soil moisture (m^3/m^3)	3-hourly 9 km soil surface (0–5 cm) moisture (m^3/m^3)
MODIS (Terra/Aqua)vegetation indices 16-day global 250 m	16-day, 250 m resolution NDVI and EVI

11.6.2.1 ERA5-Land climate reanalysis data product

An enhanced version of the ERA5 dataset, ERA5-Land, provided consistent, long-term variation in land variables at an enhanced resolution of 0.1 degree (11 km). It is produced by re-simulating the land component of the ECMWF ERA5 reanalysis data with the help of process-based modeling and ground-based observations from across the world to produce globally consistent data of climate variables on land going back decades into the past. This helps recreate past climatic conditions, which is essential for studying the evolution of climate change in the current scenario. Available at both hourly (ECMWF/ERA5_LAND/HOURLY) and daily (ECMWF/ERA5_LAND/DAILY_AGGR) aggregate, ERA5-Land provides two important parameters, hourly precipitation and daily cumulative downward shortwave radiation, respectively.

11.6.2.2 MODIS (Terra) land surface temperature product

This product provides daily 1 km land surface temperature and emissivity values for the entire globe, based on the MOD11_L2 swath product. The product is based on (1) generalized split-window LST (Wan & Dozier, 1996) and (2) day/night LST algorithm using thermal infrared (TIR), atmospheric temperature, and water vapor information. The product utilizes advanced algorithms to accurately calculate and display the generalized split-window LST and day/night LST, providing users with a comprehensive view of the thermal infrared radiation data.

11.6.2.3 SMAP L4 soil moisture product

The SMAP Level-3 (SPL3SMP_E.005) soil moisture product is available in Google Earth Engine (GEE) and provides daily composites of surface soil moisture (0–5 cm depth) for the entire world. The data were based on the Soil Moisture Active Passive (SMAP) mission with an L-band radiometer and synthetic aperture radar measuring surface soil moisture and freeze/thaw of surface ice. The data include measurements from the 6 a.m. descending pass and 6 p.m. ascending pass by the near-polar, sun-synchronous satellite at an altitude of 685 km. The data were reprojected from EASE Grid Version 2.0 to Geographic Coordinates using GDAL libraries before hosting on the GEE platform.

11.6.2.4 MODIS (Terra/Aqua) vegetation indices product

Approximately, 16-day, 250 m VIs products are available in GEE platform on a "value per pixel" basis. These products were based on the reflectance data of both Terra (MODIS/061/MOD13Q1) and Aqua (MODIS/061/MOD13Q1) after adjusting for atmospheric correction and bidirectional reflectance effects. The primary VI is the normalized difference vegetation index, which provides continuity with the NOAA-AVHRR NDVI data collection. The second VI is the EVI, which helps minimize soil background and canopy saturation effects in high-density areas. The blue band reduces atmospheric errors due to subpixel clouds and particulate matter. The Terra and Aqua data collections were merged and masked for cloudy pixels to obtain good-quality VI collection at an enhanced 8-day temporal resolution. Ancillary information was presented as additional bands of the solar zenith angle, day of year, and parent reflectance bands.

The datasets were accessed through the GEE portal and aggregated to 4-day intervals. Point sampling was performed using the region of interest (ROI) shapefile, and the sampled values were used for time-series analysis. The extracted data series was gap-filled, denoised, and filtered using a linear interpolation and sliding window algorithm. The final dataset was used for time-series plotting and analysis.

11.7 Results and discussion

The detailed examination of time-series data involving key environmental parameters—average NSR, land surface temperature (LST), precipitation (PPT), and surface soil moisture (SSM)—in conjunction with VIs such as NDVI and EVI reveals intricate relationships between climate variables and forest phenology. By aggregating all environmental parameters into 4-day averages and comparing them to the 8-day average NDVI and EVI interpolated to 4-day intervals, we can discern the nuanced impacts of climatic factors on vegetation health and dynamics.

The analysis showed that NSR significantly influenced vegetation greenness and health, as evidenced by the variations in NDVI and EVI. During periods of high NSR, there was typically an increase in NDVI and EVI, indicating enhanced photosynthetic activity and vegetative growth. This trend is particularly noticeable in the pre- and postmonsoon seasons when increased sunlight availability supports vigorous plant growth. However, despite the high NSR during the monsoon, cloud cover can reduce direct sunlight reaching the forest canopy, slightly moderating NDVI and EVI responses. The LST data revealed a complex influence on vegetation health. Higher temperatures, up to a certain threshold, can promote plant growth, as reflected by increased NDVI and EVI values. However, extreme temperatures, especially during the peak summer months, may lead to stress on vegetation, reducing greenness indices due to mechanisms such as transpiration stress and leaf senescence. This dual role of temperature underscores the importance of a balanced climate for optimal forest health. PPT and SSM are closely linked and have direct implications for vegetation health. Increased PPT, leading to higher SSM, usually results in improved NDVI and EVI scores, signifying healthy vibrant vegetation. This is particularly evident during the monsoon season, when enhanced rainfall leads to a spike in VIs, reflecting the critical role of water availability in supporting forest ecosystems. Conversely, drought conditions, indicated by reduced precipitation and SSM, were correlated with lower NDVI and EVI values, highlighting the vulnerability of forests to water stress.

Fig. 11.2 shows the seasonal variations in the chosen environmental parameters as visualized after gap-filling and smoothing of the raw time-series data. SWR decreases to a minimum of $1.25-1.42 \times 10^7$ Jm^{-2} due to sustained cloud cover during the monsoon months (June–September), increasing to a maximum of $2.25-2.41 \times 10^7$ Jm^{-2} during summer (March–May). LST follows the NSR and increases during summer to approximately 30.45°C–32.6°C and decreases to 19.27°C–20.41°C during winter. There is a visible temporal lag of approximately 5 months between the decrease in the NSR and LST due to the thermal inertia of the Earth's surface. In addition, precipitation data reveal that an increase in rainfall during October–December causes a sharp drop before both show simultaneous increases during the oncoming summer months.

Precipitation follows an Indian monsoon pattern, where yearly rainfall starts from June to September, followed by a short burst from October to December at the onset of winter. Monsoon months see approximately 8.6–13 mm of rain h^{-1}, while early winter rain ranges from 3.5 to 10 mm h^{-1}. Dry seasons show rainfall as low as 1.7–2.9 mm h^{-1}. The surface soil moisture increases to 0.47–0.51 volume fraction due to increased rainfall during monsoon months and decreases to 0.19–0.27 volume fraction during summer months. Being a moist deciduous area, the surface soil moisture content steadily follows a seasonal rainfall pattern, with a time lag of approximately 2 months.

Here, we used SWR-LST and PPT-SSM as parameter pairs and attempted to correlate them with satellite-based phenological metrics, such as NDVI and EVI. Because phenological events affect the chlorophyll content of trees, greenness indices provide sensitive information about the spatial and temporal aspects of the major phenological events of the forest flora.

Fig. 11.3 shows the seasonality and cyclic pattern of NDVI and EVI as the values of SWR and LST change over the seasons. EVI varies between 0.2 (premonsoon) to more than 0.6 (postmonsoon). NDVI varies from 0.3 (premonsoon) to almost 0.9 (postmonsoon). The green-up during the monsoon season is due to abundant rainfall and favorable temperatures (30.45°C–32.6°C) and is flanked by the dry seasons of previous and current years, where the NDVI and EVI steadily decreased from October to February–March. However, the amount of sunlight, surface temperature, precipitation, and soil moisture affect the greenness. Fig. 11.3 compares the EVI and NDVI data for the different PPT and SSM conditions.

From Fig. 11.4, a clear seasonal pattern is evident in both the precipitation and VIs, showing the direct influence of meteorological conditions on forest phenology. Precipitation peaks were consistently observed around the middle of each year, corresponding to the monsoon season, with the highest recorded at approximately 0.011 mm d^{-1} in 2022. This peak in moisture availability correlates strongly with increases in VIs, particularly EVI and NDVI, which peak around the same time. The EVI ranges from about 0.20 to 0.50, while the NDVI exhibits a broader range, peaking at approximately 0.8 and dipping to around 0.30 during drier periods.

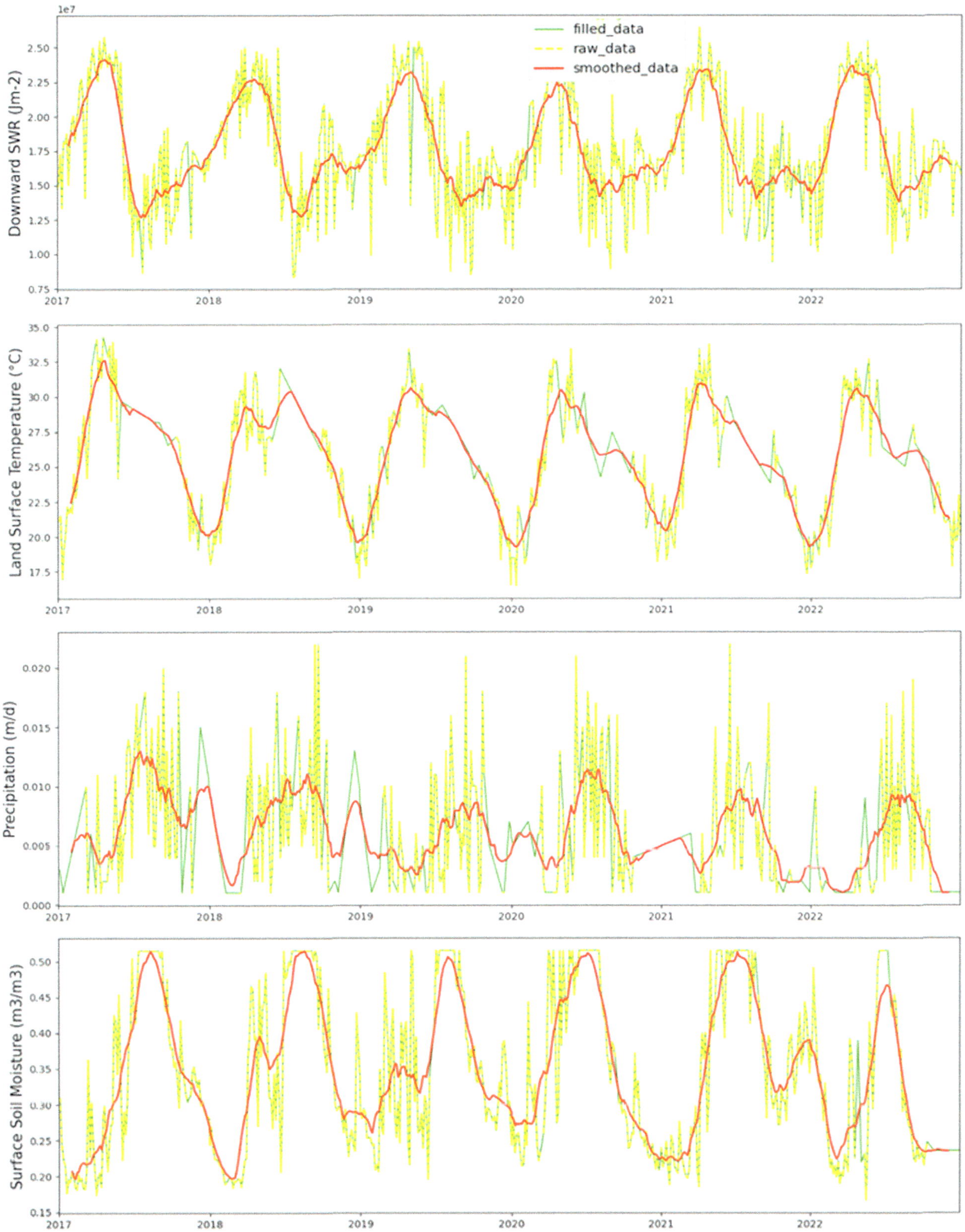

FIGURE 11.2 Gap filling and denoising of average net solar radiation, land surface temperature, precipitation and surface soil moisture (top to bottom), using linear interpolation and rolling window (n = 15) moving average filter. Similar techniques were used to preprocess phenological proxies, such as NDVI and EVI.

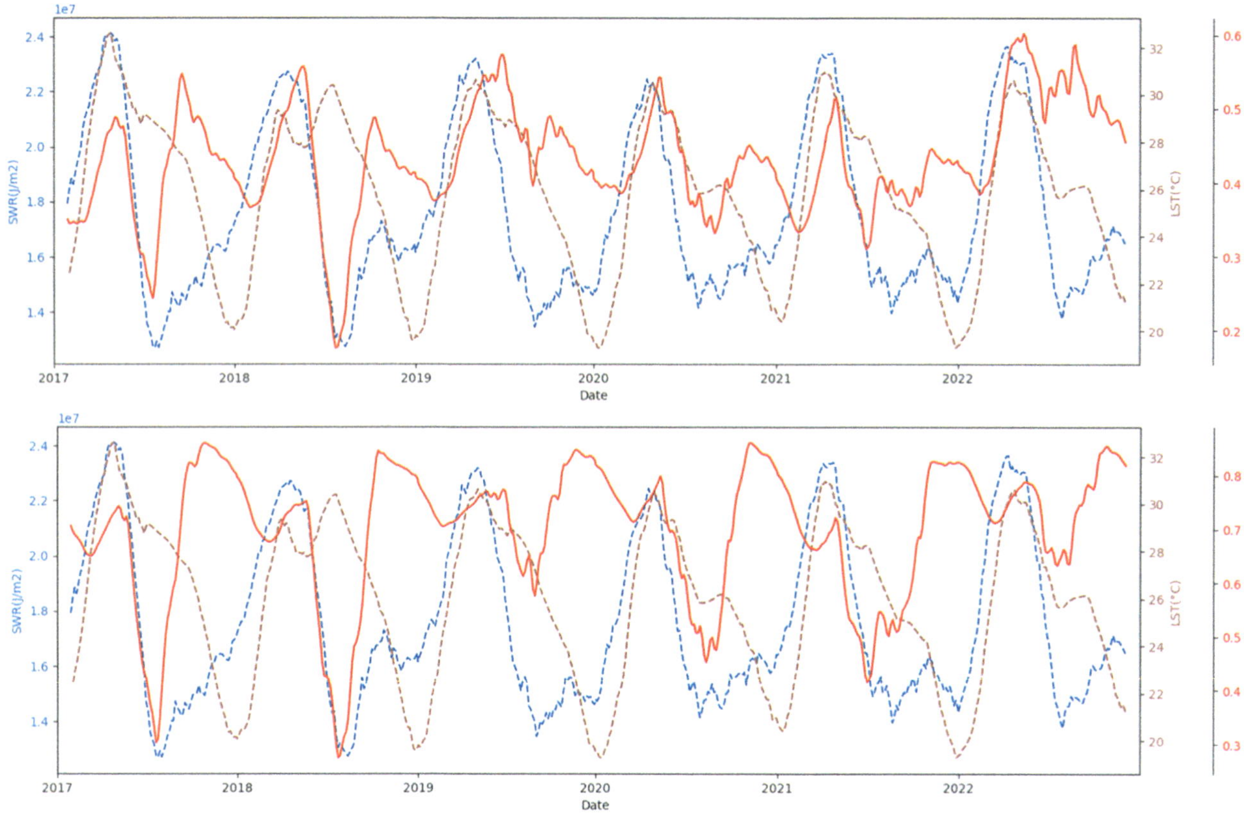

FIGURE 11.3 EVI (top) and NDVI (bottom) trends with respect to SWR (blue) and LST (brown) variations in tropical moist deciduous (TMD) forest.

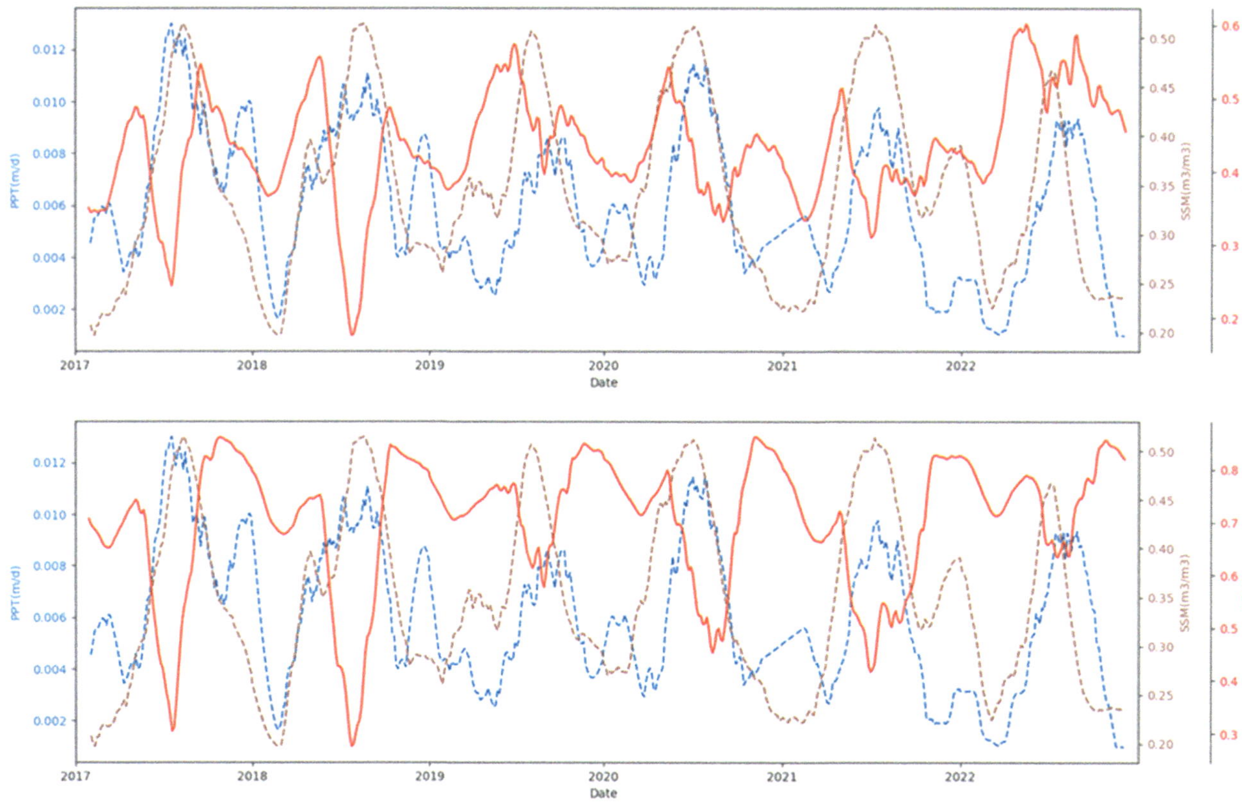

FIGURE 11.4 EVI (top) and NDVI (bottom) trends w.r.t PPT (blue) and SSM (brown) variation in a TMD forest.

The graphs reveal that both EVI and NDVI follow similar trends, with pronounced peaks during high precipitation periods, indicating enhanced vegetation health and growth. Notably, the NDVI showed a slight increase over the years, suggesting a possible gradual improvement in vegetation density or health from 2017 to 2022. This upward trend in NDVI may reflect positive outcomes from forest management practices or natural recovery processes within the studied ecosystems.

The cyclic patterns of NDVI and EVI in response to SWR and LST variations underscore the pivotal role of sunlight and temperature in driving vegetation health and productivity. During the premonsoon and summer months, the increases in SWR and LST were associated with higher NDVI and EVI values, reflecting enhanced photosynthetic activity and vegetative growth. Abundant sunlight and favorable temperatures promote forest greening, which is crucial for carbon sequestration and ecosystem services. However, the monsoon season, characterized by sustained cloud cover, decreases SWR, slightly moderating the increase in NDVI and EVI despite favorable temperature conditions. This indicates that, while temperature remains conducive to growth, reduced sunlight availability during heavy cloud cover can limit photosynthetic activity.

The precipitation pattern following the Indian monsoon system and its impact on surface soil moisture significantly affects forest phenology. The analysis showed that monsoon months with high rainfall increased soil moisture, closely followed by an increase in the NDVI and EVI values. This suggests that water availability is a critical factor for vegetation health and supports the growth and greening of forests during the monsoon season. Conversely, the decrease in soil moisture during the dry season resulted in lower NDVI and EVI values, highlighting the vulnerability of forests to water stress. The temporal lag between precipitation and changes in VIs illustrates the response time of forest ecosystems to environmental changes, with implications for understanding and managing drought effects.

The observed temporal lag between environmental conditions and vegetation responses, such as the delay in NDVI and EVI responses to changes in the SWR, LST, and soil moisture, highlights the complexity of ecosystem dynamics. The thermal inertia of the Earth's surface and the time required for ecosystems to respond to changes in water availability are crucial factors for understanding forest health and resilience. This detailed analysis of environmental parameters and their relationship with VIs provides valuable insights into the mechanisms that drive forest phenology. Understanding these dynamics is essential for developing predictive models of vegetation responses to climate change, enabling better management and conservation strategies to mitigate the impacts of environmental stressors on forest ecosystems.

11.8 Conclusions

This comprehensive study has provided invaluable insights into the intricate relationships between meteorological variables and forest phenology by analyzing EO data, meteorological measurements, and VIs. Our findings highlight the significant impact of climate variables, such as temperature, precipitation, NSR, and soil moisture, on the phenological cycles of forest ecosystems, which in turn affect their health and productivity. The analysis demonstrated that higher levels of NSR were associated with increased photosynthetic activity, as reflected by the NDVI and EVI indices. This underscores the crucial role of solar radiation in enhancing vegetative growth and health, particularly during the pre- and postmonsoon seasons. However, our study also revealed that excessive temperature can stress vegetation, reduce greenness and vitality, and emphasize the need for a balanced climate to maintain optimal forest health. Precipitation and soil moisture are closely interconnected, and their variations influence vegetation health. Increased rainfall and soil moisture significantly enhanced the NDVI and EVI values, indicating healthier and more vibrant vegetation. This relationship is particularly critical during the monsoon season and supports the robust growth of forest ecosystems. Conversely, the study noted the detrimental effects of drought conditions, where reduced precipitation and lower soil moisture were correlated with declines in VIs, highlighting the vulnerability of forests to water stress. The temporal lag observed between changes in environmental parameters and vegetation responses underscores the complexity of the ecosystem dynamics. This delay, influenced by the thermal inertia of Earth's surface and the time it takes for ecosystems to react to water availability, is crucial for understanding forest health and resilience. The use of EO data combined with advanced remote sensing technologies and data analysis methods has significantly enhanced our understanding of these dynamics. This has improved our ability to monitor and predict the impacts of climate change on forest phenology and to inform effective management and conservation strategies. These strategies are imperative to mitigate the adverse effects of environmental stressors and ensure the sustainability of forest ecosystems. As climate change continues to pose a significant threat to global ecosystems, the findings of this study highlight the importance of continued and enhanced monitoring of forest phenologies using sophisticated EO and meteorological data. Such efforts will be vital in advancing our understanding of ecosystem responses to climatic variables and guiding policymaking for forest conservation and management to combat climate change and preserve biodiversity. This study contributes to the scientific understanding of forest phenology and its drivers. This provides a crucial foundation for policymakers and forest managers to devise strategies to enhance the resilience and sustainability of forest ecosystems in the face of ongoing climate change.

Acknowledgments

We acknowledge the funding support from Indian Space Research Organisation-Geosphere Biosphere Programme (ISRO-GBP) for collaborative research, which is well coordinated by the Space Applications Centre, Ahmedabad, Gujarat. We acknowledge the Odisha State Forest, Environment, and Wildlife Department for providing institutional support to install the pheno-meteorological tower and permission to conduct this study in the Similipal Biosphere Reserve. MDB acknowledges the authority of CORAL and IIT Kharagpur for facilitating this study. AJ and MM acknowledges the fellowships received from the Ministry of Education and ISRO, respectively.

References

Atkinson, P. M., Jeganathan, C., Dash, J., & Atzberger, C. (2012). Inter-comparison of four models for smoothing satellite sensor time-series data to estimate vegetation phenology. *Remote Sensing of Environment*, *123*, 400–417.

Bajpai, O., Dutta, V., Singh, R., Chaudhary, L. B., & Pandey, J. (2020). Tree community assemblage and abiotic variables in tropical moist deciduous forest of Himalayan Terai eco-region. *Proceedings of the National Academy of Sciences, India Section B: Biological Sciences*, *90*, 873–883.

Behera, M. D., & Kushwaha, S. P. S. (2012). The charms and challenges of climate change biodiversity in a warming world. *Biodiversity and Conservation*, *21*, 1153–1158. Available from https://doi.org/10.1007/s10531-012-0281-7.

Behera, M. D., & Roy, P. S. (2002). Lidar remote sensing for forestry applications: The Indian context. *Current Science*, *83*, 1320–1328.

Berger, M., Moreno, J., Johannessen, J. A., Levelt, P. F., & Hanssen, R. F. (2012). ESA's sentinel missions in support of Earth system science. *Remote Sensing of Environment*, *120*, 84–90.

Cao, R., Chen, Y., Shen, M., Chen, J., Zhou, J., Wang, C., & Yang, W. (2018). A simple method to improve the quality of NDVI time-series data by integrating spatiotemporal information with the Savitzky-Golay filter. *Remote Sensing of Environment*, *217*, 244–257.

Cerlini, P. B., Saraceni, M., Orlandi, F., Silvestri, L., & Fornaciari, M. (2022). Phenological response to temperature variability and orography in Central Italy. *International Journal of Biometeorology*, *66*, 71–86. Available from https://doi.org/10.1007/s00484-021-02190-1.

Chen, J., Jönsson, P., Tamura, M., Gu, Z., Matsushita, B., & Eklundh, L. (2004). A simple method for reconstructing a high-quality NDVI time-series data set based on the Savitzky–Golay filter. *Remote Sensing of Environment*, *91*, 332–344.

Cleland, E. E., Chuine, I., Menzel, A., Mooney, H. A., & Schwartz, M. D. (2007). Shifting plant phenology in response to global change. *Trends in Ecology & Evolution*, *22*, 357–365. Available from https://doi.org/10.1016/j.tree.2007.04.003.

Cleveland, W. S., & Devlin, S. J. (1988). Locally weighted regression: An approach to regression analysis by local fitting. *Journal of the American Statistical Association*, *83*, 596–610.

Cui, X., Xu, G., He, X., & Luo, D. (2022). Influences of seasonal soil moisture and temperature on vegetation phenology in the Qilian mountains. *Remote Sensing*, *14*. Available from https://doi.org/10.3390/rs14153645.

Das, P., & Behera, M. D. (2019). Can the forest cover in India withstand large climate alterations? *Biodiversity and Conservation*, *28*, 2017–2033.

Dash, J., & Curran, P. J. (2004). The MERIS terrestrial chlorophyll index. *International Journal of Remote Sensing*, *25*, 5403–5413. Available from https://doi.org/10.1080/0143116042000274015.

De Frenne, P., Lenoir, J., Luoto, M., Scheffers, B. R., Zellweger, F., Aalto, J., ... De Pauw, K. (2021). Forest microclimates and climate change: Importance, drivers and future research agenda. *Global Change Biology*, *27*, 2279–2297.

De Réaumur, R. A. F. (1735). Observations du thermomere. *Memories Academie Royale Sciences Paris*, 545–576.

Diez, J. M., Ibáñez, I., Silander, J. A., Jr, Primack, R., Higuchi, H., Kobori, H., ... James, T. Y. (2014). Beyond seasonal climate: Statistical estimation of phenological responses to weather. *Ecological Applications*, *24*, 1793–1802.

Dronova, I., & Taddeo, S. (2022). Remote sensing of phenology: Towards the comprehensive indicators of plant community dynamics from species to regional scales. *Journal of Ecology*, *110*, 1460–1484. Available from https://doi.org/10.1111/1365-2745.13897.

Dugesar, V., Satish, K. V., Pandey, M. K., Srivastava, P. K., Petropoulos, G. P., Anand, A., & Behera, M. D. (2022). Impact of environmental gradients on phenometrics of major forest types of Kumaon Region of the Western Himalaya. *Forests*, *13*(12), 1973.

Fan, J., Han, F., & Liu, H. (2014). Challenges of big data analysis. *National Science Review*, *1*, 293–314.

Forrest, J., & Miller-Rushing, A. J. (2010). *Toward a synthetic understanding of the role of phenology in ecology and evolution. Philosophical Transactions of the Royal Society B: Biological Sciences*. The Royal Society.

Gao, B.-C. (1996). NDWI—A normalized difference water index for remote sensing of vegetation liquid water from space. *Remote Sensing of Environment*, *58*, 257–266.

Ghosh, S. M., & Behera, M. D. (2018). Aboveground biomass estimation using multi-sensor data synergy and machine learning algorithms in a dense tropical forest. *Applied Geography*, *96*, 29–40. Available from https://doi.org/10.1016/j.apgeog.2018.05.011.

Gitelson, A. A. (2004). Wide dynamic range vegetation index for remote quantification of biophysical characteristics of vegetation. *Journal of Plant Physiology*, *161*, 165–173.

Gonsamo, A., & Pellikka, P. (2012). The sensitivity based estimation of leaf area index from spectral vegetation indices. *ISPRS Journal of Photogrammetry and Remote Sensing*, *70*, 15–25. Available from https://doi.org/10.1016/j.isprsjprs.2012.03.009.

Gray, R. E. J., & Ewers, R. M. (2021). Monitoring forest phenology in a changing world. *Forests*, *12*. Available from https://doi.org/10.3390/f12030297.

Guo, H.-D., Zhang, L., & Zhu, L.-W. (2015). Earth observation big data for climate change research. *Advances in Climate Change Research*, *6*, 108–117. Available from https://doi.org/10.1016/j.accre.2015.09.007.

Hermance, J. F. (2007). Stabilizing high-order, non-classical harmonic analysis of NDVI data for average annual models by damping model roughness. *International Journal of Remote Sensing*, *28*, 2801–2819.

Hoegh-Guldberg, O., Jacob, D., Taylor, M., Guillén Bolaños, T., Bindi, M., Brown, S., ... Ebi, K. (2019). The human imperative of stabilizing global climate change at 1.5 C. *Science*, *365*eaaw6974.

Hoffmann, A. A., & Sgrò, C. M. (2011). Climate change and evolutionary adaptation. *Nature*, *470*, 479–485. Available from https://doi.org/10.1038/nature09670.

Holben, B. N. (1986). Characteristics of maximum-value composite images from temporal AVHRR data. *International Journal of Remote Sensing*, *7*, 1417–1434.

Hollmann, R., Merchant, C. J., Saunders, R., Downy, C., Buchwitz, M., Cazenave, A., ... Forsberg, R. (2013). The ESA climate change initiative: Satellite data records for essential climate variables. *Bulletin of the American Meteorological Society*, *94*, 1541–1552.

Huang, N. E., Shen, Z., Long, S. R., Wu, M. C., Shih, H. H., Zheng, Q., ... Liu, H. H. (1998). The empirical mode decomposition and the Hilbert spectrum for nonlinear and non-stationary time series analysis. *Proceedings of the Royal Society of London. Series A: Mathematical, Physical and Engineering sciences*, *454*, 903–995.

Huete, A., Justice, C., & van Leeuwen, W. (1999). MODIS vegetation index (MOD 13) algorithm theoretical basis document version 3. University of Arizona 1200.

Huete, A. R. (1988). A soil-adjusted vegetation index (SAVI). *Remote Sensing of Environment*, *25*, 295–309.

Jeganathan, C., Singh, B., Singh, C. P., Behera, M. D., Srivastava, S., Natesha, S. R., Singh, K., Ranjan, R., Prakash, M., Kumar, A., Pandya, M. R., Bhattacharya, B. K., Krishna, A. P., Lala, M. G. N., Rathore, V. S., Sinha, N. K., Choudhary, K., Bhuyan, M., Singh, S. S., & Sardar, P. (2024). Integrated use of field sensors, PhenoCam, and satellite data for pheno-phase monitoring in a tropical deciduous forest of Dalma Wildlife Sanctuary, Jharkhand, India: initial results from the Indian Phenology Network. *Biodiversity and Conservation*, *33*, 3539–3565. Available from https://doi.org/10.1007/s10531-024-02889-8.

Jiang, Z., Huete, A. R., Didan, K., & Miura, T. (2008). Development of a two-band enhanced vegetation index without a blue band. *Remote Sensing of Environment*, *112*, 3833–3845. Available from https://doi.org/10.1016/j.rse.2008.06.006.

Jin, H., & Eklundh, L. (2014). A physically based vegetation index for improved monitoring of plant phenology. *Remote Sensing of Environment*, *152*, 512–525.

Jönsson, P., Cai, Z., Melaas, E., Friedl, M. A., & Eklundh, L. (2018). A method for robust estimation of vegetation seasonality from Landsat and Sentinel-2 time series data. *Remote Sensing*, *10*, 635.

Jose, K., Chaturvedi, R. K., Jeganathan, C., Behera, M. D., & Singh, C. P. (2023). Plugging the Gaps in the Global PhenoCam Monitoring of Forests—The Need for a PhenoCam Network across Indian Forests. *Remote Sensing*, *15*. Available from https://doi.org/10.3390/rs15245642.

Li, D., Wu, S., Liu, L., Zhang, Y., & Li, S. (2018). Vulnerability of the global terrestrial ecosystems to climate change. *Global Change Biology*, *24*, 4095–4106.

Liu, Q., Fu, Y. H., Zhu, Z., Liu, Y., Liu, Z., Huang, M., ... Piao, S. (2016). Delayed autumn phenology in the Northern Hemisphere is related to change in both climate and spring phenology. *Global Change Biology*, *22*, 3702–3711.

Malhi, Y., Franklin, J., Seddon, N., Solan, M., Turner, M. G., Field, C. B., & Knowlton, N. (2020). Climate change and ecosystems: Threats, opportunities and solutions. *Philosophical Transactions of the Royal Society B: Biological Sciences*, *375*, 20190104. Available from https://doi.org/10.1098/rstb.2019.0104.

Menzel, A., Sparks, T. H., Estrella, N., Koch, E., Aasa, A., Ahas, R., ... Zust, A. N. A. (2006). European phenological response to climate change matches the warming pattern. *Global Change Biology*, *12*(10), 1969–1976.

Mo, L., Zohner, C. M., Reich, P. B., Liang, J., de Miguel, S., Nabuurs, G.-J., ... Crowther, T. W. (2023). Integrated global assessment of the natural forest carbon potential. *Nature*, *624*, 92–101. Available from https://doi.org/10.1038/s41586-023-06723-z.

Mohanta, M. R., Sahoo, S., & Sahu, S. C. (2021). Variation in structural diversity and regeneration potential of tree species in different tropical forest types of Similipal Biosphere Reserve, Eastern India. *Acta Ecologica Sinica*, *41*, 597–610. Available from https://doi.org/10.1016/j.chnaes.2021.08.011.

Myneni, R., Knyazikhin, Y., Glassy, J., Votava, P., & Shabanov, N. (2003). User's guide: FPAR, LAI (ESDT: MOD15A2) 8-day composite NASA MODIS land algorithm. *Terra*, 1–17.

Negi, G. C. S., Joshi, S., Singh, P., & Joshi, R. (2022). Phenological response patterns of forest communities to annual weather variability at long-term ecological monitoring sites in Western Himalaya. *Trees, Forests and People*, *8*, 100237.

Parmesan, C. (2007). Influences of species, latitudes and methodologies on estimates of phenological response to global warming. *Global Change Biology*, *13*, 1860–1872.

Piao, S., Liu, Q., Chen, A., Janssens, I. A., Fu, Y., Dai, J., ... Zhu, X. (2019). Plant phenology and global climate change: Current progresses and challenges. *Global Change Biology*, *25*, 1922–1940.

Prakash, A. J., Behera, M. D., Ghosh, S. M., Das, A., & Mishra, D. R. (2022). A new synergistic approach for Sentinel-1 and PALSAR-2 in a machine learning framework to predict aboveground biomass of a dense mangrove forest. *Ecological Informatics*, 101900. Available from https://doi.org/10.1016/j.ecoinf.2022.101900.

Prakash, A. J., Behera, M. D., Singh, C. P., Chockalingam, J., & Singh, B. (2023). Do Tropical Moist Deciduous forests fix more Carbon owing to weak Phenophases in synergy with available soil moisture? in: AGU Fall Meeting Abstracts. pp. B11L-1927.

Reddy, C. S., Padma Alekhya, V. V. L., Saranya, K. R. L., Athira, K., Jha, C. S., Diwakar, P. G., & Dadhwal, V. K. (2017). Monitoring of fire incidences in vegetation types and protected Areas of India: Implications on carbon emissions. *Journal of Earth System Science*, *126*, 1–15.

Reddy, C. S., Pattanaik, C., Dhal, N. K., & Biswal, A. K. (2006). Vegetation and floristic diversity of Bhitarkanika National Park, Orissa, India. *The Indian Forester*, *132*, 664–680.

Richardson, A. D., Keenan, T. F., Migliavacca, M., Ryu, Y., Sonnentag, O., & Toomey, M. (2013). Climate change, phenology, and phenological control of vegetation feedbacks to the climate system. *Agricultural and Forest Meteorology*, *169*, 156–173. Available from https://doi.org/10.1016/j.agrformet.2012.09.012.

Richardson, A. D., & O'Keefe, J. (2009). Phenological differences between understory and overstory: A case study using the long-term Harvard Forest records. *Phenology of Ecosystem Processes: Applications in Global Change Research*, 87–117.

Richardson, A. J., & Wiegand, C. L. (1977). Distinguishing vegetation from soil background information. *Photogrammetric Engineering and Remote Sensing*, *43*, 1541–1552.

Roerink, G. J., Menenti, M., & Verhoef, W. (2000). Reconstructing cloudfree NDVI composites using Fourier analysis of time series. *International Journal of Remote Sensing*, *21*, 1911–1917.

Rouse, J. W., Haas, R. H., Schell, J. A., & Deering, D. W. (1974). Monitoring vegetation systems in the Great Plains with ERTS. *NASA Special Publications*, *351*, 309.

Ruiz-Benito, P., Vacchiano, G., Lines, E. R., Reyer, C. P., Ratcliffe, S., Morin, X., ... Chaves, J. E. (2020). Available and missing data to model impact of climate change on European forests. *Ecological Modelling*, *416*, 108870.

Saranya, K. R. L., & Reddy, C. S. (2016). Long term changes in forest cover and land use of Similipal Biosphere Reserve of India using satellite remote sensing data. *Journal of Earth System Science*, *125*, 559–569.

Seidl, R., Rammer, W., & Spies, T. A. (2014). Disturbance legacies increase the resilience of forest ecosystem structure, composition, and functioning. *Ecological Applications*, *24*, 2063–2077. Available from https://doi.org/10.1890/14-0255.1.

Sellers, P. J., Tucker, C. J., Collatz, G. J., Los, S. O., Justice, C. O., Dazlich, D. A., & Randall, D. A. (1994). A global 1 by 1 NDVI data set for climate studies. Part 2: The generation of global fields of terrestrial biophysical parameters from the NDVI. *International Journal of Remote Sensing*, *15*, 3519–3545.

Singh, R. K., Biradar, C. M., Behera, M. D., Prakash, A. J., Das, P., Mohanta, M. R., ... Rizvi, J. (2024). Optimising carbon fixation through agroforestry: Estimation of aboveground biomass using multi-sensor data synergy and machine learning. *Ecological Informatics*, *79*, 102408. Available from https://doi.org/10.1016/j.ecoinf.2023.102408.

Spafford, L., MacDougall, A., & Steenberg, J. (2023). Climate-driven shifts in leaf senescence are greater for boreal species than temperate species in the Acadian Forest region in contrast to leaf emergence shifts. *Ecology and Evolution*, *13*, e10362.

Tucker, C. J. (1979). Red and photographic infrared linear combinations for monitoring vegetation. *Remote Sensing of Environment*, *8*, 127–150. Available from https://doi.org/10.1016/0034-4257(79)90013-0.

Vitasse, Y., Lenz, A., & Körner, C. (2014). The interaction between freezing tolerance and phenology in temperate deciduous trees. *Frontiers in Plant Science*, *5*, 541.

Walther, G.-R., Post, E., Convey, P., Menzel, A., Parmesan, C., Beebee, T. J., ... Bairlein, F. (2002). Ecological responses to recent climate change. *Nature*, *416*, 389–395.

Wan, Z., & Dozier, J. (1996). A generalized split-window algorithm for retrieving land-surface temperature from space. *IEEE Transactions on geoscience and remote sensing*, *34*(4), 892–905.

Watson, D. J. (1947). Comparative physiological studies on the growth of field crops: I. Variation in net assimilation rate and leaf area between species and varieties, and within and between years. *Annals of Botany*, *11*, 41–76.

Weiskopf, S. R., Rubenstein, M. A., Crozier, L. G., Gaichas, S., Griffis, R., Halofsky, J. E., & Whyte, K. P. (2020). Climate change effects on biodiversity, ecosystems, ecosystem services, and natural resource management in the United States. *Science of The Total Environment*, *733*, 137782. Available from https://doi.org/10.1016/j.scitotenv.2020.137782.

Xie, Y., & Wilson, A. M. (2020). Change point estimation of deciduous forest land surface phenology. *Remote Sensing of Environment*, *240*, 111698. Available from https://doi.org/10.1016/j.rse.2020.111698.

Zeng, L., Wardlow, B. D., Xiang, D., Hu, S., & Li, D. (2020). A review of vegetation phenological metrics extraction using time-series, multispectral satellite data. *Remote Sensing of Environment*, *237*, 111511. Available from https://doi.org/10.1016/j.rse.2019.111511.

Chapter 12

Earth observation for monitoring of shifting cultivation

Pulakesh Das[1], Mukunda Dev Behera[2], Saroj Kanta Barik[3] and Sujoy Mudi[2]

[1]School of Forest Resources, University of Maine, Orono, ME, United States, [2]Centre for Ocean, River, Atmosphere and Land Sciences (CORAL), Indian Institute of Technology Kharagpur, Kharagpur, West Bengal, India, [3]Centre for Advanced Studies in Botany, North-Eastern Hill University, Shillong, India

12.1 Introduction

12.1.1 Remote sensing for forest resource monitoring

Over the last five decades, earth observation satellite programs have expanded the use of remote sensing technology multiple-fold. Space research programs by the National Aeronautics and Space Administration (NASA), the Indian Space Research Organization (ISRO), the European Space Agency (ESA), and the Japan Aerospace Exploration Agency (JAXA) have enabled national to global-scale mapping and monitoring of atmospheric, oceanic, land surface, and subsurface resources and various dependent processes. Remote sensing technology provides synoptic views and consistent data for real-time and near-real-time data generation, landscape planning, management and decision-making, scientific studies, etc. Advances in data computation facilities, such as the availability of computers with large storage capabilities, have assisted in the widespread application of remote sensing technology. On the other hand, the availability of data from several satellite sensors/platforms in the public domain created a wealth of earth observation data as fine as 10 m spatial resolution data, including in countries that lack earth observation satellites. Meanwhile, commercial earth observation satellites provide user-defined imaging opportunities at a spatial resolution below 1 m. Introducing contemporary artificial intelligence (AI) techniques has opened up several horizons in geoinformatics technology and complex data processing more efficiently. WebGIS platforms, such as Google Earth Engine (GEE), consist of numerous data and allow bulk data processing with limited user-dependent computation facilities. This has reduced the need for massive data storage capacities and sophisticated systems with software for geospatial data processing. Moreover, ArcGIS Online and various webGIS platforms offer the hosting and sharing of various data and products on the web and make it available for various stakeholders' use.

Satellite remote sensing–based forest resource monitoring and characterization have gained significant attention owing to the availability of systematic spatiotemporal satellite images. Optical remote sensing data allow mapping of the forest cover area, change (deforestation and afforestation), forest type, foliar traits and biochemical properties (e.g., water, chlorophyll, and nitrogen), phenology, carbon sequestration, structural attributes (e.g., crown size), disease and pest impact, and forest fire among others (Behera, Gupta, Barik, Das, & Panda, 2018; Das et al., 2022; Ghosh, Behera, & Paramanik, 2020). On the other hand, longer-wavelength microwave (synthetic aperture radar: SAR; and light detection and ranging: LiDAR) data better predict structural attributes, including tree canopy height, canopy density, and plant biomass (Ghosh et al., 2020; Roy et al., 2021; Singh et al., 2024). Some widely used remote sensing images for forest cover monitoring are Landsat and moderate resolution imaging spectroradiometer (MODIS) data (by NASA), Sentinel-1 and Sentinel-2 data (by ESA), and linear imaging self scanner (LISS) data (by ISRO). While Landsat, Sentinel-2, and LISS data are moderate resolution data available at 5- to 24-day intervals, the MODIS data provide high-temporal frequency data (daily) at coarse spatial resolution (250 and 500 m) since 2000. Various data products have been generated using MODIS data at daily to 8-day and monthly scales, such as greenness (normalized difference vegetation index [NDVI]; enhanced vegetation index [EVI]), vegetation fraction, land surface temperature, fires (active

Sustainable Development Perspectives in Earth Observation. DOI: https://doi.org/10.1016/B978-0-443-14072-3.00013-7

fire and burned area), fraction of photosynthetically active radiation (FaPAR), leaf area index (LAI), evapotranspiration (ET), and gross primary productivity (GPP), among others, which are publicly available on multiple platforms (modis.gsfc.nasa.gov/data/dataprod/). Such vegetation parameters are also extracted from Landsat, LISS, and Sentinel data for forest mapping at a higher resolution. In addition, multiple user-ready data products have been developed using Landsat and Sentinel data for consistent forest cover and resource monitoring at the global scale, for example, Landsat data-derived tree canopy percentage, annual tree cover loss, and periodic tree cover gain data (30 m spatial resolution) by Hansen et al. (2013) and tree canopy height by Potapov et al. (2021), Sentinel-2 data-derived 10 m resolution annual land use land cover (LULC) product (livingatlas.arcgis.com/landcover/), among others.

12.1.2 Forest cover monitoring in India

In India, the Forest Survey of India (FSI) periodically maps forest and tree cover resources at 2-year intervals since 1987. FSI currently uses the LISS and Sentinel-2 images for forest and tree outside forest (TOF) cover mapping based on tree cover density. A hybrid image classification (mix of digital image classification and visual image interpretation) technique is applied for semi-automated forest cover and change monitoring. According to the latest Indian State of Forest Resources (ISFR, 2021), forest and tree cover occupy around 24.62% of the total Indian landmass, whereas 64.66% of the total geographic area in Northeast India (NEI) is forest land. Several other national-scale datasets are generated. Applying a visual image interpretation of LISS and Landsat images, Roy et al. (2015) developed national-scale LULC data for three decades, that is, 1985, 1995, and 2005, and the raster LULC data at 100 m spatial resolution is publicly available at ORNL-DAAC data portal (daac.ornl.gov/VEGETATION/guides/Decadal_LULC_India.html). The decadal LULC maps indicated a continuous gain in plantations, shrubland, and grassland and a loss in forest cover with periodic changes in shifting cultivation areas in India from 1985 to 2005. The National Remote Sensing Center (NSRC), India, also applies digitial and visual image interpretation techniques for periodic LULC map generation at various spatial and temporal scales, which are now publicly available in the Bhuvan portal (bhuvan-app1.nrsc.gov.in/thematic/thematic/index.php) (Reddy, Jha, Diwakar, & Dadhwal, 2015).

12.1.3 Shifting cultivation

Shifting or swidden cultivation is a widely traditional agricultural practice in hilly terrains mostly observed in tropical regions, that is, south and southeast Asia, central Africa, and Latin America (Jiang, Li, & Feng, 2022). This practice involves the tree or other natural vegetation slash-and-burning method, that is, the tree covers in a land parcel are removed, and the unwanted dead plant material is burned during the dry period, which clears the land and improves soil fertility. Such land parcels are used for cropping for 2–5 years, after which they are abundant for natural vegetation regeneration and reuse for shifting cultivation. The duration between successive shifting cultivation is known as the fallow period or shifting cultivation cycle. The comparison of soil organic carbon (SOC) in different soil depths and for different fallow stages in shifting cultivation lands indicated a significantly lower SOC in recent fallow than old fallow, and similarly a lower SOC in fallow shifting cultivation areas than in native forests (Nath, Sahoo, Giri, Sileshi, & Das, 2020). Thus the fallow period allows the landscape to rejuvenate the forest vegetation and replenish the soil fertility. Although this agriculture practice has been articulated as unscientific and harmful, shifting cultivation is imperative to ensure food security in hilly terrain and maintain the overall forest cover rather than permanently converting it into cropland or other land use practices. The regeneration of forest cover in the shifting cultivation areas can store more carbon than sedentary agriculture (Shaw, Natori, & Edake, 2022).

In India, shifting cultivation (locally known as "jhum") is a widespread traditional agricultural system in upland hilly terrains of NEI (Pasha, Behera, Mahawar, Barik, & Joshi, 2020; Ramakrishnan, 2007; Toky & Ramakrishnan, 1981). A significant population in India depends on such agricultural practices for sustenance (Mathur & Bhattacharya, 2023). Shifting cultivation has enabled the maintenance of high agro-biodiversity and the preservation of numerous indigenous or native crop species, including rice, millet, and yam, in NEI. In addition, the forest cover in NEI significantly contributes to the forest-based socio-economy, livelihood, and cultural diversities (Barik & Mishra, 2008). A significant proportion of forest lands in the hilly terrains of NEI are locally managed by the village communities, clans, and private individuals who supervise land use practices in the region (Duraisami et al., 2022). However, there is a significant difference in land ownership rights in various parts of NEI (Barik & Mishra, 2008). Such locally led societies enjoy traditional land use rights, protect the landscapes, and periodically allocate specific forest patches for shifting cultivation. Shifting cultivation practices cause massive deforestation, loss of mature trees, forest fragmentation, periodic alteration in the ecosystem structure, and modification of biodiversity (Das, Behera, & Murthy, 2017; Gupta, 2000;

Ramakrishnan, 2007). Various studies have indicated that population growth and food demand have reduced the shifting cultivation cycle or jhum cycle in NEI (Grogan, Lalnunmawia, & Tripathi, 2012; Ranjan & Upadhyay, 1999; Shankar & Tripathi, 1996). Such anthropogenic disturbances result in more managed and less diverse secondary forests than highly diverse primary natural-grown ones. Moreover, such anthropogenic activities have continuously modified vegetation biomass and carbon cycles in the region (Hartter, Lucas, Gaughan, & Aranda, 2008). Several attempts have been made to demotivate shifting cultivation practices through alternative agricultural practices, including agroforestry and rubber plantations in several parts of NEI (Behera, Gupta et al., 2018; Das et al., 2022; Ramakrishnan, 2007).

12.1.4 Use of remote sensing techniques

Shifting cultivation area mapping and monitoring are essential for understanding the continuous modification of forest cover, ecosystem structure, and biodiversity. Satellite remote sensing images record land surface reflectance (spectral characteristics) and backscatter values, which has enabled precise forest cover mapping and change analysis even in cloudy conditions. The revised data policy by NASA and availability of Landsat data in the public domain advanced the use of moderate resolution satellite data for shifting cultivation monitoring efforts in the last two decades, wherein a recent increasing trend in remote sensing–based studies was observed in 2008 and 2015 (Jiang et al., 2022; Mathur & Bhattacharya, 2023). Landsat data allowed land surface feature mapping since the 1980s at large spatial extents with limited ground sampling efforts. Multispectral optical remote sensing records the reflectance in multiple spectral bands, which is used to identify different features owing to their varied reflectance properties. Deforestation, followed by burning dead plant material, creates a sharp and significant decline in vegetation greenness and enables accurate shifting cultivation area mapping using multitemporal remote sensing images (Das et al., 2021). Various image-processing techniques have been developed to harness remote sensing data. Visual image interpretation, spectral indices, and principal component analysis have been widely used in shifting cultivation mapping over the past decades (Hartter et al., 2008; Pasha et al., 2020; Roy, Kaul, Sharma, Roy, & Garbyal, 1985). Several studies in India have mapped shifting cultivation areas using visual image interpretation of the landscape nationally (Lele, Joshi, & Agrawal, 2008; Reddy et al., 2015; Roy, Behera et al., 2015; Roy, Roy et al., 2015). However, most of these studies generated single-time data. Recently, Das et al. (2021) and Das et al. (2022) used Landsat images for multitemporal shifting cultivation mapping in several states in NEI. Several other studies used the LISS imagery for shifting cultivation mapping in this landscape (Bhat, Nandy, Padalia, Nath, & Pebam, 2022).

Spectral enhancements or indices have been developed to highlight various land surface features in an image (Behera, Gupta et al., 2018). Previous studies have applied spectral indices to multispectral data, such as the normalized difference vegetation index [NDVI = (NIR−Red)/(NIR + Red); NIR: near-infrared], normalized burn ratio [NBR = (NIR−SWIR2)/(NIR + SWIR2); SWIR: short-wave infrared], and normalized difference moisture index [NDMI = (NIR−SWIR1)/(NIR + SWIR1)], for shifting cultivation mapping (Key & Benson, 2006; Schneibel et al., 2017). Multitemporal NBR (dNBR = postevent−preevent NBR) was also used in such studies (Escuin, Navarro, & Fernández, 2008; Murphy, Reynolds, & Koltun, 2008; Tansey, Beston, Hoscilo, Page, & Paredes Hernández, 2008). A majority of studies have used Landsat data for shifting cultivation mapping at various spatial and temporal scales (Bhat et al., 2022; Jiang et al., 2022; Mathur & Bhattacharya, 2023; Molinario, Hansen, & Potapov, 2015), while several studies have used LISS data (Alajangi, Prasad, & Baruah, 2021; Marak et al., 2023), advanced spaceborne thermal emission and reflection radiometer (ASTER) (Stroppiana, Boschetti, Zaffaroni, & Brivio, 2009), and Sentinel-2 data (Chanu, De, Chakraborty, & Paul, 2023). Use of multiple spectral indices better identify the burned areas than a single spectral index (Das et al., 2022; Lasaponara, 2006). Multiple studies have used MODIS products due to their high temporal resolution in addition to Landsat data; for example, Hurni, Hett, Heinimann, Messerli, and Wiesmann (2013) and Li and Yang (2022) conducted annual-scale shifting cultivation monitoring in northern Laos, Kumar (2022) in Nagaland, India. Moreover, the shifting cultivation areas indicated a significant relation with topographical variables, as this practice is mostly seen in the topographically rugged terrain with a higher slope that faces the aspects with higher solar irradiance (Kumar, 2022).

Advances in AI technologies have enabled the application of various machine learning (ML) and deep learning (DL) techniques in geoinformatics. Among the various ML algorithms, the random forest (RF) indicated a higher accuracy in classification and regression analysis (Das & Pandey, 2019; Ghosh et al., 2020). RF algorithm is a decision tree-based approach that uses an ensemble of trees for voting for the most popular class, and the output class is determined by majority voting (Breiman, 2001). It uses hundreds of decision trees, overcomes overfitting and correlated variables, and provides high accuracy (Gislason, Benediktsson, & Sveinsson, 2006). Long et al. (2019) applied the RF algorithm to Landsat images for global-scale shifting cultivation mapping on the GEE platform. Their study included multitemporal data on NBR, NBR2 (normalized difference between SWIR1 and SWIR2), NDVI, NDMI, and multiple other indices and obtained an accuracy of $R^2 \geq 0.74$. However, the globally trained model could be biased due to local to landscape scale generalization errors. Hurni et al. (2013) applied the support vector machine (SVM) to MODIS EVI data for shifting cultivation mapping in northern Laos. In addition, object-based analysis has gained recent attention in image classification, including shifting cultivation monitoring (Mathur & Bhattacharya, 2023). However, there is a lack of studies on applying ML algorithms for periodic shifting cultivation mapping, especially in India. The present study employed Landsat data-derived spectral indices and shuttle radar topographic mission (SRTM) elevation data to map shifting cultivation areas using the RF algorithm in three states in NEI.

12.2 Study area

This study was conducted in three NEI states: Manipur, Meghalaya, and Nagaland. The total area of these three states is 123,204 km^2 and geographically extended between 23°49′ N and 27°2′ N, and 89°48′ E and 95°11′ E (Fig. 12.1). Forest cover is the dominant LULC in the study area, followed by cropland (Behera, Tripathi et al., 2018) (Fig. 12.1). The Indian Meteorological Department (IMD) climate data from 1986 to 2016 indicated that the average total annual precipitation was 1694, 2791, and 1852 mm in Manipur, Meghalaya, and Nagaland, respectively. The total population of Manipur in 2011 was more than 2.8 million, with more than 70% living in rural areas. The total population of Nagaland in 2011 was above 1.9 million, and more than 70% of the population lives in rural areas. Out of the above 2.9 million people in Meghalaya, around 80% live in rural areas.

12.3 Materials and method

12.3.1 Data used and preprocessing

Landsat 8 images for 2016–18 were accessed from the GEE portal. Eight Landsat image tiles were used in this study, and the least cloudy images for the driest premonsoon period (March and April) were accessed. SRTM digital elevation model (DEM) data were used as elevation data. The high-resolution Google Earth (GE; includes <1 m high-resolution Quickbird images) images were used for comparison and validation. Additionally, the course resolution MODIS burned area data (MCD64A1 version 6; spatial resolution 500 m) for 2017 and 2018 were accessed from the GEE to validate the shifting cultivation maps. The quality control layer was employed for MODIS data processing, wherein the best quality data were extracted.

12.3.2 Methodology

The false color composite (FCC) images (passing SWIR1, NIR, and red bands through red, green, and blue color channels, respectively) were used for visual image identification. In this FCC, the burned areas appear magenta, with similar

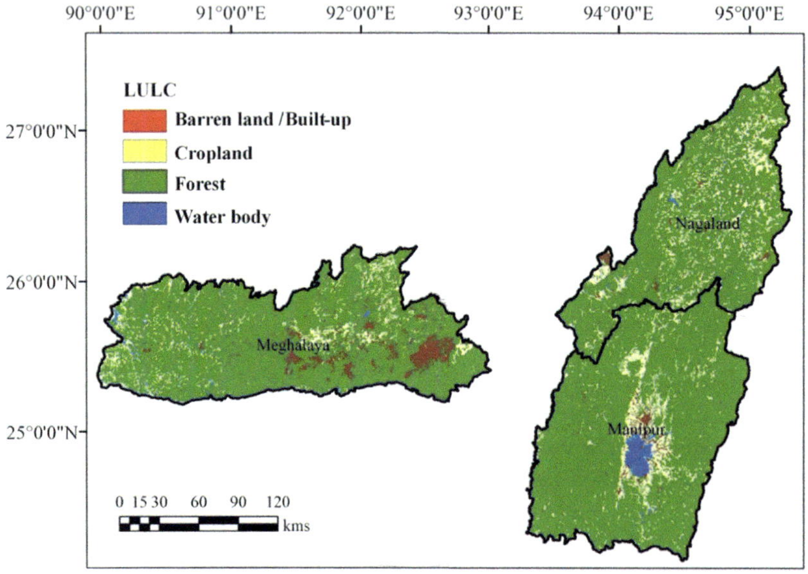

FIGURE 12.1 Land use land cover (LULC) map of Manipur, Meghalaya, and Nagaland states for year 2005. Map lines delineate study areas and do not necessarily depict accepted national boundaries. *Adapted from Behera, M. D., Tripathi, P., Das, P., Srivastava, S. K., Roy, P. S., Joshi, C., ... & Krishnamurthy, Y. V. N. (2018). Remote sensing based deforestation analysis in Mahanadi and Brahmaputra river basin in India since 1985. Journal of Environmental Management, 206, 1192–1203. https://doi.org/10.1016/j.jenvman.2017.10.015.*

reflectance in the SWIR1 and NIR bands and a relatively lower reflectance in the red band. In contrast, vegetation appears in green, bare soil appears in light magenta, and water appears in blue. The recent deforested and burned (new shifting cultivation) areas were identified through visual image interpretation by comparing last year's image, wherein vegetation cover (green pixel) was converted into barren land (magenta pixel). Seven features were identified in the satellite images: barren land, burned areas, sand, shadows, vegetation, water bodies, and clouds. Cloud and cloud shadows (shadows) were added to the image features list, as no completely cloud-free image was found, and cloud presence was widespread in satellite images in the hilly terrain of NEI. For each feature class, more than 200 reference points were identified from each image tile using visual image interpretation. Recent deforested and burned areas have more distinct spectral responses than older patches. Again, the boundary or mixed pixels are similar in the spectral response to barren land. Thus recent and pure burned areas were considered as reference, whereas old, mixed and less burned pixels were discarded.

The distinct spectral behavior of the burned area from other features was further improved by introducing spectral enhancements and indices. Three major types of indices were included: vegetation indices (NDVI, EVI, and soil adjusted vegetation index [SAVI]), vegetation water content indices (NDMI), and burned area indices (NBR and NBR2). In addition, the NDVI change from the previous year (NDVI_D) and elevation data were included as proxy variables in the shifting cultivation mapping. The NDVI_D map for 2017 was derived as NDVI (2017)−NDVI (2016), and similarly for 2018 as NDVI (2018)−NDVI (2017) (Huete, 1988; Liu & Huete, 1995; Lutes et al., 2006).

$$\mathrm{EVI} = \frac{2.5 \quad (\mathrm{NIR} - \mathrm{red})}{\mathrm{NIR} + (6 \times \mathrm{red}) - (7.5 \times \mathrm{blue}) + 1}$$

$$\mathrm{SAVI} = \frac{(1 + L) \times (\mathrm{NIR} - \mathrm{red})}{\mathrm{NIR} + \mathrm{red} + L}; L = 0.5$$

$$\mathrm{NBR2} = \frac{\mathrm{SWIR1} - \mathrm{SWIR2}}{\mathrm{SWIR1} + \mathrm{SWIR2}}$$

The RF algorithm was applied for image classification in the RStudio platform, and the "caret" and "raster" packages were used in this study. The reference data points were collected in five image tiles for 2017−18, and the developed RF model was validated using the test data points in the remaining three image tiles for 2017−18. The RF model was parameterized using "mtry" value, which regulates the number of variables used for classification in each node. The model was run using a range of "mtry" values from 1 to 10, wherein the model with the lowest error was selected as the best-performing model. The tuned and validated RF model was applied then to all eight image tiles for 2016−17 to assess the model performance. The overall and kappa accuracies were computed to determine the classification accuracy. A total of 5139 points were considered for the RF model development in 2017−18, whereas 5027 points were used for validation. In comparison, 11,157 points were considered for the accuracy assessment in 2016−17. The classified maps derived using 2017 18 images were denoted by the 2018 map and 2016−17 images by the 2017 map.

12.4 Results and discussion

The classification accuracy of the developed RF model was assessed based on the error matrix prepared using the testing or validation data points for 2017−18 (Table 12.1). An overall accuracy of 89% with a Kappa value of 0.87 was obtained, wherein the burned areas were identified with a high user's (99.86%) and producer's (80.21%) accuracy. The high user's accuracy indicates that all burned areas mapped by the RF model were validated using the reference data. However, 147 points mapped as barren land were burned areas (Table 12.1). This implies that the generated output underestimates the actual burned area in the image. The RF model validation using 2016−17 images showed good overall accuracy (90.97%) and Kappa value (0.89). It showed a low user's accuracy for barren land (67.05%) and producer's accuracy for the burned area (61.88%) (Table 12.2). However, the producer's accuracy for the burned area was 87.51%. The classification error matrix shows that approximately 440 points mapped as barren land were burned areas, and 87 points mapped as burned areas were barren land. This indicates that the developed model underestimated the burned areas rather than overestimation.

Three sites were randomly chosen for representation from Manipur, Meghalaya, and Nagaland states (Figs. 12.2−12.7). The Darengre region around the Nokrek National Park in the South Garo Hills District was studied in Meghalaya. The most recent GE imagery for this region was available for 2016 and 2017; thus, the corresponding satellite images and outputs were verified for this period. The 2016 FCC (premonsoon, March) and 2017 FCC

TABLE 12.1 The classification error matrix (validation data points of 2018).

Class	Barren land	Burned area	Cloud	Sand	Shadow	Vegetation	Water body	User's accuracy
Barren land	624	147	1	4	20	1	0	78.29
Burned area	1	689	0	0	0	0	0	99.86
Cloud	0	6	613	1	10	45	5	90.15
Sand	6	15	0	506	0	0	1	95.83
Shadow	0	2	49	0	778	198	28	73.74
Vegetation	0	0	0	0	11	738	0	98.53
Water body	0	0	0	0	4	0	524	99.24
Producer's accuracy	98.89	80.21	92.46	99.02	94.53	75.15	93.91	
Overall accuracy = 88.96%					Kappa value = 0.87			

TABLE 12.2 The classification error matrix developed using the reference data points of 2017.

Class	Barren land	Burned area	Cloud	Sand	Shadow	Vegetation	Water body	User's accuracy
Barren land	999	440	5	44	1	1	0	67.05
Burned area	87	771	21	2	0	0	0	87.51
Cloud	39	15	1688	6	2	55	0	93.52
Sand	1	0	0	862	0	0	0	99.88
Shadow	3	0	44	0	1954	0	3	97.5
Vegetation	103	20	97	0	15	2289	0	90.69
Water body	0	0	1	1	1	0	1587	99.81
Producer's accuracy	81.09	61.88	90.95	94.21	99.04	97.61	99.81	
Overall accuracy = 90.97%					Kappa value = 0.89			

(premonsoon, March) are shown in Fig. 12.2(i) and (ii). The deforested and burned areas from 2016 to 2017 are denoted as locations A–D, respectively. Locations E–G were already deforested and burned before the premonsoon season in 2016 and appeared as light magenta in the FCC image (Fig. 12.2(i) and (ii)). It can be seen that patches A–D were forests in February 2016 in the GE image and March 2016 in the FCC image, which were deforested and burned before January 2017 (GE image), thus appearing dark magenta in March 2017 (FCC image). The identified shifting cultivation patches overlaid on the 2017 FCC, which indicated that the RF model accurately classified the burned area patches; however, there could be some underestimation in the boundary of patches owing to mixed pixels. The input indices and elevation data for this region are shown in Fig. 12.3. The index maps shown in Fig. 12.3 exhibit significant differences between the burned area and other LULC features in NBR, NDVI_D, and NDMI compared to the other indices. In Nagaland, GE images of 2014 and 2017 with FCC images were studied around the Longching region, owing to the unavailability of cloud-free scenes in 2015 and 2016 (Fig. 12.4). The FCC and GE images show that locations A–D were recently burned in 2017, whereas locations E and F were old burned areas. The generated output precisely mapped the burned or shifting cultivation areas in this region (Fig. 12.4(v)). The corresponding index and elevation maps, shown in Fig. 12.5, exhibit significant differences between the burned area and other LULC features in NBR and

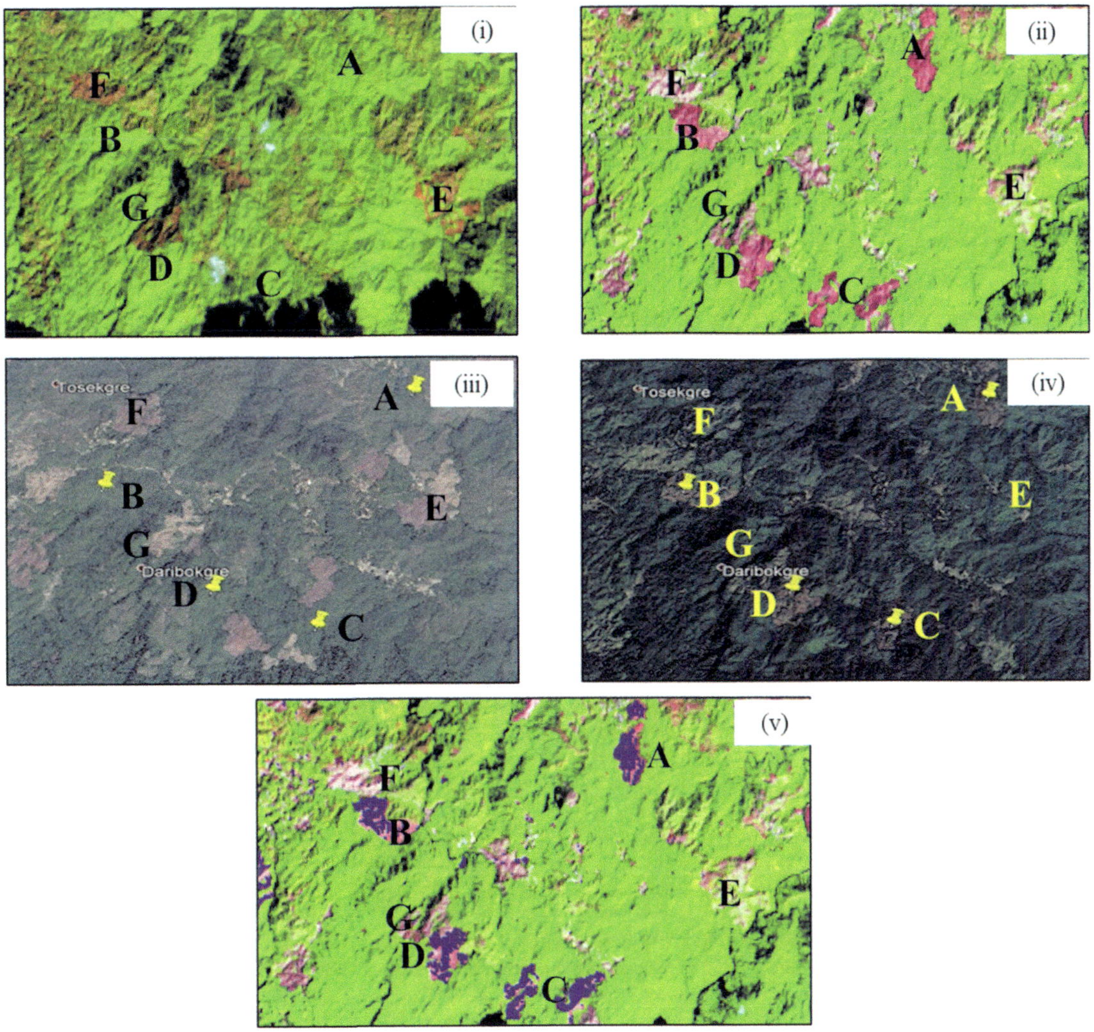

FIGURE 12.2 False color composite (FCC) map of (i) 2016, and (ii) 2017; Google Earth image of (iii) 2016 (February), and (iv) 2017 (January); (v) identified shifting cultivation areas (blue color) overlaid on FCC map of 2017. Map lines delineate study areas and do not necessarily depict accepted national boundaries.

NDMI compared to the other indices. In Manipur state, deforestation was examined near Kombiron village using the available GE images of 2016 and 2017 (Figs. 12.6 and 12.7). The highlighted patch was forest cover in 2016 (GE and FCC images) and deforested in 2017 (GE and FCC images). However, the identified area indicated an underestimation of the patch area, as seen in the FCC and GE images. The corresponding indices and elevation maps indicate a significant difference in the burned area in the NBR and NDMI compared with the other indices (Fig. 12.7). The variable importance of each input layer used in the RF model is shown in Fig. 12.8. The indices with higher importance used in the RF model validate our visual inspection, indicating the dominant contribution of NDMI, elevation, NBR, and NBR 2 than other indicators.

The maximum deforestation was identified in Manipur at 340.27 km^2 in 2017, reduced to 136.76 km^2 in 2018 (Table 12.3, Fig. 12.9). Similarly, the total deforested area identified in Meghalaya was 310.80 km^2 in 2017, which was reduced to only 70.99 km^2 in 2018, indicating a fourfold decrease in the shifting cultivation area. On the contrary, the change was relatively less in Nagaland, where 242.22 km^2 of shifting cultivation area was identified in 2017 and was reduced to 209.20 km^2 in 2018. In Meghalaya, most of the shifting cultivation patches in 2017 were identified in the East Garo and West Khasi hills, followed by the West and East Jaintia hills and Ri-Bhoi districts. In comparison, the deforested patches in 2018 were primarily concentrated in the West, Southwest, and East Khasi Hills, followed by the Ri-Bhoi District. In 2017 the majority of deforestation in Manipur was observed in the northern districts of Senapati and Ukhrul, followed by Churachandpur. However, deforestation was more dispersed in 2018. Similarly, in Nagaland,

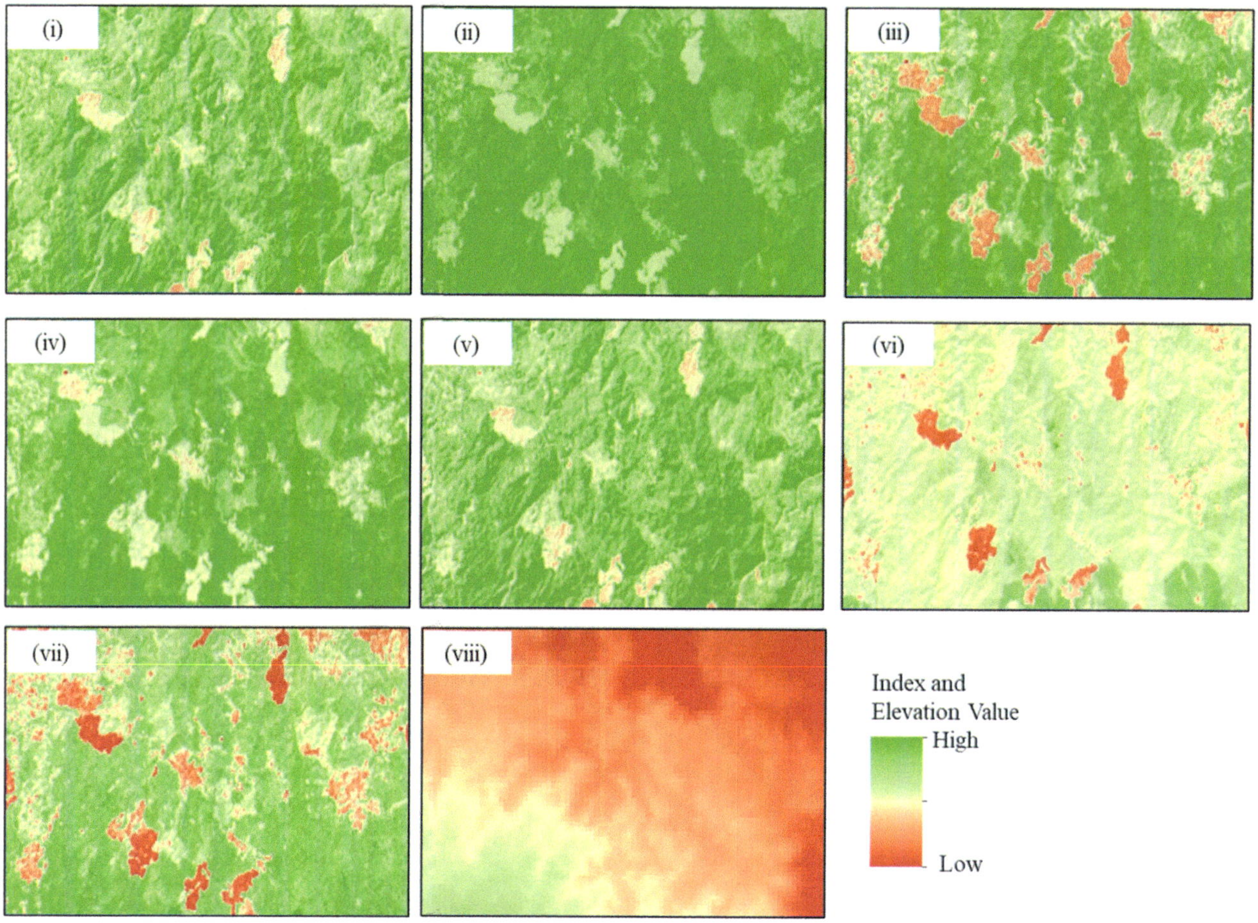

FIGURE 12.3 Maps of (i) EVI, (ii) NDVI, (iii) NBR, (iv) NBR 2, (v) SAVI, (vi) NDVI_D, (vii) NDMI, and (viii) elevation correspondsto the region shown in Fig. 12.2. Map lines delineate study areas and do not necessarily depict accepted national boundaries.

most of the deforestation in 2017 was concentrated in northeastern districts, such as Mon, Tuensand, and Longleng, followed by Phek, which was more dispersed in 2018. The spatial-temporal pattern in shifting cultivation patches identified in the study period indicated that new shifting cultivation patches were primarily created in areas adjacent to past year's shifting cultivation patches. The MODIS data-derived burned area maps of 2017 and 2018 are shown in Fig. 12.10. By comparing the RF model output with MODIS data, it can be inferred that most of the burned area identified by the MODIS data was successfully captured in the Landsat data-derived maps. However, some differences are observed in the Ri-Bhoi district of Meghalaya, which are underestimated by the MODIS data.

The variable importance chart indicated that indices such as NDMI, NBR, NBR 2, and elevation significantly contributed to identifying shifting cultivation areas compared to other input variables. Elevation data significantly reduced the mixing of cropland and barren land at lower altitudes. NDMI identifies the difference in vegetation moisture level, while NBR and NBR 2 are sensitive to charcoal deposition, and altitude (elevation) plays a crucial role in differentiating burned areas. The classification error matrices underestimated the burned areas and class mixing with barren land. The pure pixels of shifting cultivation areas were mainly utilized for training, which may have underestimated the total shifting cultivation area. Moreover, such underestimations are mainly observed around the boundaries of patches or mixed pixels. Long et al. (2019) utilized satellite images from 2014 to 2015 and validated them with the same year's data ground points and obtained a high commission, omission errors, and lowest overall accuracy of 41.23%, 57.41%, and 71.29% for Africa. Mananze, Pôças, and Cunha (2020) classified shifting cultivation areas using Landsat data with the RF algorithm and reported an overall accuracy of ~89%. Das et al. (2021) applied a decision tree-based multistep threshold (DTMT) method to Landsat imagery for shifting cultivation mapping in NEI and reported an accuracy above 85%. Kurien, Lele, and Nagendra (2019) employed multitemporal Landsat images and achieved 80% accuracy in shifting cultivation mapping in the West Garo Hills District, Meghalaya.

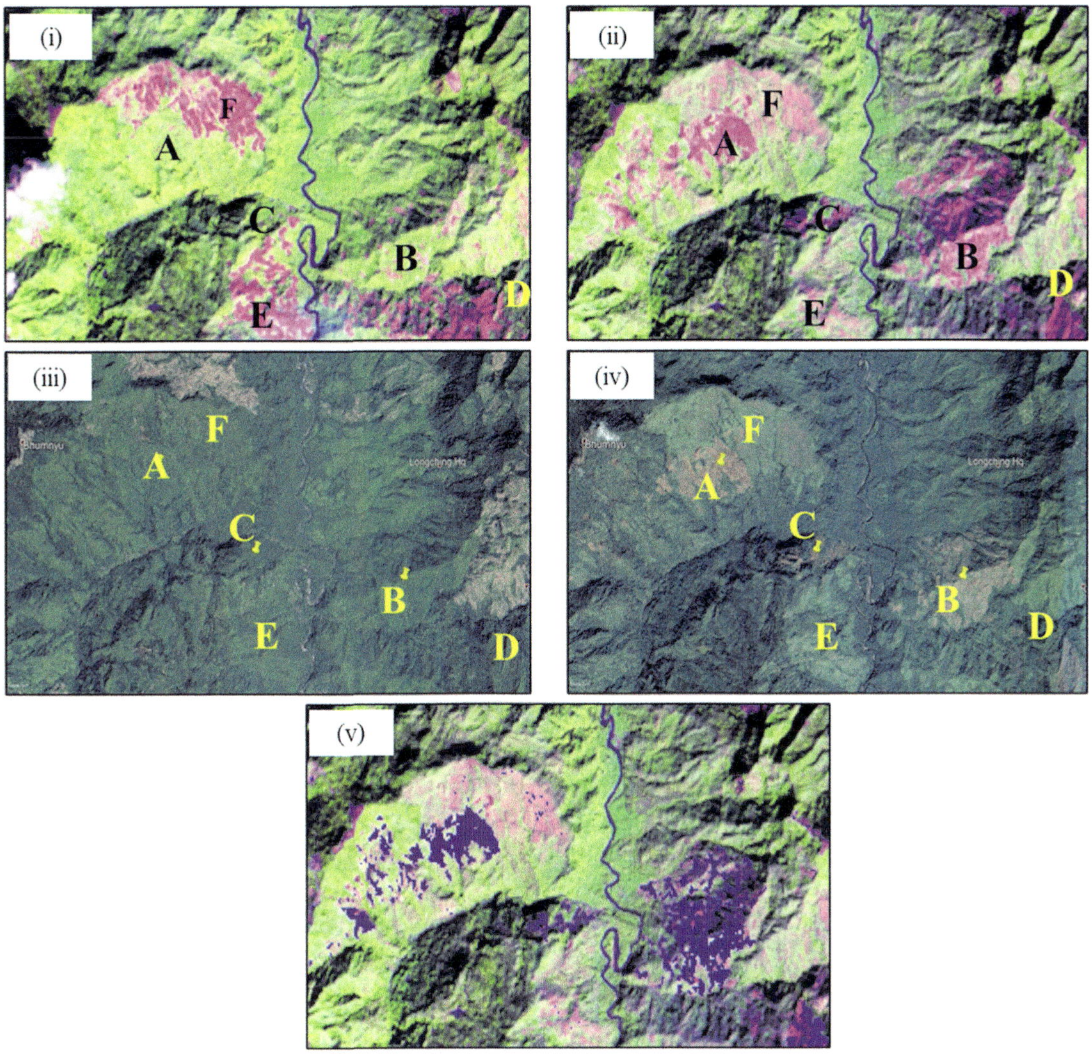

FIGURE 12.4 False color composite (FCC) map of (i) 2014, and (ii) 2017; Google Earth image of (iii) 2014 (October), and (iv) 2017 (November); (v) identified shifting cultivation areas (blue color) overlaid on FCC map of 2017. Map lines delineate study areas and do not necessarily depict accepted national boundaries.

The shifting cultivation areas identified in 2017 and 2018 indicated an overall decrease, supporting past studies in this landscape (Das et al., 2021; Das et al., 2022). However, the differences in area estimated in the current study and previous studies could be attributed to the different methods or approaches, reference data points, and indices used. The recent reduction in shifting cultivation areas could denote the gradual decrease in shifting cultivation practices in NEI and conversion to more permanent land use categories. Moreover, various governmental and nongovernmental initiatives, forest conservation, and management activities can be attributed to the reduction in shifting cultivation areas (Behera, Gupta et al., 2018; Pasha et al., 2020). Thong, Sahoo, Pebam, and Thangjam (2019) studied the rate of change in shifting cultivation areas and patch sizes in Ukhrul and Chandel, Manipur, from 1999 to 2017. They reported a reduction in the total shifting cultivation area and patch size with a jhum cycle between 9 and 14 years. Marak et al. (2023) mapped the shifting cultivation areas using Landsat images in East Garo Hills, Meghalaya, and identified ~16 km^2 of shifting cultivation areas in 2017 and 2018. They also reported a significant decreasing trend in total shifting cultivation areas from 2000 to 2018. The land use practice studied by Duraisami et al. (2022) in Mopungchuket village, Mokokchung district, Nagaland, indicated that less than 3% of the total area was used for shifting cultivation in 2010 and 2020, wherein forest and various successional stages of abandoned shifting cultivation (included scrubland and grassland) occupied ~90% of the land. They also reported an overall reduction in shifting cultivation from 2010 to 2020.

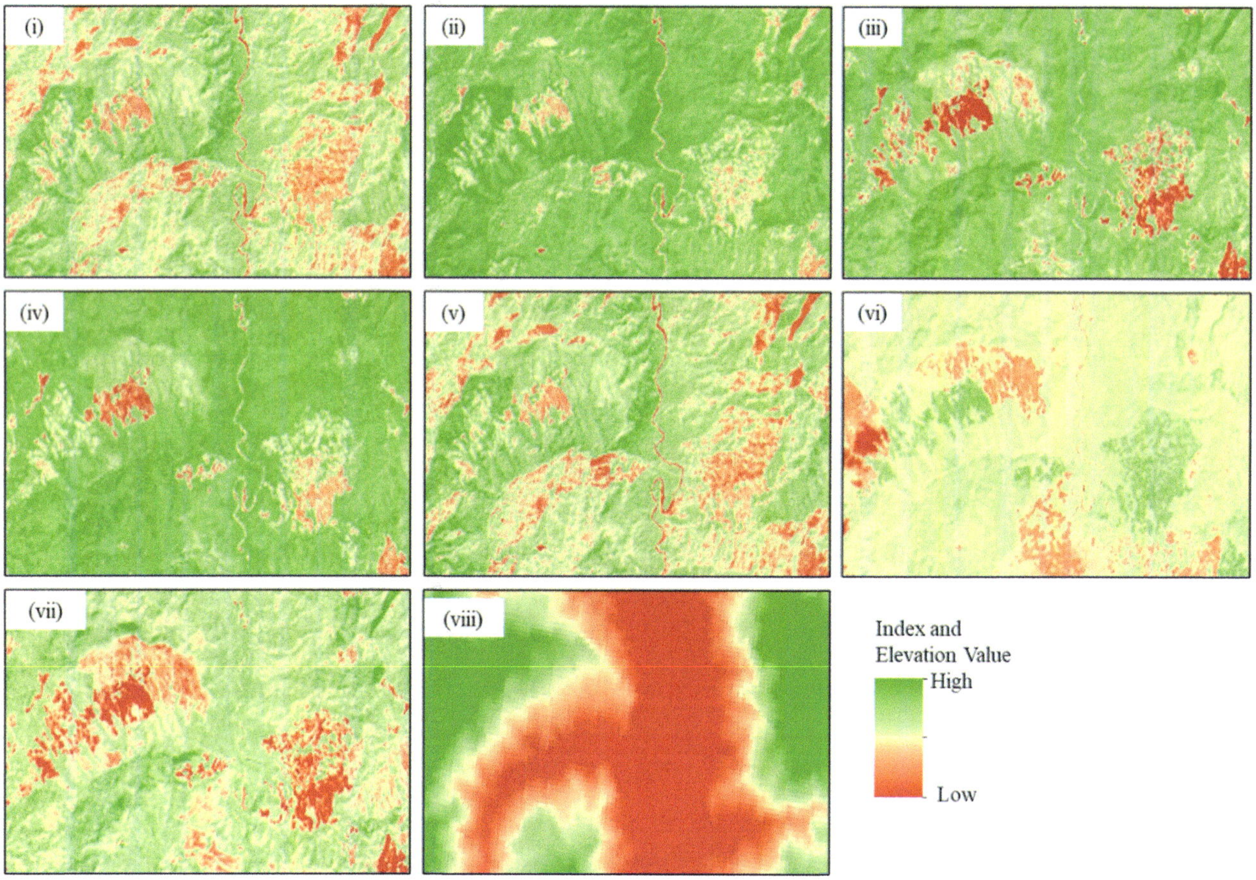

FIGURE 12.5 Maps of (i) EVI, (ii) NDVI, (iii) NBR, (iv) NBR 2, (v) SAVI, (vi) NDVI_D, (vii) NDMI, (viii) elevation correspond to the region shown in Fig. 12.4. Map lines delineate study areas and do not necessarily depict accepted national boundaries.

12.5 Way forward

The geospatial maps on forest resources have a high value in multidisciplinary use for forest conservation and management, monitoring forest structure biodiversity, ecosystem services, carbon budgeting, land use planning, agriculture, water resource management, etc. The shifting cultivation practices cause a significant change in a landscape each year, which is alternatively classified as forest and croplands based on the present land use. Thus systematic monitoring of shifting cultivation and fallow period length is essential for sustainable landscape management and developing suitable policies. Various efforts have been made to convert shifting cultivation practices to permanent agriculture, including agroforestry interventions and to reduce the dependency on shifting cultivation practices (Behera, Gupta et al., 2018; NITI Aayog, 2018). Incentivization for adopting alternative land use practices, for example, payment for ecosystem services (PES) for agroforestry, participation in restoring degraded and fallow lands, and trade of carbon credits may further reduce the shifting cultivation practices in NEI (Nath et al., 2020). A significant proportion of shifting cultivation land in Tripura state has been converted into agroforestry and rubber plantations, areca nut, cashew nut, oil palm, tea and coffee cultivation, etc. (Behera, Gupta et al., 2018; Das et al., 2022; Verma, Kumar, Chandra, & Thounaojam, 2017). Recognizing the positive outcome of promoting home gardens, NITI Aayog (2018) has recommended the inclusion of home gardens as a land use category in national LULC monitoring and allocating funds for such interventions in NEI. The tropical environmental conditions offer a high potential for rapid natural vegetation regeneration during the fallow stages in this landscape (Das & Behera, 2019; Das, Behera, Bhaskaran, & Roy, 2022). In addition to the native species, invasive species, such as *Lantana camara*, were recorded as a widely grown species in fallow shifting cultivation areas in NEI (Pasha et al., 2020; Verma et al., 2017). Thus assisted natural regeneration and farmer-managed natural regeneration via financial incentives could help in better landscape management in shifting cultivation areas. Various capacity-building programs and the promotion of increasing fallow period lengths can reduce soil erosion, replenish soil fertility better than in a shorter jhum cycle, and result in higher agronomic yield (Nath et al.,

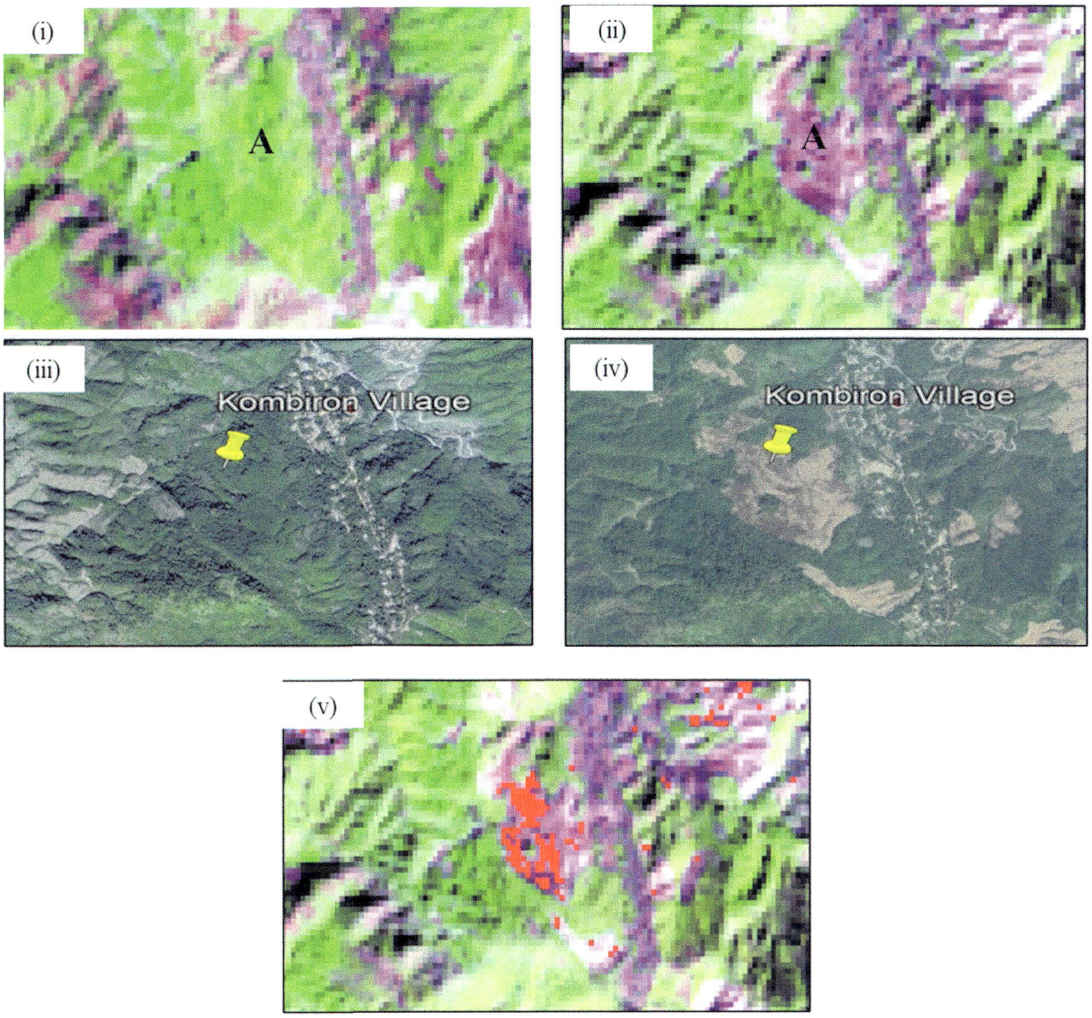

FIGURE 12.6 False color composite (FCC) map of (i) 2016, and (ii) 2017; Google Earth image of (iii) 2016 (August), and (iv) 2017 (February); (v) identified shifting cultivation areas (blue color) overlaid on FCC map of 2017. Map lines delineate study areas and do not necessarily depict accepted national boundaries.

2020). The fallow period duration directly contributes to the diversity of species recovery and aboveground plant biomass/carbon stock, and it may require around four decades to match the carbon stock in the undisturbed forests (Borah, Gilroy, Evans, & Edwards, 2022; Thong, Sahoo, Thangjam, & Pebam, 2020).

The identified burned area map in this study at a state level would provide a holistic view of the land use practice and shifting cultivation dynamics with the framework of reducing emissions from deforestation and degradation plus (REDD +) and achieving multiple sustainable development goals (SDGs) related to food security and climate change mitigation. Integration of the developed RF model into the GEE platform will generate the shifting cultivation map at an annual scale with minimal effort, reducing the cost and time of data downloading, computing facilities, processing time, and data storage devices. The adopted approach can be applied to other northeast Indian states for rapid and automated shifting cultivation area mapping. However, future research may include the 10 m resolution Sentinel-2 data and 5.8 m LISS-IV data in periodic mapping of shifting cultivation, fallow period length, change in land use practice (e.g., conversion to other land use categories such as agroforestry and permanent cropland), vegetation type, and tree growth mapping in fallow lands. The application of Sentinel-2 data will enable the identification of smaller patches in the topographically complex terrain and a higher chance of cloud-free image availability in the public domain. The red-edge spectral bands of Sentinel-2 data and high-temporal revisit time could improve vegetation type mapping compared to Landsat data. Moreover, the Sentinel-1 microwave remote sensing-based approach could be a suitable alternative for LULC and change mapping, aboveground plant biomass estimation, and vegetation growth monitoring even in cloudy

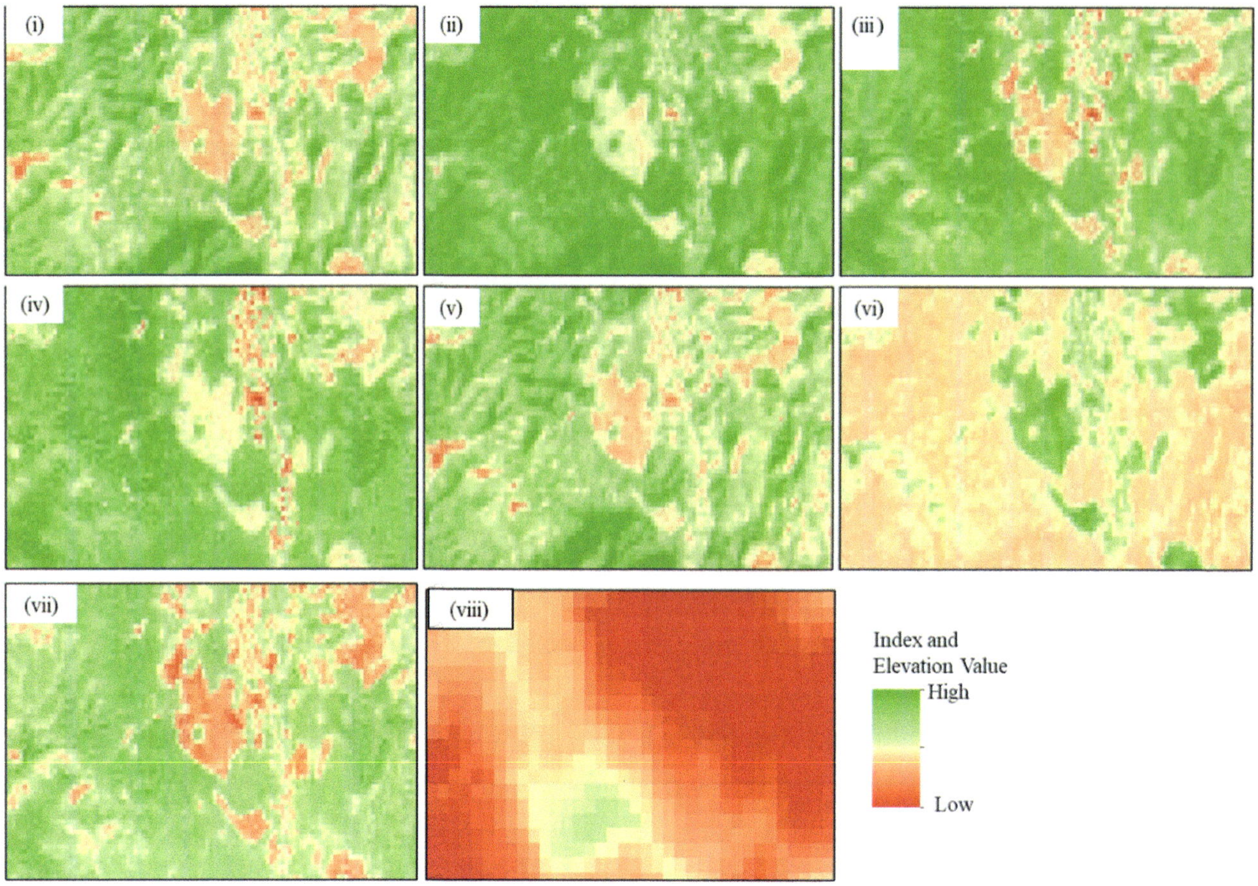

FIGURE 12.7 Maps of (i) EVI, (ii) NDVI, (iii) NBR, (iv) NBR 2, (v) SAVI, (vi) NDVI_D, (vii) NDMI, (viii) elevation correspond to the region shown in Fig. 12.6. Map lines delineate study areas and do not necessarily depict accepted national boundaries.

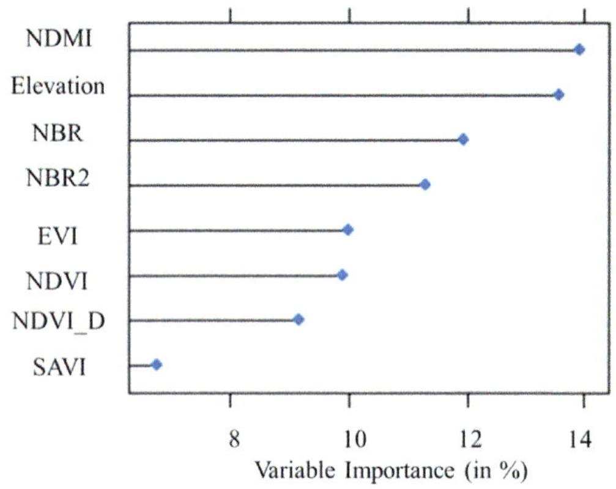

FIGURE 12.8 Variable importance plot indicating the relative contribution of each variable in identifying the shifting cultivation areas.

TABLE 12.3 The overall mapping accuracy and area statistics for three states.

State	Year	Area (km^2)
Manipur	2017	340.27
	2018	136.76
Meghalaya	2017	310.80
	2018	70.99
Nagaland	2017	242.22
	2018	209.20

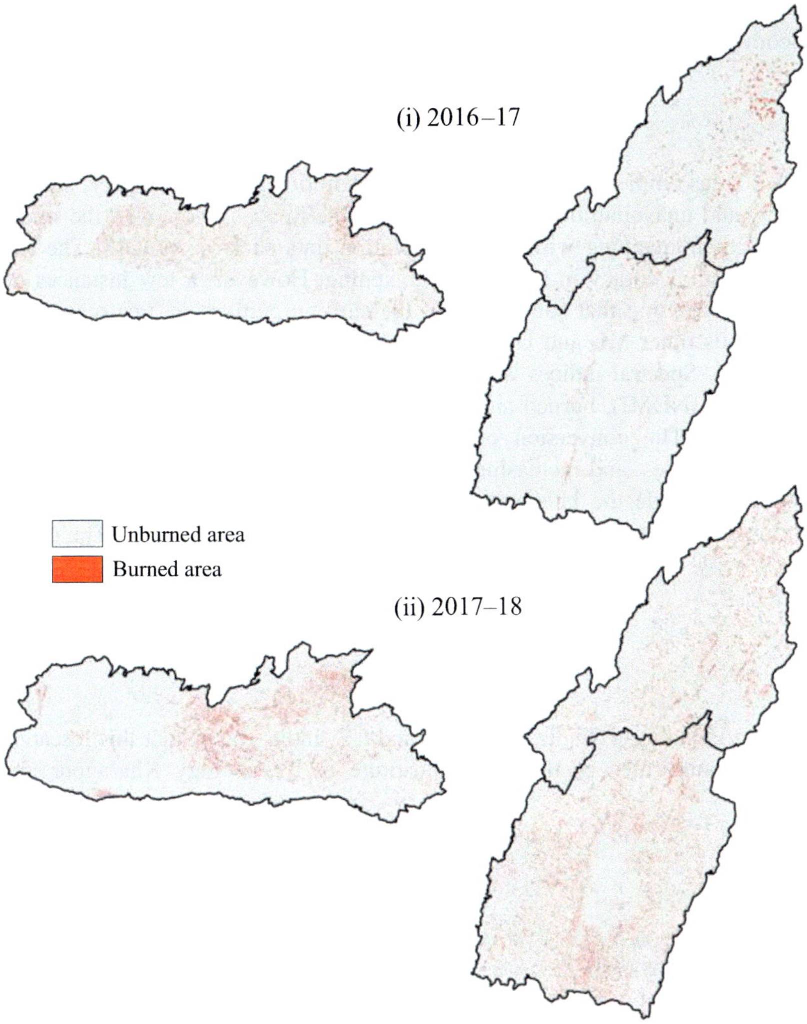

FIGURE 12.9 Deforested and burned areas identified in Manipur, Meghalaya,and Nagaland during (i) 2016–17, and (ii) 2017–18. Map lines delineate study areas and do not necessarily depict accepted national boundaries.

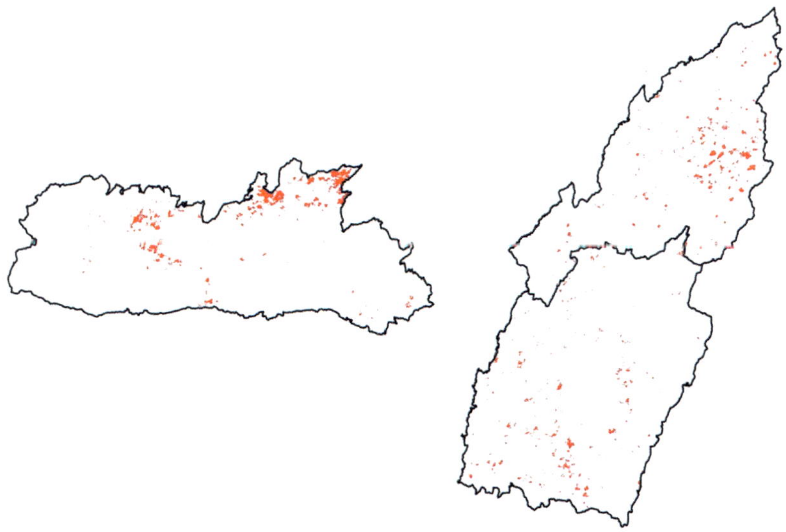

FIGURE 12.10 MODIS data-derived burned areas during 2017 and 2018 (MODIS product: MCD64A1 version 06). Map lines delineate study areas and do not necessarily depict accepted national boundaries.

conditions, which are limited in optical remote sensing data (Ghosh et al., 2020). Future studies may include multiple other ML and deep learning algorithms (e.g., convolution neural network) for shifting cultivation mapping.

12.6 Conclusions

Accurately and systematically identifying burned areas employing moderate-to-fine-resolution satellite imagery is challenging in NEI because of the rugged topography and unavailability of cloud-free satellite images. Moreover, the smaller size of the patch area imposes constraints on their mapping with coarse-resolution data such as MODIS. The RF algorithm used in the current study demonstrated high accuracy in burned area mapping. However, a few instances of misclassification were observed with barren land, indicating that some parts of the shifting cultivation area may have been underestimated. Future studies may use various other ML and DL models with extensive and multiyear training datasets, which may improve classification accuracy. Spectral indices were used as proxy variables in shifting cultivation mapping, where the indices of moisture content (NDMI), burned area (NBR and NBR2), and topography (elevation) were found to be the dominant contributors. The conversion of shifting cultivation areas to agroforestry, incentivizing farmers to preserve forest ecosystem services, and increasing the fallow period length may better protect the forest ecosystem in NEI. Future research may include the latest satellite data, from higher-resolution optical to microwave data and AI techniques for monitoring consistent shifting cultivation and allied land use practices. The geospatial shifting cultivation maps can help in sustainable land management and forest conservation and provide essential inputs with policy-level implications.

Acknowledgments

Financial assistance from the "National Mission on Himalayan Studies," MOEF, and CC, India, to conduct this research is acknowledged. The facilities provided by the authorities of the Indian Institute of Technology Kharagpur are acknowledged.

References

Alajangi, S., Prasad, N. S. R., & Baruah, A. (2021). *Monitoring and changing pattern of shifting cultivation and reclamation in hilly regions using Geospatial Technology, . Modern cartography series* (Vol. 10, pp. 465–495). Academic Press.

Barik, S. K., & Mishra, S. K. (2008). Assessment of the contribution of forests to the economy of the northeastern states of India. *International Forestry Review, 10*(2), 349–361.

Behera, M. D., Gupta, A. K., Barik, S. K., Das, P., & Panda, R. M. (2018). Use of satellite remote sensing as a monitoring tool for land and water resources development activities in an Indian tropical site. *Environmental Monitoring and Assessment, 190*, 1–21.

Behera, M. D., Tripathi, P., Das, P., Srivastava, S. K., Roy, P. S., Joshi, C., ... Krishnamurthy, Y. V. N. (2018). Remote sensing based deforestation analysis in Mahanadi and Brahmaputra river basin in India since 1985. *Journal of Environmental Management, 206*, 1192–1203. Available from https://doi.org/10.1016/j.jenvman.2017.10.015.

Bhat, Y., Nandy, S., Padalia, H., Nath, A. J., & Pebam, R. (2022). Role of geospatial technology in shifting cultivation studies in India: A systematic review. *Journal of the Indian Society of Remote Sensing, 50*(12), 2359–2379.

Borah, J. R., Gilroy, J. J., Evans, K. L., & Edwards, D. P. (2022). The value of shifting cultivation for biodiversity in Northeast India. *Diversity and Distributions, 28*(9), 1979–1992.

Breiman, L. (2001). Random forests. *Machine Learning, 45*, 5–32.

Chanu, L. K., De, A., Chakraborty, K., & Paul, S. (2023). Characterization of shifting cultivation, trends, and diversification of livelihood patterns: A case study from forest villages in Barak valley, Assam, Northeast India. *Trees, Forests and People, 14*, 100420.

Das, P., & Behera, M. D. (2019). Can the forest cover in India withstand large climate alterations? *Biodiversity and Conservation, 28*(8), 2017–2033.

Das, P., Behera, M. D., Barik, S. K., Mudi, S., Jagadish, B., Sarkar, S., & Chauhan, P. S. (2022). Shifting cultivation induced burn area dynamics using ensemble approach in Northeast India. *Trees, Forests and People, 7*, 100183.

Das, P., Behera, M. D., Bhaskaran, P. K., & Roy, P. S. (2022). Forest cover resilience to climate change over India using the MC2 dynamic vegetation model. *Environmental Monitoring and Assessment, 194*(12), 903.

Das, P., Behera, M. D., & Murthy, M. S. R. (2017). Forest fragmentation and human population varies logarithmically along elevation gradient in Hindu Kush Himalaya-utility of geospatial tools and free data set. *Journal of Mountain Science, 14*(12), 2432–2447.

Das, P., Mudi, S., Behera, M. D., Barik, S. K., Mishra, D. R., & Roy, P. S. (2021). Automated mapping for long-term analysis of shifting cultivation in Northeast India. *Remote Sensing, 13*(6), 1066.

Das, P., Panda, R. M., Dash, P., Jana, A., Jana, A., Ray, D., & Kolluru, V. (2022). Multi-decadal mapping and climate modelling indicates eastward rubber plantation expansion in India. *Sustainability, 14*(13), 7923.

Das, P., & Pandey, V. (2019). Use of logistic regression in land-cover classification with moderate-resolution multispectral data. *Journal of the Indian Society of Remote Sensing*, *47*(8), 1443–1454.

Duraisami, M., Das, P., Segarin, S., Singh, R., Jayahari, K. M., & Balasubramanian, K. (2022). *Opportunity for forest protection and landscape restoration in the Indian Himalayan Region, . Handbook of Himalayan ecosystems and sustainability* (Vol. 1, pp. 341–368). CRC Press.

Escuin, S., Navarro, R., & Fernández, P. (2008). Fire severity assessment by using NBR (normalized burn ratio) and NDVI (normalized difference vegetation index) derived from LANDSAT TM/ETM images. *International Journal of Remote Sensing*, *29*(4), 1053–1073.

Ghosh, S. M., Behera, M. D., & Paramanik, S. (2020). Canopy height estimation using sentinel series images through machine learning models in a mangrove forest. *Remote Sensing*, *12*(9), 1519.

Gislason, P. O., Benediktsson, J. A., & Sveinsson, J. R. (2006). Random forests for land cover classification. *Pattern Recognition Letters*, *27*(4), 294–300.

Grogan, P., Lalnunmawia, F., & Tripathi, S. K. (2012). Shifting cultivation in steeply sloped regions: A review of management options and research priorities for Mizoram state, Northeast India. *Agroforestry Systems*, *84*, 163–177.

Gupta, A. K. (2000). Shifting cultivation and conservation of biological diversity in Tripura, Northeast India. *Human Ecology*, *28*, 605–629.

Hansen, M. C., Potapov, P. V., Moore, R., Hancher, M., Turubanova, S. A., Tyukavina, A., & Townshend, J. (2013). High-resolution global maps of 21st-century forest cover change. *Science (New York, N.Y.)*, *342*(6160), 850–853.

Hartter, J., Lucas, C., Gaughan, A. E., & Aranda, L. L. (2008). Detecting tropical dry forest succession in a shifting cultivation mosaic of the Yucatán Peninsula, Mexico. *Applied Geography*, *28*(2), 134–149.

Huete, A. R. (1988). A soil-adjusted vegetation index (SAVI). *Remote Sensing of Environment*, *25*, 295–309.

Hurni, K., Hett, C., Heinimann, A., Messerli, P., & Wiesmann, U. (2013). Dynamics of shifting cultivation landscapes in Northern Lao PDR between 2000 and 2009 based on an analysis of MODIS time series and Landsat images. *Human Ecology*, *41*, 21–36.

ISFR. (2021). *India state of forest report 2021*. Dehradun, India: Forest Survey of India.

Jiang, N., Li, P., & Feng, Z. (2022). Remote sensing of swidden agriculture in the tropics: A review. *International Journal of Applied Earth Observation and Geoinformation*, *112*, 102876.

Key, C., & Benson, N. (2006). Landscape assessment: Ground measure of severity, the composite burn index; and remote sensing of severity, the normalized burn ratio, FIREMON: Fire effects monitoring and inventory system. U.S. Department of Agriculture, Forest Service, Rocky Mountain Research Station. p. LA-1-55.

Kumar, A. (2022). Impact of shifting cultivation on green infrastructure: A remote sensing perspective. *International Journal of Ecology and Environmental Sciences*, *48*(3), 345–356.

Kurien, A. J., Lele, S., & Nagendra, H. (2019). Farms or forests? Understanding and mapping shifting cultivation using the case study of West Garo hills, India. *Land*, *8*(9), 133.

Lasaponara, R. (2006). Estimating spectral separability of satellite derived parameters for burned areas mapping in the Calabria region by using SPOT-Vegetation data. *Ecological Modelling*, *196*(1-2), 265–270.

Lele, N., Joshi, P. K., & Agrawal, S. P. (2008). Assessing forest fragmentation in northeastern region (NER) of India using landscape matrices. *Ecological Indicators*, *8*(5), 657–663.

Li, P., & Yang, Y. (2022). Swidden agriculture landscape mapping using MODIS vegetation index time series and its spatio-temporal dynamics in northern Laos. *Remote Sensing*, *14*(23), 6173.

Liu, H. Q., & Huete, A. (1995). A feedback based modification of the NDVI to minimize canopy background and atmospheric noise. *IEEE Transactions on Geoscience and Remote Sensing*, *33*(2), 457–465.

Long, T., Zhang, Z., He, G., Jiao, W., Tang, C., Wu, B., & Yin, R. (2019). 30 m resolution global annual burned area mapping based on Landsat Images and Google Earth Engine. *Remote Sensing*, *11*(5), 489.

Lutes, D. C., Keane, R. E., Caratti, J. F., Key, C. H., Benson, N. C., Sutherland, S., & Gangi, L. J. (2006). FIREMON: Fire effects monitoring and inventory system. Gen. Tech. Rep. RMRS-GTR-164. Fort Collins, CO: US Department of Agriculture, Forest Service, Rocky Mountain Research Station. 1 CD., 164.

Mananze, S., Pôças, I., & Cunha, M. (2020). Mapping and assessing the dynamics of shifting agricultural landscapes using google earth engine cloud computing, a case study in Mozambique. *Remote Sensing*, *12*(8), 1279.

Marak, S. R., Kumar, M., Kundu, A., Dutta, D., Pebam, R., Chattoraj, K. K., & Lal, D. (2023). *Shifting cultivation in the East Garo Hills, Meghalaya (India): An Earth observation perspective. Water, land, and forest susceptibility and sustainability* (pp. 439–458). Academic Press.

Mathur, I., & Bhattacharya, P. (2023). From pixels to patterns: Review of remote sensing techniques for mapping shifting cultivation systems. *Spatial Information Research*, 1–11.

Molinario, G., Hansen, M. C., & Potapov, P. V. (2015). Forest cover dynamics of shifting cultivation in the Democratic Republic of Congo: A remote sensing-based assessment for 2000–2010. *Environmental Research Letters*, *10*(9), 094009.

Murphy, K. A., Reynolds, J. H., & Koltun, J. M. (2008). Evaluating the ability of the differenced Normalized Burn Ratio (dNBR) to predict ecologically significant burn severity in Alaskan boreal forests. *International Journal of Wildland Fire*, *17*(4), 490–499.

Nath, A. J., Sahoo, U. K., Giri, K., Sileshi, G. W., & Das, A. K. (2020). Incentivizing hill farmers for promoting agroforestry as an alternative to shifting cultivation in Northeast India. *Agroforestry for Degraded Landscapes: Recent Advances and Emerging Challenges*, *1*, 425–444, Vol.

NITI Aayog. (2018). Report of working group III shifting cultivation: Towards a transformational approach, New Delhi.

Pasha, S. V., Behera, M. D., Mahawar, S. K., Barik, S. K., & Joshi, S. R. (2020). Assessment of shifting cultivation fallows in Northeastern India using Landsat imageries. *Tropical Ecology*, *61*, 65–75.

Potapov, P., Li, X., Hernandez-Serna, A., Tyukavina, A., Hansen, M. C., Kommareddy, A., & Hofton, M. (2021). Mapping global forest canopy height through integration of GEDI and Landsat data. *Remote Sensing of Environment*, *253*, 112165.

Ramakrishnan, P. S. (2007). Traditional forest knowledge and sustainable forestry: A north-east India perspective. *Forest Ecology and Management*, *249*(1-2), 91–99.

Ranjan, R., & Upadhyay, V. P. (1999). Ecological problems due to shifting cultivation. *Current Science*, *77*(10), 1246–1250.

Reddy, C. S., Jha, C. S., Diwakar, P. G., & Dadhwal, V. K. (2015). Nationwide classification of forest types of India using remote sensing and GIS. *Environmental Monitoring and Assessment*, *187*, 1–30.

Roy, P. S., Behera, M. D., Murthy, M. S. R., Roy, A., Singh, S., Kushwaha, S. P. S., & Ramachandran, R. M. (2015). New vegetation type map of India prepared using satellite remote sensing: Comparison with global vegetation maps and utilities. *International Journal of Applied Earth Observation and Geoinformation*, *39*, 142–159.

Roy, P. S., Kaul, R. N., Sharma., Roy, M. R., & Garbyal, S. S. (1985). Forest-type stratification and delineation of shifting cultivation areas in the eastern part of Arunachal Pradesh using LANDSAT MSS data. *International Journal of Remote Sensing*, *6*(3-4), 411–418.

Roy, P. S., Roy, A., Joshi, P. K., Kale, M. P., Srivastava, V. K., Srivastava, S. K., & Kushwaha, D. (2015). Development of decadal (1985–1995–2005) land use and land cover database for India. *Remote Sensing*, *7*(3), 2401–2430.

Roy, S., Mudi, S., Das, P., Ghosh, S., Shit, P. K., Bhunia, G. S., & Kim, J. (2021). Estimating above ground biomass (agb) and tree density using sentinel-1 data. *Spatial Modeling in Forest Resources Management: Rural Livelihood and Sustainable Development*, 259–280.

Schneibel, A., Stellmes, M., Röder, A., Frantz, D., Kowalski, B., Haß, E., & Hill, J. (2017). Assessment of spatio-temporal changes of smallholder cultivation patterns in the Angolan Miombo belt using segmentation of Landsat time series. *Remote Sensing of Environment*, *195*, 118–129.

Shankar, U., & Tripathi, R. (1996). Evaluating second year cropping on jhum fallows in Mizoram, northeastern India—phytomass dynamics and primary productivity. *Journal of Biosciences*, *21*, 563–575.

Shaw, A. K., Natori, Y., & Edake, S. (2022). *A Tradition in transition: Understanding the role of shifting cultivation for sustainable development of North East India*. New Delhi: TERI.

Singh, R. K., Biradar, C. M., Behera, M. D., Prakash, A. J., Das, P., Mohanta, M. R., & Rizvi, J. (2024). Optimising carbon fixation through agroforestry: Estimation of aboveground biomass using multi-sensor data synergy and machine learning. *Ecological Informatics*, *79*, 102408.

Stroppiana, D., Boschetti, M., Zaffaroni, P., & Brivio, P. A. (2009). Analysis and interpretation of spectral indices for soft multicriteria burned-area mapping in mediterranean regions. *IEEE Geoscience and Remote Sensing Letters*, *6*(3), 499–503.

Tansey, K., Beston, J., Hoscilo, A., Page, S. E., & Paredes Hernández, C. U. (2008). Relationship between MODIS fire hot spot count and burned area in a degraded tropical peat swamp forest in Central Kalimantan, Indonesia. *Journal of Geophysical Research: Atmospheres*, *113*(D23).

Thong, P., Sahoo, U. K., Pebam, R., & Thangjam, U. (2019). Spatial and temporal dynamics of shifting cultivation in Manipur, Northeast India based on time-series satellite data. *Remote Sensing Applications: Society and Environment*, *14*, 126–137.

Thong, P., Sahoo, U. K., Thangjam, U., & Pebam, R. (2020). Pattern of forest recovery and carbon stock following shifting cultivation in Manipur, North-East India. *PLoS ONE*, *15*(10), e0239906.

Toky, O. P., & Ramakrishnan, P. S. (1981). Cropping and yields in agricultural systems of the north-eastern hill region of India. *Agro-ecosystems*, *7* (1), 11–25.

Verma, P. K., Kumar, V., Chandra, A., & Thounaojam, B. (2017). Alternatives of Shifting Cultivation in North-eastern region of India. *Report and Opinion*, *9*(12), 1–8.

Chapter 13

Role of phenology as discriminant in vegetation type mapping

A. Kurein[1], M.R. Sahu[1], Mukunda Dev Behera[1], A.J. Prakash[1], M. Mukhopadhyay[1], Soumit K. Behera[2], Saroj Kanta Barik[3], S. Gowda[4], P.C. Gogineni[4], S.K. Biswal[4] and Sudam C. Sahu[5]

[1]*Centre for Ocean, River, Atmosphere and Land Sciences (CORAL), Indian Institute of Technology Kharagpur, Kharagpur, West Bengal, India,* [2]*CSIR-National Botanical Research Institute, Lucknow, Uttar Pradesh, India,* [3]*Centre for Advanced Studies in Botany, North-Eastern Hill University, Shillong, India,* [4]*Odisha State Wildlife Department, Baripada, Odisha, India,* [5]*Department of Botany, Maharaja Sriram Chandra Bhanja Deo University, Baripada, Odisha, India*

13.1 Introduction

Vegetation is one of the primary constituents of an ecosystem which reflects the influence of environmental conditions in a quantifiable manner. Understanding the type of vegetation is crucial for effectively characterizing the landscape in terms of its structure and function (Behera, Kushwaha, & Roy, 2001). The classification and mapping of vegetation types is essential for the management of natural resources since they profoundly impact all living organisms and playsa pivotal role in shaping the global climate and terrestrial carbon cycle (Sala et al., 2000; Xiao et al., 2004). These classification schemes encompass all structural and functional aspects of vegetation in relation to regional and local climatic conditions, necessitating the inclusion of a multitude of spatially explicit inputs (Reddy, Jha, Diwakar, & Dadhwal, 2015). Despite the hierarchical organization of biological systems from the molecular to ecosystem levels, the variability in species, communities, and ecosystem distribution adds complexity to vegetation classification (Margules & Pressey, 2000). Mapping vegetation types also yields invaluable insights into the distribution of natural and human-made systems, continuously quantifying vegetation cover across local to global scales at any given time (Paramanik, Rajdeep, Behera, Bhattacharya, & Dash, 2023). Understanding the distribution of vegetation types is essential for national-level planning concerning food security, wildlife habitats, sustainable natural resource management, agroforestry, and biodiversity conservation, particularly in hotspots (Patidar & Behera, 2019). It also aids in planning protected areas and developing forest corridors. Accurate assessment of the current state of vegetation cover is crucial for developing programs to protect and restore vegetation (Egbert et al., 2002; Myers, Mittermeier, Mittermeier, Da Fonseca, & Kent, 2000). The extended lifespan of trees makes forest vegetation particularly sensitive to climate change as it limits their ability to adapt rapidly (Das, Behera, & Murthy, 2017; Das, Behera, Roy, & Barik, 2024). Land use change is projected to be the foremost driver impacting biodiversity on a global scale by 2100 when averaged across biomes. This is primarily because of its adverse effects on habitat availability and species extinction (Sala et al., 2000).

Satellite remote sensing is an efficient and economical way to map vegetation types and monitor changes in synoptic coverage (Navalgund, Jayaraman, & Roy, 2007). Accurate, geo-referenced vegetation-type data at the global, continental, and regional levels are indispensable for research on global change and modeling the Earth's system (Roy et al., 2015). Satellite sensor data offer a synoptic view of Earth, enhancing the quality, internal consistency, and reproducibility of global land and vegetation cover information. Remote sensing has been instrumental in advancing vegetation mapping and enhancing our primary understanding of terrestrial ecosystems, by elucidating the relationships between reflectance and vegetation structure (Roy et al., 2015). The reason for including topographic variables in single-date forest mapping is their correlation with species or life forms (Guisan & Zimmermann, 2000). Multitemporal remote sensing is pivotal for mapping various vegetation types (Singh et al., 2024). Analyzing time series of satellite imagery enables the integration of diverse features and spectral–temporal metrics to capture seasonal and phenological characteristics of different land use and land cover (LULC) classes (JayaPrakash et al., 2023). Applying these features and

Sustainable Development Perspectives in Earth Observation. DOI: https://doi.org/10.1016/B978-0-443-14072-3.00009-5

metrics to LULC classification significantly improves accuracy (Luo, Zhang, Wang, Men, & Liu, 2022; Zhou, Zhu, Xian, & Li, 2022).

Landsat-8 and Sentinel-2 (S-2) products are prominent among medium-resolution satellite imagery because of their high temporal resolution, short revisit time, and rich spectral configuration, making them ideal sources for time-series feature extraction (Kowalski, Okujeni, Brell, & Hostert, 2022). The standout feature of S-2 is its high spatial resolution and revisit time of five days, along with the availability of narrow red-edge bands compared to current multispectral missions (Drusch et al., 2012). Three methods are generally utilized to extract time-series features and manage missing data: (1) time-series composition, (2) spectral–temporal metrics, and (3) phenological metrics (Hermosilla, Wulder, White, Coops, & Hobart, 2018). Composition methods aggregate all images into a single image encompassing a specified timeframe (annual, seasonal, etc.). Each pixel in the resultant image denotes the reflectance of all pixels from the original image collection according to the defined rules. The primary advantage of this approach is the mitigation of atmospheric effects such as cloud and snow masking (Pflugmacher, Rabe, Peters, & Hostert, 2019). Spectral–temporal metrics yield statistical spectral information of all pixels over a temporal scale, which can include metrics based on mean, standard deviation, maximum, minimum, or percentiles of spectral information during the defined period (White, Hermosilla, Wulder, & Coops, 2022). Phenological metrics are derived from the periodicity of land cover classes, such as the start, peak, and end dates of the growing season based on change in spectral trajectory which effectively capture the phenological variation of different LULC classes or vegetation types (Salinero-Delgado et al., 2022).

Multiseason images (December, March, and May) from LISS III with a 23.5 m spatial resolution were used as the base for mapping forests and other land-use features in the Similipal Biosphere Reserve (SBR) in a previous study (Saranya & Reddy, 2016). In the present study, S-2 time-series data in conjunction with Digital Elevation Model (DEM), slope and spectral indices are used to delineate vegetation types in the study area and evaluate the performance of different data combinations in enhancing the accuracy of LULC mapping.

13.2 Materials and methods

13.2.1 Study area

The SBR, situated in the Mayurbhanj district of Odisha, India, extends between 21° 28′ and 22° 15′ North and 86° 03′ and 86° 37′ East (Fig. 13.1). Encompassing an area of approximately 4500 km^2, it comprises a core zone spanning 845 km^2 and a buffer zone spanning 1905 km^2 (Biswal, Mukherjee, Jeyaram, & Murthy, 2012). The false color composite (FCC) and Shuttle Radar Topography Mission (SRTM) derived DEM, including the core and buffer zones, are shown in Fig. 13.1; the region experiences a tropical monsoonal climate characterized by three distinct seasons: summer (March–May), monsoon (June–October), and winter (November–February). Premonsoon showers occur in May and June, whereas postmonsoon showers are experienced mostly in November and December.

Natural vegetation encompasses around 65.6% of the total geographical area (Saranya & Reddy, 2016). In semievergreen forests (SEFs), the dominant species in the top canopy include *Michelia champaca,Syzygium operculatum, Dillenia pentagyna,Syzygium cumini,* and *Litsea glutinosa* (Reddy, Pattanaik, Mohapatra, & Biswal, 2007). Moist deciduous forests (MDFs), which constitute a significant portion of vegetation cover, are primarily characterized by species such as *S. robusta, Terminalia alata, Anogeissus latifolia, Hadinia cordifolia* and *Protium serratum.* The dry deciduous forests (DDFs) are primarily composed of *S. robusta, T. alata, Buchanania lanzan, Madhuca indica,* and *Diospyros melanoxylon* (Saranya & Reddy, 2016). The SBR, a mega biodiversity zone in India, was introduced into the Man and Biosphere Programme of UNESCO in May 2009.

13.2.2 Datasets

13.2.2.1 Sentinel-2

S-2 multispectral instrument (MSI; Level-1C) data, obtained from the European Space Agency's (ESA) S-2 mission as part of the Copernicus Programme, was utilized for vegetation-type mapping. The mission is characterized by its wide swath coverage, fine spatial resolution ranging from 10 to 60 m, multispectral features comprising 13 spectral bands, and high revisit frequency. These attributes enable effective monitoring of vegetation changes throughout the growing season, forest monitoring, and detection of land cover changes. The multispectral bands used in this study include blue (B2), green (B3), red (B4), Red Edge 1 (B5), Red Edge 2 (B6), Red Edge 3 (B7), near-infrared (NIR) (B8), Red Edge 4

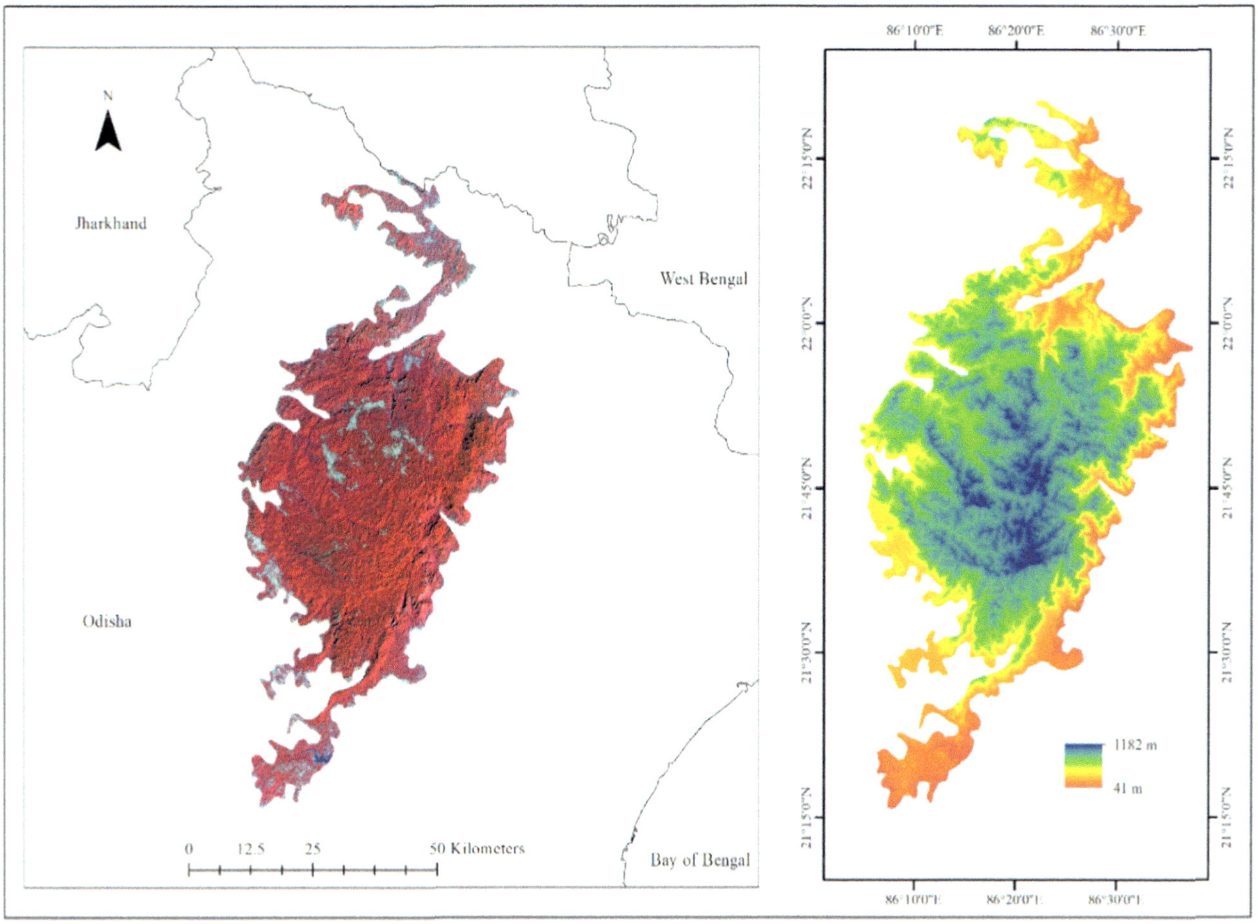

FIGURE 13.1 Location of the study area, Similipal Biosphere Reserve, Mayurbhanj district, Odisha, India; left-hand panel shows a false color composite image of the Biosphere Reserve, while the right-hand panel shows a digital elevation model wherein the elevation of the Biosphere Reserve varies between 41 and 1182 m. Map lines delineate study areas and do not necessarily depict accepted national boundaries.

(B8A), short-wave infrared (SWIR) 1 (B11), and SWIR 2 (B12). Additionally, 13 spectral indices were computed from the leaf-off season image for March 2023 (Table 13.1).

13.2.2.1.1 Normalized difference vegetation index

Normalized difference vegetation index (NDVI) is expressed as the normalized ratio between the NIR and red band reflectances. The measurements in the red and NIR spectra represent the spectral radiance (or reflectance) obtained by the sensors. Reflectance measures the proportion of the radiation reflected by an object to the radiation received. NDVI values, ranging from -1 to 1, remained consistent regardless of whether radiance, reflectance, or digital number (DN) were utilized as inputs. Negative NDVI values typically correlate with water bodies; values near zero are common for nonvegetated surfaces, such as rocks, sands, or concrete, while positive values denote vegetation, encompassing crops, shrubs, grasses, and forests (Jones and Vaughan, 2010). Higher NDVI values signify stronger indications of vigorous vegetation greenness (Fig. 13.2A).

13.2.2.1.2 Enhanced vegetation index

The enhanced vegetation index (EVI) is a modified NDVI that incorporates a soil adjustment factor (L) and two coefficients (C1 and C2) to account for atmospheric aerosol scattering by utilizing the blue band to correct the red band. EVI offers enhanced sensitivity for detecting high-biomass areas while minimizing atmospheric interference. The value ranges from -1 to 1, similar to the NDVI. Higher positive EVI values indicate healthier and more vigorous vegetation, whereas lower or negative values suggest less healthy or sparse vegetation cover (Fig. 13.2B).

TABLE 13.1 Spectral indices utilized in the study.

Indices	Formula	References
Normalized difference vegetation index (NDVI)	NIR − Red/NIR + Red	Rouse, Haas, Schell, & Deering (1973)
Red-edge NDVI (NDVI Red edge)	NIR − Red edge/NIR + Red edge	Gitelson & Merzlyak (1994)
Enhanced vegetation index (EVI)	2.5* [NIR − Red/NIR + 6* Red − 7.5* Blue + 1]	Huete, Liu, Batchily, & Van Leeuwen (1997)
Soil-adjusted vegetation index (SAVI)	(NIR − Red/NIR + Red + 0.5) * 1.5	Huete (1988)
Green chlorophyll index (CI green)	(NIR/green) − 1	Gitelson, Gritz, & Merzlyak (2003)
Red-edge chlorophyll index (CI Red edge)	(NIR/Red edge) − 1	Gitelson, Gritz, & Merzlyak (2003)
Normalized difference moisture index (NDMI)	NIR − SWIR1/NIR + SWIR1	Wilson & Sader (2002)
Modified normalized difference water index (MNDWI)	Green − SWIR1/Green + SWIR1	Xu (2006)
Normalized difference built-up index (NDBI)	SWIR1 − NIR/SWIR1 + NIR	Zha, Gao, & Ni (2003)
Normalized difference tillage index (NDTI)	SWIR1 − SWIR2/SWIR1 + SWIR2	Van Deventer, Ward, Gowda, & Lyon (1997)
Tasseled cap brightness (TCB)	(0.3037 * Blue + 0.2793 * Green + 0.4743 * Red + 0.5585 *NIR + 0.5082 * SWIR1 + 0.1863 * SWIR2)	Crist & Cicone (1984)
Tasseled cap wetness (TCW)	(0.1509 * Blue + 0.1973 * Green + 0.3279 * Red + 0.3406 *NIR + 0.7112 * SWIR1 + 0.4572 * SWIR2)	Crist & Cicone (1984)
Tasseled cap greenness (TCG)	(− 0.2848 * Blue − 0.2435 * Green − 0.5436 * Red + 0.7243 *NIR + 0.0840 * SWIR1 − 0.1800 * SWIR2)	Crist & Cicone (1984)

NIR, Near-infrared; *SWIR*, short-wave infrared.

13.2.2.1.3 Soil-adjusted vegetation index

The soil-adjusted vegetation index (SAVI) is designed to alleviate the impact of the soil background on vegetation signals by integrating a constant soil adjustment factor (L) into the denominator of the NDVI equation. The choice of L is contingent upon the reflective attributes of the soil, including its color and brightness. The vegetation density being assessed determines the specific value of the L factor 1.0 for areas with sparse vegetation, 0.5 for intermediate, and 0.25 for high density. This value typically ranges from −1 to 1 (Fig. 13.2C).

13.2.2.1.4 Normalized difference moisture index

Normalized difference moisture index (NDMI) values range from −1 to 1, with higher positive values indicative of increased moisture levels, often correlating with robust vegetative growth and adequate water availability (Fig. 13.2D). Conversely, lower or negative NDMI values signify diminished moisture content, implying drought stress or aridity. By leveraging its quantitative range, the NDMI facilitates the precise monitoring of soil moisture dynamics, drought severity, and vegetative health across diverse landscapes (Fig. 13.2D).

13.2.2.1.5 Red-edge normalized difference vegetation index

The red-edge normalized difference vegetation index (NDVI Red edge) incorporates spectral information from the red-edge region of the electromagnetic spectrum, typically between red and NIR wavelengths. This index is designed to enhance the sensitivity of the NDVI to changes in vegetation health and stress, particularly in agricultural and ecological applications. By utilizing the red-edge band, which corresponds to the transition zone between chlorophyll absorption and leaf cell structure reflectance, Red edge NDVI offers improved discrimination between vegetation types and conditions compared with traditional NDVI (Fig. 13.2E).

13.2.2.1.6 Modified normalized difference water index

Modified normalized difference water index (MNDWI) often yields higher values for surface water than NDWI because water bodies exhibit greater absorption in the SWIR band than in the NIR band. Conversely, objects such as soil, plants, or earthen constructions typically produce smaller (often negative) values in MNDWI because they primarily reflect light in SWIR rather than green. Its values typically range from -1 to 1, with higher positive values indicating the presence of water and lower or negative values suggesting land or built-up areas (Fig. 13.2F).

13.2.2.1.7 Red-edge chlorophyll index

The red-edge chlorophyll index (RECI) is specifically designed to estimate the chlorophyll content in vegetation using spectral information from the red edge and the NIR region of the electromagnetic spectrum. The RECI was formulated to quantify the relationship between the chlorophyll concentration and reflectance in the red-edge region. Higher RECI values indicated higher chlorophyll content, typically associated with healthy and vigorously growing vegetation (Fig. 13.2G). Conversely, lower RECI values suggest lower chlorophyll concentrations, which may indicate stress or senescence in the vegetation. The advantage of RECI lies in its sensitivity to subtle changes in chlorophyll concentrations, making it valuable for monitoring vegetation health, growth, and stress responses.

13.2.2.1.8 Green chlorophyll index

The green chlorophyll index (GCI) is a vegetation index used in remote sensing to assess the chlorophyll content of vegetation. Unlike traditional indices such as NDVI, which rely on NIR and red spectral bands, GCI utilizes green and NIR bands to estimate the chlorophyll concentration. High GCI values indicate higher chlorophyll content, typically observed in healthy and vigorous vegetation, whereas lower values suggest lower chlorophyll levels, indicative of stressed or senescent vegetation (Fig. 13.2H). The GCI is particularly useful in dense forests or regions with abundant vegetation cover.

13.2.2.1.9 Normalized difference built-up index

The normalized difference built-up index (NDBI) is used to delineate built-up areas and is calculated as the normalized difference between the NIR and SWIR bands. NDBI values typically range from -1 to 1, with higher positive values indicating the presence of built-up or impervious surfaces, such as buildings, roads, and pavements, while lower or negative values suggest non-built-up areas, such as vegetation or bare soil. NDBI exploits the differential reflectance properties of built-up surfaces and the natural features in the NIR and SWIR spectral bands. Built-up areas tend to exhibit higher reflectance in the SWIR band owing to their thermal properties and low moisture content, whereas natural features such as vegetation have higher reflectance in the NIR band. By subtracting NIR from SWIR reflectance and normalizing the results, NDBI enhances the contrast between built-up and non-built-up surfaces.

13.2.2.1.10 Normalized difference tillage index

The normalized difference tillage index (NDTI) is utilized to detect and assess the extent of tillage or soil disturbance in agricultural fields. It delineates areas with minimal or no disturbance to those with significant soil disturbance or tillage activity. The specific bands used in the NDTI calculation can vary depending on the sensor and its application. SWIR1 and SWIR2 are used for S-2.

13.2.2.1.11 Tasseled cap brightness

Tasseled cap transformation (TCT) is a remote-sensing method used to transform satellite imagery into a set of orthogonal components that represent specific landscape spectral and spatial characteristics. One of these components is brightness, which is one of the three primary components generated by TCT, along with greenness and wetness.

Brightness in the context of TCT represents the overall level of reflectance or radiance across the landscape. It combines information from multiple spectral bands to capture variations in surface properties such as soil, vegetation, and

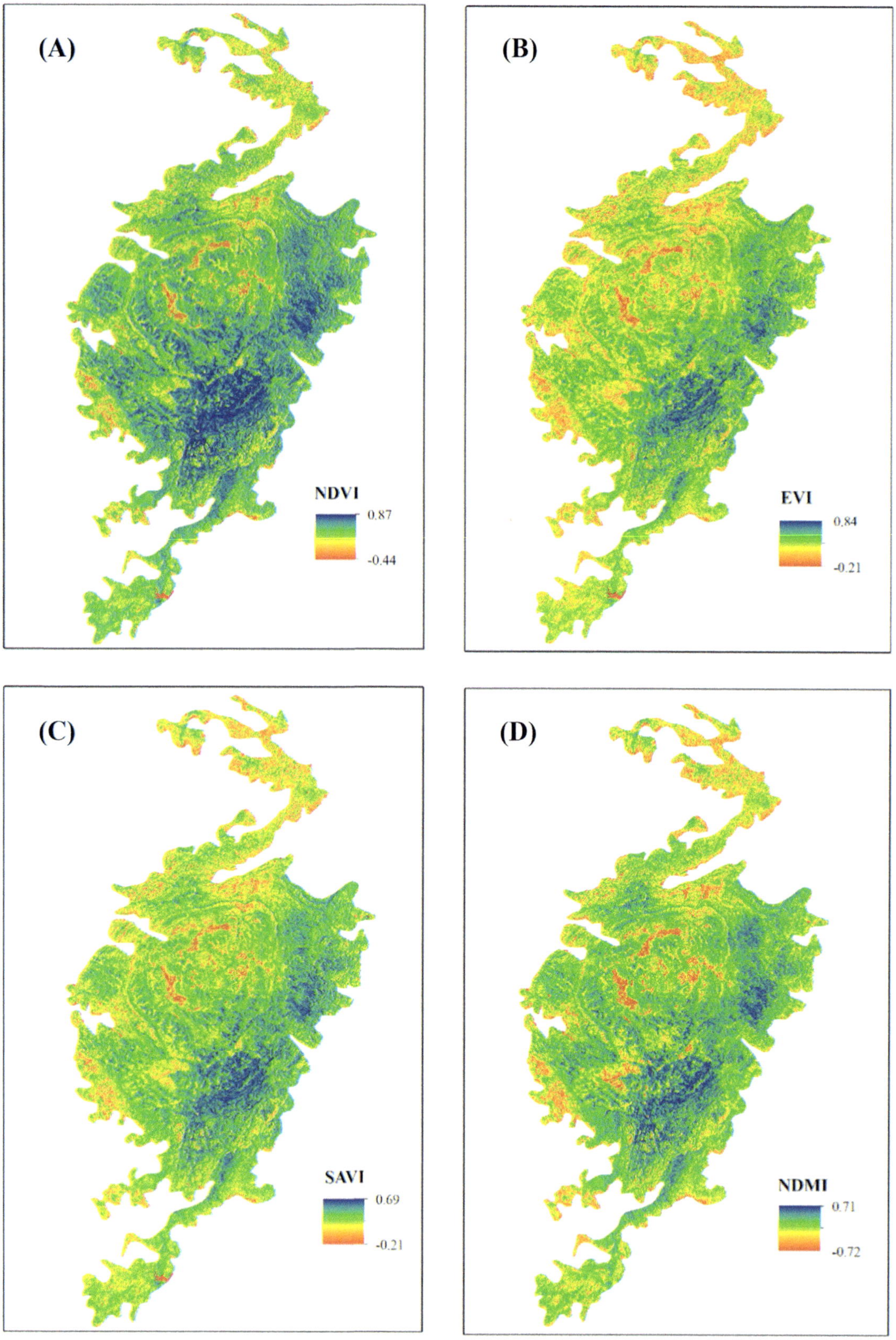

FIGURE 13.2 (A) NDVI, (B) EVI, (C) SAVI, and (D) NDMI images of SBR. (E) Red-edge NDVI, (F) MNDWI, (G) Red-edge CI, and (H) GCI images of SBR. Map lines delineate study areas and do not necessarily depict accepted national boundaries.

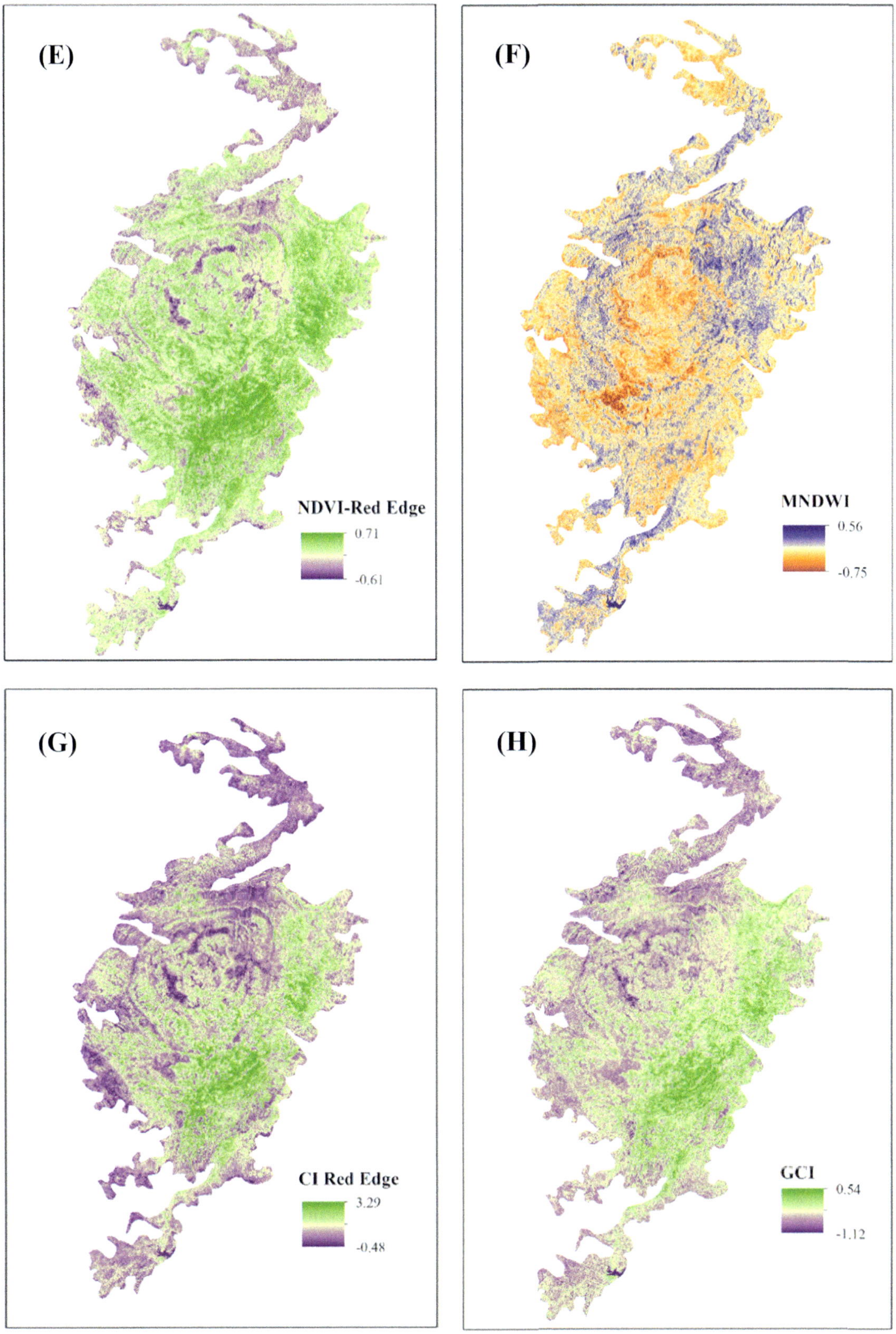

FIGURE 13.2 (Continued)

built-up areas. Higher brightness values indicate areas with higher reflectance or radiance, which may correspond to features such as bare soil, urban areas, or bright vegetation.

13.2.2.1.12 Tasseled cap wetness

Tasseled cap wetness (TCW) captures the degree of moisture content or wetness across a landscape. It combines information from multiple spectral bands to identify areas with higher moisture content, such as water bodies, wetlands, and areas with dense vegetation that retain moisture. Higher values of TCW indicate wetter conditions, whereas lower values suggest drier areas.

13.2.2.1.13 Tasseled cap greenness

Tasseled cap greenness (TCG) quantifies the vigor and health of vegetation across the landscape. It combines information from multiple spectral bands to identify dense and healthy vegetation areas. Higher values of TCG indicate areas with lush vegetation cover, whereas lower values suggest sparse or stressed vegetation.

13.2.2.2 Topographic variables

The elevation and slope were derived from the SRTM, DEM, with a resolution of 1 arcsecond (approximately 30 m). This high-resolution dataset is a product of collaboration among the National Aeronautics and Space Administration (NASA), National Geospatial-Intelligence Agency (NGA), and Italian and German space agencies.

13.2.3 Methodology

The study utilizes the S-2 image collection (L2A surface reflectance) spanning from January to December 2023, with cloud cover less than 10%, acquired via the Google Earth Engine (GEE) platform. The images are generated after applying quality bit masks to identify cloud and cirrus pixels within the quality assessment band.

To explore various data combinations for generating vegetation-type maps, four approaches are tested: (1) single image from the leaf-off season (spectral bands only), (2) single image along with topographic variables and spectral indices, (3) S-2 time series (January–December 2023), and (4) S-2 time series along with topographic variables. May–June and November–early January were considered the start and end of the growing season for the dominant vegetation types in the study area (Fig. 13.1). The phenological difference between the vegetation types is observed most during the summer period (March–early April) when sal and tropical sal-mixed DDFs remained leafless with no evidence of foliage, whereas tropical sal-mixed MDFs appeared in bright red-brown tinged tones in FCC images (Reddy et al., 2015). Tropical SEFs have a predominance of evergreen broad-leaved trees, and the canopy is never without green foliage. The FCC image shows these forests with a bright red tone or pinkish red with a medium to rough texture across all seasons (Reddy et al., 2015). Barren lands and settlements mostly appear in cyan shades, water in blue tones, and agriculture fields in pink shades.

A survey is carried out by reviewing the published literature to collect data on the distribution of vegetation types in the study area. The Random Forest classifier is trained using samples collected by leveraging existing LULC maps, high-resolution imagery from Google Earth, and FCC images of the S-2 time series. The phenological disparity between the dominant vegetation types is most pronounced in the images from the leaf-off season (March–April). The study focused on mapping the major vegetation types in the study area, which included tropical sal-mixed MDF, tropical sal-mixed DDF, tropical SEF, and sal. Other land cover classes include scrub, barren land, water bodies, settlements, and agricultural lands (Fig. 13.3).

13.2.3.1 Random Forest

Random Forest (RF) is an ensemble machine learning algorithm that improves prediction and classification accuracy without overfitting the data (Breiman, 2001). Classification and regression trees are the foundation of the RF. The reference dataset is randomly split into training (70%) and validation (30%) datasets to develop and validate the RF classification model, respectively. The training dataset is used to train the RF model, whereas the validation dataset is used for accuracy assessment (Adam, Mutanga, Odindi, & Abdel-Rahman, 2014). The 70/30 split proportion is common in machine learning, balancing model development, and evaluation (Dobbin & Simon, 2011).

Model accuracy assessment is conducted using the validation dataset by employing standard performance metrics. These metrics include the error matrix, overall accuracy (OA), producer accuracy (PA), user accuracy (UA), and kappa

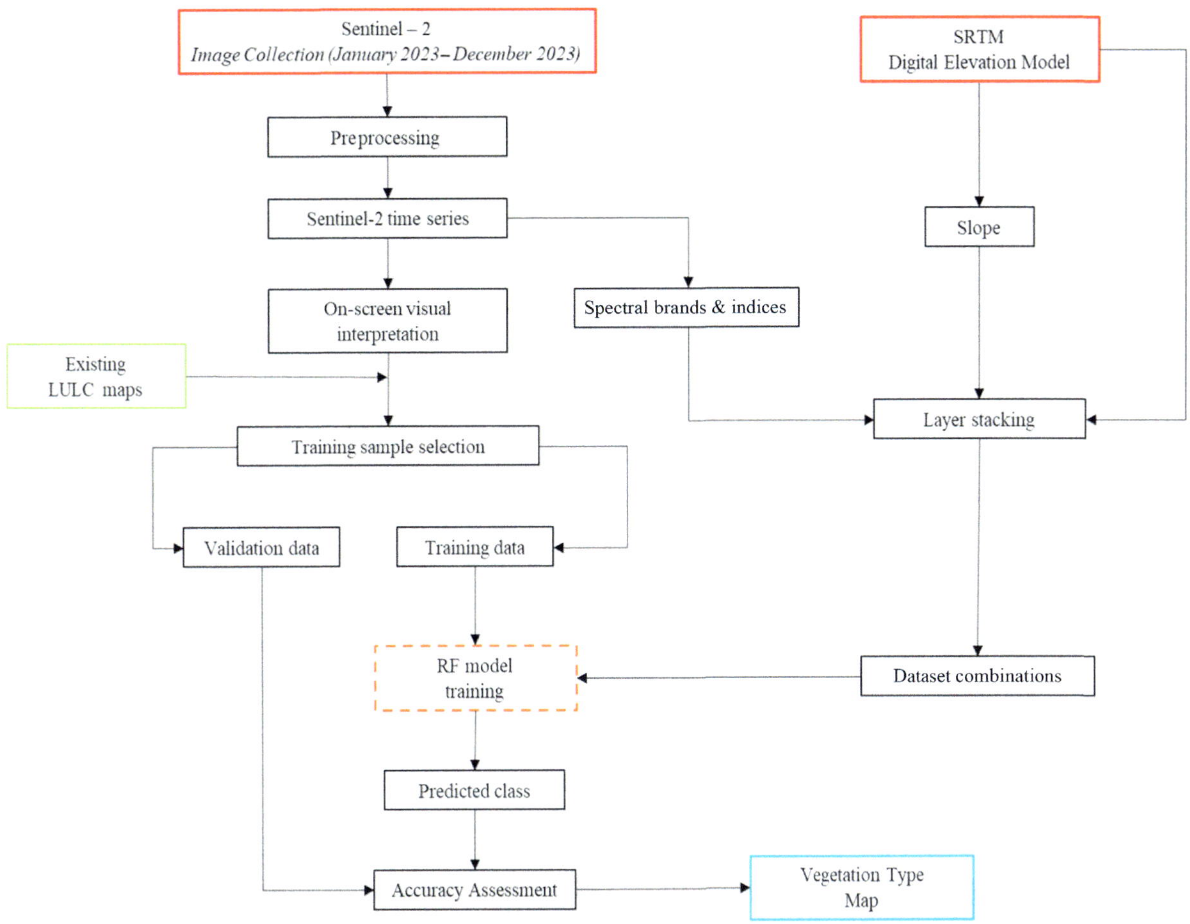

FIGURE 13.3 Methodology flowchart. Map lines delineate study areas and do not necessarily depict accepted national boundaries.

(k) coefficient (Congalton, 1991). The error matrix quantitatively compares the similarities between the samples classified by the model and validation data. OA represents the proportion of correctly classified samples relative to their input labels, while the k-coefficient provides a more robust measure of performance, mitigating the tendency of OA to overestimate the actual performance (Schulz, Holtgrave, & Kleinschmit, 2021). PA indicates the proportion of correctly classified samples within a specific category, whereas UA represents the proportion of correctly classified samples out of the total samples classified into that category (Olofsson, Foody, Stehman, & Woodcock, 2013). Furthermore, variable importance analysis is conducted to assess the significance of each of the predictor variables in the optimal RF model. This statistical measure quantifies the weight of each predictor variable contributing to the model's predictive performance, aiding in the interpretation of model outcomes and identifying influential predictors.

LULC mapping for the nine classes is carried out by randomly splitting the reference samples into training (70%) and validation (30%) datasets. The training and validation data applied for classifying and assessing the accuracy of the land-cover maps were consistent across the four datasets. Alongside the spectral bands of the S-2 images, additional variables such as spectral indices and topographic variables (DEM and slope) are included to test their potential contribution to map accuracy.

13.3 Results

13.3.1 Overall accuracy with and without auxiliary variables across datasets

All datasets considered could generate vegetation-type maps with an OA exceeding 86%. The classification accuracy achieved using a single image during the leaf-off season is 86.6%. Upon incorporating spectral indices and topographic

TABLE 13.2 Overall accuracy (%) and kappa coefficient from different data combinations.

Data	Overall accuracy (OA) %	Kappa coefficient (k)
Dataset 1 (D1) *(Single image – leaf-off season)*	86.6	0.84
Dataset 2 (D2) *(D1 + spectral indices + topographic variables)*	94.3	0.93
Dataset 3 (D3) *(S-2 time series)*	96.3	0.95
Dataset 4 (D4) *(D3 + topographic variables)*	97.1	0.96

variables along with spectral bands, the accuracy increases to 94.3%, showing a notable improvement of 7.7%. Dataset 3, which included S-2 time-series images covering various time windows, such as the start of the growing season, end of the growing season, and leaf-off season, approaches an accuracy of 96.3%. The addition of topographic variables (DEM and slope) to Dataset 3 further boosted the OA. The highest accuracy was observed with S-2 time-series data combined with topographic variables, achieving 97.1% OA, and k value was 0.96 (Table 13.2).

Among the variables, the blue band, SWIR1, and SWIR2 stood out as the most significant in Dataset 1. Elevation consistently ranked highest in importance across all combinations involving variables beyond the spectral bands. Dataset 2 highlighted elevation, slope, and NDTI as significant factors. This signifies the role of NDTI in effectively delineating agricultural fields from barren land. Dataset 4, which attained the highest OA, emphasized the importance of spectral bands from November, December, March, April, and May, underscoring the significance of the growing season time frame, duration of growth, and leaf-shedding period in providing spectral cues for vegetation classification. The high variable importance score of the GCI in Dataset 2 signifies the role of the GCI in delineating vegetation types during the dry season (Fig. 13.4).

13.3.2 Effect of different dataset combinations on vegetation-type classification accuracy

The major vegetation types in the study area are identified and mapped, and the results indicate that the tropical sal-mixed MDF dominates SBR, followed by tropical sal-mixed DDF, SEF, and sal, which is in agreement with the vegetation-type survey carried out in Simlipal by Reddy et al. (2007).

The UA and PA for almost all classes show significant improvement while transitioning the training dataset from a single image to time-series data (Table 13.3). Tropical sal-mixed DDF and sal forests are known for their unique seasonal pattern of partial-to-complete leaf loss during the dry season (typically from February to April) (Reddy et al., 2015). Considering these notable phenological dynamics within the major vegetation types in the study area, time-series data proved instrumental in capturing and leveraging such temporal nuances.

The vegetation-type maps produced using datasets D1 and D2 exhibit distinct differences compared to those generated using D3 and D4. Due to the dominance of vegetation in the area, the most pronounced disparities in the classified maps are observed for dry deciduous, sal, and SEFs. One of the most interesting findings is the noticeable shift in the classification accuracy, based on a single image (D1) to multitemporal images (D3), emphasizing the importance of incorporating phenological cycles across different forest types, as depicted in the classified outputs. Furthermore, the significant role of topographic variables like elevation in enhancing the accuracy of vegetation-type mapping, along with other spectral information from time-series data, is evident from the results (Fig. 13.5).

Similarly, a significant enhancement in classification accuracy is observed while delineating agriculture, barren lands, scrubs, and settlements. This indicates that the improvement in accuracy due to the utilization of time-series data extends beyond forest types to encompass nonforest classes. The RF model distinguishes between agriculture and barren land when multitemporal data, instead of a single satellite image, is used along with topographical variables. D4 exhibits higher OA, UA, and PA compared to D2, which utilized spectral bands, indices, and topographic variables (Table 13.4).

From the classified map that attained the highest OA, the areas covered by various vegetation types in the study area were calculated. Tropical sal-mixed MDF covered the major area among the forest types (1602.6 km^2), followed by tropical sal-mixed DDF (609.8 km^2), tropical SEF (231.3 km^2), and sal (80.3 km^2), which is consistent with the findings of Saranya & Reddy (2016) (Fig. 13.6).

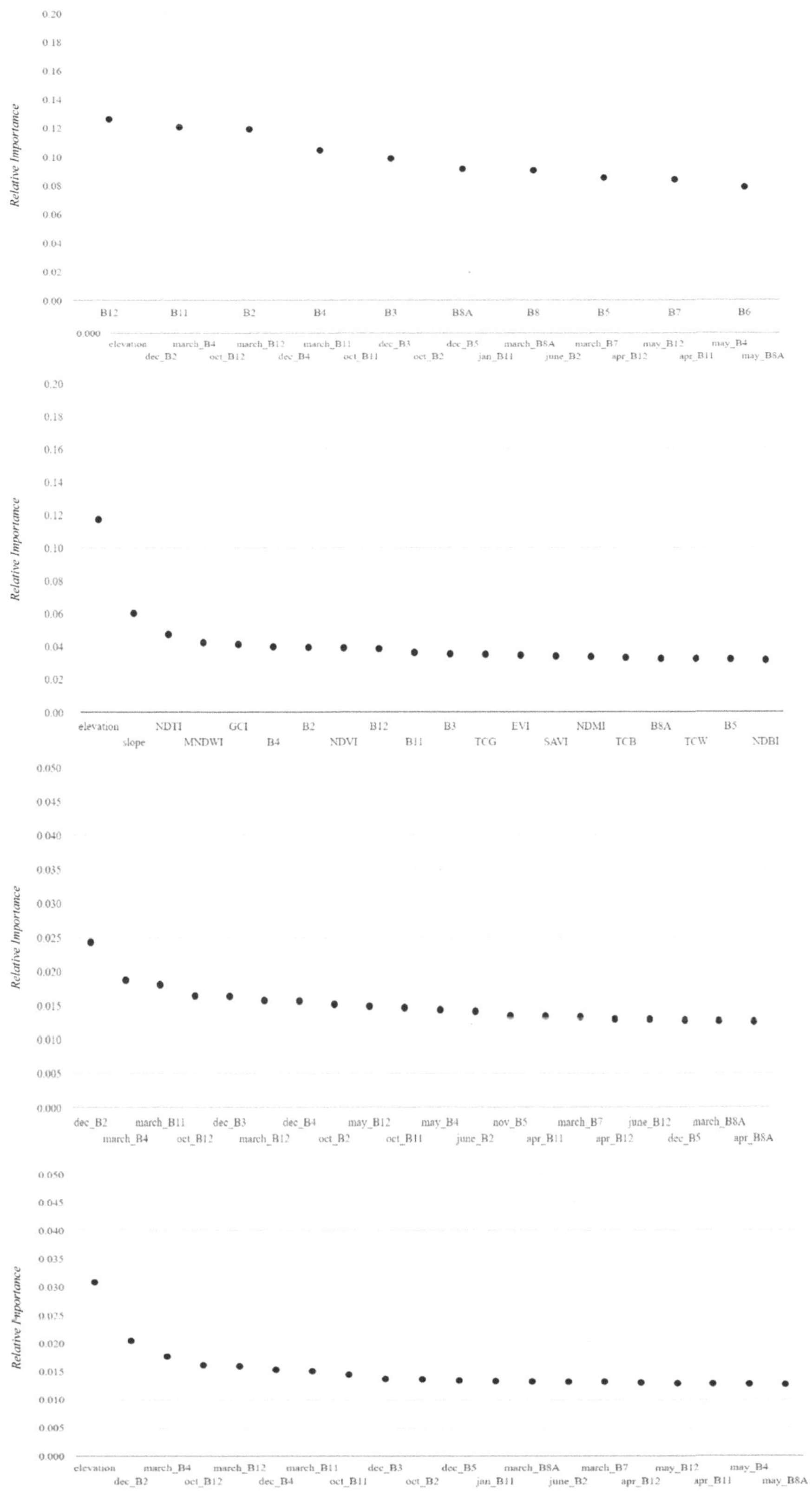

FIGURE 13.4 D1–3: Variable importance in the Random Forest models trained on the three (D1—upper, D2—middle, D3—lower) datasets. D4: Variable importance in the Random Forest models trained on the datasets.

TABLE 13.3 User accuracy and producer accuracy of different classes based on the four datasets.

Data	%	Agriculture	Barren land	Scrub	Settlement	Water	Sal	Dry deciduous	Moist deciduous	Semievergreen
D1	PA	87.1	85.4	79.1	29	99.1	96.3	75	89	95.8
	UA	79.8	87.5	83.3	50	97.5	87.3	78.5	91.5	95.6
D2	PA	98.6	95.3	94.1	46.8	100	99.3	89.2	93.6	95.2
	UA	92.1	96.1	92.4	100	98	96.9	90	91.3	97.2
D3	PA	98.7	98.9	99.1	88.8	100	99.3	91.2	78.3	99.2
	UA	100	96.1	94.3	100	100	98.6	100	97	88.6
D4	PA	100	96.5	98.4	90	100	99.4	88.3	92.4	97.4
	UA	99.2	97.6	94.6	97.2	100	98.4	96.3	90.7	97.4

PA, Producer accuracy; *UA*, user accuracy.

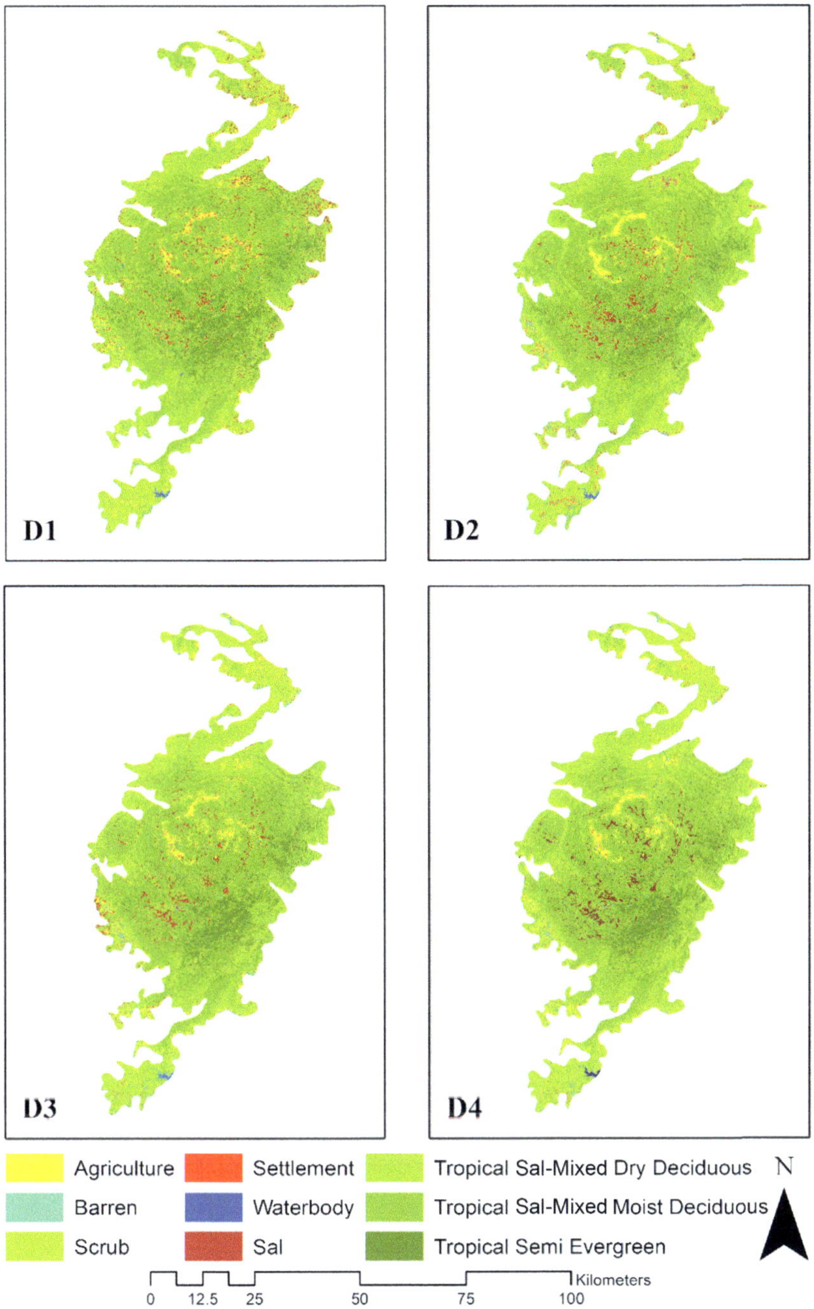

FIGURE 13.5 Vegetation-type-map-generated single image (D1 and D2) and multitemporal image collection (D3 and D4). Map lines delineate study areas and do not necessarily depict accepted national boundaries.

13.4 Discussion

In the Eastern Ghats of Odisha, the dominant forest types are tropical moist deciduous, tropical dry deciduous, and semi-evergreen, as identified by Champion & Seth (1968). The RF model used for classification was trained and validated against the same reference dataset to ensure comparability across the four classified maps. This study compared the accuracy of vegetation-type mapping when the datasets were changed from a single image to multitemporal data, with a prime focus on phenological dynamics. When vegetation responds noticeably to moisture stress by shedding leaves, the dry season of the year will stand out when spectral separability is considered across the time-series images. Therefore, in areas like the SBR where tropical sal-mixed dry deciduous and MDFs dominate (Reddy et al., 2007, 2015), the leaf-off season

TABLE 13.4 Confusion matrix for accuracy assessment (Dataset 4).

D4	Agriculture	Barren land	Scrub	Settlement	Water	Sal	Dry deciduous	Moist deciduous	Semievergreen
Agriculture	134	0	0	0	0	0	0	0	0
Barren land	0	83	3	0	0	0	0	0	0
Scrub	0	0	124	0	0	0	2	0	0
Settlement	1	2	1	36	0	0	0	0	0
Water	0	0	0	0	57	0	0	0	0
Sal	0	0	0	0	0	187	0	1	0
Dry deciduous	0	0	3	1	0	3	53	0	0
Moist deciduous	0	0	0	0	0	0	0	49	4
Semievergreen	0	0	0	0	0	0	0	4	154

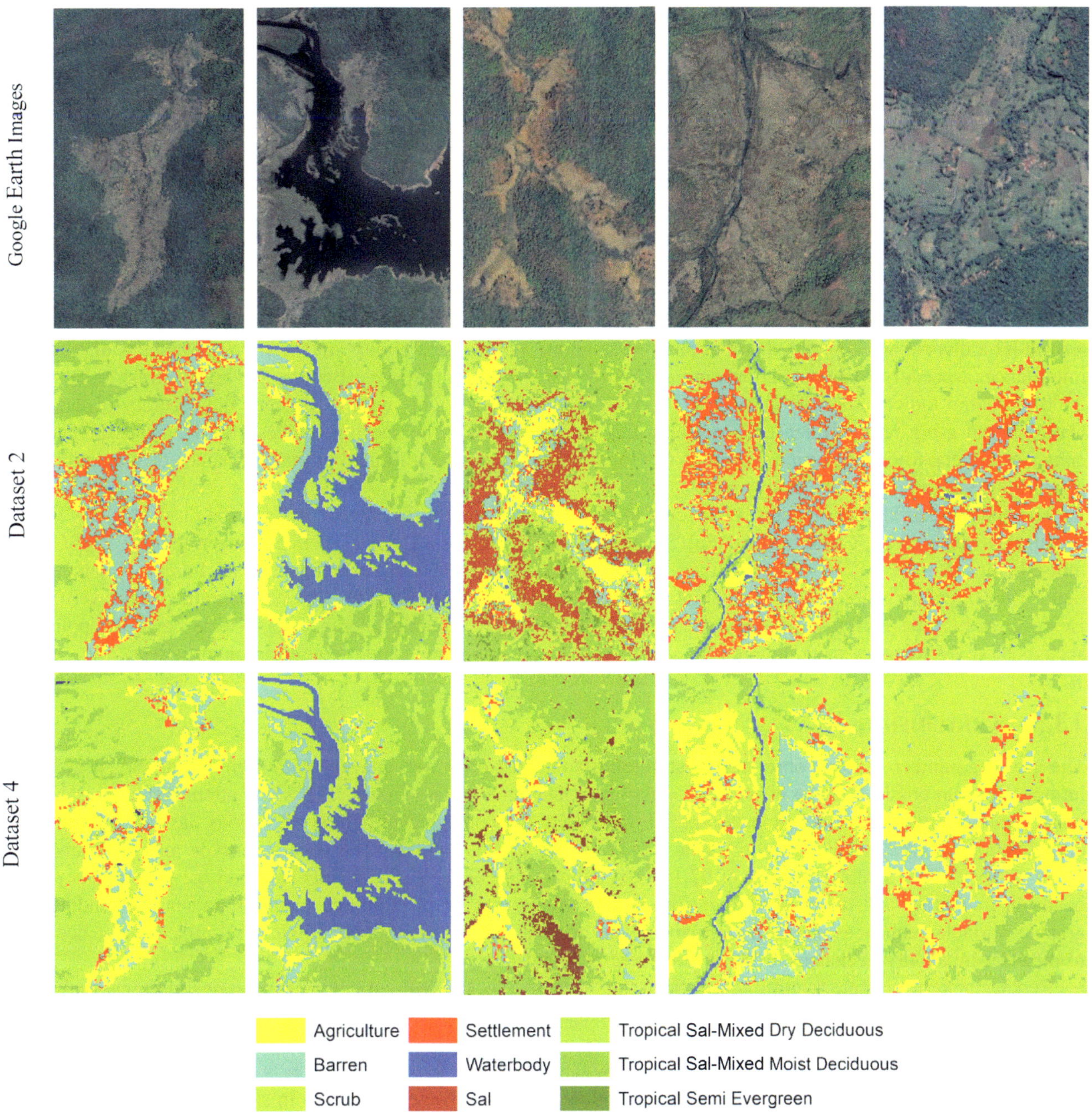

FIGURE 13.6 Disagreement among various classes in classified map derived from D2 to D4. Map lines delineate study areas and do not necessarily depict accepted national boundaries.

will contribute substantially while extracting phenological characteristics of vegetation. Hence, it is important to study how the classified output accuracy varies with the dry season image and time-series data, and assess the role of auxiliary variables in boosting model performance.

While an image from the dry "leaf-off" season is useful for spectral unmixing of pixels belonging to vegetation types having contrasting phenological cycles (e.g., MDF vs DDF), incorporating multitemporal images (after appropriate preprocessing) provides extended temporal data spanning over multiple seasons which are required to distinguish spectrally similar vegetation types. The fact that anthropogenic features are either spectrally invariant or have dissimilar spectral trajectories compared to the natural environment around them, also aids in their proper segregation. Topographical data and spectral indices also add layers of auxiliary environmental gradient information that helps to train a robust classification model and optimize its accuracy.

With the availability of GEE, the prevailing approach for filling gaps in cloudy images is to utilize median metrics, which is known as the temporal aggregation method. Carrasco, O'Neil, Daniel Morton, and Rowland (2019) underscore the merits of this approach, particularly its ability to substantially reduce data volume, thus enabling a faster and more efficient analysis. Here, the time-series data are used in place of median images to capture the nuanced shifts in vegetation phenology, particularly at the onset of the growing season and leaf-off period, when the major vegetation types in SBR exhibit variations in the timing of growth initiation and leaf shedding.

The classified map with the highest classification accuracy among the datasets achieves an output consistent with that of the vegetation-type maps generated by Roy et al. (2015) and Saranya & Reddy (2016). However, the extent and distribution of tropical SEFs in the SBR are different in the existing reference maps that should be studied and mapped precisely with vegetation surveys and multisensor data synergy. Regardless of the dataset used, elevation comes out to be an important variable when classifying LULC (Noi Phan, Kuch, & Lehnert, 2020). The elevation gradient efficiently separates scrublands, settlement areas, and different forest types when used as model inputs. In a tropical forest characterized by a diverse and complex distribution of vegetation, spectral separability among various classes is of paramount importance, especially when optical data are utilized for classification purposes (Behera, Srivastava, Kushwaha, & Roy, 2000). Spectral indices, such as NDVI, EVI, and SAVI, have been widely used to enhance the accuracy of land cover classification using remotely sensed images. In this study, including the spectral indices and S-2 bands of Dataset 1 enhances the classification accuracy from 86.6% to 94.3%. This can be attributed to the dominance of vegetation as the primary land-cover type in the study area.

The spectral-based classification is expected to be different from the field-based forest type classification (3C/2e—Moist Peninsular Sal; 3C/DSI—Moist Sal Savannah; 2B/3C—Orissa Tropical Semievergreen Forest; 2B/3C—very moist Peninsular Sal; 3C/3C—Moist Mixed Deciduous Forests; 5B/Cic—Dry Peninsular Sal Forests: Champion and Seth, 1968). Future studies should attempt to improve classification accuracy by corroborating field verification. However, a high-resolution vegetation-type map could serve as a valuable reference for a wide range of applications such as biodiversity assessment, monitoring, carbon stock estimation, and habitat-conservation initiatives.

13.5 Conclusions

The growing number of Earth observation satellite missions that offer free imagery with high spatial and spectral resolutions opens up possibilities for efficiently assessing vegetation over large areas. This study attempted to map the dominant vegetation types in the SBR using the RF algorithm and compared the effect of various datasets on the overall classification accuracy using freely available S-2 images acquired via GEE. The vegetation-type map generated using S-2 time-series data achieved an OA of 97.1% and kappa value of 0.96. The results revealed the potential of spectral bands in extracting and mapping various forest types by leveraging phenological patterns and how including topographic variables enhances model performance. The dominant forest types were tropical sal-mixed moist deciduous, tropical sal-mixed dry deciduous, tropical semievergreen, and sal.

Replacement of a single satellite image from the "leaf-off" season with an annual time-series multiseasonal data as the sole training data for the RF model provides a quantum leap in the classification OA by 9.7% and k-coefficient by 0.11 (D3–D1), surpassing the single image–auxililary variables combination in D2. Addition of topographical data in D4 marginally contributes to increase the statistics by 0.8% and 0.01%, respectively. It is clear that multitemporal phenological data provide the strongest discriminant for classification of vegetation types in a diverse and complex ecosystem such as SBR. Remote-sensing images by virtue of their repetitive data capture well captures the phenological varioans, permitting discrimination in lieu of variation in forest type, species composition, topography, and moisture content. Therefore, multivariate statistical approaches should be utilized through advanced machine learning approaches to study and understand the phenological variations in distribution in forest types.

Integrating data from Synthetic Aperture Radar (SAR), Light Detection and Ranging (LiDAR), and optical sources offers a promising approach for investigating and improving the accuracy of vegetation-type mapping, thereby offering more comprehensive insight into ecosystem dynamics. This spatially explicit classification of vegetation types in the SBR can be beneficial for studies related to carbon stocks, climate-vegetation modeling, LULC change detection, and vegetation-type responses to climatic variability.

Acknowledgments

We thank the Department of Biotechnology (DBT), Government of India, for research funding in the form of a project "Biotechnological Interventions for Management of Protected Areas." We acknowledge the Odisha State Forest, Environment, and Wildlife Department for their kind support and permission to conduct this study in the Similipal Biosphere Reserve. MDB acknowledges the authority of CORAL and IIT Kharagpur for facilitating this study. We thank Dr. A Sashney, Director, CSIR-NBRI, for his kind coordination.

References

Adam, E., Mutanga, O., Odindi, J., & Abdel-Rahman, E. M. (2014). Land-use/cover classification in a heterogeneous coastal landscape using RapidEye imagery: Evaluating the performance of random forest and support vector machines classifiers. *International Journal of Remote Sensing*, *35*(10), 3440–3458. Available from https://doi.org/10.1080/01431161.2014.903435.

Behera, M. D., Kushwaha, S. P. S., & Roy, P. S. (2001). Forest vegetation characterization & mapping using IRS-1C satellite images in eastern Himalayan region. *Geocarto International*, *16*(3), 53–62.

Behera, M. D., Srivastava, S., Kushwaha, S. P. S., & Roy, P. S. (2000). Stratification and mapping of *Taxus baccata* L. bearing forests in Tale valley using remote sensing and GIS. *Current Science*, *78*(8), 1008–1013.

Biswal, A., Mukherjee, S., Jeyaram, A., & Murthy, K. (2012). Man in biosphere reserve: A remote sensing study in Similipal, Orissa. *The International Archives of the Photogrammetry, Remote Sensing and Spatial Information Sciences*, *38*, 82–86.

Breiman, L. (2001). Random forests. *Machine Learning*, *45*, 5–32.

Carrasco, L., O'Neil, A. W., Daniel Morton, R., & Rowland, C. S. (2019). Evaluating combinations of temporally aggregated Sentinel-1, Sentinel-2 and Landsat 8 for land cover mapping with Google Earth Engine. *Remote Sensing*, *11*(3). Available from https://doi.org/10.3390/rs11030288.

Congalton, R. G. (1991). A review of assessing the accuracy of classifications of remotely sensed data. *Remote Sensing of Environment*, *37*.

Champion, H. G., & Seth, S. K. (1968). *A Revised Survey of the Forest Types of India* (pp. 404–pp). New Delhi: Manager of Publication, Government of India.

Crist, E. P., & Cicone, R. C. (1984). A physically-based transformation of Thematic Mapper data—The TM Tasseled Cap. *IEEE Transactions on Geoscience and Remote Sensing*, *3*, 256–263.

Das, P., Behera, M. D., & Murthy, M. S. R. (2017). Forest Fragmentation and Human Population varies logarithmically along elevation gradient in Hindu Kush Himalaya-Utility of Geospatial tools and free data set. *Journal of Mountain Science*, *14*(12), 2432–2447.

Das, P., Behera, M. D., Roy, P. S., & Barik, S. K. (2024). Predicting tipping points of forest cover resilience as a response to precipitation – A step towards understanding climate resilience. *Biodiversity and Conservation*. Available from https://doi.org/10.1007/s10531-024-02804-1.

Dobbin, K. K., & Simon, R. M. (2011). Optimally splitting cases for training and testing high dimensional classifiers. *BMC Medical Genomics*, *4*. Available from https://doi.org/10.1186/1755-8794-4-31.

Drusch, M., Del Bello, U., Carlier, S., Colin, O., Fernandez, V., Gascon, F., ... Bargellini, P. (2012). Sentinel-2: ESA's optical high-resolution mission for GMES operational services. *Remote Sensing of Environment*, *120*, 25–36. Available from https://doi.org/10.1016/j.rse.2011.11.026.

Egbert, S. L., Park, S., Price, K. P., Lee, R.-Y., Wu, J., & Nellis, M. D. (2002). Using conservation reserve program maps derived from satellite imagery to characterize landscape structure. *Computers and Electronics in Agriculture*, *37*. Available from http://www.elsevier.com/locate/compag.

Gitelson, A., & Merzlyak, M. N. (1994). Quantitative estimation of chlorophyll-a using reflectance spectra: Experiments with autumn chestnut and maple leaves. *Journal of Photochemistry and Photobiology B: Biology*, *22*(3), 247–252.

Gitelson, A. A., Gritz, Y., & Merzlyak, M. N. (2003). Relationships between leaf chlorophyll content and spectral reflectance and algorithms for non-destructive chlorophyll assessment in higher plant leaves. *Journal of Plant Physiology*, *160*(3), 271–282.

Guisan, A., & Zimmermann, N. E. (2000). Predictive habitat distribution models in ecology. *Ecological Modelling*, *135*. Available from http://www.elsevier.com/locate/ecolmodel.

Hermosilla, T., Wulder, M. A., White, J. C., Coops, N. C., & Hobart, G. W. (2018). Disturbance-informed annual land cover classification maps of Canada's forested ecosystems for a 29-year landsat time series. *Canadian Journal of Remote Sensing*, *44*(1), 67–87. Available from https://doi.org/10.1080/07038992.2018.1437719.

Huete, A. R. (1988). A soil-adjusted vegetation index (SAVI). *Remote Sensing of Environment*, *25*(3), 295–309.

Huete, A. R., Liu, H. Q., Batchily, K. V., & Van Leeuwen. (1997). A comparison of vegetation indices over a global set of TM images for EOS-MODIS. *Remote Sensing of Environment*, *59*(3), 440–451.

Jaya Prakash, A., Shubham Kumar., Behera, M. D., Das, P., Kumar, A., & Srivastava, P. K. (2023). Impact of extreme weather events on crop land inundation over Indian Subcontinent. *Environmental Monitoring and Assessment*, *195*, 50. Available from https://doi.org/10.1007/s10661-022-10553-3.

Jones, H. G., & Vaughan, R. A. (2010). *Remote sensing of vegetation: Principles, techniques, and applications*. USA: Oxford University Press.

Kowalski, K., Okujeni, A., Brell, M., & Hostert, P. (2022). Quantifying drought effects in Central European grasslands through regression-based unmixing of intra-annual Sentinel-2 time series. *Remote Sensing of Environment*, *268*. Available from https://doi.org/10.1016/j.rse.2021.112781.

Luo, C., Zhang, X., Wang, Y., Men, Z., & Liu, H. (2022). Regional soil organic matter mapping models based on the optimal time window, feature selection algorithm and Google Earth Engine. *Soil and Tillage Research*, *219*. Available from https://doi.org/10.1016/j.still.2022.105325.

Margules, C. R., & Pressey, R. L. (2000). Systematic conservation planning. *Nature*, *405*(6783), 243–253. Available from https://doi.org/10.1038/35012251.

Myers, N., Mittermeier, R. A., Mittermeier, C. G., Da Fonseca, G. A. B., & Kent, J. (2000). Biodiversity hotspots for conservation priorities. *Nature*, *403*. Available from http://www.nature.com.

Navalgund, R. R., Jayaraman, V., & Roy, P. S. (2007). Remote sensing applications: An overview. *Current Science*, *93*(12). Available from https://www.jstor.org/stable/24102069.

Noi Phan, T., Kuch, V., & Lehnert, L. W. (2020). Land cover classification using Google Earth Engine and random forest classifier—the role of image composition. *Remote Sensing*, *12*(15). Available from https://doi.org/10.3390/RS12152411.

Olofsson, P., Foody, G. M., Stehman, S. V., & Woodcock, C. E. (2013). Making better use of accuracy data in land change studies: Estimating accuracy and area and quantifying uncertainty using stratified estimation. *Remote Sensing of Environment*, *129*, 122–131. Available from https://doi.org/10.1016/j.rse.2012.10.031.

Paramanik, S., Rajdeep, N., Behera, M. D., Bhattacharya, B. K., & Dash, J. (2023). Species-level classification of mangrove forest using AVIRIS-NG hyperspectral imagery. *Remote Sensing Letters*, *14*(5), 522–533.

Patidar, N., & Behera, M. D. (2019). How significantly do land use and land cover (LULC) changes influence the water balance of a river basin? A study in Ganga river basin, India. *Proceedings of the National Academy of Sciences, India Section A*, *89*, 353–365.

Pflugmacher, D., Rabe, A., Peters, M., & Hostert, P. (2019). Mapping pan-European land cover using Landsat spectral-temporal metrics and the European LUCAS survey. *Remote Sensing of Environment*, *221*, 583–595. Available from https://doi.org/10.1016/J.RSE.2018.12.001.

Reddy, C. S., Jha, C. S., Diwakar, P. G., & Dadhwal, V. K. (2015). Nationwide classification of forest types of India using remote sensing and GIS. *Environmental Monitoring and Assessment*, *187*(12), 1–30. Available from https://doi.org/10.1007/s10661-015-4990-8.

Reddy, C. S., Pattanaik, C., Mohapatra, A., & Biswal, A. K. (2007). Phytosociological observations on tree diversity of tropical forest of Similipal Biosphere Reserve, Orissa, India. *Taiwania*, *52*(4).

Rouse, J., Haas, R., Schell, J., & Deering, D. (1973). Monitoring vegetation systems in the great plains with ERTS. *Third ERTS Symposium, NASA*, *1*, 309–317.

Roy, P. S., Behera, M. D., Murthy, M. S. R., Roy, A., Singh, S., Kushwaha, S. P. S., … Ramachandran, R. M. (2015). New vegetation type map of India prepared using satellite remote sensing: Comparison with global vegetation maps and utilities. *International Journal of Applied Earth Observation and Geoinformation*, *39*, 142–159. Available from https://doi.org/10.1016/j.jag.2015.03.003.

Sala, O. E., Stuart Chapin, F. I. I. I., Armesto, J. J., Berlow, E., Bloomfield, J., Dirzo, R., … Leemans, R. (2000). Global biodiversity scenarios for the year 2100. *Science*, *287*(5459), 1770–1774.

Salinero-Delgado, M., Estévez, J., Pipia, L., Belda, S., Berger, K., Gómez, V. P., & Verrelst, J. (2022). Monitoring cropland phenology on Google Earth Engine using gaussian process regression. *Remote Sensing*, *14*(1). Available from https://doi.org/10.3390/rs14010146.

Saranya, K. R. L., & Reddy, C. S. (2016). Long term changes in forest cover and land use of Similipal Biosphere Reserve of India using satellite remote sensing data. *Journal of Earth System Science*, *125*(3), 559–569.

Schulz, C., Holtgrave, A. K., & Kleinschmit, B. (2021). Large-scale winter catch crop monitoring with Sentinel-2 time series and machine learning—An alternative to on-site controls? *Computers and Electronics in Agriculture*, *186*. Available from https://doi.org/10.1016/j.compag.2021.106173.

Singh, B., Jeganathan, C., Rathore, V. S., Atkinson, P. M., Behera, M. D., Singh, C. P., … Roy, P. S. (2024). The relationship between central Indian terrestrial vegetation and monsoon rainfall distributions in different hydroclimatic extreme years using time-series satellite data. *Theoretical and Applied Climatology*, *155*, 45–69.

Van Deventer., Ward, A. D., Gowda, P. H., & Lyon, J. G. (1997). Using thematic mapper data to identify contrasting soil plains and tillage practices. *Photogrammetric Engineering and Remote Sensing*, *63*, 87–93.

White, J. C., Hermosilla, T., Wulder, M. A., & Coops, N. C. (2022). Mapping, validating, and interpreting spatio-temporal trends in post-disturbance forest recovery. *Remote Sensing of Environment*, *271*, 112904. Available from https://doi.org/10.1016/J.RSE.2022.112904.

Wilson, E. H., & Sader, S. A. (2002). Detection of forest harvest type using multiple dates of Landsat TM imagery. *Remote Sensing of Environment*, *80*(3), 385–396.

Xiao, X., Zhang, Q., Braswell, B., Urbanski, S., Boles, S., Wofsy, S., … Ojima, D. (2004). Modeling gross primary production of temperate deciduous broadleaf forest using satellite images and climate data. *Remote Sensing of Environment*, *91*(2), 256–270. Available from https://doi.org/10.1016/j.rse.2004.03.010.

Xu, H. (2006). Modification of normalised difference water index (NDWI) to enhance open water features in remotely sensed imagery. *International Journal of Remote Sensing*, *27*(14), 3025–3033.

Zha, Y., Gao, J., & Ni, S. (2003). Use of normalized difference built-up index in automatically mapping urban areas from TM imagery. *International Journal of Remote Sensing*, *24*(3), 583–594.

Zhou, Q., Zhu, Z., Xian, G., & Li, C. (2022). A novel regression method for harmonic analysis of time series. *ISPRS Journal of Photogrammetry and Remote Sensing*, *185*, 48–61. Available from https://doi.org/10.1016/j.isprsjprs.2022.01.006.

Chapter 14

Satellite-based monitoring of agriculture resilience succeeding watershed development interventions

Abigith Baby[1], Pulakesh Das[2] and Stephanie Katsir[1]

[1]*Deutsche Gesellschaft für Internationale Zusammenarbeit (GIZ), New Delhi, India,* [2]*School of Forest Resources, University of Maine, Orono, ME, United States*

14.1 Introduction

India, the world's most populous nation, possesses only 4% of global freshwater resources (WRIS, 2023). The majority of Indian farmers rely on monsoon rainfall to support their agriculture and to replenish their water supplies. Although the "Green Revolution" in India resulted in increased crop production, it also increased the usage of fertilizers and pesticides and had a detrimental effect on soil health. The sustainability of agriculture and maintaining soil health is one of the major crises India faces today (Prajapati, Tripathi, & Mishra, 2010). India is highly dependent on agriculture, employing 54.6% of the total workforce (Office of the Registrar General & Census Commissioner, 2011). India is expected to produce 345 million tons (Mt) of food grains annually to feed the growing population by 2030 (Indian Council for Agricultural Research, 2011), against the production of about 265 Mt in 2013–14. Farm holdings increased from 71 to 138 million during the same period from 1970–71 to 2010–11, mainly due to the progressive fragmentation of land holdings (Ganeshamurthy, 2014). Thus, it is essential to promote policies and initiatives that benefit smallholding and subsistent Indian farmers. India's average annual temperature increased by about 0.7°C during the period between 1901 and 2018 (MoES, 2020). The Intergovernmental Panel on Climate Change (IPCC) projects that the global surface temperature may rise between 1.0°C and 5.7°C by 2100 (IPCC, 2021). The global hydrological system, ecosystems, sea level, food output, and related processes are predicted to be severely impacted by this unprecedented temperature surge and precipitation change. In addition to climate change, the increased anthropogenic activities and land use practices modify the hydrological conditions and agricultural productivity in different river basins (Das et al., 2018). India's agriculture is heavily reliant on seasonal weather conditions, which have been experiencing multiple extreme weather events such as erratic rainfall, frequent droughts, and floods, causing crop damage and yield reduction in different Indian agro-ecological zones (Pandey & Srivastava, 2019; Prakash et al., 2023; Mudi, Prakash, & Das, 2022). In addition, the excessive use of pesticides and fertilizers has resulted in agricultural land degradation, a major issue that must be addressed in the coming years (Times of India, 2022). India is challenged by widespread land degradation, and climate change exacerbates the risk of lower yields and crop failure, leaving smallholders, marginal farmers, and dependent socioeconomy vulnerable (Matin & Behera, 2019).

Drought is a natural hazard that differs from other hazards since it has a slow onset, evolves over months or even years, and affects a large spatial extent. Depending on the severity and duration, it causes significant alteration to hydrological parameters and damage to agricultural productivity (Pandey & Srivastava, 2019). Its onset, end, and severity are often challenging to determine. In India, the likelihood of drought varies from occurring every 2 years in western Rajasthan to every 15 years in Assam (Shah & Mishra, 2020). High population rise, agricultural output demand, and various agroclimatic conditions combine to put excessive pressure on the land and increase the risk of land degradation in India. Land degradation can lead to the removal of topsoil due to runoff, which causes soil organic matter loss and decreased soil water-holding capacity. Thus, the soil becomes more prone to drying out during periods of low rainfall, exacerbating the impacts of drought. A recent national assessment has reported that 29.77% of India's total geographic

Sustainable Development Perspectives in Earth Observation. DOI: https://doi.org/10.1016/B978-0-443-14072-3.00020-4

area is undergoing land degradation (Space Applications Centre, 2021). Water erosion has been responsible for the most significant amount of land degradation in India (11.01% in 2018–19, 10.98% in 2011–13, and 10.83% in 2003–05), followed by vegetation degradation (9.15% in 2018–19, 8.91% in 2011–13, and 8.60% in 2003–05), and wind erosion (5.46% in 2018–19, 5.55% in 2011–13, and 5.58% in 2003–05) (Space Applications Centre, 2021).

Maharashtra is one the most important states, with a significant population dependent on agriculture and allied socioeconomy. Maharashtra experienced multiple droughts in the past decade in 2011–12, 2012–13, 2014–15, and 2018–19 (Chatterjee & Katewa, 2024). Maharashtra is the third most vulnerable state to extreme events such as cyclones, floods, and droughts after Assam and Andhra Pradesh (Mohanty & Wadhawan, 2021). Several governmental and nongovernmental initiatives have been taken to check land degradation, mitigate drought impact, and develop sustainable agricultural practices in India. The Soil Protection and Rehabilitation for Food Security in India (ProSoil) project implemented by Deutsche Gesellschaft für Internationale Zusammenarbeit (GIZ) GmbH and the National Bank for Agriculture and Rural Development adopts sustainable approaches for protecting and rehabilitating degraded soils. The project aims to protect and rehabilitate 155,000 ha of land by December 2024 through soil protection interventions. The ProSoil is working on soil protection and water conservation demonstration projects in Maharashtra and Madhya Pradesh. The ProSoil project intervention has been classified and explained as sustainable land management (SLM) measures comprising practices to conserve the soil and water resources in and around the project areas. The major objective of SLM measures is to achieve maximum production without losing many resources. The ProSoil project interventions by GIZ have been implemented over many districts in Maharashtra and Madhya Pradesh from 2015 to 2023. In the initial phase, the project focused on efficient water management through soil protection, water conservation, and rehabilitation of land activities. The activities under the ProSoil interventions are listed in Table 14.1. The project aims to improve water availability to increase the cropping area and develop resilient agriculture practices to combat drought.

Systematic monitoring and evaluation (M&E) of landscape developmental projects at a landscape level are required to ensure the activities meet the targeted objectives. Monitoring and quantifying the project outcomes are also necessary to understand the effectiveness of the various interventions and to take the learning forward for future projects. The commonly used tools for M&E are field data collection, surveys, interviews, and change analysis using geospatial technologies. The M&E using geospatial technology is becoming significant owing to the availability of higher-resolution satellite images, gridded climate data, and other relevant data in the public domain. Geospatial techniques have been widely used for drought detection and decision-making based on the spatio-temporal drought conditions map. Several studies have used satellite images and derived indicators to monitor land and water resource developmental activities. For example, Behera, Gupta, Barik, Das, & Panda (2018) used the Landsat and Sentinel-2 optical satellite images to assess vegetation growth and surface water change due to the project intervention by Indo-German Development Cooperation in two districts in Tripura, India. They derived spectral indices from satellite images as normalized difference water index to identify the surface water area change due to the creation of various water-harvesting structures. They also derived the enhanced vegetation index (EVI) to monitor vegetation cover change due to various plantation activities. The satellite image-derived indicators such as vegetation health index (VHI), temperature condition index (TCI), vegetation condition index (VCI), and normalized difference vegetation index (NDVI) are also used for agricultural drought detection (Kloos et al., 2021). Bento et al. (2018) studied the VHI, VCI, and TCI from 1982 to 2009 and linked them with the Standardized Precipitation-Evapotranspiration Index (SEPI) to assess agriculture drought impact in the Euro-Mediterranean region (Bento et al., 2018). They used the Global Inventory Modelling and Mapping Studies (GIMMS) data available at 8-km spatial and bi-monthly temporal scale for deriving VCI (to compute VHI) and 0.5°C scale hourly LST data for deriving the TCI. They have reported that VCI is more evident in water-scarce regions, and the TCI is more useful where solar radiation is the major limiting factor. Kundu et al. (2016a,b) employed the National Oceanic and Atmospheric Administration (NOAA) Advanced Very High Resolution Radiometer (AVHRR) data to derive VHI, VCI, and TCI to assess agricultural monitoring in August and September (peak Kharif season) in India (Kundu et al., 2016a,b). Their study identified the different drought conditions experienced in several Indian states in 2009. Yoon et al. (2020) studied agricultural drought in East Asia using remote sensing inputs to derive the evaporative stress index (a drought index based on evapotranspiration) and compared it with VHI, EVI, and leaf area index (LAI) on a monthly scale . Wu et al. (2015) used Moderate Resolution Imaging Spectroradiometer (MODIS) vegetation and LST data for drought monitoring in the US corn belt areas (Wu et al., 2015). Several studies have applied the MK test for trend analysis; for example, the hydrological drought trend analyzed in Turkey from 1970 to 2017 (Yilmaz, 2019), in arid and semiarid regions of Iran from 1975 to 2005 (Dashtpagerdi et al., 2015).

Several developmental interventions have been implemented to address the agriculture drought and improve climate resilience in the Bhokardan block, Jalna district, Maharashtra. The current study aims to assess the effectiveness of soil

TABLE 14.1 ProSoil SLM interventions.

SLM measures	ProSoil interventions
Efficient water management	• Drip irrigation • Sprinklers • Well digging • Farm ponds • Water-harvesting structures
Soil fertility management	• Demonstrations/application of biochar, BioPROM, compost, green manure, amrutpani, and other bioresource interventions • Soil sampling and testing lab kits, • Farmers using urban city compost as fertilizers • Mulching demonstrations projects • Soil testing of farms and distribution of findings • Soil health card and farmer's education about it • Crop rotation
Soil protection and water conservation	• Bunding • Construction of check dams, bunds, gully plugs • Nalla deepening/desiltation, contour bund
Rehabilitation of land activities	• Plantations for erosion check • Glyricidia plantation on bunds and wasteland
Crop management	• Seed treatment before sowing • Single sapling with fixed spacing • Seed rate management • Sustainable crop intensification (SCI), System of rice intensification (SRI)/Sustainable rice technologies (SRT) application practiced • Demonstrations on various Practices of organic productions (POPs) like trap crop, micronutrients, intercropping, and crop diversification • Integrated pest and disease management practices • Demonstration on fodder crops on bunds • Silage preparation • Knowledge of farmer about fodder production by hydroponic technology • Knowledge of farmers for certification of organic produce
Knowledge management, integration into institutional framework, dissemination, and networking	Localized quality advisory to farmers on crop diversification, intercropping, symptom identification for crop-specific nutrient deficiency, local pest threats and diseases, integrated pest, and disease management practices

protection, water conservation, and water management initiatives of the ProSoil SLM initiatives through climate resilience due to drought conditions. The study employed various satellite data and derived proxies for long-term crop health monitoring and recent cropping intensity change assessment.

14.2 Study area

The study was conducted in the Bhokardan block, located in Jalna district, in the Marathwada region of Maharashtra. Maharashtra is the third-largest state in India, wherein about 54% of the population resides in rural areas (Office of the Registrar General & Census Commissioner, 2011). The total area of the Jalna district is 7719 km^2 (WOTR, 2021). Most agricultural activity in this region is primarily reliant on the monsoon rainfall. Black soil is the dominant soil type in the area, with different textures and densities, including light, medium, and heavy soils. The Bhokardan block is a drought-prone area with a cluster of about 69.89 km^2 (WOTR, 2021). Rain-fed agriculture, livestock raising, and labor work for agricultural and other activities comprise the population's main occupations. There are nine Agro-Climatic Zones in Maharashtra. Bhokardan block comes under the "assured rainfall zone" where the average annual rainfall varies between 700 and 900 mm. Most of this area is a Kharif tract suited for single rainfed crops. Soybean, cotton, and maize are the main crops cultivated during the Kharif season, while wheat and chickpea are grown during the Rabi season. Rain is necessary to sow Kharif crops, particularly in June and July (Chatterjee & Katewa, 2024), whereas Rabi crops are cultivated using irrigated water.

TABLE 14.2 The details of the ProSoil interventions area.

Block	Bhokardan
District	Jalna
Villages	Chandai Tepli, Chandai Thombri, Chincholi, Deulgaon Tad, Pimpalgaon Barao, Banegaon, Thigalkheda, and Palaskheda Thombari
Project area	6989 ha (69.89 km^2)
Number of households	2127

The areas of interest considered for the study are as follows:

- Bhokardan block: The complete administrative boundary of Bhokardan block, consisting of 157 villages in Jalna district, was considered for landscape-level analysis. This majorly consists of areas without ProSoil Interventions.
- ProSoil intervention area: The ProSoil intervention area is a cluster of eight villages in the Bhokardan block, where the different soil and water resource developmental activities were implemented (Table 14.2).

The majority of farmers in the study area are small and marginal landowners. The field visit in this region identified the issues related to soil erosion, the decline in yield due to monocropping, and chemical fertilizers. The interaction with farmers also indicated a reduction of crop yield due to the overuse of chemical fertilizers, monoculture agriculture practices, and crops. At the start of the ProSoil project, the majority of the farmers used flood irrigation, and only a few farmers used drip irrigation or sprinkler systems. Moreover, a significant population in this region relocate to urban areas for three to four months for work. The migration time may go more than six months during a drought year.

14.3 Materials and methodology

14.3.1 Data used

The publicly available satellite data and derived products were used in this study. The high-resolution Sentinel-2 optical data (10 and 20 m spatial resolution) and coarse resolution MODIS (500 m) data were accessed from the Googe Earth Engine (GEE) platform. The coarse resolution monthly MODIS vegetation parameters, such as NDVI, LAI, and LST products, were accessed from 2001 to 2023 for long-term analysis. The MODIS data were preprocessed using the quality control data layer and multiplied with the scale-factor values. The high-resolution Sentinel-2 data available since 2015 were accessed, wherein the cloud-free images were considered. The cloud-free image tiles were visually shortlisted to avoid misclassification due to cloud-cover presence in the satellite images.

In addition, the Environmental Systems Research Institute (ESRI) 2020 Global land use land cover (LULC) map, generated using the Sentinel-2 data, was used in the study (Fig. 14.1 d). This map was generated by a deep learning model trained on samples from more than 20,000 sites distributed across the world's major biomes and more than 5 billion manually labeled Sentinel-2 pixels (Karra et al., 2021). The data included 10 land cover classes as follows: water, tree, grass, flooded vegetation, crops, scrub/shrub, built area, bare ground, snow/ice, and clouds.

The rainfall data from 2001 to 2023 were obtained from the Indian Meteorological Department (IMD), which was accessed from the India Water Resources Information System (WRIS, 2023). The reference data on the cropping cycle and agricultural-related issues faced in the region were collected by interacting with the farmers in the study area.

14.3.2 Methodology

The study aims to analyze the improvement of the watershed's overall health in the ProSoil intervention area and compare it with the surrounding region (Bhokardan block), where no such interventions were implemented. The ProSoil project interventions in this area were started in 2015. Long-term and consistent MODIS data were employed to analyze the vegetation conditions since 2001 and recent changes. The change in vegetation conditions in the agricultural fields was assessed using vegetation health indicators as follows.

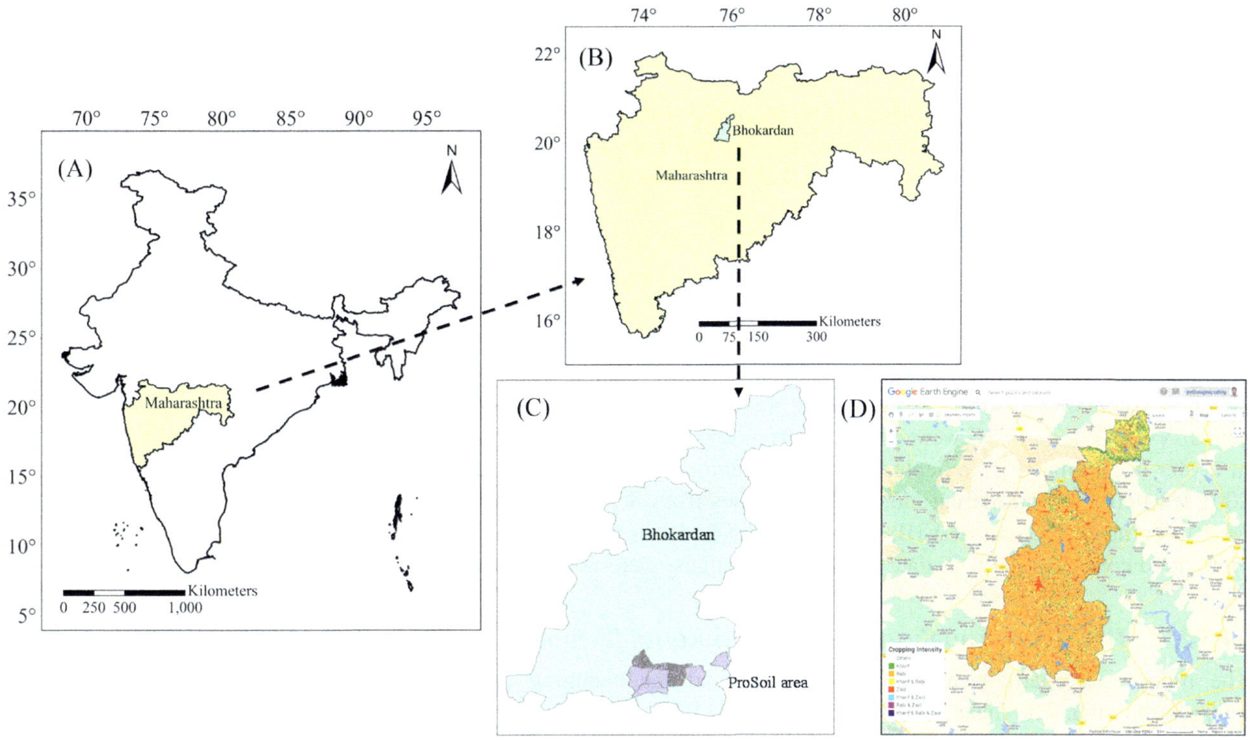

FIGURE 14.1 Study area location of (A) Maharashtra, (B) Bhokardan block, (C) PoSoil intervention area, and (D) Land use land cover map. Map lines delineate study areas and do not necessarily depict accepted national boundaries. *Courtesy: Deutsche Gesellschaft für Internationale Zusammenarbeit (GIZ)*

TABLE 14.3 Vegetation health index threshold for drought classification (Kogan, 2002).

Drought class	Vegetation health index
Extreme drought	<10
Severe drought	≥ 10 and <20
Moderate drought	≥ 20 and <30
Mild drought	≥ 30 and <40
No drought	≥ 40

VHI: Using a combined estimation of the vegetation and thermal conditions, the VHI represents the overall health of the vegetation. It is computed employing the VCI and TCI as follows:

$$\text{VHI} = \alpha\text{VCI} + (1 - \alpha)\text{TCI}$$

where α is a coefficient determining the contribution of the two indices and is generally taken as 0.5. The five classes of the VHI are used for the vegetative drought classifications (Table 14.3) (Kogan, 2002). A lower VHI value indicates higher drought intensity and vice versa.

VCI: It estimates the vegetation condition based on the best and worst vegetation vigor in different years over a specific period, providing a better vegetation condition than annual NDVI (Bajgiran, Darvishsefat, Khalili, & Makhdoum, 2008). In order to indicate the proportional changes in vegetation conditions from extremely bad to optimal, the VCI (expressed in percentage) rescales vegetation dynamics between 0 and 100. The VCI is derived using NDVI as follows:

$$\text{VCI} = 100\left[(\text{NDVI} - \text{NDVI}_{\min})/(\text{NDVI}_{\max} - \text{NDVI}_{\min})\right]$$

where NDVI, $NDVI_{min}$, and $NDVI_{max}$ are the seasonal average of smoothed (no-noise) monthly NDVI and its multiyear absolute minimum and maximum values, respectively (Dutta et al., 2015). VCI value is lower than 35, indicating a severe drought condition.

TCI determines the stresses caused by excessive wetness and temperature-related vegetation, estimated using the LST. The TCI is a measure of how much the thermal condition has changed in relation to the brightness temperature (BT) gleaned from the thermal bands (Kundu et al., 2016a,b).

$$TCI = 100\left[(BT_{max} - BT)/(BT_{max} - BT_{min})\right]$$

where BT, BT_{min}, and BT_{max} are the seasonal average of monthly brightness temperature and its multiyear absolute minimum and maximum values, respectively.

The monthly VCI, TCI, and VHI were used in this study to monitor and evaluate the crop health or vegetation drought in the Bhokardan block and the ProSoil intervention area. The Kharif season is considered from June to October, whereas the Rabi season is from November to February, and the Zaid season is from March to May. The annual VCI, TCI, and VHI were computed from 2001 to 2022 using the monthly data, wherein the vegetation condition in a year is calculated from September in the previous year to May in the current year. For example, the mean VCI of 2001 is calculated considering monthly VCI values from September 2000 to May 2001. The annual TCI and VCI were used to compute the annual VHI, which was masked using the croplands identified in the ESRI LULC map. Due to high cloud cover, the MODIS data from June to August were not considered.

14.3.2.1 Time-series analysis using Mann Kendall trend test

The time-series charts were created to understand the changing pattern and the recent changes in the project-intervention area. The mean values of NDVI, LAI, VHI, and LST data were extracted for eight villages and the entire Bhokardan block from 2001 to 2023 for trend analysis. The Mann Kendall (MK) test is a statistical test commonly used for trend analysis in climatology (Mavromatis & Stathis, 2011). The advantage of using the MK test is that the data do not need to be normalized because it is a nonparametric test (Dashtpagerdi et al., 2015; Yilmaz, 2019). The MK test has a low sensitivity to abrupt breaks in inhomogeneity of the time-series data (Tabari et al., 2011). The significance level is derived during the trend analysis. The MK test was performed using Addinsoft's XLSTAT 2012.

Rainfall is a major water source for agriculture in Maharashtra, and it is essential to analyze the relationship between rainfall and VHI. Thus, the relationship between VHI and rainfall was examined using the Spearman rank correlation test.

14.3.2.2 Cropping intensity mapping

The seasonal crop cover change assessment is an important parameter for studying cropping intensity change due to increased irrigation water availability. The seasonal cropping area change was studied over the three seasons: Kharif, Rabi, and Zaid from 2016 to 2023. The monthly cloud-free Sentinel-2 images were used to generate the monthly NDVI images from September to May, which were masked using the cropland area identified in the ESRI LULC map. The monthly NDVI images were used to map the seasonal cropping area by applying a threshold (0.3) to NDVI images. The Kharif season was identified using the September and October images, while November and February images were used to map the Rabi season crop area, and March to May images were used for Zaid-cropping area mapping.

The flowchart of the methodology developed is shown in Fig. 14.2.

14.4 Results

14.4.1 Bhokardan block

14.4.1.1 Vegetation conditions change

The mean monthly NDVI, LAI, and LST time series generated from September 2001 to March 2023 are shown in Fig. 14.3. The monthly MODIS NDVI values varied between 0.18 and 0.75. The NDVI was minimum in May when there was no cultivation (fallow period) and maximum in September–October (the peak growing period) in the Kharif season. A second peak, mostly observed in January, indicates the growing period in the Rabi season, which has mostly been observed since 2009. The trend analysis from 2001 to 2023 indicated a significant positive slope (Sen slope = 0.02; $P < .05$), showing an increasing trend. This explains that crop or vegetation cover is increasing overall in

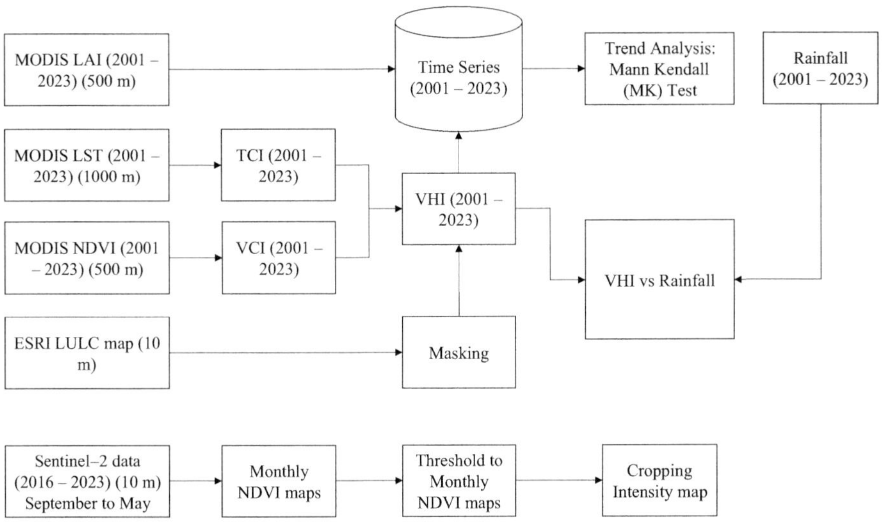

FIGURE 14.2 Data processing flowchart.

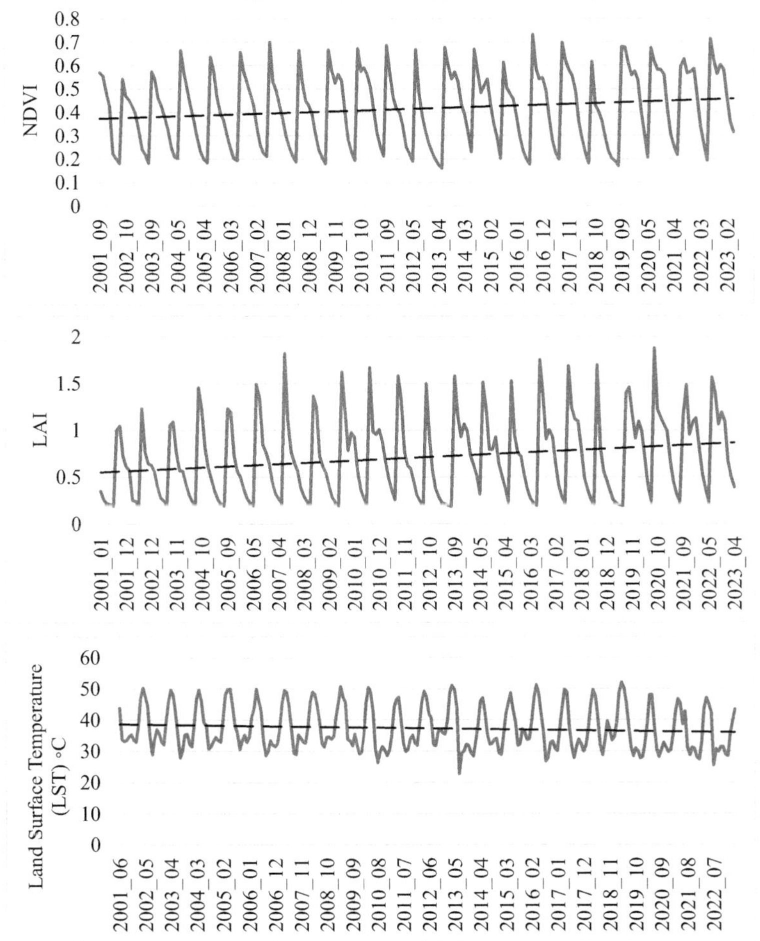

FIGURE 14.3 Monthly mean NDVI, LAI, and LST (top to bottom) from 2001 to 2023 for the Bhokardan block.

the region. The LAI chart generated from the MODIS dataset varies between 0.19 and 1.8 (Fig. 14.3). The minimum and maximum were observed in March–May (fallow period) and September–October (peak growing season in Kharif), respectively. The second peak was also observed in LAI time-series data during the Rabi season, as seen in NDVI. The trend analysis indicated a positive, that is, increasing trend (Sen slope = 0.001; $P < .05$), showing that LAI,

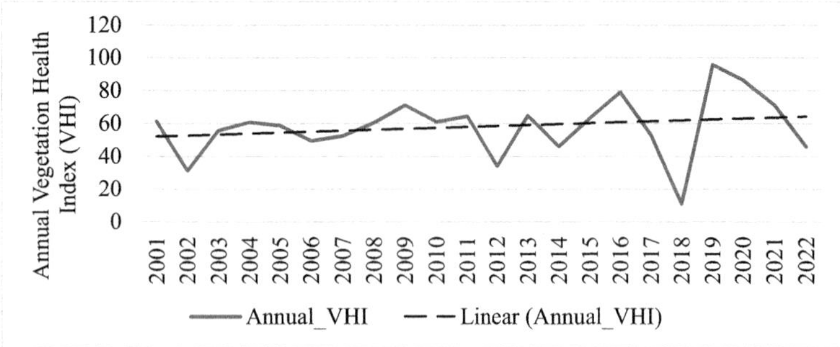

FIGURE 14.4 Annual VHI from 2001 to 2022 for the Bhokardan Block.

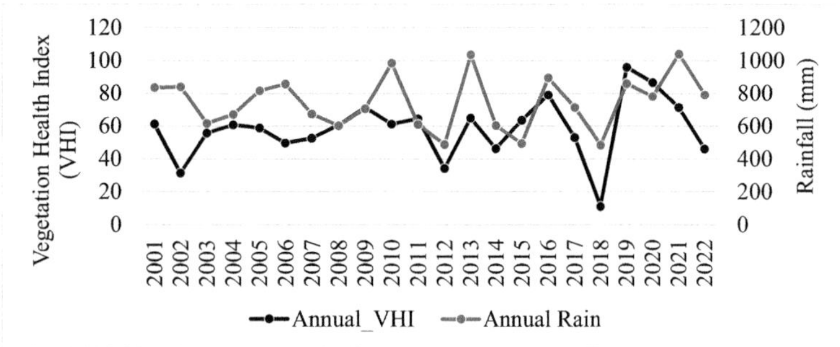

FIGURE 14.5 VHI versus Rainfall variation from 2001 to 2022–23 in the Bhokardan block.

which is a proxy for canopy density, increased in the region. The LST chart generated from the MODIS dataset showed a significant decreasing trend (Sen slope = −0.011; $P < .05$) (Fig. 14.3). It showed that LST has reduced in the agricultural lands, which could be attributed to the increased greenness and lower moisture stress in the leaf canopy.

The annual VHI time series generated from 2001 to 2023 is shown in Fig. 14.4, which showed an increasing trend with a dip in 2002, 2012, 2018, and 2022. The VHI was below 20 in 2018, indicating a severe drought year. Similarly, droughts were identified in 2002 and 2012, when the VHI was reduced to below 40. The Spearman correlation coefficient test that derives the relationship between rainfall and VHI (r = 0.43; $P < .05$) showed a positive relation (Fig. 14.5). The positive correlation indicates that the VHI was high in normal rainfall years and reduced in the rainfall-deficit (drought) years.

14.4.1.2 Cropping intensity change

The cropping intensity was identified from 2016 to 2023 (Fig. 14.6). The cropping areas in different seasons were identified using the GEE platform. The total cropping area shows an increasing trend, with a significant rise in the Rabi and Zaid-cropping area. Although the majority of the interventions were implemented starting in 2015, the change in cropping intensity was compared between 2016–17 and 2019–23, as the high-resolution Sentinel-2 were mostly available since 2016 (Fig. 14.6). However, the cropping area data for 2018 were omitted as it was a major drought year. The seasonal crop cover area showed that there was an increment in the Kharif-cropping area by 0.52%, the Rabbi cropping area by 3.46%, and the Zaid-cropping area by 349.01% from 2019 to 2023 compared to the baseline average from 2016 to 2018.

14.4.1.3 Succeeding ProSoil intervention area

14.4.1.3.1 Vegetation conditions change

The NDVI value for the ProSoil intervention area varied between 0.15 and 0.74 (Fig. 14.7). The value was minimal in May during the fallow period and generally maximum in September–October, which is the peak growing time of the Kharif season. From 2010 onward, a second peak was observed mostly in January, which indicates the cropping in the Rabi season. The MK trend test analysis of NDVI for 2001–23 indicated a significant positive value (Sen slope = 0.02, $P < .05$). This shows an overall increase in cropping (vegetation) in the region. The MODIS LAI value varies between 0.2 and 2.3, and the time-series change showed an increasing trend, as observed in NDVI (Fig. 14.7). The minimum LAI value was observed in March–May (fallow period) and the maximum in September–October (i.e., in the Kharif growing season). The MK test of LAI from 2001 to 2023 showed an increasing positive trend (Sen slope = 0.001; $P < .05$).

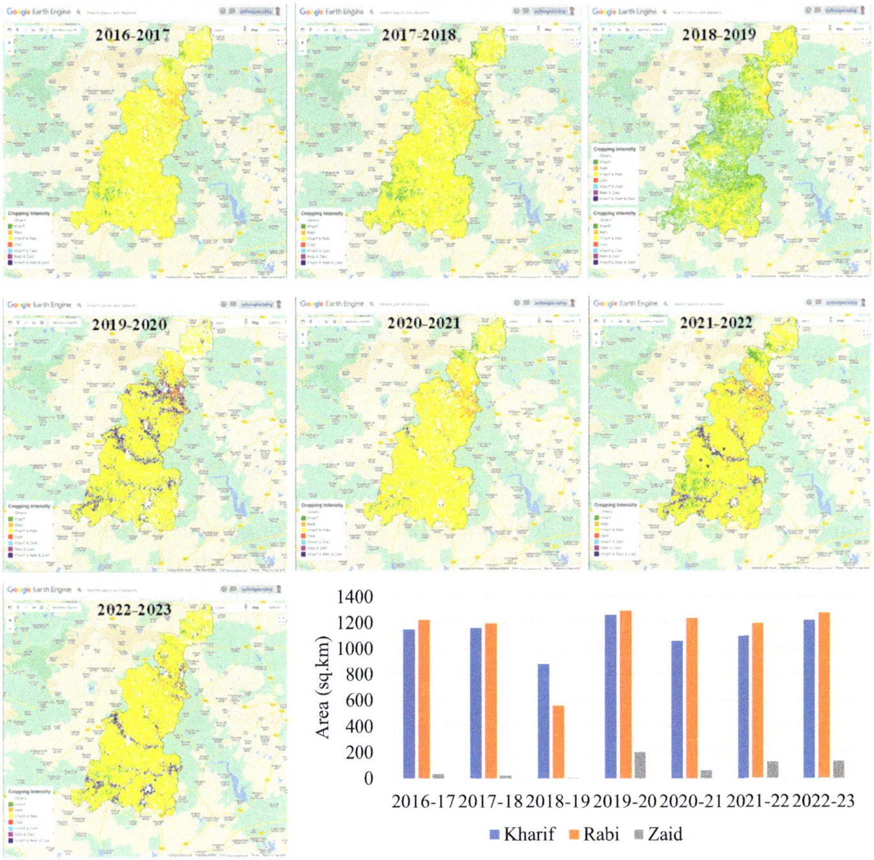

FIGURE 14.6 Seasonal cropping map and area identified from 2016 to 2023. Map lines delineate study areas and do not necessarily depict accepted national boundaries.

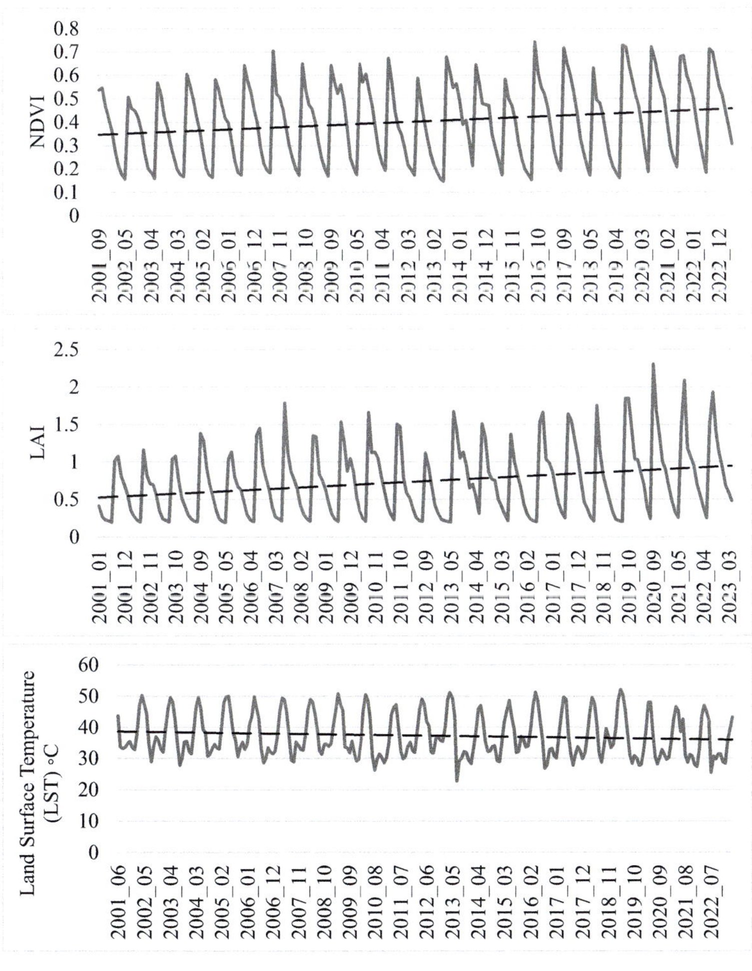

FIGURE 14.7 Monthly mean NDVI, LAI, and LST data (top to bottom) from 2001 to 2023 for the ProSoil cluster.

This shows that the LAI, a proxy for vegetation canopy density, was increasing in the region. The MODIS LST dataset showed a significant decreasing trend (Sen slope = −0.011; $P < .05$), which could be attributed to the increased crop cover vegetation (Fig. 14.7).

The annual VHI time series from 2001 to 2023 showed a higher value than 40, which indicates that the ProSoil intervention area did not experience deteriorated crop health during the study period (Fig. 14.8). The ProSoil intervention area where water-conservation interventions are carried out did not show any drought impact even during the confirmed drought years of 2012, 2015, and 2018, as observed in the Bhokardan block. The Spearman correlation coefficient between rainfall and VHI ($r = 0.119$, $P > .05$) showed an insignificant relationship which means that the VHI was less dependent on rainfall variation (Fig. 14.9).

14.4.1.4 Cropping intensity change

The cropping area in different seasons was identified from 2016 to 2023 (Fig. 14.10). The total cropping area shows a minor increasing trend; however, an overall increment was observed in the Zaid-cropping area. The cropping area in different seasons was compared between the average from 2016 to 2018 and 2019 to 2023. For consistency, the data for 2018 were omitted, similar to the Bhokardan block. The data showed a marginal change in the Kharif-cropping area, reduced by −0.32%, an increase in the Rabbi cropping area by 1.17%, and a significant increment in the Zaid-cropping area by 167.92% in 2019–23 compared to 2016–18 (Fig. 14.10).

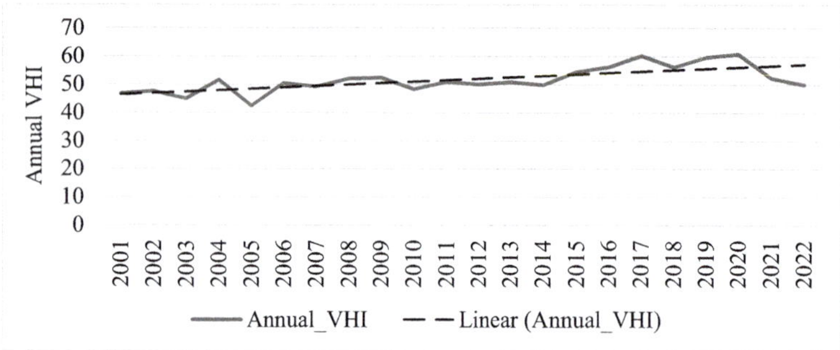

FIGURE 14.8 Annual VHI time series from 2001 to 2023 for the ProSoil intervention area.

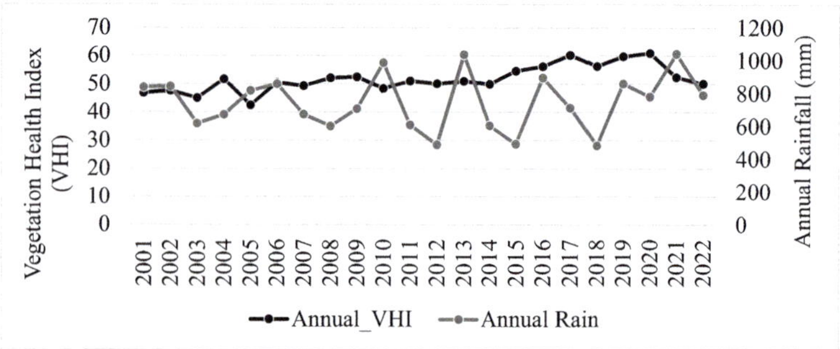

FIGURE 14.9 VHI versus Rainfall variation from 2001 to 2022 in the ProSoil intervention area.

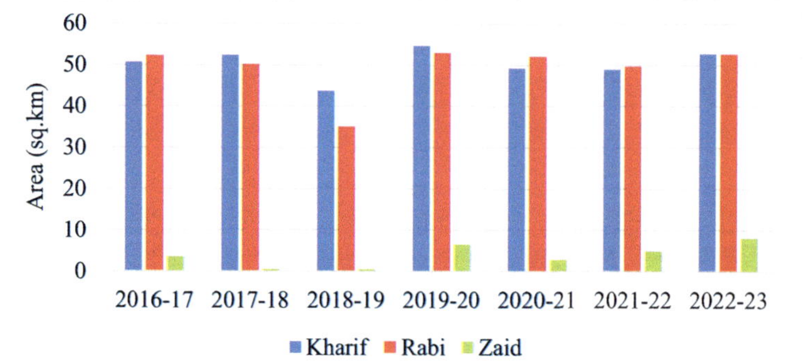

FIGURE 14.10 Cropping area in different seasons from 2016 to 2023 in the ProSoil intervention area.

The canopy density (LAI), greenness (NDVI), and (LST) were compared before the intervention in the ProSoil area (i.e., from 2001 to 2014) and the start of interventions (i.e., 2015 to 2022). A significant rise in canopy density (19.79%) and greenness (10.47%), as well as a reduction in LST (1.25%), were observed in the Bhokardan block. However, a higher rise in LAI (31%) and NDVI (15.5%) and a reduction in LST (4.60%) were recorded in the ProSoil intervention area.

14.5 Discussion

The long-term MODIS data-derived greenness (NDVI), canopy density (LAI), and temperature (LST) were used in the study to compute VHI and examine the vegetation health from 2001 to 2023. Although a significant annual change was less prominent separately by NDVI, LAI, and LST, the VHI captured the annual change in vegetation health in terms of rainfall deficit or meteorological drought. The lowest VHI value, ≈ 20, recorded in 2018 indicated the most drought-affected year in the last two decades in the Bhokardan Block. Moreover, the VHI was low in 2002 and 2014, indicating poor crop health due to rainfall deficit. Past studies have reported the drought in this region in 2002 and 2012 (Chavadekar & Kashid, 2018) and 2018 (Gangan, 2019), which was well identified in the VHI time-series data. Aswathi et al. (2018) studied the meteorological and agricultural drought in Maharashtra, India, from 2001 to 2015 using multiple indicators and satellite data, including MODIS-based VHI assessment (Aswathi et al., 2018). They reported a severe agricultural drought in 2002 compared to a less severe drought in 2010. C.M. et al. (2023) studied the meteorological, hydrological, and agricultural droughts from 2001 to 2017 in Maharashtra and Madhya Pradesh, India (Arun Murali, Chowdary, Kesarwani, & Neeti, 2023). They employed MODIS data for VHI mapping in different cropping seasons and reported 2002 as one of the most severe drought years with reduced crop production. Moreover, their agricultural drought frequency map shows that the current study area is among the most drought-prone areas in Maharashtra. The VHI-based crop health assessment demonstrates that the satellite data-derived vegetation health indicators could detect deteriorated crop health due to drought in the region. The trend analysis of vegetation greenness (NDVI) and canopy density (LAI) showed a consistent increase (positive slope), which was also captured in the VHI time series in the Bhokardan block. The high positive correlation ($r = 0.43$; $P < .05$) between VHI and rainfall demonstrated crop health's dependency on the region's rainfall deficit.

Similarly, the long-term greenness and canopy density showed a positive trend, that is, a continuous improvement in crop health in the ProSoil intervention area. A similar trend was also observed in time-series VHI data, which showed improved crop health in recent years. Unlike Bhokardan, the reduction in VHI was not observed in the ProSoil intervention area, even in the most rainfall-deficit year, as in 2018. The mean VHIs from 2001 to 2014 were close to 50, which increased to above 55 from 2015 to 2022. The VHI time-series data showed a relatively stable value in the ProSoil intervention area, which has been less prone to rainfall deficit or drought than the Bhokardan block. The lower relation between VHI and rainfall in this area compared to the Bhokardan block demonstrated a lower dependency on rainfall for crop growth. This could be attributed to the increased drought resilience in the ProSoil intervention area due to the various water resource developmental activities. The higher increase in canopy density and greenness in the intervention area (ProSoil) than in its surrounding area (Bhokardan block) demonstrated the notable outcome of the soil and water-harvesting structures that improved the overall crop health. Moreover, the increased water availability for irrigation during the dry period increased the Zaid-cropping area by 167.92%.

Various studies have reported an increasing trend of greenness in India. Kuttippurath and Kashyap (2023) studied the greenness change in India from 2000 to 2019, including MODIS data and observed an increase in NDVI and LAI by 10% and 11%, respectively (Kuttippurath & Kashyap, 2023). Their study reported that vegetation change due to agricultural activities is two times higher than in forests, wherein the Zaid cropping dominated the greenness change. They have reported that the increased irrigation facilities have led to a higher cropping area in recent years. Park et al. (2023) studied the MODIS LAI data for 19 years and highlighted that the dry season irrigation activities played an important role in increased greenness in India (Park et al., 2023). They attributed that the CO_2 fertilization heightened in leaf area changes compared to land-management activities. A satellite data-based global assessment by Chen et al. (2019) articulated that the greening in India was dominantly contributed by croplands (82%), with an estimated food production surge of 35% since 2000. They also reported an increased leaf area of >11% from 2000 to 2017.

The current study was suitable for landscape-level analysis using coarse resolution (500 m) MODIS data. However, it is difficult to assess farm-level crop health using MODIS data. Alternatively, efforts can be made to use Landsat data (30 m spatial resolution), which has been available for more than 30 years. Although the Sentinel-2 data have 10 m spatial resolution bands, these data have been available since 2015. The farm-level crop health analysis would require other remote sensing data, such as PlanetScope with 3 m spatial resolution or other very high resolution satellite data (<1 m spatial resolution). Furthermore, the current study has mainly focused only on the effectiveness of water-conservation interventions. Future studies may examine satellite data to assess the impact of soil-conservation activities. The current study employed greenness, canopy density, and land surface temperature parameters, which can be replaced by estimating and analyzing the crop yield.

14.6 Conclusion

The positive trend in long-term (from 2001 to 2022) satellite data-derived greenness and canopy density proxies explained an overall improved crop health. Although deteriorated crop health was seen in the nearby area during rainfall-deficit years, such changes were not observed in the ProSoil intervention area. The high-resolution cropping intensity map generated using the Sentinel-2 data exhibited an increased cropping area during the Zaid season. The unaltered crop health during rainfall-deficit years and drastic improvement in cropping areas in the Zaid season demonstrated the improvement in the region's irrigated water supply. The farmers are utilizing the water from wells, lakes, nallas, check dams, etc., constructed as part of the ProSoil project for irrigation activity. The linear regression performed to understand the relationship between VHI and rainfall demonstrated a higher dependency on rainfall for crop growth in the Bhokardan block while a lower dependency in the ProSoil intervention area. The study demonstrated a robust method for crop health monitoring using coarse and moderate-resolution satellite remote sensing. The study showed the importance of sustainable agricultural practices and water-harvesting structures in reducing the impact of droughts or developing drought-resilient agriculture practices. Similar studies in the other ProSoil intervention areas, such as Madhya Pradesh, can validate the applicability of the approach developed in the current study. Future studies may include very high resolution satellite data, other parameters (proxy to soil health), crop yield, etc., for monitoring various ProSoil interventions in different landscapes.

Acknowledgments

The book chapter has been developed from the outcomes of master's thesis of the author developed during his internship with Deutsche Gesellschaft für Internationale Zusammenarbeit (GIZ) GmbH. The authors wish to express gratitude to Navin Horo and Indraneel Ghosh from the Soil Protection and Rehabilitation for Food Security (ProSoil) project team, GIZ, who gave valuable insights about the project and were instrumental in shaping the research methodology initially. The authors also thank the entire GIZ team for their mentorship and support during the internship experience.

Referances

Arun Murali, C. M., Chowdary, V. M., Kesarwani, M., & Neeti, N. (2023). Integrated drought monitoring and assessment using multi-sensor and multi-temporal earth observation datasets: A case study of two agriculture-dominated states of India. *Environmental Monitoring and Assessment*, *195*(1), 1. Available from https://doi.org/10.1007/s10661-022-10550-6, https://link.springer.com/article/10.1007/s10661-022-10550-6#article-info.

Aswathi, P. V., Nikam, B. R., Chouksey, A., & Aggarwal, S. P. (2018). Assessment and monitoring of agricultural droughts in Maharashtra using meteorological and remote sensing based indices. *ISPRS Annals of the Photogrammetry, Remote Sensing and Spatial Information Sciences*, *IV–5*, 253–264. Available from https://doi.org/10.5194/isprs-annals-IV-5-253-2018.

Bajgiran, P. R., Darvishsefat, A. A., Khalili, A., & Makhdoum, M. F. (2008). Using AVHRR-based vegetation indices for drought monitoring in the Northwest of Iran. *Journal of Arid Environments*, *72*(6), 1086–1096. Available from https://doi.org/10.1016/j.jaridenv.2007.12.004.

Behera, M. D., Gupta, A. K., Barik, S. K., Das, P., & Panda, R. M. (2018). Use of satellite remote sensing as a monitoring tool for land and water resources development activities in an Indian tropical site. *Environmental Monitoring and Assessment*, *190*(7). Available from https://doi.org/10.1007/s10661-018-6770-8.

Bento, V. A., Gouveia, C. M., DaCamara, C. C., & Trigo, I. F. (2018). A climatological assessment of drought impact on vegetation health index. *Agricultural and Forest Meteorology*, *259*, 286–295. Available from https://doi.org/10.1016/j.agrformet.2018.05.014.

Chatterjee, S., & Katewa, N. (2024). *Droughts: Learning continuity and recovery* (No. 1466). ADBI Working Paper. https://www.adb.org/sites/default/files/publication/984656/adbi-wp1466.pdf. (Accessed 3 June 2023).

Chavadekar, A., & Kashid, S. S. (2018). *Historical drought analysis of Maharashtra state by using SPI Index. Techno-societal 2016* (pp. 1097–1104). Springer International Publishing. Available from https://doi.org/10.1007/978-3-319-53556-2_109.

Chen, C., Park, T., Wang, X., Piao, S., Xu, B., Chaturvedi, R. K., … Myneni, R. B. (2019). China and India lead in greening of the world through land-use management. *Nature Sustainability*, *2*(2), 122–129. Available from https://doi.org/10.1038/s41893-019-0220-7.

Das, P., Behera, M. D., Patidar, N., Sahoo, B., Tripathi, P., Behera, P. R., … Krishnamurthy, Y. V. N. (2018). Impact of LULC change on the runoff, base flow and evapotranspiration dynamics in eastern Indian river basins during 1985–2005 using variable infiltration capacity approach. *Journal of Earth System Science*, *127*(2)). Available from https://doi.org/10.1007/s12040-018-0921-8.

Dashtpagerdi, M. M., Kousari, M. R., Vagharfard, H., Ghonchepour, D., Hosseini, M. E., & Ahani, H. (2015). An investigation of drought magnitude trend during 1975–2005 in arid and semi-arid regions of Iran. *Environmental Earth Sciences*, *73*(3), 1231–1244. Available from https://doi.org/10.1007/s12665-014-3477-1.

Dutta, D., Kundu, A., Patel, N. R., Saha, S. K., & Siddiqui, A. R. (2015). Assessment of agricultural drought in Rajasthan (India) using remote sensing derived Vegetation Condition Index (VCI) and Standardized Precipitation Index (SPI). *The Egyptian Journal of Remote Sensing and Space Science*, *18*(1), 53–63. Available from https://doi.org/10.1016/j.ejrs.2015.03.006.

Ganeshamurthy. (2014). Towards a national soil policy for food and nutritional security. *Indian Journal of Fertilizers*, *10*(12), 62–67.

Gangan, S. P. (2019). Drought. *Hindustan Times*. https://www.hindustantimes.com/mumbai-news/90-drought-in-marathwada-100-apathy-claim-villagers/story-dMTmZERpB3ryjlV1qU4cUM.html.

Indian Council for Agricultural Research. (2011). *Vision 2030*.

IPCC. (2021). *Climate change 2021, the physical science basis summary for policymakers, working group I sixth assessment report*. https://www.ipcc.ch/report/ar6/wg1/downloads/report/IPCC_AR6_WGI_SPM_final.pdf.

Karra, K., Kontgis, C., Statman-Weil, Z., Mazzariello, J. C., Mathis, M., & Brumby, S. P. (2021). Global land use/land cover with Sentinel 2 and deep learning. In *2021 IEEE international geoscience and remote sensing symposium IGARSS*, (pp. 4704–4707). IEEE. Available from https://doi.org/10.1109/IGARSS47720.2021.9553499.

Kloos, S., Yuan, Y., Castelli, M., & Menzel, A. (2021). Agricultural drought detection with MODIS based vegetation health indices in Southeast Germany. *Remote Sensing*, *13*(19), 3907. Available from https://doi.org/10.3390/rs13193907.

Kogan, F. (2002). World droughts in the new millennium from AVHRR-based vegetation health indices. *Eos, Transactions American Geophysical Union*, *83*(48), 557–563. Available from https://doi.org/10.1029/2002EO000382.

Kundu, A., Denis, D.R.P., & Dutta, D. (2016a). Spatial pattern of agricultural drought using NOAA-AVHRR derived vegetation indices. In: Remote Sensing for Natural Resources Monitoring and Management (pp. 272–281). India: Scientific Publishers.

Kundu, A., Dwivedi, S., & Dutta, D. (2016b). Monitoring the vegetation health over India during contrasting monsoon years using satellite remote sensing indices. *Arabian Journal of Geosciences*, *9*(2), 144. Available from https://doi.org/10.1007/s12517-015-2185-9.

Kuttippurath, J., & Kashyap, R. (2023). Greening of India: Forests or croplands? *Applied Geography*, *161*, 103115. Available from https://doi.org/10.1016/j.apgeog.2023.103115.

Matin, S., & Behera, M. D. (2019). Studying evidence of land degradation in the Indian Ganga River Basin—A Geoinformatics approach. *Environmental Monitoring and Assessment*, *191*(3).

Mavromatis, T., & Stathis, D. (2011). Response of the water balance in Greece to temperature and precipitation trends. *Theoretical and Applied Climatology*, *104*, 13–24. Available from https://doi.org/10.1007/s00704-010-0320-9.

MoES, G. (2020). In R. Krishnan, J. Sanjay, C. Gnanaseelan, M. Mujumdar, A. Kulkarni, & S. Chakraborty (Eds.), *Assessment of climate change over the Indian region*. Springer Singapore. Available from https://doi.org/10.1007/978-981-15-4327-2.

Mudi, S., Prakash, A. J., & Das, P. (2022). *Flood hazard mapping in Assam using sentinel-1 SAR data*. Geospatial technology for environmental hazards. *Advances in Geographic Information Science* (pp. 459–473). Cham: Springer.

Mohanty, A., & Wadhawan, S. (2021). *Mapping India's climate vulnerability: A district level assessment*. New Delhi. Council on Energy, Environment and Water. https://www.ceew.in/sites/default/files/ceew-study-on-climate-change-vulnerability-index-and-district-level-risk-assessment.pdf. (Accessed 3 June 2023).

Office of the Registrar General & Census Commissioner, I (2011). *Census*. Census of India. https://censusindia.gov.in/census.website/.

Pandey, V., & Srivastava, P. K. (2019). Integration of microwave and optical/infrared derived datasets for a drought hazard inventory in a sub-tropical region of India. *Remote Sensing*, *11*(4), 439.

Park, T., Gumma, M. K., Wang, W., Panjala, P., Dubey, S. K., & Nemani, R. R. (2023). Greening of human-dominated ecosystems in India. *Communications Earth & Environment*, *4*(1), 419. Available from https://doi.org/10.1038/s43247-023-01078-9.

Prajapati, R. K., Tripathi, S. K., & Mishra, R. M. (2010). Impact of climate change on agriculture: An Indian perspective. *Current World Environment*, *5*(1), 39. Available from http://dx.doi.org/10.12944/CWE.5.1.06. Accessed 03.06.23.

Prakash, A. J., Kumar, S., Behera, M. D., Das, P., Kumar, A., & Srivastava, P. K. (2023). Impact of extreme weather events on cropland inundation over Indian subcontinent. *Environmental Monitoring and Assessment*, *50*(195(1)).

Shah, D., & Mishra, V. (2020). Drought onset and termination in India. *Journal of Geophysical Research: Atmospheres*, *125*(15).

Space Applications Centre. (2021). *Desertification and land degradation atlas of India (Assessment and analysis of changes over 15 years based on remote sensing)*. Ahmedabad, India: Space Applications Centre, ISRO. https://vedas.sac.gov.in/static/atlas/dsm/DLD_Atlas_SAC_2021.pdf. (Accessed 3 June 2023).

Tabari, H., Marofi, S., Aeini, A., Talaee, P. H., & Mohammadi, K. (2011). Trend analysis of reference evapotranspiration in the western half of Iran. *Agricultural and Forest Meteorology*, *151*(2), 128–136. Available from https://doi.org/10.1016/j.agrformet.2010.09.009.

Times of India. (2022). *Excessive use of pesticides leads to soil degradation. Times of India*. https://timesofindia.indiatimes.com/city/patna/excessive-use-of-pesticides-leads-to-soil-degradation-say-experts/articleshow/95990295.cms.

WRIS. (2023). *India water resources information system*. IMD. Available from https://indiawris.gov.in/wris/#/rainfall.

Wu, D., Qu, J., & Hao, X. (2015). Agricultural drought monitoring using MODIS-based drought indices over the Corn Belt. *International Journal of Remote Sensing*, *36*, 5403–5425.

Yilmaz, B. (2019). Analysis of hydrological drought trends in the gap region (Southeastern Turkey) by Mann-Kendall test and innovative Şen method. *Applied Ecology and Environmental Research*, *17*(2), 3325–3342. Available from https://doi.org/10.15666/aeer/1702_33253342.

Yoon, D. H., Nam, W. H., Lee, H. J., Hong, E. M., Feng, S., Wardlow, B. D., ... Kim, D. E. (2020). Agricultural drought assessment in East Asia using satellite-based indices. *Remote Sensing*, *12*(3), 444. Available from https://doi.org/10.3390/rs12030444.

WOTR (2021). Impact of land use and land cover changes on the land surface temperature: A case study of Bhokardan, Maharashtra. https://wotr.org/2021/05/17/impact-of-lulc-changes-lst-casestudy-bhokardan/. (Accessed 3 June 2023).

Chapter 15

Assessment of agroforestry land use systems for sustainable agriculture development: geospatial perspective using AI

Shivam Trivedi[1], Vinod P.V.[1], Rama Subramoniam S.[1], Chandrasekaran B.[1], Nagashree M.K.[1], Nagajothi K.[1], Divya Shah[2], R. Hebbar[1], C.S. Jha[3], S.K. Srivastav[4], R.B. Sinha[2] and Prakash Chauhan[5]

[1]*Regional Remote Sensing Centre-South, National Remote Sensing Centre (NRSC), Indian Space Research Organisation (ISRO), ISITE Campus, Bengaluru, Karnataka, India,* [2]*Food and Agriculture Organisation of United Nations, New Delhi, India,* [3]*Agricultural Sciences and Applications Group, National Remote Sensing Centre, Indian Space Research Organization, Hyderabad, Telangana, India,* [4]*Regional Centres, NRSC, Indian Space Research Organisation (ISRO), New Delhi, India,* [5]*National Remote Sensing Centre, Indian Space Research Organization, Hyderabad, Telangana, India*

15.1 Introduction

Agroforestry is a sustainable and promising, age-old practice of cultivating agricultural crops and trees together on the same land parcel, sometimes with livestock. Such efficient, diversified, and intensive land management systems create numerous ecological, social, and economic benefits, in addition to reducing pressure on forests (FAO, 2005; ICRAF, 1993; Nair, 1985; van Noordwijk, 2021; Wilson & Lovell, 2016). The tree components in agroforestry land use systems include a wide and diverse range of tree formations, from individual scattered trees on farmlands to plantations crops, tree clusters, home gardens, and shaded perennial, multistrata tree crop associations. Distribution of agroforestry land use systems in tropical and temperate regions is mainly governed by the variations in their climatic and socioeconomic characteristics. Globally, the total area under agroforestry land use systems is estimated to be approximately 1.6 billion hectares, with nearly 78% in tropics and 22% in temperate regions (Nair, Kumar, & Vimala, 2021; Zomer et al., 2014). Asia Pacific region, especially countries like India, Indonesia, Malaysia, Sri Lanka, and Bangladesh, harbor several traditional and modern agroforestry land use systems abundantly, under various agro-ecological environments, including multifunctional home gardens (Ahmad, Uddin, Goparaju, Rizvi, & Biradar, 2020; Cannell, 1987; Shin et al., 2020).

In India, multifunctional agroforestry is an ancient traditional practice and plays a significant role in such a vast agrarian country (Fig. 15.1). In the first updated Nationally Determined Contributions under Paris Agreement (2021–30), submitted to the United Nations Framework Convention on Climate Change (UNFCCC), India aims to create an additional carbon sink of 2.5–3 billion tonnes of CO_2 equivalent through additional forest and tree cover by 2030 (NDC, 2022). Importantly, agroforestry practices have emerged as one of the strategies envisaged to achieve this goal by increasing the tree cover (National Agroforestry Policy (NAP), India, 2014; Zomer et al., 2016).

15.1.1 Agroforestry as a multifunctional, sustainable agriculture solution

The concept of diversified use of land with a blend of both agronomic and perennial species is gaining traction as an appropriate response to the multiple challenges arising from ecological and economic domains (Nair et al., 2021; Udayakumar, Lai, Ussiri, Kumar, & Clay, 2021). Agroforestry practices bring real benefits for livelihood of farming communities in various ways, while supporting mitigation and adaptation to climate change, as well as increasing agrobiodiversity. In the context of climate change, mitigation refers to reducing or minimizing climate change, while adaptation implies adapting to life in a changing climate. With the intentional integration of woody perennials (such as trees, shrubs, palms, bamboo, etc.) into crop and animal farming systems, the potential of agroforestry land use systems

Sustainable Development Perspectives in Earth Observation. DOI: http://dx.doi.org/10.1016/B978-0-443-14072-3.00010-1

FIGURE 15.1 Agroforestry land use systems in India. Traditional farmland with Maize crop and trees, plowing of field using bullocks in Dharwad, Karnataka (top); perennial coffee plantations under shade trees in Hassan, Karnataka (bottom). *Photo credit: RRSC-South, NRSC, ISRO.*

as a promising adaptation strategy to mitigate global ecological challenges and achievement of a wide range of Sustainable Development Goals (SDGs) is well established (Dhyani et al., 2021; FAO, ICRAF, 2019; Lasco, Delfino, Catacutan, Simelton, & Wilson, 2014; Uthappa et al., 2016). Climate adaptation is facilitated through main-streaming adaptation actions into national/regional development planning, integrated water resources management, agroforestry practices and coastal reforestation of mangroves, etc. (Puthalpet, 2019; Quandt, Neufeldt, & Gorman, 2023).

Agroforestry is economically, a risk-negotiating tool at the farm level as it diversifies agricultural production-matrix and its ecological benefits contributes to an array of ecosystem services. The Millenium Ecosystem Assessment, 2005 carried out by United Nations categorized these services into four broad groups, which include provisioning, regulatory, cultural, and supporting services (Nair et al., 2021). Provisioning services provide products from ecosystems, such as genetic resources, food, fiber, fodder, and freshwater, while regulating services provide benefits from ecosystem processes such as regulation of climate, water, and some human diseases. Cultural services refer to nonmaterial benefits from ecosystems through spiritual enrichment, cognitive development, and recreation. Supporting services imply biomass production, soil formation, nutrient cycling, etc. that are necessary for producing all other ecosystem services (Barrios et al., 2017; MA, 2005; Santos, Crouzeilles, & Sansevero, 2019; Torralba, Fagerholm, Burgess, Moreno, & Plieninger, 2016; van Noordwijk, 2021). These services facilitate development of sustainable agriculture, which include agricultural production, rural economy, ecological, and environmental sustainability within the agricultural systems.

15.1.2 Contribution of agroforestry practices in achieving sustainable development goals

Agroforestry practices are at the forefront of policymaking as a key approach to help achieve the Global Goals. These practices align very well with SDGs of United Nations under 2030 Agenda, highlighting its multifaceted contribution to sustainable development perspective. It is a traditionally and culturally acceptable practice in a large number of countries, improving rural livelihoods, mitigating poverty, and enhancing the farmers' income (Ahmad, Goparaju, & Qayum, 2017; Basu, 2014; FAO, 2004; Puri & Nair, 2004; Waldron et al., 2017; Zomer et al., 2016). In parallel, agroforestry resources have the potential to mitigate the impacts of climate change through moderation of microclimate and conservation of natural resources in the short term, while sequestering carbon in the long term (Dhyani, 2014; National Agroforestry Policy (NAP), India, 2014). Agroforestry land use systems can contribute toward achieving at least 9 out of 17 SDGs (https://sdgs.un.org/goals), either directly through strong impact potential or indirectly by supporting the cause of other goals as documented through evidence-based information (Jamnadass et al., 2011; Rolo, 2014;

Montagnini & Metzel, 2017; Waldron et al., 2017; Karlsson, 2018; Andersson, 2018; Dhyani et al., 2021; Nair et al., 2021; Ntawuruhunga, Ngowi, Mangi, Salanga, & Shikuku, 2023; SDG, 2023) (Fig. 15.2).

The various ways in which agroforestry brings broader tangible and intangible environmental and societal benefits, in the context of specific SDGs are as follows:

SDG 1. No Poverty: *End poverty in all its forms everywhere*—Poverty reduction by enhancing income of smallholder farmers by diversifying their income sources through production of various crops, timber, and nontimber products.

SDG 2. Zero Hunger: *End hunger, achieve food security and improved nutrition and promote sustainable agriculture*—Hunger alleviation through food and nutritional security by providing a diverse range of food products.

SDG 13. Climate Action: *Take urgent action to combat climate change and its impacts*—Enhancing resilience to climate change through diversified and adaptive land use systems; mitigating climate change by sequestering carbon in trees and soil.

SDG 15. Life on Land: *Protect, restore and promote sustainable use of terrestrial ecosystems, sustainably manage forests, combat desertification, and halt and reverse land degradation and halt biodiversity loss*—Promoting biodiversity conservation by providing habitat for various species and enhancing sustainable land management.

Additionally, agroforestry systems also support in achieving SDGs 3, 5, 6, 7, and 12. These specifically include good health and well-being, by improving overall nutritional health and supply of medicinal plants for traditional medicine (SDG 3); improving gender equality by empowering women, through income-generating opportunities and involvement in decision-making (SDG 5); increasing access to clean water, contributing to water conservation through improved watershed management (SDG 6), sustainable energy solutions through alternative methods, reducing dependencies on fossil fuels (SDG 7), and responsible agricultural production through sustainable use of resources; promoting organic matter and nutrients recycling (SDG 12).

15.1.3 Agroforestry land use systems in India

Trees and forests are not only critical components of ecology, but are also considered as integral to the Indian culture. Agroforestry has been practiced in India since time immemorial and also an inherent part of India's evergreen revolution, being a nature-based sustainable solution. India is a vast, developing country with a population of 1.44 billion and agriculture is the mainstay of Indian economy. About 70% of rural households depend on agriculture for their livelihood. Majority of the farmers are either small or marginal (https://www.fao.org/india/fao-in-india/india-at-a-glance/en/), with lower per capita income. Panchayati Raj, which is the system of local self-governance of villages in rural India under Ministry of Panchayati Raj, Government of India. It enables overall development of the rural community, with Gram Panchayat (GP) as its basic functional unit. Rural Area Development Plan Formulation and Implementation (RADPFI) Guidelines, 2021 highlight the significance of spatial development planning for rural areas at grassroots or GP level (RADPFI, 2021).

Agroforestry land use systems are emerging as promising solutions for development of sustainable agriculture, while immensely contributing to upliftment of farming community as discussed in Section 1.2. Prioritization of GPs for

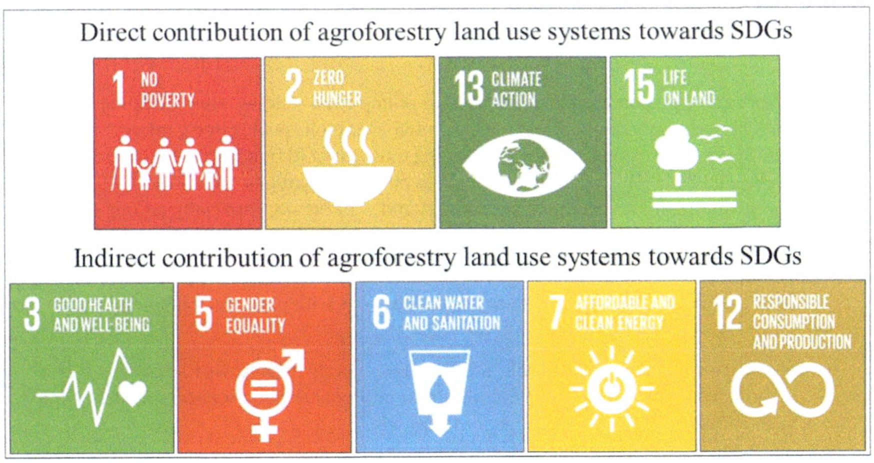

FIGURE 15.2 Contribution of agroforestry land use systems in achieving SDGs under 2030 Agenda for Sustainable Development. *Adapted from https://sdgs.un.org/goals.*

intensification of agroforestry land use systems under Gram Panchayat Spatial Development Planning (GPSDP) for land resources will facilitate increase in tree cover, as well as overall socioeconomic development of the farming community (RRSC-South, 2021; Trivedi et al., 2024a). Agroforestry systems can be classified on the basis of several factors, viz., structure, function, physiognomy, floristics, ecological considerations, socioeconomic factors, history, etc. (Dwivedi, 2022). Based on structural classification, referring to composition, stratification, and dimension of the crop, the most widely recognized classification is based on the composition (Gholz, 1988; Nair, 1985; Sinclair, 1999). Predominant Indian agroforestry land use systems are briefly described in Table 15.1.

These agroforestry practices are prevalent in diverse agro-ecological regions (AER) of the country in various proportions. AER map of India was prepared in 1992 with 20 regions, while revised edition was brought out in 2015 (Singh, Chatterji, Chattaraj, & Butte, 2018). These 20 regions were grouped into arid, semiarid, subhumid, humid, and coastal ecosystems in the revised map. The spatial distribution and extent of multistrata agroforestry land use systems in these varied AER across the vast expanse of Indian croplands is not well understood. The reasons can be attributed to the inherent complexities in accurately mapping agroforestry components and trees outside forest (TOF) in mixed tree and agricultural cropping systems, including scattered individual farm trees (FSI, 2020; FAO, 2013; Rizvi et al., 2020). The National Agroforestry Policy, initiated by Government of India, highlights various impediments to the promotion of agroforestry and also suggests ways forward to promote agroforestry systems in the country. One such way is to develop a sound spatial database and reliable baseline information system (National Agroforestry Policy (NAP), India, 2014).

The National Agroforestry Policy of India initiated in 2014 is a comprehensive policy framework designed to improve agricultural livelihoods by maximizing agricultural productivity for mitigating climate change and to bring in coordination among different schemes, programs, and policies of Government of India (National Agroforestry Policy (NAP), India, 2014). Scientific database on different types of agroforestry systems prevalent in India is very important for planners and decision-makers to implement government sponsored schemes at GP level. Although forest resources are routinely assessed at state and national level, reliable database on agroforestry and trees growing outside forest lands are limited. Additionally, TOFs, which are inherent part of agroforestry land use systems, are a key contributor in the carbon balance sheet prepared and consolidated by Ministry of Environment, Forest and Climate Change (MoEFCC), Government of India. A current lack of robust monitoring mechanisms has contributed to an insufficient grasp of distribution of agroforestry land use systems in relation to management practices, as well as their vulnerability to climate change and diseases (Brandt et al., 2024).

15.1.4 Role of Earth Observation in mapping agroforestry resources

During the past decade, Earth Observation Systems have shown significant advances globally, resulting in an unprecedented improvement in data quality. In the formative years, low- and medium-resolution satellite images were utilized for various geospatial applications including regional scale and landscape level vegetation mapping, forest cover and type mapping, etc. (IRS LISS-I, -II, and -III). Medium-resolution satellite data (Indian Remote Sensing, Linear Imaging

TABLE 15.1 Predominant Indian agroforestry land use systems.

S.No.	Agroforestry land use systems and components	Description
1.	Agri-silvicultural systems (trees combined with crops)	Introduction of woody species in the form of improved fallows, alley cropping with trees as part of hedges in linear pattern, multilayer multispecies tree gardens, multipurpose trees on crop lands including those on field bunds; plantation crop combinations as block plantations; kitchen and home gardens, trees in soil conservation and reclamation; shelterbelts and windbreaks, especially around coastal regions; live hedges and fuelwood production through firewood species.
2.	Silvopastoral systems (trees and pastures or animals)	Trees on rangeland or pastures including scattered individual trees; protein banks on farmland (protein rich tree fodder); plantation crops with pastures and animals/cattle under shade of trees.
3.	Agro-silvopastoral systems (animals, trees and crops)	Home gardens involving animals, multipurpose woody hedgerows for mulching, green manure, etc.; apiculture with trees for honey production; aqua-forestry with trees lining lakes, ponds, tree leaves as forage for fishponds.

Self-Scanning Sensor, i.e. IRS LISS-III, 23.5 m) were used to assess percentage tree cover at pixel level, under various categories such as tree groups and cropland. With the availability of improved sensors such as LISS-IV (5.8 m), a wide range of vegetation-mapping applications were carried out at local/community level, with better accuracy. LISS-IV (5.8 m) images provided relatively better results for delineation of scattered trees, linear patterns, and block plantations with reasonably good accuracy (Rizvi et al., 2020). These datasets, coupled with Cartosat-1 (2.5 m), were also used for studies such as site-suitability for the introduction of various vegetation categories including horticulture plantations, commercial plantations, forest plantations, and TOFs (Hebbar et al., 2014, 2019; Pujar et al., 2016; Ray et al., 2016; Rizvi et al., 2020; Ravishankar et al., 2022).

Forest Survey of India followed a grid-based approach (5x5 km), utilizing 10 m Sentinel-2 images for unsupervised classification of TOF in selected grids. The classified image is visually analyzed and edited to arrive at three classes in areas under TOF (block, linear, and scattered), thereby estimating about 29.38 million ha area in the country under TOF through extrapolation technique (FSI, 2019; Rizvi et al., 2020). Recent years have witnessed a paradigm shift in sensor technology with availability of high to very high resolution satellite (VHRS) images (1 m and better pixel resolution) with improved spatial, spectral, and temporal resolutions. This has enabled spatial demarcation of perennial vegetation types at microlevel, including scattered individual trees in the farmlands. Recent studies have demonstrated the potential of VHRS images in mapping TOF and other tree components at various scales (Brandt & Stolle, 2020; Brandt et al., 2020; Vinod, Trivedi, Hebbar, & Jha, 2023).

15.1.5 Advances in image processing and analysis techniques

Machine learning (ML) techniques have been extensively used for vegetation analysis from satellite images of varying resolutions. Early methods for classifying individual pixels from coarse-resolution satellite images, such as k-Nearest Neighbors and Decision Trees, relied heavily on the spectral characteristics of the pixels. These techniques have evolved significantly over time, moving from per-pixel-based methods for low-resolution data to object-based image analysis (OBIA) techniques for medium- and high-resolution images, and more recently, to advanced deep learning (DL) techniques. Friedl and Brodley (1997) highlighted the effectiveness of decision tree classification techniques for land cover classification using low-resolution images. Subsequently, other techniques like Random Forests (Pal, 2005) and support vector machines (Huang, Davis, & Townshend, 2002) demonstrated their potential in land cover classification and vegetation analysis using low- and medium-resolution images. However, these per-pixel techniques proved less effective for medium- and high-resolution images where spatial context becomes critical. With the advent of medium- and high-resolution satellite imagery, the focus shifted toward OBIA, which groups pixels into meaningful objects or segments and classifies them based on spectral, spatial, and contextual information. These OBIA techniques have been successfully applied to numerous remote sensing problems, including tree crown extraction and vegetation analysis (Blaschke, 2010; Hebbar et al., 2014; Vinod et al., 2023).

Recent advancements in ML, particularly DL techniques like Convolutional Neural Networks (CNNs), have revolutionized satellite image analysis. CNNs have been widely adopted for tasks such as land cover classification. Liu and Chen (2015) showcased the effectiveness of scalable deep CNNs for tree crown extraction, tree health estimation, and vegetation analysis in high-resolution satellite images. Semantic segmentation, a popular DL technique where each pixel is assigned a class label, has become a cornerstone in satellite image analysis. Shelhamer, Long, and Darrell (2017) employed Fully Convolutional Networks for the semantic segmentation of satellite images for vegetation analysis.

The U-Net architecture (Ronneberger, Fischer, & Brox, 2015), specifically designed for semantic segmentation, has gained wide adoption in satellite image analysis due to its encoder–decoder structure combined with skip connections, resulting in precise image segmentation. Vinod et al. (2023) utilized a modified U-Net architecture along with OBIA for classifying TOF in urban environments. Other CNN-based architectures like SegNet and DeepLab (Chen, Papandreou, Kokkinos, Murphy, & Yuille, 2018) have been successfully applied to various tasks, including agricultural field segmentation and urban area classification. Transfer learning, which leverages pretrained models on large datasets, has further enhanced the performance of DL-based methods in satellite image analysis. By fine-tuning pretrained models, researchers have achieved improved results with smaller datasets and reduced training times. Popular pretrained models include ImageNet-trained CNNs like Visual Geometry Group (VGG), ResNet, and Inception, which have been adapted for various satellite image analysis tasks. In the broader context of Earth Observation Systems for vegetation studies, particularly agroforestry mapping, these advancements in ML and DL techniques provide powerful tools for more accurate and efficient analysis. These methodologies not only improve our understanding of land cover and vegetation health but also support sustainable management practices in agroforestry and other vegetation-related domains.

15.2 Assessment of agroforestry land use systems—a case study in Karnataka, India

15.2.1 Background

One of the key requirements for implementation of National Agroforestry Policy of India is a robust, geospatial baseline database of agroforestry land use systems. Currently, details on agroforestry components in the country are primarily available in nonspatial format. It is estimated that around 29 Mha of the 142 Mha of the net sown area (NSA) in the country falls under the category of TOF (FSI, 2021). Horticultural and commercial plantations alone contribute more than 12 Mha and a significant portion of this has been mapped/inventoried using geospatial techniques under different flagship projects jointly undertaken by Indian Space Research Organization and Department of Agriculture and Cooperation (Ray et al., 2016).

According to TOF classification (FSI, 2020), trees within agricultural lands (based on land use) are categorized under agroforestry. Most of the earlier studies have used medium-resolution data for mapping TOF and agroforestry systems, which is not adequate to address the large diversity of the agroforestry systems of India and thus, in order to get a quantitative and reliable estimate, higher spatial resolutions (1 m or better) are required for inventory of major components of the agroforestry systems, including individual scattered trees on the farmlands. However, reliable spatial database on area under agroforestry land use systems is not available in India. In this background, the importance of holistic baseline spatial database on the existing agroforestry land use systems has been perceived (Nair & Dagar, 1991; Rizvi, Handa, & Arunachalam, 2022).

In this context, the present research was carried out under Technical Cooperation Programme of Food and Agriculture Organization (FAO), United Nations, New Delhi, India with National Remote Sensing Centre (NRSC), Indian Space Research Organization (ISRO) (Trivedi et al., 2024b). This study aimed to develop a novel, robust and scalable geospatial approach for the assessment of area under agroforestry systems in two selected pilot study districts, utilizing VHRS images and Artificial Intelligence techniques (RRSC-South, 2022). This standard procedure could be used for upscaling to arrive at regional/national level agroforestry estimates, prioritization for agroforestry intensification, as well as suitability analysis for sustainable farming under climate change scenarios. This chapter discusses the procedure adopted for assessment of tree components in agroforestry land use systems using VHRS images and DL for prioritization of agroforestry systems at grassroots level.

15.2.2 Materials and methods

15.2.2.1 Study areas

As part of the pilot studies, two districts Dharwad and Hassan, from Karnataka in southern India, representing diverse agro-ecological regimes, were considered (Fig. 15.3). These districts were selected in consultation with National Rainfed Area Authority (NRAA), Ministry of Agriculture and Farmers Welfare, Government of India for assessment of agroforestry land use systems. The agro-ecological characteristics of the study areas are provided in Table 15.2.

15.2.2.1.1 Dharwad district (Karnataka, India)

Current-day Dharwad district is a remnant of the erstwhile undivided Dharwad district, which was divided into three districts in 1997, with Haveri and Gadag being the other two districts. The district is situated in the western region of the northern half of Karnataka. Dharwad district may be divided into three natural regions, viz., the malnad, semimalnad (transitional region), and the plains. The district encompasses an area of 4265 km^2, lying between the latitudinal parallels of 15°02′ and 15°51′ N and longitudes of 73°43′ and 75°35′ E. The district lies approximately 800 m above the mean sea level. It is bounded on the north by Belgaum district, on the east by Gadag district, on the south by Haveri district, and on the west by Uttara Kannada district. The district now comprises a total of eight taluks, namely, Dharwad (district headquarters), Hubballi, Hubballi Nagara, Kalaghatagi, Navalgund, Kundgol, Alnavar, and Annigere. Hubballi is unequivocally considered as the commercial hub of northern Karnataka. There are 144 GPs in the district. District receives an annual average rainfall of 787 mm with the distribution being uneven; Kalghatagi taluk and Alnavar area in Dharwad taluk, in particular receive more rainfall than other taluks of the district. These regions, on an average, receive moderate to heavy rainfall and have dense vegetation. The minimum and maximum temperatures recorded in the district are 12°C and 40°C, respectively.

It is a unique district, in the sense that the western half is red-soil-dominant region, while the eastern half is predominantly black soil. The Western Ghats lining the western border of the district have dense forest cover (in Dharwad and Kalghatagi taluks), and the perennial tree cover progressively tapers toward the eastern edge, except for the horticulture

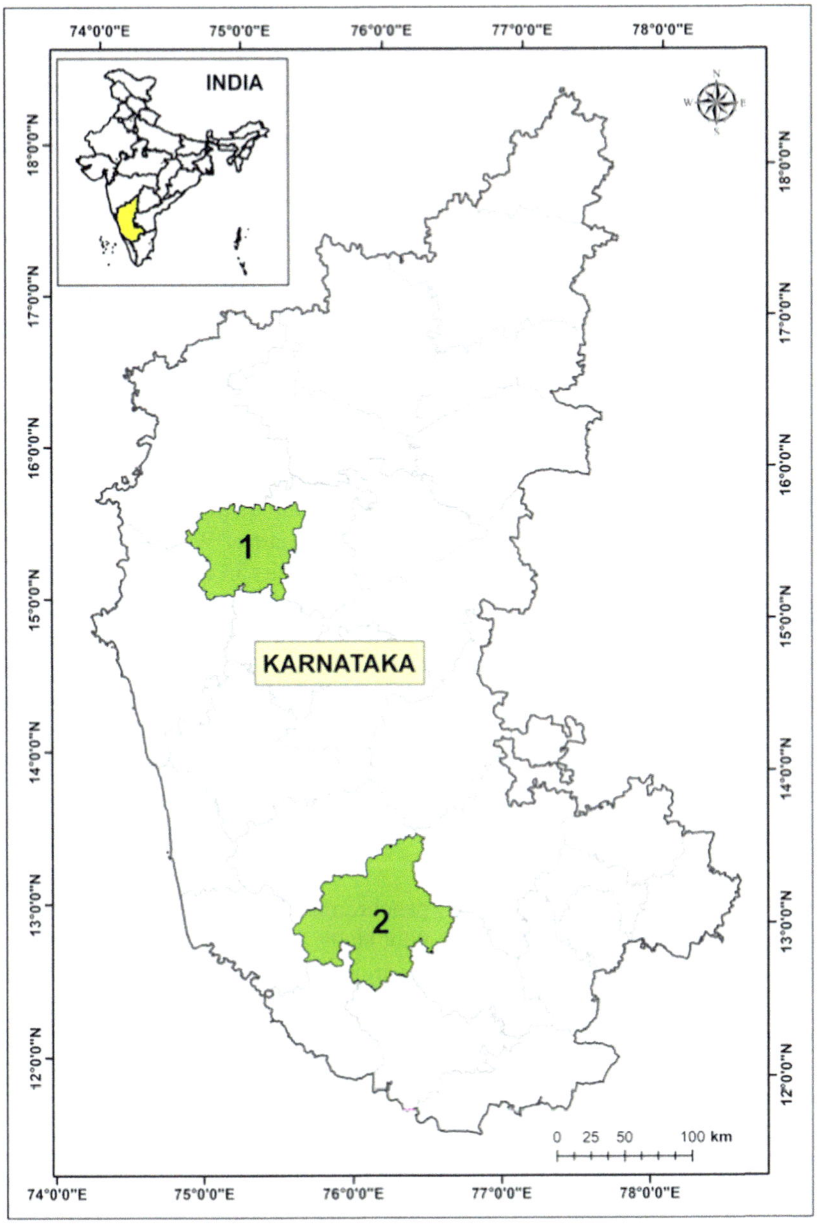

FIGURE 15.3 Study location map. Map lines delineate study areas and do not necessarily depict accepted national boundaries. Pilot districts in Karnataka, India—1. Dharwad 2. Hassan.

TABLE 15.2 Details of study areas and corresponding agro-ecological regions.

S.No.	District (State)	Total Geographic Area (TGA) (sq.km)	Agro-ecological region (AER)	Dominant soil type	Details of AER
1	Dharwad (Karnataka)	4265	8 Deccan Plateau	Mixed red and black soils	Hot semiarid eco-region; Length of growing period (LGP) 150–180 and 120–150 days
2	Hassan (Karnataka)	6845	9 Deccan Plateau	Red loamy soils	Hot semiarid eco-region; LGP 150–180 (inclusion of 180–210 days)
TOTAL		**11,110**			

plantations, which are again mostly confined to red-soil regions. Dominant crops in black-soil regions are cotton, wheat, jowar (rabi), pulses/oilseeds, maize, and chillies; red-soil regions are predominantly suitable for cultivation of paddy. Mango is the most dominant horticulture plantations in western parts of Dharwad with few patches of coconut and arecanut (*Areca catechu*) plantations, apart from few tree species on farm land and roadsides like neem (*Azadirachta indica*), Pongamia (*Pongamia pinnata*), *Acacia* spp., and others. This district is considered the capital of the cultural and the educational activities in Karnataka.

15.2.2.1.2 Hassan district (Karnataka, India)

Hassan district is located in the south-western part of Karnataka, between 12°13′ and 13°33′ N latitudes and 75°33′ and 76°38′ E longitudes. The district has a total area of 6845 km^2. The district is divided into three distinct geomorphic units, that is, the western hilly terrains (part of the Western Ghats), the central transition zone and the eastern *Maidan* (plain) region. Hassan is divided into eight taluks, namely, Hassan (district headquarters), Alur, Arkalgud, Arsikere, Belur, Chennarayapatna, Holenarsipur, and Sakleshpur. The climate in Hassan is tropical. Average rainfall is 1211 mm and the average day-time maximum temperature ranges between 27°C and 34°C, and the winters are cool and chilly. The soils of the district display a wide diversity and are quite fertile. The main soil types are red soil, red sandy soil, mixed soil, and silty clay soil. Around 70% of the population depends on agriculture and the district has a strong agriculture base, facilitating agro and food-processing industries. Pristine evergreen forests, as part of Western Ghats, form the extreme western region of the district. Hassan district is specifically chosen for its diversity in terms of commercial coffee and tea plantations under evergreen shade trees dotting the western portion of the district, and the coconut-dominant horticulture plantations forming the major agroforestry component more specifically in/along the valleys. Arecanut forms the second important horticulture plantation, present either as a mixed understorey in coconut plantations or as pure stands. Further, horticulture/commercial plantations occupy almost one-third of the geographical area of the district. Irrigated paddy, ragi, and other cereals and pulses are the major agricultural crops grown in Hassan. Hemavathi reservoir located in the southern part of Hassan meets the critical irrigation and drinking water needs of the district; Yagachi and Vatehole reservoirs also supplement the requirements.

15.2.2.2 Satellite and ancillary datasets used

Submeter VHRS datasets of 0.6 m (Cartosat-2 series multispectral (MX) and panchromatic (PAN) merged) and 0.7 m (KOMPSAT 3/3 A, merged data) acquired during 2018–21 were used for the pilot study areas. Ancillary data in terms of administrative boundaries, Survey of India (SOI) topo-sheets, and digital elevation model (DEM) were used (SOI, 2023). Ancillary datasets were also referred for interpretation of satellite data, improved stratification of forest/nonforest regions, segregation of agricultural areas into required categories to improve the quality of deliverables.

15.2.2.3 Methodology

The methodology for generation of agroforestry land use system map consists of two main stages, namely: (1) extraction of tree components from VHRS data using DL techniques and (2) integration of tree components with LULC map (Fig. 15.4). The first stage involves creation of tree-component layer (up to individual scattered trees in farmlands) using semantic segmentation-based DL model. It involves several steps such as satellite data preprocessing, creation of annotated training samples, DL model training and validation, and inference. This is an iterative process involving model fine-tuning in order to get a reasonably good output.

In the second stage, the tree-component map generated in the first step is integrated with harmonized classes derived from 1:10,000 LULC map using spatial analysis to arrive at the integrated agroforestry land use map. Further, prioritization of GPs was also carried out for selection of panchayats with marginal of tree components for promoting agroforestry land use systems at grassroots levels.

The details of the steps involved are described below.

15.2.2.3.1 Extraction of tree components using deep learning technique

15.2.2.3.1.1 Satellite data processing

15.2.2.3.1.1.1 Orthorectification. Orthorectification corrects satellite imagery for positional accuracy and terrain relief, addressing errors such as terrain relief, scale variation, sensor orientation, and internal issues. This process is essential when multiple images need to be mosaicked to create seamless data for study areas and is crucial for Geographic Information System (GIS) and other mapping applications. Orthorectification was performed using

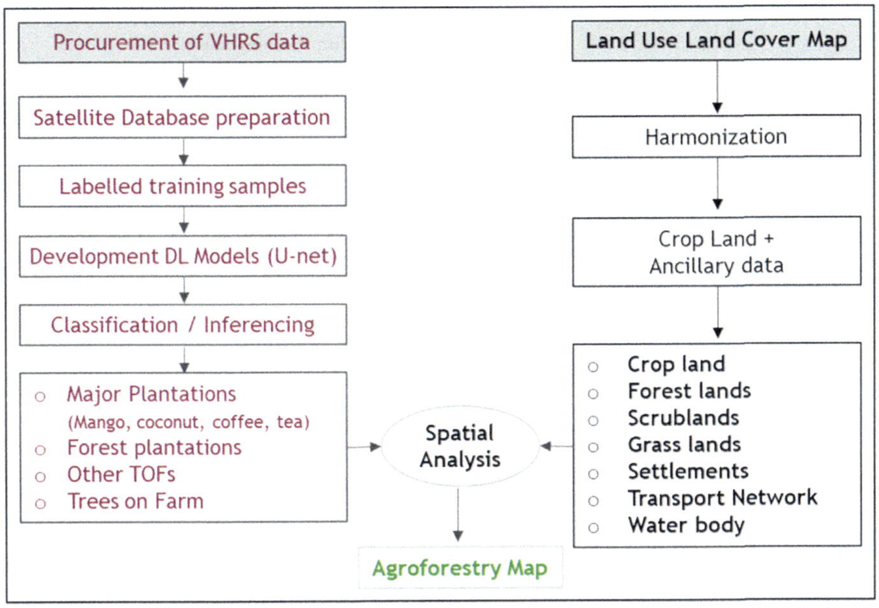

FIGURE 15.4 Overall methodology flowchart for generation of integrated agroforestry land use systems map.

reference ortho-data and a DEM with Universal Transverse Mercator (UTM) projection and World Geodetic System (WGS) 84 spheroid. The process included collecting and refining control/tie points from reference images and applying contrast enhancement to generate a seamless mosaic and improve image interpretability.

15.2.2.3.1.1.2 ***Data fusion.*** PAN and MX datasets from the Cartosat-2 series and KOMPSAT 3/3 A satellites were merged using a standard data-fusion technique. This integration combines the spectral information from multispectral data with the spatial detail from panchromatic data, resulting in a high-resolution multispectral product. After quality checking, these datasets were mosaicked at the district level.

15.2.2.3.1.2 Development of deep learning models In order to develop the DL model for tree-component extraction from the VHRS data, a transfer learning based approach, which employed popular semantic segmentation architecture, namely U-Net, was used. The U-Net architecture consists of an encoder–decoder structure (Ronneberger et al., 2015). The encoder creates the feature map of the image through a series of convolution and max pooling operations, which are then passed to a decoder with a series of upsampling, max pooling and skip connection operations, and finally softmax based prediction to generate final tree-component output. The U-Net model with a pretrained ResNet34 backbone has been used for training the VHRS data. The various steps involved in the trained DL model development are described below.

Training data preparation

Generation of accurately labeled training samples is an important step for development of a robust DL model for this study. About 3486 multiclass-labeled training samples representing the overall variability of the tree components were prepared from VHRS images (Fig. 15.5), utilizing field information from a total of 788 geotagged ground truth (GT) locations. These training samples consisting of image patches and their corresponding masks were used for training the DL model. Standard data augmentation techniques, such as image flipping, rotation, cropping, and resizing, were applied to prevent model overfitting. These samples were then split into training and validation sets, ensuring balanced class distribution in both subsets.

Model training

To train the U-Net-based model, the training samples were divided into training and validation samples (2789 samples for training, i.e., 80% and 697 samples for validation, which is 20%). The U-Net model was fine-tuned in two stages, with binary cross entropy (BCE) as the loss function and batch gradient descent algorithm for model optimization. Model fine-tuning was conducted in two stages:

1. Stage 1: The last layer of the pretrained model was modified and retrained using a technique called one cycle learning (Smith, 2018) with a single learning rate (LR). In this stage, only the parameters of the last layer were updated.
2. Stage 2: Discriminative LRs (i.e., lower LR for later layers and higher LR for earlier layers) were used to fine-tune the entire model, updating the parameters of all layers. This two-stage learning process results in a more robust model compared to training only the last layer.

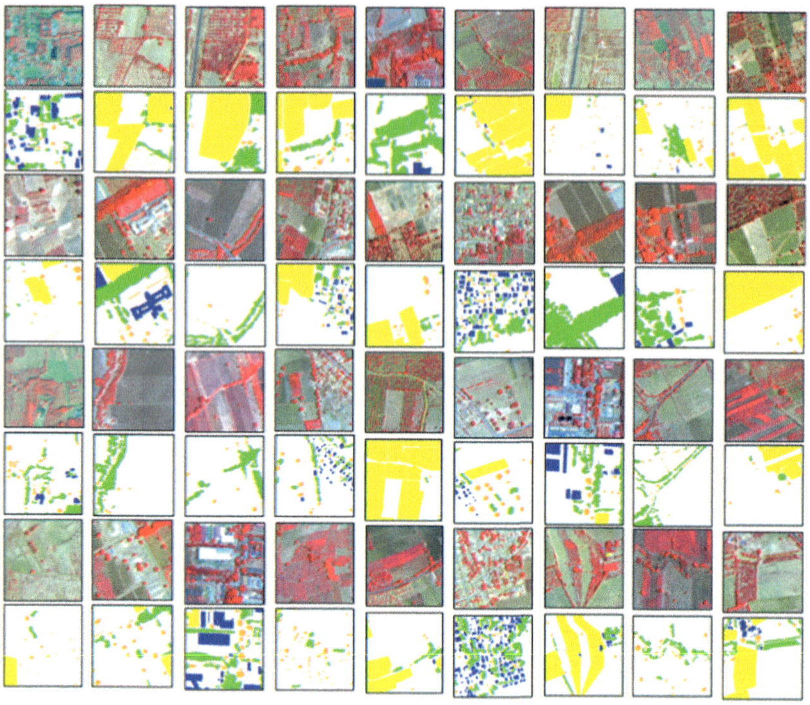

FIGURE 15.5 Multiclass-labeled training samples for deep learning model.

During training, a BCE loss function was optimized using a batch gradient descent algorithm. The model was validated using the validation dataset, and the corresponding losses were calculated for both training and validation data. Training continued for a specified number of epochs until the training and validation losses reached acceptable minimum values, indicating reasonably good model performance. The trained model was then saved for inference and classifying tree components from VHRS data for the entire study region.

Inference and output generation

During inference, the satellite images consisting of various tree classes were classified using the trained model. The classified outputs were geo-referenced, mosaicked, and converted into vector format for integration with LULC maps at a 1:10,000 scale. Vector operations like smoothening functions were used for removing small polygons and outliers.

15.2.2.3.2 Integration of deep learning outputs with land use land cover

Spatial analysis was conducted in a GIS environment to create an integrated agroforestry land use system map for various study areas and spatial extents of classes. This process involved multiple steps. First, the DL model-based classification identified different categories, such as trees on farmland and dominant plantations including mango, coconut, coffee, and tea, as well as agricultural plantations, depending on the diversity in the study areas. The remaining land use categories were generated through a two-step process using a legacy 1:10,000 scale LULC map produced under the "Space based Information Support for Decentralized Planning" (SIS-DP) initiative (Bhuvan Panchayat, 2023; NRSC, 2020a, 2020b) along with the DL output. Initially, the different Level-3/Level-4 classes from the LULC map were consolidated into broader categories to align with the agroforestry land use map. After creating these harmonized classes, a defined GIS procedure was followed to integrate them with the DL-based output classes. This integration resulted in the agroforestry land use map, providing microlevel details down to individual farm trees. Finally, majority filtering and dissolving operations were performed to smooth the output for generating a comprehensive and integrated agroforestry map for both the study areas.

15.2.2.3.3 Prioritization of Gram Panchayats for intensification of agroforestry land use systems

The integrated agroforestry land use system maps, generated in the previous step were further analyzed for prioritization of GPs for promoting agroforestry practices. The integrated agroforestry land use layer was intersected with GP boundary of respective districts, Dharwad and Hassan. Class-wise statistics were then generated for each GP. Area under agriculture plantations, mango plantations and agriculture TOF, in the case of Dharwad and agriculture plantations, coconut plantations and agriculture TOF, in the case of Hassan, was added to arrive at total area under agroforestry tree components.

The proportion of these agroforestry tree components within NSA was calculated and expressed as percentage for each district, at GP level. Accordingly, all GPs were divided into four priority classes based on proportion of existing agroforestry tree components, into very high (<5%), high (5%–15%), medium (15%–30%), and low (<30%), indicating higher priority for GPs with lesser area under trees within NSA. GPs dominated with forests/forest plantations and also those dominated by perennial coffee plantations, existing mainly under natural shade trees were not considered for prioritization. Remaining areas including built-up class were categorized under settlement-dominated regions and not included in the analysis.

15.2.3 Results and discussion

This section presents the results of DL model training and inference, integration of DL outputs with LULC map, and prioritization of GPs for agroforestry intensification.

15.2.3.1 Deep learning model training

In stage 1, the head layer of the U-Net model was fine-tuned with the following hyperparameters: batch size: 8, LR: 0.000158, weight decay: 0.01, momentum: (0.95, 0.85, 0.95), and epoch: 10. "Ranger" was used for model optimization. In stage 2, the following hyperparameters were used: LR: (0.0000003.962e − 7, 3.962e − 5), weight decay: 0.01, batch size: 8, epoch: 12, and momentum: (0.95, 0.85, 0.95). In this stage, all the layers were fine-tuned.

DL model training and validation curves were obtained at different training stages of the training process (Fig. 15.6). During the first stage of the training wherein the last layer of the U-Net model with pretrained ResNet50 was retrained, the training and validation losses were relatively high at 1.910 and 0.462, respectively, and gradually decreased over the epochs, reaching smaller training and validation losses of 0.237 and 0.239 at the end of the 10th epoch. The model validation accuracy at this point was found to be 93.45%. In the second stage, the training and validation losses were found to be 0.1716 and 0.1823, respectively, at the beginning and decreased to 0.1428 and 0.1762 toward the end of the training epochs with model accuracy slightly increasing from 93.45% to 93.72%. As observed in Fig. 15.6, the training and validation loss curves were not smooth in this stage, mainly due to changes in parameters of all layers in contrast to the first-stage training, where only the last layer was getting modified. To get a smoother loss curve, the number of training samples may be increased and a larger batch size may be used. But since the losses were in decreasing trend, it did not affect the overall performance of the model.

The model validation estimated an accuracy of 93.7% and Intersection of Union (IoU) of 0.73, which shows the robustness of the model for classification of various tree components. The inference results demonstrate the excellent performance of the model in classifying different tree components such as individual trees on farmland, agriculture, and horticulture plantations including mango, coconut, tea, and coffee (Fig. 15.7).

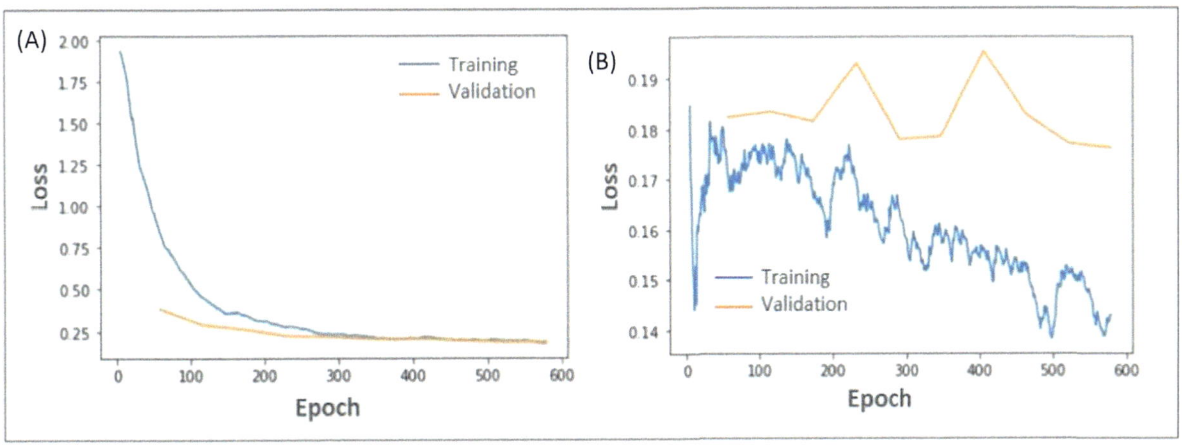

FIGURE 15.6 Deep learning model development. Loss versus epochs/batches processed: (A) training and validation losses after head layer training; (B) training and validation losses after training using discriminative learning rate.

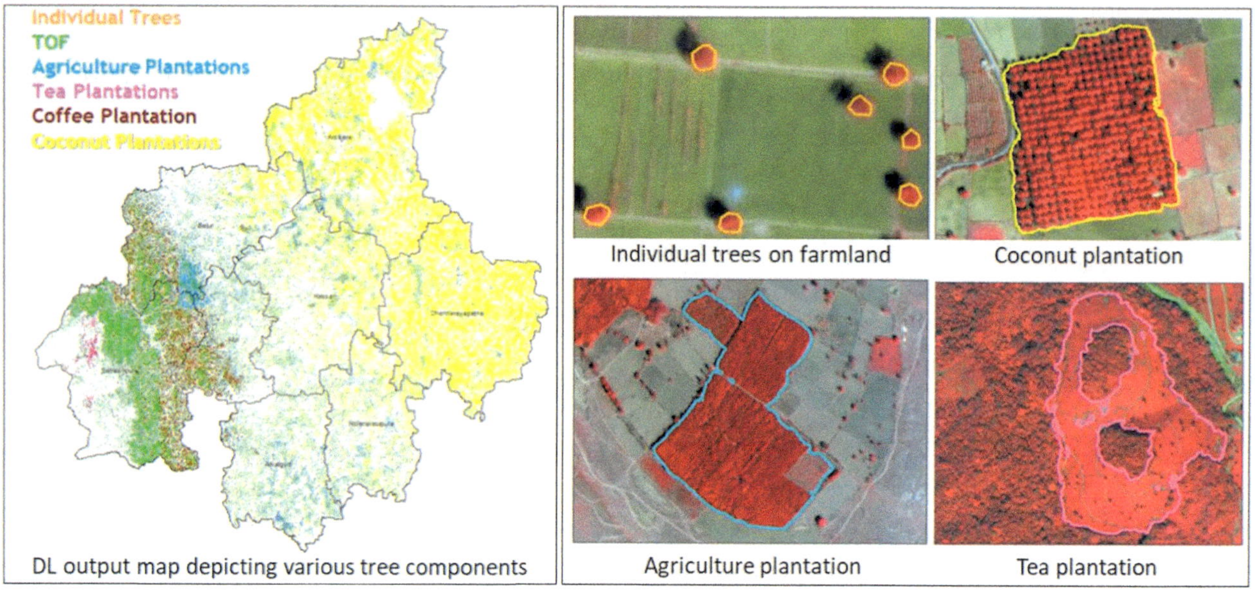

FIGURE 15.7 **Hassan district. Map lines delineate study areas and do not necessarily depict accepted national boundaries.** Deep learning output map and zoomed views for different tree components overlaid on False Colour Composite (FCC) image, depicting microlevel information.

15.2.3.2 Integrated agroforestry land use map

The local knowledge and long experience of domain experts helped in the accuracy assessment of agroforestry land use map qualitatively utilizing GT information and existing plantation crop maps. Then, a statistical approach was employed for accuracy assessment, quantitatively using independent samples, identified using legacy field database collected as part of other national/regional level projects (Chaudhari et al., 2019; Hebbar et al., 2014, 2019; Ray et al., 2016). For this purpose, 1248 GT locations were marked on VHRS data independently (based on field data and interpretation knowledge), objectively, representing all classes and diversities. Overall classification accuracy was estimated to be 86.5%, with a kappa coefficient of 0.847.

The integrated agroforestry land use map for Dharwad district (Fig. 15.8) indicate that the agriculture cropland category is the dominant class (318,241.6 ha) in the district. Agriculture plantations (dominated by mango plantations) accounted for 15,491 ha and the Agriculture TOF class accounted for 4635.2 ha (Table 15.3). As observed during the GT collection, *Prosopis juliflora*, commonly known as Bellary Jali, is the major component of TOF as most dominant perennial scrub vegetation, almost omnipresent in the black-soil region, especially alongside drainages, roads, and wastelands. Both the classes put together occupy spatial extent of around 6.0% of cumulative agriculture area (including agriculture plantations) in the district. There is also a considerable area of 14,814.7 ha under scrubland and forest land (33,679.5 ha). Apart from this, urban vegetation also covered an extent of around 2722.7 ha. Based on the outcome, agri-silviculture (agriculture, plantations, and TOF including scattered individual trees) is the dominant agroforestry system, followed by the TOF inside scrubland. In a nutshell, it is very clear, that there is scope for promotion of agroforestry, not only in the croplands, but also to a certain extent even in the scrublands to improve the tree cover of the district.

Agroforestry systems are well known in Hassan district which exhibits maximum diversity in terms of agroforestry land use system components (Fig. 15.9). The western part of the district is dominated by commercial plantations like coffee and tea while eastern part is well known for horticulture plantations, namely, coconut and arecanut, and to a lesser extent, some fruit crops. In view of large diversity, training labels were generated independently for dominant commercial and other agricultural plantations and the integrated map includes three different plantation categories, namely, coffee plantations, tea plantations, and coconut-dominant plantations, apart from other agriculture plantations.

The satellite data analysis indicated that the western four taluks are dominated by coffee plantations and to lesser extent by tea while coconut is cultivated throughout the district. The agriculture plantations are dominated by arecanut plantations and other fruit crops. Although pure stands of coconut are present in the district, multitier cropping systems with a combination of coconut, areacanut, banana (*Musa* spp.), and other spices is very common in the district indicating diversity and profitable agroforestry land use systems already present in the district. In such complex systems,

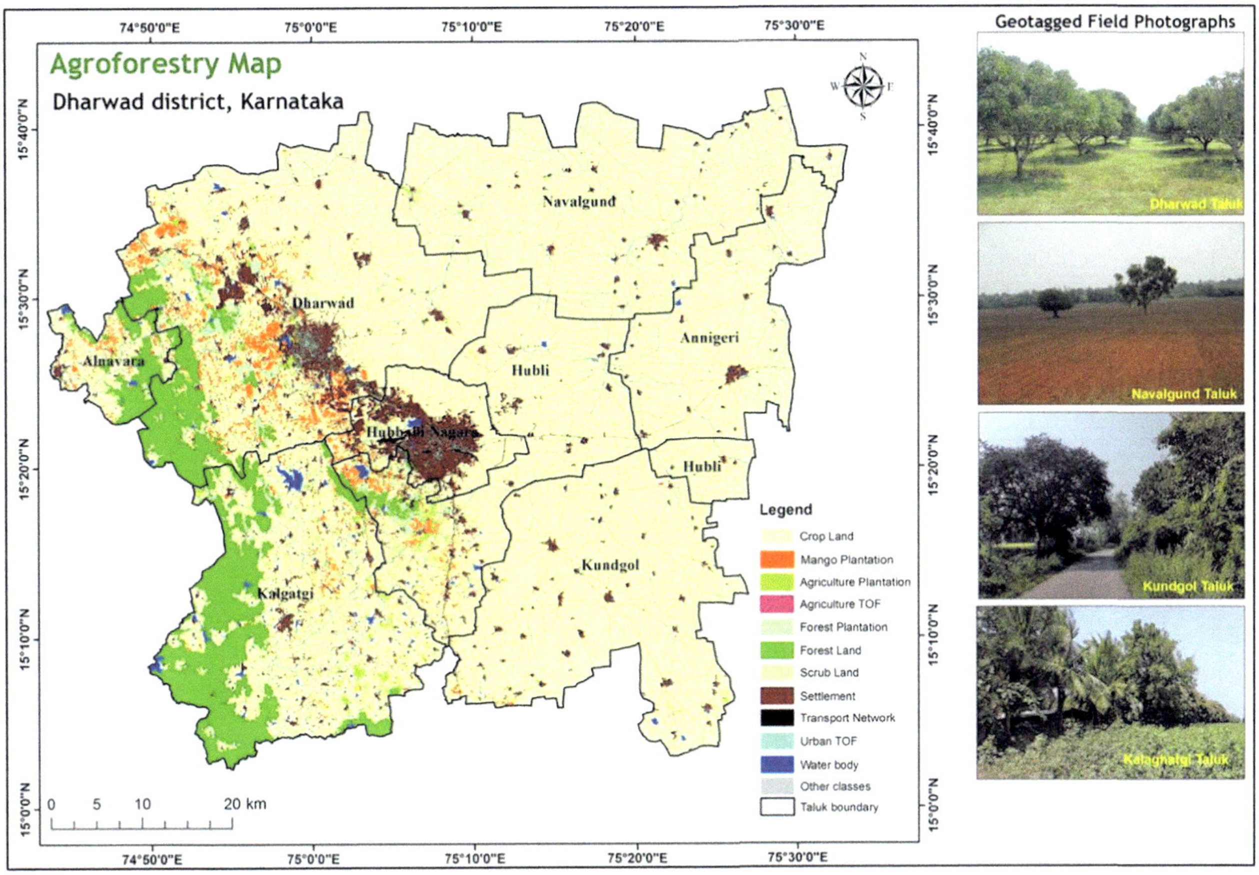

FIGURE 15.8 Integrated agroforestry land use map of Dharwad district, Karnataka, India. Map lines delineate study areas and do not necessarily depict accepted national boundaries.

TABLE 15.3 Area estimates of agroforestry components and other land use in pilot districts.

S.No.	District	Dharwad		Hassan	
	Class name	Area (ha)	% TGA	Area (ha)	% TGA
1	Crop land	318,241.6	74.8	284,735.4	41.8
2	Mango plantation	10,821.2	2.5	–	–
3	Coconut plantation	–	–	129,319.8	19.1
4	Tea plantation	–	–	420.6	0.1
5	Coffee plantation	–	–	67,280.3	9.9
6	Agriculture plantation	4669.8	1.1	2607.3	0.4
7	Agriculture TOF	4635.2	1.1	9671.7	1.4
8	Forest plantation	3143.1	0.7	8938.4	1.3
9	Forest land	33,679.5	7.9	48,059.9	7
10	Scrub land	14,814.7	3.5	47,324.1	6.9
11	Grass land	–	–	5812.1	0.9
12	Built-up	23,155.1	5.4	33,129.0	4.9

(Continued)

TABLE 15.3 (Continued)

S.No.	District	Dharwad		Hassan	
	Class name	Area (ha)	% TGA	Area (ha)	% TGA
13	Urban TOF	2722.7	0.6	3109.5	0.5
14	Water body	8828.6	2.1	37,881.1	5.6
15	Other classes	757.2	0.2	3517.1	0.5
	TOTAL	**425,468.7**		**681,806.3**	

TOF, Trees outside forest.

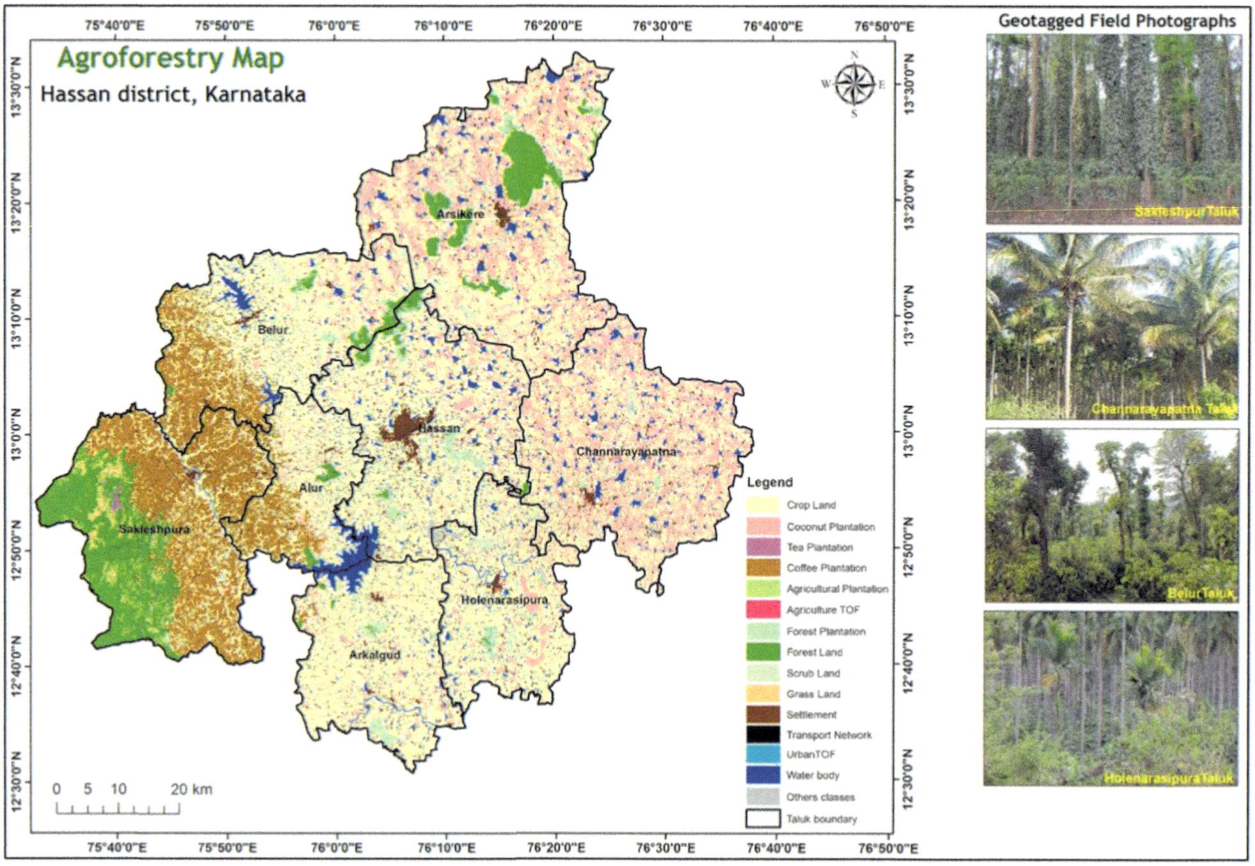

FIGURE 15.9 Integrated agroforestry land use map of Hassan district, Karnataka, India. Map lines delineate study areas and do not necessarily depict accepted national boundaries.

discrimination of each plantation types is quite challenging. The VHRS image with DL enabled classification of major plantation components like coffee (67,280.3 ha), tea (420.6 ha), coconut (129,319.8 ha), forest (48,059.9 ha), and other plantations (2607.3 ha) along with individual trees (9671.7 ha). Forest mask was used for segregating forest lands from other vegetation types.

The agroforestry land use map indicates that the Western Ghats region (western part of Hassan district) is dominated by forest lands with evergreen forest. In the lower regions adjacent to Western Ghats, coffee plantations are predominantly cultivated with variety of shade trees, namely, silver oak (*Grevillea robusta*), arecanut, and mixed forest. Rest of the district is dominated by a mix of agriculture crops and plantations. Thus, Hassan district represents a combination of agroforestry systems including agri-silviculture, horti-silviculture, silvi-pastures, and forest plantations (Table 15.3).

15.2.3.3 Prioritization of Gram Panchayats for intensification of agroforestry components

The two districts, Dharwad and Hassan in Karnataka, considered for prioritization toward intensification of agroforestry practices, are diverse and contrasting in nature, in terms of existing agroforestry tree components within NSA (Fig. 15.10).

In the case of Dharwad, 144 GPs show a decreasing trend in terms of number of GPs to be considered for agroforestry intensification/expansion, implying that a large number of GPs are categorized in very high and high priority zones, mainly located in eastern half of the district (Table 15.4). The eastern half of the district has black cotton soil and dominated by cropland, dotted with individual scattered trees. The western half of the district is dominated by forests, forest plantations, and agriculture/horticulture plantations, mainly mango (Fig. 15.11).

Hassan district has various plantations, including large patches of coffee in the western part of the district (adjacent to reserve forest) and coconut toward north and northeastern region. Other agriculture plantations in farmlands and few hundred hectares of tea plantations also exist. These trees along with agriculture TOF (or trees on farmland) constitute a significant area within NSA. Out of 267 GPs in Hassan, there is an increasing trend in terms of number of GPs to be considered for agroforestry expansion, which means that very few GPs require attention in terms of agroforestry intensification (Table 15.4). Majority of GPs are falling in low- and medium-priority category (Fig. 15.12).

This exercise provides an actionable dataset for implementation at grassroots level and agroforestry intensification may be taken up in very high and high priority classes, followed by medium and low priority, especially in Dharwad district. In the remaining GPs categorized as forest/forest plantation dominated, coffee dominated, and settlement dominated, there is little scope for expansion activities.

The western part of Dharwad is predominantly paddy-growing region, where tree shade might adversely impact the yield of primary crops. Taking into consideration the interaction of woody and nonwoody components, characterization of soil and climatic factors, suitable plantation crops maybe selected to increase the tree cover, while mitigating the impact of future climate change.

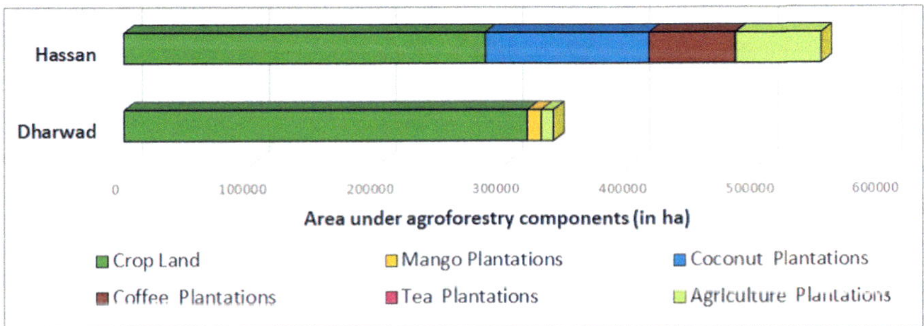

FIGURE 15.10 Distribution of various agroforestry components in Dharwad and Hassan.

TABLE 15.4 Gram Panchayat level prioritization for agroforestry intensification.

S.No.	Priority class/dominant land use	Proportion of agroforestry components in NSA (%)	Number of GPs	
			Dharwad	Hassan
1	Very high	<5	74	0
2	High	5–15	25	38
3	Medium	15–30	18	77
4	Low	>30	6	104
5	Forest/forest plantation	–	21	7
6	Coffee plantation dominated	–	0	41
	Total	**100**	**144**	**267**

NSA, Net sown area.

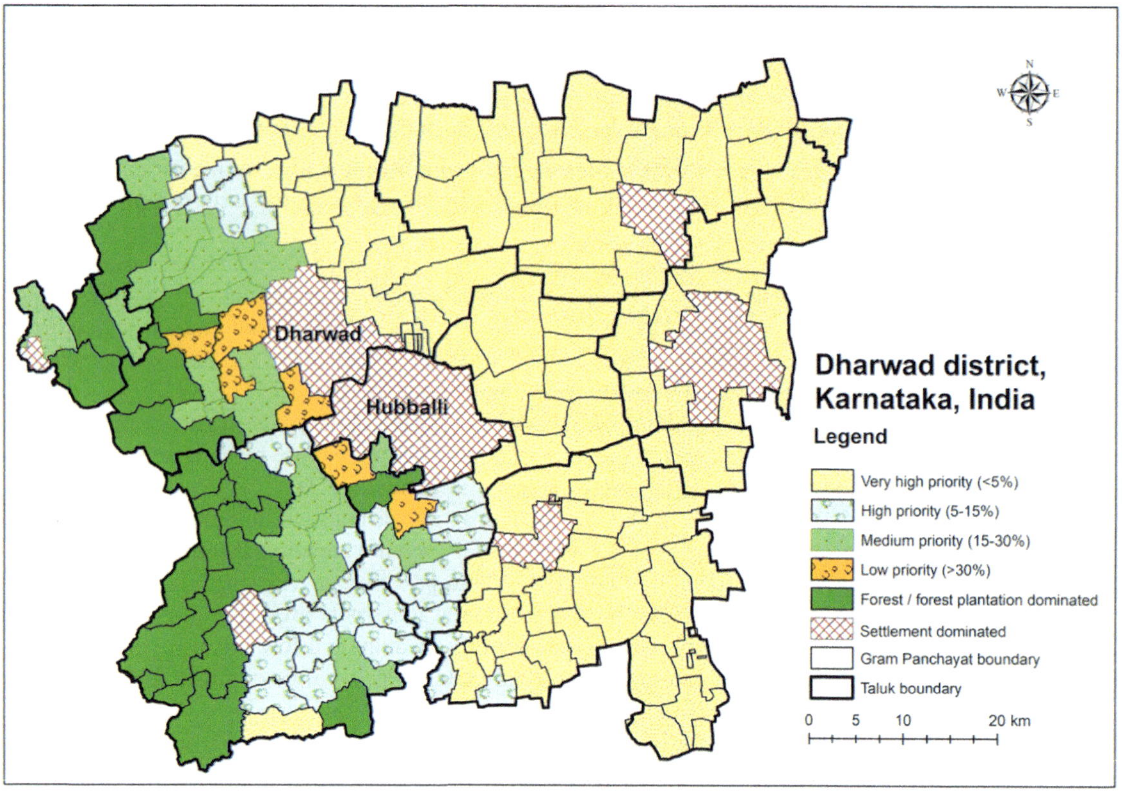

FIGURE 15.11 Prioritization for agroforestry at Gram Panchayat level in Dharwad district. Map lines delineate study areas and do not necessarily depict accepted national boundaries.

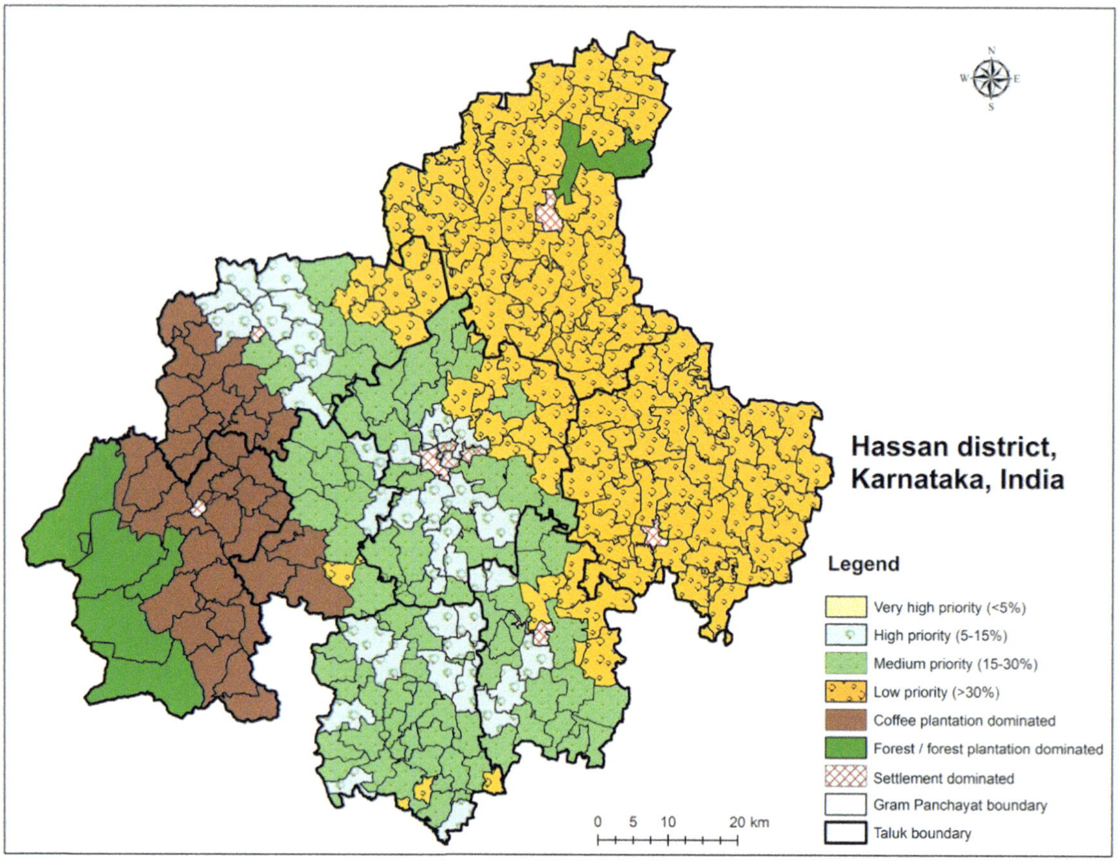

FIGURE 15.12 Prioritization for agroforestry at Gram Panchayat in Hassan district. Map lines delineate study areas and do not necessarily depict accepted national boundaries.

15.3 Conclusions and recommendations

Agroforestry land use systems support farmers' livelihoods while reducing the pressure on forests, and adoption of these practices can be a safe bet for carbon sequestration, climate change mitigation while achieving multiple SDGs. The present research was carried out to evolve a standard geospatial approach/procedure for assessment of agroforestry land use systems using DL techniques and submeter VHRS images to create a detailed geospatial database of agroforestry land use system components, generating information at microlevel. This database will help assess the extent of existing agroforestry systems and to plan for its expansion at large scales. The outcome of this research aligns with the National Agroforestry Policy strategy, supporting farmers in adopting agroforestry practices for increased income generation, food and nutritional security and climate change mitigation.

The methodology developed in this study reduces the large uncertainties in the area estimates of agroforestry land use systems, compared to the conventional approaches. Hence, significant improvement in the methodology with better accuracy and precision for the India's National Communication (NATCOM) and Biennial Update Report (BUR) submitted to UNFCCC is a very important contribution from a large tropical, forested country like India (MoEFCC, 2021, 2023). This methodology can be upscaled to regional/national level with locale specific customization, particularly in developing countries where rainfed farming continues to be mainstay of livelihoods of large segment of population.

In view of the multifaceted role and global importance of agroforestry land use systems, the following recommendations are made:

1. Generation of national level agroforestry baseline database using geospatial techniques at periodic intervals through augmentation of training samples for regionally/nationally relevant DL models, in view of large diversity and heterogeneity of agroforestry tree components across the country in various agro-climatic regions.
2. Identification of dominant tree species with higher ecological, economic, and cultural values for enhanced benefits and better resource management.
3. Prioritization of GPs for intensification of agroforestry land use systems and implementation for very high and high priority regions for increasing tree cover.
4. Characterization of agroforestry land use systems in terms of pedoclimatic parameters for site-suitability analysis.
5. Development of agroforestry hotspots as biodiversity and ecotourism attractions to create public awareness and promote nature-based sustainability solutions.
6. Bridging the gap between scientific knowledge on agroforestry land use systems and the extent to which it is applied or transferred to the practitioners.
7. Enabling policy implementation addressing the agroforestry sector as a whole, with tangible and intangible benefits reaching the grassroots level, up to smallholder farmers.

Acknowledgments

Authors express their sincere gratitude to team members from Food and Agriculture Organization (FAO), United Nations, New Delhi, India and National Rainfed Area Authority (NRAA), Ministry of Agriculture and Farmers Welfare, Government of India for their involvement and contribution in the study. The support received from State Forest Department, Karnataka for in-season field data collection and contribution from scientists, other technical and administrative staff of NRSC and Regional Centres is thankfully acknowledged.

Author contributions

Conceptualization and methodology: HR, ST, VPV, RSS, JCS; **Software development**: VPV, CB, NK; **Validation**: NMK, RSS, NK; **Data analysis**: RSS, ST, NK, VPV, CB; **Field data:** ST, CB, NMK; **Data curation:** ST, RSS, NK, VPV, CB, NMK; **Writing (original draft):** ST, VPV; **Writing (reviewing):** HR, NK; **Supervision:** HR, JCS, SRB, SKS, PC; **Funding acquisition:** DS, SRB.

Funding

This study was funded by Food and Agriculture Organisation (United Nations), New Delhi under Technical Cooperation Programme (TCP/IND/3710) with National Remote Sensing Centre, Indian Space Research Organisation, Government of India.

Declaration of competing interest

The authors declare no competing interests.

References

Ahmad, F., Goparaju, L., & Qayum, A. (2017). Agroforestry suitability analysis based upon nutrient availability mapping: A GIS based suitability mapping. *AIMS Agriculture and Food*, *2*(2), 201–220. Available from https://doi.org/10.3934/agrfood.2017.2.201.

Ahmad, F., Uddin, M. M., Goparaju, L., Rizvi, J., & Biradar, C. (2020). Quantification of the land potential for scaling agroforestry in South Asia. *Journal of Cartography and Geographic Information*, *70*, 71–89. Available from https://doi.org/10.1007/s42489-020-00045-0.

Andersson, L. (2018). *Achieving the Global Goals through agroforestry*. Stockholm: Agroforestry Network. ISBN: 978-91-985041-1-8. Agroforestry_8-siding_oktober.indd (viagroforestry.org).

Barrios, E., Valencia, V., Jonsson, M., Brauman, A., Hairiah, K., Mortimer, P. E., & Okubo, S. (2017). Contribution of trees to the conservation of biodiversity and ecosystem services in agricultural landscapes. *International Journal of Biodiversity Science, Ecosystem Services & Management*, *14*(1), 1–16. Available from https://doi.org/10.1080/21513732.2017.1399167.

Basu, J. P. (2014). Agroforestry, climate change mitigation and livelihood security in India. *The New Zealand Journal of Forestry Science*, *44* (Suppl.1), S11. Available from https://doi.org/10.1186/1179-5395-44-S1-S11.

Bhuvan Panchayat (2023). *Bhuvan Panchayat* 3.0, National Remote Sensing Centre (NRSC), Indian Space Research Organisation (ISRO). https://bhuvan-panchayat3.nrsc.gov.in/index.html.

Blaschke, T. (2010). Object-based image analysis for remote sensing. *ISPRS Journal of Photogrammetry and Remote Sensing*, *65*(1), 2–16. Available from https://doi.org/10.1016/j.isprsjprs.2009.06.004.

Brandt, J., & Stolle, F. (2020). A global method to identify trees inside and outside of forests with medium resolution satellite imagery. *Remote Sensing of Environment*. Available from https://arxiv.org/abs/2005.08702.

Brandt, M., Gominski, D., Reiner, F., Karirya, A., Guthula, V. B., Ciais, P. . . ., & Fensholt, R. (2024). Severe decline in large farmland trees in India over the past decade. *Nature Sustainability*. Available from https://doi.org/10.1038/s41893-024-01356-0.

Brandt, M., Tucker, C. J., Kariryaa, A., Rasmussen, K., Abel, C., Small, J. . . ., & Fensholt, R. (2020). An unexpectedly large count of trees in the West African Sahara and Sahel. *Nature*, *587*, 78–82. Available from https://doi.org/10.1038/s41586-020-2824-5.

Cannell, M. G. R. (1987). Agroforestry – A decade of development. *International Center for Research in Agroforestry, ICRAF*, *24*, 393.

Chaudhari, K., Nishant, N., Upadhyay, G., More, R., Singh, N., Vyas, S. P., & Bhattacharya, B. K. (2019). Crop inventory of orchard crops in India using remotely sensed data. *The international archives of the photogrammetry, remote sensing and spatial information sciences, XLII-3/W6, 2019 ISPRS-GEOGLAM-ISRS Joint int. workshop on "earth observations for agricultural monitoring", 18–20 February 2019, New Delhi, India*. isprs-archives-XLII-3-W6-269-2019.pdf (copernicus.org).

Chen, L. C., Papandreou, G., Kokkinos, I., Murphy, K., & Yuille, A. L. (2018). DeepLab: Semantic image segmentation with deep convolutional nets, atrous convolution, and fully connected CRFs. *IEEE Transactions on Pattern Analysis and Machine Intelligence*, *40*(4), 834–848. Available from https://doi.org/10.1109/TPAMI.2017.2699184.

Dhyani, S. K. (2014). National Agroforestry Policy 2014 and the need for area estimation under agroforestry. *Current Science*, *107*(1), 9–10. Available from https://www.currentscience.ac.in/Volumes/107/01/0009.pdf.

Dhyani, S. K., Murthy, I. K., Kadaverugu, R., Dasgupta, R., Kumar, M., & Adesh Gadpayle, K. (2021). Agroforestry to achieve global climate adaptation and mitigation targets: Are South Asian countries sufficiently prepared? *Forests*, *12*, 303. Available from https://doi.org/10.3390/f12030303.

Dwivedi, A. P. (2022). *Agroforestry: Principles and practices* (pp. 1–355). Oxford & IBH Publishing Company.

FAO. (2004). *Improvement of rural livelihoods: The role of agroforestry*. .. . Available from https://www.fao.org/forestry/7459-03b98002736c1812f203700d24b933c91.pdf.

FAO. (2005). FAO environment and natural resources service series, No. 8. In A. Di Gregorio & Louisa J.M. Jansen (Eds.), *Land cover classification system (LCCS), Version 2: Classification concepts and user manual*. Rome: Food and Agriculture Organisation (United Nations).fao.org.

FAO. (2013). Towards the assessment of trees outside forests. *Forest resources assessment, working paper 183. A thematic report prepared in the framework of The Global Forest Resources Assessment 2010* (p. 335). https://www.fao.org/3/aq071e/aq071e00.htm.

FAO, ICRAF. (2019). Agroforestry and tenure. *Forestry working paper no. 8* (p. 40). fao.org.

Friedl, M. A., & Brodley, C. E. (1997). Decision tree classification of land cover from remotely sensed data. *Remote Sensing of Environment*, *61*(3), 399–409. Available from https://doi.org/10.1016/S0034-4257(97)00049-7.

FSI. (2019). *India state of forest report* (Vols. 1 & 2) (p. 16). Forest Survey of India.

FSI. (2020). Trees outside forests resources in India. forest survey of India technical information series. *Ministry of Environment, Forest and Climate Change, Government of India*, *2*(1), 1–30.

FSI. (2021). *India State of Forests Report-2021, forest survey of India* (Vol. 17) (pp. 209–220). Ministry of Environment, Forest and Climate Change, Government of India, ISBN: 978-81-9500073-1-8. https://fsi.nic.in/forest-report-2021.

Gholz, H. L. (1988). Agroforestry – Introduction. In H. L. Gholz (Ed.), *Agroforestry: Realities, possibilities and potentials* (p. 227). Dordrecht, The Netherlands: Martinus Nijhoff Publishers.

Hebbar, R., Ravishankar, H. M., Trivedi, S., Manjula, V. B., Kumar, N. M., Mukharib, D. S., . . . Ganesh Raj K. (2019). National level inventory of coffee plantations using high resolution satellite data. *ISPRS international archives of the photogrammetry, remote sensing and spatial information sciences, XLII-3/W6* (pp. 293–298). https://doi.org/10.5194/isprs-archives-XLII-3-W6-293-2019.

Hebbar, R., Ravishankar, H. M., Trivedi, S., R. Subramoniam, S, U., Raj, Dadhwal, V. K. (2014) Object-oriented classification of high-resolution data for inventory of horticultural crops. In *The international archives of the photogrammetry, remote sensing and spatial information sciences. XL-8, 2014 ISPRS technical commission VIII symposium, 09–12 December 2014*. isprsarchives-XL-8-745-2014.pdf (copernicus.org).

Huang, C., Davis, L. S., & Townshend, J. R. G. (2002). An assessment of support vector machines for land cover classification. *International Journal of Remote Sensing*, *23*(4), 729–745. Available from https://doi.org/10.1080/01431160110040323.

ICRAF. (1993). A global agenda by Helen Van Houten. *Annual report 1993, International Centre for Research in Agroforestry* (pp. 1–208).

Jamnadass, R. H., Dawson, I. K., Franzel, S., Leakey, R. R. B., Mithofer, D., Akinnifesi, F. K., & Tchoundjeu, Z. (2011). Improving livelihoods and nutrition in Sub-Saharan Africa through the promotion of indigenous and exotic fruit production in smallholders' agroforestry systems. *International Forestry Review*, *13*(3)). Available from https://doi.org/10.1505/146554811798293836.

Karlsson, L. (2018) *Scaling up agroforestry—Potentials, challenges and barriers, a review of environmental, social and economic aspects at the farmer, community and landscape levels*. Stockholm: Agroforestry Network. ISBN: 978-91-985041-0-1. Mall brevpapper. hemsida.eu.

Lasco, R. D., Delfino, R. J. P., Catacutan, D. C., Simelton, E. S., & Wilson, D. M. (2014). Climate risk adaptation by smallholder farmers: The roles of trees and agroforestry. *Current Opinion in Environmental Sustainability*, *6*, 83–88. Available from https://doi.org/10.1016/j.cosust.2013.11.013, ISSN 1877-3435.

Liu, T., & Chen, W. (2015). A scalable deep convolutional neural network for high-resolution imagery scene classification. *IEEE Geoscience and Remote Sensing Letters*, *12*(8), 1621–1625.

MA. (2005). *Millennium ecosystem assessment*. Washington, DC: Island Press.

MoEFCC. (2021). *India: Third biennial update report to the United Nations framework convention on climate change* (pp. 1–479). Ministry of Environment, Forest and Climate Change, Government of India. Available from https://moef.gov.in/wp-content/uploads/2024/01/INDIA_BUR-3.pdf.

MoEFCC. (2023). *India: Third National communication and initial adaptation communication to the United Nations framework convention on climate change* (pp. 1–613). Ministry of Environment, Forest and Climate Change, Government of India. Available from https://moef.gov.in/wp-content/uploads/2024/01/India-TNC-IAC-revised.pdf.

Montagnini, F., & Metzel, R. (2017). The contribution of agroforestry to sustainable development goal 2: End hunger, achieve food security and improved nutrition, and promote sustainable agriculture. *Agroforestry* (p. 12). https://doi.org/10.1007/978-3-319-69371-2_2.

Nair, P. K. R. (1985). Classification of agroforestry systems. *Agroforestry Systems*, *3*, 97–128.

Nair, P. K. R., & Dagar, J. C. (1991). An approach to developing methodologies for evaluating agroforestry systems in India. *Agroforestry Systems*, *16*, 55–81. Available from https://doi.org/10.1007/BF00053197.

Nair, P. K. R., Kumar B, M., & Vimala, N. (2021). *An introduction to agroforestry: Four decades of scientific developments* (2nd ed.) (pp. 1–666). Springer. ISBN: 978-3-030-75357-3, . https://doi.org/10.1007/978-3-030-75358-0.

National Agroforestry Policy (NAP), India. (2014). *National agroforestry policy* 2014.pdf (agricoop.nic.in).

NDC. (2022). *India's updated first nationally determined contribution under Paris agreement (2021–2030) – Working towards climate justice* (pp. 1–3), August 2022 Submission to UNFCCC.

NRSC. (2020a). *Technical guidelines under SIS-DP update project* (pp. 1–110). National Remote Sensing Centre, ISRO, NRSC-RC-DELHI-OCT2020-TR0001686-V1.0.

NRSC. (2020b). *Revised legend for LULC thematic map under 'SIS-DP Update' project*. National Remote Sensing Centre, ISRO, NRSC-NR-DIR-AUG-2020-TR-1371-V2.0.

Ntawuruhunga, D., Ngowi, E., Mangi, H., Salanga, R., & Shikuku, K. (2023). Geospatial characterization of climate-smart agroforestry in two contrasting physiographic zones of Rwanda. *Trees, Forests and People*, *14*100446. Available from https://doi.org/10.1016/j.tfp.2023.100446.

Pal, M. (2005). Random forest classifier for remote sensing classification. *International Journal of Remote Sensing*, *26*(1), 217–222. Available from https://doi.org/10.1080/01431160412331269698.

Pujar, G. S., Dadhwal, V. K., Murthy, M. S. R., Trivedi, Shivam, Reddy, P. M. ..., & Jha, C. S. (2016). Geospatial approach for national level TOF assessment using IRS high resolution imaging: Early results. *Journal of the Indian Society of Remote Sensing*. Available from https://doi.org/10.1007/s12524-015-0476-y.

Puri, S., & Nair, P. K. R. (2004). Agroforestry research for development in India: 25 years of experiences of a national program. *Agroforestry Systems*, *61*, 437–452. Available from https://doi.org/10.1023/B:AGFO.0000029014.66729.e0.

Puthalpet, J. R. (2019). *The daunting climate change: Science, impacts, adaptation & mitigation strategies, policy responses* (pp. 1–523). BS Publications, ISBN: 978-93-8830-583-9.

Quandt, A., Neufeldt, H., & Gorman, K. (2023). Climate change adaptation through agroforestry: Opportunities and gaps. *Current Opinion in Environmental Sustainability*, *60*, 101244. Available from https://doi.org/10.1016/j.cosust.2022.101244.

RADPFI. (2021). *Rural area development plan formulation and implementation guidelines, technical guidelines report* (pp. 1–266). Ministry of Panchayati Raj. https://cdnbbsr.s3waas.gov.in/s316026d60ff9b54410b3435b403afd226/uploads/2023/05/2023053052.pdf.

Ravishankar, H. M., Trivedi, S., Rama Subramoniam, S., Mohammed Ahamed, J., Nagashree, T. R., Manjula, V. B. ..., & Dadhwal, V. K. (2022). Geospatial applications in inventory of horticulture plantations. In C. S. Jha, A. Pandey, V. Chowdary, & V. Singh (Eds.), *Geospatial technologies for resources planning and management. Water science and technology library*. Cham: Springer, Vol. 115https://doi.org/10.1007/978-3-030-98981-1_12.

Ray, S. S., Mamatha, S., Manjunath, K. R., U., Raj, Seshasai, M. V. R., Singh K.,K., ... Saxena, M. (2016) *CHAMAN—A national level programme for horticulture assessment and development* (pp. 1–6). NNRMS e-Bulletin, Department of Space, Government of India, May 2016. nnrms_bulletin_may_2016.pdf (isro.gov.in).

Rizvi, R. H., Handa, A. K., & Arunachalam, A. (2022). Agroforestry in India: Area estimates and methods. *Current Science*, *123*(6), 743–744. Available from https://www.currentscience.ac.in/Volumes/123/06/0743.pdf.

Rizvi, R. H., Handa, A. K., Sridhar, K. B., Kumar, A., Bhaskar, S., Chaudhari, S. K., ... Dongre, G. (2020). *Mapping agroforestry and trees outside forest*. Jointly published by the ICAR, Central Agroforestry Research Institute (CAFRI), Jhansi and World Agroforestry (ICRAF), South Asia Regional Programme, New Delhi. B20033.pdf (worldagroforestry.org).

Rolo, V. (2014). Agroforestry for sustainable food production. *Sustainability*, *14*, 10193. Available from https://doi.org/10.3390/su141610193.

Ronneberger, O., Fischer, P., & Brox, T. (2015). U-Net: Convolutional networks for biomedical image segmentation. *Medical image computing and computer-assisted intervention—MICCAI 2015* (pp. 234-241). https://doi.org/10.48550/arXiv.1505.04597.

RRSC-South. (2021). *Gram Panchayat Spatial Development Planning (GPSDP) Karnataka State Report* (pp. 1−79). Regional Remote Sensing Centre-South, NRSC, ISRO. NRSC-RC-REGBANG-RRSC-BANG-MAR2021-TR-0001825-V1.0.

RRSC-South. (2022). *Spatial inventory of agroforestry resources using geospatial technology and AI with very high resolution images, Technical Report* (pp. 1−118). Regional Remote Sensing Centre-South, NRSC, ISRO. NRSC-RRSC-BANG-APR2022-TR-000199-V1.0.

Santos, P. Z. F., Crouzeilles, R., & Sansevero, J. B. B. (2019). Can agroforestry systems enhance biodiversity and ecosystem service provision in agricultural landscapes? A meta-analysis for the Brazilian Atlantic Forest. *Forest Ecology and Management*, *433*, 140−145. Available from https://doi.org/10.1016/j.foreco.2018.10.064, ISSN 0378-1127.

SDG. (2023). https://unstats.un.org/sdgs/report/2023/The-Sustainable-Development-Goals-Report-2023.pdf.

Shelhamer, E., Long, J., & Darrell, T. (2017). Fully convolutional networks for semantic segmentation. *IEEE Transactions on Pattern Analysis and Machine Intelligence*, *39*(4), 640−651.

Shin, S., Soe, K. T., Lee, H., Kim, T. H., Lee, S., & Park, M. S. (2020). A systematic map of agroforestry research focusing on ecosystem services in the Asia-Pacific region. *Forests*, *11*(4), 368. Available from https://doi.org/10.3390/f11040368.

Sinclair, F. L. (1999). A general classification of agroforestry practice. *Agroforestry Systems*, *46*(2), 161−180.

Singh, S. K., Chatterji, S., Chattaraj, S., & Butte, P. S. (2018). *ICAR-NBSS&LUP Technologies, ICAR—The National Bureau of Soil Survey and Land Use Planning (NBSS & LUP). Technical report, NBSS Publ. No.176* (p. 90). Avialble from https://doi.org/10.13140/RG.2.2.23495.11687.

Smith, L. N. (2018). *A disciplined approach to neural network hyperparameters: Part 1-learning rate, batch size, momentum, and weight decay*. https://doi.org/10.48550/arXiv.1803.09820.

SOI. (2023). *Survey of India open series maps*. Survey of India, Ministry of Science and Technology, Government of India.

Torralba, M., Fagerholm, N., Burgess, P. J., Moreno, G., & Plieninger, T. (2016). Do European agroforestry systems enhance biodiversity and ecosystem services? A meta-analysis. *Agriculture, Ecosystems & Environment*, *230*, 150−161. Available from https://doi.org/10.1016/j.agee.2016.06.002, ISSN 0167-8809.

Trivedi, S., Rama Subramoniam, S., Manjula, V. B., Ravindranath, S., Shetty, D., & Hebbar, R. (2024a). Application of high resolution satellite images for comprehensive spatial development planning at grassroot level *IEEE xplore (2023 IEEE InGARSS)* (pp. 1−4). Available from https://doi.org/10.1109/InGARSS59135.2023.10490422.

Trivedi, S., Vinod, P. V., Chandrasekaran, B., Nagashree, M. K., Rama Subramoniam, S., Manjula, VB., ... Chauhan, P. (2024b) Geospatial assessment of agroforestry land use systems using very high-resolution satellite images and artificial intelligence. Agroforestry Systems, https://doi.org/10.1007/s10457-024-01042-2.

Udayakumar, S., Lai, L., Ussiri, D. A. N., Kumar, S., & Clay, S. (2021). Role of integrated crop-livestock systems in improving agriculture production and addressing food security—A review. *Journal of Agriculture and Food Research*, 5100190. Available from https://doi.org/10.1016/j.jafr.2021.100190, ISSN 2666-1543.

Uthappa, A. R., Chavan, S. B., Handa, A. K., Newaj, R., Kumar, D., Sridhar, K. B., & Chaturvedi, O. P. (2016). Agroforestry – A sustainable solution to address climate change challenges. In S. K. Gupta, P. Panwar, & R. Kaushal (Eds.), *Agroforestry for increased production and livelihood security* (pp. 1−22). New India Publishing Agency.

van Noordwijk, M. (2021). Agroforestry-based ecosystem services. *Land*, *10*, 770. Available from https://doi.org/10.3390/land10080770.

Vinod, P. V., Trivedi, S., Hebbar, R., & Jha, C. S. (2023). Assessment of Trees Outside Forest (TOF) in urban landscape using high-resolution satellite images and Deep Learning techniques. *Journal of the Indian Society of Remote Sensing*, *51*(3), 549−564. Available from https://doi.org/10.1007/s12524-022-01646-0.

Waldron, A., Garrity, D., Malhi, Y., Girardin, C., Miller, D. C., & Seddon, N. (2017). Agroforestry can enhance Food Security while meeting other Sustainable Development Goals. *Tropical Conservation Science*, *10*. Available from https://doi.org/10.1177/1940082917720667.

Wilson, M., & Lovell, S. (2016). Agroforestry – The next step in sustainable and resilient agriculture. *Sustainability*, *8*(6), 574. Available from https://doi.org/10.3390/su8060574.

Zomer, R., Trabucco, A., Coe, R., Place, F., van Noordwijk, M., & Xu, J. (2014). *Trees on farms: An update and reanalysis of agroforestry's global extent and socio-ecological characteristics* (pp. 1−33). Working Paper 179, World Agroforestry Center. WP14064.pdf worldagroforestry.org.

Zomer, R. J., Neufeldt, H., Xu, J., Ahrends, A., Bossio, D., Trabucco, A., ... Wang, M. (2016). Global tree cover and biomass carbon on agricultural land: The contribution of agroforestry to global and national carbon budgets. *Nature Scientific Reports*, *6*, 29987, https://doi.org/10.1038/srep29987.

Chapter 16

Earth observation based mapping and site-suitability analysis of coffee agro-ecosystem in the face of climate change

Rama Subramoniam S.[1], Nagashree M.K.[1], Shivam Trivedi[1], D.S. Mukharib[2], Kishore Mote J.[2], R. Hebbar[1], S.K. Srivastav[3] and Prakash Chauhan[4]

[1]*Regional Remote Sensing Centre-South, National Remote Sensing Centre (NRSC), Indian Space Research Organisation (ISRO), ISITE Campus, Bengaluru, Karnataka, India,* [2]*Coffee Board, Ministry of Commerce, Government of India, Bengaluru, Karnataka, India,* [3]*Regional Centres, NRSC, Indian Space Research Organisation (ISRO), New Delhi, India,* [4]*National Remote Sensing Centre, Indian Space Research Organization, Hyderabad, Telangana, India*

16.1 Introduction

Coffee is the world's second-most consumed beverage and second-most traded commodity in world, playing a significant role in global economy, employing more than125 million people (Bernardes, Moreira, Adami, Giarolla, & Rudorff, 2012). Arabica (*Coffea arabica*), which contributes 60% of traded coffee, and Robusta (*Coffea canephora*), which contributes the remaining 40%, are major two species (International Coffee Organization ICO Trade Statistics, 2018) even though the taxonomy of the genus *Coffea* presents 130 species (Davis, Bridson, Stoffelen, & Govaeirts, 2006). Coffee production in the world is about 10.0 million tonnes and harvested area is 11.0 M ha. The South America produce over 55.5% of the whole world's production, followed by Asia (31.9%) (FAOSTAT, 2021). India ranks 7th in the world in terms of area and production. Indian coffee is known for its reputation as the best shade-grown coffee in the world. Coffee is grown in open area, without any shade coupled with extensive cultivation practices mostly practiced in Brazil and Vietnam and highly amenable for mechanization of farm operations. These countries are facing local and global impacts of climate change over the decades as fluctuations in climate and agronomic parameters could affect the growth of coffee.

Sustainable coffee agro-ecosystems may provide several ecological services, such as maintaining biodiversity, soil fertility, and C sequestration, and regulating pests and diseases (Kath et al., 2022). Indian coffee is grown under two-tier shade in a more sustainable way for centuries. These plantations are grown on undulating terrains of varying slope, which gives less scope for mechanization. Indian coffee is unique as it is grown in forests under natural shade and raised in virgin jungles and is hand-picked and sun-dried. This uniqueness makes it more labor-intensive and provides employment to the native populations of hilly regions. Climate change has impacted the coffee production significantly in India, with increasing temperatures and unpredictable weather patterns. This can lead to reduction in yield and increases vulnerability to pests and diseases, particularly for both varieties. Deterioration of soil health, biodiversity conservation (flora, fauna, and shade trees), and pollution are serious environmental challenges to this ecosystem functioning (Damodaran, 2002).

The traditional regions in the south form the cradle of Indian coffee. The country share of world coffee production is 3.67%, and in global exports it is 4.53%. The crop estimates for the year 2017–18 is around 316,000 million tonnes (Coffee Board, 2018). Karnataka, Kerala, Tamil Nadu, partly in Andhra Pradesh, Orissa, and marginal extent in northeastern states are major coffee-cultivation states. These areas have diverse climatic conditions, which are suited for cultivation of coffee. Some regions with high elevations are ideal for Arabica coffee and warm humid conditions are suited for Robusta coffee (Coffee Board, 2018). The coffee-production systems may play an important role in species conservation and other environmental services (Moguel & Toledo, 1999). Coffee cultivation in traditional systems helps in conservation of biodiversity. Shade trees in coffee eco-system control the soil erosion on a sloping terrain, enrich the soil fertility, and protect the coffee from seasonal temperature fluctuations (Komar, 2006; Perfecto, Rice, Greenberg, & Van der Voort, 1996). A strong foundation was laid for sustainability of coffee sector and India has become one of the few countries in the world to grow all of its coffee under natural shade (Anonymous, 2014).

Sustainable Development Perspectives in Earth Observation. DOI: https://doi.org/10.1016/B978-0-443-14072-3.00011-3

High-resolution satellite data along with advanced data analytics techniques enables mapping and monitoring of natural resources (Miguel, Komar, Kevin, & Hugo, 2012). Information from satellite data depends on spatial, spectral, and temporal resolutions from various sensors (Bernardes et al., 2012). Due to complex cultivation practices the application of remote sensing data for inventory of coffee plantations is limited (Cordero-Sancho & Sader, 2007; Miguel et al., 2012). Object-oriented classification techniques have shown some good results for classification of high-resolution data (Bock, Xofis, Mitchley, Rossner, & Wissen, 2005; Gamanya, Maeyer, & Dapper, 2007; Mueller, Segl, & Kaufmann, 2004). Hebbar et al. (2014) and (2019) demonstrated object-based image analysis (OBIA) methodology for classification of high-resolution satellite images for inventory of horticulture plantations. Use of high-resolution data for demarcation of banana orchards using a combination of OBIA and visual interpretation techniques (Kasper, Stuart, Christian, Seonaid, & Lisa, 2009).

The accurate spatial database and geographic location of these two dominant coffee types are not available. The spatial database is essential for improved plantation management, yield estimation, and pest and disease forecasting, and for providing agro-advisory services. The discrimination of coffee types (Arabica and Robusta) is possible based on the region having suitable bio-climatic and topographic regimes. Major parameters which decide location and spatial extent of coffee types include elevation, slope, aspect, rainfall, and temperature. The shade tree and its density also vary with different coffee types and Arabica coffee has higher shade density. Arabica have been traditionally raised at higher altitudes (1000 m above MSL) while Robusta on lower altitudes. Arabica coffee is characterized by small, dark leaves as opposed to broad, pale green leaves of Robusta. Arabica plants grows 2.5–4.5 m high and its beans are oval in shape. Robusta can grow up to 6 m high and the beans have circular appearance. Arabica are self-pollinated and Robusta cross-pollinated. Both require to be raised as silvi-horticultural crops under shade in Indian conditions to avoid enervation and dieback due to overbearing under sun-exposed conditions. Most common shade species employed are Dadap and silver oak during early stages and silver oak, *Ficus/Albizzia* spp., jack fruit sp., apart from other preexisting mixed forest/jungle species, during later stages. The crop is a short-day plant and its floral commencement takes place during 8–11 hours of daylight (between September and December). Water, either as premonsoon showers or through sprinkler irrigation, is essential for ideal flowering. Flowering occurs during February–March/April and the Arabica comes to fruition in 8–9 months and Robusta 2 months later. Average productivity of Arabica is about 643 kg ha^{-1} while Robusta is about 1035 kg ha^{-1}, in India (Coffee Board, 2019).

Land-suitability study for agriculture is an important technique in deciding agricultural cropping pattern, monitoring, planning, and advisory services. Site-suitability analysis is an assessment of an area to determine how appropriate it is for a particular use of the land in a particular location and its suitability for a particular crop. Land-suitability tools is applied to identify better management practices in agricultural fields (Hidayat, Afriliana, Gusmini, & Harada, 2020; Neupane & Sharma, 2022). Policymakers and planners use Geographic Information System (GIS) data to decide on the best regions for a location-specific agricultural planning by combining data on soils, topography, and climatic parameters to determine the size and location of suitable areas for introducing crops. The land ownership, transport, infrastructure, distance to market centers, etc. could be considered for optimal site selection. Advanced GIS tools brought about new approaches for spatial modeling of factors affecting crop suitability and coupling of this type of model with GIS and remote sensing. The integration of RS-GIS, Fuzzy-logic, and application of multicriteria evaluation using analytical hierarchy process (AHP) could provide a superior database and guide decision-makers for considering crop land substitution in order to achieve better agricultural production (Salas Lopez et al., 2020). Different multicriteria analysis techniques/tools to develop land-suitability maps have been extensively reviewed by Singha and Swain (2016). Remote sensing and GIS along with multicriteria modeling were used to analyze topography, soil, land cover, and climatic factors for site-suitability analysis for cultivation of coffee in Kenya (Frankline et al., 2016). In a similar study, GIS was used for a weighted overlay analysis to assess the major production zones of Arabica coffee and their qualitative productivity indices (Nzeyimana, Hartemink, & Geissen, 2014).

Global warming is predicted such that water shortages and increasing temperatures will negatively affect the coffee production at lower elevations and vice versa. Impacts of climate change will not only be threat to small-scale growers but also all stakeholders associated with coffee sector. It is possible to withstand the negative effects of climate change by various adaptation and mitigation practices, namely, reforestation, shade density, crop improvement, intercropping, and other agronomic practices (Rudragouda, D'souza, & Somashekhargouda, 2017). Comprehensive accomplishment of these practices supports to improve the impact of climate change. Studies predict that suitable coffee-cultivation area could drop by around 50% by 2050 for both Arabica and Robusta, owing to climate change. Systematic studies are required to assess impact of climate change in the context of shade-grown Indian coffee plantations for formulating suitable adaptation and management strategies. This requires timely and effective agronomic changes to reduce the potential risks and ensure the long-term viability of coffee production (Poltronieri & Rossi, 2016). Similar to all other agro-systems, the consequences of climate change on coffee production are also associated to optimal management of crop, soil, water, and nutrient to mitigate risks such as salinity, drought, biodiversity decline, suitability losses, change of species seed availability, and abiotic and biotic stress.

As weather patterns shift and global temperatures rise, long-term studies are essential to understand the dynamics of coffee agro-ecosystems as well as to evaluate the effects of adaptation and climate change on coffee quality and yields in controlled experiments. Rising temperatures are projected to reduce coffee yields in low-altitude regions (Adhikari, Elizabeth, Russell Paterson, & Mark, 2020; Bunn, Laderach, Ovalle Rivera, & Kirschke, 2015). Intergovernmental Panel on Climate Change (IPCC) reports show that climate change will reduce the yields on average and decrease coffee-suitable land by 2050 (Bilen et al., 2023). According to Global Climate Risk Index, India is among those countries most affected by climate change, particularly southern states, where majority of coffee is grown. The major coffee-producing countries also facing local and global climate change due to fluctuations in rainfall, temperature, relative humidity, soil nutrients, moisture, sunlight, aeration, etc. Soil temperature could impact the growth and productivity. Adoption of mitigation strategies to reduce global warming is important to overcome the impact of climate variations on coffee quality and production (Nagaraj & Mote, 2020).

The vulnerability of coffee-growing areas is affected by several factors, namely, land size, infrastructure, income level, labor availability, etc. (IPCC, 2022). Coffee growers must implement integrated management practices on their farms to ensure the long-term viability of coffee production to overcome these hindrances (Velmourougane & Bhat, 2017). Decision-makers and stakeholders require consolidated scientific evidence to design and implement successful mitigation and adaptation practices and resilience strategies for coffee production in environmentally sensitive areas. United Nations Framework Convention on Climate Change (UNFCCC) (URT, 2003) projected that there is no adverse impact on coffee production with a rise in temperature up to 2°C. Rainfall variability was identified as the decisive factor and an increase in precipitation could increase coffee yield around 17% (CIAT, 2019; GIZ, 2010; Simms, 2005). This will limit the production of coffee due to several factors including availability of land, ownership, and social preference (Adhikari, Nejadhashemi, & Woznicki, 2015).

Detailed inventory of coffee plantations is extremely essential due to nonavailability of reliable regional and Indian national geospatial database on coffee plantations after deregulation. Collaborative study between ISRO and Coffee Board was initiated for inventory of coffee plantations in six traditional coffee-growing talukas using high-resolution satellite data (RRSC-S, 2016). This study was upscaled to national-level project GeoCUP based on the encouraging results of the pilot study. Geodatabase on potential areas toward expansion plans to introduce coffee plantations in non-traditional regions would serve as an important input. The suitability of two dominant coffee types (Arabica and Robusta) for particular region depends on pedoclimatic parameters. The shade tree type and density vary with coffee types particularly higher shade density for Arabica. Major parameters which decide location and spatial extent of coffee types are elevation, slope, aspect, rainfall, and temperature have used or suitability analysis. The discrimination of coffee types in major growing regions have suitable bioclimatic and topographic regimes is studied. Indian national-level coffee map, coffee type map, and suitability analysis are the major inputs essential for analyzing coffee-growing environment and generating inputs for studying climate change impact assessment. This chapter deals with following specific objectives:

- Inventory of coffee plantations at national level using high-resolution satellite images
- Discrimination of two dominant coffee varieties, Robusta and Arabica
- Site-suitability analysis for expansion of coffee plantations

16.2 Materials and methods

16.2.1 Satellite and ancillary data

16.2.1.1 National-level coffee mapping

Multiresolution remote sensing data, namely, Resourcesat-1 and -2 and Cartosat-1, have been used. Multidate IRS LISS-III data (24 m) were used to study the phenological characterization of different vegetation categories as coffee is cultivated along with forest and shade trees. A total of 19 LISS-III images were used for phenological characterization of different vegetation classes while 57 Resourcesat-LISS-IV scenes (5.0 m) and 168 Cartosat-1 scenes (2.5 m) acquired during 2016–18 periods have been utilized for mapping.

The following ancillary data from different sources and digital devices were also used:

- Cartosat-1 stereo pairs and Digital Elevation Model (DEM) (10 m)

- Spatial database on transport network, built-up, administrative boundaries, etc.
- Coffee Board statistics (area and production)
- Survey of India toposheets (1:50,000 scale)
- Reserved forest boundary
- Handheld Global Positioning System (GPS) device for geo-tagging of sample sites
- Image processing and GIS software packages

16.2.1.2 Discrimination of coffee types

The existing coffee map, Sentinel-1 temporal data and very high resolution data along with other ancillary data have been used. The input layers required including pedoclimatic parameters are presented in Table 16.1.

16.2.1.3 Site-suitability analysis

The DEM, Sentinel-1 data, and land use land cover (LULC) map along with other ancillary data have been used. Ten major parameters were used in the analysis. The details of the parameters used for the study is provided in Table 16.2. Major selection criteria, namely, elevation, aspect, slope, rainfall (quantum/pattern), soil depth, etc., were considered to determine the suitability under different LULC categories in general and under wasteland areas, in particular.

TABLE 16.1 Typical pedoclimatic parameters for Arabica and Robusta coffee types.

S. No	Factors	Arabica	Robusta
1	Soils	Deep, fertile, high OM, well drained, and slightly acidic (pH: 6.0–6.5)	
2	Slopes	Gentle to moderate slopes	Gentle slopes to fairly level fields
3	Elevation	1000–1500 m	500–1000 m
4	Aspect	North, east, and N-E aspects	North, east, and N-E aspects
5	Temperature	15°C–25°C; cool, equable	20°C–30°C; hot, humid
6	Relative humidity	70%–80%	80%–90%
7	Annual precipitation	1600–2500 mm	1000–2000 mm
8	Blossom showers	March–April (25–40 mm)	February–March (25–40 mm)
9	Backing showers	April–May (50–75 mm)	March-April (50–75 mm)

TABLE 16.2 Site-suitability parameters selected for analysis.

S. No.	Criteria	Range/Quantum	Comments
1	Annual rainfall	1600–2500	Average of last 5 years
2	Blossom showers	>25 mm (Mar–Apr)	Average for 2 months
3	Backing showers	>50 mm (Apr–May)	Average for 2 months
4	Elevation	1000–1500 m	DEM (10 m)
5	Slope	<15.0%	From DEM
6	Aspect	North, east, and N-E	From DEM
7	LULC map	Scrubland, waste land, grassland	1:50,000 scale
8	Temperature	15°C–40°C	Gridded data
9	Soil map	pH acidic, depth: deep	1:250,000 scale
10	Relative humidity	70%–80%	

LULC, Land use land cover; *DEM*, Digital Elevation Model.

16.2.2 Study area

16.2.2.1 National-level coffee mapping

Three major coffee-growing regions in the country are (1) traditional areas in the southern states (Karnataka, Kerala, and Tamil Nadu); (2) nontraditional coffee-growing areas (Andhra Pradesh and Odisha in the Eastern Ghats), and (3) northeastern region (Assam, Manipur, Meghalaya, Mizoram, Tripura, Nagaland, and Arunachal Pradesh). The northeastern states and Eastern Ghats are recently developed new areas of coffee cultivation (Fig. 16.1).

16.2.2.2 Discrimination of coffee types

Three sites have been selected representing the diversity in coffee cultivation for developing the methodology and one site for validation. The study area consisted of Belur Taluk, Karnataka, India which consisted of equal proportions of

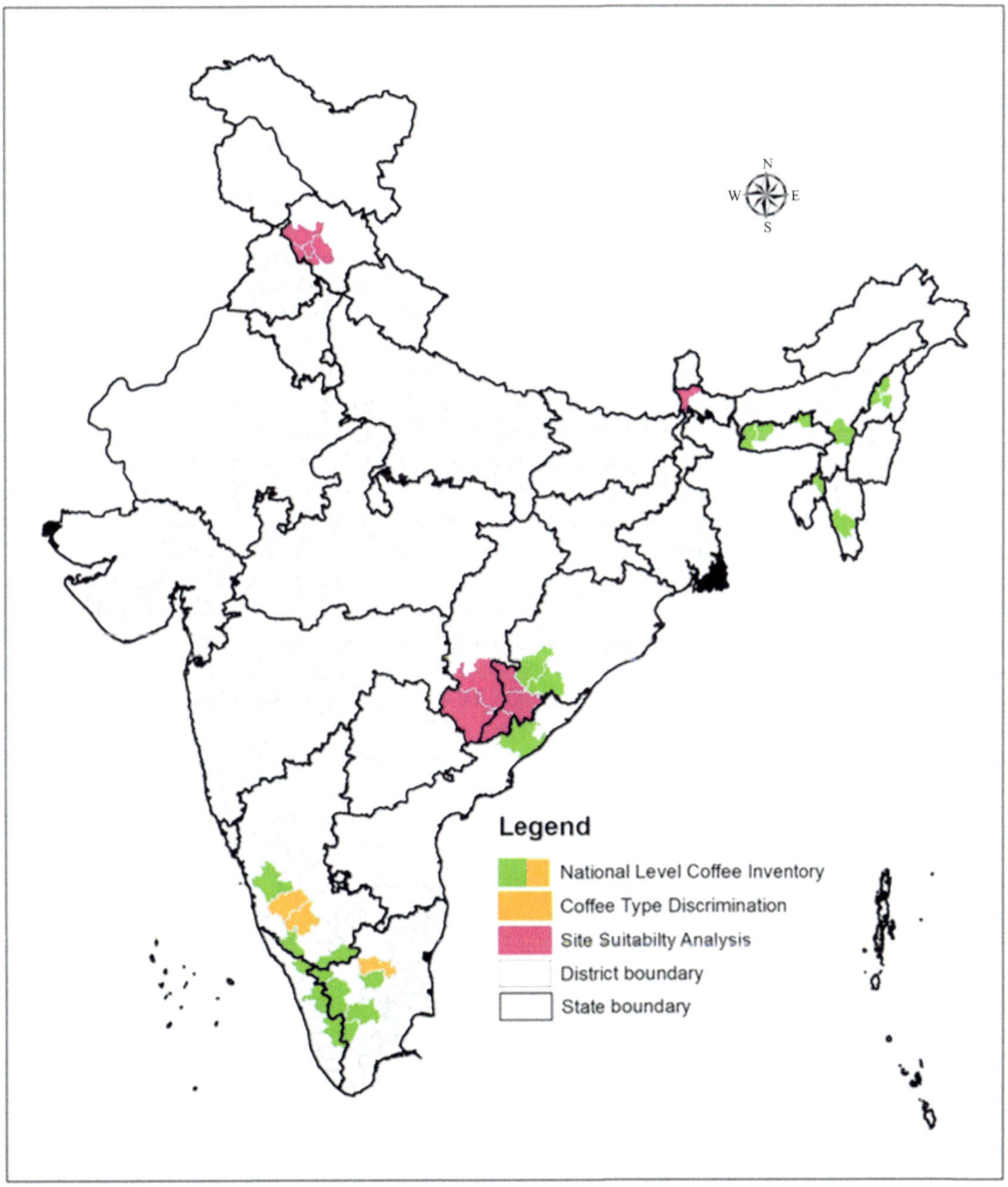

FIGURE 16.1 Study area map for Coffee inventory, discrimination, and suitability analysis. Map lines delineate study areas and do not necessarily depict accepted national boundaries.

both coffee types while Virajpet Taluk, Karnataka dominant with Robusta and Yercaud Taluk, Tamil Nadu is Arabica dominant. Chikkamagaluru district is selected as validation site. The study area map is depicted in Fig. 16.1.

16.2.2.3 Site-suitability analysis

The study districts are Mandi, Kangra, Una, Hamirpur, and Bilaspur of Himachal Pradesh; Koraput, Malkangiri, and Nabrangapur of Odisha; Dantewada and Bastar of Chhattisgarh; and Darjeeling in West Bengal.

16.2.3 Methodology

16.2.3.1 National-level coffee mapping

The overall methodology consisted of selection of optimum satellite data, datasets processing, and field data collection and analysis for delineation of coffee plantations. Data processing involved orthorectification, radiometric correction, computing normalised difference vegetation index (NDVI), and merging of multisource datasets. Orthorectification was carried out utilizing DEM. LISS-IV data were used for generation of vegetation indices which were in turn utilized in classification using 2.5 m composite images. A total of 9387 ground truth samples (including 2242 geotagged field photographs) representing the overall variability of coffee plantations was collected. Hybrid approach was employed for generating the spatial extent of coffee plantations at taluk levels, which consisted of digital classification and visual interpretation techniques.

The spectral signature of coffee plantations is complex because of varying shade tree types, density, and terrain conditions. Digital classification/visual interpretation approaches were adopted to achieve acceptable mapping accuracy. The study taluks were grouped into two types, namely, homogeneous and contiguous patches of coffee plantations with uniform spectral signature and heterogeneous areas having sparse distribution of coffee plantations with mixed spectral signature. Heterogeneous and smallholding areas were delineated using onscreen visual interpretation at 1:5000 scale using tone, texture, shape, and terrain characteristics.

OBIA method was explored in the case of homogeneous areas covering about 40% of the study area. For OBIA method, Cartosat-1 and LISS-IV data with NDVI was used for multiresolution segmentation. Scale parameter of 20 was selected after optimizing the segmentation parameters such as scale, shape, and compactness. Training samples were generated for coffee and other vegetation categories based on the field data. Mean, standard deviation, NDVI, geometry, and other contextual parameters were analyzed for classification using SVM algorithm. Feature space optimization function was used for selection of image features for classification. The object-based classified output showed about 75.0% accuracy and required further improvement to achieve acceptable accuracy. Heterogeneous areas showed very lower accuracy and postclassification, updating map was tedious and time consuming. Therefore, onscreen visual interpretation was done using GIS. The procedure adopted for inventory coffee plantations is depicted in Fig. 16.2.

For visual interpretation, prefield-interpreted coffee plantation maps were updated using the ground truth information through onscreen interpretation at 1:5000 scale using tone, texture, terrain, contextual, and other associated image parameters. Field information and expert knowledge of Coffee Board officials played an important role in achieving reasonable accuracy. Postclassification/interpretation accuracy was estimated using 6152 samples spread across coffee-growing regions using field validation. Taluk-/district-wise coffee acreage statistics were extracted and aggregated at different administrative levels.

16.2.3.2 Coffee type discrimination

The two main varieties of coffee, namely, Arabica and Robusta, are grown in India. Arabica is grown in higher altitudes than Robusta. The cool and equable temperature, ranging between 15°C and 25°C, is suitable for Arabica while for Robusta, hot and humid climate with temperature ranging from 20°C to 30°C is suitable. Thus, temperature and rainfall conditions along with terrain are considered to be important factors in defining extent of these coffee types. Major parameters which decide location and spatial extent of coffee types include elevation, slope, aspect, rainfall, temperature, and vegetation indices. During the fieldwork, information from sufficient number of sample sites representing the two coffee types in the selected study areas have been collected in a predefined format. The flowchart showing the methodology and workflow is shown in Fig. 16.3.

Terrain characteristics like elevation and slope were derived from DEM derived from Cartosat-1 Stereo pairs. Soil texture was generated from 1:50,000 soil map. Climatic parameters in rainfall and temperature in spatial domain were derived using 5-year climatic database. Coffee type maps have been generated through spatial analysis of pedoclimatic parameters and coffee map. Area statistics of coffee type map were generated for the selected study areas.

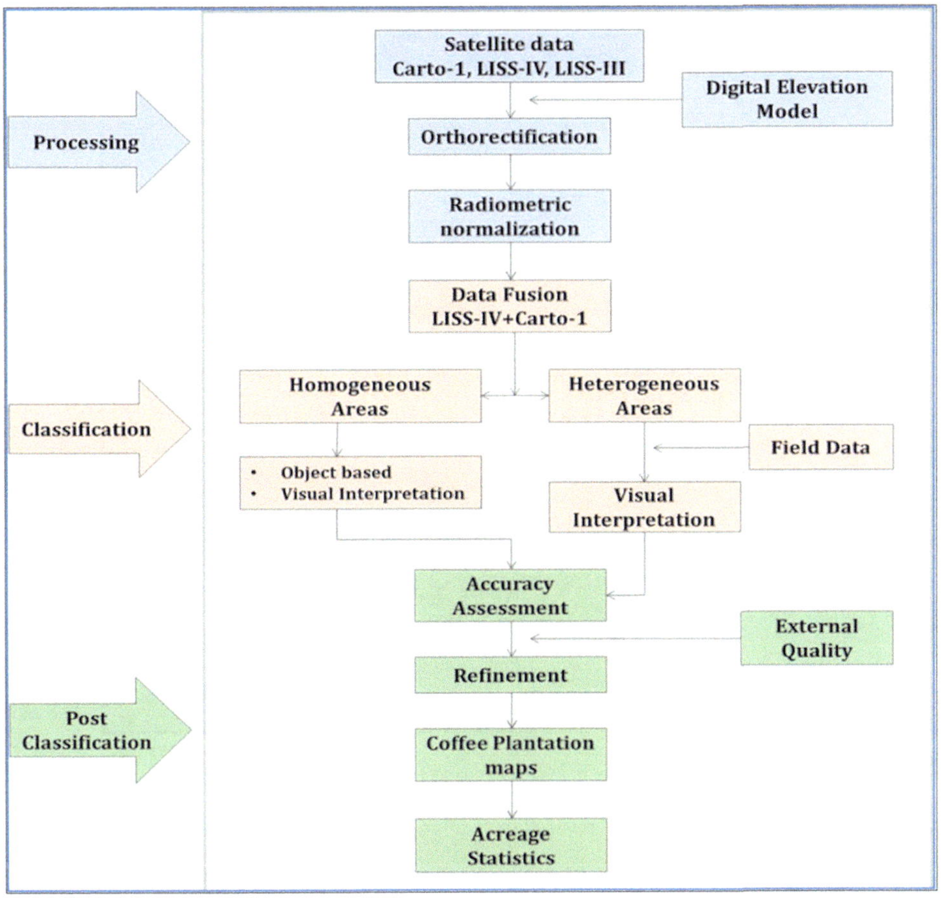

FIGURE 16.2 Flowchart for National level coffee inventory.

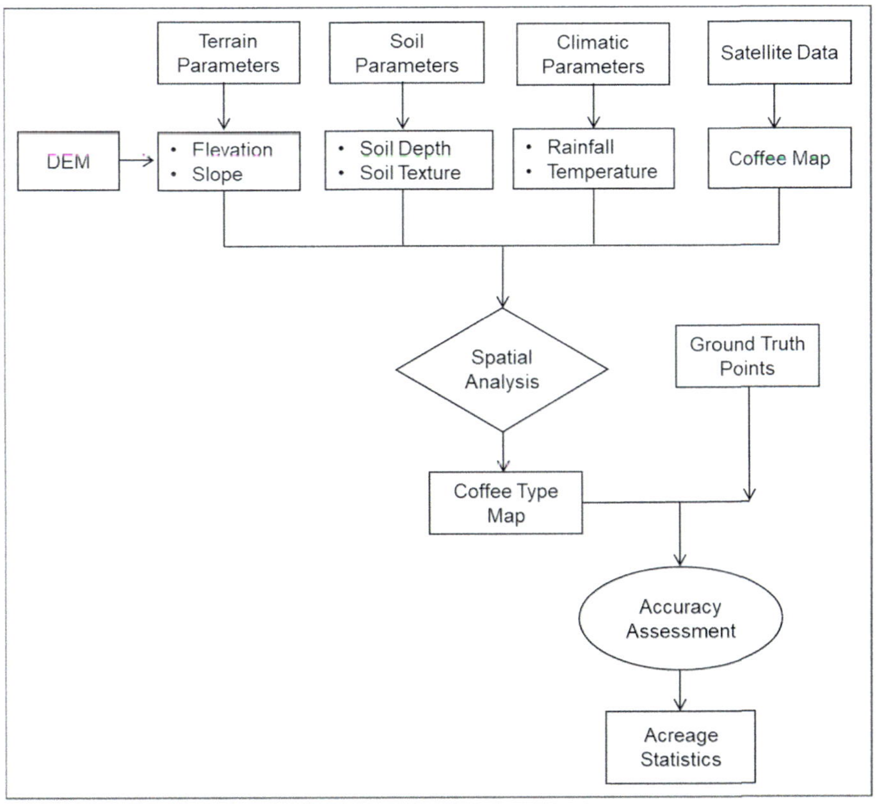

FIGURE 16.3 Methodology flowchart for coffee type discrimination.

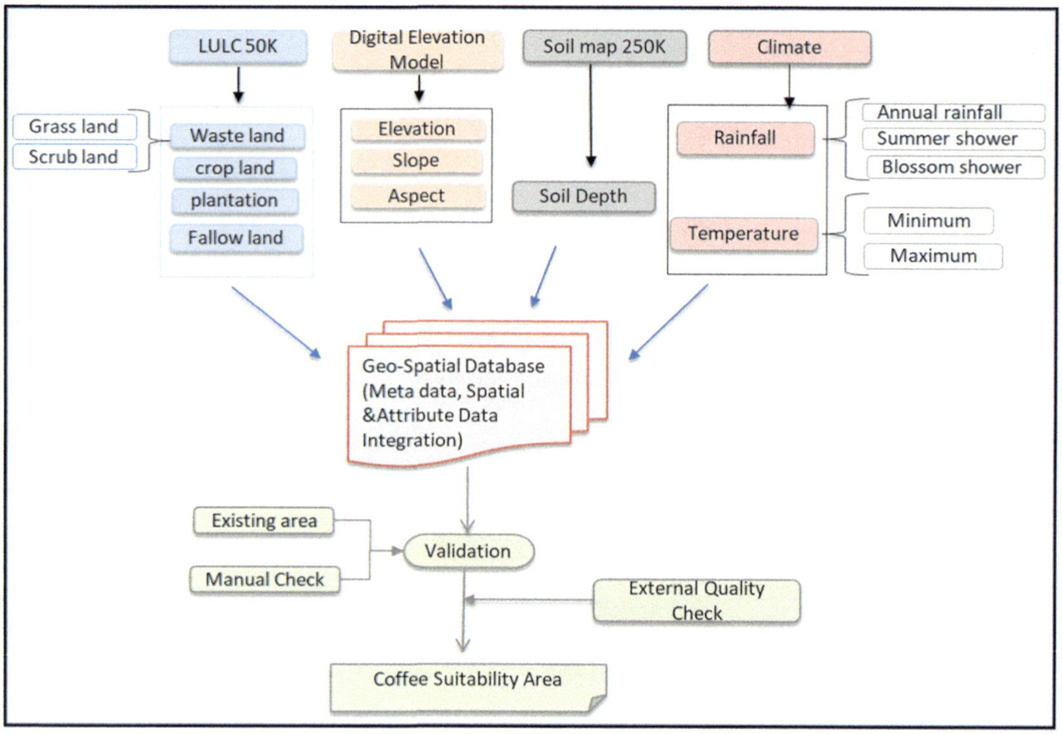

FIGURE 16.4 Flowchart depicting methodology for site-suitability analysis.

16.2.3.3 Site-suitability analysis

For introduction and expansion of coffee plantations in newer or nontraditional regions of the country, site-suitability analysis was taken up in 11 districts spread across 4 states based on pedoclimatic requirements. The essential parameters for quality coffee production include altitude (400–1700 m), annual temperature (16°C–24°C), annual precipitation (1000–1400 mm), gently sloping terrain (5%–15%), and moderately drained to well-drained soils with slightly acidic soils (Raghuramulu, 2019). In the present study, suitability analysis in selected districts was carried out using pedoclimatic parameters, specifically for Arabica coffee (Fig. 16.4). Arabica coffee requires higher altitude of 1000–1500 m and southern and western aspects should be avoided especially at lower elevations. Analysis of potential areas of coffee plantation involves geospatial techniques for data processing, thematic map preparation, spatial analysis generation of maps, and statistical outputs.

16.3 Results and discussion

16.3.1 Spectral signature of coffee plantations

Spectral response of coffee and other vegetation types was studied in depth. Coffee plantations manifested in different tones of red and mottled texture due to evergreen nature of coffee and shade trees. Signatures of monoshaded coffee plantations were distinguishable from other plantation categories, namely, coconut, areca nut, silver oak, and acacia. Typical spatial/spectral pattern of coffee plantation types (Arabica and Robusta) is depicted in Fig. 16.5. Coffee plantation appeared as brownish to deep red with mottled appearance due to the presence of interspersed shade trees. Coffee plantations typically exist at higher altitudes in India and across all slope categories except in very steep slopes (Vieira, Alves, Lacerda, Veiga, & Epiphanio, 2006). Delineation of coffee plantations was quite challenging in such cases as the signature of coffee plantations under mixed shade trees is confusing with forest.

16.3.1.1 National-level coffee mapping

Temporal LISS-III images were used for phenological characterization of coffee and other competing crops for acquisition of suitable satellite datasets. Cartosat-1 (2.5 m) and LISS-IV multispectral (5.0 m) datasets of February–April months were used for delineation. OBIA approach showed promising results in homogenous and contiguous large areas.

FIGURE 16.5 Spatial/spectral pattern of coffee plantations on high-resolution satellite images.

It showed lower accuracy in heterogeneous areas due to mixed signature and varying texture. The visual interpretation technique is the appropriate method for mapping of coffee plantations to achieve acceptable accuracy of better than 85%. However, manual interpretation is tedious, time consuming, and requires human and computer resources. Depending on the spectral characteristics and distribution of coffee plantations and other competing crops, digital or visual interpretation technique was followed.

A comparative analysis was carried out between visual interpretation and OBIA-based classification for accuracy assessment for one test area. The analysis indicated that the commission error was 14.2% which was mainly due to spectral similarity of forest and coffee plantations under mixed shade trees. The misinterpretation along the field boundary/fringes was observed because of mixed pixels contributing omission error of 10.4% with overall accuracy was 75.4% from OBIA method. An accuracy of 75%–80% for mapping of coffee plantations due to similar signature of coffee and other vegetation types has been reported by earlier studies (Bouma & Kobryn, 2004; Lisa, Lincoln, & Chris, 2018; Miguel et al., 2012). OBIA approach adopted in the heterogeneous areas showed very poor classification accuracy of about 65.3% and postclassification refinement was time consuming and tedious. Hence, hybrid approach using both OBIA and visual interpretation techniques were used in this study.

Field verification of the interpreted maps was carried out for the accuracy assessment after classification. The mapping accuracy achieved was better than 90.0%. Classified coffee maps at taluk level were spatially integrated at regional levels (taluk and district) to generate coffee map at Indian national level as depicted in Fig. 16.6. The state-wise acreage was estimated from the consolidated maps (Table 16.3). Total area under coffee plantations was about 4.45 lakh ha. The coffee area estimated from remote sensing is comparable with the official estimate at India level. However, some deviations were observed at disaggregated levels (taluk and district levels).

16.3.1.1.1 Traditional regions

Traditional regios comprised southern states of Karnataka, Kerala, and Tamil Nadu contributing nearly 91.5% of area under coffee plantations. Coffee plantations in these regions are more contiguous, occupying a large extent. Karnataka leads in total area of coffee cultivation, accounting for about 68.2% of the national acreage, followed by Kerala (15.3%) and Tamil Nadu (8.2%).

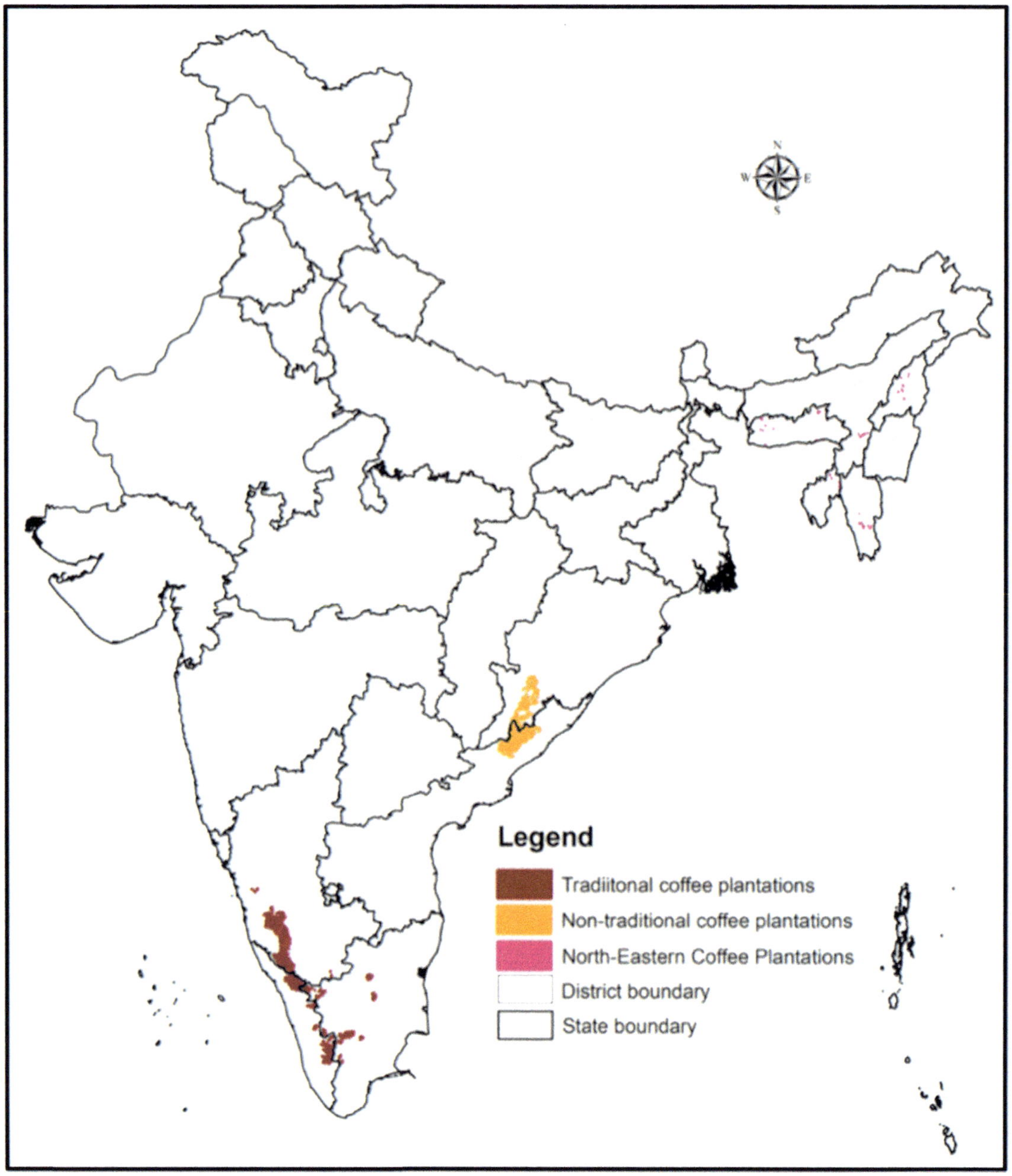

FIGURE 16.6 Spatial distribution of coffee plantations at national level. Map lines delineate study areas and do not necessarily depict accepted national boundaries.

16.3.1.1.2 Nontraditional regions

These regions accounted for about 8.0% and are mainly located in states of Andhra Pradesh and Odisha. Pedoclimatic and terrain conditions are more favorable for growing Arabica coffee in such regions. Coffee plantations were introduced here by Integrated Tribal Development Agency for providing livelihood to tribal community. New areas of coffee cultivation have come up in this region and are less contiguous, with moderate density. Most of the coffee plantations in Andhra Pradesh are distributed in and around Araku valley in Alluri Sitharamaraju district (earlier Vishakhapatnam), contributing about 7.5% area. Coffee area in Odisha state accounts for only 0.8% and distributed in Koraput, Kalahandi, and Rayagada districts.

16.3.1.1.3 Northeastern (N-E) regions

Coffee plantations exist in small and scattered plots in northeastern states of Assam, Tripura, Meghalaya, Manipur, Mizoram, Nagaland, and Arunachal Pradesh. This region contributes lesser than 0.5% of national coffee area.

TABLE 16.3 State-level coffee acreage estimates at national level.

S. No.	Coffee-growing region/State	Area under coffee plantations (ha)
A	**Traditional coffee-growing region**	
1	Karnataka	300,943
2	Tamil Nadu	36,960
3	Kerala	67,472
	Subtotal	405,375
B	**Nontraditional coffee-growing region**	
4	Andhra Pradesh	35,423
5	Odisha	3897
	Subtotal	39,320
C	**Northeastern coffee-growing region**	
6	Assam, Arunachal Pradesh, Mizoram, Meghalaya, Nagaland, Tripura	674
Total		**445,369**

TABLE 16.4 Coffee type statistics for study sites and validation site.

Test/Validation sites	Coffee type statistics (ha)			Accuracy (%)
	Arabica	Robusta	Mixed	
Belur taluk	6758.0	4698.0	3811.0	83.6
Yercaud taluk	8483.2	1.2	–	95.0
Virajpet taluk	–	62,110.0	–	95.7
Chikkamagaluru district	38,042.3	918.7	64,017.2	92.0

The analysis indicates that coffee mapping in the N-E states is quite challenging due to sparse distribution and mixed and terrace cultivation practiced in the region.

16.3.2 Coffee type discrimination

The coffee plantation map generated using the satellite data was used as an input for discrimination of coffee types in the selected study areas. The pedoclimatic parameters along terrain conditions were integrated coffee map to characterize bioclimatic regimes. The coffee type map was generated based on the bioclimatic regimes for suitable for the particular coffee type. Coffee type statistics for the study areas along with accuracy has been shown in Table 16.4. The analysis indicated that both Arabica and Robusta are cultivated in Belur taluk due to suitable climatic and terrain conditions in the taluk. In Yercaud taluk, terrain conditions and pedoclimatic parameters are ideal for Arabica cultivation with an accuracy of 95%. About 99% coffee area in the taluk is covered by Arabica, and few traces of Robusta are also cultivated in the lower altitudes. Virajpet taluk is suited for Robusta coffee because of its low elevation and moderate rainfall. As per ground information, the entire taluk is dominated by Robusta with small patches of Arabica in few places. Chikkamagaluru district comprising Chikkamagaluru, Koppa, Mudigere, and N.R. Pura taluks is considered for validation of coffee types. Chikkamagaluru taluk is dominated by Arabica and N.R. Pura is dominated by Robusta. Mudigere and Koppa taluks have mix of both Arabica and Robusta.

Chikkamagaluru taluk receives normal average rainfall of 1925 mm. Black soil is found around Baba Budan Hills whereas red and gravel soil are found in the southern parts. Coffee plantations are Arabica dominant, existing in high-altitude regions, followed by Robusta in the lower altitudes. The coffee plantations are done mainly under silver oak,

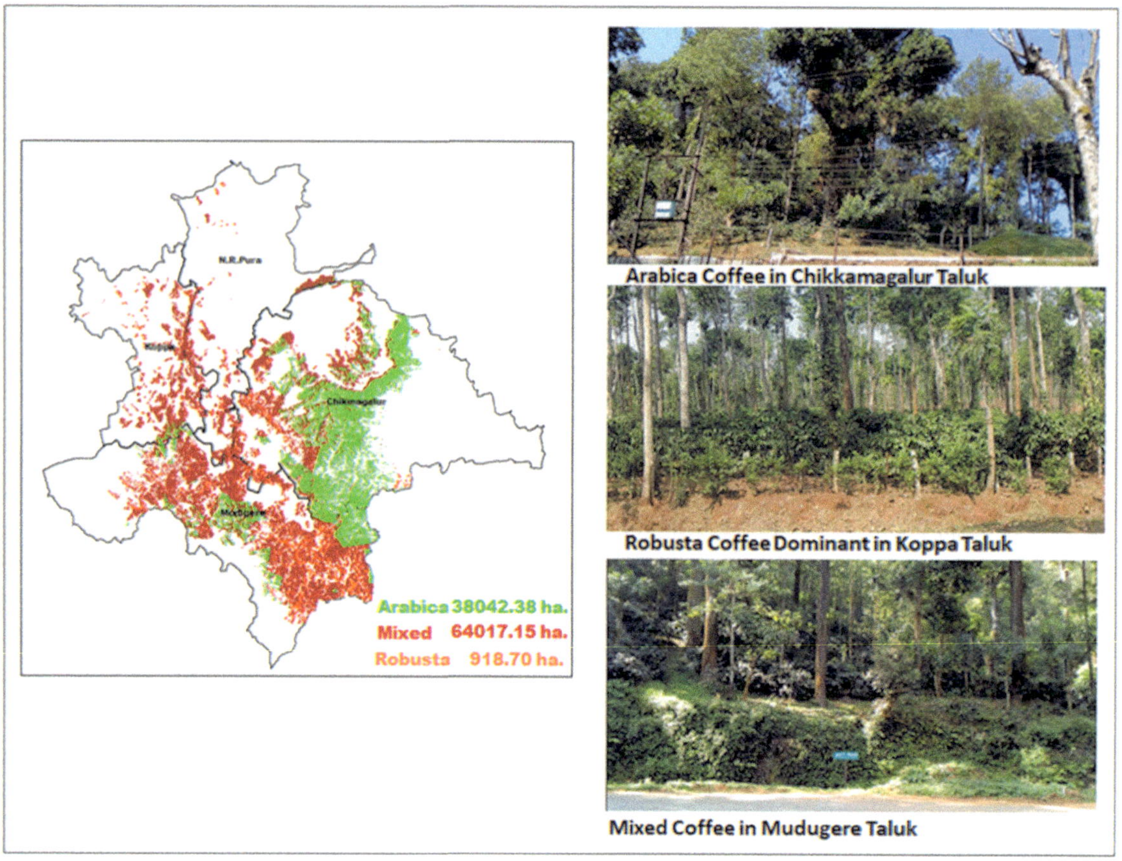

FIGURE 16.7 Coffee type map of four taluks, Chikkamagaluru district, Karnataka. Map lines delineate study areas and do not necessarily depict accepted national boundaries.

mixed shade of jungle trees, and areca nut plantations. N.R. Pura Taluk receives good rainfall during the monsoon season, and robusta is concentrated toward southern region of the taluk. Mudigere taluk has mixed proportion of Arabica and Robusta toward eastern, northeastern, and central region of the taluk. Coffee plantations are done mainly under silver oak and diversified with pepper vines. Few coffee plantations are dominated with Arabica in Koppa taluk with an elevation of about 570 m above mean sea level (MSL) Coffee plantations are mostly Robusta and partly Arabica cultivated under silver oak, mixed shade of jungle trees and areca nut plantations. The final coffee type map of Chikkamagaluru is shown in Fig. 16.7. The results indicated that the total Coffee cultivation, Arabica is found to be 38,042.38 ha, Robusta 918.70 ha, and mixed 64,017.15 ha. The results revealed that the overall accuracy was 92%.

16.3.3 Site-suitability analysis

The site-suitability analysis plays an important role for expansion of coffee plantations to meet the growing demand for coffee beans. In this context, the site-suitability analysis was carried out for 11 selected districts in the nontraditional regions using bioclimatic regimes. Site-suitability analysis was taken up in three districts in Odisha, namely, Nabrangapur, Koraput, and Malkangiri. The spatial distribution of suitable areas for coffee cultivation in the three districts of Orissa is depicted in Fig. 16.8. The study indicated that the pedoclimatic conditions in the state are ideally suitable for cultivation of Arabica coffee. Significant amount of contiguous land is available for coffee cultivation in Odisha especially in Koraput (77,978 ha) and Malkangiri (23,263 ha) followed by Nabrangapur (7767 ha).

In the case of Himachal Pradesh, the analysis revealed that about 9152 ha of coffee-suitable areas is available surrounding Hamirpur district of Himachal Pradesh. In Chhattisgarh, site-suitability analysis was taken up in two districts, namely Bastar and Dantewada. The analysis indicated that availability of coffee-suitable areas in the two districts is very limited (843 ha) and suitable areas are distributed throughout the district, indicting limited scope for expansion in the districts. In West Bengal, site-suitability analysis was taken up in Darjeeling district and the results indicated marginal area of 735 ha is available for coffee cultivation in Darjeeling (Table 16.5).

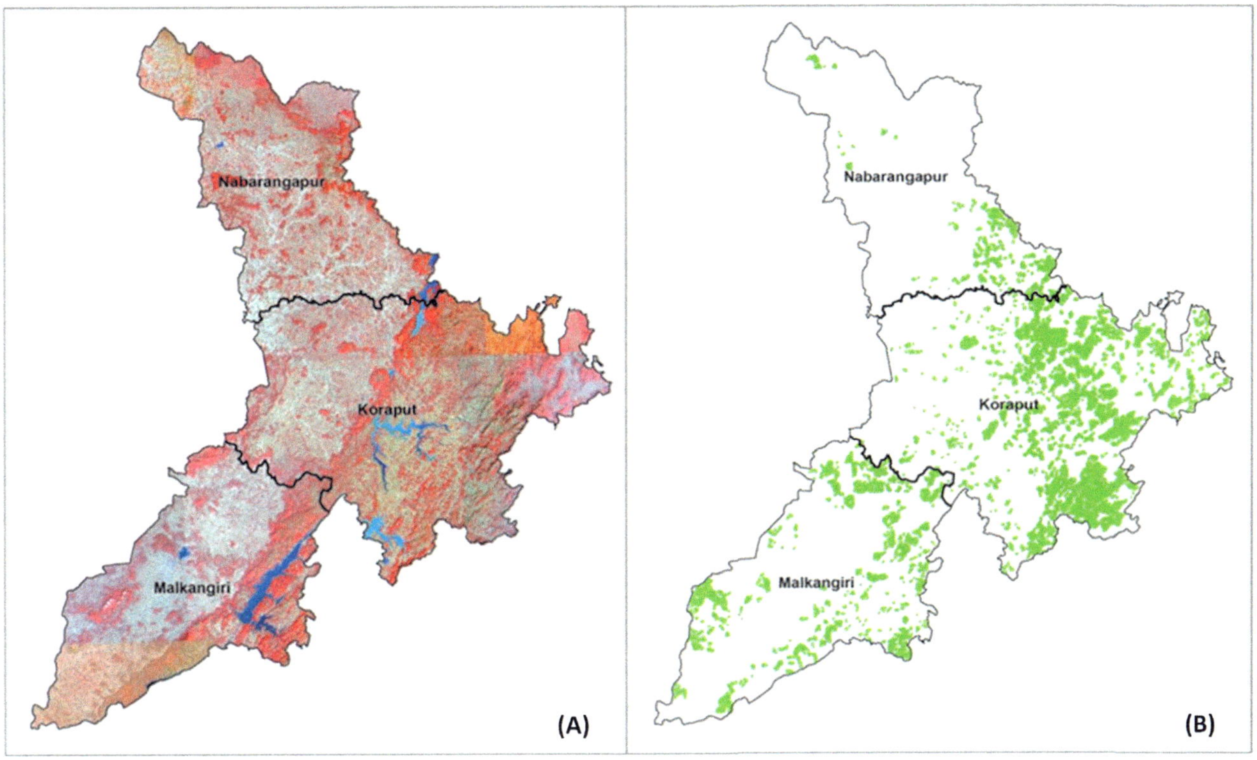

FIGURE 16.8 Spatial distribution of coffee-suitable areas in three districts of Odisha. Map lines delineate study areas and do not necessarily depict accepted national boundaries.

TABLE 16.5 District-wise suitable areas for coffee cultivation.

S. No.	State	District	Suitable area (ha)
1	Himachal Pradesh	Hamirpur	3098
		Kangra	3421
		Mandi	884
		Una	941
		Bilaspur	838
		Subtotal	***9182***
2	Chhattisgarh	Dantewada	709
		Bastar	134
		Subtotal	***843***
3	West Bengal	Darjeeling	**735**
4	Odisha	Koraput	77,978
		Malkangiri	23,263
		Nabrangapur	7767
		Subtotal	***109,008***
	TOTAL		**119,768**

16.4 Conclusions

The potential applications of geospatial technology has been showcased for inventory of coffee plantations in diverse traditional and nontraditional coffee-growing regions with an acceptable overall mapping accuracy. LISS-IV + Cartosat-1 data (2.5 m) were used for area-assessment coffee plantations, achieving better than 90.0% overall accuracy at the national level. Total area under coffee plantations was about 4.45 lakh ha and interpreted coffee maps at taluk level were spatially integrated at the national level. This study is first of its kind in the country to utilize geospatial techniques to create coffee map at the national level for generating geospatial database. This database can be updated regularly as per the need of Coffee Board and shall go a long way in meeting operational requirements for planning, management practices, developmental, and extension activities.

In the context of coffee, research studies for discrimination of coffee types are limited. Long-term climatic parameters like precipitation and temperature and terrain characteristics like elevation and slope were integrated with coffee map for generating spatial extent of coffee types for Belur and Virajpet taluks of Karnataka and Yercaud taluk of Tamil Nadu as study sites. Chikkamagaluru district has been selected as validation site. Total coffee area in Belur taluk is 15,270 ha in which Arabica coffee (6758 ha), Robusta (4698 ha), and mixed (3811 ha) and Virajpet taluk have 62,110 ha Robusta coffee estimated from this study. Coffee type map showed reasonable accuracy ranging from 83.6% to 95.7% from the test sites. Field validation of coffee type map was done for four taluks of Chikkamagaluru district as validation site showed reasonable accuracy of 92%. The analysis indicated promising results for discrimination of two dominant coffee varieties, and methodology is being extended to other coffee-growing regions of the country for generating coffee type map for the country.

The scope of the study also included the site-suitability analysis for identification of potential areas for expansion of coffee cultivation in nontraditional coffee-growing regions. The suitability criteria included 10 parameters such as climate (rainfall, blossom showers, temperature, etc.), soils (depth and texture), and terrain parameters (elevation, slope, and aspect). Spatial modeling using GIS was used for integration of different parameters for site-suitability analysis. The study indicated that about 1.20 lakh ha of suitable lands are mapped in 11 study areas which are the potential areas for introduction of coffee plantations. The climatic and terrain conditions are favorable for cultivation of Arabica coffee in hilly tribal belts of Odisha and contiguous areas are available in Koraput, Malkangiri, and Nabrangapur districts of Odisha for possible expansion. In the case of Odisha, substantial amount of contiguous land is available for coffee cultivation especially in Koraput (77,978 ha) and Malkangiri (23,263 ha) followed by Nabrangapur (7767 ha). The analysis in three study districts in Chhattisgarh and West Bengal indicated limited scope for expansion of coffee cultivation in the region. However, further characterization of these lands at microlevel/field level could be taken up by the Coffee Board using field survey to collect relevant information about the ownership details, land holding, farmer details, soils and terrain conditions, and other collateral information for field/farmer level implementation. In general, the suitable areas were located in the neighborhood of exiting coffee areas and hence, farmers in these regions could be easily motivated to take up coffee cultivation based on economics and success story of existing coffee planters.

Climate change affects coffee production and livelihoods of coffee-growing community. Changes in temperature and rainfall patterns, as well as extreme weather events, can impact production cycles and incidence of pests and diseases. Given the sensitivity of coffee quality to environmental variation, evidence-based innovations are required to enhance the sustainability and resilience of coffee sector in the context of global change. In this regard, characterization coffee agro-ecosystem has been studied in detail using geospatial technology. Further work is in progress to study the impact of climate change in coffee agro-ecosystem and carbon sequestration studies. Establishment of eddy-covariance flux tower at Central Coffee Research Institute for Carbon dynamics at microlevel.

16.5 Future scope

The future of Indian coffee in the face of climate change is a complex interplay of ecological, economic, and social factors. While significant challenges loom, there are also opportunities for adaptation and building resilience. The shift toward climate-resilient agriculture, leveraging Information and Communication Technology (ICTs) for agro-advisory services, and creating off-farm employment opportunities are essential steps toward sustaining India's coffee industry. Understanding the intricate dynamics of climate change's impact on coffee production is critical for formulating effective adaptation strategies and ensuring the sector's long-term viability. After successfully accomplishing the challenging task of mapping of coffee plantations, related aspects of coffee cultivation are required to be addressed at country level, namely, discrimination of coffee types, shade tree types and density, production estimation, pests and diseases, etc.

Further research utilizing hyperspectral, microwave techniques along with advanced data analysis techniques need to be explored for shaded Indian coffee plantations.

Overall, we recommend the following research actions for advancing the evidence on climate effects on coffee quality toward strengthening the sustainability and resilience of the coffee agro-ecosystems:

- Identification of dominant shade tree species and extending within selected coffee agro-ecosystems (Arabica and Robusta) from UAV and LiDAR data.
- Carbon sequestration studies (above-ground carbon stock estimation) using Ground, UAV, and LiDAR.
- Long-term research to capture variability of environmental exposure and impact on coffee quality over time.
- Given the northward shift of agriculture with climate change, research on climate change and coffee quality at different elevations.
- More focussed research is essential to study climate adaptation strategies in coffee agro-ecosystems as well as post-harvest technologies to mitigate climate change impact.

Acknowledgments

Authors express their sincere gratitude to Secretary, Coffee Board and the concerned authorities of Coffee Board for providing valuable guidance and support for carrying out the study. Contribution of project team of GeoCUP national project and support from other technical and administrative staff members of RRSC-South, NRSC is also duly acknowledged.

Author contributions

Conceptualization and Methodology: HR, RSS, NMK, ST; **Data Analysis:** RSS, NMK, ST; **Validation:** RSS, NMK, ST, MDS, KMJ; **Field Data:** MDS, KMJ, RSS, NMK; **Data Curation:** RSS, ST, NMK; **Writing (original draft):** RSS, NMK; **Writing (reviewing):** HR; **Supervision**: SKS, PC.

Funding

This study was funded by Coffee Board and National Remote Sensing Centre, Indian Space Research Organisation, Government of India.

Declaration of competing interest

The authors declare no competing interests.

References

Adhikari, M., Elizabeth, L. I., Russell Paterson, M., & Mark, A. M. (2020). A review of potential impacts of climate change on coffee cultivation and mycotoxigenic. *Fungi Microorganisms*, *8*, 1625. Available from https://doi.org/10.3390/microorganisms8101625.

Adhikari, U., Nejadhashemi, A. P., & Woznicki, S. A. (2015). Climate change and Eastern Africa: A review of impact on major crops. *Food and Energy Security*, *4*, 110–132. Available from https://doi.org/10.1002/fes3.61.

Anonymous. (2014). *Coffee guide*. Karnataka, India: Central Coffee Research Institute, Coffee Research Station, Government of India, Chikkamagaluru district.

Bernardes, T., Moreira, M. A., Adami, M., Giarolla, A., & Rudorff, B. F. T. (2012). Monitoring biennial bearing effect on coffee yield using MODIS remote sensing imagery. *Special Issue on Advances in Remote Sensing of Agriculture*, 2492–2509.

Bilen, C., El Chami, D., Mereu, V., Trabucco, A., Marras, S., & Spano, D. (2023). A systematic review on the impacts of climate change on coffee agro systems. *Plants*, *12*, 102. Available from https://doi.org/10.3390/plants12010102.

Bock, M., Xofis, P., Mitchley, J., Rossner, G., & Wissen, M. (2005). Object-oriented methods for habitat mapping at multiple scales – Case studies from Northern Germany and Wye Downs, UK. *Journal for Nature Conservation*, *13*, 75–89.

Bouma, G. A., & Kobryn, H. T. (2004). Change in vegetation cover in East Timor, (1989-1999). *Natural Resources Forum*, *28*, 1–12.

Bunn, C., Laderach, P., Ovalle Rivera, O., & Kirschke, D. (2015). A bitter cup: Climate change profile of global production of Arabica and Robusta coffee. *Climatic Change*, 89–101.

CIAT. (2019). In A. Eitzinger, L. Eugene, A. Jarvis, A. Martínez, & C. Navarrete (Eds.), *Disaster and climate risk assessment for PITAG regions in Haiti. International Center for Tropical Agriculture* (p. 105). Cali, Colombia: CIAT Publication (Technical Report, No. 476).

Coffee Board. (2018). Database on coffee. *Annual report of Coffee Board of India.* http://www.indiacoffee.org.

Coffee Board. (2019). Database on coffee. *Annual report of Coffee Board of India.* http://www.indiacoffee.org.

Cordero-Sancho, S., & Sader, S. A. (2007). Spectral analysis and classification accuracy of coffee crops using Landsat and a topographic-environmental model. *International Journal. of Remote Sensing, 28*, 1577–1593.

Damodaran, A. (2002). Conflict of trade-facilitating environmental regulations with biodiversity concerns: The case of coffee-farming units in India. *World Development, 30*(7), 1123–1135.

Davis, A. P., Bridson, D. M., Stoffelen, P., & Govaeirts, R. (2006). An annotated taxonomic conspectus of the genus Coffea (Rubiaceae). *Botanical Journal of the Linnean Society, 152*, 465–512.

FAOSTAT. (2021). *Food and Agriculture Organisation of the United Nations Statistical Database*. Rome: Statistical Division, FAO.

Frankline, R., & Charles, C. M. (2016). GIS based suitability analysis for coffee farming in Kenya, (2016). *International Journal of Geomatics and Geosciences, 6*(3), 1722–1733.

Gamanya, R., Maeyer, P. D., & Dapper, M. D. (2007). An automated satellite image classification design using object-oriented segmentation algorithms: A move towards standardization. *Expert Systems with Applications, 32*, 616–624.

GIZ. (2010). *Climate change adaptation and mitigation in the Kenyan coffee sector.* Eschborn, Germany: Deutsche Gesellschaft für Internationale Zusammenarbeit (GIZ) GmbH.

Hebbar, R., Ravishankar, H. M., Trivedi, S., Manjula, V. B., Kumar, N. M., Mukharib, D. S., Mote, J. K., ... Ganesh Raj, K. (2019). National level inventory of coffee plantations using high resolution satellite data. ISPRS international archives of the photogrammetry. *Remote Sensing and Spatial Information Sciences*, 293–298. Available from https://doi.org/10.5194/isprs-archives-XLII-3-W6-293-2019, XLII-3/W6.

Hebbar, R., Ravishankar, H. M., Trivedi, S., Rama Subramoniam, S., Raj, U., & Dadhwal, V. K. (2014). Object oriented classification of high-resolution data for inventory of horticulture crops. *The international archives of the photogrammetry, remote sensing and spatial information sciences* (Vol. XL-VIII). *ISPRS Technical Commission VIII Symposium*, , 09–12 Dec. 2014, Hyderabad, India.

Hidayat, E., Afriliana, A., Gusmini., & Harada, H. (2020). Land suitability evaluation of coffee in Tokunoshima Island, Japan. *Journal of Applied Agricultural Science and Technology, 4*(2), 146–154.

International Coffee Organization (ICO) Trade Statistics (2018).

IPCC. (2022). In H. O. Portner, A. Alegria, M. Craig, D. Roberts, M. Tignor, S. Langsdorf, & K. Mintenbeck (Eds.), *Climate change: Impacts, adaptation and vulnerability. Contribution of working group II to the sixth assessment report of the intergovernmental panel on climate change* (p. 3056). New York, NY: Cambridge University Press.

Kasper, J., Stuart, P., Christian, W., Seonaid, P., & Lisa, N. (2009). Mapping banana plantations from object-oriented classification of SPOT-5 imagery. *PE& RS, 75*(9), 1069–1081.

Kath, J., Byrareddy, V., Davis, A. P., Craparo, A., Fong, Y., King, R., Nquyen-Huy, T., van Asten, P. J. A., Marcussen, T., Mushtaq, S., et al. (2022). Vapour pressure deficit determines critical thresholds for global coffee production under climate change. *Nat. Food., 3*, 2022.

Komar, O. (2006). Ecology and conservation of birds in coffee plantations: A critical review. *Bird Conservation International, 16*, 1–23.

Lisa, C. K., Lincoln, P., & Chris, B. (2018). Using Google Earth engine to map complex shade-grown coffee landscapes in northern Nicaragua. *Remote Sensing, 10*(952), 1–19. Available from https://doi.org/10.3390/rs10060952.

Miguel, A. O. H., Komar, O., Kevin, P. P., & Hugo, J. V. (2012). Mapping coffee plantations with Landsat imagery: An example from El Salvador. *International Journal of Remote Sensing, 33*(1), 220–242.

Moguel, P., & Toledo, V. M. (1999). Review: Biodiversity conservation in traditional coffee systems of Mexico. *Conservation Biology, 13*, 11–21.

Mueller, M., Segl, K., & Kaufmann, H. (2004). Edge- and region-based segmentation technique for the extraction of large, man-made objects in high-resolution satellite imagery. *Pattern Recognition, 37*, 1619–1628.

Nagaraj, G., & Mote, K. (2020). Impact of climate change on coffee production: An overview. *Journal of Pharmacognosy and Phytochemistry, 9*(3), 1850–1858.

Neupane, K., & Sharma, T. P. P. (2022). Land suitability analysis for coffee production in Hilly Districts of Lumbini Province. *The Geographic Base*, 1–14. Available from https://doi.org/10.3126/tgb.v9i1.55421.

Nzeyimana, I., Hartemink, A. E., & Geissen, V. (2014). GIS-based multi-criteria analysis for arabica coffee expansion in Rwanda. *PLoS One, 9*(10), e107449. Available from https://doi.org/10.1371/journal.pone.0107449.

Perfecto, I., Rice, R., Greenberg, R., & Van der Voort, M. E. (1996). Shade coffee: A disappearing refuge for biodiversity. *Bioscience, 46*(8), 598–608.

Poltronieri, P., & Rossi, F. (2016). Challenges in specialty coffee processing and quality assurance. *Challenges, 7*, 19.

Raghuramulu, Y. (2019). Climate resilient technologies for sustainable coffee production. *Indian Coffee, LXXXIII*(5), 4–12.

RRSC-S. (2016). *Inventory of coffee plantations using Geo-spatial technology: Pilot study in traditional coffee growing regions of India* (NRSC-RC-APR-2016-TR-842, pp. 1–150).

Rudragouda, C., D'souza, G. F., & Somashekhargouda, P. (2017). Climate change in coffee tracts of Western Ghats of India. *International Journal of Agriculture Sciences, 9*(52), 4893–4897.

Salas Lopez, R., Gamez Fernandez, D., Silva Lopez, J. O., Rojas Briceno, N. B., Oliva, M., Terrones Murga, R. E., et al. (2020). Land suitability for coffee (*Coffea arabica*) growing in Amazonas, Peru: Integrated use of AHP, GIS and RS. *ISPRS International Journal of Geo-Information, 9*, 673. Available from https://doi.org/10.3390/ijgi9110673.

Simms, A. (2005). *Africa: Up in smoke? The second report from the Working Group on Climate Change and Development (Series Up)*. London, United Kingdom: New Economics Foundation.

Singha, C., & Swain, C. K. (2016). Land suitability evaluation criteria for agricultural crop selection: A review. *Agricultural Reviews*, *37*(2), 125–132. Available from https://doi.org/10.18805/ar.v37i2.10737.

URT. (2003). *Initial national communication under the United Nations Framework Convention on Climate Change (UNFCCC)*. Dar es Salaam, Tanzania: United Republic of Tanzania, Vice President' s Office.

Velmourougane, K., & Bhat, R. (2017). Sustainability challenges in the coffee plantation sector. In R. Bhat (Ed.), *Sustainability challenges in the agrofood sector* (pp. 616–642). New York, NY: John Wiley & Sons Ltd. (1st ed.).

Vieira, T. G. C., Alves, H. M. R., Lacerda, M. P. C., Veiga, R. D., & Epiphanio, J. C. N. (2006). Crop parameters for the evaluation of the spectral response of coffee (Coffea arabica) areas within the state of Minas Gerais, Brazil, Coffee Science. *Lavras*, *1*(2), 111.

Chapter 17

Spatio-temporal assessment of surface dynamics of high-altitude wetlands using Earth-Observation Datasets

Mukunda Dev Behera[1], K.G. Yadhukrishna[1], A. Raj[1], Ishita S. Srivastava[2], Pulakesh Das[3], A. Kashyap[1] and R. Joshi[4]
[1]*Centre for Ocean, River, Atmosphere and Land Sciences (CORAL), Indian Institute of Technology Kharagpur, Kharagpur, West Bengal, India,* [2]*School of Environmental Science and Engineering, Indian Institute of Technology Kharagpur, Kharagpur, West Bengal, India,* [3]*School of Forest Resources, University of Maine, Orono, ME, United States,* [4]*Sikkim Regional Centre, G.B. Pant National Institute of Himalayan Environment, Gangtok, Sikkim, India*

17.1 Introduction

High-altitude wetlands (HAWs) refer to wetlands such as areas of swamp, marsh, meadow, fen, peatland, or waterbody above 3000 m altitude (Chatterjee et al., 2010). They are characterized by high ultraviolet radiation and low oxygen partial pressure, with plant and animal species having specific tolerances of harsh conditions, including low temperatures (O'Neill, 2019). These wetlands are often glacial and include water bodies like lakes, ponds, and rivers (Begam & Sen, 2019). A wide range of biodiversity, including several endemic and endangered species of plants, birds, and mammals, is sustained by HAWs. The term "high altitude" is independent of the surrounding topography, meaning these wetlands can exist even in low-relief regions. HAWs also influence the downstream region economy and thereby people's knowledge, traditions, and socio-economy. Wetlands have a significant impact on the hydrological cycle because they affect surface runoff, river discharge, groundwater replenishment, evaporation, etc. The HAWs serve as natural reservoirs, regulating water flow and supporting downstream water needs (Bullock & Acreman, 2003). The Indian Himalayan Region (IHR), located in the northern parts of India, provides various ecosystem services to millions of individuals residing in this region. IHRs are notably sensitive to climate change and human disturbances (Das, Behera, Bhaskaran, & Roy, 2022; Murthy, Das, & Behera, 2016). Majority of the wetlands in HAW in IHR are smaller than 10 ha (Panigrahy, 2017). The temperature climate with seasonal precipitation and melting of snow and glaciers in the Himalayan region feed the HAWs in this region. These HAWs have a significant impact on the regional climatic conditions and regional biodiversity, hydrology, and carbon storage (Blackwell & Pilgrim, 2011). These wetlands serve as important breeding grounds for migratory birds.

Developing spatio-temporal data on wetland's water spread and water quality is a prerequisite for wetland management and conservation. Precise monitoring is crucial for high-altitude wetlands due to their significance in supporting biodiversity, regulating water, and sequestering carbon. The conventional techniques for mapping and monitoring wetland zones, such as field surveys and sampling, are typically labor-intensive, costly, time-consuming, and frequently unable to identify changes over large areas. The surface reflectance measured in satellite remote sensing enables the mapping of land surface features and their properties at a large spatial extent in a definite interval (Roy & Behera, 2003). Satellite data provide repetitive coverage, allowing for seasonal or annual monitoring of wetlands at different temporal and spatial scales. The free access to past satellite imageries allows monitoring changes since 1907s. Sentinel-2 data is one of the highest-resolution data publicly available for monitoring waterbodies, estimating chlorophyll content, colored dissolved organic matter, and dissolved organic carbon assessing water quality in inland and coastal waterbodies, detecting algal blooms and aquatic ecosystem health (Behera, Gupta, Barik, Das, & Panda, 2018; Hossen, Mahmod, Negm, & Nakamura, 2022; Kislik, Dronova, Grantham, & Kelly, 2022; Toming et al., 2016).

Various techniques have been developed for land surface monitoring, including waterbodies, such as visual image interpretation and digital or automated image classification (Behera et al., 2018). The spectral image enhancement technique is one of the robust approaches for water body mapping. This approach involves multiple spectral bands and

Sustainable Development Perspectives in Earth Observation. DOI: https://doi.org/10.1016/B978-0-443-14072-3.00022-8

highlights different land surfaces in an image. The spectral indices minimize atmospheric effects, background soil reflectance for vegetation, sensor calibration, different environmental conditions, and illumination. Therefore the use of spectral indices became an important tool for multitemporal change detection studies (Behera et al., 2018). The normalized index limits the spectral index values between −1 and +1 and allows easy interpretation of images based on the varied reflectance properties of the objects. Normalized difference vegetation index (NDVI) is a widely used index which enhances the vegetation greenness and water body in images based on their varied reflectance between near-infrared (NIR) and red bands. The higher reflectance of vegetation in the NIR band and lower in the red band causes a positive value for vegetation, whereas a lower reflectance in the NIR band and moderate reflectance in the red bands induce a negative value for the water body.

In addition, several spectral indices have been developed for water body and water quality mapping. The normalized difference water index (NDWI) is a popular method for determining moisture content and identifying open water features (Das, Pandey, & Dutta, 2021; Talukdar & Pal, 2019). It is particularly useful in distinguishing between water bodies and other land surface features. Behera, Chitale, Shaw, Roy, and Murthy (2012) applied NDWI and NDVI to Landsat and LISS-III data to understand the seasonal dynamics of Samaspur wetlands in Raebareli, Uttar Pradesh, India. Their study highlighted the construction of the Sharda Canal and other human activities, such as agricultural activities, that modified the water spread and water quality. Several other studies have used NDWI for surface and change monitoring using Landsat and Sentinel-2 optical data (Behera et al., 2018; Panigrahy, 2017; Srivastava & Agarwal, 2020). The modified normalized difference water index (MNDWI) outperforms the NDWI in extracting surface water from urban environments because it improves water bodies while suppressing built-up features more efficiently. Normalize difference turbidity index (NDTI) is a useful spectral index for determining the amount of silt in rivers and monitoring their quality (Das et al., 2021). NDTI shows a good agreement ($R^2 = 0.9$) with water turbidity in lakes (Bid & Siddique, 2019). Singh, Bhardwaj, and Verma (2020) employed NDTI for water turbidity assessment in Harike wetland, Punjab, India. Rawat and Singh (2024) applied multiple spectral indices including NDWI and NDTI to Sentinel-2 imagery for water spread and quality monitoring in the Nainital lake (>1900 m a.m.s.l), Nainital district, Uttarakhand, India, from 2017 to 2022. The Google Earth Engine (GEE) is a Web Geographic Information System platform that provides several datasets, including satellite images, topography, and climate, and allows their processing online. The user can download the processed data without downloading the satellite images, which reduces the storage space required for satellite data processing. Thus it does not require any geospatial software in the user's system. Such advantages have dramatically increased the use of the GEE platform in geospatial research (Das et al., 2021).

Various global databases are developed on wetlands (Zhang, et al., 2022). Sustainable Wetlands Adaptation and Mitigation Program portals show India has the third highest wetland area (2.7 million ha) after Brazil and Indonesia (https://www2.cifor.org/global-wetlands/ accessed on September 1, 2024). Zhang, et al. (2022) integrated Landsat optical and Sentinel-1 microwave data for global scale wetland mapping by applying the random forest machine learning algorithm in the GEE platform. They employed multiple spectral indices, including NDVI and NDWI, and reported ~86% accuracy, validating with more than 25,000 data points. In India, the National Remote Sensing Centre (NRSC), Hyderabad, uses satellite data to produce a national database on land use land cover (LULC) maps at periodic intervals. These data also classify the waterbody extent; however, the database does not produce LULC data every year and does not study seasonal variations or water quality. Although several studies were conducted for waterbody area mapping and quality assessment in India, limited studies assessed HAWs.

17.2 Study area

Sikkim is one of the smallest Indian states in the North-Eastern Himalayan region. It is located between the Lesser and Greater Himalayas and exhibits faults and folds in younger mountain systems. Geographically, it is located between 27°5′ and 28°10′N and between 88°01′ and 88°58′E. The highest mountain peak, Kangchenjunga, is located in Sikkim. Sikkim is characterized by its diverse climatic zones that range from subtropical valleys to alpine heights. HAWs above ~3000 m.a.s.l in the Sikkim Himalayas inhabit glacially carved valleys, cirques, and basin depressions framed by towering peaks and ridges. Gurudongmar Lake Complex, Tso Lhamu, and Tsomgo Lake are the few important lakes of Sikkim situated above 3500 m altitude, wherein the Tsomgo Lake, Gurudongmar Lake Complex, Khecheopalri Lake, Tamzey Lake Complex, and Tembava Lake complex were included under the National Wetland Conservation Program (Srivastava & Agarwal, 2020). The climatology in this region is defined by cold temperatures, substantial precipitation, and pronounced seasonality, with monsoon rains replenishing the wetlands during the summer. The landforms undergo slow sedimentation processes, with organic matter accumulation driving the formation of peatlands. Freeze-thaw cycles influence soil dynamics and hydrology, shaping diverse microenvironments. The weather has become unpredictable and

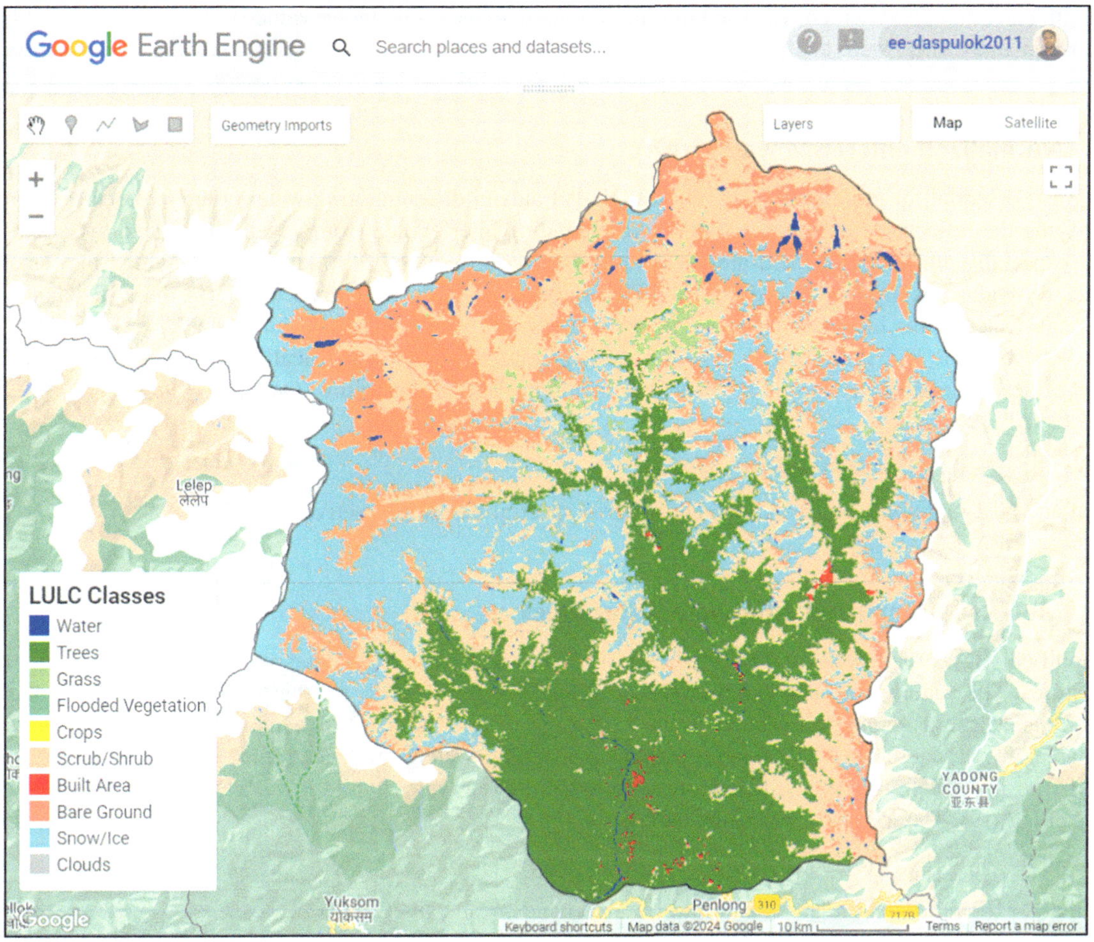

FIGURE 17.1 The Environmental Systems Research Institute (ESRI) land use land cover (LULC) map of the year 2020 was accessed from Google Earth Engine (GEE). Map lines delineate study areas and do not necessarily depict accepted national boundaries.

inconsistent, while global warming in recent decades has caused rapid snow melting and sources of HAW. Alpine grasses, sedges, and mosses foster rich biodiversity and serve as critical habitats for endemic and migratory fauna (Fig. 17.1).

17.3 Data and methodology

17.3.1 Data used

This study used Sentinel-2 satellite imagery from 2016 to 2023 to map the wetlands over 0.1 ha using the GEE platform. With two satellites (2A and 2B), it has 5 days of temporal frequency in the equator. The onboard multispectral instrument collects data in 13 spectral bands in visual NIR range at 10, 20, and 60 m spatial resolutions, which allows detailed examination of a variety of land surface properties, including vegetation condition and water quality assessment (Behera et al., 2018; Ghosh, Mishra, & Gitelson, 2016; Varghese, Behera, & Behera, 2023). The cloud-free Sentinel-2 images for summer, monsoon, and postmonsoon (winter) were accessed for seasonal waterbody change monitoring. The Shuttle Radar Topographic Mission (SRTM) digital elevation model (DEM) data at 30 m resolution was obtained from the United States Geological Survey. The DEM data were also accessed from the GEE platform.

17.3.2 Methodology

Multiple indices were derived using cloud-free Sentinel-2 imagery to study the surface water dynamics in the study area (detailed below). The SRTM DEM was used to identify the high-altitude areas above 3000 m. A binary mask was created with elevation 3000, which was used to clip the index images. The index values in three different seasons (summer, monsoon and winter) were analyzed from 2016 to 2023 to understand the seasonal and annual variability.

17.3.3 Normalized difference water index

The NDWI is essential for monitoring waterbodies. NDWI aids in differentiating water bodies from vegetation and soil by amplifying the reflectance of water features while reducing interference from vegetation and soil (Gao, 1996). This ability is especially crucial in elevated areas where marshes are frequently limited in size, divided into smaller parts, and affected by changes in seasons and climate. The waterbodies show lower reflectance in the NIR band than in the visible range, wherein the highest reflectance is observed in the blue-green transition zone. Therefore with the higher gradient between NIR and green, NDWI index values greater than 0.5 are typically associated with aquatic bodies. NDWI is widely employed in various applications, including wetland mapping (Ouma & Tateishi, 2006). NDWI has been calculated as:

$NDWI = \frac{Green - NIR}{Green + NIR}$ (McFeeters, 1996) (17.1)

Green and NIR represent the green and near-infrared spectral bands, respectively.

17.3.4 Modified normalized difference water index

It is a modified NDWI that uses a mid-infrared (MIR) band instead of an NIR band and can considerably improve open water features. The waterbodies also exhibit a lower reflectance in the MIR, similar to the NIR band. On the contrary, the built-up areas show a higher reflectance in the MIR band rather than the NIR band. Thus MNDWI better segregates built-up areas from waterbodies and is more suited for mapping waterbodies near densely populated areas. Similar to NDWI, the positive MNDWI values indicate the presence of open water, while lower values represent land or vegetation (O'Neill, 2019). The MNDWI was calculated as:

$MNDWI = \frac{Green - MIR}{Green + MIR}$ (Xu, 2006) (17.2)

Green and MIR represent the green and mid-infrared bands, respectively.

17.3.5 Normalized difference turbidity index

NDTI is used to estimate the turbidity in the water bodies and their temporal changes (Garg, Aggarwal, & Chauhan, 2020; Khan & Baig, 2017). The amount of total suspended sediments and other contaminants in the water is correlated with turbidity, which measures water transparency. Clear water has a higher reflectance in the blue-green transition zone, wherein soil reflectance increases with increasing wavelength in the visible spectrum. Therefore clear water shows higher reflectance in the green spectrum than the red spectrum, while turbid water shows higher reflectance in the red spectrum than in green. The NDTI was derived as:

$NDTI = \frac{Red - Green}{Red + Green}$ (Lacaux, Tourre, Vignolles, Ndione, & Lafaye, 2007) (17.3)

where green and red represent the green and red bands, respectively.

TABLE 17.1 Summary of indices used in this study.

Index	Formula	Spectral range (μm)	Index with mid-band spectra (μm)	References
NDWI	$\frac{Green - NIR}{Green + NIR}$	0.60–0.70 0.80–1.10	$\frac{0.65 - 0.95}{0.65 + 0.95}$	McFeeters (1996)
MNDWI	$\frac{Green - MIR}{Green + MIR}$	0.52–0.60 1.55–1.75	$\frac{0.56 - 1.65}{0.56 + 1.65}$	Xu (2006)
NDTI	$\frac{Red - Green}{Red + Green}$	0.61–0.68 0.50–0.59	$\frac{0.64 - 0.54}{0.64 + 0.54}$	Lacaux et al. (2007)

NDTI values near zero indicate high turbidity when the red and green reflectance values are similar. A high negative value indicates clear water, as low-turbidity water shows higher green reflectance than red. Other elements that can affect NDTI include sun glint, shallow water, and exposed clouds, all of which increase red reflectance (Table 17.1).

17.4 Results

The derived spectral indices (NDWI, MNDWI, and NDTI) for summer, monsoon, and winter from 2016 to 2023 are shown in Figs. 17.2–17.9.

Index value ranges in 2016:

The NDTI values varied between 0.4 to −0.5, 0.9 to −0.2, and 0.6 to −0.6 for the summer, monsoon, and winter seasons, respectively, for 2016. Similarly, the NDWI values varied between 0.4 to −0.8, 0.8 to −0.4, and −0.6 to −0.4 for the summer, monsoon, and winter seasons. In contrast, the MNDWI values varied between 0.9 to −0.7, 0.9 to −0.4, and 0.2 to −0.6 for the summer, monsoon, and winter seasons, respectively (Fig. 17.2).

Index value ranges in 2017:

The NDTI values ranged from 0.7 to −0.5, 0.4 to −0.8, and 0.7 to −0.4 for the summer, monsoon, and winter seasons, respectively, in 2017. The NDWI values ranged from −0.6 to 0.5 for the summer season, from −0.6 to 0.9 for the monsoon season, and from −0.6 to 0.2 for the winter season in 2017. On the other hand, the MNDWI values ranged from −0.6 to 0.6 for the summer season, from 0.4 to 0.8 for the monsoon season, and from −0.7 to 0.4 for the winter season in 2017 (Fig. 17.3).

Index value ranges in 2018:

The NDTI values ranged from 0.4 to −0.6, 0.8 to −0.4, and 0.6 to −0.2 for the summer, monsoon, and winter seasons, respectively, in 2018. The NDWI values ranged from 0.5 to −0.6, 0.9 to 0.2, and 0.6 to −0.4, while the MNDWI values ranged from 0.4 to −0.6, 0.8 to −0.2, and 0.3 to −0.7, respectively (Fig. 17.4).

Index value ranges in 2019:

The NDTI values ranged from 0.5 to −0.3, 0.6 to −0.7, and 0.8 to −0.7 in the summer, monsoon, and winter seasons of 2019 (Fig. 17.5), the NDWI values ranged from 0.3 to −0.4, 0.9 to 0.4, and 0.6 to −0.6, respectively, while the MNDWI values varied from 0.3 to −0.3, 0.7 to −0.4, and 0.4 to −0.4, respectively.

Index value ranges in 2020:

The NDTI values for the 2020 summer, monsoon, and winter seasons ranged from 0.3 to −0.4, 0.6 to −0.7, and 0.8 to −0.5, respectively. The NDWI values for the 2020 summer, monsoon, and winter seasons were 0.2 to −0.3, 0.7 to −0.5, and 0.4 to −0.6, respectively, whereas the MNDWI values were 0.4 to −0.6, 0.7 to 0.4, and 0.6 to −0.8, respectively (Fig. 17.6).

Index value ranges in 2021:

In contrast, NDTI values ranged from 0.5 to −0.4, 0.6 to −0.6, and 0.8 to −0.4 in the summer, monsoon, and winter seasons of 2021 (Fig. 17.7). The NDWI values ranged from 0.4 to −0.4, 0.8 to −0.3, and 0.7 to −0.4, respectively. In contrast, the MNDWI ranged from 0.3 to −0.5, 0.5 to −0.6, and 0.9 to −0.4, respectively.

Index value ranges in 2022:

In 2022 the NDTI values ranged from 0.7 to −0.5, 0.8 to −0.7, and 0.4 to −0.4 for the summer, monsoon, and winter seasons, respectively. The NDWI values for the summer, monsoon, and winter seasons of 2022 ranged from 0.5 to −0.3, 0.7 to −0.3, and 0.4 to −0.5, respectively, whereas the MNDWI values varied from 0.4 to −0.3, 0.8 to 0.6, and 0.7 to −0.7, respectively (Fig. 17.8).

Index value ranges in 2023:

NDTI values ranged from 0.3 to −0.4, 0.9 to −0.5, and 0.6 to −0.4 for the summer, monsoon, and winter seasons, respectively. For the summer, monsoon, and winter seasons of 2023, the NDWI values ranged from 0.4 to −0.3, 0.8 to −0.4, and 0.6 to −0.4, respectively, whereas the MNDWI values ranged from 0.3 to −0.3, 0.9 to 0.7, and 0.6 to −0.5, respectively (Fig. 17.9).

The spatio-temporal extent of NDWI across the Sikkim Himalayas varies from high values in the northern part to low values in the southern part. The spatio-temporal maps from 2016 to 2023 show lower NDWI values during the summer. This could be attributed to lower water levels in wetlands due to dry summer (i.e., lower precipitation). Moreover, higher summer temperatures lead to high evaporation, reducing water levels (Chatterjee et al., 2010). Comparatively, monsoon precipitation leads to increased water levels and a higher NDWI value in the wetlands. Compared to summer, a lower monsoon temperature and higher humidity lead to reduced evaporation rates. The snowfall and low-temperature freeze waterbodies in the HAW in the winter, leading to lower NDWI. However, the unfrozen waterbodies indicated a higher NDWI value. Similarly, MNDWI decreases in the summer due to reduced water levels

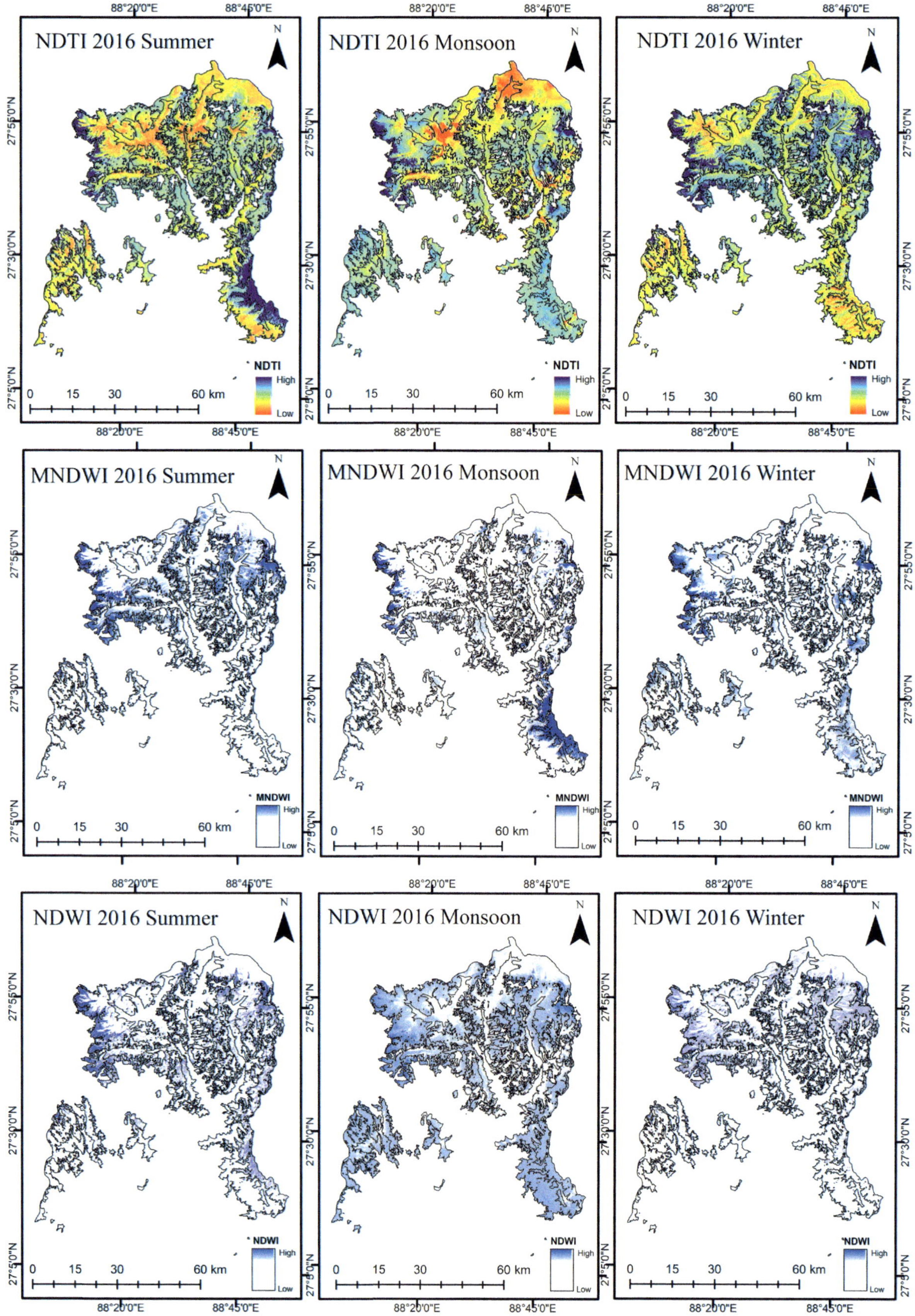

FIGURE 17.2 Seasonal distribution of NDTI, MNDWI, and NDWI for 2016 derived from Sentinel-2. Map lines delineate study areas and do not necessarily depict accepted national boundaries.

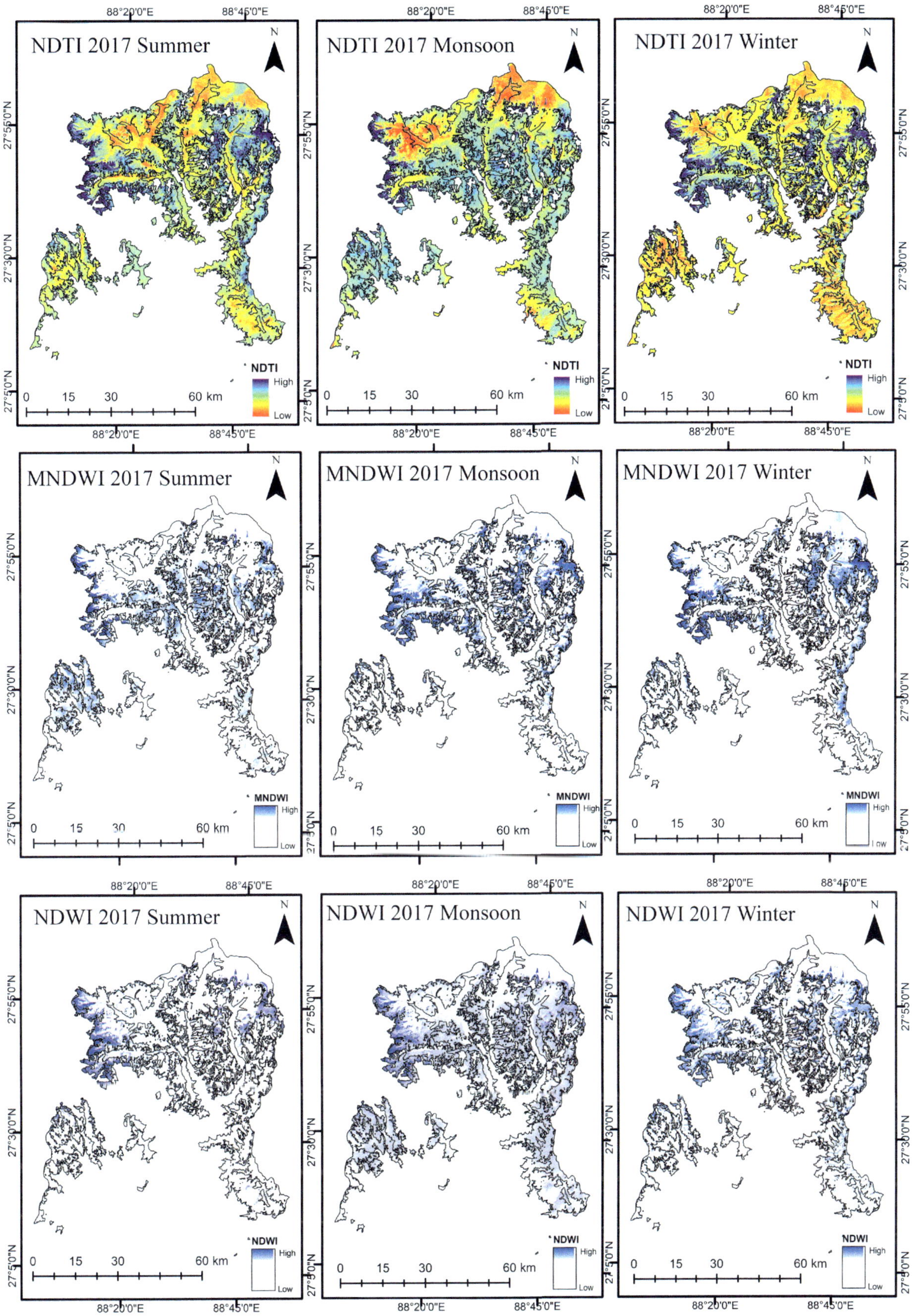

FIGURE 17.3 Seasonal distribution of NDTI, MNDWI, and NDWI for 2017 derived from Sentinel-2. Map lines delineate study areas and do not necessarily depict accepted national boundaries.

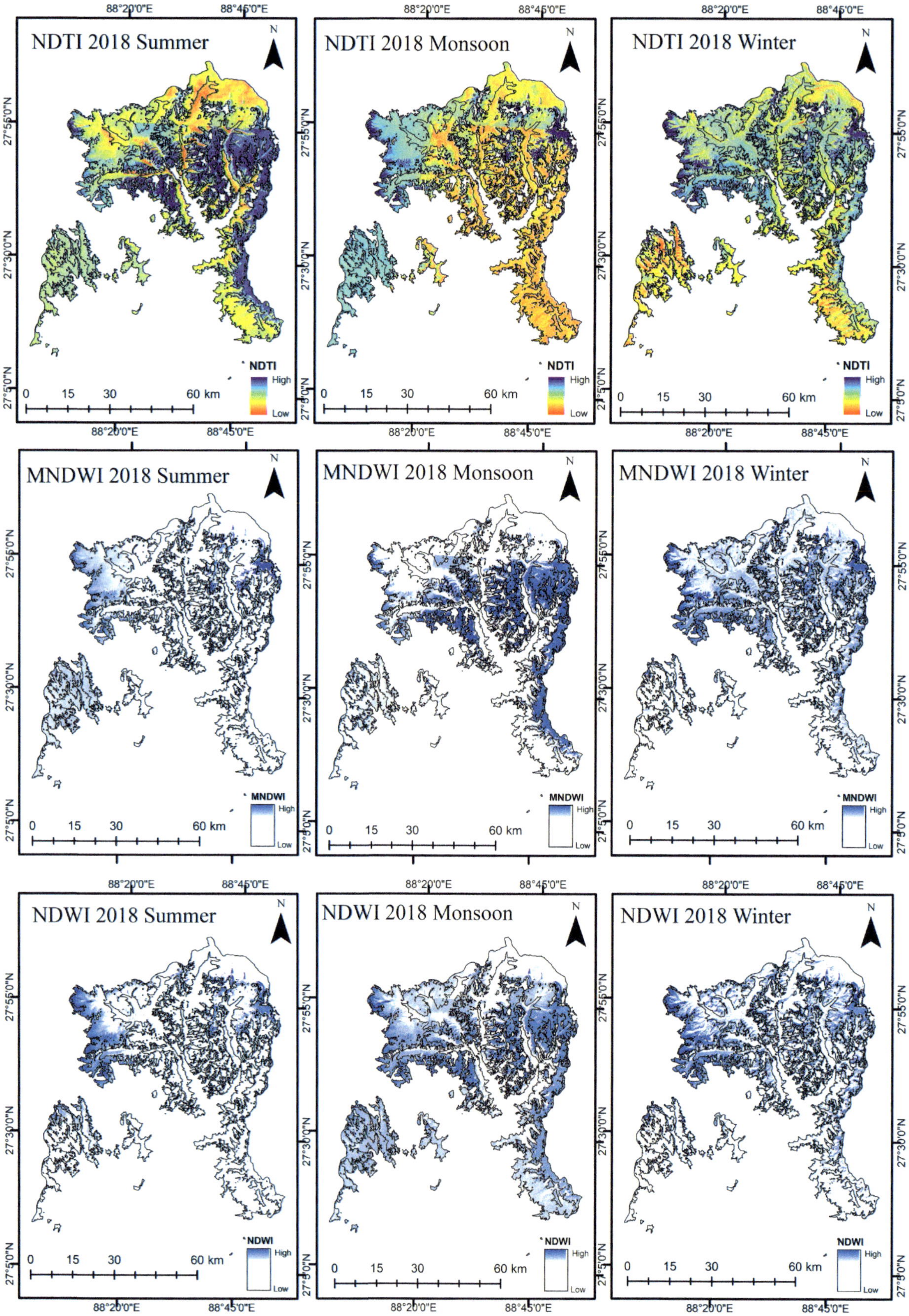

FIGURE 17.4 Seasonal distribution of NDTI, MNDWI, and NDWI for 2018 derived from Sentinel-2. Map lines delineate study areas and do not necessarily depict accepted national boundaries.

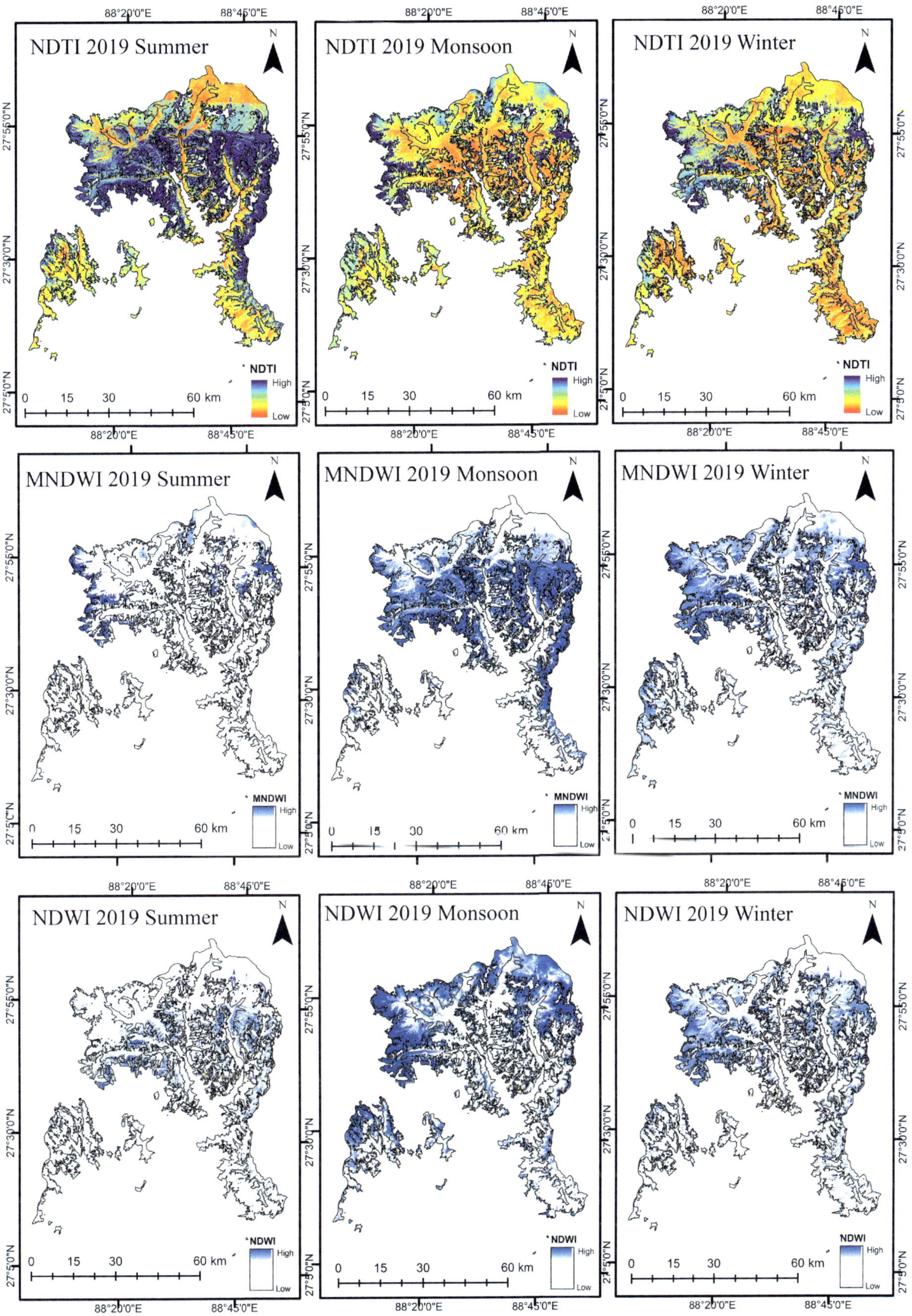

FIGURE 17.5 Seasonal distribution of NDTI, MNDWI, and NDWI for 2019 derived from Sentinel-2. Map lines delineate study areas and do not necessarily depict accepted national boundaries.

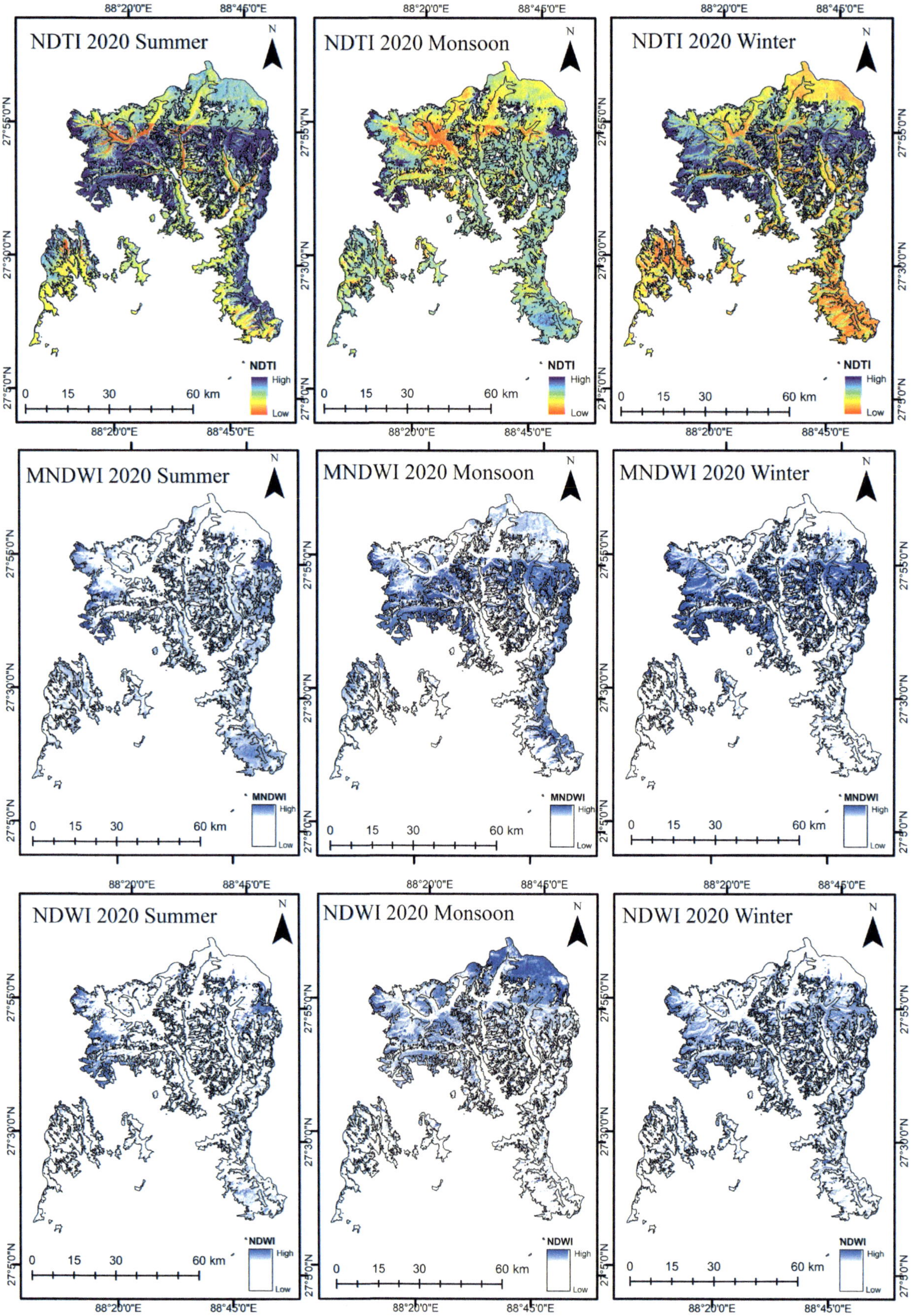

FIGURE 17.6 Seasonal distribution of NDTI, MNDWI, and NDWI for 2020 derived from Sentinel-2. Map lines delineate study areas and do not necessarily depict accepted national boundaries.

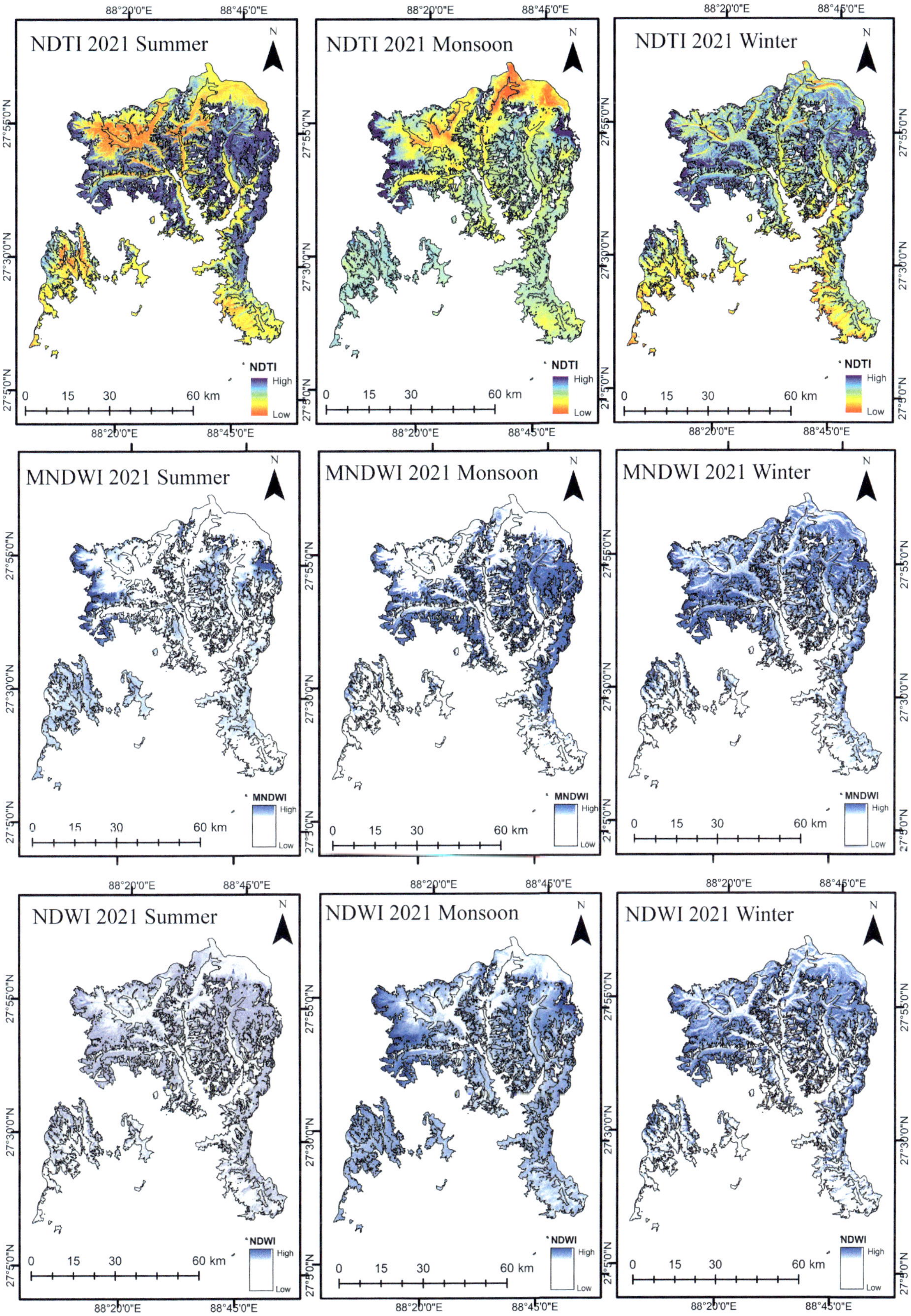

FIGURE 17.7 Seasonal distribution of NDTI, MNDWI, and NDWI for 2021 derived from Sentinel-2. Map lines delineate study areas and do not necessarily depict accepted national boundaries.

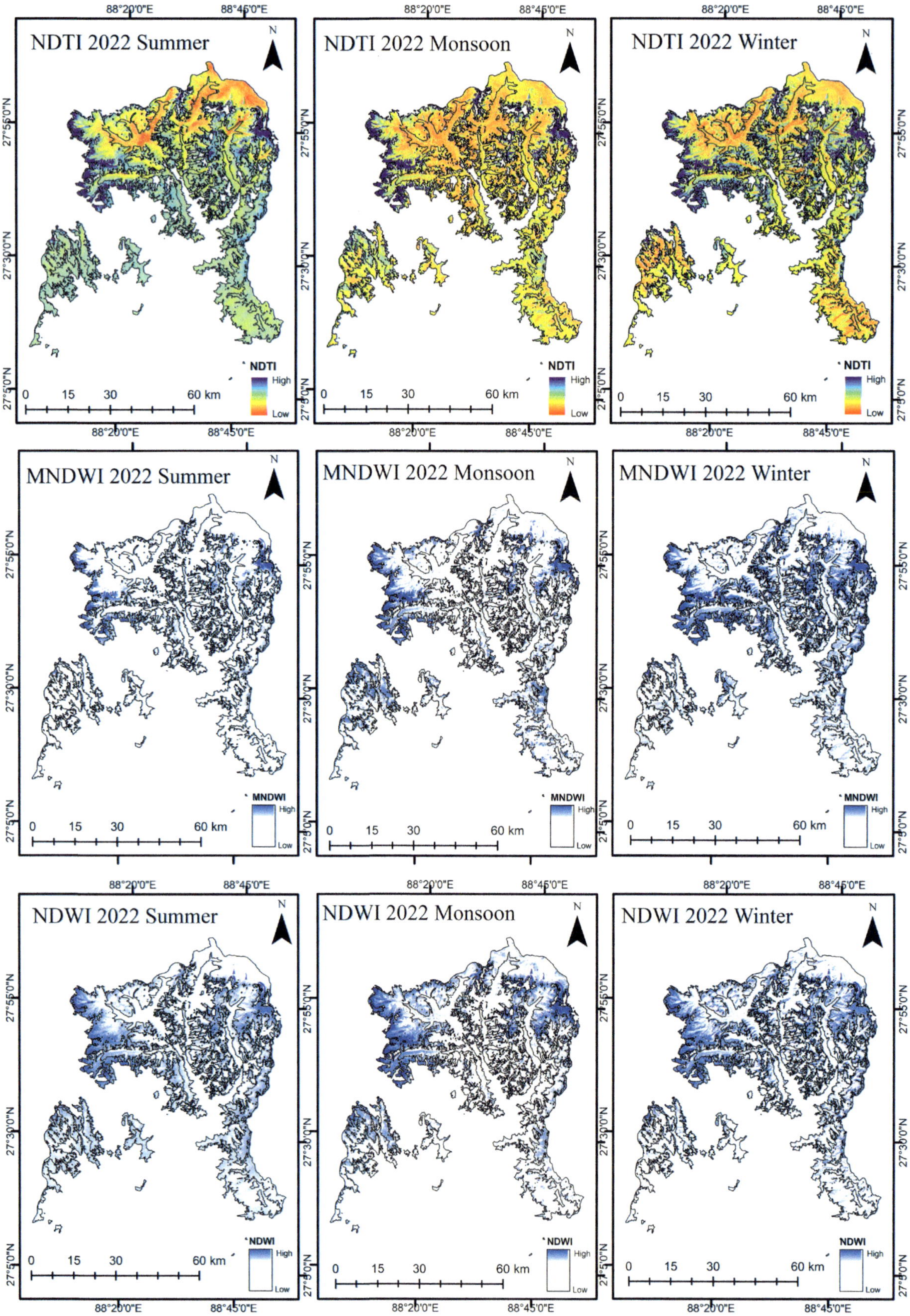

FIGURE 17.8 Seasonal distribution of NDTI, MNDWI, and NDWI 2022 derived from Sentinel-2. Map lines delineate study areas and do not necessarily depict accepted national boundaries.

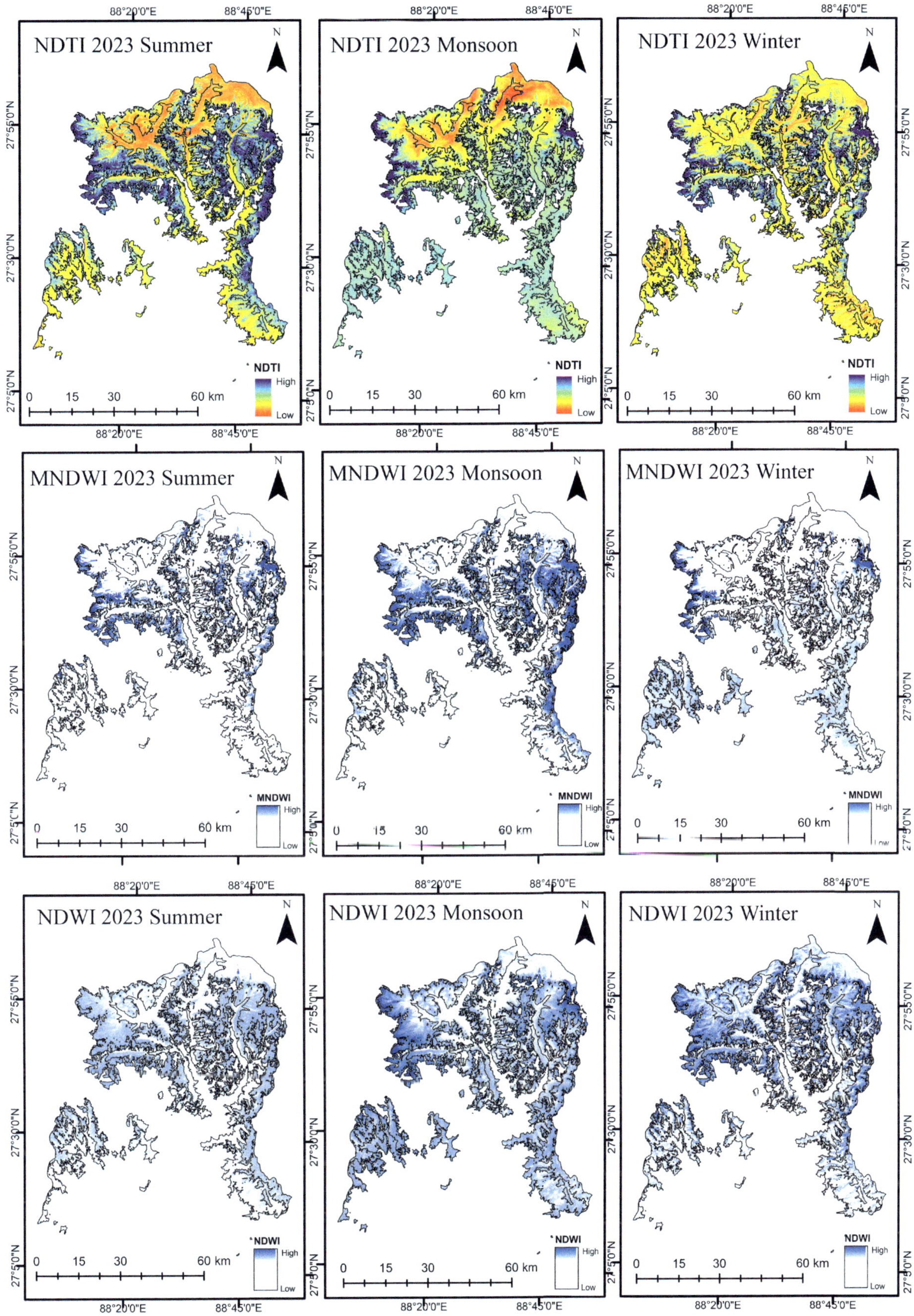

FIGURE 17.9 Seasonal distribution of NDTI, MNDWI, and NDWI for 2023 derived from Sentinel-2. Map lines delineate study areas and do not necessarily depict accepted national boundaries.

in wetlands. Comparatively, a higher MNDWI value in monsoon is due to the higher water levels. Again, the change in surface reflectance due to snow cover and frozen waterbodies reduces the MNDWI values in the winter, while a higher MNDWI value is observed in unfrozen wetlands.

Lower precipitation and higher temperatures with evaporation in summer generally lead to lower water levels, which in turn reduces sediment load and potentially reduces the water turbidity. Therefore the NDTI in the summer indicates clear water. However, short-term spikes caused by localized storms can cause an increase in NDTI in the summer. In monsoons, higher NDTI values could be attributed to the higher turbidity due to heavy precipitation, high runoff, and sediment load. Comparatively, lower NDTI values in winter indicate clearer water due to snow cover and freezing conditions with less sediment suspension.

17.5 Discussion

The NDWI variation across the season suggests a higher water spread in the monsoon season, followed by a moderate variation in the winter season and a lower water spread in summer. Past studies showed that seasonal precipitation significantly impacts the water body, primarily regulating water spread and modifying the NDWI and MNDWI values. Srivastava and Agarwal (2020) studied seasonal water spread and observed higher areas in the postmonsoon season than in premonsoon areas. At the same time, the reduced precipitation and runoff cause sediment resuspension and potentially clear the water, that is, reducing the NDTI values. The short-term spikes caused by localized storms can cause an increase in NDTI in the summer. Moreover, the snow melts water, and evaporation has contrasting positive and negative effects on the water spread during summer (Begam & Sen, 2019). The higher precipitation and lower evaporation due to high humidity in the monsoon season cause the highest water spread, followed by surface runoff. Such events expand the surface water area and increase the water turbidity, leading to higher NDWI, MNDWI, and NDTI values. On the contrary, the lower temperature and snowfall events freeze the water bodies during winter, leading to lower NDWI, while the unfrozen waterbodies indicate higher NDWI values. Again, the water turbidity was observed lower, indicated by a lower NDTI value in the winter compared to summer. Rawat and Singh (2024) employed NDWI and NDTI for water spread and quality monitoring in the Nainital lake (~1900 m a.m.s.l) from 2017 to 2022. They reported a significant decrease in NDWI values in summer and winter, compared to a higher value in monsoon. On the contrary to this study, they observed a higher NDTI value in the summer due to eutrophication under warm climate conditions, while a lower NDTI value in monsoon indicated an influx of water through precipitation.

The National Wetland Inventory and Assessment (NWIA) was conducted by Space Applications Centre, Ahmedabad, India (National Wetland Atlas, 2011). The NWIA in Sikkim was conducted using LISS-III and LISS-IV data at 1:25,000 scale in 2009, which indicated lower open water in premonsoon and higher in the postmonsoon season. Ashok, Rani, and Jayakumar (2021) studied the Renuka wetland, a Ramsar site in Sirmaur District, Uttarakhand, in the Western IHR using NDWI and NDWI derived from Landsat 7 and 8 images in the GEE platform from 2013 to 2019. Their study indicated a prominent seasonal variation in the NWDI values, as observed in the current study. Bid and Siddique (2019) applied NDTI to Landsat images for water quality monitoring in Panchet Hill Dam in the Jharkhand and West Bengal border, India. They reported higher turbidity in the monsoon season, followed by moderate turbidity in the postmonsoon season and lowest turbidity in the premonsoon season.

In the HAWs of Sikkim, the seasonal study of NDTI, MNDWI, and NDWI values provides important information about the dynamics and quality of water. In the preview of climatic and environmental changes, these findings highlight the significance of ongoing monitoring and adaptive management for preserving water supplies and wetland ecosystems. Begam and Sen (2019) studied the moraine-dammed glacial lakes (MDGLs) and mapped the HAW above 3800 m in Nepal, Bhutan, and the Indian state of Sikkim. They observed a more than 40% increase in MDGLs from 1990 to 2015, wherein most of the wetlands indicated an increasing trend in the water surface area or lateral expansion. The rise in temperature and melting of glaciers cause HAW expansion and may trigger glacier lake outbursts, which can cause flooding downstream and loss of life and properties.

Past studies in the Western Himalayan region reported a reduction in wetland areas, accompanied by a loss of forest cover and an increase in settlement areas (Romshoo & Rashid, 2014). Several studies indicated a warmer and wetter climate in the Himalayan region in future, which may alter the water cycle and wetland's dynamics and functionality (Das et al., 2022). The rise in precipitation may increase water availability and turbidity, while the temperature rise will accompany glacier melting, increased evaporation, and reduced dissolved oxygen, among others (Tse-Ring, Sharma, Chettri, & Shrestha, 2010). Periodic wetland area and water quality assessment is important to develop suitable conservation and management practices, which will help protect the flora and fauna (more than 800 species) (O'Neill, 2019).

The HAW mapping is challenging due to topographic complexity, such as the hill shade (Mandal, Bandyopadhyay, & Bhadra, 2023). In addition, the smaller size, cloud presence, and class mixing with snow cover and barren land limits water body mapping (Behera et al., 2018). The use of Sentinel-2 data with a higher spatial resolution (10 and 20 m bands) allows for the mapping of smaller water bodies than Landsat data. Moreover, the Sentinel-2 has the advantage of higher temporal frequency (5 days) than Landsat data with 16 days, which improves the chances of getting cloud-free imageries. Future studies may include the Indian Remote Sensing LISS-IV satellite data with 5.6 m spatial resolution, which was made publicly available recently. The microwave data operates a higher wavelength (cm scale), which can penetrate the clouds and thus has a benefit than optical data hindered by the cloud presence. Therefore studies may include microwave or synthetic aperture radar data (e.g., Sentinel-1) for water body mapping under cloudy conditions (Behera et al., 2022).

17.6 Conclusion

This research highlights the significant seasonal variations in water dynamics and quality in the HAWs of Sikkim using indices such as NDWI, MNDWI, and NDTI. Lower NDWI, MNDWI, and NDPI values indicate that summertime brings lower water levels and clearer water, while localized storms might produce momentary spikes in turbidity. Significant rises in these indices during the monsoon season indicate saturated conditions and high-water levels, along with increased turbidity, as shown by elevated NDTI values brought on by heavy precipitation and sediment-carrying runoff. Winter offers a more varied picture, with lower index values in frozen parts and comparatively higher values in unfrozen areas, as well as lower NDTI values indicating clearer water conditions. These results highlight the need for ongoing observation and flexible management to guarantee the IHR's resilience to the effects of climate change, the preservation of ecosystem health, and the sustainable use of water resources.

The work shows how useful remote sensing indices are in high-altitude areas, providing a quick and easy way to observe and evaluate these ecosystems. The results show how important monsoon precipitation is for replenishing water bodies, even though there are difficulties because of the high turbidity levels. The unpredictability of wintertime highlights the need for more research on how freezing and thawing cycles affect water quality. Future investigations should focus on long-term monitoring to understand patterns and changes in water dynamics, evaluate the impact of human activity, examine the connection between biodiversity and water quality, and use climate models to forecast future effects on these delicate ecosystems. This study offers a basis for protecting Sikkim's essential HAWs in the face of climatic and environmental difficulties. This study can be further extended by investigating hydroclimatic variables, like snow cover, snow albedo, surface temperature, and precipitation, that affect these wetlands as these are snow-driven lakes, and if water is reduced at high altitudes, it will affect low wetland communities.

Acknowledgment

The authors thank the project titled "Multidimensional Assessment of Ecological Dynamics and Ecosystem Health of Selected High-Altitude Wetlands of the Indian Himalayan Region (IHR) for Effective Conservation and Management Planning" funded under the National Mission on Himalayan Studies (NMHS), Ministry of Environment and Forest and Climate Change, Government of India.

References

Ashok, A., Rani, H. P., & Jayakumar, K. V. (2021). Monitoring of dynamic wetland changes using NDVI and NDWI based landsat imagery. *Remote Sensing Applications: Society and Environment*, *23*. Available from https://doi.org/10.1016/j.rsase.2021.100547.

Begam, S., & Sen, D. (2019). Mapping of moraine dammed glacial lakes and assessment of their areal changes in the central and eastern Himalayas using satellite data. *Journal of Mountain Science*, *16*(1), 77–94. Available from https://doi.org/10.1007/s11629-018-5023-1.

Behera, M. D., Chitale, V. S., Shaw, A., Roy, P. S., & Murthy, M. S. R. (2012). Wetland monitoring, serving as an index of land use change-a study in Samaspur Wetlands, Uttar Pradesh, India. *Journal of the Indian Society of Remote Sensing*, *40*, 287–297.

Behera, M. D., Gupta, A. K., Barik, S. K., Das, P., & Panda, R. M. (2018). Use of satellite remote sensing as a monitoring tool for land and water resources development activities in an Indian tropical site. *Environmental Monitoring and Assessment*, *190*, 1–21. Available from https://doi.org/10.1007/s10661-018-6770-8.

Behera, M. D., Prakash, J., Paramanik, S., Mudi, S., Dash, J., Varghese, R., & Srivastava, P. K. (2022). Assessment of tropical cyclone amphan affected inundation areas using sentinel-1 satellite data. *Tropical Ecology*, 1–11. Available from https://doi.org/10.1007/s42965-021-00187-w.

Bid, S., & Siddique, G. (2019). Identification of seasonal variation of water turbidity using NDTI method in Panchet Hill Dam, India. *In Modeling Earth Systems and Environment*, *5*(Issue 4), 1179–1200. Available from https://doi.org/10.1007/s40808-019-00609-8, Springer Science and Business Media Deutschland GmbH.

Blackwell, M. S. A., & Pilgrim, E. S. (2011). Ecosystem services delivered by small-scale wetlands. *Hydrological Sciences Journal, 56*. Available from https://doi.org/10.1080/02626667.2011.630317.

Bullock, A., & Acreman, M. (2003). The role of wetlands in the hydrological cycle. *Hydrology and Earth System Sciences, 7*. Available from https://doi.org/10.5194/hess-7-358-2003.

Chatterjee, A., Blom, E., Gujja, B., Jacimovic, R., Beevers, L., O'Keeffe, J., & Biggs, T. (2010). WWF initiatives to study the impact of climate change on Himalayan high-altitude wetlands (HAWs). *Mountain Research and Development, 30*(1), 42–52.

Das, P., Behera, M. D., Bhaskaran, P. K., & Roy, P. S. (2022). Forest cover resilience to climate change over India using the MC2 dynamic vegetation model. *Environmental Monitoring and Assessment, 194*(12), 903. Available from https://doi.org/10.1007/s10661-022-10545-3.

Das, P., Pandey, V., & Dutta, D. (2021). *Land surface water resource monitoring and climate change. Mapping, Monitoring, and Modeling Land and Water Resources* (pp. 311–326). CRC Press. Available from https://doi.org/10.1201/9781003181293.

Gao, B. (1996). NDWI—A normalized difference water index for remote sensing of vegetation liquid water from space. *Remote Sensing of Environment, 58*(3), 257–266. Available from https://doi.org/10.1016/S0034-4257(96)00067-3.

Garg, V., Aggarwal, S. P., & Chauhan, P. (2020). Changes in turbidity along Ganga River using Sentinel-2 satellite data during lockdown associated with COVID-19. *Geomatics, Natural Hazards and Risk, 11*(1), 1175–1195. Available from https://doi.org/10.1080/19475705.2020.1782482.

Ghosh, S., Mishra, D. R., & Gitelson, A. A. (2016). Long-term monitoring of biophysical characteristics of tidal wetlands in the northern Gulf of Mexico – A methodological approach using MODIS. *Remote Sensing of Environment, 173*, 39–58. Available from https://doi.org/10.1016/j.rse.2015.11.015.

Hossen, H., Mahmod, W. E., Negm, A., & Nakamura, T. (2022). Assessing water quality parameters in burullus lake using sentinel-2 satellite images. *Water Resources, 49*(2), 321–331. Available from https://doi.org/10.1134/S0097807822020087.

Khan, H., & Baig, S. (2017). High altitude wetlands of the HKH region of Northern Pakistan – Status of current knowledge, challenges and research opportunities. *In Wetlands, 37*(2), 371–380. Available from https://doi.org/10.1007/s13157-016-0868-y.

Kislik, C., Dronova, I., Grantham, T. E., & Kelly, M. (2022). Mapping algal bloom dynamics in small reservoirs using Sentinel-2 imagery in Google Earth Engine. *Ecological Indicators, 140*, 109041. Available from https://doi.org/10.1016/j.ecolind.2022.109041.

Lacaux, J. P., Tourre, Y. M., Vignolles, C., Ndione, J. A., & Lafaye, M. (2007). Classification of ponds from high-spatial resolution remote sensing: Application to Rift Valley Fever epidemics in Senegal. *Remote sensing of environment, 106*(1), 66–74. Available from https://doi.org/10.1016/j.rse.2006.07.012.

Mandal, S., Bandyopadhyay, A., & Bhadra, A. (2023). Performance evaluation of MBT, OBIA and PBRF method for high-altitude water body mapping: A case study of Mago River Basin of Arunachal Pradesh. *Journal of the Indian Society of Remote Sensing, 51*(11), 2187–2200.

McFeeters, S. K. (1996). The use of the Normalized Difference Water Index (NDWI) in the delineation of open water features. *International Journal of Remote Sensing, 17*(7), 1425–1432.

National Wetland Atlas. (2011). Sikkim, SAC/EPSA/ABHG/NWIA/ATLAS/36/2011 Space Applications Centre (ISRO), Ahmedabad, India, p. 84

Murthy, M. S. R., Das, P. D., & Behera, M. D. (2016). Road accessibility, population proximity and temperature increase are major drivers of forest cover change in the Hindu Kush Himalayan Region. *Current Science, 111*(10), 1599–1602.

O'Neill, A. (2019). Evaluating high-altitude Ramsar wetlands in the Eastern Himalayas. *Global Ecology and Conservation, 20*, e00715. Available from https://doi.org/10.1016/j.gecco.2019.e00715.

Ouma, Y. O., & Tateishi, R. (2006). A water index for rapid mapping of shoreline changes of five East African Rift Valley lakes: An empirical analysis using Landsat TM and ETM + data. *International Journal of Remote Sensing, 27*(15), 3153–3181. Available from https://doi.org/10.1080/01431160500309934.

Panigrahy, S. (2017). Mapping of wetlands using satellite remote sensing data: Indian experience. *Wetland Science: Perspectives From South Asia*, 423–448. Available from https://doi.org/10.1007/978-81-322-3715-0_22.

Rawat, K. S., & Singh, S. K. (2024). Monitoring water spread and aquatic vegetation using earth observational data for Nani-High Altitude Lake (N-HAL) of Uttarakhand State, India. *Journal of Engineering Research, 12*(1), 64–74. Available from https://doi.org/10.1016/j.jer.2023.10.014.

Romshoo, S. A., & Rashid, I. (2014). Assessing the impacts of changing land cover and climate on Hokersar wetland in Indian Himalayas. *Arabian Journal of Geosciences, 7*, 143–160. Available from https://doi.org/10.1007/s12517-012-0761-9.

Roy, P. S., & Behera, M. D. (2003). Wetland mapping: A remote sensing perspective. In J. Parikh, & H. Datye (Eds.), *Sustainable management of wetlands: Biodiversity and beyond* (pp. 99–116). New Delhi: Sage Publishers, ISBN: 0-7610-9602-8.

Singh, S., Bhardwaj, A., & Verma, V. K. (2020). Remote sensing and GIS based analysis of temporal land use/land cover and water quality changes in Harike wetland ecosystem, Punjab, India. *Journal of Environmental Management, 262*, 110355. Available from https://doi.org/10.1016/j.jenvman.2020.110355.

Srivastava, P., & Agarwal, N. (2020). Geo-information tools in implementing nature-based solutions from high altitude wetlands: From mapping to decision-making support for disaster risk reduction. *Nature-based Solutions for Resilient Ecosystems and Societies*, 185–196. Available from https://doi.org/10.1007/978-981-15-4712-6_11.

Talukdar, S., & Pal, S. (2019). Effects of damming on the hydrological regime of Punarbhaba river basin wetlands. *Ecological Engineering, 135*, 61–74. Available from https://doi.org/10.1016/j.ecoleng.2019.05.014.

Toming, K., Kutser, T., Laas, A., Sepp, M., Paavel, B., & Nõges, T. (2016). First experiences in mapping lake water quality parameters with Sentinel-2 MSI imagery. *Remote Sensing, 8*(8), 640. Available from https://doi.org/10.3390/rs8080640.

Tse-Ring, K., Sharma, E., Chettri, N., & Shrestha, A. B. (2010). *Climate change vulnerability of mountain ecosystems in the Eastern Himalayas*. International Centre for Integrated Mountain Development (ICIMOD).

Varghese, R., Behera, S. K., & Behera, M. D. (2023). Significant inverse influence of tropical Indian Ocean SST on SIF of Indian vegetation during the summer monsoon onset phase. *Remote Sensing*, *15*(7), 1756. Available from https://doi.org/10.3390/rs15071756.

Xu, H. (2006). Modification of normalised difference water index (NDWI) to enhance open water features in remotely sensed imagery. *International Journal of Remote Sensing*, *27*(14), 3025–3033. Available from https://doi.org/10.1080/01431160600589179.

Zhang, X., Liu, L., Zhao, T., Chen, X., Lin, S., Wang, J., & Liu, W. (2022). GWL_FCS30: Global 30 m wetland map with fine classification system using multi-sourced and time-series remote sensing imagery in 2020. *Earth System Science Data Discussions*, *2022*, 1–31. Available from https://doi.org/10.5194/essd-15-265-2023.

Section C

Ocean-atmosphere dynamics and climate change

Chapter 18

The Antarctic climate system: recent changes and their drivers

Pranab Deb, Sai Prabala Swetha Chittella, Arnab Sen and Sanjiban Dutta
Centre for Ocean, River, Atmosphere and Land Sciences (CORAL), Indian Institute of Technology Kharagpur, Kharagpur, West Bengal, India

Antarctica, the fifth-largest continent on Earth, is predominantly covered by a massive ice sheet, reaching depths of over 4 km, concealing a hidden landscape of mountains, valleys, and plains formed over hundreds of thousands of years by accumulated and compressed snow (Fig. 18.1). From the towering Transantarctic Mountains to the Ellsworth Mountains in West Antarctica, the Antarctic ice sheet features distinct mountain systems beneath its surface. Flowing from the continent's center toward the surrounding ocean, the ice gives rise to numerous glaciers extending into the sea, where coastal ice chunks break away to form drifting icebergs. With the exception of coastal peaks, only the highest Antarctic mountains emerge above the icecap. In winter, the surrounding sea freezes for hundreds of kilometers offshore, forming pack ice that undergoes constant changes in shape and distribution due to winds and currents. Situated in the high latitudes of the Southern Hemisphere, Antarctica experiences temperatures below freezing year-round. In winter, the continent is shrouded in complete darkness for 24 hours a day, while summer months witness minimal sunset. Antarctica is encircled by the Southern Ocean, featuring the Antarctic Circumpolar Current as the largest current globally. The Antarctic Convergence, where cold Antarctic waters meet warmer northern waters, fosters a zone with elevated nutrient concentrations, supporting abundant marine plant and animal life. Antarctica hosts various cryospheric features, including ice sheets, floating ice shelves along the coastline, and expansive sea ice detached from the land.

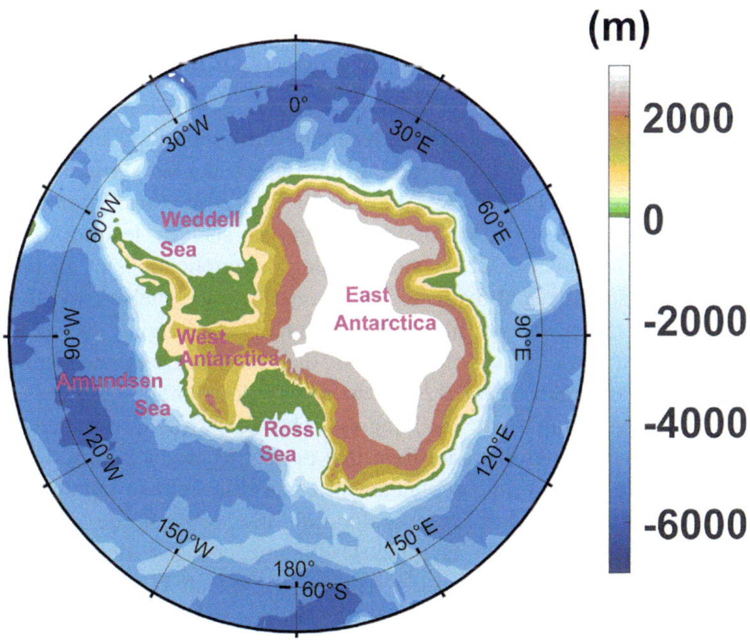

FIGURE 18.1 Map of Antarctica with surface elevation in the background (shaded, in m). Map lines delineate study areas and do not necessarily depict accepted national boundaries.

Sustainable Development Perspectives in Earth Observation. DOI: https://doi.org/10.1016/B978-0-443-14072-3.00005-8

The latter plays a crucial role in supporting wildlife like krill and emperor penguin. Notably, ongoing research indicates that climate change is impacting the timing and extent of sea ice in the region. Icebergs, formed by the calving of mainland glaciers and ice sheets, float in the surrounding ocean. Remarkably, despite their massive size, about 90% of an iceberg's mass is submerged underwater due to the slight difference in density between ice and water. In recent decades, extensive field expeditions and satellite data have allowed scientists to map the hidden topography beneath the ice, unveiling Antarctica's geological structure.

18.1 Early expeditions, discovery, and treaty

The history of Antarctic exploration is rich with tales of courage, perseverance, and scientific discovery. The first confirmed sighting of Antarctica is attributed to the Russian expedition led by Fabian Gottlieb von Bellingshausen and Mikhail Lazarev in 1820. However, the continent's existence was hinted at as early as the ancient Greeks, who hypothesized the existence of a landmass in the Southern Hemisphere to balance the known world. Throughout the 19th and early 20th centuries, expeditions from various nations, including Britain, France, Norway, and Australia, braved the harsh Antarctic conditions in pursuit of geographical knowledge and scientific understanding. Notable explorers like Ernest Shackleton, Robert Falcon Scott, and Roald Amundsen became household names as they pushed the boundaries of human endurance in their quests to reach the South Pole. Ultimately, it was Roald Amundsen and his team of Norwegian explorers who first reached the South Pole on December 14, 1911, marking a monumental achievement in the annals of exploration. These expeditions laid the foundation for modern Antarctic research and continue to inspire explorers and scientists to this day.

The Antarctic Treaty System, established in 1961, serves as the cornerstone of international cooperation for the peaceful and scientific management of Antarctica. Initially motivated by the success of the International Geophysical Year (IGY) in 1957–1958, which saw the establishment of numerous scientific stations across the continent, the treaty dedicates Antarctica to peaceful scientific investigation and suspends all territorial claims. Signed by 12 nations in 1959 and later joined by a total of 54 countries (until January 2022), the treaty prohibits military activity and nuclear testing, promotes scientific collaboration, and sets aside disputes over territorial sovereignty. Additionally, a protocol added in 1991 prohibits mineral exploration until at least 2048, with a unanimous agreement required for any changes before then. As a result, Antarctica remains a demilitarized, nuclear-free zone, fostering international scientific cooperation and environmental protection in a region untouched by war.

Several aspects of Antarctic climate have undergone dramatic changes in recent decades. The following sections discuss these changes and the underlying processes.

18.2 Antarctic temperature trends and drivers

Air temperature plays a crucial role in influencing crucial aspects of Antarctica's dynamics and stability. Over the past few decades, elevated air temperatures have been identified as a key driver behind the unprecedented shrinking and thinning of floating ice shelves surrounding the continent (Paolo, Fricker, & Padman, 2015). These warmer temperatures, particularly notable during the austral summer, trigger intense surface melting on the ice shelves, leading to the formation of surface meltwater ponds (Johnson, Hock, & Fahnestock, 2022). Understanding the implications of air temperature variations is therefore paramount in assessing the stability and resilience of Antarctica's ice shelves and their impact on the broader climate system. However, studying those changes in air temperature is challenging due to the harsh weather conditions and the presence of only a few in situ observations. Most of the automatic weather stations are located in the coastal region. Only two stations hold long records for the continent and deep interior. Moreover, instrument issues and data discontinuities make the investigation more challenging. The Scientific Committee on Antarctic Research undertook the Reference Antarctic Data for Environmental Research project to produce station data based on six hourly synoptic observations (Turner et al., 2004). Also, the advancement of satellite remote sensing has given researchers an opportunity for detailed exploration of the spatial and temporal variabilities of Antarctic temperatures. Over the modern satellite era, Antarctica has experienced zonally asymmetric climatic changes. Before 2000, rapid warming was observed over the western Antarctic and Antarctic Peninsula regions (Bromwich et al., 2013). On the contrary, the east Antarctic region experienced a mild cooling trend (Turner et al., 2005). Several factors are responsible for the variability of Antarctic surface air temperature.

18.3 Southern Annular Mode

One significant contributor to Antarctic temperature modification is the positive trend observed in the Southern Annular Mode (SAM), which is attributed to radiative forcing associated with greenhouse gas increase and stratospheric ozone loss (Arblaster & Meehl, 2006; Thompson et al., 2011). SAM signifies the latitudinal movement of the Southern Hemisphere westerly winds encircling Antarctica, influencing weather and climate patterns across the region (see Fig. 18.2). Positive SAM phases entail strengthened westerly winds converging toward Antarctica, resulting in cooler temperatures over the continent and enhanced rainfall in select areas. Conversely, negative SAM phases feature weakened westerly winds extending toward the equator, leading to warmer temperatures in Antarctica and reduced precipitation in certain zones. SAM's impact extends beyond polar regions, affecting weather patterns in mid-latitude regions and influencing global climate variability. Understanding SAM dynamics is crucial for anticipating climate fluctuations and extreme weather events in the Southern Hemisphere. During positive SAM phases, the intensified polar vortex and displaced jet stream result in strengthened prevailing eastward atmospheric flow. SAM also modulates geostrophic flow around Antarctica, bringing anomalous warm northwesterly winds to the Peninsula region, causing warming there while inducing cooling over the continent. Intriguingly, atmospheric internal variability, such as Atlantic multidecadal oscillation (AMO)-induced tropical Atlantic sea surface temperature warming, can drive a positive SAM response, strengthening the Amundsen Sea Low (ASL) during austral winter and thus impacting West Antarctica's climate (Li, Gerber, Holland, & Yoo, 2015).

18.4 Tropics-Antarctic teleconnection

Tropics-Antarctic teleconnections have also been found to be important for understanding recent climatic changes around Antarctica (Li et al., 2021; Orr et al., 2023; Yuan, Kaplan, & Cane, 2018). Tropical sea surface temperature (SST) variability modulates the Antarctic climate through Rossby wave dynamics (Li et al., 2021). SST anomalies associated with the El Nino-Southern Oscillation (ENSO) have been linked to anomalous patterns of sea ice and surface temperature across the Antarctic Peninsula, known as the Antarctic Dipole (Yuan & Martinson, 2001; Yuan, 2004).

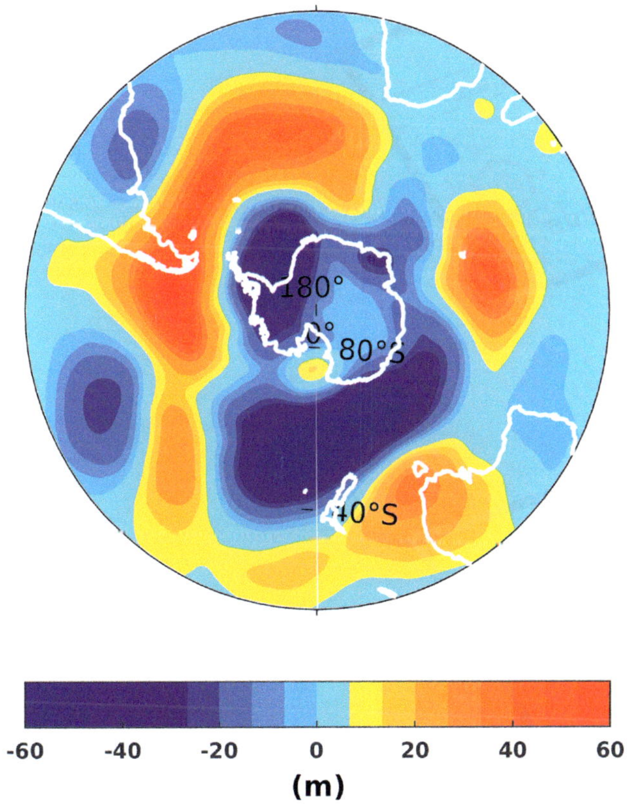

FIGURE 18.2 Spatial structure of southern annular mode (SAM) in austral summer (December-January-February) shown as the regression of SAM index on to 500 hPa geopotential height during summer, 1979–2018.

Recent studies have demonstrated the role of the Interdecadal Pacific Oscillation (IPO) and the South Pacific convergence zone in modulating sea level pressure and surface wind fields around Antarctica (Clem et al., 2020; Meehl, Arblaster, Bitz, Chung, & Teng, 2016). Atlantic multidecadal variability contributes to Antarctic Peninsula warming by reducing the surface pressure in the Amundsen Sea (Li, Holland, Gerber, & Yoo, 2014; Simpkins, McGregor, Taschetto, Ciasto, & England, 2014). SST anomalies over the tropical Indian Ocean have also been found to have significantly contributed to sea ice changes around Antarctica (Purich & England, 2019; Wang et al., 2019; Yu, Zhong, Vihma, Sui, & Sun, 2022).

During austral fall, the role of the extratropical Rossby wave train forced by the tropical Pacific SST anomalies has been found important for the warming trend over the Peninsula (Ding & Steig, 2013). Since the late 20th century, this zonal asymmetric surface air temperature trend has been reversed. During austral summer, warming over west Antarctica and Antarctic Peninsula is absent (Turner et al., 2016), and inland Antarctica experienced rapid warming (Clem et al., 2020). Since 2001, the Antarctic ozone hole has exhibited a sign of recovery, which caused the lower polar stratosphere and warming and the weakening of the polar vortex (Zambri, Solomon, Thompson, & Fu, 2021). The phase change of the decadal variabilities of the tropical ocean is also important for the reversed temperature trend. In the early 2000s, the IPO entered its negative phase. The coupling of negative IPO and positive SAM during the 21st century has resulted in the advection of warm and moist air from the South Atlantic to the interior of Antarctica, causing regional climatic change (Clem et al., 2020).

A general depiction of extratropical Rossby wave generated by heating anomaly near the tropics is provided in Fig. 18.3.

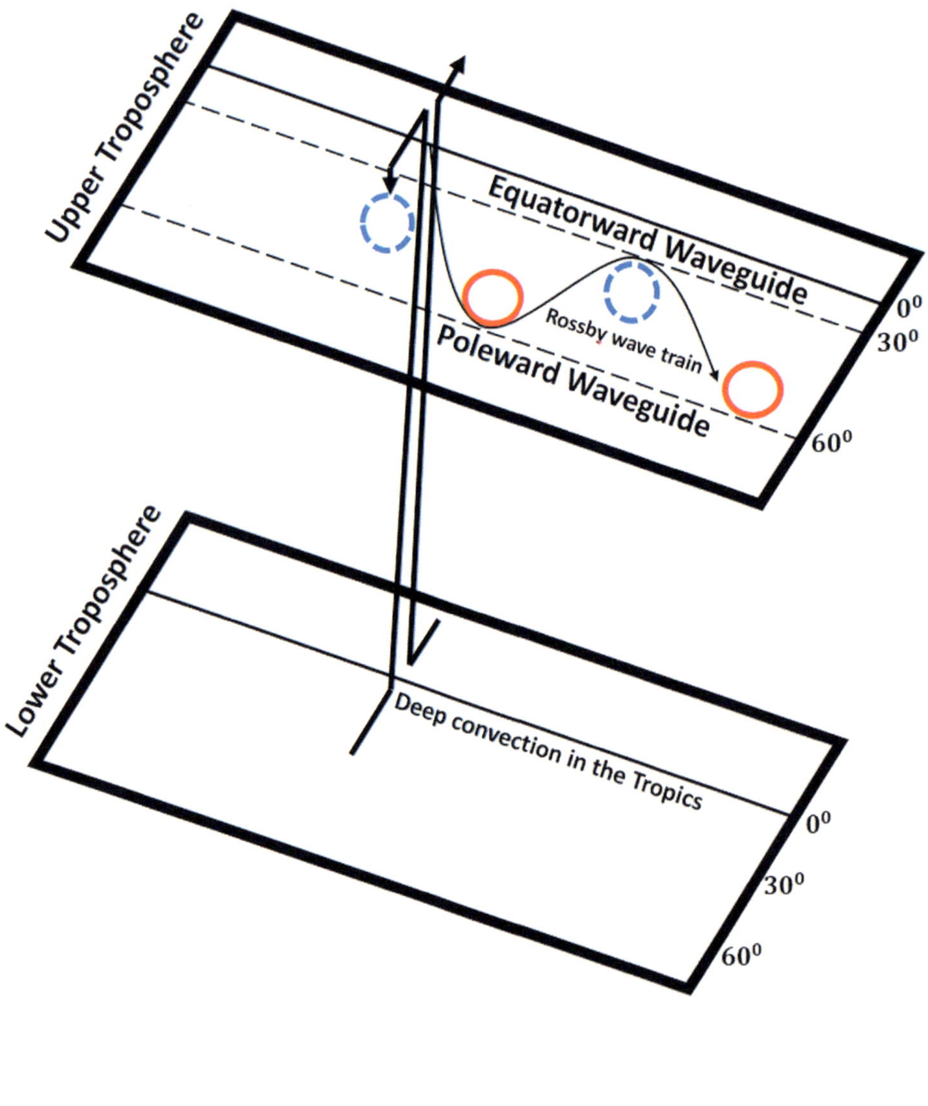

FIGURE 18.3 Schematic depiction of Rossby wave generation and its propagation to the Southern high latitudes.

18.5 Antarctic sea ice variability during 1979 to present

In polar oceans, reduced sunlight and cold temperatures cause seawater to freeze, resulting in the formation of a thin and intermittent layer called sea ice. As temperatures drop below freezing, seawater solidifies into ice, creating a barrier between the ocean surface and the atmosphere. This process occurs seasonally in polar regions due to variations in sunlight and temperature. Fig. 18.4 shows the seasonal sea ice cover surrounding Antarctica. Sea ice serves as a crucial boundary between the ocean surface and the atmosphere. Situated between the Antarctic continent and the sub-Antarctic zone, Antarctic sea ice exhibits substantial seasonal variations, expanding from approximately 1.8 million km^2 at the southern summer's conclusion to about 20 million km^2 by the southern winter's close (Rana, Dash, Routray, & Pandey, 2011). This dynamic coverage significantly influences ocean and atmospheric conditions, exerting far-reaching effects on the global climate system. Key to its impact is feedback mechanisms, with the ice-albedo feedback being particularly notable (Curry, Schramm, & Ebert, 1995; Hall, 2004). This positive feedback process amplifies initial warming or cooling, resulting in a reduction or increase in sea ice extent (SIE), subsequently affecting solar energy reflection and absorption. Antarctic albedo (Brandt, Warren, Worby, & Grenfell, 2005; Laine, 2008; Shao & Ke, 2015; Zhou, Zhang, & Zheng, 2019) plays a vital role, reflecting about 98% of incoming solar radiation on freshly fallen snow compared to 10%–15% on open water. The interaction between sea ice and the atmosphere involves comparatively lower heat exchange compared to open ocean areas (Rana et al., 2011). The sea ice zone significantly influences cloud formation, atmospheric stability, and rainfall patterns over the Southern Ocean and Antarctica (Shao & Ke, 2015). Moreover, Antarctic sea ice serves a critical role in the formation of Antarctic Bottom Water, integral to global thermohaline circulation or the ocean's conveyor belt (Jacobs, 2004; Orsi, Johnson, & Bullister, 1999). The cold, dense, and saline waters formed in the Southern Ocean sink and initiate a deep southward flow as the Antarctic Bottom Water, influencing the lower limb of thermohaline circulation. This water gradually resurfaces in other oceans, impacting weather patterns, contributing to regional climate variations, and influencing Earth's overall climate by exchanging heat and gases between the atmosphere and ocean. Beyond its climatic functions, sea ice variability also has large impacts

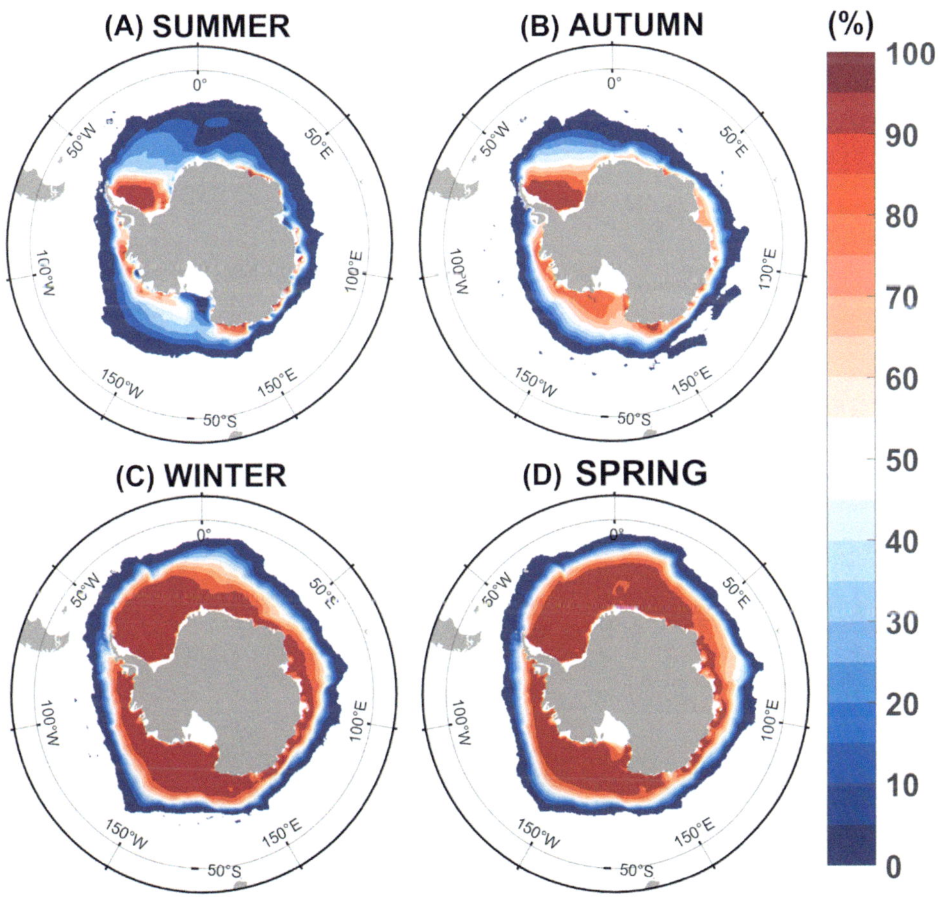

FIGURE 18.4 Maps of Southern Hemisphere sea ice concentrations for (A) summer, (B) autumn, (C) winter and (D) spring, averaged over the years 1979–2022, as derived from the NOAA/NSIDC Climate Data Record of Passive Microwave Sea Ice Concentration Version 4. Map lines delineate study areas and do not necessarily depict accepted national boundaries.

on local ecosystems (Ainley et al., 2003). The Antarctic sea-ice zone provides habitat for various biota species (Garrison, 1991) and supports scientific research initiatives.

18.5.1 Sea ice increase during 1979–2016

Antarctic SIE has been consistently monitored by satellites for the past four decades. Unlike the Arctic, the overall Antarctic sea ice cover exhibited an increase during 1979–2016 in all seasons (Fig. 18.5). The reliable record of passive microwave satellite estimates of SIE, starting in late 1978 with data from the Special Sensor Microwave Imager, initially showed a slight decrease in Antarctic SIE during the first decade. Subsequently, several studies highlighted a statistically significant overall increase in extent (Cavalieri et al., 1997; Comiso & Nishio 2008; Zwally et al., 2002). However, it was often emphasized that this general increase obscured substantial regional variations, notably a decrease in the Bellingshausen Sea and an increase in the Ross Sea (Comiso et al., 2011; Parkinson & Cavalieri 2012; Stammerjohn et al., 2012). Despite elevated greenhouse gas concentrations, the anticipated decline in SIE in both polar regions is not evident. Compounding the puzzle is the discrepancy in the majority of coupled climate models, which, when subjected to observed forcings over recent decades, project a decline in Antarctic sea ice similar to that of the Arctic (Eisenman et al., 2011). This incongruity suggests the potential omission of certain processes in the current generation of climate models or that the observed trend, albeit over a relatively short period since 1979, may push the limits of the simulations produced by these models.

The weakly positive trend in Antarctic SIE has led to various proposed mechanisms. A freshened ocean surface, reducing the upwelling of warmer subsurface waters, may contribute to increased Antarctic sea ice cover. The cooling of surface waters in the Southern Ocean, attributed to the observed strengthening of westerly winds and an upward trend in the SAM, is another potential factor for the positive trend in Antarctic SIE. The intensified westerly winds have a

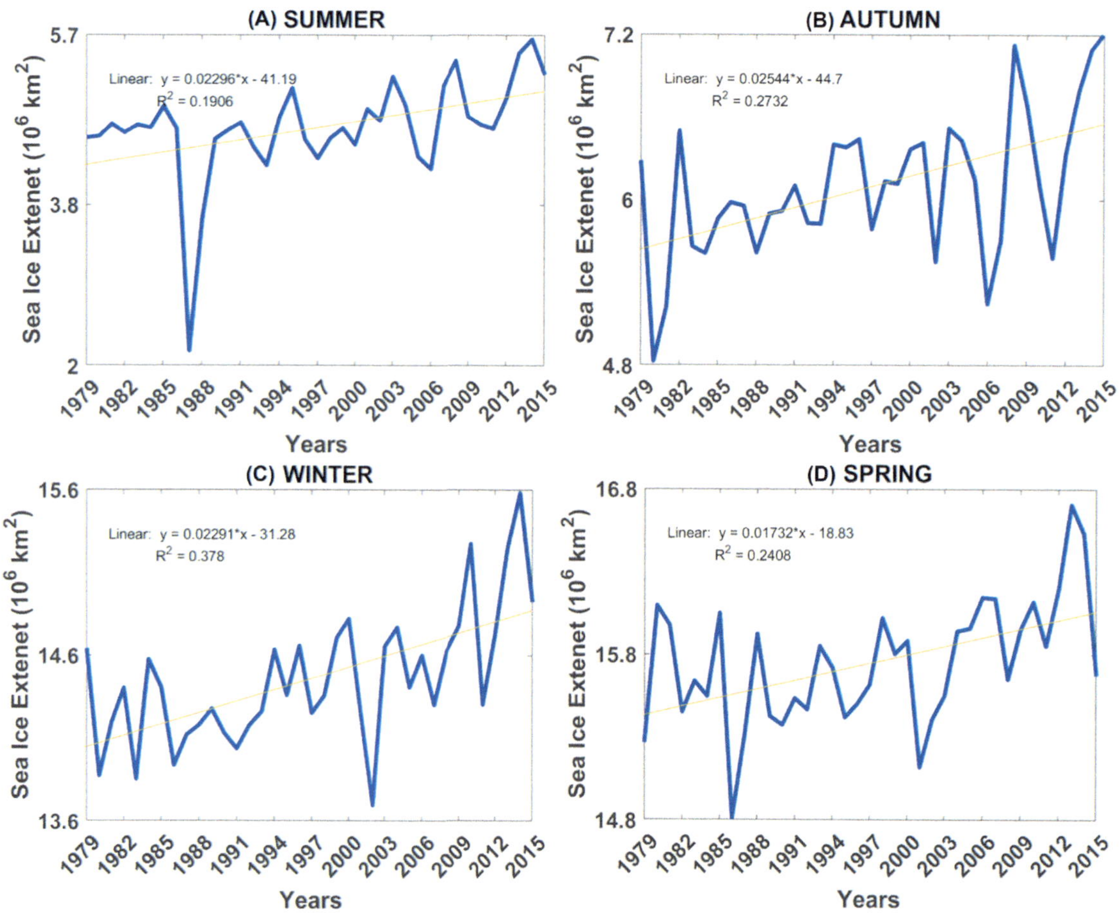

FIGURE 18.5 Trends of Antarctic sea ice area between 1979 and 2015 for (A) summer, (B) autumn, (C) winter and (D) spring seasons, as derived from the NOAA/NSIDC Climate Data Record of Passive Microwave Sea Ice Concentration Version 4.

short-term cooling effect, leading to sea ice expansion by pushing cold surface waters northward. However, over longer timescales, this northward flow may upwell warm deep waters, resulting in reduced sea ice. The SAM typically exhibits symmetric wind anomalies, with a notable nonannular component in the Amundsen Sea region. Contrasting SIE trends in the Ross (increasing) and Bellingshausen (decreasing) Seas suggest a link with the ASL. ASL is a dominant climatological feature and is the deepest of three low-pressure areas near Antarctica. In its usual position, it pushes cold air over the Ross Sea, expanding sea ice northward. On its eastern side, warm air limits sea ice in the Bellingshausen and Amundsen Seas. The ASL has deepened in recent decades, aligning with the dipole of ice loss/increase across West Antarctica. Model results by Turner and Coauthors, (2009) propose that stratospheric ozone loss deepened the ASL, strengthening southerly winds over the Ross Sea and contributing to increased SIE. The intrinsic variability of the ASL, influencing the strength of southerly wind over the Ross Sea, results from various contributing factors, including tropical forcing (Deb et al., 2018), the SAM phase, and regional flow-orography interactions. Additionally, atmosphere–ocean feedback plays a role (Goosse & Zunz, 2014).

The SAM and ASL are responsive to broader atmospheric patterns associated with alterations in tropical regions. ENSO-induced Rossby wave patterns can elevate pressure over the Amundsen Sea and influence sea ice dynamics and thermodynamics. The AMO (reflective of North Atlantic Ocean SSTs) and the IPO (tied to western Pacific SST variations) also exercise control over these tropical SSTs, thereby influencing the atmospheric conditions surrounding Antarctica. During a positive IPO phase, the encircling winds weaken, exerting a subsiding influence on the atmosphere over the Amundsen and Bellingshausen seas. Conversely, a negative IPO phase strengthens the winds, intensifying the ASL. The AMO exerts analogous effects. The positive phase of the AMO since the late 1970s and the negative phase of the IPO since the early 2000s produced two outcomes: deepening the ASL and fortifying the SAM. These atmospheric alterations facilitated an augmentation in the overall sea ice cover around Antarctica until approximately 2015. Some research suggests that the strengthening of westerlies during the negative IPO phase (2000–2014) led to a decade-scale warming of the ocean subsurface, creating conditions conducive to sea ice melting (Eayrs, Li, Raphael, & Holland, 2021). The influence of tropics-Antarctic teleconnection on ASL (Hoskins & Karoly, 1981; Li et al., 2015) contrasts with the proposed influences of increased SSTs in the southwestern tropical and subtropical Pacific.

Furthermore, the mass loss of the West Antarctic Ice Sheet, particularly in the Amundsen Sea Embayment area, has led to coastal water freshening. Bintanja et al. (2013) suggested that fresher water impedes upward heat flux, contributing to a greater extent of ice in the Ross Sea. Interestingly, wind-driven changes in sea ice advection were found to dominate trends around much of West Antarctica, with wind-driven thermodynamic changes prevailing elsewhere (Holland & Kwok, 2012). Overall, due to the inadequate representation of tropical-high latitude teleconnections (as well as other key processes) in many climate models, the current generation of climate models struggles to simulate the observed sea ice changes since 1979. The discrepancy between climate models and observations likely stems from climate forcing, rather than sea ice physics.

18.5.2 Recent Antarctic sea ice decline—evidence and theories

The sea ice growth rate surged nearly fivefold between 2000 and 2014, reaching a record high of 12.8 million km^2 in 2014. However, this was followed by an unexpected reversal in trend, leading to a decline in SIE from the spring of 2016, reaching record lows in 2017 and 2023 (NASA Earth Observatory, 2023) (also see Fig. 18.6). The current trend indicates a decline of 0.11% per decade from 1979 to 2022 (Suryawanshi, Jena, Bajish, & Anilkumar, 2023), with the lowest extent recorded on February 21, 2023, measuring 1.79 million km^2 (NASA Earth Observatory, 2023). The rapid decline in Antarctic sea ice since 2016 is a significant departure from the long-term trend, with a 3-year loss equivalent to the typical sea ice loss in the Arctic over 30 years. The Weddell Sea and Ross Sea areas experienced pronounced regional declines, with notable occurrences of open-water areas known as polynyas (Eayrs et al., 2021; Suryawanshi et al., 2023). West Antarctica exhibited a substantial decrease in sea ice and remained ice-free in 2022 (Suryawanshi et al., 2023). Several factors contribute to this sea ice decline. The intensification of atmospheric zonal waves transporting warm and moist air toward polar regions played a role. Changes in the ZW3 index, associated with El Niño events, contributed to the sea ice reduction, particularly over the Weddell Sea and Ross Sea (Suryawanshi et al., 2023). In addition, an unusual warming of the Southern Ocean mixed layer (by around 1°C) significantly influenced the decline. Furthermore, the transition of IPO from a negative to a positive phase in the spring of 2016 coincided with the initial decrease, while a subsequent negative IPO and a positive SAM primarily influenced the post-2016 decline (Suryawanshi et al., 2023). Additional factors contributing to the decline include intense storm events, heightened meridional heat exchange, a robust Indian Ocean Dipole, the Madden-Julian Oscillation, weakened La Niña conditions, and a negative phase of the SAM (Meehl et al., 2019). Also, anomalously active planetary wave patterns during winter,

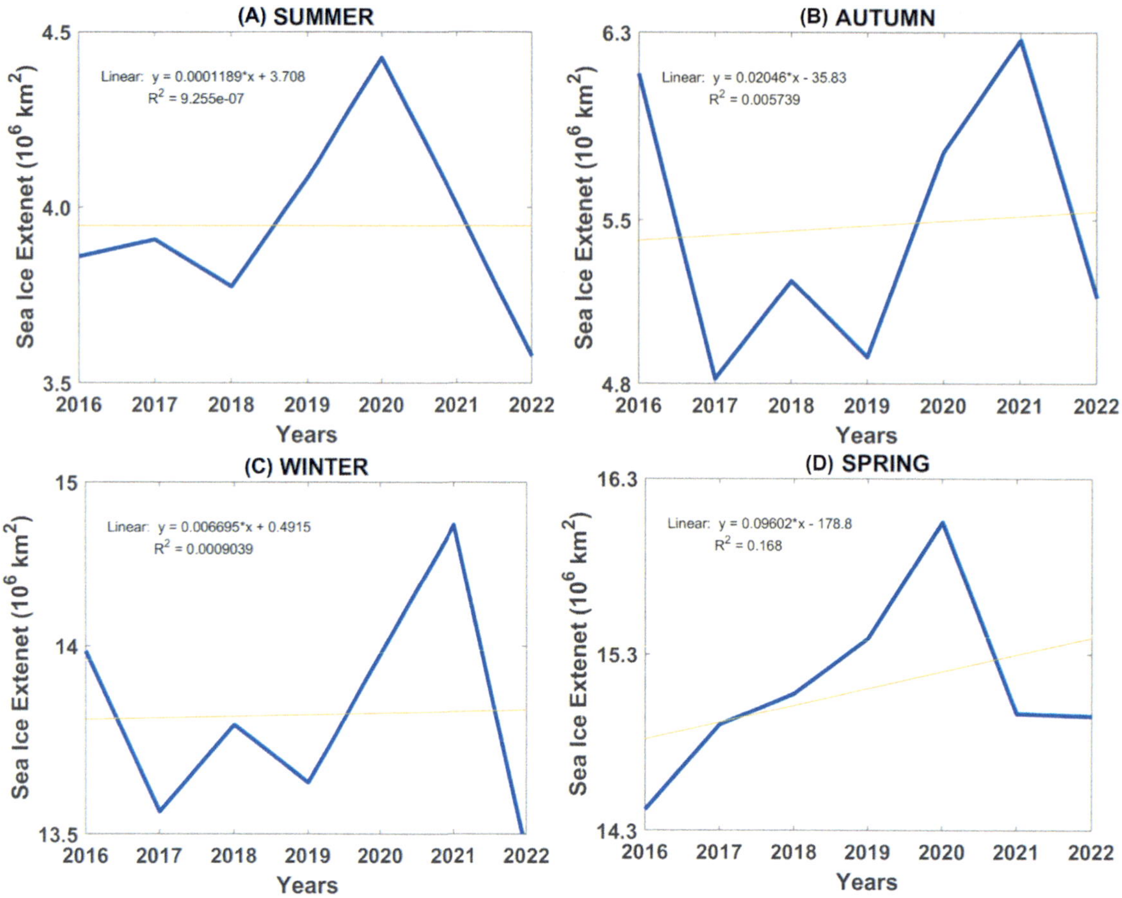

FIGURE 18.6 Seasonal trend of Antarctic sea ice area between 2016 and 2022 for (A) summer, (B) autumn, (C) winter and (D) spring seasons, as derived from the NOAA/NSIDC Climate Data Record of Passive Microwave Sea Ice Concentration Version 4.

coupled with anticyclones in midlatitudes, disrupted the polar vortex, leading to elevated ozone levels over the Antarctic region in 2017 and exacerbating the decrease in sea ice. Human-induced factors like greenhouse gas emissions may have played a role in the recent rapid sea ice decline. Anthropogenic warming in the tropical Indian Ocean could be connected to anomalies observed in 2016, with potential implications for future investigations into the extent of human influence on tropical teleconnections (Eayrs et al., 2021).

18.6 Stability of Antarctic ice sheet: mass balance and accumulation

18.6.1 Surface mass balance

An Antarctic surface mass balance (SMB) is the overall change in the amount of ice and snow that the continent receives or loses. It is essential to monitor SMB in order to quantify the ice sheet's response to global warming and its role in the increase of sea levels (Cogley et al., 2010). Snowfall, melting, sublimation, and wind-driven erosion are the key factors that contribute to the net change in surface mass, which is then used to determine the SMB. The formula that is commonly used to calculate SMB is expressed as follows: $SMB = P - M - E$, where P represents precipitation (snowfall), M represents surface melting, and E represents sublimation and wind-driven erosion. A positive SMB in this equation means that mass is being added (accumulation), while a negative SMB means that mass is being removed (ablation). Regional variations in SMB provide insights into the complex dynamics of Antarctica's ice sheet, highlighting the importance of accurate measurements and modeling for climate predictions and assessments. In Antarctica, accumulation is primarily the result of solid precipitation, such as snow, and occasionally liquid precipitation; wind-driven processes also influence accumulation rates (Mottram et al., 2021). Ablation comprises the mechanisms responsible for mass loss, primarily evaporation, sublimation due to high winds, and run-off (Agosta et al., 2019; Das et al.,

2013; Scambos et al., 2012). According to a recent research by Lenaerts, Medley, van den Broeke, and Wouters (2019), runoff is a relatively minor contributor.

Various approaches exist for estimating SMB—the standard method of measuring SMB is with stakes (Black & Budd, 1964; Dibb & Fahnestock, 2004; McConnell, Bales, & Davis, 1997). Installing the stakes vertically above the surface, taking an initial reading of the height between the surface and the top of the stake, repeating the measurement, and calculating the difference between the initial and final readings yields an estimate. In accumulation zones, this is used to assess accumulation, while in ablation zones, it is used to estimate ice ablation (Braithwaite, 1995; Van de Wal, Greuell, van den Broeke, Reijmer, & Oerlemans, 2005). The other measurements are obtained from snow pits, firn or ice core records, ground-penetrating radars, automatic weather stations, GPS Interferometric Reflectometry, Airborne and Satellite Altimetry, and Precipitation Radar such as CloudSat. All of these methods are described in detail in Lenaerts et al. (2019). As the observations are restricted to a few locations due to Antarctica's complex topography, they are unable to reproduce the high-accumulation marginal zones of the ice sheet, as well as the coastal and inland blue ice regions (Eisen et al., 2008). And also interpolation techniques used to interpolate these observations (Arthern, Winebrenner, & Vaughan, 2006; Vaughan, Bamber, Giovinetto, Russell, & Cooper, 1999) are subject to significant limitations (Genthon, Krinner, & Castebrunet, 2009; Magand, Picard, Brucker, Fily, & Genthon, 2008; Picard, Arnaud, Domine, & Fily, 2009). This implies that we have to rely on modeling techniques to estimate SMB in these regions. Previous research conducted by Rignot, Velicogna, van den Broeke, Monaghan, and Lenaerts (2011), Gardner et al. (2018), and Shepherd et al. (2018) has utilized the results obtained from regional climate models to assess the SMB. The study conducted by Mottram et al. (2021) involved a comparative analysis of various regional climate models through the simulation of SMB. They conducted a comparison between the modeled SMB and a substantial dataset of observations. Despite the dataset being biased due to undersampling, it reveals that several biases in SMB are shared among the models.

18.6.2 Recent trends in surface mass balance

According to recent studies, there are notable trends in the SMB. Shepherd et al. (2019) conducted a comprehensive analysis utilizing 25 years of satellite radar altimeter observations and a regional climate model. Their findings indicate that East Antarctica contributed -1.1 ± 0.4 mm and West Antarctica contributed $+5.7 \pm 0.8$ mm to the global sea level between the years 1992 and 2017. Rignot et al. (2019) employed a SMB model to assess the changes in total mass loss. Their findings indicated an increase in total mass loss from 40 ± 9 Gt y^{-1} during the period of 1979–1990 to 50 ± 14 Gt y^{-1} in 1989–2000, further escalating to 166 ± 18 Gt y^{-1} in 1999–2009 and ultimately reaching 252 ± 26 Gt y^{-1} in 2009–2017. The primary sources of mass loss in Antarctica are mostly concentrated in the Amundsen and Bellingshausen Sea sectors, as well as in Wilkes Land in West Antarctica, and the western and northeastern regions of the Antarctic Peninsula in East Antarctica. It has been discovered that the observed mass loss in Antarctica primarily occurs in regions adjacent to warm, circumpolar deep water (CDW). Their finding aligns with the hypothesis that intensified polar westerlies facilitate the transportation of CDW toward Antarctica, resulting in the melting of its floating ice shelves, destabilization of glaciers, and subsequent rise in sea level.

18.6.3 Accumulation

The significance of accumulation has been identified as the dominant driver of SMB (Shepherd & Wingham, 2007; Wingham, Shepherd, Muir, & Marshall, 2006), with the contribution of runoff to the SMB being relatively negligible (Lenaerts et al., 2019). Accumulation, the process by which precipitation and snowfall are accumulated and stored, serves as the primary driver of ice sheet growth, representing the net gain of mass from snowfall. Accumulation not only acts as a counterbalance to potential sea level rise by adding mass to the ice sheet but also preserves a historical record of past climate conditions in layers of snow. Analyzing ice cores extracted from Antarctica provides insights into ancient climates, atmospheric composition, and temperature. Moreover, accumulation influences ice flow dynamics, playing a crucial role in modeling ice sheet behavior and predicting its impact on sea levels. Changes in accumulation patterns also serve as indicators of climate change, with variations affecting freshwater input to the Southern Ocean, ocean circulation, and marine ecosystems. Accurate estimation of accumulation is essential for predicting the future behavior of Antarctica in the context of ongoing climate change. The absence of in situ data gives rise to significant disparities in the recorded precipitation levels among various reanalyzes conducted on the continent (Behrangi et al., 2016). Due to the challenging climatic conditions and restricted accessibility of Antarctica, unmanned instrumentation is employed to minimize human involvement and facilitate automated measurement of precipitation, which also has been constrained (Tang et al., 2018). In order to estimate precipitation using conventional techniques, it is important to consider the impact of high wind speeds on the measurement process. Specifically, the deflection of snow particles across the gauge during such conditions can result in a significant underestimation of the collected precipitation, hence posing challenges to accurate measurement (Folland, 1988; Kochendorfer et al., 2017). As a result of the complexities inherent in conducting in situ observations, current research has primarily depended on models at the global and regional scales to ensure the most precise results. Based on a modeling study performed by Bromwich, Guo, Bai, and Chen (2004), it was predicted that the mean annual precipitation in Antarctica would surpass 800 mm. Turner et al. (2019) determined in a recent study that extreme precipitation events comprise over 40% of the annual total precipitation.

A large fraction of the accumulated snowfall takes place in the form of extreme precipitation events. Extremes can be defined as infrequent occurrences that take place within a specific location and temporal context, as outlined by the Intergovernmental Panel on Climate Change's Fifth Assessment Report (IPCC AR5). Various methodologies have been employed in past studies to delineate severe precipitation occurrences, typically characterized by daily precipitation exceeding the 90th, 95th, or 99th percentile of the entire time series of daily values (Swetha Chittella, Deb, & Melchior van Wessem, 2022; Turner et al., 2019; Vavrus, Notaro, & Lorenz, 2015). The contribution of extreme precipitation over West Antarctica is shown in Fig. 18.7. A study conducted by Noone, Turner, and Mulvaney (1999) found that short occurrences of maritime air incursions have the potential to cause increased precipitation events. This effect is particularly prominent when these incursions happen during periods characterized by strong meridional flow patterns and amplified mid-tropospheric planetary waves, as noted by Hirasawa, Nakamura, and Yamanouchi (2000) and Massom et al. (2004). In accordance with Zhu and Newell (1998), the occurrence of extreme precipitation events is linked to atmospheric rivers in coastal areas. Subsequently, these events also have a significant role in generating heavy precipitation in the interior of Antarctica (Genthon, Krinner, & Deque, 1998; Gorodetskaya et al., 2014). In their study, Marshall, Thompson, and van den Broeke (2017) determined that these extreme events can be attributed to SAM or the Pacific-South American patterns. The latter pattern is closely associated with the variability of ENSO.

18.6.4 Drivers of accumulation and precipitation

Atmospheric Circulation and Moisture Transport: The patterns of atmospheric circulation and the movement of moisture from lower latitudes have a significant impact on the accumulation and precipitation that occurs in Antarctica. The Southern Ocean, which encircles Antarctica, serves a critical function. A high-pressure system generated by cold, dense air over the continent redirects damp air masses from the Southern Ocean into Antarctica's interior. Orographic uplift transpires as these air masses cross the East Antarctic ice sheet as they approach the landmass. As the air that has been elevated descends, condensation occurs, which subsequently results in the deposition of snow. The precise atmospheric circulation patterns, including the location and intensity of the ASL and the West Antarctic Low, have a substantial influence on the volume and dispersion of precipitation throughout distinct Antarctic regions.

Climate Change and Temperature Trends: The effects of climate change on Antarctica's accumulation and precipitation are obvious. An elevation in temperature has the potential to augment the atmospheric capacity to retain moisture, which might result in an overabundance of precipitation. Nevertheless, a complex relationship exists between temperature and precipitation. Elevated temperatures may lead to increased liquid precipitation in certain areas, thereby

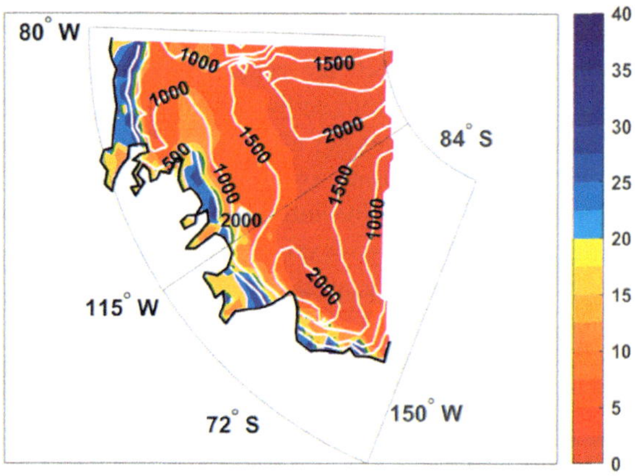

FIGURE 18.7 Precipitation contribution (mm day^{-1}) from all extreme precipitation events (EPEs) at each grid for time period 1979–2016 over west Antarctica using RACMO model daily data (Swetha Chittella et al., 2022).

potentially diminishing its capacity to contribute to the mass of ice sheets. Moreover, elevated temperatures have the potential to induce greater surface melting, thereby influencing the SMB as a whole. It is of the utmost importance to monitor these temperature-related variations in order to comprehend the dynamic evolution of the Antarctic ice sheet and its impact on the rise in sea level.

Interactions Between the Ocean and Atmosphere and Sea Ice: The precipitation over Antarctica is impacted by the heat and moisture exchange between the atmosphere and the ocean, which is influenced by the presence and variability of sea ice in the Southern Ocean. The regional climate is influenced by alterations in heat flux and moisture availability caused by variations in SIE. Furthermore, the interactions between the ocean and atmosphere, which encompass El Niño and the SAM, have the capacity to alter patterns of precipitation. As an illustration, El Niño is characterized by SSTs in the equatorial Pacific that are higher than average, which has far-reaching effects on atmospheric circulation patterns worldwide, including those that extend to Antarctica. It is imperative to comprehend these interactions in order to model and predict with precision the factors that influence accumulation and precipitation in Antarctica, as well as their reaction to larger-scale climate dynamics.

18.7 Antarctic ice shelves and their importance

The Antarctic ice sheet, Earth's largest freshwater reservoir with a global sea-level equivalent of ~58 m (Morlighem et al., 2020), is experiencing an accelerated loss of mass, contributing significantly to rising global sea levels (Rignot et al., 2019; Shepherd et al., 2019). Land ice, encompassing ice sheets and glaciers, has emerged as the primary contributor to the recent surge in global mean sea level over the period 2006–2018 (Fox-Kemper et al., 2021). Over the last decade, Antarctic ice sheet (AIS) mass loss has tripled, primarily attributed to heightened ice discharge into the Amundsen Sea (Shepherd et al., 2018). The potential mass loss of the AIS is projected to exceed that of Greenland contributing to sea-level rise by 2100 (Pörtner et al., 2022). Geological data and models project the AIS to be a predominant driver of sea level rise over millennia (Bamber et al., 2019).

The primary mechanism behind Antarctic ice mass loss lies in ice dynamics (Blankenship et al., 1993; Pritchard et al., 2012). The dynamic contribution of the AIS, particularly through increased ice discharge into the Amundsen Sea, plays a crucial role in determining the pace of future sea level changes. Key processes behind the accelerated ice loss are the enhanced melting of ice shelves due to warmer ocean water (basal melt) and increased calving (iceberg formation) triggered by basal melt and/or surface melt (Pritchard et al., 2012; Van de Wal et al., 2005).

The discharge rate of Antarctica's ice into the ocean is strongly influenced by fringing ice shelves (encompassing ~74% of Antarctica's coastline) that provide resistive "buttressing" forces on upstream grounded ice. These ice shelves are essential because (1) they have experienced retreat and thinning in key regions during recent decades (Paolo et al., 2015; Shepherd, Wingham, & Rignot, 2004), (2) their buttressing effect moderates the grounded ice sheet's contribution to global sea level rise (Rignot et al., 2004; Reese, Gudmundsson, Levermann, & Winkelmann, 2018), (3) ice melting induces ocean freshening, influencing ocean circulation patterns, and (4) changes in their mass contribute modestly ($49 \pm 8\ \mu m\ yr^{-1}$) to the rate of sea level rise due to steric effects (Shepherd et al., 2010). A schematic diagram showing the orientation of ice shelf and grinding line in West Antarctica is shown in Fig. 18.8.

18.7.1 Thinning of the ice shelves around Antarctica

The floating ice shelves along Antarctica's coastline have experienced unprecedented thinning and shrinking in recent decades (Paolo et al., 2015). Over the past three decades, satellite observations have documented significant changes in Antarctic ice shelves, including retreat (Cook & Vaughan, 2010; Scambos et al., 2009), thinning (Paolo et al., 2015), and disintegration (Rott, Skvarca, & Nagler, 1996) in key sectors. These changes are closely linked to atmospheric and oceanic processes, operating on various timescales (Adusumilli, Fricker, Medley, Padman, & Siegfried, 2020; Paolo et al., 2018). Ice shelves undergo mass gain through ice influx across the grounding line and local surface accumulation via snowfall, while mass loss occurs through calving, basal melt, and minor surface effects like surface melting, wind scour, sublimation, and surface runoff. Despite satellite-derived estimates of ice shelf thickness, previous studies faced challenges in directly measuring basal melt rates (Christianson et al., 2016; Dutrieux et al., 2014; Jacobs et al., 2013; Jenkins et al., 2018), hindering our understanding of driving mechanisms and temporal variations in basal melt rates.

Antarctic Peninsula ice shelves have experienced an average retreat of 18% over the last 50 years, with catastrophic collapses observed in the Larsen-A, Larsen-B, and Wilkins ice shelves in 1995, 2002, and 2008, respectively. The Larsen-C Ice Shelf lost a significant portion in July 2017, prompting questions about the remaining ice shelf's stability and the triggering mechanism (Hogg & Gudmundsson, 2017). Potential drivers include crevasse propagation due to

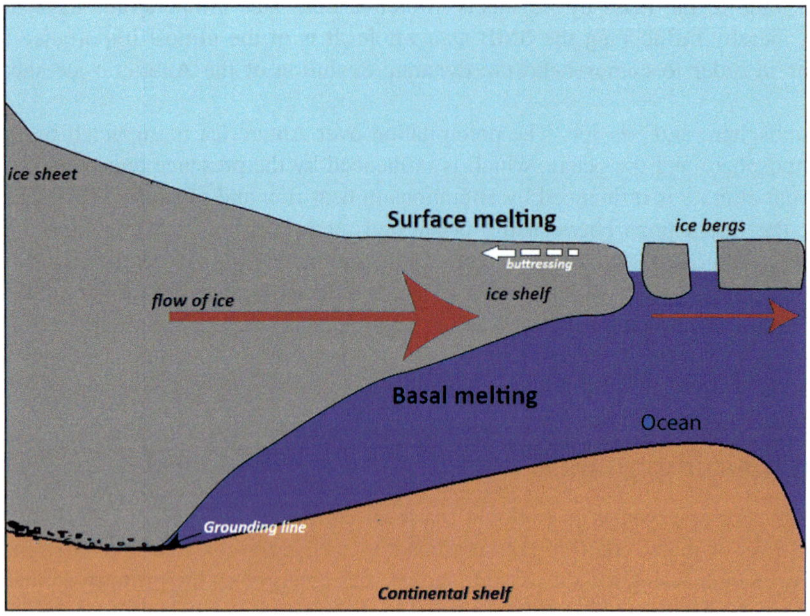

FIGURE 18.8 A schematic diagram showing the orientation of ice shelve and grounding line in West Antarctica. *Courtesy of Ronja Reese and Maria Zeitz.*

atmospheric warming, leading to meltwater ponding, or warmer ocean water causing thinning through increased submarine basal melt rates. In the Amundsen Sea Embayment, ice shelves at the terminus of the Pine Island and Thwaites Glaciers have exhibited substantial thinning for over two decades, linked to the long-term influence of warm ocean water eroding the ice shelf base (Depoorter et al., 2013; Paolo et al., 2018). Recent decades have witnessed rapid thinning of ice shelves in West Antarctica's Amundsen Sea Embayment, with indications of similar changes in some East Antarctic basins (Khazendar et al., 2013; Roberts et al., 2018). Even in East Antarctica's Totten and Cook regions, where surface melt is minimal, ice shelf thinning has been observed, hinting at ocean-driven processes (Rignot, Jacobs, Mouginot, & Scheuchl, 2013). A recent study utilizes 26 years (1992–2017) of satellite-derived data, including ice shelf thickness, flow, and basal melt rates, to reveal an overall pattern of thinning across Antarctica (Paolo et al., 2023). In agreement with earlier studies, the highest rates of ice shelf thinning are found for those ice shelves located in the Amundsen and Bellingshausen sectors. Interestingly, a slowdown in thinning rates was noted from around 2008, particularly in the Amundsen, Bellingshausen, and Wilkes sectors, which is attributed to modulation in external ocean forcing, altered in West Antarctica by negative feedbacks between ice shelf thinning rates and grounded ice flow, and subice shelf cavity geometry and basal melting. The slight reduction in basal melting rates over the observational period suggests that long-term changes in ocean conditions predominantly drive high thinning rates. Thus it is important to consider complex feedbacks between thinning and ice advection and between ice shelf draft and basal melt rates when making causal inferences from ice shelf observations (Paolo et al., 2023). Overall the existing studies suggest a potential reduction in buttressing across Antarctica, initiating a process of runaway retreat in inherently unstable regions due to marine ice sheet instability (Joughin, Smith, & Medley, 2014; Rignot, Mouginot, Morlighem, Seroussi, & Scheuchl, 2014; Weertman, 1974). Consequently, understanding the mechanisms governing ice shelf thickness changes becomes crucial for predicting Antarctica's future contribution to global sea level rise.

18.7.2 Surface melting across the Antarctic ice shelves

Ice shelves in the Antarctic can melt either from below due to increased incursions of relatively warm water (Pritchard et al., 2012) or from above due to rising air temperatures (Scambos, Hulbe, Fahnestock, & Bohlander, 2000). Despite surface melt representing a growing component of the mass budget due to atmospheric warming (DeConto & Pollard, 2016), its extent is relatively limited, encompassing only 11% of the total surface area from 2000 to 2009 (Trusel, Frey, & Das, 2012). However, this area is still comparable to the entire surface area of the Greenland Ice Sheet (1.7 million km^2). The presence of substantial surface meltwater ponds is widespread across many Antarctic ice shelves during austral summertime in response to intense or prolonged surface melting (Arthur, Stokes, Jamieson, Carr, & Leeson, 2020; Banwell et al., 2021; Dell et al., 2020; Kingslake, Ely, Das, & Bell, 2017; Luckman et al., 2014; Stokes, Sanderson, Miles, Jamieson, & Leeson, 2019). Vertical fracturing (hydrofracturing) can occur if meltwater enters downward, enlarging fractures in the ice and potentially triggering catastrophic collapse (Banwell & Macayeal, 2015; Banwell, MacAyeal, & Sergienko, 2013; Banwell, Willis, Macdonald, Goodsell, & MacAyeal, 2019; MacAyeal, Scambos, Hulbe, & Fahnestock, 2003; Scambos et al., 2000). This phenomenon has been observed on several ice shelves on the Antarctic Peninsula in recent decades (Doake, Corr, Rott, Skvarca, & Young, 1998; Glasser et al., 2011; Rott et al., 1996; Scambos et al., 2000, 2009).

The thinning and loss of ice shelves can diminish the backstress they provide on grounded ice, leading to glacier acceleration and thinning (Dupont & Alley, 2005). Surface meltwater reaching the bed has resulted in increased sliding, accelerating Antarctic Peninsula outlet glaciers (Tuckett et al., 2019). Overall, the thinning and/or collapse of ice shelves therefore cause acceleration of glacier flow into the ocean, resulting in increased rates of ice sheet mass loss and consequent sea level rise (Pritchard et al., 2012; Rignot et al., 2019). The increasing melt in the coming decades is anticipated to result in mass loss through associated runoff, and surface melt is also linked to dynamic ice loss mechanisms (Kittel et al., 2021).

18.7.3 Processes driving surface melt

Local-scale processes associated with warm signatures and ice shelf surface melting encompass foehn winds (Datta et al., 2019; Elvidge, Renfrew, King, Orr, & Lachlan-Cope, 2016; Gilbert, Orr, Renfrew, King, & Lachlan-Cope, 2022; Orr et al., 2008; Orr et al., 2021; Zou, Bromwich, Montenegro, Wang, & Bai, 2021), katabatic winds (Bromwich, Carrasco, & Stearns, 1992; Coggins, McDonald, & Jolly, 2014; Heinemann, Glaw, & Willmes, 2019; Lenaerts et al., 2017; Parish & Bromwich, 1989), and barrier winds (Coggins et al., 2014; Orr et al., 2004, 2014). Foehn events, in particular, are known for inducing extreme temperature increases within a few hours, with temperature spikes in Antarctica exceeding 10K (Bozkurt, Rondanelli, Marín, & Garreaud, 2018; Laffin et al., 2021; Speirs, Steinhoff, McGowan, Bromwich, & Monaghan, 2010).

At the synoptic scale, warm air intrusions and atmospheric rivers contribute to rapid temperature increases of over 10K, sometimes coinciding with foehn events (Bozkurt et al., 2018; Clem et al., 2020; Djoumna & Holland, 2021; Nicolas & Bromwich, 2011; Nicolas et al., 2017; Scott, Nicolas, Bromwich, Norris, & Lubin, 2019; Wille et al., 2019, 2022). Furthermore, synoptic-scale depressions circulating around Antarctica can lead to sudden warming over coastal locations (Orr et al., 2014; Simmonds & Keay, 2000).

Air temperature and melt-related aspects, such as liquid-water production, exhibit a strong correlation (Hock, 2005; Ohmura, 2001). These processes occur in response to a positive surface energy balance when the air/surface temperature exceeds the threshold of 273.15K (0°C), the melting point of ice/snow (Banwell et al., 2021; Barrand et al., 2013; Laffin et al., 2021; Tedesco, Abdalati, & Zwally, 2007; Trusel et al., 2012). However, local snowpack properties can lead to melting even at air temperatures below 273.15K. For instance, larger snow grains in the snowpack enhance solar radiation penetration and absorption into the subsurface, providing more energy for melt despite subfreezing air temperatures (Liston & Winther, 2005; Nicolas et al., 2017; Tedesco, 2009). The relationship between air temperatures above the melt threshold and the amount of meltwater produced is not strictly linear (Abram et al., 2013; Jakobs, Reijmer, van den Broeke, Van de Berg, & Van Wessem, 2021; Trusel et al., 2015). Melt production is amplified in regions with lower surface albedo due to increased shortwave radiation absorption, augmenting the energy available for surface melt (Kingslake et al., 2017; Lenaerts et al., 2017; Liston & Winther, 2005). This nonlinear relation is further enhanced by melt–albedo feedback, where melting enlarges snow grain size, reducing albedo and intensifying melt. This feedback is most pronounced for low melt rates (<100 mm w.e. yr^{-1}) but weaker for higher melt rates (>200 mm w.e. yr^{-1}) required for ice shelf hydrofracturing (Jakobs et al., 2021).

Passive microwave observations from radiometers have played a key role in quantifying melt extent and duration across the entire Antarctic ice sheet. These observations have facilitated the assessment of trends and climate teleconnections (Picard, Fily, & Gallée, 2007; Tedesco & Monaghan, 2009; Torinesi, Fily, & Genthon, 2003; Zwally & Fiegles, 1994). A study using passive microwave measurements, spanning July 1979 to June 2020, revealed surface meltwater presence across Antarctic ice shelves. The Antarctic Peninsula exhibits the highest annual melt days, peaking at 89 days. Overall, negligible linear trends were detected in yearly melt days. High annual melt years were noticed in Dronning Maud Land and Wilkes Land associated with negative SAM mode. The Antarctic Peninsula and Bellingshausen Sea sector of West Antarctica are other regions with instances of high melt years. Regions with prolonged melt show reduced winter microwave emissions due to firn structural changes, notably observed during the austral summers of 1991–1992 and 2015–2016, particularly impacting the Ross Ice Shelf. From 2001 onwards, the advent of rapid-repeat scatterometers like QuikSCAT and the Advanced SCATterometer has significantly contributed to quantifying melt extent and duration across the entire continent. Despite the advancements, it is noteworthy that the most pronounced melt events, particularly during the 1990s, were not covered by recent active microwave records. Recent innovations, such as high-resolution, rapid-repeat Sentinel-1 SAR imagery, have provided near-daily melt detection since 2014, offering valuable insights into more contemporary melt patterns (Johnson, Fahnestock, & Hock, 2020).

A recent study Orr et al. (2023) introduced a surface "melt potential" index (MPI) over Antarctic ice shelves that describes the frequency (MPI-freq; %) and intensity (MPI-int; K) of daily maximum summer temperatures (from high-resolution regional climate models) exceeding a melt threshold of 273.15K. This index helped identify which ice shelves are vulnerable to melt-induced hydrofracture. For example, MPI is highest for Antarctic Peninsula ice shelves (MPI-freq 23%–35%, MPI-int 1.2–2.1K), lowest (2%–3%, <0K) for the Ronne–Filchner and Ross ice shelves, and around 10%–24% and 0.6–1.7K for the other West and East Antarctic ice shelves. Hotspots of MPI are apparent over many ice shelves (Fig. 18.8), and they also show a decreasing trend in MPI-freq (Fig. 18.8). The high melt events over West Antarctic ice shelves resemble the central Pacific El Niño pattern with a stationary Rossby wave and a strong anticyclone over the high-latitude South Pacific. Interannual changes in summer melt across West Antarctic ice shelves was also found to be influenced by the canonical El Niño events (Deb et al., 2018). By contrast, the East Antarctic circulation comprises a zonally symmetric negative SAM pattern with a strong regional anticyclone on the plateau and enhanced coastal easterlies/weakened Southern Ocean westerlies. However, the decadal, multi-decadal, and long-term trends in surface melt across Antarctic ice shelves remain poorly understood, highlighting a critical research gap that future studies must address.

References

Abram, N. J., Mulvaney, R., Wolff, E. W., Triest, J., Kipfstuhl, S., Trusel, L. D., ... Arrowsmith, C. (2013). Acceleration of snow melt in an Antarctic Peninsula ice core during the twentieth century. *Nature Geoscience*, *6*(5), 404–411.

Adusumilli, S., Fricker, H. A., Medley, B., Padman, L., & Siegfried, M. R. (2020). Interannual variations in meltwater input to the Southern Ocean from Antarctic ice shelves. *Nature Geoscience*, *13*(9), 616–620.

Ainley, et al. (2003). *Sea Ice: An Introduction to its Physics, Chemistry, Biology and Geology*. Thomas, D. N., & Dieckmann, G. S. (Eds.), Available from https://doi.org/10.1002/9780470757161

Agosta, C., Amory, C., Kittel, C., Orsi, A., Favier, V., Gallée, H., ... Fettweis, X. (2019). Estimation of the Antarctic surface mass balance using the regional climate model MAR (1979–2015) and identification of dominant processes. *The Cryosphere*, *13*(1), 281–296.

Arblaster, J. M., & Meehl, G. A. (2006). Contributions of external forcings to southern annular mode trends. *Journal of Climate*, *19*(12), 2896–2905.

Arthern, R. J., Winebrenner, D. P., & Vaughan, D. G. (2006). Antarctic snow accumulation mapped using polarization of 4.3-cm wavelength microwave emission. *Journal of Geophysical Research: Atmospheres*, *111*(D6).

Arthur, J. F., Stokes, C., Jamieson, S. S., Carr, J. R., & Leeson, A. A. (2020). Recent understanding of Antarctic supraglacial lakes using satellite remote sensing. *Progress in Physical Geography: Earth and Environment*, *44*(6), 837–869.

Bamber, J. L., Oppenheimer, M., Kopp, R. E., Aspinall, W. P., & Cooke, R. M. (2019). Ice sheet contributions to future sea-level rise from structured expert judgment. *Proceedings of the National Academy of Sciences USA*, *116*(23), 11195–11200. Available from https://doi.org/10.1073/pnas.1817205116.

Banwell, A. F., Datta, R. T., Dell, R. L., Moussavi, M., Brucker, L., Picard, G., ... Stevens, L. A. (2021). The 32-year record-high surface melt in 2019/2020 on the northern George VI Ice Shelf, Antarctic Peninsula. *The Cryosphere*, *15*(2), 909–925.

Banwell, A. F., & Macayeal, D. R. (2015). Ice-shelf fracture due to viscoelastic flexure stress induced by fill/drain cycles of supraglacial lakes. *Antarctic Science*, *27*(6), 587–597.

Banwell, A. F., MacAyeal, D. R., & Sergienko, O. V. (2013). Breakup of the Larsen B Ice Shelf triggered by chain reaction drainage of supraglacial lakes. *Geophysical Research Letters*, *40*(22), 5872–5876.

Banwell, A. F., Willis, I. C., Macdonald, G. J., Goodsell, B., & MacAyeal, D. R. (2019). Direct measurements of ice-shelf flexure caused by surface meltwater ponding and drainage. *Nature Communications*, *10*(1), 730.

Barrand, N. E., Vaughan, D. G., Steiner, N., Tedesco, M., Kuipers Munneke, P., Van Den Broeke, M. R., & Hosking, J. S. (2013). Trends in Antarctic Peninsula surface melting conditions from observations and regional climate modeling. *Journal of Geophysical Research: Earth Surface*, *118*(1), 315–330.

Behrangi, A., Christensen, M., Richardson, M., Lebsock, M., Stephens, G., Huffman, G. J., et al. (2016). Status of high-latitude precipitation estimates from observations and reanalyses. *Journal of Geophysical Research: Atmospheres*, *121*(9), 4468–4486.

Bintanja, R., van Oldenborgh, G. J., Drijfhout, S. S., Wouters, B., & Katsman, C. A. (2013). Important role for ocean warming and increased ice-shelf melt in Antarctic sea-ice expansion. *Nature Geoscience*, *6*, 376–379. Available from https://doi.org/10.1038/NGEO1767.

Black, H. P., & Budd, W. (1964). Accumulation in the region of Wilkes, Wilkes Land, Antarctica. *Journal of Glaciology*, *5*(37), 3–15.

Blankenship, D. D., Bell, R. E., Hodge, S. M., Brozena, J. M., Behrendt, J. C., & Finn, C. A. (1993). Active volcanism beneath the West Antarctic ice sheet and implications for ice-sheet stability. *Nature*, *361*(6412), 526–529.

Bozkurt, D., Rondanelli, R., Marín, J. C., & Garreaud, R. (2018). Foehn event triggered by an atmospheric river underlies record-setting temperature along continental Antarctica. *Journal of Geophysical Research: Atmospheres*, *123*(8), 3871–3892.

Braithwaite, R. J. (1995). Positive degree-day factors for ablation on the Greenland ice sheet studied by energy-balance modelling. *J. Glaciol., 41* (137), 153–160.

Brandt, R. E., Warren, S. G., Worby, A. P., & Grenfell, T. C. (2005). Surface albedo of the Antarctic sea ice zone. *Journal of Climate, 18*(17), 3606–3622.

Bromwich, D. H., Carrasco, J. F., & Stearns, C. R. (1992). Satellite observations of katabatic-wind propagation for great distances across the Ross Ice Shelf. *Monthly Weather Review, 120*(9), 1940–1949.

Bromwich, D. H., Guo, Z., Bai, L., & Chen, Q. S. (2004). Modeled Antarctic precipitation. Part I: Spatial and temporal variability. *Journal of Climate, 17*(3), 427–447.

Bromwich, D. H., Nicolas, J. P., Monaghan, A. J., Lazzara, M. A., Keller, L. M., Weidner, G. A., & Wilson, A. B. (2013). Central West Antarctica among the most rapidly warming regions on Earth. *Nature Geoscience, 6*(2), 139–145.

Cavalieri, D. J., Gloersen, P., Parkinson, C., Comiso, J., & Zwally, H. J. (1997). Observed hemispheric asymmetry in global sea ice changes. *Science, 278*, 1104–1106. Available from https://doi.org/10.1126/science.278.5340.1104.

Christianson, K., Bushuk, M., Dutrieux, P., Parizek, B. R., Joughin, I. R., Alley, R. B. ,&, ... Kim, T. W. (2016). Sensitivity of Pine Island Glacier to observed ocean forcing. *Geophysical Research Letters, 43*(20), 10–817.

Clem, K. R., Fogt, R. L., Turner, J., Lintner, B. R., Marshall, G. J., Miller, J. R., & Renwick, J. A. (2020). Record warming at the South Pole during the past three decades. *Nature Climate Change, 10*(8), 762–770.

Coggins, J. H., McDonald, A. J., & Jolly, B. (2014). Synoptic climatology of the Ross Ice Shelf and Ross Sea region of Antarctica: k-means clustering and validation. *International Journal of Climatology, 34*(7), 2330–2348.

Cogley, J. G., Arendt, A. A., Bauder, A., Braithwaite, R. J., Hock, R., Jansson, P., ... Zemp, M. (2010). Glossary of glacier mass balance and related terms. (IHP-VII Technical Documents in Hydrology). *International Hydrological Programme*. Available from https://research.manchester.ac.uk/en/publications/glossary-of-glacier-mass-balance-and-related-terms.

Comiso, J. C., Kwok, R., Martin, S., & Gordon, A. (2011). Variability and trends in sea ice and ice production in the Ross Sea. *Journal of Geophysical Research, 116*, C04021. Available from https://doi.org/10.1029/2010JC006391.

Comiso, J. C., & Nishio, F. (2008). Trends in the sea ice cover using enhanced and compatible AMSR-E, SSM/I, and SMMR data. *Journal of Geophysical Research, 113*, C02S07. Available from https://doi.org/10.1029/2007JC004257.

Cook, A. J., & Vaughan, D. G. (2010). Overview of areal changes of the ice shelves on the Antarctic Peninsula over the past 50 years. *The Cryosphere, 4*(1), 77–98.

Curry, J. A., Schramm, J. L., & Ebert, E. E. (1995). Sea ice-albedo climate feedback mechanism. *Journal of Climate, 8*(2), 240–247.

Das, I., Bell, R. E., Scambos, T. A., Wolovick, M., Creyts, T. T., Studinger, M. ,, ... van den Broeke, M. R. (2013). Influence of persistent wind scour on the surface mass balance of Antarctica. *Nature Geoscience, 6*(5), 367–371.

Datta, R. T., Tedesco, M., Fettweis, X., Agosta, C., Lhermitte, S., Lenaerts, J. T. M., & Wever, N. (2019). The effect of Foehn-induced surface melt on firn evolution over the northeast Antarctic peninsula. *Geophysical Research Letters, 46*(7), 3822–3831.

Deb, P., Orr, A., Bromwich, D. H., Nicolas, J. P., Turner, J., & Hosking, J. S. (2018). Summer drivers of atmospheric variability affecting ice shelf thinning in the Amundsen Sea Embayment, West Antarctica. *Geophysical Research Letters, 45*(9), 4124–4133.

DeConto, R. M., & Pollard, D. (2016). Contribution of Antarctica to past and future sea-level rise. *Nature, 531*(7596), 591–597.

Dell, R., Arnold, N., Willis, I., Banwell, A., Williamson, A., Pritchard, H., & Orr, A. (2020). Lateral meltwater transfer across an Antarctic ice shelf. *The Cryosphere, 14*(7), 2313–2330.

Depoorter, M. A., Bamber, J. L., Griggs, J. A., Lenaerts, J. T., Ligtenberg, S. R., van den Broeke, M. R., & Moholdt, G. (2013). Calving fluxes and basal melt rates of Antarctic ice shelves. *Nature, 502*(7469), 89–92.

Dibb, J. E., & Fahnestock, M. (2004). Snow accumulation, surface height change, and firn densification at Summit, Greenland: Insights from 2 years of in situ observation. *Journal of Geophysical Research: Atmospheres, 109*(D24).

Ding, Q., & Steig, E. J. (2013). Temperature change on the Antarctic Peninsula linked to the tropical Pacific. *Journal of Climate, 26*(19), 7570–7585.

Djoumna, G., & Holland, D. M. (2021). Atmospheric rivers, warm air intrusions, and surface radiation balance in the Amundsen Sea Embayment. *Journal of Geophysical Research: Atmospheres, 126*(13), e2020JD034119.

Doake, C. S. M., Corr, H. F. J., Rott, H., Skvarca, P., & Young, N. W. (1998). Breakup and conditions for stability of the northern Larsen Ice Shelf, Antarctica. *Nature, 391*(6669), 778–780.

Dupont, T. K., & Alley, R. B. (2005). Assessment of the importance of ice-shelf buttressing to ice-sheet flow. *Geophysical Research Letters, 32*(4).

Dutrieux, P., De Rydt, J., Jenkins, A., Holland, P. R., Ha, H. K., Lee, S. H., ... Schröder, M. (2014). Strong sensitivity of Pine Island ice-shelf melting to climatic variability. *Science, 343*(6167), 174–178.

Eayrs, C., Li, X., Raphael, M. N., & Holland, D. M. (2021). Rapid decline in Antarctic sea ice in recent years hints at future change. *Nature Geoscience, 14*(7), 460–464.

Eisen, O., Frezzotti, M., Genthon, C., Isaksson, E., Magand, O., van den Broeke, M. R., ... Vaughan, D. G. (2008). Ground-based measurements of spatial and temporal variability of snow accumulation in East Antarctica. *Reviews of Geophysics, 46*(2).

Eisenman, I., Schneider, T., Battisti, D. S., & Bitz, C. M. (2011). Consistent changes in the sea ice seasonal cycle in response to global warming. *J. Climate, 24*, 5325–5335. Available from https://doi.org/10.1175/2011JCLI4051.1.

Elvidge, A. D., Renfrew, I. A., King, J. C., Orr, A., & Lachlan-Cope, T. A. (2016). Foehn warming distributions in nonlinear and linear flow regimes: A focus on the Antarctic Peninsula. *Quarterly Journal of the Royal Meteorological Society, 142*(695), 618–631.

Folland, C. K. (1988). Numerical models of the raingauge exposure problem, field experiments and an improved collector design. *Quarterly Journal of the Royal Meteorological Society*, *114*, 1485–1516.

Fox-Kemper, B., Hewitt, H. T., Xiao, C., Aðalgeirsdóttir, G., Drijfhout, S. S., Edwards, T. L., … Yu, Y. (2021). *Ocean, Cryosphere and Sea Level Change. In Climate Change 2021: The Physical Science Basis. Contribution of Working Group I to the Sixth Assessment Report of the Intergovernmental Panel on Climate Change [Masson-Delmotte, V., P. Zhai, A. Pirani, S.L. Connors, C. Péan, S. Berger, N. Caud, Y. Chen, L. Goldfarb, M.I. Gomis, M. Huang, K. Leitzell, E. Lonnoy, J.B.R. Matthews, T.K. Maycock, T. Waterfield, O. Yelekçi, R. Yu, and B. Zhou (eds.)]* (pp. 1211–1362). Cambridge, United Kingdom and New York, NY, USA: Cambridge University Press. Available from https://doi.org/10.1017/9781009157896.011.

Gardner, A. S., Moholdt, G., Scambos, T., Fahnstock, M., Ligtenberg, S., Van Den Broeke, M., & Nilsson, J. (2018). Increased West Antarctic and unchanged East Antarctic ice discharge over the last 7 years. *The Cryosphere*, *12*(2), 521–547.

Garrison, D. L. (1991). Antarctic sea ice biota. *American Zoologist*, *31*(1), 17–34.

Genthon, C., Krinner, G., & Castebrunet, H. (2009). Antarctic precipitation and climate-change predictions: Horizontal resolution and margin vs plateau issues. *Annals of Glaciology*, *50*(50), 55–60.

Genthon, C., Krinner, G., & Deque, M. (1998). Intra-annual variability of Antarctic precipitation from weather forecasts and high-resolution climate models. *Annals of Glaciology*, *27*, 488–494.

Gilbert, E., Orr, A., Renfrew, I. A., King, J. C., & Lachlan-Cope, T. (2022). A 20-year study of melt processes over larsen C ice shelf using a high-resolution regional atmospheric model: 2. Drivers of surface melting. *Journal of Geophysical Research: Atmospheres*, *127*(8), e2021JD036012.

Glasser, N. F., Scambos, T. A., Bohlander, J., Truffer, M., Pettit, E., & Davies, B. J. (2011). From ice-shelf tributary to tidewater glacier: Continued rapid recession, acceleration and thinning of Röhss Glacier following the 1995 collapse of the Prince Gustav Ice Shelf, Antarctic Peninsula. *Journal of Glaciology*, *57*(203), 397–406.

Goosse, H., & Zunz, V. (2014). Decadal trends in the Antarctic sea ice extent ultimately controlled by ice-ocean feedback. *Cryosphere*, *8*, 453–470. Available from https://doi.org/10.5194/tc-8-453-2014.

Gorodetskaya, I. V., Tsukernik, M., Claes, K., Ralph, M. F., Neff, W. D., & van Lipzig, N. P. M. (2014). The role of atmospheric rivers in anomalous snow accumulation in East Antarctica. *Geophysical Research Letters*, *41*, 6199–6206.

Hall, A. (2004). The role of surface albedo feedback in climate. *Journal of Climate*, *17*(7), 1550–1568.

Heinemann, G., Glaw, L., & Willmes, S. (2019). A satellite-based climatology of wind-induced surface temperature anomalies for the Antarctic. *Remote Sensing*, *11*(13), 1539.

Hirasawa, N., Nakamura, H., & Yamanouchi, T. (2000). Abrupt changes in meteorological conditions observed at an inland Antarctic station in association with wintertime blocking. *Geophysical Research Letters*, *27*(13), 1911–1914.

Hock, R. (2005). Glacier melt: A review of processes and their modelling. *Progress in Physical Geography*, *29*(3), 362–391.

Hogg, A. E., & Gudmundsson, G. H. (2017). Impacts of the Larsen-C Ice Shelf calving event. *Nature Climate Change*, *7*(8), 540–542.

Holland, P. R., & Kwok, R. (2012). Wind-driven trends in Antarctic sea-ice drift. *Nature Geoscience*, *5*, 872–875. Available from https://doi.org/10.1038/ngeo1627.

Hoskins, B. J., & Karoly, D. J. (1981). The Steady Linear Response of a Spherical Atmosphere to Thermal and Orographic Forcing. *Journal of the Atmospheric Sciences*, *38*, 1179–1196. Available from http://doi.org/10.1175/1520-0469(1981)038 < 1179:TSLROA > 2.0.CO;2.

Jacobs, S., Giulivi, C., Dutrieux, P., Rignot, E., Nitsche, F., & Mouginot, J. (2013). Getz Ice Shelf melting response to changes in ocean forcing. *Journal of Geophysical Research: Oceans*, *118*(9), 4152–4168.

Jacobs, S. S. (2004). Bottom water production and its links with the thermohaline circulation. *Antarctic Science*, *16*(4), 427–437.

Jakobs, C. L., Reijmer, C. H., van den Broeke, M. R., Van de Berg, W. J., & Van Wessem, J. M. (2021). Spatial variability of the snowmelt-albedo feedback in Antarctica. *Journal of Geophysical Research: Earth Surface*, *126*(2), e2020JF005696.

Jenkins, A., Shoosmith, D., Dutrieux, P., Jacobs, S., Kim, T. W., Lee, S. H., … Stammerjohn, S. (2018). West Antarctic Ice Sheet retreat in the Amundsen Sea driven by decadal oceanic variability. *Nature Geoscience*, *11*(10), 733–738.

Johnson, A., Fahnestock, M., & Hock, R. (2020). Evaluation of passive microwave melt detection methods on Antarctic Peninsula ice shelves using time series of Sentinel-1 SAR. *Remote Sensing of Environment*, *250*, 112044.

Johnson, A., Hock, R., & Fahnestock, M. (2022). Spatial variability and regional trends of Antarctic ice shelf surface melt duration over 1979–2020 derived from passive microwave data. *Journal of Glaciology*, *68*(269), 533–546.

Joughin, I., Smith, B. E., & Medley, B. (2014). Marine ice sheet collapse potentially under way for the Thwaites Glacier Basin, West Antarctica. *Science*, *344*(6185), 735–738.

Khazendar, A., Schodlok, M. P., Fenty, I., Ligtenberg, S. R. M., Rignot, E., & Van den Broeke, M. R. (2013). Observed thinning of Totten Glacier is linked to coastal polynya variability. *Nature Communications*, *4*(1), 2857.

Kingslake, J., Ely, J. C., Das, I., & Bell, R. E. (2017). Widespread movement of meltwater onto and across Antarctic ice shelves. *Nature*, *544*(7650), 349–352.

Kittel, C., Amory, C., Agosta, C., Jourdain, N. C., Hofer, S., Delhasse, A., … Fettweis, X. (2021). Diverging future surface mass balance between the Antarctic ice shelves and grounded ice sheet. *The Cryosphere*, *15*(3), 1215–1236.

Kochendorfer, J., Rasmussen, R., Wolff, M., Baker, B., Hall, M. E., Meyers, T., et al. (2017). The quantification and correction of wind-induced precipitation measurement errors. *Hydrology and Earth System Sciences*, *21*, 1973–1989.

Laffin, M. K., Zender, C. S., Singh, S., Van Wessem, J. M., Smeets, C. J. P. P., & Reijmer, C. H. (2021). Climatology and evolution of the Antarctic Peninsula föhn wind-induced melt regime from 1979–2018. *Journal of Geophysical Research: Atmospheres*, *126*(4), e2020JD033682.

Laine, V. (2008). Antarctic ice sheet and sea ice regional albedo and temperature change, 1981–2000, from AVHRR Polar Pathfinder data. *Remote Sensing of Environment*, *112*(3), 646–667.

Lenaerts, J. T., Medley, B., van den Broeke, M. R., & Wouters, B. (2019). Observing and modeling ice sheet surface mass balance. *Reviews of Geophysics*, *57*(2), 376–420.

Lenaerts, J. T. M., Lhermitte, S., Drews, R., Ligtenberg, S. R. M., Berger, S., Helm, V., … Eijkelboom, M. (2017). Meltwater produced by wind–albedo interaction stored in an East Antarctic ice shelf. *Nature Climate Change*, *7*(1), 58–62.

Li, X., Cai, W., Meehl, G. A., Chen, D., Yuan, X., Raphael, M., … Wang, G. (2021). Tropical teleconnection impacts on Antarctic climate changes. *Nature Reviews Earth & Environment*, *2*(10), 680–698.

Li, X., Gerber, E. P., Holland, D. M., & Yoo, C. (2015). A Rossby wave bridge from the tropical Atlantic to West Antarctica. *Journal of Climate*, *28*(6), 2256–2273.

Li, X., Holland, D. M., Gerber, E. P., & Yoo, C. (2014). Impacts of the north and tropical Atlantic Ocean on the Antarctic Peninsula and sea ice. *Nature*, *505*(7484), 538–542.

Liston, G. E., & Winther, J. G. (2005). Antarctic surface and subsurface snow and ice melt fluxes. *Journal of Climate*, *18*(10), 1469–1481.

Luckman, A., Elvidge, A., Jansen, D., Kulessa, B., Munneke, P. K., King, J., & Barrand, N. E. (2014). Surface melt and ponding on Larsen C Ice Shelf and the impact of föhn winds. *Antarctic Science*, *26*(6), 625–635.

MacAyeal, D. R., Scambos, T. A., Hulbe, C. L., & Fahnestock, M. A. (2003). Catastrophic ice-shelf break-up by an ice-shelf-fragment-capsize mechanism. *Journal of Glaciology*, *49*(164), 22–36.

Magand, O., Picard, G., Brucker, L., Fily, M., & Genthon, C. (2008). Snow melting bias in microwave mapping of Antarctic snow accumulation. *The Cryosphere*, *2*(2), 109–115.

Marshall, G. J., Thompson, D. W. J., & van den Broeke, M. R. (2017). The signature of Southern Hemisphere atmospheric circulation patterns in Antarctic precipitation. *Geophysical Research Letters*, *44*(11), 580–11, 589.

Massom, R. A., Pook, M. J., Comiso, J. C., Adams, N., Turner, J., Lachlan-Cope, T. A., & Gibson, T. (2004). Precipitation over the interior East Antarctic ice sheet related to mid-latitude blocking-high activity. *Journal of Climate*, *17*, 1914–1928.

McConnell, J. R., Bales, R. C., & Davis, D. R. (1997). Recent intra-annual snow accumulation at South Pole: Implications for ice core interpretation. *Journal of Geophysical Research: Atmospheres*, *102*(D18), 21947–21954.

Meehl, G. A., Arblaster, J. M., Bitz, C. M., Chung, C. T., & Teng, H. (2016). Antarctic sea-ice expansion between 2000 and 2014 driven by tropical Pacific decadal climate variability. *Nature Geoscience*, *9*(8), 590–595.

Meehl, G. A., Arblaster, J. M., Chung, C. T., Holland, M. M., DuVivier, A., Thompson, L., … Bitz, C. M. (2019). Sustained ocean changes contributed to sudden Antarctic sea ice retreat in late 2016. *Nature Communications*, *10*(1), 14.

Morlighem, M., Rignot, E., Binder, T., Blankenship, D., Drews, R., Eagles, G., … Goel, V. (2020). Deep glacial troughs and stabilizing ridges unveiled beneath the margins of the Antarctic ice sheet. *Nature Geoscience*, *13*(2), 132–137.

Mottram, R., Hansen, N., Kittel, C., van Wessem, J. M., Agosta, C., Amory, C., … … Souverijns, N. (2021). What is the surface mass balance of Antarctica? An intercomparison of regional climate model estimates. *The Cryosphere*, *15*(8), 3751–3784.

Nicolas, J. P., & Bromwich, D. H. (2011). Climate of West Antarctica and influence of marine air intrusions. *Journal of Climate*, *24*(1), 49–67.

Nicolas, J. P., Vogelmann, A. M., Scott, R. C., Wilson, A. B., Cadeddu, M. P., Bromwich, D. H., … Powers, H. H. (2017). January 2016 extensive summer melt in West Antarctica favoured by strong El Niño. *Nature Communications*, *8*(1), 15799.

Noone, D., Turner, J., & Mulvaney, R. (1999). Atmospheric signals and characteristics of accumulation in Dronning Maud Land, Antarctica. *Journal of Geophysical Research*, *104*(D16), 19,191–19,211.

Ohmura, A. (2001). Physical basis for the temperature-based melt-index method. *Journal of Applied Meteorology and Climatology*, *40*(4), 753–761.

Orr, A., Cresswell, D., Marshall, G. J., Hunt, J. C., Sommeria, J., Wang, C. G., & Light, M. (2004). A 'low-level' explanation for the recent large warming trend over the western Antarctic Peninsula involving blocked winds and changes in zonal circulation. *Geophysical Research Letters*, *31*(6).

Orr, A., Deb, P., Clem, K. R., Gilbert, E., Bromwich, D. H., Boberg, F., … Zou, X. (2023). Characteristics of Surface "Melt Potential" over Antarctic Ice Shelves based on Regional Atmospheric Model Simulations of Summer Air Temperature Extremes from 1979/80 to 2018/19. *Journal of Climate*, *36*(10), 3357–3383. Available from https://doi.org/10.1175/JCLI-D-22-0386.1.

Orr, A., Kirchgaessner, A., King, J., Phillips, T., Gilbert, E., Elvidge, A., … Webster, S. (2021). Comparison of kilometre and sub-kilometre scale simulations of a foehn wind event over the Larsen C Ice Shelf, Antarctic Peninsula using the Met Office Unified Model (MetUM). *Quarterly Journal of the Royal Meteorological Society*, *147*(739), 3472–3492.

Orr, A., Marshall, G. J., Hunt, J. C., Sommeria, J., Wang, C. G., Van Lipzig, N. P., … King, J. C. (2008). Characteristics of summer airflow over the Antarctic Peninsula in response to recent strengthening of westerly circumpolar winds. *Journal of the Atmospheric Sciences*, *65*(4), 1396–1413.

Orr, A., Phillips, T., Webster, S., Elvidge, A., Weeks, M., Hosking, S., & Turner, J. (2014). Met Office Unified Model high-resolution simulations of a strong wind event in Antarctica. *Quarterly Journal of the Royal Meteorological Society*, *140*(684), 2287–2297.

Orsi, A. H., Johnson, G. C., & Bullister, J. L. (1999). Circulation, mixing, and production of Antarctic Bottom Water. *Progress in Oceanography*, *43*(1), 55–109.

Paolo, F. S., Fricker, H. A., & Padman, L. (2015). Volume loss from Antarctic ice shelves is accelerating. *Science)*, *348*(6232), 327–331.

Paolo, F. S., Gardner, A. S., Greene, C. A., Nilsson, J., Schodlok, M. P., Schlegel, N. J., & Fricker, H. A. (2023). Widespread slowdown in thinning rates of West Antarctic ice shelves. *The Cryosphere*, *17*(8), 3409–3433.

Paolo, F. S., Padman, L., Fricker, H. A., Adusumilli, S., Howard, S., & Siegfried, M. R. (2018). Response of Pacific-sector Antarctic ice shelves to the El Niño/Southern oscillation. *Nature Geoscience*, *11*(2), 121–126.

Parkinson, C. L., & Cavalieri, D. J. (2012). Antarctic sea ice variability and trends, 1979–2010. *Cryosphere*, *6*, 871–880. Available from https://doi.org/10.5194/tc-6-871-2012.

Parish, T. R., & Bromwich, D. H. (1989). Instrumented aircraft observations of the katabatic wind regime near Terra Nova Bay. *Monthly Weather Review*, *117*(7), 1570–1585.

Picard, G., Arnaud, L., Domine, F., & Fily, M. (2009). Determining snow specific surface area from near-infrared reflectance measurements: Numerical study of the influence of grain shape. *Cold Regions Science and Technology*, *56*(1), 10–17.

Pritchard, H., Ligtenberg, S. R., Fricker, H. A., Vaughan, D. G., van den Broeke, M. R., & Padman, L. (2012). Antarctic ice-sheet loss driven by basal melting of ice shelves. *Nature*, *484*(7395), 502–505.

Purich, A., & England, M. H. (2019). Tropical teleconnections to Antarctic sea ice during austral spring 2016 in coupled pacemaker experiments. *Geophysical Research Letters*, *46*(12), 6848–6858.

Picard, G., Fily, M., & Gallée, H. (2007). Surface melting derived from microwave radiometers: A climatic indicator in Antarctica. *Annals of Glaciology*, *46*, 29–34.

Pörtner, H. O., Roberts, D. C., Adams, H., Adelekan, I., Adler, C., Adrian, R., ... Kerr, R. B. (2022). Technical summary. *Climate Change*, 37–118.

Rana, P. K., Dash, M. K., Routray, A., & Pandey, P. C. (2011). Prediction of sea ice edge in the Antarctic using GVF Snake model. *Journal of the Geological Society of India*, *78*, 99–108.

Reese, R., Gudmundsson, G. H., Levermann, A., & Winkelmann, R. (2018). The far reach of ice-shelf thinning in Antarctica. *Nature Climate Change*, *8*(1), 53–57.

Rignot, E., Casassa, G., Gogineni, P., Krabill, W., Rivera, A., & Thomas, R. (2004). Accelerated ice discharge from the Antarctic Peninsula following the collapse of Larsen B ice shelf. *Geophysical Research Letters*, *31*, L18401. Available from https://doi.org/10.1029/2004GL020697.

Rignot, E., Jacobs, S., Mouginot, J., & Scheuchl, B. (2013). Ice-shelf melting around Antarctica. *Science*, *341*(6143), 266–270.

Rignot, E., Mouginot, J., Morlighem, M., Seroussi, H., & Scheuchl, B. (2014). Widespread, rapid grounding line retreat of Pine Island, Thwaites, Smith, and Kohler glaciers, West Antarctica, from 1992 to 2011. *Geophysical Research Letters*, *41*(10), 3502–3509.

Rignot, E., Mouginot, J., Scheuchl, B., Van Den Broeke, M., Van Wessem, M. J., & Morlighem, M. (2019). Four decades of Antarctic Ice Sheet mass balance from 1979–2017. *Proceedings of the National Academy of Sciences*, *116*(4), 1095–1103.

Rignot, E., Velicogna, I., van den Broeke, M. R., Monaghan, A., & Lenaerts, J. T. (2011). Acceleration of the contribution of the Greenland and Antarctic ice sheets to sea level rise. *Geophysical Research Letters*, *38*(5).

Roberts, J., Galton-Fenzi, B. K., Paolo, F. S., Donnelly, C., Gwyther, D. E., Padman, L., ... Payne, A. J. (2018). Ocean forced variability of Totten Glacier mass loss. *Geological Society, London, Special Publications*, *461*(1), 175–186.

Rott, H., Skvarca, P., & Nagler, T. (1996). Rapid collapse of northern Larsen ice shelf, Antarctica. *Science*, *271*(5250), 788–792.

Scambos, T., Fricker, H. A., Liu, C. C., Bohlander, J., Fastook, J., Sargent, A., ... Wu, A. M. (2009). Ice shelf disintegration by plate bending and hydro-fracture: Satellite observations and model results of the 2008 Wilkins ice shelf break-ups. *Earth and Planetary Science Letters*, *280*(1-4), 51–60.

Scambos, T. A., Frezzotti, M., Haran, T., Bohlander, J., Lenaerts, J. T. M., Van Den Broeke, M. R., ... Neumann, T. (2012). Extent of low-accumulation, wind glaze, areas on the East Antarctic plateau: Implications for continental ice mass balance. *Journal of Glaciology*, *58*(210), 633–647.

Scambos, T. A., Hulbe, C., Fahnestock, M., & Bohlander, J. (2000). The link between climate warming and break-up of ice shelves in the Antarctic Peninsula. *Journal of Glaciology*, *46*(154), 516–530.

Scott, R. C., Nicolas, J. P., Bromwich, D. H., Norris, J. R., & Lubin, D. (2019). Meteorological drivers and large-scale climate forcing of West Antarctic surface melt. *Journal of Climate*, *32*(3), 665–684.

Shao, Z. D., & Ke, C. Q. (2015). Spring–summer albedo variations of Antarctic sea ice from 1982 to 2009. *Environmental Research Letters*, *10*(6), 064001.

Shepherd, A., Gilbert, L., Muir, A. S., Konrad, H., McMillan, M., Slater, T., ... Engdahl, M. E. (2019). Trends in Antarctic Ice Sheet elevation and mass. *Geophysical Research Letters*, *46*(14), 8174–8183.

Shepherd, A., Ivins, E., Rignot, E., Smith, B., Van Den Broeke, M., Velicogna, I., ... Wouters, B. (2018). Mass balance of the Antarctic Ice Sheet from 1992 to 2017. *Nature*, *558*(7709), 219–222.

Shepherd, A., & Wingham, D. (2007). Recent sea-level contributions of the Antarctic and Greenland ice sheets. *Science*, *315*(5818), 1529–1532.

Shepherd, A., Wingham, D., & Rignot, E. (2004). Warm ocean is eroding West Antarctic ice sheet. *Geophysical Research Letters*, *31*(23).

Shepherd, A., Wingham, D., Wallis, D., Giles, K., Laxon, S., & Sundal, A. V. (2010). Recent loss of floating ice and the consequent sea level contribution. *Geophysical Research Letters*, *37*(13).

Simmonds, I., & Keay, K. (2000). Mean Southern Hemisphere extratropical cyclone behavior in the 40-year NCEP–NCAR reanalysis. *Journal of Climate*, *13*(5), 873–885.

Simpkins, G. R., McGregor, S., Taschetto, A. S., Ciasto, L. M., & England, M. H. (2014). Tropical connections to climatic change in the extratropical Southern Hemisphere: The role of Atlantic SST trends. *Journal of Climate*, *27*(13), 4923–4936.

Speirs, J. C., Steinhoff, D. F., McGowan, H. A., Bromwich, D. H., & Monaghan, A. J. (2010). Foehn winds in the McMurdo Dry Valleys, Antarctica: The origin of extreme warming events. *Journal of Climate*, *23*(13), 3577–3598.

Stammerjohn, S., Massom, R., Rind, D., & Martinson, D. (2012). Regions of rapid sea ice change: an inter-hemispheric seasonal comparison. *Geophysical Research Letters*, *39*. Available from https://doi.org/10.1029/2012gl050874.

Stokes, C. R., Sanderson, J. E., Miles, B. W., Jamieson, S. S., & Leeson, A. A. (2019). Widespread distribution of supraglacial lakes around the margin of the East Antarctic Ice Sheet. *Scientific Reports*, *9*(1), 13823.

Suryawanshi, K., Jena, B., Bajish, C.C. and Anilkumar, N., 2023. Recent Decline in Antarctic Sea Ice Cover From 2016 to 2022: Insights From Satellite Observations, Argo Floats, and Model Reanalysis.

Swetha Chittella, S. P., Deb, P., & Melchior van Wessem, J. (2022). Relative contribution of atmospheric drivers to "extreme" snowfall over the Amundsen Sea Embayment. *Geophysical Research Letters*, *49*(16), e2022GL098661.

Tang, M. S. Y., Chenoli, S. N., Colwell, S., Grant, R., Simms, M., Law, J., et al. (2018). Precipitation instruments at Rothera Station, Antarctic Peninsula: A comparative study. *Polar Research*, *37*(1), 1503906.

Tedesco, M. (2009). Assessment and development of snowmelt retrieval algorithms over Antarctica from K-band spaceborne brightness temperature (1979–2008). *Remote Sensing of Environment*, *113*(5), 979–997.

Tedesco, M., Abdalati, W., & Zwally, H. J. (2007). Persistent surface snowmelt over Antarctica (1987–2006) from 19.35 GHz brightness temperatures. *Geophysical Research Letters*, *34*(18).

Tedesco, M., & Monaghan, A. J. (2009). An updated Antarctic melt record through 2009 and its linkages to high-latitude and tropical climate variability. *Geophysical Research Letters*, *36*(18).

Thompson, D. W., Solomon, S., Kushner, P. J., England, M. H., Grise, K. M., & Karoly, D. J. (2011). Signatures of the Antarctic ozone hole in Southern Hemisphere surface climate change. *Nature Geoscience*, *4*(11), 741–749.

Torinesi, O., Fily, M., & Genthon, C. (2003). Variability and trends of the summer melt period of Antarctic ice margins since 1980 from microwave sensors. *Journal of Climate*, *16*(7), 1047–1060.

Trusel, L. D., Frey, K. E., & Das, S. B. (2012). Antarctic surface melting dynamics: Enhanced perspectives from radar scatterometer data. *Journal of Geophysical Research: Earth Surface*, *117*(F2).

Trusel, L. D., Frey, K. E., Das, S. B., Karnauskas, K. B., Kuipers Munneke, P., Van Meijgaard, E., & Van Den Broeke, M. R. (2015). Divergent trajectories of Antarctic surface melt under two twenty-first-century climate scenarios. *Nature Geoscience*, *8*(12), 927–932.

Tuckett, P. A., Ely, J. C., Sole, A. J., Livingstone, S. J., Davison, B. J., Melchior van Wessem, J., & Howard, J. (2019). Rapid accelerations of Antarctic Peninsula outlet glaciers driven by surface melt. *Nature Communications*, *10*(1), 4311.

Turner, J., Colwell, S. R., Marshall, G. J., Lachlan-Cope, T. A., Carleton, A. M., Jones, P. D., … Iagovkina, S. (2004). The SCAR READER project: Toward a high-quality database of mean Antarctic meteorological observations. *Journal of Climate*, *17*(14), 2890–2898.

Turner, J., Colwell, S. R., Marshall, G. J., Lachlan-Cope, T. A., Carleton, A. M., Jones, P. D., … Iagovkina, S. (2005). Antarctic climate change during the last 50 years. *International Journal of Climatology*, *25*(3), 279–294.

Turner, J., & Coauthors. (2009). Non-annular atmospheric circulation change induced by stratospheric ozone depletion and its role in the recent increase of Antarctic sea ice extent. *Geophysical Research Letters*, *36*, L08502. Available from https://doi.org/10.1029/2009GL037524.

Turner, J., Lu, H., White, I., King, J. C., Phillips, T., Hosking, J. S., … Deb, P. (2016). Absence of 21st century warming on Antarctic Peninsula consistent with natural variability. *Nature*, *535*(7612), 411–415.

Turner, J., Phillips, T., Thamban, M., Rahaman, W., Marshall, G. J., Wille, J. D., … Lachlan-Cope, T. (2019). The dominant role of extreme precipitation events in Antarctic snowfall variability. *Geophysical Research Letters*, *46*(6), 3502–3511.

Vaughan, D. G., Bamber, J. L., Giovinetto, M., Russell, J., & Cooper, A. P. R. (1999). Reassessment of net surface mass balance in Antarctica. *Journal of Climate*, *12*(4), 933–946.

Vavrus, S. J., Notaro, M., & Lorenz, D. J. (2015). Interpreting climate model projections of extreme weather events. *Weather and Climate Extremes*, *10*, 10–28.

Van de Wal, R. S. W., Greuell, W., van den Broeke, M. R., Reijmer, C. H., & Oerlemans, J. (2005). Surface mass-balance observations and automatic weather station data along a transect near Kangerlussuaq, West Greenland. *Annals of Glaciology*, *42*, 311–316.

Wang, G., Hendon, H. H., Arblaster, J. M., Lim, E. P., Abhik, S., & van Rensch, P. (2019). Compounding tropical and stratospheric forcing of the record low Antarctic sea-ice in 2016. *Nature Communications*, *10*(1), 13.

Weertman, J. (1974). Stability of the junction of an ice sheet and an ice shelf. *Journal of Glaciology*, *13*(67), 3–11.

Wille, J. D., Favier, V., Dufour, A., Gorodetskaya, I. V., Turner, J., Agosta, C., & Codron, F. (2019). West Antarctic surface melt triggered by atmospheric rivers. *Nature Geoscience*, *12*(11), 911–916.

Wille, J. D., Favier, V., Jourdain, N. C., Kittel, C., Turton, J. V., Agosta, C., … Amory, C. (2022). Intense atmospheric rivers can weaken ice shelf stability at the Antarctic Peninsula. *Communications Earth & Environment*, *3*(1), 90.

Wingham, D. J., Shepherd, A., Muir, A., & Marshall, G. J. (2006). Mass balance of the Antarctic ice sheet. *Philosophical Transactions of the Royal Society of London Series A*, *364*(1844), 1627–1635.

Yu, L., Zhong, S., Vihma, T., Sui, C., & Sun, B. (2022). The impact of the Indian Ocean Basin mode on Antarctic sea ice concentration in interannual time scales. *Geophysical Research Letters*, *49*(11), e2022GL097745.

Yuan, X. (2004). ENSO-related impacts on Antarctic sea ice: A synthesis of phenomenon and mechanisms. *Antarctic Science*, *16*(4), 415–425.

Yuan, X., Kaplan, M. R., & Cane, M. A. (2018). The interconnected global climate system—A review of tropical–polar teleconnections. *Journal of Climate*, *31*(15), 5765–5792.

Yuan, X., & Martinson, D. G. (2001). The Antarctic dipole and its predictability. *Geophysical Research Letters*, *28*(18), 3609–3612.

Zambri, B., Solomon, S., Thompson, D. W., & Fu, Q. (2021). Emergence of Southern Hemisphere stratospheric circulation changes in response to ozone recovery. *Nature Geoscience*, *14*(9), 638–644.

Zhou, C., Zhang, T., & Zheng, L. (2019). The characteristics of surface albedo change trends over the Antarctic sea ice region during recent decades. *Remote Sensing*, *11*(7), 821.

Zhu, Y., & Newell, R. E. (1998). A proposed algorithm for moisture fluxes from atmospheric rivers. *Monthly Weather Review*, *126*(3), 725–735.

Zou, X., Bromwich, D. H., Montenegro, A., Wang, S. H., & Bai, L. (2021). Major surface melting over the Ross Ice Shelf part I: Foehn effect. *Quarterly Journal of the Royal Meteorological Society*, *147*(738), 2874–2894.

Zwally, H. J., & Fiegles, S. (1994). Extent and duration of Antarctic surface melting. *Journal of Glaciology*, *40*(136), 463–475.

Zwally, H. J., Comiso, J. C., Parkinson, C. L., Cavalieri, D. J., & Gloersen, P. (2002). Variability of the Antarctic sea ice cover. *Journal of Geophysical Research*, *107*, 1029–1047. Available from https://doi.org/10.1029/2000JC000733.

Chapter 19

Sea level change in Asia

Dhrubajyoti Samanta
Earth Observatory of Singapore, Nanyang Technological University, Singapore

19.1 Introduction

Sea level rise is a crucial indicator of anthropogenic climate change (Fox-Kemper et al., 2021). Consequently, it is a subject of great interest to both researchers and policymakers. Changes in global mean sea level (GMSL) result from several contributing factors involving ocean, cryosphere, and solid Earth, making this a uniquely complex phenomenon (Stammer, Cazenave, Ponte, & Tamisiea, 2013). Observations from satellite altimetry and tide gauge measurements reveal that GMSL has increased by approximately 110 mm since 1993 until 2023 (e.g., Cazenave & Moreira, 2022; Guérou et al., 2023; Hamlington et al., 2024), primarily due to the thermal expansion of ocean waters and increasing ocean mass due to land ice melt (Fox-Kemper et al., 2021). Global climate models project a continued acceleration in GMSL during the 21st century unless substantial reductions in greenhouse gas emissions are achieved (Hermans et al., 2021; Horton et al., 2020; Nerem et al., 2018; Oppenheimer et al., 2019). However, sea level rise is not spatially uniform across the globe (Fasullo & Nerem, 2018; Hamlington et al., 2020; Stammer et al., 2013), owing to various regional to local scale processes, such as ocean dynamics, vertical land motion, and land water storage (Church et al., 2013; Couldrey et al., 2021; Horton et al., 2018). The Intergovernmental Panel on Climate Change's Sixth Assessment Report (IPCC AR6) highlights that regional deviations in sea level projections for the 21st century can reach up to $\pm 20\%$ from the GMSL in many coastal regions, emphasizing the importance of understanding regional to local sea level dynamics and their relevance to coastal communities (Fox-Kemper et al., 2021; Hamlington et al., 2020).

Advances in understanding the processes driving GMSL changes and their regional variability have led to significant improvements in models used for projection of future climate changes. To assess the accuracy and reliability of our understanding about sea level change, an examination of the sea level budget is indispensable. Closure of the sea level budget implies that the observed changes of GMSL equal the sum of observed or estimated contributions (e.g., Horwath et al., 2021). Due to improvements in sea level observational systems, the contributing processes to sea level change are now well known, and closing the GMSL budget is no longer challenging (Frederikse et al., 2020). Satellite observations, along with in situ measurements, have also led to substantial improvement in understanding processes causing regional differences in sea level changes (Hamlington et al., 2020). Importantly, the drivers of GMSL changes have a regional signature and typically vary over time. Therefore, separating the contributors spatially and temporally is essential when planning for a particular time horizon to mitigate the impact of sea level changes (Hamlington et al., 2020).

Asia has a population of more than 4.82 billion (https://www.worldometers.info/world-population/asia-population/) accounting for $\sim 60\%$ of the world's population. South and Southeast Asia have very highly dense coastal populations vulnerable to sea level rise (Fig. 19.1A). Notably, in most Asian coasts, the sea level rise trend in the last 30 years is higher than the global average (Fig. 19.1B), suggesting a higher flooding risk in the region.

Despite the abundance of original research and comprehensive reviews on sea level studies (e.g., Hamlington et al., 2020; Kopp et al., 2015; Slangen et al., 2017), there is a need for a synthesis of the current state of knowledge in Asia, especially following the publication of IPCC AR6. In this review, the author summarizes aspects of contemporary Asian sea level change and its projections, underlying mechanisms, and coastal impacts. The author also discusses the implications and requirements for adaptation and mitigation pathways. Finally, the author includes two annexes that may be useful for new researchers in the field of sea level studies.

Sustainable Development Perspectives in Earth Observation. DOI: https://doi.org/10.1016/B978-0-443-14072-3.00015-0

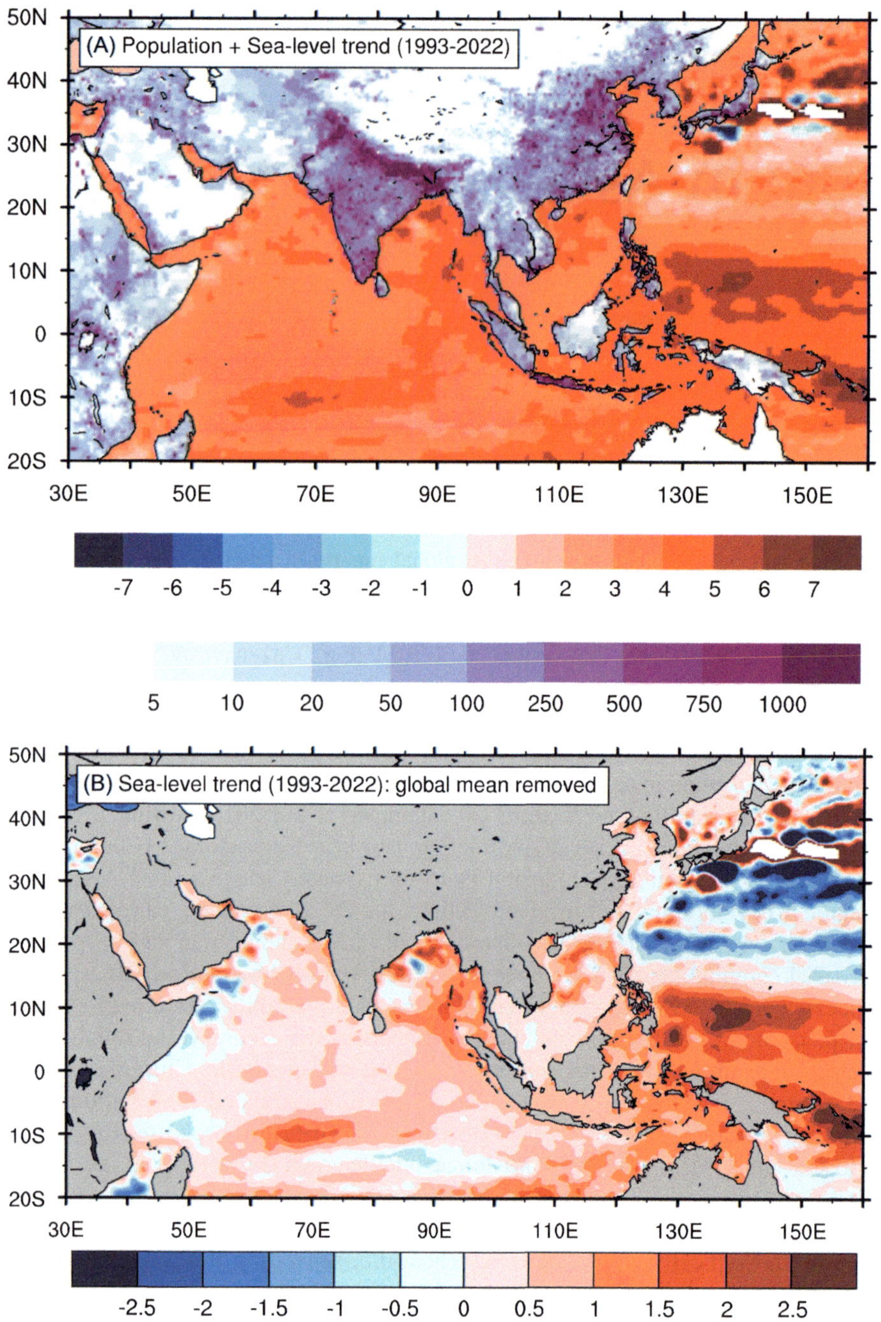

FIGURE 19.1 Association of higher sea level trend (in mm/year) and population density (in person/km^2) over Asian region. (A) Sea level trend from satellite observations during 1993–2022 (shaded over oceans) and 2020 population density (shaded over land); (B) sea level trend during 1993–2022 after removing global mean sea level trend (3.4 mm/year). Map lines delineate study areas and do not necessarily depict accepted national boundaries. *Sea level data are is from two-satellite merged constellation (Copernicus Climate Change Service [(C3S])) available from https://cds.climate.copernicus.eu/cdsapp#!/dataset/satellite-sea-level-global?tab=form. Gridded population data areis available from https://sedac.ciesin.columbia.edu/data/set/gpw-v4-population-density-rev11/data-download.*

19.2 Drivers of sea level change

Sea level change is governed by a complex interplay of global and regional processes, each contributing to its variability and impacts. During the 20th century sea level rise was predominantly driven by thermosteric effects, which are the expansion of seawater due to the warming of ocean temperatures (Fox-Kemper et al., 2021). However, in the 21st century, sea level rise is projected to be primarily controlled by barystatic sea level, resulting from the addition of water to the oceans from the melting of the ice sheets and glaciers (Fox-Kemper et al., 2021).

Understanding the implications of a warming climate on coastal regions requires examining both GMSL and regional relative sea level. Regional relative sea level change is influenced by a multitude of physical processes, which are spatially and temporally variable, deviating from the long-term rate of GMSL rise. Essentially, regional relative sea level is a combination of regional and global drivers.

$$\text{Regional relative sea level} = \text{sterodynamic variability} + \text{short-term effects} + \text{glaciers} \\ + \text{land water storage} + \text{ice sheets} + \text{subsidence}$$

Regional sterodynamic sea level is the combination of global thermosteric sea level and regional dynamic sea level. In other words, the regional sterodynamic sea level is the sum of the regional thermosteric sea level, regional halosteric sea level, and regional manometric sea level (Gregory et al., 2019; Jin, Köhl, Jungclaus, & Stammer, 2023). Dynamic sea level involves variability in the ocean's circulation and density. Multiscale climate variability can modulate the sterodynamic sea level at different time scales. Short-term effects encompass variations in sea level occurring over periods ranging from minutes to weeks. These arise from processes like storm surges, wave runup, wave setup, astronomical tides, and freshwater input. Glaciers outside the ice sheets contribute $\sim 1\%$ of the total ice trapped on land. These glacier-related changes are expressed regionally through gravitational, rotational, and deformational (GRD) alternations, characterized by unique fingerprints or patterns. Land water storage pertains to changes in the transfer of water between land and ocean. It includes variability in groundwater withdrawal, water impoundment, and the global water cycle. Regionally, these changes manifest through distinctive GRD patterns or fingerprints. The Greenland and Antarctic ice sheets significantly contribute to the GMSL rise. The regional manifestations of changes induced by ice sheets are identified through GRD patterns or fingerprints. Subsidence refers to the movement of the coastal land, which can also include uplift. It is a response to various physical processes, including glacial isostatic adjustment, groundwater, and hydrocarbon withdrawal, and even tectonic activity. Subsidence changes can exhibit substantial variability over time and space, and partially attributable to anthropogenic activities.

On a regional scale, dynamic sea level plays a pivotal role in shaping regional sea level (e.g., Gregory et al., 2019). However, IPCC AR6 expressed medium confidence in our understanding of the oceanic processes driving dynamic sea level changes (Fox-Kemper et al., 2021). Dynamic sea level represents sea level change relative to the geoid and is determined by the dynamic equilibrium between ocean circulation and density distribution (Gregory et al., 2019). Dynamic sea level simulation in climate models relies on the ocean's dynamic responses to changing radiative forcing, as simulated by global climate models (Lyu, Zhang, & Church, 2020). The foundation of GMSL projections lies in ensembles of complex global climate models from CMIP5 (Taylor et al., 2012) and CMIP6 (Eyring et al., 2016). The global mean dynamic sea level is considered to be zero. Researchers have investigated multiple factors influencing dynamic sea level, such as changes in heat flux (Couldrey et al., 2021; Todd et al., 2020), upper-ocean stratification (Little, Horton, Kopp, Oppenheimer, & Yip, 2015), internal variability, and external forcings (Ferrero, Tonelli, Marcello, & Wainer, 2021).

Regionally, sea level can vary up to $\pm 20\%$ (Fox-Kemper et al., 2021) from GMSL due to various geophysical processes. The Indian Ocean sea level changes are affected by natural modes of climate variability as well as anthropogenic influence (e.g., Han et al., 2010). Nidheesh et al. (2013) argued that thermosteric sea level is the major contributor to the spatial patterns of sea level variability in the Indian Ocean, except for the Bay of Bengal and the south-east Indian Ocean, where the halosteric component is important (e.g., Llovel & Lee, 2015). In the south-east Indian Ocean, halosteric sea level rise is $\sim 6.41 \pm 0.62$ mm/year, relative to 3.72 ± 1.04 mm/year contribution from the thermosteric component (Llovel & Lee, 2015). The redistribution of ocean mass or manometric sea level serves as a counterbalance to the contrasting sea level changes induced by steric effects, occurring both in the deep South China Sea and on shallow continental shelves (Thompson, Jevrejeva, Zachariah, Faller, & Tkalich, 2023). Indeed, using regional ocean model simulations from 1961 to 2022, Thomson et al. (2023) argued that about 80% of the sea level changes over the South China Sea shelves are from manometric sea level.

19.3 Contemporary sea level change

The GMSL rose faster in the 20th century than in any earlier century over the last three millennia (Fox-Kemper et al., 2021). There have been a substantial number of studies of 20th century GMSL rise based on analysis of past tide gauge data. Those studies (e.g., Church et al., 2001; Frederikse et al., 2020) showed a GMSL rise of ~ 1.5 mm/year. This rise has accelerated recently, particularly in the altimetry era (e.g., Hamlington et al., 2024). Accurate assessments of present-day GMSL changes and understanding its driving factors are essential. Estimated GMSL rates are shown in Table 19.1 which displays the total and individual contributions at the global scale.

On a regional scale, earlier studies (e.g., Unnikrishnan & Shankar, 2007) showed a rate of sea level rise of about 1.06–1.75 mm/year in the Indian Ocean during 1874–2004, consistent with the global rising trend of 1.7 mm/year

TABLE 19.1 Estimation of global mean sea level (GMSL).

Period	Estimate and their uncertainty (mm/year)	References	Data source
Total sea level			
1993–2020 1900–2018	3.34 ± 0.40 1.56 ± 0.30	Frederikse et al. (2020)	Satellite altimetry Tide gauges
Steric sea level			
2005–19 1900–2018	1.10 ± 0.20 0.52 ± 0.18	Roemmich and Gilson (2009) Frederikse et al. (2020)	Argo floats In situ observations
Ocean mass			
2002–20 1900–2018	2.10 ± 0.30 1.00 ± 0.30	Watkins et al. (2015) Frederikse et al. (2020)	GRACE/GRACE-FO Satellite and in situ observations
Greenland mass loss			
2002–20 1900–2020	0.78 ± 0.06 0.44 ± 0.04	Watkins et al. (2015) Frederikse et al. (2020)	GRACE/GRACE-FO Satellite and in situ observations
Antarctic mass loss			
2002–20 1900–2020	0.41 ± 0.11 0.08 ± 0.08	Watkins et al. (2015) Frederikse et al. (2020)	GRACE/GRACE-FO Satellite and in situ observations
Glacier mass loss			
2002–16 1900–2018	0.57 ± 0.10 0.70 ± 0.18	Wouters et al. (2019) Frederikse et al. (2020)	GRACE Satellite and in situ observations
Land water storage			
2002–14 1900–2018	− 0.33 ± 0.16 − 0.21 ± 0.13	Reager et al. (2016) Frederikse et al. (2020)	GRACE Satellite and in situ observations

Uncertainty estimates indicates the 90% confidence interval.

estimated for 1880–2009 (Church & White, 2011). However, Han et al. (2010) showed a higher sea level rise trend along the Indian Ocean coasts since the 1960s after corrections for vertical land movement from glacial isostatic adjustment and barometric pressure correction, except for the fall at Zanzibar.

During the era of satellite altimetry, the fastest sea level rise on a basin scale occurred in the western Pacific Ocean (Fig. 19.1; Fox-Kemper et al., 2021). In the past 30 years (1993–2022), the sea level rise rate in almost the entire Asian region has been higher than GMSL rise (Fig. 19.1B). The east coast of Japan has the highest sea level trend globally in the satellite era, reaching up to 9–10 mm/year. The east coast of Philippines also has ~7 mm/year trend over the same period, whereas the rest of the Asian coast has a sea level rise trend of 3–6 mm/year from 1993 to 2022 (Fig. 19.1). The present rate of sea level rise over the entire Asian region is higher than previous reviews on the Indian Ocean sea level trend (e.g., Swapna et al., 2020). Observations also demonstrated that regional high tide flooding events were more than eight times per year during 1995–2014 compared to five times per year during 1960–1980 (Fox-Kemper et al., 2021).

Understanding all the contributing factors to sea level changes and closing its budget is an important aspect of impact assessment and resource management. Frederikse et al. (2020) reconstructed GMSL since 1900 and compared it to the sum of contributing processes. They successfully showed the closure of 20th century sea level budget, implying that no additional unknown processes, such as large-scale deep-ocean thermal expansion or additional mass loss from the Antarctic ice sheet are required to explain the observed GMSL. Their demonstration of GMSL budget closure provides a consistent baseline against which semiempirical and process-based sea level projections can be benchmarked.

However, on the regional scale, such as over the Asian region, sea level budget closure is rare and challenging. That is one of the causes of inconsistent sea level rise trends between tide gauge records and satellite altimetry trends, such as in the discussion by Swapna et al. (2020). While the long-pending challenge of closing the GMSL budget is closed by Frederikse et al. (2020), closing regional sea level budgets is still challenging due to inadequate understanding of regional to local scale contributing factors. In a process-based assessment of sea level rise in the northwestern Pacific marginal seas, Cha et al. (2023) reveal that changes in ocean circulation have led to significant sterodynamic sea level changes along continental shelves, driven by ocean mass redistribution. They highlight that ocean mass distribution between the deep ocean and shallow marginal seas plays an important role in driving regional sea level rise and variability. Sea level budget can be reasonably close to observations in the Yello and East China Seas (Cha et al., 2023). Yet, challenges persist in the South China Sea, mainly attributable to insufficient observational data and inconsistencies in different datasets in the region. Nonetheless, with continued efforts to enhance satellite and in situ measurements, a better understanding, and prediction of sea level changes, especially in the Asian region is feasible.

19.3.1 Observation methods of sea level change

The observation of sea level change builds on the fundamental physical principles established in earlier centuries, as well as in modern near-global observations. Regional sea level changes are resolved by both tide gauge and satellite altimetry observations (Hamlington et al., 2020). Until 1993, sea level observations were primarily based on a network of tide gauge stations, some with over 100 years of continuous records. However, in many regions, especially in Asia, these records are either short or discontinuous due to several challenges, such as funding limitations, institutional coordination, and public awareness and engagement. It is noteworthy that the tide gauge measurements are referenced to a local vertical datum, which is a fixed reference point on the land. This reference point can shift over time due to vertical land motion. In contrast, satellite altimeters measure geocentric sea level, relative to the Earth's center of the mass. The differences between these two sea level measurements can be crucial for acheiving a comprehensive understanding and accurate projections of sea level change. Since 1992, changes in GMSL and regional patterns have been meticulously monitored using satellite altimeters. TOPEX/Poseidon was the first satellite altimeter to provide accurate, near-global sea level measurements every 10 days. With the success of the Sentinel-6 Michael Freilich satellite, the satellite altimetry record of sea level change has now surpassed 30 years in length (Hamlington, Willis, & Vinogradova, 2023). Observations indicate that the GMSL rise trend of 3.4 mm/year is higher than pre-satellite era estimates.

To gain a comprehensive understanding of the factors contributing to GMSL rise, various components of the sea level observation network are utilized. The GRACE and GRACE-FO satellites play a crucial role in this by measuring mass-driven GMSL changes. Such as ice melt and variations in land water storage. Additionally, Argo profiling floats monitor ocean temperature and salinity changes, enabling estimates of thermosteric and halosteric sea level changes. By combining the GMSL estimates from GRACE/GRACE-FO and Argo with satellite altimetry data, researchers can validate their understanding of GMSL contributions and improve the accuracy of sea level budgets.

Efforts to enhance satellite observations of sea level changes focus on measuring smaller scales and obtaining data closer to the coast (Hamlington et al., 2023). For example, ICESat-2 extends the records of regional glacier mass change after ICESat and operation ICEBridge. The NASA-ISRO synthetic aperture radar (NISAR) provides all-weather, day/night imaging of the Earth repeated four to six times per month, resolving coastal inundation with a horizontal resolution of 3–10 m. NISAR is immensely helpful for monitoring changes in Antarctic and Greenland ice sheet flow. Moreover, NISAR data, when linked with extreme sea level data from tide gauges and satellites, can help improve digital elevation models and hydrodynamic flood models, as well as for new estimates of coastal vertical land motion at high resolution. The biggest step forward in sea level observation using satellites occurred with the launch of surface water and ocean topography (SWOT) mission in December 2022. The SWOT provides a better understanding of changing regional hydrology such as the inclusion of human activities and water consumption. As SWOT provides near-coast measurements of sea level, it fills the gap between the open-ocean measurements of altimetry and the coastal measurements of tide gauge (Hamlington et al., 2020). As the satellite-based sea level records continue to lengthen, the ability to isolate natural variability from long-term sea level rise continues to improve (Hamlington et al., 2023). The international partnership supporting altimeter records, such as a commitment to launch Sentinel-6B and beyond, will provide a multidecadal and multigenerational look at rising seas.

The development of new technologies, such as global navigation satellite system (GNSS), provides important information on vertical land motion, mitigating the limitations of tide gauge measurements. The Global Extreme Sea Level Analysis (GESLA) project further expands the availability of extreme sea level data from tide gauges worldwide (Woodworth et al., 2016).

Despite these advancements, the brevity of the instrumental records limits analyses of low-frequency sea level variability, especially in multidecadal to centennial time scales. The paleo sea level reconstructions offer valuable insihgts into global and regional sea level change, particularly over data-sparse regions. For example, using coral microatolls Majewski et al. (2022) extended sea level records in Southeast Asia. Chua et al. (2021) reconstructed the Holocene sea level records in Singapore based on $\sim$40 m long sediment core. Shaw (2023) provided a comprehensive review of quaternary relative sea level changes in the tropics. Moreover, studies on sea level surveys based on experts' judgments provide an additional way of understanding sea level changes (e.g., Horton et al., 2020). However, these paleo and survey-based methods are beyond the primary focus of this discussion.

19.3.2 Temporal variability in contemporary sea level changes and climate variabilities

Sea level over the Asian region shows large regional variability in all temporal scales, interannual to decadal and multidecadal scales. In addition to the long-term trend, sea level in the Indian Ocean exhibits substantial temporal variability (e.g., Swapna et al., 2020). Steric effects are the primary contributor to sea level changes in the Indian Ocean, which is linked with interannual modes of climate variability such as the El Niño Southern Oscillation (ENSO) and the Indian Ocean Dipole (IOD). Decadal sea level variability is driven mainly by the surface wind forcing over the Indo-Pacific Ocean (e.g., Han et al., 2019; Nidheesh, Lengaigne, Vialard, Unnikrishnan, and Dayan, 2013). The decadal sea level variability is particularly noticeable in the southwestern tropical Indian Ocean over the thermocline ridge region or ocean upwelling region (Vialard et al., 2009), predominantly driven by Pacific Decadal Oscillation (Deepa et al., 2019). In the southern Indian Ocean, the dominance of decadal ENSO is reported in the literature (Swapna et al., 2020), with Indonesian Throughflow as a significant conduit linking the Pacific and Indian Oceans (e.g., Samanta, Goodkin, & Karnauskas, 2021). The Indonesian Throughflow is often referred to as the oceanic bridge between these oceans and plays a key role in sea level variability in the southeastern Indian Ocean (e.g., Deepa et al., 2019; Han et al., 2010). Swapna et al. (2020) emphasized that the representation of decadal sea level variability in the Indian Ocean in observation-based sea level products, including reanalysis and reconstructions, varies among products due to limited observational data in the region. For example, in the Southeast Asian seas particularly in the shallow ocean, ocean observations are historically very limited or even absent in some regions. Therefore a cautious interpretation should be made, considering the limitations of observational data sampling.

Over a multidecadal timescale, sea level variability in the Indian Ocean is primarily influenced by thermosteric changes driven by surface winds (Swapna et al., 2020) and is correlated with the strength of the summer monsoon circulation. In addition to ENSO and IOD on an interannual scale, Southeast Asia is subject to strong seasonal variability in sea level corresponding to two monsoon seasons. For example, the Gulf of Thailand experiences ~40 cm of sea level rise in the winter (northeast) monsoon season compared to summer (southwest) monsoon. Qu, Jevrejeva, Williams, and Moore (2022) showed in addition to the South China Sea, the Bohai, Yellow, and East China Seas are also subject to strong seasonal variability in sea level. They further showed that the seasonal cycle accounts for 37%–94% of monthly sea level variability in their studied 81 tide gauge records in the region. Qu et al. (2022) found that northern China, Taiwan, Korea, Japan, and Philippines experience dominant influence from inverse barometer effects that are linked with seasonality along with a steric contribution. Furthermore, seasonal monsoons associated with barotropic response and freshwater exchange play a critical role in eastern and southern Chinese coasts, east Malaysia, Vietnam, and the Gulf of Thailand. Seasonal cycle amplitudes, corrected for the inverse barometer effect, are more pronounced in continental shelf areas compared to the deep ocean, with the maximum observed in the Gulf of Thailand (Qu et al., 2022).

The connection between long-term changes in wind patterns and anthropogenic climate change in the Indo-Pacific region has been a topic of discussion among researchers. Although there is consensus that sea level trends in the region over the past two decades are closely linked to intensified wind stress, questions have been raised regarding whether the persistence of these recent wind patterns is associated with anthropogenic factors (Stammer et al., 2013). Merrifield and Maltrud (2011) argued that during 1990–2010, sea level trends increased in the western tropical Pacific Ocean due to the intensified trade winds and Pacific Ocean circulation, indicating a response to anthropogenic global warming rather than natural climate variability. Similarly, Han et al. (2010) proposed that changes in regional sea levels in the Indian Ocean since 1960 are driven by altered wind patterns associated with enhanced Hadley and Walker circulations. However, reaching such conclusions requires caution due to reported variations in wind regime and sea level over a multidecadal timescale in the Indo-Pacific region, including a trend reversal in the early 1990s (Stammer et al., 2013). This shift was communicated between the Indian and Pacific Ocean basins through wave transmission across the Indonesian archipelago (Stammer et al., 2013).

19.4 Attribution of historical sea level changes to anthropogenic forcings

The knowledge of drivers of past climate changes enhances confidence in climate projections. However, our limited knowledge of historical changes in response to anthropogenic forcing poses a major barrier to sea level projections (Samanta, Jevrejeva et al., 2024). Global and regional sea level changes are influenced by a variety of anthropogenic and natural forcings (e.g., Dangendorf et al., 2015; Slangen et al., 2014). The IPCC AR6 highlights evidence of anthropogenic influence on the two primary drivers of sea level rise since 1950; the thermal expansion of ocean water (thermosteric effect) and the melting of ice sheets and glaciers (Fox-Kemper et al., 2021).

Prior research on attributing sea level change to anthropogenic forcing predominantly focuses on GMSL. Tokarska, Hegerl, Schurer, Ribes, and Fasullo (2019) showed that GMSL along with ocean warming is detectable over the historical period and attributable to anthropogenic forcings. On a longer timescale, Jevrejeva, Grinsted, and Moore (2009) assessed the influence of various natural and anthropogenic forcings on GMSL over the past 1000 years based on a statistical model and argued that in the past 200 years, GMSL rise has been mostly associated with anthropogenic forcings. Similarly, using optimal fingerprinting, Slangen et al. (2014) showed that anthropogenic forcings are necessary to account for the observed magnitude of global mean thermosteric sea level rise during 1957–2005. Expanding on these results, Marzeion, Cogley, Richter, and Parkes (2014) reported a significant increase in the anthropogenic contribution to glacial mass loss over the recent decades. Dangendorf et al. (2015) provided compelling evidence of a substantial anthropogenic footprint on GMSL rise since 1900 by examining the spectral properties of both the slowly varying volumetric component and the rapidly changing atmospheric component. From a different perspective, Walker, Kopp, Little, and Horton (2022) utilized proxy sea level records from the Common Era (0–2000 CE) to demonstrate that the time of emergence of sea level rise surpassing preindustrial rates aligns closely with early ocean warming and glacier melting.

Detecting anthropogenic influences on a regional scale sea level is more challenging than on a global scale due to the different expressions of natural climate variability (Hamlington et al., 2020; Richter et al., 2020). Natural climate variability can emerge over multiple timescales without the impact of external forces from anthropogenic activities. For example, the regional sea level in the tropical Pacific Ocean is predominantly influenced by decadal-scale natural variability rather than external forcing (Meyssignac, Salas y Melia, Becker, Llovel, & Cazenave, 2012), particularly the Pacific decadal oscillation (Hamlington et al., 2014; Scafetta, 2014). However, a forced signal can be detected at all latitudes of each major ocean basin (Marcos et al., 2017; Richter et al., 2020). Despite a substantial number of global scale studies, it is worth noting that regional sea level changes can diverge from the global average which holds relevance when considering the impacts on local communities (Fox-Kemper et al., 2021).

To address the gap of regional sea level attribution, Samanta, Vairagi et al. (2024) employed a large ensemble climate model simulation from the detection and attribution of model intercomparison project (DAMIP) under CMIP6 to examine spatiotemporal patterns of sea level changes in the Indo-Pacific warm pool region. This region mostly covers the highly populated Asian coasts. They emphasized the predominant influence of greenhouse gas-induced sea level changes in the historical period (1950–2014). However, they noted that the time of emergence of anthropogenic signals in sea level is not uniform across the Asian region. For example, the time of emergence of greenhouse gas signals in the western Indian Ocean is ~1925, whereas it is ~1965 in the eastern Indian Ocean. Samanta, Vairagi et al. (2024) further discovered the emergence of a positive Indian Ocean dipole-like pattern in recent decades in dynamic and dynamic sea level changes. They argued that sea level rise in deeper oceans of the Asian region is primarily governed by changes in ocean temperature (termed thermosteric) and salinity (termed halosteric). Conversely, greenhouse gas-induced manometric sea level rise is predominant in the continental shelves and shallow regions such as maritime Southeast Asia (Fig. 19.2; Samanta, Vairagi et al., 2024). With the simultaneous increase in greenhouse gas forcing and decrease in aerosol forcing in the 21st century (Forster et al., 2023), manometric sea level further may emerge as a stronger driver for sea level rise in the shallow oceans and continental shelves.

19.5 Sea level projections

The GMSL will continue to rise to 2100 and beyond, as all contributors to GMSL will remain active throughout this century (Fox-Kemper et al., 2021). For example, deep-ocean heat uptake and mass loss of the Greenland and Antarctic ice sheets will persist for several thousands of years (Fox-Kemper et al., 2021).

Projecting GMSL and regional sea level changes requires integrating various components influenced by natural and anthropogenic forcings (Fox-Kemper et al., 2021). Climate models, specifically global atmosphere-ocean-land-sea-ice are employed to assess the sea level changes arising from ocean warming, changes in ocean circulation, and water density. Additionally, other models (e.g., statistical models,) are used to evaluate variations in the mass of glaciers, ice sheets, and land water, which directly affect the ocean mass. These mass changes drive both GMSL and distinct regional patterns in sea level change, shaped by variations in Earth's shape, gravitational fields, and rotational dynamics in response to these significant mass transfers (Church & Zhang, 2023).

Sea level projections in IPCC AR6 and other studies are framed using socio-economic pathways (SSPs; O'Neill et al., 2016), which span low to high emissions scenarios and associated socio-economic conditions. These SSPs reflect scenarios from SSP1-1.9 to SSP5-8.5, with SSP5-8.5 representing the worst-case scenario with unchecked greenhouse

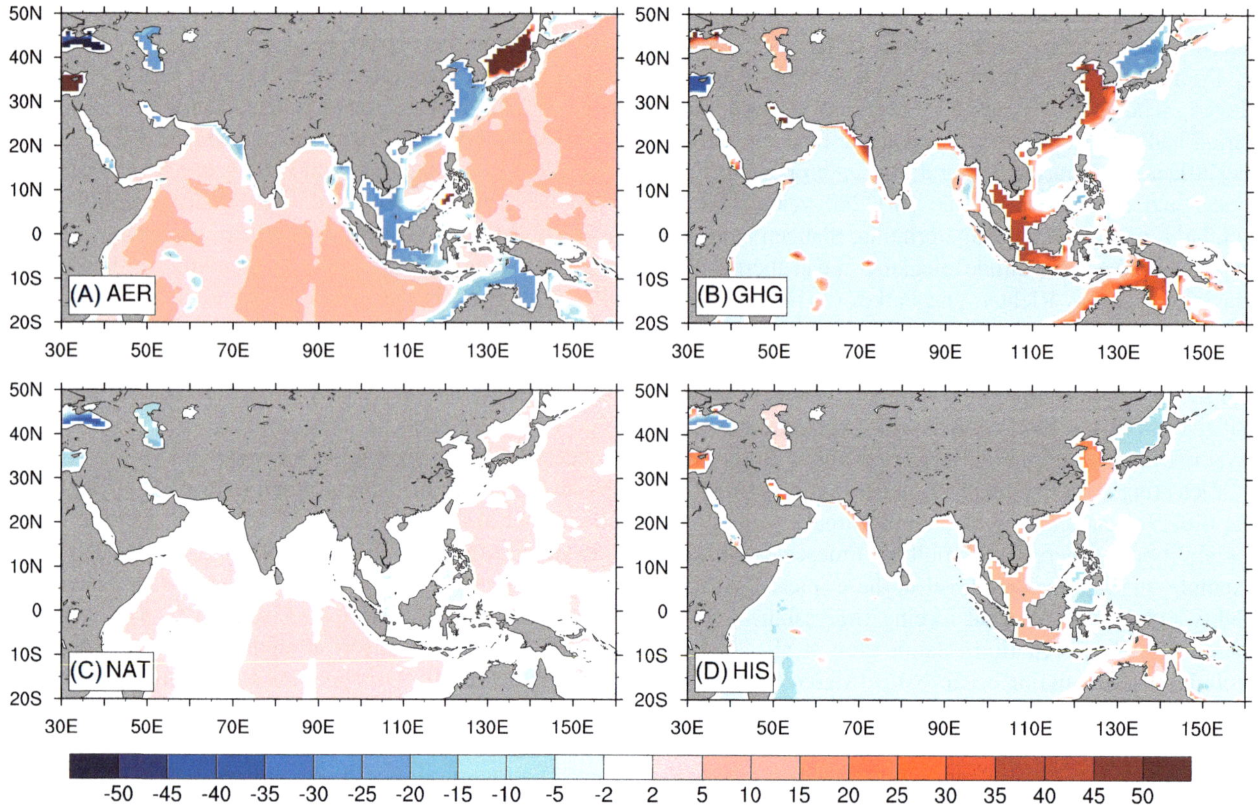

FIGURE 19.2 Manometric sea level change (1985–2014 minus 1950–1979; in mm) from ensemble mean (N = 30) CanESM from DAMIP simulation for (A) aerosol-only (AER), (B) greenhouse gas-only (GHG), (C) natural-only (NAT), and (D) historical (HIS) simulations. Historical manometric sea level change is the manifestation of the net impact of aerosol and greenhouse gas forcings. Map lines delineate study areas and do not necessarily depict accepted national boundaries.

gas emissions. The likely ranges in the IPCC AR6 are similar to those in AR5. The IPCC AR6 projections are relative to 1995–2014 and include a lower greenhouse gas scenario (SSP1-1.9) consistent with a global mean surface temperature increase of 1.4 °C by 2081–2100. To generate sea level projections, IPCC AR6 adopts multiple methods. For example, thermal expansion is based on a two-layer emulator with climate sensitivity and expansion coefficients calibrated to emulate CMIP6 models. Glacial isostatic adjustment and other vertical land motion from a spatiotemporal statistical model of tide gauge data (Kopp et al., 2014). Dynamic sea level changes are from CMIP6 model outputs. Whereas, the Greenland ice sheet component is from the emulated Ice Sheet Model Intercomparison Project (ISMIP6) of CMIP6 up to 2100, and the parametric model fit to ISMIP6 thereafter. The Antarctic ice sheet component is derived from emulated ISMIP6 and Linear Antarctic Respose Model Intercomparison Project simulations for medium confidence processes. Whereas, low confidence processes are based on single ice-sheet-model ensemble simulations and structured expert judgement (see Fox-Kemper et al., 2021 Table 9.7 for details).

IPCC AR6 suggests a GMSL rise ranking from 0.38 m in SSP1-1.9 to 1 m in SSP5-8.5 to ~2 m in SSP5-8.5 low confidence scenario (Fig. 19.3). The SSP5-8.5 low confidence scenario involves the uncertainty in ice sheet processes, and projects GMSL to rise by 5 m by 2150. Median projected relative sea level changes suggests about 60% (SSP1-1.9) to 70% (SSP5-8.5) of the global coastal sea level and has a projected 21st-century median within ± 20% of GMSL increase (Fox-Kemper et al., 2021). Alternative perspectives, such as temperature-based projections (e.g., Vermeer & Rahmstorf, 2009) links GMSL variation on decadal to century-scale global mean temperature.

Regional sea level projections, derived from global climate models (e.g., CMIP6), show greater-than-average rise near poleward boundaries of the subtropical gyres (Church & Zhang, 2023). Projections in IPCC AR6, including vertical land motions estimates (Kopp et al., 2014), are available for global coastal tide gauge locations for the global ocean from NASA sea level projection toolbox (https://sealevel.nasa.gov/ipcc-ar6-sea-level-projection-tool. However, IPCC AR6 (Fox-Kemper et al., 2021) cautioned that there is only "low to medium confidence" in vertical land motion and glacial isostatic adjustment assumptions, necessitating more detailed regional-to-local scale analysis. Tay et al. (2022)

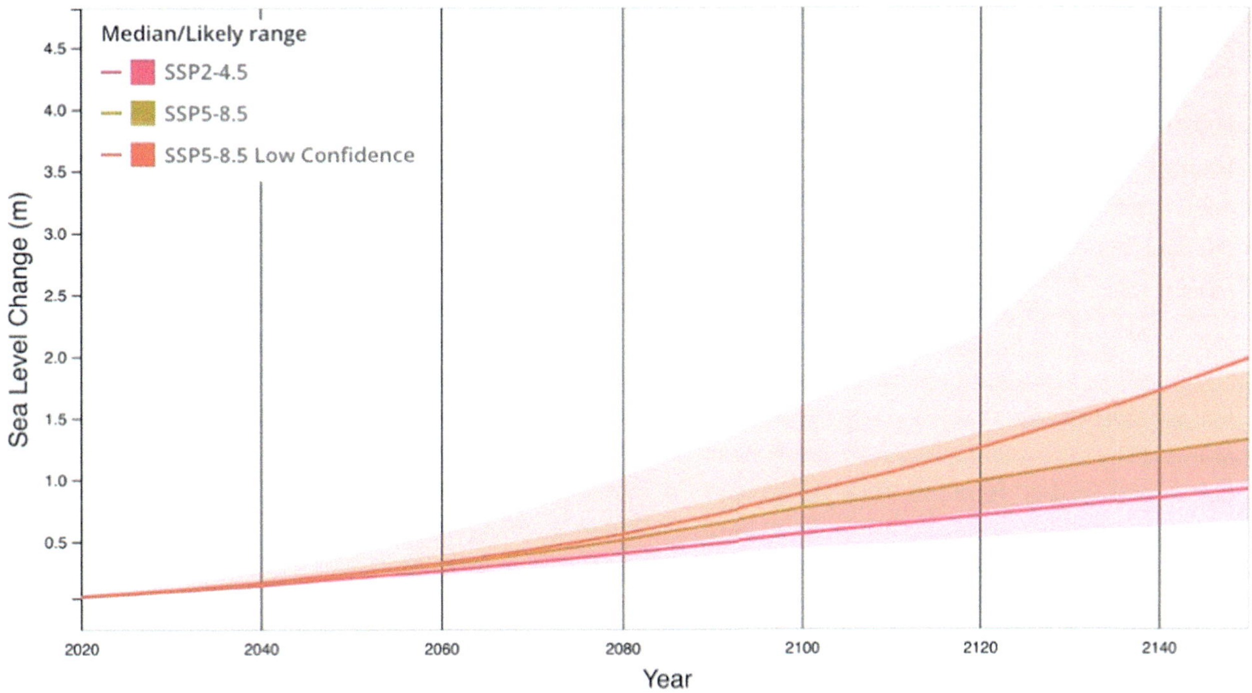

FIGURE 19.3 Global mean Sea level projection for different scenarios relative to 1995–2014 reference. Shaded regions indicate 17th to 83rd percentile ranges. *The figure is updated based on the data available from https://sealevel.nasa.gov/ipcc-ar6-sea level-projection-tool?type = global.*

emphasized significant subsidence in major Asian cities, highlighting their vulnerability to future sea level rise. Table 19.2 provides the projected sea level rise for major cities in Asia, based on IPCC AR6.

IPCC sea level projections primarily rely on large scale (>100 km) CMIP6 model simulations. However, these models inadequately represents oceanic circulation in regions with complex geography, such as Southeast Asia (Samanta et al., 2021). To address this, on a regional scale, Samanta, Jevrejeva et al. (2024) used High Resolution Model Intercomparison Project models (Haarsma et al., 2016) from CMIP6 to evaluate dynamic sea level projection in Southeast Asia under shared socioeconomic pathways five (SSP5-8.5) and diagnose the role of regional ocean heat transport. They found a projected sterodynamic sea level rise up to 0.45 m over Southeast Asia by 2100. They argued the important role of ocean heat and mass transport changes and Ekman dynamics in shaping dynamic sea level pattern over Southeast Asia. Jin et al. (2023) further highlighted the role of weakening Indonesian throughflow on sea level projection in Southeast Asia by 2100.

Government agencies in Asia, such as Singapore's third National Climate Change Study (V3), provide additional insights into regional sea level projection in Asia. The V3 report projected a sea level rise of 0.57 ± 0.04 m under SSP2-4.5 and 0.79 ± 0.04 m under SSP5-8.5 by the end of 2100 (Moise et al., 2024). The report further argued sea level could rise to 2.12 m at Sultan Shoal, Singapore under SSP5-8.5 by 2150.

On the regional and local scale, the effects of changing sea levels are most observed through the impacts of extreme events. The magnitude of these extreme events is shaped by various processes spanning a wide range of spatial scales. These include global factors like large-scale climate variability, as well as local features such as tides and storm surges (Church & Zhang, 2023). If the contributing factors remain constant, extreme sea levels that previously occurred once per century will occur annually or more frequently at about 19%–31% of tide gauges by 2050 and at about 82% by 2100 under the SSP5-8.5 scenario. Such extreme sea level events will occur ~20–30 times more frequently by 2050 and up to 530 times more frequently by 2100 relative to the recent past, as referred from SSP1-2.6, SSP2-4.5, and SSP5-8.5. Jevrejeva, Williams, Vousdoukas, and Jackson (2023) projected that by 2100, nearly 45% of the global coastline will experience extreme sea level exceeding the global mean extreme level of 4.2 m under high tide and low probability sea level rise scenario (95th percentile). They further showed that extreme sea level could reach 9–10 m for the East China Sea and Japan by 2100.

The sea level projections beyond 2150 is subject to significant uncertainties, particularly regarding the responses of the Greenland and Antarctic ice sheets to dynamic processes such as marine ice sheet instability (MISI) and marine ice

TABLE 19.2 Examples of projected sea level rise (in m) in Asian cities in 2100 for SSP scenarios in which there is medium confidence.

Projected sea level rise (in m) in 2100 in Asian cities based on SSP scenarios			
Location	SSP2-4.5	SSP5-8.5	SSP5-8.5 Low confidence
Aden, Yemen	0.54	0.77	0.89
Bhaunagar, India	0.71	0.95	*1.07*
Mumbai, India	0.46	0.70	0.82
Cochin, India	0.58	0.82	0.95
Chennai, India	0.45	0.69	0.82
Diamond Harbour, India	0.64	0.86	0.98
Male, Maldives	0.62	0.86	0.99
Khepupara, Bangladesh	*1.05*	*1.25*	*1.40*
Rangoon, Mynmar	0.71	0.93	*1.05*
Fort Phrachula Chomkalo, Thailand	*1.68*	*1.89*	**2.02**
Tanjong Pagar, Singapore	0.54	0.76	0.90
Danang, Vietnam	0.63	0.84	0.97
Manila, Philippines	*1.12*	*1.33*	*1.46*
Xiemen, China	0.59	0.81	0.94
Jeju, South Korea	0.82	*1.08*	*1.21*
Pohang, South Korea	0.80	*1.05*	*1.18*
Toba, Japan	**2.36**	**2.61**	**2.75**
Hiroshima, Japan	0.81	*1.06*	*1.19*
Yokosuka, Japan	0.89	*1.16*	*1.29*
Kushiro, Japan	*1.21*	*1.48*	*1.61*

The median values are mentioned in the table. The cells reaching 1.00 m are given in italic and exceeding 2.0 is in bold. Projections are relative to the 1995–2014 baseline and the data are from NASA sea level projection toolbox (https://sealevel.nasa.gov/ipcc-ar6-sea level-projection-tool).

cliff instability (MICI) (Fox-Kemper et al., 2021). As a result, IPCC AR6 sea level projection beyond 2150 carry low confidence, reflecting limited evidence and low agreement in assessments.

19.6 Uncertainties in sea level projections

The future change in sea levels is characterized by a blend of quantifiable and unquantifiable uncertainties (Kopp et al., 2023). Effectively communicating both types of uncertainty poses a significant challenge in translating sea level science into actionable insights for long-term coastal planning. Lessons from earlier IPCC reports emphasize the need to communicate these uncertainties clearly and effectively (Kopp et al., 2023). IPCC AR6 builds on this approach, striveing to represent the uncertainty in sea level projections including careful emphasis on the "likely" range. Both the Special Report on the Ocean and Cryosphere in a Changing Climate (SROCC) and IPCC AR6 offer guidance on how various users with different risk tolerances can make use of sea level projection. This includes risk-neutral approaches, focusing on likely outcomes, as well as risk-averse high-end outcomes (Kopp et al., 2023). However, global survey shows planners consistently use widely varying sea level rise projections for coastal projections (Hirschfeld et al., 2023). Uncertainties in projected sea level rise vary across spatiall and temporal scales. For instance, considerable uncertainties persist regarding the contributions of the Antarctic ice sheet to sea level change (e.g., Church & Zhang, 2023; Fox-Kemper et al., 2021; Oppenheimer et al., 2019). Additionally, the regional distribution of dynamic (sterodynamic) sea level changes (Gregory et al., 2019).

A key limitation in climate model-based sea level projections such as those using CMIP6 models, is the absence of barystatic components in the model simulations. IPCC AR6 addresses this by combining different components for sea level projections. As the CMIP6 models do not encompass all these processes, various methods are employed to estimate the contribution from different components in sea level projections. Processes related to ice sheets, such as the Antarctica ice sheet, are characterized by deep uncertainty stemming from limited evaluation data, inadequate understanding of processes, and high sensitivity to uncertain boundary conditions and parameters (Fox-Kemper et al., 2021). A higher GMSL rise before 2100 is possible due to the earlier-than-projected abrupt and widespread onset of MISI and MICI around Antarctica, and faster changes in surface mass balance and discharge from Greenland (Fox-Kemper et al., 2021). Improving the representation of important physical processes in these climate models related to polar climate has the potential to reduce uncertainty in sea level projections (e.g., Li, DeConto, & Pollard, 2023; Park et al., 2023; Topál et al., 2022).

On the regional to local scale, extreme sea level constructed along coastlines with coastal megacities (such as Mumbai, Singapore, and Tokyo) and settlements near large river deltas (such as the Mekong delta, Ganges delta) often underestimate risks. This underestimations stems from the omission of local processes, such as subsidence from groundwater extraction (e.g., in Jakarta, Bangkok) and land subsidence (e.g., Tay et al., 2022) in sea level projection from CMIP6 models.

Furthermore, it is essential to acknowledge that global climate models exhibit various mean-state biases, particularly over the Indo-Pacific region (e.g., Samanta, Karnauskas, et al., 2018; Samanta, Hameed, et al., 2018; Samanta, Karnauskas, & Goodkin, 2019) that may impact sea level simulations. For example, Lyu et al. (2020) have argued that CMIP6 models underestimate the dynamic sea level rise in the western Pacific Ocean due to cold-tongue sea surface temperature bias in the tropical Pacific Ocean.

Additionally, dynamic sea level projections are constrained by the relatively low spatial resolution of global climate models, which is around 110 km in the ocean components. While computationally efficient compared to eddy-resolving models, this resolution is insufficient for capturing key oceanic processes in many regions, including Southeast Asia (Lee, Fournier, Gordon, & Sprintall, 2019; Samanta et al., 2021; Wyrtki, 1961). Moreover, a few studies also demonstrated that low-resolution global climate models (such as in CMIP6) underestimate dynamic sea level projection (Samanta, Jevrejeva et al., 2024; Wickramage, Köhl, Jungclaus, & Stammer, 2023) compared to the lower-resolution version of the same model. Wickramage et al. (2023) emphasized the importance of careful evaluation of sea level projection from low-resolution global climate models due to the absence of proper representation of small-scale oceanic eddies, such as those over the eddy-active Kursohio region.

19.7 Impact of sea level change on coastal community and economy

Rising sea levels are a significant concern as they lead to the submergence of low-lying wetlands and dry land, shoreline erosion, increased risk of coastal flooding, and the intrusion of saltwater into estuaries and nearby groundwater aquifers. Moreover, rising sea levels threaten coastal infrastructure, making it more vulnerable to storm-related damage. Like other low-lying global coastal regions, Asian coastal communities, particularly in tropical areas, are already feeling the impact of sea level rise. The hazards of rising seas affect not only small islands but also continental regions, such as Sundarbans in India and Bangladesh where coastal settlements face submergence (Minunno, Andersson, & Morrison, 2023). Shoreline retreat due to combined effect of relative sea level rise and subsidence due to groundwater extraction is evident in Asia, for example, in Chao-Pharya River delta in Thailand (Uehara, Sojisuporn, Saito, & Jarupongsakul, 2010).

Land subsidence (Tay et al., 2022) is an additional factor that makes sea level rise impact more concerning. For example, land subsidence in Jakarta (Indonesia) and Manila (Philippines) already led to more frequent flooding during high tide events. The increasing flood impacts from sea level rise are among the major reasons behind the Indonesian government's decision to shift their capital from Jakarta to Nusantara in East Kalimantan (https://www.ikn.go.id/en/high-investor-interest-in-the-development-of-the-nusantara-capital-1). Accelerated and prolonged exploration of sea level rise along Asian coasts poses potentially catastrophic risks to the growing population.

For many years, coastal ecosystems have been recognized as crucial for the well being and subsistence of millions of people (Saintilan, Horton, & Guntenspergen, 2023). However, these important ecosystems face an uncertain future because of anthropogenic climate change and rising sea levels. Paleo and instrumental observational data from Saintilan et al. (2023) suggest that due to sea level rise the eventual retreat of mangroves and tidal marshes is likely (4 mm/year) or very likely (7 mm/year). Whereas, reef islands face likely increase in island instability by relative sea level rise of 7 mm/year.

Under 1.5 °C scenario, the likely ($p >= 0.66$) GMSL rise rate at 2080–2100 is between 2.4 and 6.4 mm/year. This corresponds to a relative sea level rise of 4 to7 mm/year along coastlines, similar to contemporary mangrove development in Southeast Asia (Saintilan et al., 2023). Subsidence in Asian megacities (e.g., Tay et al., 2022) can cause the relative sea level rise rate to exceed 7 mm/year. Whereas, under 3 °C warming scenario, nearly the entire Asian coastline will be exposed to $>=7$ mm/year of relative sea level (Saintilan et al., 2023). This will also impact along with regions of other tropical and subtropical coastlines where abundant mangroves and coral reefs are located.

Sea level rise will pose major sustainability challenges for deltas and their inhabitants. The increased delta flooding in deltas, such as in the Sundarbans and Mekong Delta, heightens risks, including salinization of arable lands and land loss. These changes will further reduce the available habitable areas, potentially triggering conflicts over livelihoods. Historically sea level rise caused forced human migration in South Asia (Kim et al., 2023). However, the projected increase in population may pose further challenges for such migrations in the future.

19.8 Implications for adaptation and mitigation pathways

Understanding multifaceted processes and their regional variations is important for accurately addressing the impacts of sea level changes and for developing effective mitigation and adaptation strategies. However, combating sea level rise is heavily dependent on concerted global action. Until now only a few countries are 2°C compatible with most countries remaining in the insufficient or critically insufficient categories, which impedes the ability to address rapidly rising sea level through emission mitigation pathways (Minunno et al., 2023). Given that insufficiency of current mainstream mitigation measures to effectively slow down sea level rise, more radical measures are deemed necessary.

There has been a consensus that reducing heat-trapping greenhouse gas emissions is the most optimal solution for limiting the magnitude and rate of sea level rise (Jevrejeva et al., 2009; Lockley et al., 2020; Moore et al., 2010; Yue, Jevrejeva, Qu, Zhao, & Moore, 2023). Several studies suggested the option of geoengineering to mitigate unprecedented sea level rise (e.g., Minunno et al., 2023; Moore et al., 2010) such as sulfur dioxide injection into the stratosphere, mirrors in space, biochar, afforestation, and bioenergy with carbon sequestration (Moore et al., 2010). Moore et al. (2010) showed that the most reliable and environmentally desirable approach to mitigating sea level rise involves utilizing bioenergy with carbon sequestration. Using five Earth system model simulations of two solar geoengineering methods (solar dimming and stratospheric sulfate aerosol injection), Yue et al. (2023) demonstrated that global mean thermosteric sea level could be reduced by 11.2–12.6 cm (36–41%) relative to SSP5-8.5 by 2080-2099, effectively restoring levels to SSP2-4.5. However, these methods do not perfectly restore regional sea level rise patterns. They further showed the east coast of Japan as the region of the most sterodynamic sea level reduction within Asia from geoengineering solutions.

However, these options progressively reduce solar radiation by 1 W m^{-2} per decade until 2100 and have the potential to limit or even reverse rising sea levels. For instance, an aerosol injection system that consistently reduces radiative forcing by 4 W m^{-2} (comparable to the cooling effect of a volcanic eruption like Mount Pinatubo in 1991) could potentially delay in sea level rise by 40–80 years. However, the aerosol injection strategies may not be cost-effective unless they can be maintained continuously, and the harm caused by the climate response to the aerosols remains below approximately 0.6% of the Global World Product. Moore et al. (2010) further emphasized that geoengineering cannot be the solution to sea level rise, because most extreme geoengineering approaches will not stop sea level rise due to anthropogenic climate change. Further, the approach such as injecting sulfur dioxide into the atmosphere can alter global rainfall and weather patterns.

The committed loss of coastal habitats under high-warming scenarios should not be a barrier to conservation and restoration efforts (Saintilan et al., 2023). The mangroves show landward tendencies below 7 mm/year of sea level rise rate. However, the Asian coastlines pose a barrier to such landward retreats or coastal ecosystems (Saintilan et al., 2023). The implications of the gap between the Paris Agreement (2°C with an aim of 1.5°C) and pathways of current policies (2.4°C–3.5°C by 2081–2100) are significant for coastal ecosystems. The thermal inertia of ocean waters is likely to drive Antarctic ice instability, therefore, a higher sea level rise rate is likely to continue for centuries. The boundaries for a safe operating space for coastal ecosystems are approaching as coastal ecosystems represent several tipping elements for climate change impacts that are vital for human well-being (Saintilan et al., 2023), not only in Asia but also worldwide. These suggest the urgent need to contribute toward the preservation, restoration, and landward accommodation of coastal blue carbon ecosystems.

As the sea level rise, coastal flood protection infrastructure is an important alternative to relocating to inland regions. Depending on the future concentration of greenhouse gas and polar ice sheet melting (Fox-Kemper et al., 2021), a global two meters rise in sea level is not unexpected within one century to as long as two millennia (Le Cozannet et al., 2023). In addition to climate mitigation efforts aimed at slowing down the rate of sea level rise, it is imperative to start an adaptation strategy to 2 m of sea level rise without delay. The burden of present coastal flood risk and future sea level rise falls disproportionately on tropical regions, especially in Asia. Indeed Becker, Karpytchev, and Hu (2023) showed that several Asian megacities (Chennai, Kolkata, Yangon, Bangkok, Ho Chi Minh City, and Manila) could be the new sea level rise hotspot by 2100.

The escalating rise of sea level on Asian coasts and beyond necessitates a shift in our mindset, requiring the establishment of a strategic vision for vulnerable coastal areas and the identification of realistic pathways to realize this vision. This approach can aid in mitigating future damages and losses, preventing the entrenchment of unsustainable situations, proactively addressing emerging issues, and capitalizing the opportunities as they arise (Le Cozannet et al., 2023). Coastal adaptation must take a central role in coastal development on highly populated Asian coasts, especially concerning long-lasting critical infrastructure. Coastal adaptation requires robust support from climate services, which bridges the gap between scientific knowledge, policy decisions, and adaptation practices.

19.9 Conclusions

Sea level rise is one of the most severe, and costly risk of anthropogenic climate change, with far-reaching and irreversible consequences for many coastal zones in Asia and beyond. Since 1900 GMSL has risen at a rate of 1.7 mm/year — an order of magnitude faster than in previous centuries, resulting in an increased severity of extreme sea level events. During the satellite era (since 1993), the GMSL rise rate further accelerated up to 3.4 mm/year. Regional sea level changes can deviate up to $\pm 20\%$ from the GMSL due to varying physical processes and can occur accross a wide range of timescales. Contemporary regional sea level changes results from the combined influences of natural climate variability and anthropogenic forcings. As climate change continues, the anthropogenic impact on regional sea level will dominate toward the end of the 21st century. With increasing greenhouse gas forcing and decreasing aerosol forcings, manometric sea level is going to be a key component of the coastal and shallow oceans of Asian seas.

The impact of sea level rise on coastal populations, infrastructure, ecosystems, economies, and livelihoods are already disproportionately affecting the densely populated Asian coasts and are projected to intensify in the coming decades. Appropriately managing the sea level rise risk requires planners to be cognizant of both structural uncertainty and quantifiable uncertainty in the projection of future sea level change. In parallel to the multiple modeling approaches, a flexible, open-source platform, such as framework for assessing changes to sea level (FACTS) can support specific assessment processes, particularly that of the IPCC AR6 (Kopp et al., 2023). The implementation of appropriate mitigation and adaptation measures must start now.

Acknowledgment

The author acknowledges Dr. Stephen Chua from Earth Observatory of Singapore, Nanyang Technological University for useful feedback on the initial draft. This is Earth Observatory of Singapore contribution number 561.

Annex 1 Selected acronyms and associated links

IPCC AR6: Intergovernmental panel on climate change sixth assessment report (https://www.ipcc.ch/assessment-report/ar6/)

GMSL: Global mean sea level

GRACE: The gravity recovery and climate experiment (https://www.jpl.nasa.gov/missions/gravity-recovery-and-climate-experiment-grace)

GRACE-FO: GRACE follow on

ICESat-2: Ice, cloud and land elevation satellite 2 (https://icesat-2.gsfc.nasa.gov/)

NISAR: NASA-ISRO synthetic aperture radar (https://nisar.jpl.nasa.gov/)

SWOT: Surface water and ocean topography (https://swot.jpl.nasa.gov/mission/overview/)

GNSS: Global navigation satellite system (https://www.gps.gov/systems/gnss/)

Annex 2 Terminologies in sea level studies (see Gregory et al., 2019 for details)

Barystatic sea level rise: global mean sea level rise due to changes in the mass of the ocean, but not its density.

Global mean sea level rise: global mean of relative sea level change due to change in the volume of the ocean.

Ocean dynamic sea level: mean sea level above geoid due to ocean dynamics.

Relative sea level change: relative sea level change or rise for the change in mean sea level relative to land.

Extreme sea level: the occurrence of exceptionally high local sea surface due to short-term phenomena (e.g., storm surges, tides, and waves), or extreme coastal water level.

Steric sea level change: part of relative sea level change due to change in ocean density, assuming the local mass of ocean per unit area is constant; It consists of thermosteric (due to thermal expansion) and halosteric (due to salinity/density change) sea level changes.

Manometric sea level: nonsteric sea level change due to change in time-mean local mass of the ocean per unit area; associated with ocean bottom pressure change.

Geocentric sea level change: change in local mean sea level relative to the terrestrial reference frame.

Sterodynamic sea level change: relative sea level change due to changes in ocean density and circulation, with inverse barometer (IB) correction; considered as the sum of ocean dynamic sea level plus global mean thermosteric sea level rise.

Vertical land movement (VLM): the change in the height of the sea floor or the land surface.

Glacial isostatic adjustment (GIA): changes in Earth Gravity, Earth Rotation, and viscoelastic solid-Earth Deformation (GRD) due to ongoing changes in solid Earth caused by past changes in land ice.

Marine ice sheet instability (MISI): a potentially catastrophic process by which ice sheets resting below sea level on a downward-slopping bedrock (such as over the West Antarctic) become unstable and experience rapid and irreversible runaway retreat. This can happen if ocean temperature rises and melt the ice near the grounding line.

Marine ice cliff instability (MICI): a potentially catastrophic process by which ice cliffs at the terminus of the marine-based ice sheets become unstable and collapse under their weight. Ice cliffs are formed when ice shelves, which are floating extensions of ice sheets, break off.

Annex 3 Relevant websites for sea level studies

https://psmsl.org/ Global tide gauge data at monthly and annual resolution from Permanent Service for Mean Sea Level

https://uhslc.soest.hawaii.edu/datainfo/ Research quality global tide gauge data at hourly and monthly resolution from University of Hawaii Sea Level Center

https://cds.climate.copernicus.eu/cdsapp#!/dataset/satellite-sea-level-global?tab = overview Sea level gridded data from satellite observations based on two satellite (Copernicus Climate Change Service (C3S))

https://climate.esa.int/en/projects/sea-level/data/ Different altimeter bases sea level products from European Space Agency Climate Change Initiative (ESA CCI)

https://data.marine.copernicus.eu/products?facets = mainVariables%7ESea + surface + height--areas%7EGlobal + Ocean Various satellite based and reanalysis sea level products, including 0.083° × 0.083° spatial resolution reanalysis product since 1993

https://esgf-index1.ceda.ac.uk/projects/cmip6-ceda/ Global climate model data from Coupled Model Intercomparison Project Phase 6 (CMIP6) simulations

https://www.sonel.org/-Vertical-land-movements-.html High-quality continuous measurements of sea and land levels at the coast from tide gauges (relative sea level) and from modern geodetic techniques

https://doi.pangaea.de/10.1594/PANGAEA.889923 A global GPS dataset for testing and improving modelled GIA uplift rates

https://gesla787883612.wordpress.com/ Global extreme sea level analysis (GESLA)

https://sealevel.nasa.gov/ipcc-ar6-sea-level-projection-tool NASA sea level projection tool

https://sealevel.nasa.gov/data_tools/16 NASA sea level evaluation and assessment tool

https://argo.ucsd.edu/data/ Argo data products for temperature, salinity observations

https://www.ncei.noaa.gov/access/global-ocean-heat-content/ Global ocean heat and salt content from NOAA

References

Becker, M., Karpytchev, M., & Hu, A. (2023). Increased exposure of coastal cities to sea-level rise due to internal climate variability. *Nature Climate Change*, *13*(4), 367–374.

Cazenave, A., & Moreira, L. (2022). Contemporary sea-level changes from global to local scales: A review. *Proceedings of the Royal Society A*, *478* (2261), 20220049.

Cha, H., Moon, J. H., Kim, T., & Song, Y. T. (2023). A process-based assessment of the sea-level rise in the northwestern Pacific marginal seas. *Communications Earth and Environment*, *4*(1), 300.

Chua, S., Switzer, A. D., Li, T., Chen, H., Christie, M., Shaw, T. A., & Horton, B. P. (2021). A new Holocene sea-level record for Singapore. *The Holocene*, *31*(9), 1376–1390.

Church, J. A., Gregory, J. M., Huybrechts, P., Kuhn, M., Lambeck, K., Nhuan, M. T., … Woodworth, P. L. (2001). Changes in sea level. In J. T. Houghton, Y. Ding, D. J. Griggs, M. Noguer, P. J. Van der Linden, X. Dai, K. Maskell, & C. A. Johnson (eds): *Climate Change 2001: The Scientific Basis: Contribution of Working Group I to the Third Assessment Report of the Intergovernmental Panel* (pp. 639–694).

Church, J. A., Clark, P. U., Cazenave, A., Gregory, J. M., Jevrejeva, S., Levermann, A., … Payne, A. J. (2013). Sea level change, in Climate Change 2013: The Physical Science Basis. In T. F. Stocker, D. Qin, G.-K. Plattner, M. Tignor, S. K. Allen, J. Boschung, & P. M. Midgley (Eds.), *Contribution of working group I to the fifth assessment report of the intergovernmental panel on climate change* (pp. 1137–1216). Cambridge, U. K., and New York: Cambridge Univ. Press.

Church, J. A., & White, N. J. (2011). Sea-level rise from the late 19th to the early 21st century. *Surveys in Geophysics*, *32*, 585–602.

Church, J. A., & Zhang, X. (2023). *Sea-level rise and estuaries. Climate change and estuaries* (1st, p. 20). CRC Press.

Couldrey, M. P., Gregory, J. M., Dias, F. B., Dobrohotoff, P., Domingues, C. M., Garuba, O., & Zanna, L. (2021). What causes the spread of model projections of ocean dynamic sea-level change in response to greenhouse gas forcing? *Climate Dynamics*, *56*(1), 155–187.

Dangendorf, S., Marcos, M., Müller, A., Zorita, E., Riva, R., Berk, K., & Jensen, J. (2015). Detecting anthropogenic footprints in sea-level rise. *Nature Communications*, *6*(1), 1–9.

Deepa, J. S., Gnanaseelan, C., Mohapatra, S., Chowdary, J. S., Karmakar, A., Kakatkar, R., & Parekh, A. (2019). The tropical Indian Ocean decadal sea level response to the Pacific decadal oscillation forcing. *Climate Dynamics*, *52*, 5045–5058. Available from https://doi.org/10.1007/s10712-016-9386-y.

Eyring, V., Bony, S., Meehl, G. A., Senior, C. A., Stevens, B., Stouffer, R. J., & Taylor, K. E. (2016). Overview of the Coupled Model Intercomparison Project Phase 6 (CMIP6) experimental design and organization. *Geoscientific Model Development*, *9*(5), 1937–1958.

Fasullo, J. T., & Nerem, R. S. (2018). Altimeter-era emergence of the patterns of forced sea-level rise in climate models and implications for the future. *Proceedings of the National Academy of Sciences*, *115*(51), 12944–12949.

Ferrero, B., Tonelli, M., Marcello, F., & Wainer, I. (2021). Long-term regional dynamic sea level changes from CMIP6 projections. *Advances in Atmospheric Sciences*, *38*(2), 157–167.

Forster, P. M., Smith, C. J., Walsh, T., Lamb, W. F., Lamboll, R., Hauser, M., & Zhai, P. (2023). Indicators of Global Climate Change 2022: Annual update of large-scale indicators of the state of the climate system and human influence. *Earth System Science Data*, *15*(6), 2295–2327.

Fox-Kemper, B., Hewitt, H. T., Xiao, C., Aðalgeirsdóttir, G., Drijfhout, S. S., Edwards, T. L., … Yu, Y. (2021). Ocean, Cryosphere and Sea Level Change. In V. Masson-Delmotte, P. Zhai, A. Pirani, S. L. Connors, C. Péan, S. Berger, N. Caud, Y. Chen, L. Goldfarb, M. I. Gomis, M. Huang, K. Leitzell, E. Lonnoy, J. B. R. Matthews, T. K. Maycock, T. Waterfield, O. Yelekçi, R. Yu, & B. Zhou (Eds.), *Climate change 2021: The physical science basis. Contribution of working group I to the sixth assessment report of the intergovernmental panel on climate change* (pp. 1211–1362). Cambridge, United Kingdom and New York, NY, USA: Cambridge University Press. Available from https://doi.org/10.1017/9781009157896.011.

Frederikse, T., Landerer, F., Caron, L., Adhikari, S., Parkes, D., Humphrey, V. W., … Wu, Y. H. (2020). The causes of sea-level rise since 1900. *Nature*, *584*(7821), 393–397. Available from https://www.nature.com/articles/s41586-020-2591-3.

Gregory, J. M., Griffies, S. M., Hughes, C. W., Lowe, J. A., Church, J. A., Fukimori, I., & Van de Wal, R. S. (2019). Concepts and terminology for sea level: Mean, variability and change, both local and global. *Surveys in Geophysics*, *40*(6), 1251–1289.

Guérou, A., Meyssignac, B., Prandi, P., Ablain, M., Ribes, A., & Bignalet-Cazalet, F. (2023). Current observed global mean sea level rise and acceleration estimated from satellite altimetry and the associated measurement uncertainty. *Ocean Science*, *19*(2), 431–451.

Haarsma, R. J., Roberts, M. J., Vidale, P. L., Senior, C. A., Bellucci, A., Bao, Q., & von Hardenberg, J. (2016). High resolution model intercomparison project (HighResMIP v1. 0) for CMIP6. *Geoscientific Model Development*, *9*(11), 4185–4208.

Hamlington, B. D., Bellas-Manley, A., Willis, J. K., Fournier, S., Vinogradova, N., Nerem, R. S., … Kopp, R. (2024). The rate of global sea level rise doubled during the past three decades. *Communications Earth & Environment*, *5*(1), 601.

Hamlington, B. D., Gardner, A. S., Ivins, E., Lenaerts, J. T., Reager, J. T., Trossman, D. S., et al. (2020). Understanding of contemporary regional sea-level change and the implications for the future. *Reviews of Geophysics*, *58*(3). Available from https://doi.org/10.1029/2019rg000672, e2019RG000672.

Hamlington, B. D., Strassburg, M. W., Leben, R. R., Han, W., Nerem, R. S., & Kim, K. Y. (2014). Uncovering an anthropogenic sea-level rise signal in the Pacific Ocean. *Nature Climate Change*, *4*(9), 782–785.

Hamlington, B. D., Willis, J. K., & Vinogradova, N. (2023). The emerging golden age of satellite altimetry to prepare humanity for rising seas. *Earth's Future*, *11*(11), e2023EF003673.

Han, W., Meehl, G. A., Rajagopalan, B., Fasullo, J. T., Hu, A., Lin, J., & Yeager, S. (2010). Patterns of Indian Ocean sea-level change in a warming climate. *Nature Geoscience*, *3*(8), 546–550.

Han, W., Stammer, D., Thompson, P., Ezer, T., Palanisamy, H., Zhang, X., … Yuan, D. (2019). Impacts of basin-scale climate modes on coastal sea level: a review. *Surveys in geophysics*, *40*, 1493–1541.

Hermans, T. H., Gregory, J. M., Palmer, M. D., Ringer, M. A., Katsman, C. A., & Slangen, A. B. (2021). Projecting global mean sea-level change using CMIP6 models. *Geophysical Research Letters*, *48*(5), e2020GL092064.

Hirschfeld, D., Behar, D., Nicholls, R. J., Cahill, N., James, T., Horton, B. P., & Yokoki, H. (2023). Global survey shows planners use widely varying sea-level rise projections for coastal adaptation. *Communications Earth and Environment*, *4*(1), 102.

Horton, B. P., Khan, N. S., Cahill, N., Lee, J. S., Shaw, T. A., Garner, A. J., & Rahmstorf, S. (2020). Estimating global mean sea-level rise and its uncertainties by 2100 and 2300 from an expert survey. *NPJ Climate and Atmospheric Science*, *3*(1), 18.

Horton, B. P., Kopp, R. E., Garner, A. J., Hay, C. C., Khan, N. S., Roy, K., & Shaw, T. A. (2018). Mapping sea-level change in time, space, and probability. *Annual Review of Environment and Resources*, *43*, 481–521.

Horwath, M., Gutknecht, B. D., Cazenave, A., Palanisamy, H. K., Marti, F., Marzeion, B., & Benveniste, J. (2021). Global sea-level budget and ocean-mass budget, with focus on advanced data products and uncertainty characterisation. *Earth System Science Data Discussions*, *2021*, 1–51.

Jevrejeva, S., Grinsted, A., & Moore, J. C. (2009). Anthropogenic forcing dominates sea-level rise since 1850. *Geophysical Research Letters*, *36*(20).

Jevrejeva, S., Moore, J. C., & Grinsted, A. (2012). Sea level projections to AD2500 with a new generation of climate change scenarios. *Global and Planetary Change*, *80*, 14–20.

Jevrejeva, S., Williams, J., Vousdoukas, M. I., & Jackson, L. P. (2023). Future sea level rise dominates changes in worst case extreme sea levels along the global coastline by 2100. *Environmental Research Letters*, *18*(2), 024037.

Jin, Y., Köhl, A., Jungclaus, J., & Stammer, D. (2023). Mechanisms of projected sea-level trends and variability in the Southeast Asia region based on MPI-ESM-ER. *Climate Dynamics*, 1–16.

Kim, H. L., Li, T., Kalsi, N., Nguyen, H. T. T., Shaw, T. A., Ang, K. C., & Horton, B. P. (2023). Prehistoric human migration between Sundaland and South Asia was driven by sea-level rise. *Communications Biology*, *6*(1), 150.

Kopp, R. E., Garner, G. G., Hermans, T. H., Jha, S., Kumar, P., Reedy, A., & Smith, C. (2023). The Framework for Assessing Changes To Sea-level (FACTS) v1. 0: a platform for characterizing parametric and structural uncertainty in future global, relative, and extreme sea-level change. *Geoscientific Model Development*, *16*(24), 7461–7489.

Kopp, R. E., Hay, C. C., Little, C. M., & Mitrovica, J. X. (2015). Geographic variability of sea-level change. *Current Climate Change Reports*, *1*, 192–204. Available from https://link.springer.com/article/10.1007/s40641-015-0015-5.

Kopp, R. E., Horton, R. M., Little, C. M., Mitrovica, J. X., Oppenheimer, M., Rasmussen, D. J., & Tebaldi, C. (2014). Probabilistic 21st and 22nd century sea-level projections at a global network of tide-gauge sites. *Earth's Future*, *2*(8), 383–384.

Kopp, R. E., Oppenheimer, M., O'Reilly, J. L., Drijfhout, S. S., Edwards, T. L., Fox-Kemper, B., & Xiao, C. (2023). Communicating future sea-level rise uncertainty and ambiguity to assessment users. *Nature Climate Change*, 1–13.

Le Cozannet, G., Nicholls, R. J., Durand, G., Slangen, A., Lincke, D., & Chapuis, A. (2023). Adaptation to multi-meter sea-level rise should start now. *Environmental Research Letters*.

Lee, T., Fournier, S., Gordon, A. L., & Sprintall, J. (2019). Maritime continent water cycle regulates low-latitude chokepoint of global ocean circulation. *Nature Communications*, *10*(1), 1–13.

Li, D., DeConto, R. M., & Pollard, D. (2023). Climate model differences contribute deep uncertainty in future Antarctic ice loss. *Science Advances*, *9* (7), eadd7082.

Little, C. M., Horton, R. M., Kopp, R. E., Oppenheimer, M., & Yip, S. (2015). Uncertainty in twenty-first-century CMIP5 sea level projections. *Journal of Climate*, *28*(2), 838–852.

Llovel, W., & Lee, T. (2015). Importance and origin of halosteric contribution to sea level change in the southeast Indian Ocean during 2005–2013. *Geophysical Research Letters*, *42*(4), 1148–1157.

Lockley, A., Wolovick, M., Keefer, B., Gladstone, R., Zhao, L. Y., & Moore, J. C. (2020). Glacier geoengineering to address sea-level rise: a geotechnical approach. *Advances in Climate Change Research*, *11*(4), 401–414.

Lyu, K., Zhang, X., & Church, J. A. (2020). Regional dynamic sea level simulated in the CMIP5 and CMIP6 models: Mean biases, future projections, and their linkages. *Journal of Climate*, *33*(15), 6377–6398.

Marcos, M., Marzeion, B., Dangendorf, S., Slangen, A. B., Palanisamy, H., & Fenoglio-Marc, L. (2017). Internal variability versus anthropogenic forcing on sea level and its components. *Integrative study of the mean sea level and its components*, 337–356.

Marzeion, B., Cogley, J. G., Richter, K., & Parkes, D. (2014). Attribution of global glacier mass loss to anthropogenic and natural causes. *Science*, *345*(6199), 919–921.

Majewski, J. M., Meltzner, A. J., Switzer, A. D., Shaw, T. A., Li, T., Bradley, S., … Horton, B. P. (2022). Extending Instrumental Sea-Level Records Using Coral Microatolls, an Example From Southeast Asia. *Geophysical Research Letters*, *49*(4), e2021GL095710.

Merrifield, M. A., & Maltrud, M. E. (2011). Regional sea level trends due to a Pacific trade wind intensification. *Geophysical Research Letters*, *38*(21).

Meyssignac, B., Salas y Melia, D., Becker, M., Llovel, W., & Cazenave, A. (2012). Tropical Pacific spatial trend patterns in observed sea level: internal variability and/or anthropogenic signature? *Climate of the Past*, *8*(2), 787–802.

Minunno, R., Andersson, N., & Morrison, G. M. (2023). A systematic literature review considering the implementation of planetary geoengineering techniques for the mitigation of sea-level rise. *Earth-Science Reviews*, 104431.

Moise, et al. (2024). Singapore's 3rd national climate change report, accessed from https://www.mss-int.sg/V3-climate-projections.

Moore, J. C., Jevrejeva, S., & Grinsted, A. (2010). Efficacy of geoengineering to limit 21st century sea-level rise. *Proceedings of the National Academy of Sciences*, *107*(36), 15699–15703.

Moore, J. C., Yue, C., Chen, Y., Jevrejeva, S., Visioni, D., Uotila, P., & Zhao, L. (2024). Multi-model simulation of solar geoengineering indicates avoidable destabilization of the West Antarctic ice sheet. *Earth's Future*, *12*(6), e2024EF004424.

Nerem, R. S., Beckley, B. D., Fasullo, J. T., Hamlington, B. D., Masters, D., & Mitchum, G. T. (2018). Climate-change–driven accelerated sea-level rise detected in the altimeter era. *Proceedings of the National Academy of Sciences*, *115*(9), 2022–2025. Available from https://doi.org/10.1073/pnas.1717312115.

Nidheesh, A. G., Lengaigne, M., Vialard, J., Unnikrishnan, A. S., & Dayan, H. (2013). Decadal and long-term sea level variability in the tropical Indo-Pacific Ocean. *Climate dynamics*, *41*, 381–402.

O'Neill, B. C., Tebaldi, C., Van Vuuren, D. P., Eyring, V., Friedlingstein, P., Hurtt, G., & Sanderson, B. M. (2016). The scenario model intercomparison project (ScenarioMIP) for CMIP6. *Geoscientific Model Development*, *9*(9), 3461–3482.

Oppenheimer, M., Glavovic, B., Hinkel, J., van de Wal, R., Magnan, A. K., Abd-Elgawad, A., & Sebesvari, Z. (2019). *Sea level rise and implications for low lying islands, coasts and communities*.

Park, J. Y., Schloesser, F., Timmermann, A., Choudhury, D., Lee, J. Y., & Nellikkattil, A. B. (2023). Future sea-level projections with a coupled atmosphere-ocean-ice-sheet model. *Nature Communications*, *14*(1), 636.

Qu, Y., Jevrejeva, S., Williams, J., & Moore, J. C. (2022). Drivers for seasonal variability in sea level around the China seas. *Global and Planetary Change*, *213*, 103819.

Reager, J. T., Gardner, A. S., Famiglietti, J. S., Wiese, D. N., Eicker, A., & Lo, M. H. (2016). A decade of sea level rise slowed by climate-driven hydrology. *Science (New York, N.Y.)*, *351*(6274), 699–703.

Richter, K., Meyssignac, B., Slangen, A. B., Melet, A., Church, J. A., Fettweis, X., & Champollion, N. (2020). Detecting a forced signal in satellite-era sea-level change. *Environmental Research Letters*, *15*(9), 094079.

Roemmich, D., & Gilson, J. (2009). The 2004–2008 mean and annual cycle of temperature, salinity, and steric height in the global ocean from the Argo Program. *Progress in oceanography*, *82*(2), 81–100.

Saintilan, N., Horton, B., & Guntenspergen, G. (2023). Widespread retreat of coastal habitat is likely at warming levels above 1.5 C. *Nature*.

Samanta, D., Goodkin, N. F., & Karnauskas, K. B. (2021). Volume and heat transport in the South China Sea and Maritime Continent at present and the end of the 21st century. *Journal of Geophysical Research: Oceans*, e2020JC016901. Available from https://doi.org/10.1029/2020JC016901.

Samanta, D., Hameed, S. N., Jin, D., Thilakan, V., Ganai, M., Rao, S. A., & Deshpande, M. (2018). Impact of a narrow coastal Bay of Bengal sea surface temperature front on an Indian summer monsoon simulation. *Scientific Reports*, *8*(1), 1–12.

Samanta, D., Jevrejeva, S., Karnauskas, K. B., Palanisamy, H., Goodkin, N. F., Chew, L. Y., ... Horton, B. P. (2024). Dynamic sea level in Southeast Asia by the end of 21st century. *Journal of Geophysical Research Oceans*. (under review).

Samanta, D., Karnauskas, K. B., & Goodkin, N. F. (2019). Tropical Pacific SST and ITCZ biases in climate models: Double trouble for future rainfall projections? *Geophysical Research Letters*, *46*(4), 2242–2252.

Samanta, D., Karnauskas, K. B., Goodkin, N. F., Coats, S., Smerdon, J. E., & Zhang, L. (2018). Coupled model biases breed spurious low-frequency variability in the tropical Pacific Ocean. *Geophysical Research Letters*, *45*(19), 10–609.

Samanta, D., Vairagi, V., Richter, K., McDoagh, E. L., Karnauskas, K. B., Goodkin, N. F., ... Horton, B. P. (2024). The role of anthropogenic forcings on historical sea-lvel change in the Indo-Pacific warm pool region. *Earth's Future*. Available from https://doi.org/10.1029/2023EF003684.

Scafetta, N. (2014). Multi-scale dynamical analysis (MSDA) of sea level records versus PDO, AMO, and NAO indexes. *Climate Dynamics*, *43*(1-2), 175–192.

Shaw, T. (2023). Late quaternary relative sea-level changes in the tropics. In Scott A. Elias, & Cary J. Mock (Eds.), *Encyclopedia of quaternary science* (2nd ed., pp. 495–502). Elsevier, ISBN 9780444536426.

Slangen, A. B. A., Adloff, F., Jevrejeva, S., Leclercq, P. W., Marzeion, B., Wada, Y., & Winkelmann, R. (2017). A review of recent updates of sea-level projections at global and regional scales. *Integrative Study of the Mean Sea level and Its Components*, 395–416.

Slangen, A. B., Church, J. A., Zhang, X., & Monselesan, D. (2014). Detection and attribution of global mean thermosteric sea-level change. *Geophysical Research Letters*, *41*(16), 5951–5959.

Stammer, D., Cazenave, A., Ponte, R. M., & Tamisiea, M. E. (2013). Causes for contemporary regional sea level changes. *Annual Review of Marine Science*, *5*(1), 21–46. Available from https://doi.org/10.1146/annurev-marine-121211-172406.

Swapna, P., Ravichandran, M., Nidheesh, G., Jyoti, J., Sandeep, N., Deepa, J. S., Unnikrishnan, A. S. (2020). Sea-level rise. Government of India, pp. 175–189.

Tay, C., Lindsey, E. O., Chin, S. T., McCaughey, J. W., Bekaert, D., Nguyen, M., & Hill, E. M. (2022). Sea-level rise from land subsidence in major coastal cities. *Nature Sustainability*, 1–9.

Taylor, K. E., Stouffer, R. J., & Meehl, G. A. (2012). An overview of CMIP5 and the experiment design. *Bulletin of the American meteorological Society*, *93*(4), 485–498.

Thompson, B., Jevrejeva, S., Zachariah, J., Faller, D. G., & Tkalich, P. (2023). Impact of mass redistribution on regional sea level changes over the South China Sea shelves. *Geophysical Research Letters*, *50*(23), e2023GL105740.

Todd, A., Zanna, L., Couldrey, M., Gregory, J., Wu, Q., Church, J. A., & Zhang, X. (2020). Ocean-only FAFMIP: Understanding regional patterns of ocean heat content and dynamic sea level change. *Journal of Advances in Modeling Earth Systems*, *12*(8), e2019MS002027.

Tokarska, K. B., Hegerl, G. C., Schurer, A. P., Ribes, A., & Fasullo, J. T. (2019). Quantifying human contributions to past and future ocean warming and thermosteric sea-level rise. *Environmental Research Letters*, *14*(7), 074020.

Topál, D., Ding, Q., Ballinger, T. J., Hanna, E., Fettweis, X., Li, Z., & Pieczka, I. (2022). Discrepancies between observations and climate models of large scale wind-driven Greenland melt influence sea-level rise projections. *Nature Communications*, *13*(1), 6833.

Uehara, K., Sojisuporn, P., Saito, Y., & Jarupongsakul, T. (2010). Erosion and accretion processes in a muddy dissipative coast, the Chao Phraya River delta, Thailand. *Earth Surface Processes and Landforms*, *35*(14), 1701–1711.

Unnikrishnan, A. S., & Shankar, D. (2007). Are sea-level-rise trends along the coasts of the north Indian Ocean consistent with global estimates? *Global and Planetary Change*, *57*(3-4), 301–307.

Vermeer, M., & Rahmstorf, S. (2009). Global sea level linked to global temperature. *Proceedings of the National Academy of Sciences*, *106*(51), 21527–21532.

Vialard, J., Duvel, J. P., Mcphaden, M. J., Bouruet-Aubertot, P., Ward, B., Key, E., ... Kennan, S. (2009). Supplement to cirene: Air—sea interactions in the seychelles—chagos thermocline ridge region. *Bulletin of the American Meteorological Society*, *90*(1), ES1–ES4.

Walker, J. S., Kopp, R. E., Little, C. M., & Horton, B. P. (2022). Timing of emergence of modern rates of sea-level rise by 1863. *Nature Communications*, *13*(1), 1–8.

Watkins, M. M., Wiese, D. N., Yuan, D. N., Boening, C., & Landerer, F. W. (2015). Improved methods for observing Earth's time variable mass distribution with GRACE using spherical cap mascons. *Journal of Geophysical Research: Solid Earth*, *120*(4), 2648–2671.

Wickramage, C., Köhl, A., Jungclaus, J., & Stammer, D. (2023). Sensitivity of MPI-ESM sea level projections to its ocean spatial resolution. *Journal of Climate*, *36*(6), 1957–1980.

Woodworth, P. L., Hunter, J. R., Marcos, M., Caldwell, P., Menéndez, M., & Haigh, I. (2016). Towards a global higher-frequency sea level dataset. *Geoscience Data Journal*, *3*(2), 50–59.

Wouters, B., Gardner, A. S., & Moholdt, G. (2019). Global glacier mass loss during the GRACE satellite mission (2002–2016). *Frontiers in Earth Science, 7*, 96.

Wyrtki, K. (1961). *Physical oceanography of the Southeast Asian waters* (Vol. 2). University of California, Scripps Institution of Oceanography.

Yue, C., Jevrejeva, S., Qu, Y., Zhao, L., & Moore, J. C. (2023). Thermosteric and dynamic sea level under solar geoengineering. *NPJ Climate and Atmospheric Science*, *6*(1), 135.

Chapter 20

An overview of atmospheric carbon dioxide and global warming interlinks

Chiranjit Das[1], Prabir K. Patra[2,3] and Ravi Kumar Kunchala[1]

[1]*Centre for Atmospheric Sciences, Indian Institute of Technology Delhi, New Delhi, NCT of Delhi, India,* [2]*Japan Agency for Marine-Earth Science and Technology (JAMSTEC), Yokohama, Kanagawa, Japan,* [3]*Research Institute for Humanity and Nature, Kyoto, Kyoto Prefecture, Japan*

20.1 Introduction

The Earth's atmosphere is composed of nitrogen (N_2; 78.08%), oxygen (O_2; 20.95%), argon (Ar; 0.93%), and trace gases (0.04%) that include both inert gases, such as neon, helium, and krypton, as well as greenhouse gases (GHGs) and reactive species. GHGs maintain a habitable temperature for earth's living organisms; however, the recent dramatic increase in their concentration presents numerous direct and indirect climate threats experienced in the present years. Among various GHGs, including carbon dioxide (CO_2), methane (CH_4) and nitrous oxide (N_2O), CO_2 stands the most significant anthropogenic GHG in the atmosphere and will be primarily emphasized in this chapter in terms of its contribution to global warming. According to the Intergovernmental Panel on Climate Change (IPCC) 6th Assessment Report (AR6), atmospheric concentrations of CO_2, CH_4, and N_2O have increased by 47%, 156%, and 23%, respectively, since the preindustrial level (i.e., 1750), primarily as a result of human activities (Canadell et al., 2021). This increase in weighted-mean greenhouse gas concentration leads to a rise in global-mean surface air temperature (GSAT) by trapping excess heat through absorption of the earth's outgoing longwave radiation (Liou, 1980).

The relationship between atmospheric CO_2 and global warming is well-established and supported by multiple lines of evidence. The increase in CO_2 has been accompanied by a 1.07°C rise in the average surface temperature of the Earth during the 2000–2019 period compared to the 1850–1900 period (direct influence). Further, the rising CO_2 also indirectly affects in the form of melting terrestrial permafrost, sea level rise, and ocean acidification, which can further contribute to global warming (IPCC, 2021). Studies show the increase in anthropogenic CO_2 emission is partly removed by natural carbon sinks present in both land and ocean, with an estimated cumulative removal of 54% of total carbon emission in 2000–2019 or 56% over the period of 1960–2019, leaving about 46% or 44%, respectively, in the atmosphere (IPCC, 2021). The amount that stays in the atmosphere is often referred to as the "airborne fraction."

The primary land sink in the biosphere occurs through the photosynthesis process by plants and in soil through the formation of soil organic matter of dead plants, animals, and microorganisms. On the other hand, the ocean acts as a net sink, facilitating different processes like mixing/circulation (physical process), phytoplankton photosynthesis (biological process), and chemical processes such as bicarbonate formation, which leads to ocean acidification (Orr et al., 2005; Sabine et al., 2004). Strengthening of this sink capacity partly slowed the rate of increase in global temperature, removing more CO_2 from the atmosphere through CO_2 fertilization effects in the terrestrial biosphere and ocean. Although regional heterogeneity in sink response to climate change is observed, in the tropics, a declining trend in sink strength is reported, whereas mid-latitude and boreal forests are turning into net stronger sinks of CO_2 (Hubau et al., 2020; Keeling, Chin, & Whorf, 1996). However, global total land and ocean sinks respond slower than the CO_2 concentration or emission rise. Therefore, we must reduce fossil fuel emissions to meet the Paris Agreement commitments and avoid a 2°C increase in Earth's surface air temperature by 2100.

Since emissions and global warming are inextricably linked, researchers emerged with the concept of a "carbon budget" or an estimation of a "remaining carbon budget" to determine the maximum amount of cumulative CO_2 emissions that would be acceptable while maintaining to stay under a global warming level. The remaining carbon budget for reaching 1.5°C, 1.7°C, and 2°C warming is now estimated at 380, 730, and 1230 $GtCO_2$, which could be reached in only 9, 18,

Sustainable Development Perspectives in Earth Observation. DOI: https://doi.org/10.1016/B978-0-443-14072-3.00016-2

and 30 years if CO_2 emissions remain at 2022 levels (Friedlingstein et al., 2022). The estimates are also subject to uncertainty due to uncertainty in the budget component estimates of fossil CO_2 emission (Liu et al., 2015; Saeki & Patra, 2017), net land use change fluxes (Bastos et al., 2021; Houghton et al., 2012), ocean (Aumont et al., 2001; Gloege et al., 2021), and land sink fluxes (Cox et al., 2013; Walker et al., 2021). Nevertheless, to avoid the consequences of climate change, immediate, rapid, and large-scale reductions in emissions are required to meet the net-zero emission commitment that demands alternative climate-friendly energy solutions (IPCC, 2021). In this context, renewable energy growth could help reduce dependency on traditional fossil energy sources to meet the energy requirement, emission reduction, and limit global warming (Gielen et al., 2019). This could be achieved through reducing solar power and battery storage prices (Bloomberg New Energy Finance BNEF, 2021). Additionally, fulfilling ambitions like cutting down CO_2 emissions by 50% by the year 2050 to minimize the possibility of crossing the 2°C limit (Meinshausen et al., 2009).

Atmospheric CO_2 emissions due to human activities and its complex and ever-changing link with global warming have significant implications for the earth's temperature and its inhabitants' well-being. In this context, to determine the emission sources and sinks pattern (location and time evolution) response to global warming, a chemistry transport model, or "top-down," is generally employed, which links the atmospheric concentration with emission. This earth observation-based modeling methodology was first introduced in the 1980s to understand the global pattern of sources and sinks of CO_2 (Heimann & Keeling, 1986; Tans, Conway, & Nakazawa, 1989). Other than this, the "bottom-up" modeling approach is also widely used to interpret the carbon source-sink pattern through observations based on the land and ocean over the earth's surface (Huntzinger et al., 2012; Matsumoto et al., 2004). A detailed discussion of this modeling is not possible to cover here; however, readers will find many aspects of carbon cycle science through this chapter. In this chapter, the readers will be introduced to different CO_2 observational systems (surface, aircraft, and satellite-based), variability of CO_2 at different time and space scales, main components of the global CO_2 budget, the role of different atmospheric constituents for global warming, future projection of carbon sink changes (response and feedback) under different emission scenarios, and lastly, the mitigation techniques.

20.2 Observations of atmospheric CO_2

The importance of CO_2 in global warming has been well documented since the 19th century; using the basic principles of physical chemistry, Svante Arrhenius was the first to calculate the extent to which earth's surface temperature would increase if CO_2 increased in the atmosphere through the greenhouse effect (Arrhenius, 1896). In the late 1950s, the high-quality in situ measurement of CO_2 was first conducted by Charles David Keeling of the Scripps Institution of Oceanography at the South Pole, Antarctica, and soon after at Mauna Loa (MLO), Hawaii (Keeling, 1960). Since then, the network of CO_2 measurement has been expanded to hundreds of sites globally, coordinated mainly by the World Meteorological Organization (WMO), and the collected measurements are made accessible through the World Data Centre for Greenhouse Gases (WDCGG). In situ air sampling programs are also maintained from mobile platforms, such as ships and aircraft. Later, in the 2000s, to improve the spatial and temporal coverage of CO_2 measurements for a better understanding of the global pattern of CO_2, space-based satellite remote sensing technology was used to supplement the limited spatial coverage of the in situ networks.

20.2.1 In situ measurements

Surface-based in situ CO_2 measurements are made in ambient air near the earth's surface over land or ocean, usually from a tower or mast mounted over a hundred meters above the surface. These CO_2 dry-air mole fraction measurements are expressed in units of parts per million by volume, or ppm refers to the number of CO_2 molecules per 10^6 dry-air molecules. The first direct measurement of atmospheric CO_2 was started at Mauna Loa station, Hawaii, and the South Pole Observatory (SPO) in the late 1950s (Keeling et al., 2001). Later, the surface in situ network is expanded globally and routinely takes continuous or weekly CO_2 measurements at WMO Global Atmospheric Watch (GAW) and other regionally representative stations network under NOAA Earth System Research Laboratories (NOAA ESRL; https://www.esrl.noaa.gov/gmd/ccgg) and Integrated Carbon Observation System (ICOS; https://www.icos-ri.eu/home). More details about the operational sites and measurement techniques at the networks can be accessed using the website links.

Additionally, vertical profiles of CO_2 are measured in the troposphere and lower stratosphere using aircraft. NOAA routinely collects airborne CO_2 profiles every 2–3 weeks from sites mainly in North America (https://gml.noaa.gov/dv/data/) (Sweeney et al., 2015). Tohoku University has been running aircraft and stratospheric balloon measurement programs over Japan since the 1980s (http://caos.sakura.ne.jp/tgr/) (Nakazawa, Miyashita, Aoki, & Tanaka, 1991). Other notable aircraft

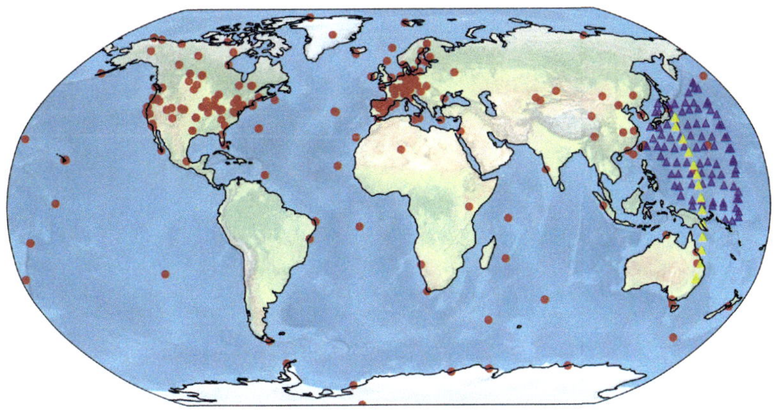

FIGURE 20.1 Spatial distribution of in situ observation locations (ground-based network and on mobile platforms such as aircraft and ships). Most sites are operated to collect CO_2 observation, and data are submitted to the WDCGG (World Data Centre for Greenhouse Gases), an effort under the WMO Global Atmosphere Watch. Here, ground-based site information taken from NOAA's obspack_co2_1_GLOBALVIEWplus_v6.1 product (https://gml.noaa.gov/ccgg/obspack/), Keifu Maru (KEF) ship CO_2 measurements locations from Japan Meteorological Agency (https://gaw.kishou.go.jp), and aircraft air-borne CO_2 sampling locations from CONTRAIL aircraft (https://www.cger.nies.go.jp/contrail/) between Japan and Australia. Map lines delineate study areas and do not necessarily depict accepted national boundaries.

measurements are Comprehensive Observation Network for Trace Gases by AIrLiner (CONTRAIL) (Machida et al., 2008), HIAPER Pole-to-Pole (HIPPO) (Wofsy, 2011), In-Service Aircraft for a Global Observing System (IAGOS) (Petzold et al., 2015), and Atmospheric Tomography (ATom) Mission (Thompson, Chevallier, Maksyutov, Patra, & Bowman, 2022; Thompson, Wofsy, et al., 2022). Further, partial CO_2 pressure in seawater is measured using research and commercial ship cruises, for example, the SOCAT database (Bakker et al., 2016), NIES (Tohjima et al., 2012), and the Global Ocean Data Analysis Project (GODAP) (Olsen et al., 2020). These data products enabled robust estimation of distribution, trends, and variability of net air–sea CO_2 fluxes (Gregor, Kok, & Monteiro, 2018; Takahashi et al., 2009) (Fig. 20.1).

20.2.2 Remote sensing measurements

The advancement of space-based satellite remote sensing technology enabled us to observe CO_2 at very high resolution with greater spatiotemporal coverage compared to surface in situ networks. Satellite collects CO_2 at column average dry air mixing ratio, that is, mixing ratio of columnar CO_2 abundance to molecular O_2 abundance, termed as XCO_2. These satellites retrieve spectral information of reflected sunlight within CO_2 and O_2 bands, which are later analyzed to estimate the spatiotemporal global distribution of XCO_2. The first space-based instrument to facilitate different trace gases, including CO_2, was the Scanning Imaging Absorption SpectroMeter for Atmospheric CartograpHY (SCIAMACHY), which was a boarded European Space Agency (ESA) Environmental Satellite (ENVISAT) satellite in 2002 and was operational till 2012 (Buchwitz et al., 2005). Later, Japan's Greenhouse gas Observing SATellite "IBUKI" (GOSAT) was launched in 2009 to retrieve information of two most important greenhouse gases, CO_2 and CH_4, at high precision, enabling estimation of regional scale sources and sinks by inverse modeling (Hamazaki, Kaneko, Kuze, & Kondo, 2005). GOSAT's Thermal and Near-Infrared Sensor for Carbon Observation Fourier-Transform Spectrometer (TANSO-FTS) retrieves XCO_2 from reflected sunlight from the Earth's surface at 1.6 and 2.0 μm with 0.2 cm^{-1} spectral resolution (Kuze, Suto, Nakajima, & Hamazaki, 2009) (Fig. 20.2).

GOSAT is followed by NASA's Orbiting Carbon Observatory 2 (OCO-2) satellite, launched in 2014, having higher spatial resolution and retrieval accuracy than prior carbon-measuring satellites (Crisp, 2015). A monthly snapshot of February, 2015 of GOSAT and OCO-2 soundings could be visualized in Fig. 20.3. The disparity in the number of footprints between the two satellites clearly illustrates the high data density of OCO-2, with approximately 100 times as many samples per day or month recorded by OCO-2 as compared to GOSAT (Crisp et al., 2022; Das, Kunchala, Chandra, Chhabra, & Pandya, 2023). This is attributable to the OCO-2's powerful high-resolution grating spectrometer instrument capable of high precision and accurate measurement of CO_2 with an increasing number of cloud-free soundings to capture precise regional scale CO_2 variation (O'Dell et al., 2018). The OCO-2 series measurements are made at the same wavelength ranges used in GOSAT, detected by diffraction grating spectrometers on satellites (O'Dell et al., 2018). Most recently, GOSAT-2 and OCO-3 were launched in 2018 and 2019, respectively, to provide additional spatiotemporal

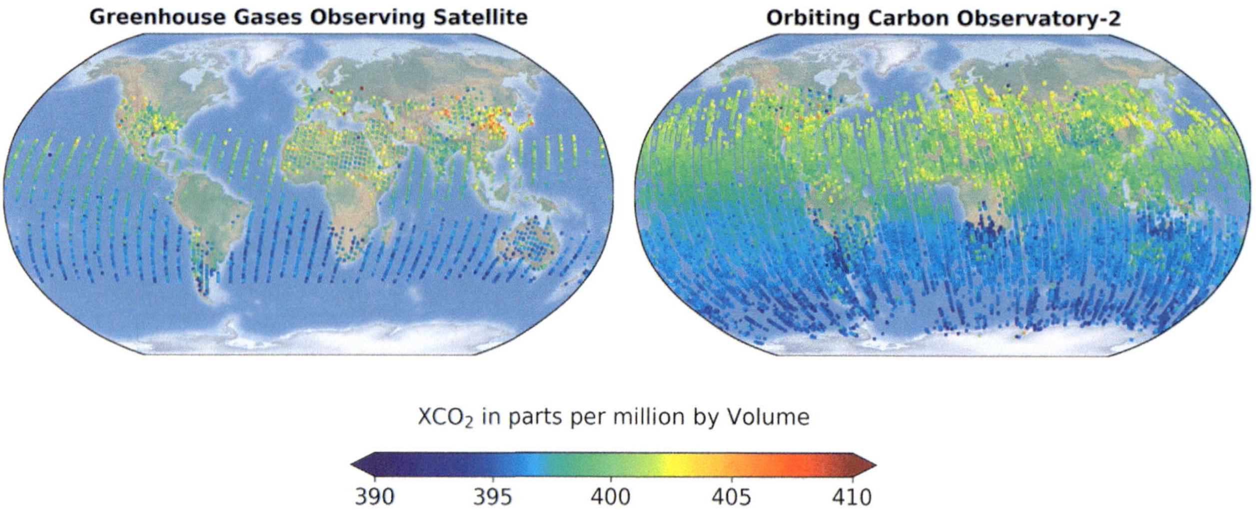

FIGURE 20.2 Monthly column-averaged CO_2 or XCO_2 distribution from GOSAT and OCO-2 satellite observation between February 1 to February 28 of 2015 (GOSAT: https://data2.gosat.nies.go.jp; OCO-2: https://co2.jpl.nasa.gov/). Map lines delineate study areas and do not necessarily depict accepted national boundaries.

coverage and resolution. GOSAT-2 is capable of locating carbon sources of combustion because it is equipped with a new CO channel, and OCO-3 is now able to take XCO_2 observation of a small area (100 × 100 km) like a large urban region (Thompson, Chevallier, et al., 2022; Thompson, Wofsy, et al., 2022). These satellite retrievals provide global XCO_2 distribution with greater integrated details of regional carbon source-sink mechanisms and time evolution.

Apart from space-based measurements of XCO_2, an upward-looking remote sensing technique was also adopted at Total Carbon Column Observing Network (TCCON) sites covering latitudes from 45°S to 80°N using FTS (Wunch et al., 2011). More details of TCCON sites can be accessed at https://tccondata.org/. The TCCON-retrieved XCO_2 forms the basis of validations of satellite XCO_2 measurements and acts as a "bridge" to the transfer standard between satellite and in situ networks (Inoue et al., 2016; O'Dell et al., 2018; Wunch et al., 2011). Because the TCCON instruments are detecting direct solar radiation, even though in identical wavelength ranges, the spectra are less disturbed by the airmass factor in comparison to that experienced by the reflected sunlight for the satellites.

20.3 Atmospheric CO_2 long-term variability

Thus far, the discussion has been focused on the preliminary aspects of in situ networks and recent satellite observation, aimed at comprehending the long-term variability of atmospheric CO_2, as these measurements provide unparalleled insight into carbon source-sink activity. Time series of CO_2, at monthly average time, are shown for two representative stations of the northern (MLO) and southern hemisphere (SPO), respectively, and global mean CO_2 variability from satellite measurements (GOSAT and OCO-2). The results clearly demonstrate CO_2 concentration increases with time recorded in both satellite and in situ measurements (Fig. 20.3A). The longest-running observations from MLO and SPO indicate a significant increase of 100 ppm in CO_2 concentration during the period from 1958 to 2022. A positive difference in the yearly mean CO_2 concentrations between MLO and SPO stations serves as evidence of a higher rate of increase in CO_2 concentration in the northern hemisphere compared to the southern hemisphere. This can largely be attributed to the fact that most anthropogenic emission sources are located in the northern hemisphere, and also the net land and ocean uptakes are smaller than anthropogenic emissions.

The CO_2 growth rate (ppm yr^{-1}), which denotes the rate of increase in CO_2 concentration over time, has exhibited an upward trend, rising from approximately 1 ppm yr^{-1} in 1958 to nearly 2.5 ppm yr^{-1} in 2022, primarily because of the increase in anthropogenic emissions (Fig. 20.3B). The CO_2 growth rate also exhibits interannual variability, owing to the interannual variability of land-atmosphere CO_2 exchange over highly climate-sensitive tropical regions, which is largely controlled by the El Niño-Southern Oscillation (ENSO) events in the tropical Pacific. This interannual variability of CO_2 growth rate generally shows an increase (decrease) during the warm (cold) phase of ENSO, which is mainly controlled by tropical land CO_2 fluxes (Canadell et al., 2021; Das et al., 2022). The satellite retrievals of XCO_2 from GOSAT (2009–2022) and OCO-2 (2014–2022) showed similar concentration and growth rate variability as in situ measurements.

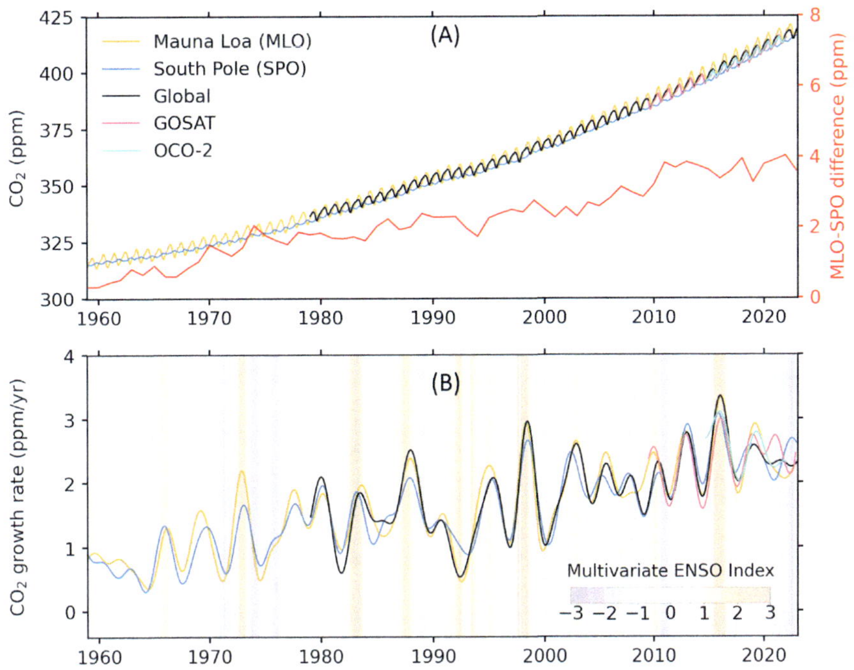

FIGURE 20.3 Time series of CO_2 concentration (top panel; A) and growth rate (bottom panel; B) using in situ observations (Mauna Loa, South Pole, and Global mean) and satellite observations (GOSAT, and OCO-2) till December 2022 (MLO, SPO: https://scrippsco2.ucsd.edu:, Global: https://gml.noaa.gov, GOSAT: https://data2.gosat.nies.go.jp, OCO-2: https://co2.jpl.nasa.gov). The growth rate (ppm/yr) is calculated by taking time derivatives on the long-term trend (LTR) (d [LTR]/dt) of concentration time series following Thoning, Tans, and Komhyr (1989). *This figure is inspired from Figure 5.6 of Canadell, J. G., Monteiro, P. M. S., Costa, M. H., da Cunha, L. C., Cox, P. M., Eliseev, A. V., et al. (2021). Global carbon and other biogeochemical cycles and feedbacks. In* Climate Change 2021: The physical science basis. Contribution of working group I to the sixth assessment report of the Intergovernmental Panel on Climate Change. *Cambridge University Press, Cambridge, United Kingdom and New York, NY, USA, pp. 673–816, https://doi.org/10.1017/9781009157896.007.*

20.4 Atmospheric CO_2 budget

20.4.1 Global carbon source and sinks balance

The global anthropogenic carbon emission, contributions from fossil fuels (FF) and land use changes (LUC), has exhibited an almost linear increase from approximately 4 GtC yr^{-1} in 1960 to a level of ~12 GtC yr^{-1} in 2019 (Canadell et al., 2021). This significant tripling of the emission rate over a period of sixty years is mainly because of human consumption of fossil fuels, which constitutes the primary source of total anthropogenic emissions. Among the various types of fossil fuels (coal, oil, and gas), the emission from coal-based fossil fuels is the dominant contributor of FF emission (Friedlingstein et al., 2022). The second largest contributor to anthropogenic emissions is land use change (LUC)-based emission, where deforestation is the main dominant emission source compared to other LUC based sources like organic soil and other transitions. The LUC also consists of sink components, such as afforestation and regrowth of natural vegetation after harvesting. Despite these sink components, the overall net effect of LUC remains a carbon source, as gross emissions outweigh gross removals (Friedlingstein et al., 2022).

The carbon sinks are integral for CO_2 removal from the atmosphere and a significant component of the CO_2 budget. CO_2 uptake through photosynthesis processes by vegetation and soil carbon storage in the form of organic matter serves as primary carbon sinks. The ocean acts as a persistent sink with CO_2 uptake facilitated by phytoplankton photosynthesis, air–sea gas exchange, and deep ocean CO_2 storage through the ocean currents. The components of the global CO_2 budget include FF and LUC emissions, which are partitioned into the atmosphere in terms of atmospheric CO_2 burden, land sink, and ocean sink (Fig. 20.4A). A time series of combined partitioned components and their relative contributions to carbon budgets for periods 1960–2021 is shown in Fig. 20.4(B). It is observed that emission components of carbon budgets have grown tremendously since 1960, largely dominated by FF emission, and further land sink and atmospheric CO_2 burden showing the large interannual variability, while the variability in ocean sink is minimal (Bousquet et al., 2000).

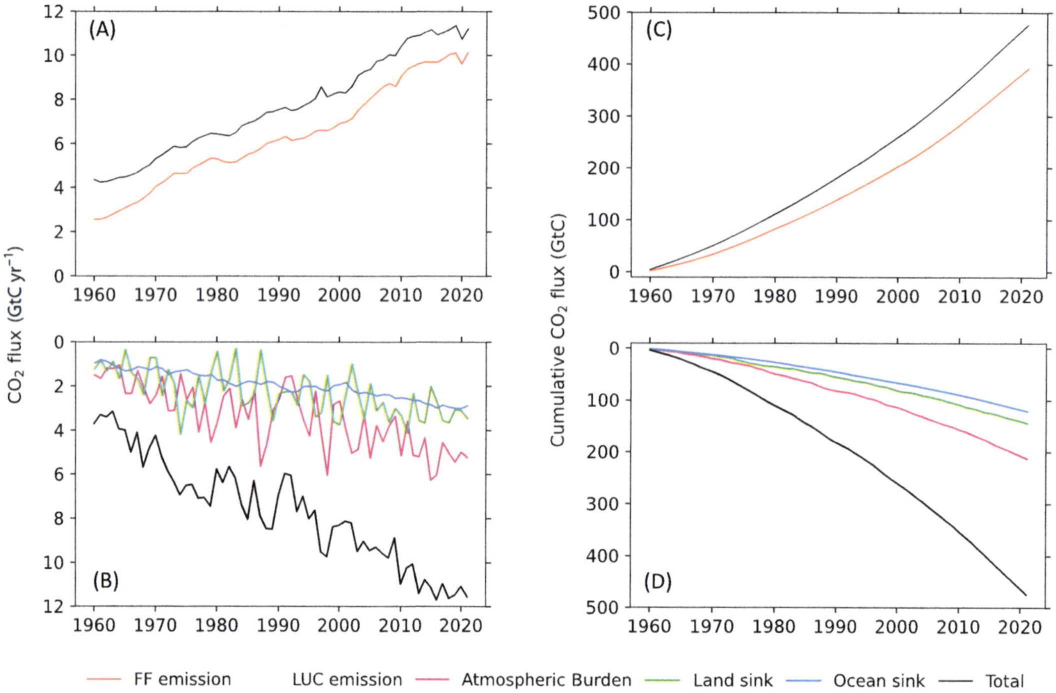

FIGURE 20.4 Time series of annual CO_2 fluxes (in GtC yr^{-1}) of global carbon budget components represented for anthropogenic emissions (fossil fuel and land-use change emission), their partitioning into atmospheric burden, land sink and ocean sink (left column; panels A and B) as well as. The cumulative fluxes (estimated by summing all prior annual fluxes since the year 1960) of the mentioned components and their total presented in units of GtC (right column; panels C and D) for 1960–2021. *Annual mean CO_2 flux data are taken from Friedlingstein, P., O'Sullivan, M., Jones, M. W., Andrew, R. M., Gregor, L., Hauck, J. et al. (2022). Global carbon budget 2022.* Earth System Science Data, 14*(11), 4811–4900 https://doi.org/10.5194/essd-14-4811–2022.*

The budget mismatch between total emissions and partitioned fluxes is considered a net budget imbalance (B_{IM} = anthropogenic emission-partitioned emission). The substantial semidecadal variability in this budget imbalance, with minimal change over time, is unlikely to be solely attributed to emissions, which generally exhibit an increasing trend. Therefore, errors in land and ocean sink estimation are the most probable possibility for this budget imbalance (Friedlingstein et al., 2022). Cumulative flux of the constituting components of the global carbon budget since 1960 is also exhibited (Fig. 20.4C and D). Total cumulative CO_2 emission flux reached nearly 500 GtC with contributions of approximately 400 GtC and 100 GtC from FF and LUC, respectively. On the other hand, the cumulative flux of partitioned components is dominated by atmospheric burden, then the land and ocean uptakes.

20.4.2 Airborne fraction

Along with the increase of CO_2 concentration or total anthropogenic emissions, carbon sinks also continued to grow; otherwise, the CO_2 growth rate would have been twice as much as observed (Figs. 20.3 and 20.4), and its feedback on the earth's temperature would have reached an uninhabitable situation. Strengthening of sink is evident in fraction of total emission that stays in the atmosphere or airborne fraction (AF), mainly responsible for global warming, which has remained constant at 44% over the past 60 years despite its large year-to-year variability, which coincides with ENSO events (Ballantyne, Alden, Miller, Tans, & White, 2012; Canadell et al., 2021; Patra, Maksyutov, & Nakazawa, 2005; Raupach, Canadell, & Le Quéré, 2008). During El Niño (La Niña) episodes, we observe a large (small) fraction of CO_2 staying in the atmosphere, suggesting natural carbon sinking strength on land and ocean becomes less (more) effective in these conditions (Fig. 20.5). Remarkably, there are some volcanic eruption years when AF is low (not necessarily coinciding with La Niña years), implying high sink efficiency because eruption-injected sulfate aerosols in the atmosphere have a cooling effect in the atmosphere through blocking incoming solar radiation and effectively increase the photosynthetic activity under diffuse radiation (Gu et al., 2003). Further, during major drought years (like the drought of 1988, 1997, and 2015), we see high AF, which indicates low sink efficiency due to suppressed photosynthetic activity resulting from high water stress in dry conditions, as well as increased forest mortality and induced forest wildfires (Brando et al., 2019; Patra, Ishizawa, Maksyutov, Nakazawa, & Inoue, 2005).

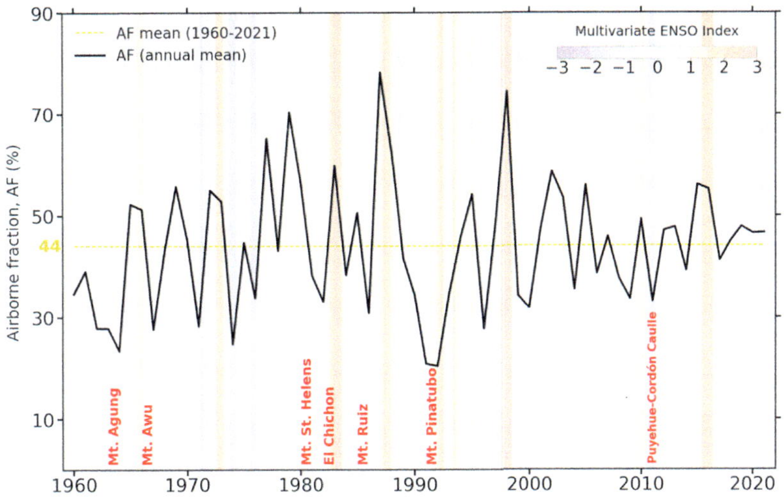

FIGURE 20.5 Time series of airborne fraction (AF) in % with Multivariate ENSO index (shaded background). Occurrences of major volcanic eruptions are represented by text in red color font along the x-axis. Annual mean CO_2 flux data are taken from Friedlingstein et al. (2022), ENSO index is taken from https://psl.noaa.gov/enso/mei. AF is calculated by taking the ratio of CO_2 growth rate to the sum of anthropogenic emission from fossil fuel and land use change. *This figure is inspired from figure 5.7 of Canadell, J. G., Monteiro, P. M. S., Costa, M. H., da Cunha, L. C., Cox, P. M., Eliseev, A. V., et al. (2021). Global carbon and other biogeochemical cycles and feedbacks. In* Climate Change 2021: The physical science basis. Contribution of working group I to the sixth assessment report of the Intergovernmental Panel on Climate Change. *Cambridge University Press, Cambridge, United Kingdom and New York, NY, USA, pp. 673–816, https://doi.org/10.1017/9781009157896.007.*

20.5 Role of atmospheric constituents in global-mean surface air temperature change

20.5.1 Role of CO_2, other greenhouse gases and aerosols

In this section, we discuss the roles of various atmospheric constituents, such as aerosols (particulate matter) and other GHGs, in addition to CO_2, in changing the surface air temperature. The effects of atmospheric compositions on earth's climate are studied using the Coupled Model Intercomparison Project (CMIP); the latest version 6 (CMIP6) was adopted for the IPCC AR6 report (IPCC, 2021) for preparation of the overall climate change assessment (Fig. 20.6). Methane (CH_4), Nitrous Oxide (N_2O), and halogenated gases are the other GHGs that have a warming effect on the global climate. CH_4 is a more powerful warming agent than CO_2 because its global warming potential (GWP) is about 29 (81) times higher than CO_2 in a 100 (20) year time scale (Forster et al., 2021). However, the concentration or emission of CH_4 is less than CO_2, therefore its overall warming effect is still less than CO_2 (Fig. 20.6C). The natural sources of methane emission are wetlands (180 $TgCH_4\ yr^{-1}$) and other natural sources (36 $TgCH_4\ yr^{-1}$) contributed mainly by freshwater (lakes and rivers) and geological onshore/offshore processes during 2010–2019. The anthropogenic CH_4 emission sources are agriculture and waste (221 Tg $CH_4\ yr^{-1}$), fossil fuel (106 $TgCH_4\ yr^{-1}$), and biomass burning and biofuel (30 $TgCH_4\ yr^{-1}$) (Canadell et al., 2021; Chandra et al., 2021; Saunois et al., 2019). Another important greenhouse gas that has received a lot of attention is N_2O, notable for its GWP of 273 on 20- and 100-year timescale, long lifetime of 116 years (Forster et al., 2021), and its role in stratospheric ozone loss (Ravishankara, Daniel, & Portmann, 2009). It is mainly emitted from natural soils ($\sim$7.0 TgN yr^{-1}) and via oceanic exchanges ($\sim$2.9 TgN yr^{-1}) under naturally occurring anaerobic conditions and released due to human activities such as use of synthetic fertilizers and manures ($\sim$4.0 TgN yr^{-1}) and industrial processes ($\sim$2.5 TgN yr^{-1}) during 2010–2019 (Patra et al., 2022; Thompson et al., 2019; Tian et al., 2020).

The aerosols, on the other hand, offset a great fraction of the global warming caused by the GHGs. Aerosols are suspended solid or liquid particles present in the atmosphere, interact with solar and terrestrial radiations directly or indirectly, affect earth's radiation balance, and have the potential to perturb global and regional climates (Charlson et al., 1992). The primary anthropogenic aerosol compounds are sulfate, nitrate, black carbon, and organic carbon. Sulfate (SO_4^-) aerosols are the main compounds that cool the lower atmosphere by reflecting the incoming solar radiation. SO_4^- is released directly to the atmosphere or produced secondarily from both the natural (volcanic eruptions) and anthropogenic (combustion of coal and sulfate-based fossil fuels) sources as sulfur dioxide (SO_2). The manifestation of less solar energy reaching to the earth's surface due to the sulfate aerosols is known as the solar dimming effect when looked up from ground (Stanhill & Cohen, 2001). Black carbon (BC) is another important aerosol compound that warms the planet through

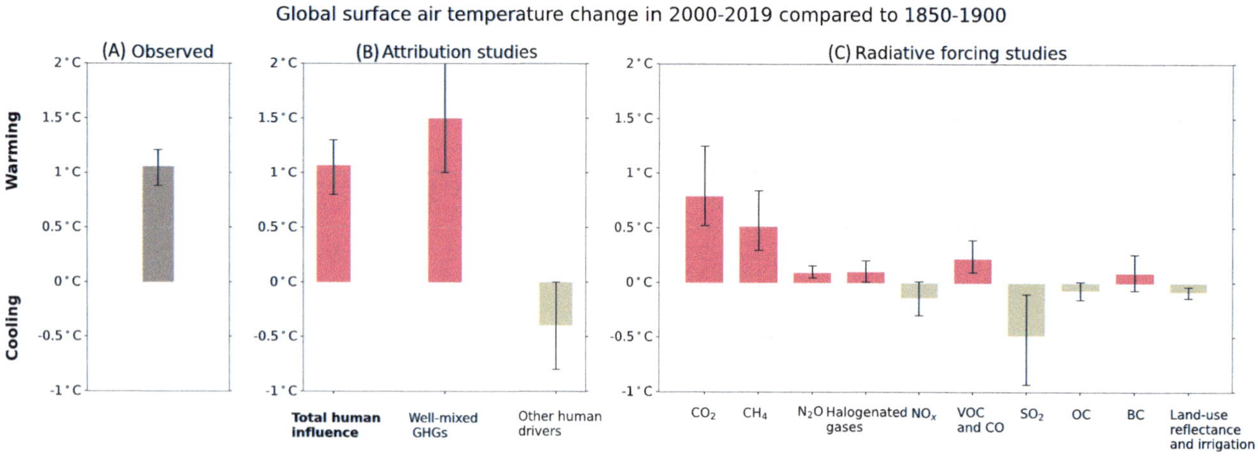

FIGURE 20.6 Illustration of observed warming (A), contribution of well mixed GHGs and other human drivers based on attribution approach (B), and CO_2 and individual constituents (other GHGs and aerosols) role using radiative forcing studies (C) to the increase of GSAT during 2000–2019 compared to 1850–1900. NOx, VOC, CO, and OC are nitrogen oxides, volatile organic compounds, carbon monoxide, and organic carbon, respectively. Whiskers at the top of the bars represent the possible range. *Figure data taken from IPCC. (2021).: Summary for Policymakers. In:* Climate Change 2021: The Physical science basis. Contribution of working group I to the sixth assessment report of the Intergovernmental Panel on Climate Change. *Cambridge University Press, Cambridge, United Kingdom and New York, NY, USA, pp. 3 – 32, https://doi.org/10.1017/9781009157896.001.; Summary for Policymakers.*

absorption of solar radiation, results in heating of the atmosphere (Fig. 20.6C) and is also responsible for snow and ice melting when deposited on the surface, which has positive feedback on global warming as it changes earth's surface albedo. A strong reduction in the role of black carbon in the global warming effect is estimated in AR6 (IPCC, 2021), compared to AR5 (IPCC, 2013), by the earth system models. The aerosols have strong impacts on the cloud formation and warm rainfall processes and thus a much wider impact on future climate change if atmospheric aerosol composition is modified such as those occurred during the social restrictions to combat the COVID-19 pandemic (Khatri et al., 2021; Szopa et al., 2021). The effect of the aerosol and aerosol composition changes due to COVID-19, however, showed negligible effect on the global climate as per the simulations using the CMIP6 (Szopa et al., 2021). The CMIP6 models use simplified representations of atmospheric aerosol emission, production, and growth/decay mechanisms for computational efficiency, which is required for millennium-scale simulations.

Observed warming or increase in GSAT between 2010 and 2019 in comparison to 1850–1900 is observed in the 0.9°C–1.2°C range (Fig. 20.6A). Attribution studies considering synthesized information from climate models and observation highlighted that total human-caused warming of 2010–2019 is between 0.8°C and 1.3°C, or the best estimate of 1.07°C relative to 1850–1900. The well-mixed GHGs contributed warming in the range of 1°C–2°C, which is partly offset by aerosols cooling in the range of 0°C to –0.8°C, a minimal contribution of ±0.1°C by the natural drivers (solar and volcanic), and ±0.2°C constitute internal variability of the earth's climate system (Fig. 20.6B). Further details on warming estimation by radiative forcing and climate sensitivity are given as the contributions of CO_2, non-CO_2 GHGs, and various aerosol species (Fig. 20.6C). Both CO_2 and N_2O have a long lifetime of the order of 100 years, while CH_4 is a short-lived climate forcer (SLCF) of a chemical lifetime of 9.1 years (Szopa et al., 2021). Due to the shorter lifetime and very high GWP in 20-year integration time, tackling CH_4 emissions can have a rapid impact on controlling global warming at the intermediate time horizon. The effect of CO_2 or N_2O emission reductions on GSAT change takes much longer to be observed due to their long residence time of more than 100 hundred years (Chandra & Niwa, 2022). Ultimately, the total GHG emissions must be reduced to near zero by around 2050 to achieve the ambitious climate goal set by the Paris Agreement for limiting the GSAT change below 2°C by the end of the century. Notably, the aerosol cooling of approximately –0.5°C is dominated by the sulfur dioxide, which if reduced along with the reduction in air pollution, could exert additional stress on the GHG emission mitigation targets.

20.5.2 Assessing CO_2 emissions and global-mean surface air temperature increase

The process of emission with corresponding increase in GSAT depends on an intertwined time-evolving mechanism depending on the radiative forcing of CO_2, sink carbon uptake, and ocean heat uptake (Canadell et al., 2021).

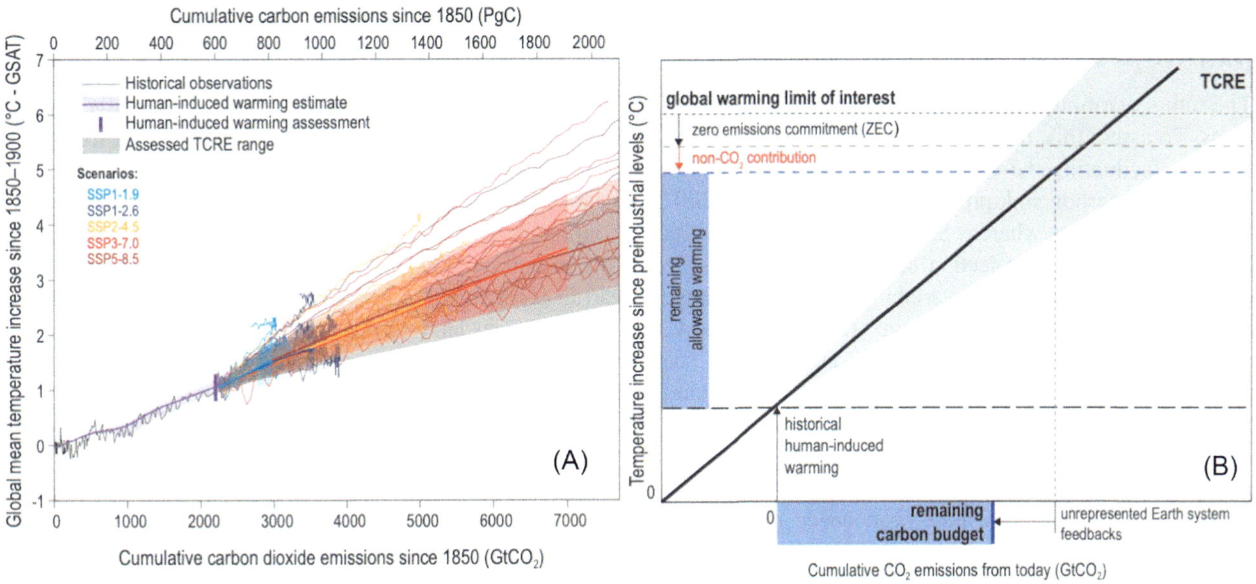

FIGURE 20.7 Representation of covarying relationship between cumulative CO_2 emission and global mean temperature increase from historical observations, human-induced warming estimation, and assessment, assessed range of TCRE as well as under different SSP scenarios since 1850–1900 (left panel; A). Illustration of the constituting components for estimating the remaining carbon budget (right panel; B). *Figure taken from figure 5.31 of Canadell, J. G., Monteiro, P. M. S., Costa, M. H., da Cunha, L. C., Cox, P. M., Eliseev, A. V., et al. (2021). Global carbon and other biogeochemical cycles and feedbacks. In* Climate Change 2021: The physical science basis. Contribution of working group I to the sixth assessment report of the Intergovernmental Panel on Climate Change. *Cambridge University Press, Cambridge, United Kingdom and New York, NY, USA, pp. 673–816, https://doi.org/10.1017/9781009157896.007 (Nonexclusive copyright permission granted by IPCC Legal Office, 23 October 2024).*

To determine the effect of cumulative CO_2 emission on GSAT, a transient climate response to cumulative emissions of CO_2 (TCRE) is formulated (Fig. 20.7) (Allen et al., 2009). The estimate of TCRE ranges in between 1.0° and 2.3°C (best estimate 1.65°C) per 1000 PgC emissions (IPCC, 2021), which is consistent with the total of 650 ± 65 PgC of anthropogenic CO_2 emitted in the period 1850–2019. The studies with historical observations demonstrated a near-constant increase of TCRE, suggesting an approximate near-linear constant relationship between cumulative CO_2 emission and increase in GSAT (Fig. 20.7). The recent IPCC AR6 report emphasized that GSAT will continue to rise until at least this mid-21st century under all projected Shared Socio-economic Pathways (SSPs) scenarios and consists of five different socio-economic pathways, namely, SSP1, SSP2, SSP3, SSP4, and SSP5 (Riahi et al., 2017). This SSP scenario as compared to Representative Concentration Pathways (RCPs) used in IPCC AR5 is much more robust because it considers the future population change, economic development, equality, and education effects on greenhouse emissions for analyzing future climate change, adaptation, and mitigation. Roughly TCRE appears to be independent across a range of different pathways; therefore, it is utilized as a central metric in climate change assessment and remaining carbon budget estimation. The assessed TCRE range under different SSP scenarios shows a large spread among them; overall, it has been observed that the GSAT increases with cumulative CO_2 emissions (Fig. 20.7A).

Nevertheless, to halt the accelerating warming, mitigation actions are of utmost importance. Consequently, knowledge of how much carbon emission is still left to limit the warming below 1.5°C in line with the Paris Agreement is important. This encourages the estimation of the remaining carbon budget for different global warming levels, which is a complex task involving assessment of various contributing components like TCRE, accurate estimation of historical human-induced warming, warming contribution from non-CO_2 agents (CH_4, N_2O), and zero emission commitment (ZEC). The estimation considering possible uncertainties of TCRE and non-CO_2 warming in the remaining carbon budget suggested that it is highly unlikely that the remaining carbon budget is effectively zero for limiting warming to 1.5°C by 2100. However, there remains significant uncertainty in the earth system feedback component, which could reduce the remaining carbon budget to a small, yet not zero value (Canadell et al., 2021). Remaining carbon budgets (starting from 1 January 2020) for limiting warming to 1.5°C and 2.0°C are 140 and 370 PgC, respectively, based on the 50th percentile of TCRE (IPCC, 2021). These remaining carbon budgets may vary by an estimated ± 60 PgC depending on how successfully future non-CO_2 emissions can be reduced.

20.6 Future projection of carbon sink under different shared socio-economic pathways scenarios

The future warming will largely depend on how carbon sinks in land and ocean will respond to future anthropogenic emissions. To assess this, the recent IPCC AR6 report considers different concentration-driven projection scenarios for the 21st century, facilitating the CMIP6 Earth System Models (ESMs) to evaluate the range of potential responses of the ocean and net land carbon sink presented in Fig. 20.8. The SSP1-2.6 and SSP2-4.5 are representative of future concentration stabilization scenarios, whereas SSP3-7.0 and SSP5-8.5 present continuous increase in concentration scenarios. CMIP6 ESM historical and future projection suggested less variability in ocean carbon sink, unlike the net land carbon sink, which exhibited higher interannual variability as well as a large spread in cumulative land carbon uptake among projection scenarios in the rest of the 21st century (Fig. 20.8B and C). Both ocean and land carbon sinks are projected to grow until about 2050 under SSP2-4.5, SSP3-7.0, and SSP5-8.5 scenarios. The oceanic sinks are strongest over the Southern Ocean and North Atlantic under all scenarios, whereas the land carbon sinks are regions of present-day forests (Canadell et al., 2021). This increase in terrestrial carbon sinks is mainly the effect of CO_2 fertilization and increased water-use efficiency of plants, helping the removal of atmospheric CO_2. However, after 2050, ocean carbon sink projects decelerating growth for SSP3-7.0 and SSP5-8.5 scenarios, mainly because increased warming of ocean water reduces ocean carbon buffering capacity, an effect that is more pronounced in the SSP1-2.6 scenario when CO_2 starts declining (Terhaar, Frölicher, & Joos, 2022). Similar time dynamics of land carbon sinks have been observed post-2050 due to net carbon loss (respiration > photosynthesis) and forest dieback in a warmed world (Allen et al., 2010; Doughty et al., 2023).

Fig. 20.9 bar chart displays the projected partitioned proportions (in %) of cumulative CO_2 emission removed by land and ocean sinks and the airborne fraction by 2100 under five different SSP scenarios. Similar to sink temporal dynamics explained in Fig. 20.8, overall, it has been observed that under low emission scenarios (SSP1-1.9 and SSP1-2.6), land and ocean uptake proportion increases; however, scenarios with higher cumulative emission (SSP2-4.5, SSP3-7.0, and SSP5-8.5) projected a lesser removal fraction of emissions by sink implies a larger proportion of CO_2 will reside in the atmosphere (Fig. 20.9). Ocean acts as a major heat and carbon sink; it has taken up nearly 90% of the excess heat and 23% ± 5% carbon since the 1970s. However, recent warming has led to ocean stratification and less ventilation, resulting in a cumulative reduction of ocean carbon sinks globally. In contrast, there is medium confidence that land sinks will transform into sources under a high emission (SSP5-8.5) scenario (Canadell et al., 2021). This increase in global warming will further enhance the chances of carbon release from the permafrost, projected to be 3–41 PgC per 1°C of warming by 2100 (Canadell et al., 2021). These experiment scenarios are especially useful in understanding or determining the necessary level of emission mitigation required and the extent to which natural land and ocean sinks can aid in achieving climate goals.

In 2015, through the United Nations Framework Convention on Climate Change (UNFCCC) Paris agreement, a Nationally Determined Contributions (NDCs) were introduced. It represents the specific commitments or steps a country undertakes for emission reduction and adaptation to climate change to stay below 2°C global warming (United Nations

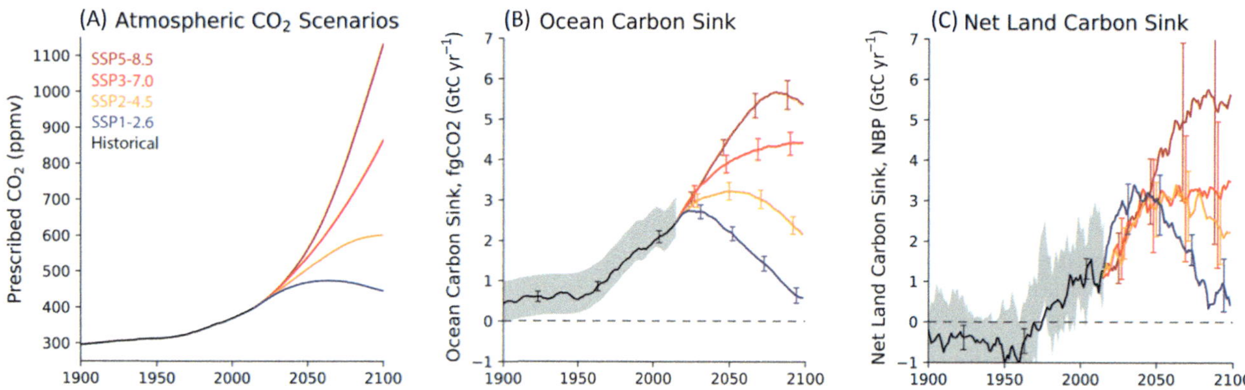

FIGURE 20.8 Time evolution of prescribed atmospheric CO_2 concentration scenarios (A), ocean carbon sink (B), and net land carbon sink (C) for different concentration driven SSP scenarios in CMIP6 ESM model for 1900–2100. *The figure taken from Fig. 5.25 (a,b,c) of Canadell, J. G., Monteiro, P. M. S., Costa, M. H., da Cunha, L. C., Cox, P. M., Eliseev, A. V., et al. (2021). Global carbon and other biogeochemical cycles and feedbacks. In* Climate Change 2021: The physical science basis. Contribution of working group I to the sixth assessment report of the Intergovernmental Panel on Climate Change. *Cambridge University Press, Cambridge, United Kingdom and New York, NY, USA, pp. 673–816, https://doi.org/10.1017/9781009157896.007 (Nonexclusive copyright permission granted by IPCC Legal Office, October 24, 2024).*

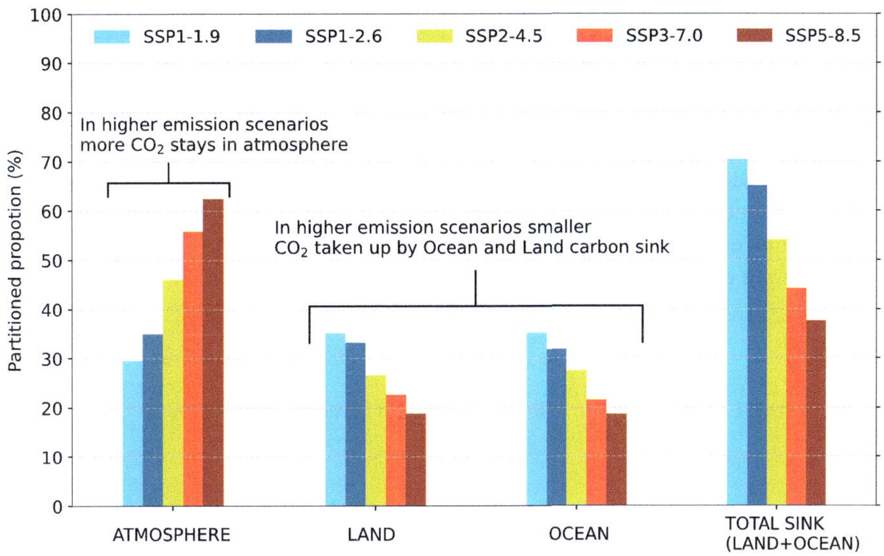

FIGURE 20.9 The bar chart represents the projected partitioned proportion of cumulative CO_2 emissions staying in the atmosphere, removed by land and ocean sinks by 2100 at five different SSP scenarios. *Figure data taken from Canadell, J. G., Monteiro, P. M. S., Costa, M. H., da Cunha, L. C., Cox, P. M., Eliseev, A. V., et al. (2021). Global carbon and other biogeochemical cycles and feedbacks. In* Climate Change 2021: The physical science basis. Contribution of working group I to the sixth assessment report of the Intergovernmental Panel on Climate Change. *Cambridge University Press, Cambridge, United Kingdom and New York, NY, USA, pp. 673–816, https://doi.org/10.1017/9781009157896.007 and IPCC., (2021).: Summary for Policymakers. In:* Climate Change 2021: The Physical science basis. Contribution of working group I to the sixth assessment report of the Intergovernmental Panel on Climate Change. *Cambridge University Press, Cambridge, United Kingdom and New York, NY, USA, pp. 3 – 32, https://doi.org/10.1017/9781009157896.001; Summary for Policymakers.*

Framework Convention on Climate Change, 2016). Even though we start drawing down the CO_2 emissions, the impact on climate will be seen after many years. The concentration will slowly start to decrease if we reach net emissions, either zero when emitted CO_2 is completely removed by natural and human processes, termed "net zero emission," or negative when CO_2 removal is greater as compared to the amount emitted, termed "net negative emission" (Rogelj et al., 2018). This itself will take around five to ten years to observe concentration slow down even in high mitigation scenarios. Consequently, it would be reflected in the deceleration of global warming after decades, which may not be detected at first because of natural climate variability and may be observed after a few decades. Therefore, mitigating climate change or meeting Paris goals would require rapid reduction of emissions along with supplemented scaling of carbon dioxide removal (CDR) approach through conventional (afforestation, reforestation, and forest management) and novel (Bioenergy with Carbon Capture and Storage (BECCS), biochar, and direct air carbon capture and storage (DAACS)) methods (Canadell et al., 2021; Smith et al., 2023). Conventional CDR based on land largely contributes to the current overall CDR ($\sim$0.5 GtC yr^{-1}), with a very small fraction of 0.5 MtC yr^{-1} coming from novel CDR. Still, there is a large gap between the current or proposed CDR and the amount needed to meet the Paris temperature goal; therefore, rapid acceleration in CDR is an utmost priority (Smith et al., 2023). Meanwhile, it is essential to be cautious while deploying CDR techniques for CO_2 sequestration, as it may lead to partial offsetting by releasing CO_2 from carbon sink stores (Canadell et al., 2021). A recent assessment reveals that past and current pledges countries took are still far behind the required implementation and more reduction for a better and sustainable future (Rogelj et al., 2023).

20.7 Summary and conclusions

This chapter has discussed the linkage of atmospheric CO_2 with global warming by reviewing various research articles and the latest climate assessment reports, such as the global carbon budget and IPCC AR6. We discussed the various surface and aircraft-based in situ observation sites, which cover limited geographical regions. On the other hand, space-based remote sensing used to measure CO_2 concentration provides comprehensive coverage of nearly the entire globe, offering a more detailed spatiotemporal variation of the concentration. Further, we explained the advantage of long-term CO_2 recorded through in situ measurement that integrates information of carbon source-sink variability over a long period. This chapter also provided a comprehensive overview of the main emission sources (fossil and land-use change) and sinks (vegetation in land, phytoplankton, and air-sea gas exchange in the ocean) of CO_2, as well as their balance through the atmospheric CO_2 budget. We discussed the global airborne fraction and the interannual variability of growth rates, which are influenced by climate phenomena such as ENSO. We also highlighted the role of CO_2, non-CO_2 climate factors and aerosol compounds on climate, where CO_2 stands out as the main agent for global warming. Then we introduced the readers to the quasilinear relationship between cumulative CO_2 emission and temperature change, or TCRE, leveraged for assessing the remaining carbon budget for different warming levels, an important calculation for designing future climate policy. The response of land and ocean sinks in the 21st century under five prescribed SSP scenarios shows a larger interannual variability in land than ocean carbon sinks, where both showed strong increases till 2050. Finally, we highlighted the importance of cautious implementation of CDR techniques along with reduction of CO_2 emissions to avoid dire consequences of climate change for a comprehensive understanding in this chapter.

Acknowledgments

C. Das thanks the Director, IIT Delhi, for providing the fellowship for smoothly carrying out this work.

References

Allen, C. D., Macalady, A. K., Chenchouni, H., Bachelet, D., McDowell, N., Vennetier, M., ... Cobb, N. (2010). A global overview of drought and heat-induced tree mortality reveals emerging climate change risks for forests. *Forest Ecology and Management*, *259*(4). Available from https://doi.org/10.1016/j.foreco.2009.09.001.

Allen, M. R., Frame, D. J., Huntingford, C., Jones, C. D., Lowe, J. A., Meinshausen, M., & Meinshausen, N. (2009). Warming caused by cumulative carbon emissions towards the trillionth tonne. *Nature*, *458*(7242). Available from https://doi.org/10.1038/nature08019.

Arrhenius, S. (1896). On the influence of carbonic acid in the air upon the temperature of the ground. *The London, Edinburgh, and Dublin Philosophical Magazine and Journal of Science*, *41*(251), 237–276. Available from https://doi.org/10.1080/14786449608620846.

Aumont, O., Orr, J. C., Monfray, P., Ludwig, W., Amiotte- Suchet, P., & Probst, J.-L. (2001). Riverine-driven interhemispheric transport of carbon. *Global Biogeochemical Cycles*, *15*, 393–405. Available from https://doi.org/10.1029/1999GB001238.

Bakker, D. C. E., Pfeil, B., Landa, C. S., Metzl, N., O'Brien, K. M., Olsen, A., ... Xu, S. (2016). A multi-decade record of high-quality fCO_2 data in version 3 of the surface ocean CO_2 atlas (SOCAT). In. *Earth System Science Data*, *8*(Issue 2). Available from https://doi.org/10.5194/essd-8-383-2016.

Ballantyne, A. P., Alden, C. B., Miller, J. B., Tans, P. P., & White, J. W. C. (2012). Increase in observed net carbon dioxide uptake by land and oceans during the past 50 years. *Nature*, *488*(7409). Available from https://doi.org/10.1038/nature11299.

Bastos, A., Hartung, K., Nützel, T. B., Nabel, J. E. M. S., Houghton, R. A., & Pongratz, J. (2021). Comparison of uncertainties in land-use change fluxes from bookkeeping model parameterisation. *Earth System Dynamics*, *12*, 745–762. Available from https://doi.org/10.5194/esd-12-745-2021.

Bloomberg New Energy Finance (BNEF). (2021). *2H 2021 LCOE Update*. https://www.bnef.com/.

Bousquet, P., Peylin, P., Ciais, P., Le Quéré, C., Friedlingstein, P., & Tans, P. P. (2000). Regional changes in carbon dioxide fluxes of land and oceans since 1980. *New Series*, *290*(5495).

Brando, P. M., Paolucci, L., Ummenhofer, C. C., Ordway, E. M., Hartmann, H., Cattau, M. E., ... Balch, J. (2019). Droughts, wildfires, and forest carbon cycling: A pantropical synthesis. *Annual Review of Earth and Planetary Sciences*, *47*. Available from https://doi.org/10.1146/annurev-earth-082517-010235.

Buchwitz, M., de Beek, R., Burrows, J. P., Bovensmann, H., Warneke, T., Notholt, J., ... Schulz, A. (2005). Atmospheric methane and carbon dioxide from SCIAMACHY satellite data: Initial comparison with chemistry and transport models. *Atmospheric Chemistry and Physics*, *5*(4). Available from https://doi.org/10.5194/acp-5-941-2005.

Canadell, J. G., Monteiro, P. M. S., Costa, M. H., da Cunha, L. C., Cox, P. M., Eliseev, A. V., ... Zickfeld, K. (2021). Global carbon and other biogeochemical cycles and feedbacks. In *Climate Change 2021: The physical science basis. Contribution of working group I to the sixth assessment report of the Intergovernmental Panel on Climate Change*. Cambridge University Press, Cambridge, United Kingdom and New York, NY, USA, pp. 673–816, Available from https://doi.org/10.1017/9781009157896.007.

Chandra, N., & Niwa, Y. (2022). How quickly would we see the effects of changing greenhouse gas emissions? *Frontiers for Young Minds*, *10*, 695527. Available from https://doi.org/10.3389/frym.2022.695527.

Chandra, N., Patra, P. K., Bisht, J. S. H., Ito, A., Umezawa, T., Saigusa, N., ... Canadell, J. G. (2021). Emissions from the oil and gas sectors, coal mining and ruminant farming drive methane growth over the past three decades. *Journal of the Meteorological Society of Japan*, *99*(2). Available from https://doi.org/10.2151/jmsj.2021-015.

Charlson, R. J., Schwartz, S. E., Hales, J. M., Cess, R. D., Coakley, J. A., Hansen, J. E., & Hofmann, D. J. (1992). Climate forcing by anthropogenic aerosols. *Science (New York, N.Y.)*, *255*(5043). Available from https://doi.org/10.1126/science.255.5043.423.

Cox, P. M., Pearson, D., Booth, B. B., Friedlingstein, P., Hunting- ford, C., Jones, C. D., & Luke, C. M. (2013). Sensitivity of tropical carbon to climate change constrained by carbon dioxide variability. *Nature*, *494*, 341–344. Available from https://doi.org/10.1038/nature11882.

Crisp, D. (2015). Measuring atmospheric carbon dioxide from space with the Orbiting Carbon Observatory-2 (OCO-2). *Earth Observing Systems*, 9607. Available from https://doi.org/10.1117/12.2187291.

Crisp, D., Dolman, H., Tanhua, T., McKinley, G. A., Hauck, J., Bastos, A., ... Aich, V. (2022). How well do we understand the land-ocean-atmosphere carbon cycle? *Reviews of Geophysics*, *60*(Issue 2). Available from https://doi.org/10.1029/2021RG000736.

Das, C., Kunchala, R. K., Chandra, N., Chhabra, A., & Pandya, M. R. (2023). Characterizing the regional XCO_2 variability and its association with ENSO over India inferred from GOSAT and OCO-2 satellite observations. *Science of The Total Environment*, *902*, 166176. Available from https://doi.org/10.1016/j.scitotenv.2023.166176.

Das, C., Kunchala, R. K., Chandra, N., Chmura, L., Nęcki, J., & Patra, P. K. (2022). Meridional propagation of carbon dioxide (CO_2) growth rate and flux anomalies from the tropics due to ENSO. *Geophysical Research Letters*, *49*(19). Available from https://doi.org/10.1029/2022GL100105.

Doughty, C. E., Keany, J. M., Wiebe, B. C., Rey-Sanchez, C., Carter, K. R., Middleby, K. B., … Fisher, J. B. (2023). Tropical forests are approaching critical temperature thresholds. *Nature*, *621*(7977), 105–111. Available from https://doi.org/10.1038/s41586-023-06391-z.

Forster, P. T., Storelvmo, K., Armour, W., Collins, J. -L., Dufresne, D., Frame, D. J., … Zhang, H. (2021). The Earth's energy budget, climate feedbacks, and climate sensitivity. In *Climate Change 2021: The physical science basis. Contribution of working group I to the sixth assessment report of the Intergovernmental Panel on Climate Change*. Cambridge University Press, Cambridge, United Kingdom and New York, NY, USA, pp. 923–1054. Available from https://doi.org/10.1017/9781009157896.009.

Friedlingstein, P., O'Sullivan, M., Jones, M. W., Andrew, R. M., Gregor, L., Hauck, J., … Zheng, B. (2022). Global Carbon Budget 2022. *Earth System Science Data*, *14*(11), 4811–4900. Available from https://doi.org/10.5194/essd-14-4811-2022.

Gielen, D., Boshell, F., Saygin, D., Bazilian, M. D., Wagner, N., & Gorini, R. (2019). The role of renewable energy in the global energy transformation. *Energy Strategy Reviews*, *24*. Available from https://doi.org/10.1016/j.esr.2019.01.006.

Gloege, L., McKinley, G. A., Landschützer, P., Fay, A. R., Frölicher, T. L., Fyfe, J. C., … Takano, Y. (2021). Quantifying errors in observationally based estimates of ocean carbon sink variability. *Global Biogeochemical Cycles*, *35*, e2020GB006788. Available from https://doi.org/10.1029/2020GB006788.

Gregor, L., Kok, S., & Monteiro, P. M. S. (2018). Interannual drivers of the seasonal cycle of CO_2 in the Southern Ocean. *Biogeosciences*, *15*(8). Available from https://doi.org/10.5194/bg-15-2361-2018.

Gu, L., Baldocchi, D. D., Wofsy, S. C., W Munger, J., Michalsky, J. J., Urbanski, S. P., & Boden, T. A. (2003). Response of a deciduous forest to the Mount Pinatubo eruption: Enhanced photosynthesis. *Science (New York, N.Y.)*, *299*(5615). Available from https://doi.org/10.1126/science.1078366.

Hamazaki, T., Kaneko, Y., Kuze, A., & Kondo, K. (2005). Fourier transform spectrometer for Greenhouse Gases Observing Satellite (GOSAT). *Enabling Sensor and Platform Technologies for Spaceborne Remote Sensing*, 5659. Available from https://doi.org/10.1117/12.581198.

Heimann, M., & Keeling, C. D. (1986). Meridional eddy diffusion model of the transport of atmospheric carbon dioxide: 1. Seasonal carbon cycle over the tropical Pacific Ocean. *Journal of Geophysical Research: Atmospheres*, *91*(D7). Available from https://doi.org/10.1029/jd091id07p07765.

Houghton, R. A., House, J. I., Pongratz, J., Van Der Werf, G. R., Defries, R. S., Hansen, M. C., … Ramankutty, N. (2012). Carbon emissions from land use and land-cover change. *Biogeosciences*, *9*(12). Available from https://doi.org/10.5194/bg-9-5125-2012.

Hubau, W., Lewis, S. L., Phillips, O. L., Affum-Baffoe, K., Beeckman, K. H., Cuni-Sanchez, A., … Zemagho, L. (2020). Asynchronous carbon sink saturation in African and Amazonian tropical forests. *Nature*, *579*(7797), 80–87. Available from https://doi.org/10.1038/s41586-020-2035-0.

Huntzinger, D. N., Post, W. M., Wei, Y., Michalak, A. M., West, T. O., Jacobson, A. R., … Cook, R. (2012). North American Carbon Program (NACP) regional interim synthesis: Terrestrial biospheric model intercomparison. *Ecological Modelling*, 232. Available from https://doi.org/10.1016/j.ecolmodel.2012.02.004.

Inoue, M., Morino, I., Uchino, O., Nakatsuru, T., Yoshida, Y., Yokota, T., … Tanaka, T. (2016). Bias corrections of GOSAT SWIR XCO_2 and XCH_4 with TCCON data and their evaluation using aircraft measurement data. *Atmospheric Measurement Techniques*, *9*(8). Available from https://doi.org/10.5194/amt-9-3491-2016.

IPCC. (2013). Climate Change 2013: The physical science basis. *Contribution of working group I to the fifth assessment report of the Intergovernmental Panel on Climate Change*. Cambridge University Press, Cambridge, United Kingdom and New York, NY, USA, p. 1535, Available from https://doi.org/10.1017/CBO9781107415324.

IPCC. (2021). Summary for Policymakers. In *Climate Change 2021: The Physical science basis. Contribution of working group I to the sixth assessment report of the Intergovernmental Panel on Climate Change*. Cambridge University Press, Cambridge, United Kingdom and New York, NY, USA, pp. 3 – 32, Available from https://doi.org/10.1017/9781009157896.001.

Keeling, C. D. (1960). The concentration and isotopic abundances of carbon dioxide in the atmosphere. *Tellus*, *12*(2), 200–203. Available from https://doi.org/10.1111/j.2153-3490.1960.tb01300.x.

Keeling, C. D., Chin, J. F. S., & Whorf, T. P. (1996). Increased activity of northern vegetation inferred from atmospheric CO_2 measurements. *Nature*, *382*(6587), 146–149. Available from https://doi.org/10.1038/382146a0.

Keeling, C. D., Piper, S. C., Bacastow, R. B., Wahlen, M., Whorf, T. P., Heimann, M., & Meijer, H. A. (2001). UC San Diego SIO reference title exchanges of atmospheric CO_2 and $^{13}CO_2$ with the terrestrial biosphere and oceans from 1978 to 2000. I. Global aspects publication date exchanges of atmospheric CO_2 and $^{13}CO_2$ with the terrestrial biosphere and oceans from 1978 to 2000. I. Global Aspects.

Khatri, P., Hayasaka, T., Holben, B., Tripathi, S. N., Misra, P., Patra, P. K., … Dumka, U. C. (2021). Aerosol loading and radiation budget perturbations in densely populated and highly polluted Indo-Gangetic Plain by COVID-19: Influences on cloud properties and air temperature. *Geophysical Research Letters*, *48*(20). Available from https://doi.org/10.1029/2021GL093796.

Kuze, A., Suto, H., Nakajima, M., & Hamazaki, T. (2009). Thermal and near infrared sensor for carbon observation Fourier-transform spectrometer on the Greenhouse Gases Observing Satellite for greenhouse gases monitoring. *Applied Optics*, *48*(35). Available from https://doi.org/10.1364/AO.48.006716.

Liou, K. N. (1980). An introduction to atmospheric radiation. *In"l Geophys. Ser.*, 26, p. 392.

Liu, Z., Guan, D., Wei, W., Davis, S. J., Ciais, P., Bai, J., … He, K. (2015). Reduced carbon emission estimates from fossil fuel combustion and cement production in China. *Nature*, *524*, 335–338. Available from https://doi.org/10.1038/nature14677.

Machida, T., Matsueda, H., Sawa, Y., Nakagawa, Y., Hirotani, K., Kondo, N., … Ogawa, T. (2008). Worldwide measurements of atmospheric CO_2 and other trace gas species using commercial airlines. *Journal of Atmospheric and Oceanic Technology*, *25*(10), 1744–1754. Available from https://doi.org/10.1175/2008JTECHA1082.1.

Matsumoto, K., Sarmiento, J. L., Key, R. M., Aumont, O., Bullister, J. L., Caldeira, K., ... Orr, J. C. (2004). Evaluation of ocean carbon cycle models with data-based metrics. *Geophysical Research Letters*, *31*(7). Available from https://doi.org/10.1029/2003GL018970.

Meinshausen, M., Meinshausen, N., Hare, W., Raper, S. C. B., Frieler, K., Knutti, R., ... Allen, M. R. (2009). Greenhouse-gas emission targets for limiting global warming to 2°C. *Nature*, *458*(7242). Available from https://doi.org/10.1038/nature08017.

Nakazawa, T., Miyashita, K., Aoki, S., & Tanaka, M. (1991). Temporal and spatial variations of upper tropospheric and lower stratospheric carbon dioxide. *Tellus B*, *43*(2). Available from https://doi.org/10.1034/j.1600-0889.1991.t01-1-00005.x.

O'Dell, C. W., Eldering, A., Wennberg, P. O., Crisp, D., Gunson, M. R., Fisher, B., ... Velazco, V. A. (2018). Improved retrievals of carbon dioxide from Orbiting Carbon Observatory-2 with the version 8 ACOS algorithm. *Atmospheric Measurement Techniques*, *11*(12). Available from https://doi.org/10.5194/amt-11-6539-2018.

Olsen, A., Lange, N., Key, R. M., Tanhua, T., Bittig, H. C., Kozyr, A., ... Woosley, R. J. (2020). An updated version of the global interior ocean biogeochemical data product, GLODAPv2.2020. *Earth System Science Data*, *12*(Issue 4). Available from https://doi.org/10.5194/essd-12-3653-2020.

Orr, J. C., Fabry, V. J., Aumont, O., Bopp, L., Doney, S. C., Feely, R. A., ... Yool, A. (2005). Anthropogenic ocean acidification over the twenty-first century and its impact on calcifying organisms. *Nature*, *437*(7059). Available from https://doi.org/10.1038/nature0409.

Patra, P. K., Dlugokencky, E. J., Elkins, J. W., Dutton, G. S., Tohjima, Y., Sasakawa, M., ... Takigawa, M. (2022). Forward and inverse modelling of atmospheric nitrous oxide using MIROC4-atmospheric chemistry-transport model. *Journal of the Meteorological Society of Japan*, *100*(2). Available from https://doi.org/10.2151/jmsj.2022-018.

Patra, P. K., Maksyutov, S., & Nakazawa, T. (2005a). Analysis of atmospheric CO_2 growth rates at Mauna Loa using CO_2 fluxes derived from an inverse model. *Tellus, Series B: Chemical and Physical Meteorology*, *57*(5). Available from https://doi.org/10.1111/j.1600-0889.2005.00159.x.

Patra, P. K., Ishizawa, M., Maksyutov, S., Nakazawa, T., & Inoue, G. (2005b). Role of biomass burning and climate anomalies for land-atmosphere carbon fluxes based on inverse modeling of atmospheric CO_2. *Global Biogeochemical Cycles*, *19*(3), 1–10. Available from https://doi.org/10.1029/2004GB002258.

Petzold, A., Thouret, V., Gerbig, C., Zahn, A., Brenninkmeijer, C. A. M., Gallagher, M., ... Gallagher, M. (2015). Global-scale atmosphere monitoring by in-service aircraft-- Current achievements and future prospects of the European Research Infrastructure IAGOS. *Tellus, Series B: Chemical and Physical Meteorology*, *6*(1), 1–24. Available from https://doi.org/10.3402/tellusb.v67.28452.

Raupach, M. R., Canadell, J. G., & Le Quéré, C. (2008). Anthropogenic and biophysical contributions to increasing atmospheric CO_2 growth rate and airborne fraction. *Biogeosciences*, *5*, 1601–1613. Available from https://doi.org/10.5194/bg-5-1601-2008.

Ravishankara, A. R., Daniel, J. S., & Portmann, R. W. (2009). Nitrous oxide (N_2O): The dominant ozone-depleting substance emitted in the 21st century. *Science (New York, N.Y.)*, *326*(5949). Available from https://doi.org/10.1126/science.1176985.

Riahi, K., van Vuuren, D. P., Kriegler, E., Edmonds, J., O'Neill, B. C., Fujimori, S., ... Tavoni, M. (2017). The Shared Socioeconomic Pathways and their energy, land use, and greenhouse gas emissions implications: An overview. *Global Environmental Change*, *42*. Available from https://doi.org/10.1016/j.gloenvcha.2016.05.009.

Rogelj, J., Fransen, T., den Elzen, M. G. J., Lamboll, R. D., Schumer, C., Kuramochi, T., ... Portugal-Pereira, J. (2023). Credibility gap in net-zero climate targets leaves world at high risk. *Science (New York, N.Y.)*, *380*(6649). Available from https://doi.org/10.1126/science.adg6428.

Rogelj, J., Shindell, D., Jiang, K., Fifita, S., Forster, P., Ginzburg, V., ... Vilariño, M. V. (2018). Mitigation pathways compatible with 1.5°C in the context of sustainable development. In *Global warming of 1.5°C. IPCC special report global warming of 1.5°C.*

Sabine, C. L., Feely, R. A., Gruber, N., Key, R. M., Lee, K., Bullister, J. L., ... Rios, A. F. (2004). *The Oceanic Sink for anthropogenic CO_2*. Available from https://doi.org/10.1126/science.1097. https://www.science.org.

Saeki, T., & Patra, P. K. (2017). Implications of overestimated anthropogenic CO_2 emissions on East Asian and global land CO_2 flux inversion. *Geoscience Letters*, *4*(Issue 1). Available from https://doi.org/10.1186/s40562-017-0074-7.

Saunois, M., Stavert, A. R., Poulter, B., Bousquet, P., Canadell, J. G., Jackson, R. B., ... Zhuang, Q. (2019). *The global methane budget 2000–2017. Earth system science data discussions*. https://doi.org/10.5194/essd-2019-128.

Smith, S. M., Geden, O., Nemet, G., Gidden, M., Lamb, W.F., Powis, C., ... Minx, J. C. (2023). *The state of carbon dioxide removal–* 1st ed. Available at: https://www.stateofcdr.org.

Stanhill, G., & Cohen, S. (2001). Global dimming: A review of the evidence for a widespread and significant reduction in global radiation with discussion of its probable causes and possible agricultural consequences. *Agricultural and Forest Meteorology*, *107*(Issue 4). Available from https://doi.org/10.1016/S0168-1923(00)00241-0.

Sweeney, C., Karion, A., Wolter, S., Newberger, T., Guenther, D., Higgs, J. A., ... Tans, P. P. (2015). Seasonal climatology of CO_2 across North America from aircraft measurements in the NOAA/ESRL global greenhouse gas reference network. *Journal of Geophysical Research: Atmospheres*, *120*(10), 5155–5190. Available from https://doi.org/10.1002/2014JD022591.

Szopa, S., Naik, V., Adhikary, B., Artaxo, P., Berntsen, T., Collins, W. D., S.., ... Zanis, P. (2021). *Short-lived climate forcers. Climate Change 2021: The physical science basis. Contribution of Working Group I to the sixth assessment report of the Intergovernmental Panel on Climate Change* (pp. 817–922). Cambridge, United Kingdom and New York, NY, USA: Cambridge University Press. Available from https://doi.org/10.1017/9781009157896.008.

Takahashi, T., Sutherland, S. C., Wanninkhof, R., Sweeney, C., Feely, R. A., Chipman, D. W., ... de Baar, H. J. W. (2009). Climatological mean and decadal change in surface ocean pCO_2, and net sea-air CO_2 flux over the global oceans. *Deep-Sea Research Part II: Topical Studies in Oceanography*, *56*(8–10), 554–577. Available from https://doi.org/10.1016/j.dsr2.2008.12.009.

Tans, P. P., Conway, T. J., & Nakazawa, T. (1989). Latitudinal distribution of the sources and sinks of atmospheric carbon dioxide derived from surface observations and an atmospheric transport model. *Journal of Geophysical Research: Atmospheres (1984–2012)*, *94*(D4), 5151–5172. Available from https://doi.org/10.1029/JD094iD04p05151.

Terhaar, J., Frölicher, T. L., & Joos, F. (2022). Observation-constrained estimates of the global ocean carbon sink from Earth system models. *Biogeosciences*, *19*, 4431–4457. Available from https://doi.org/10.5194/bg-19-4431-2022.

Thompson, C. R., Wofsy, S. C., Prather, M. J., Newman, P. A., Hanisco, T. F., Ryerson, T. B., ... Zeng, L. (2022). The NASA Atmospheric Tomography (ATom) mission: Imaging the chemistry of the global atmosphere. *Bulletin of the American Meteorological Society*, *103*(3). Available from https://doi.org/10.1175/BAMS-D-20-0315.1.

Thompson, R. L., Chevallier, F., Maksyutov, S., Patra, P. K., & Bowman, K. (2022). Chapter 4 – Top-down approaches. In B. Poulter, J. G. Canadell, D. J. Hayes, & R. L. Thompson (Eds.), *Balancing Greenhouse Gas Budgets* (pp. 87–155). Elsevier. Available from https://doi.org/10.1016/B978-0-12-814952-2.00008-3.

Thompson, R. L., Lassaletta, L., Patra, P. K., Wilson, C., Wells, K. C., Gressent, A., ... Canadell, J. G. (2019). Acceleration of global N_2O emissions seen from two decades of atmospheric inversion. *Nature Climate Change*, *9*(12). Available from https://doi.org/10.1038/s41558-019-0613-7.

Thoning, K. W., Tans, P. P., & Komhyr, W. D. (1989). Atmospheric carbon dioxide at Mauna Loa observatory 2. Analysis of the NOAA GMCC data, 1974–1985. *Journal of Geophysical Research*, *94*(D6), 8549–8565. Available from https://doi.org/10.1029/jd094id06p08549.

Tian, H., Xu, R., Canadell, J. G., Thompson, R. L., Winiwarter, W., Suntharalingam, P., ... Yao, Y. (2020). A comprehensive quantification of global nitrous oxide sources and sinks. *Nature*, *586*(7828). Available from https://doi.org/10.1038/s41586-020-2780-0.

Tohjima, Y., Minejima, C., Mukai, H., Machida, T., Yamagishi, H., & Nojiri, Y. (2012). Analysis of seasonality and annual mean distribution of atmospheric potential oxygen (APO) in the Pacific region. *Global Biogeochemical Cycles*, *26*(4). Available from https://doi.org/10.1029/2011GB004110.

United Nations Framework Convention on Climate Change. (2016). *NDC registry*. Retrieved from http://unfccc.int/focus/ndc_registry/items/9433.php.

Walker, A. P., De Kauwe, M. G., Bastos, A., Belmecheri, S., Georgiou, K., Keeling, R. F., ... Zuidema, P. A (2021). Integrating the evidence for a terrestrial carbon sink caused by increasing atmospheric CO_2. *New Phytologist*, *229*(Issue 5). Available from https://doi.org/10.1111/nph.16866.

Wofsy, S. C. (2011). HIAPER Pole-to-Pole Observations (HIPPO): Fine-grained, global-scale measurements of climatically important atmospheric gases and aerosols. *Philosophical Transactions of the Royal Society A*, *369*, 2073–2086. Available from https://doi.org/10.1098/rsta.2010.0313.

Wunch, D., Toon, G. C., Blavier, J. F. L., Washenfelder, R. A., Notholt, J., Connor, B. J., ... Wennberg, P. O. (2011). The total carbon column observing network. *Philosophical Transactions of the Royal Society A: Mathematical, Physical and Engineering Sciences*, *369*(1943), 2087–2112. Available from https://doi.org/10.1098/rsta.2010.0240.

Chapter 21

Advances in Earth Observation Satellites for global air quality monitoring

Jayanarayanan Kuttippurath and Vikas Kumar Patel
Centre for Ocean, River, Atmosphere and Land Sciences (CORAL), Indian Institute of Technology Kharagpur, Kharagpur, West Bengal, India

21.1 Introduction

Air pollution refers to the intrusion of chemical, physical, or biological substances into the indoor or outdoor environment that changes the inherent properties of the atmosphere. It is caused by multiple factors, such as household combustion devices, motor vehicles, industries, and forest fires. The key pollutants that are of significant public health concern are particulate matter ($PM_{2.5}$), carbon monoxide (CO), ozone (O_3), nitrogen dioxide (NO_2), and sulfur dioxide (SO_2). Both outdoor and indoor air pollution have been identified as significant contributors to respiratory and various other diseases, and thus, they play a crucial role in the occurrence of illness and death. As per the World Health Organization (WHO), a vast majority of the global population (about 99% in areas beyond the air quality criterion) is exposed to air pollution (WHO, 2022). Policies aimed at mitigating air pollution present a mutually beneficial approach that addresses climate and health concerns. These policies would help to reduce the adverse health effects associated with air pollution and decrease GreenHouse Gas (GHG) emissions, and thereby aid in the near and long-term efforts to combat climate change.

As per WHO, the annual premature mortality due to outdoor air pollution exceeded 3.2 million individuals in the year 2020 (WHO, 2021). The primary contributors consist of Particulate Matter (PM) with sizes 2.5 μm or smaller, commonly referred to as $PM_{2.5}$. These particles have the ability to deeply infiltrate the respiratory, cardiovascular, and circulatory systems; resulting in the development of many diseases and malignancies (WHO, 2022). However, the global average estimates presuppose the uniformity of these particles across different regions. $PM_{2.5}$ comprises a mixture of many chemical substances, including hydrocarbons, salts, and other compounds emitted by cars, cooking stoves, and industrial activities (Alves et al., 2023). Additionally, $PM_{2.5}$ also contains naturally occurring constituents such as dust particles and microbes. One prominent cause of $PM_{2.5}$ in Asia is the emission of soot resulting from domestic heating and cooking activities (Weagle et al., 2018). The presence of desert dust significantly contributes to the exacerbation of air pollution in the northern Africa, Middle East, and central Asia, and identifying its most hazardous source remains ambiguous (Weagle et al., 2018).

Over the past decade, satellite data have increasingly been applied to public health and air quality analyzes (Holloway et al., 2021). Investments in earth observation satellites (EOS) and similar initiatives have supported these applications to spur the effective utilization of satellite data for health and air quality research. Though health-relevant, near-surface concentrations of pollutants may be inferred from satellite measurements, the successful interpretation of these data for health exposure requires understanding of their strengths and limitations (Holloway et al., 2021). Health applications require linking of satellite data to surface-level air quality observations. Nevertheless, the geographical coverage of ground-based observations is constrained due to their high initial cost and the expenses associated with maintaining a network. Therefore these stations are commonly positioned in city areas to monitor air pollution effectively, although their deployment in rural regions is highly warranted (Pathak and Kuttippurath, 2022). The limited spatial coverage of stations hinders the assessment of air pollution exposure in rural areas and poses challenges in evaluating emission sources outside the urban regions, such as agriculture and forest fires (Ravishankara et al., 2007).

Many countries have a dearth or absence of publicly accessible monitoring data, and even common air pollutants are inadequately documented. A recent study found that a total of 141 countries lack a systematic monitoring system

Sustainable Development Perspectives in Earth Observation. DOI: http://dx.doi.org/10.1016/B978-0-443-14072-3.00021-6

for $PM_{2.5}$, but 23 have more than three sensors per million inhabitants (Martin et al., 2019). de Sherbinin, Levy, Zell, Weber, and Jaiteh (2014) have emphasized the potential of well-designed satellite-based indicators in providing valuable insights into global air pollution levels, which can be utilized for decision-making processes and public outreach initiatives. One of the main benefits of satellite data is their extensive spatial coverage, which fills the existing spatial gaps in the current ground-based monitoring networks. Satellite data provide valuable insights into the spatial distribution of pollutants, which help to identify areas with their highest concentrations (Anenberg et al., 2019; Dutta et al., 2021). Furthermore, these data allow us to track changes in pollution over the past several decades, providing its temporal evolution and spatial extent. Additionally, satellite observations enable us to trace the transport of pollutants on the regional and intercontinental scales (Kibirige et al., 2023; Vijayaraghavan et al., 2008). However, there are differences between satellite and surface observations, and certain instruments encounter difficulty in accurately accounting for variations in instrument performance during their lifetime (e.g., Kuttippurath et al., 2018).

One of the primary limitations associated with satellite measurements in the context of health applications pertains to the absence of surface-level information (Holloway et al., 2021). Satellite data generally provide information regarding the chemical composition of atmospheric column above the Earth's surface. However, further analytical procedures are necessary to establish a connection between the atmospheric column and air quality at the surface Diao et al., 2019; Shaddick et al., 2018). In recent decades, multiple advancements have been made in satellite observations, such as satellites like Advanced Very High-Resolution Radiometer (AVHRR) or Total Ozone Mapping Spectrometer (TOMS) were launched in late 1970s to monitor air quality. Later, multiple satellite instruments with latest technology were deployed to explore the air pollution sources and pollutant concentrations across the globe with high spatial resolution. Some of them are EOS Terra and Aqua with MODerate resolution Imaging Spectroradiometer (MODIS) and Multiangle Imaging SpectroRadiometer (MISR) instruments to measure aerosols in the atmosphere.

21.2 Satellite measurements of aerosols and particulate matter

Aerosol optical depth (AOD) refers to the aggregated quantification of diverse aerosols, such as urban haze, smoke particles, desert dust, and sea salt, dispersed throughout an atmospheric column Kaufman et al., 2001; Kuttippurath et al., 2020; Rosenfeld et al., 2014; Satheesh et al., 2006; Suresh Babu et al., 2013). It represents the extent to which these aerosols impede the passage of light from the Earth's surface to the uppermost layer of atmosphere. Aerosol ambient extinction coefficient inside a vertical column spanning from the Earth's surface to the top of the atmosphere is frequently employed in various assessments. Therefore the measurement of AOD has emerged as a crucial parameter for investigating the influence of aerosols on human health and climate, particularly in relation to the fine PM. Two predominant methods for monitoring the AOD are ground-based sun photometers and satellites. Ground-based measurements of AOD offer the advantage of providing real-time data, and are valuable for the calibration of AOD instruments aboard satellites and for enhancing the accuracy of remote sensing retrieval algorithms (Wei et al., 2019). The AERosol RObotic NETwork (AERONET) is a prominent terrestrial-based network for monitoring AOD. It encompasses a global network of more than 500 aerosol monitoring stations for the past 25 years (Smirnov et al., 2002). However, the operation of AERONET is associated with significant costs and its coverage is limited, which is particularly evident in less populated regions. In contrast, satellite remote sensing can effectively monitor AOD across the regions; including the areas that are difficult to access.

The availability of satellite-derived AOD emerged in the 1980s because of the availability of a range of EOS instruments. These include optical instruments such as MODIS and MERIS (MEdium Resolution Imaging Spectrometer), multiangle instrument MISR, polarization instrument Advanced Earth Observing Satellite (ADEOS) Polarization and Directionality of the Earth's Reflectance (POLDER)/Polarization and Anisotropy of Reflectances for Atmospheric Science coupled with Observations from a LIDAR (PARASOL), and lidar instrument Cloud-Aerosol Lidar and Infrared Pathfinder Satellite Observations (CALIPSO). The agreement between AOD derived from different instruments is limited due to various factors. These include the complexity of the Earth's surface conditions, the uncertainty associated with the cloud mask method (Kacenelenbogen et al., 2014), variations in calibration methods (Ignatov & Stowe, 2002), and the utilization of different retrieval algorithms (Kokhanovsky et al., 2007).

TOMS performed aerosol measurements aboard different satellite platforms (e.g., Nimbus-7 and Earth Probe) from 1979 to 2005. Similarly, AVHRR onboard the National Oceanic and Atmospheric Administration (NOAA)-6 satellite launched in 1978 also measured AOD across the globe. Furthermore, like TOMS or NOAA-6, Sea-viewing Wide Field-of-view Sensor (SeaWiFS) was deployed into orbit in 1997 to monitor aerosols over both ocean and land at a spatial resolution of 13.5 km. However, understanding of aerosols has been highly facilitated after the deployment of the Terra (1999) and Aqua (2002) satellites, which are equipped with the MODIS and MISR instruments, respectively. MODIS data are available globally in daily temporal resolution. The MODIS Level 2 AOD data are available at 1 km spatial resolution, which would help to identify the sources and to monitor air quality at small spatial scales. Fig. 21.1 depicts the distribution of AOD as obtained from the MODIS Terra and Aqua satellites. The ability of MODIS to detect and record areas of high AOD is readily visible (e.g., Sanatan et al., 2024). For example, the Indo-Gangatic Plain (IGP) and East China (EC) exhibit high values of AOD, ranging from 0.64 to 0.8. High AOD is also found in desert regions such as the Sahara, Taklamakan, and Thar, about 0.8. Similarly, it is shown that both SeaWiFS and MISR effectively capture AOD, but MISR tends to underestimate as compared to MODIS and SeaWiFS (e.g., Kuttippurath & Raj, 2021). The SeaWiFS and MISR were decommissioned in 2010 and 2017, respectively, and MODIS on Terra and Aqua satellites have been operating effectively to date. In addition, MODIS exclusively captures the vertical column of aerosols in the atmosphere. In the year 2006, CALIPSO emerged as a significant advancement in the field of aerosol studies, as it enabled the acquisition of vertical profiles of atmospheric aerosols (Winker et al., 2010). The CALIPSO satellite was developed through a collaborative effort between National Aeronautics and Space Administration (NASA) and the French space agency Centre National d'Etudes Spatiales (CNES) to fulfill the requirement of observing aerosols and clouds from space. CALIPSO is equipped with cloud-aerosol lidar with orthogonal polarization (CALIOP), a lidar that is sensitive to polarization and operates at two different wavelengths (Winker et al., 2010). Additionally, CALIPSO has two passive sensors that operate in the visible and thermal infrared spectral areas. CALIOP is the first lidar instrument

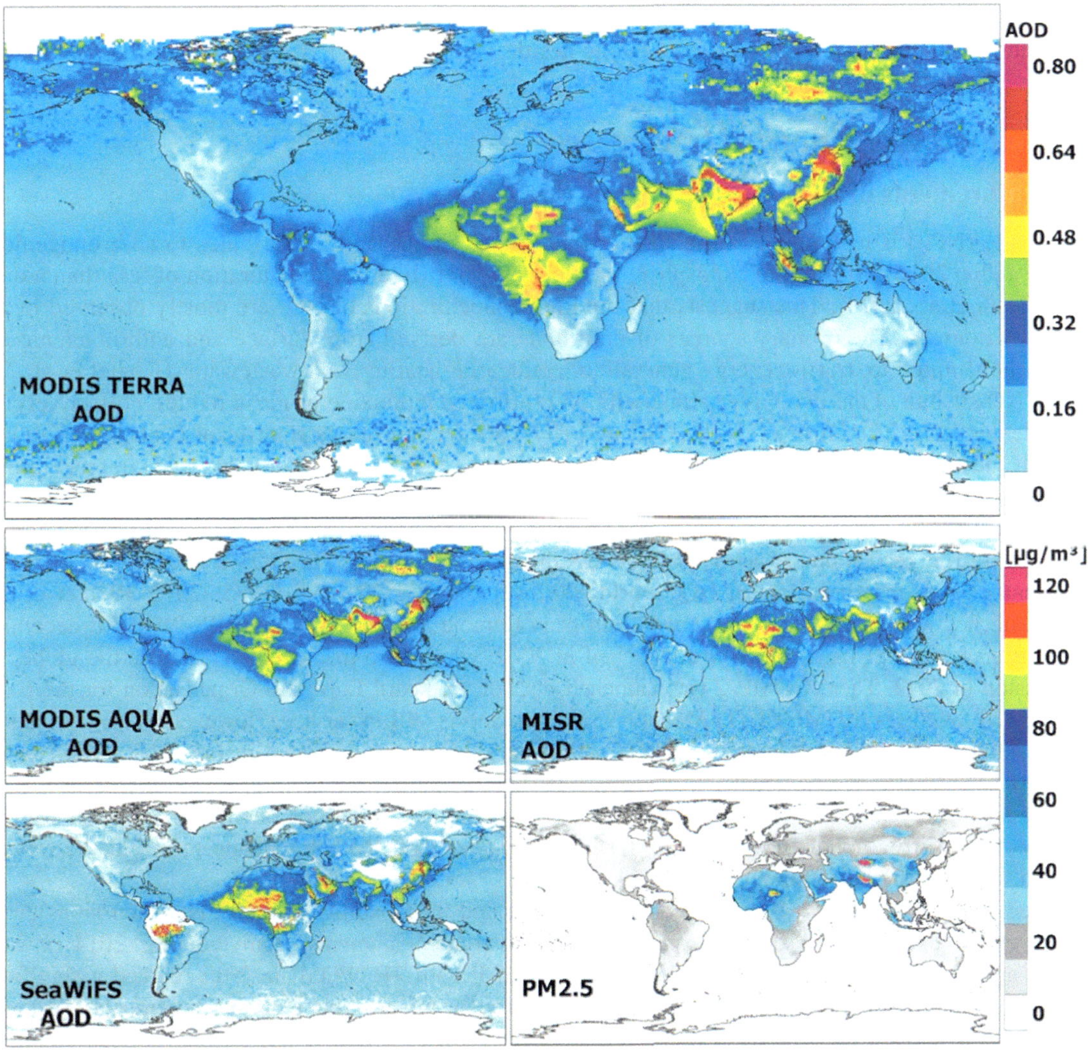

FIGURE 21.1 Distribution of AOD from MODIS Terra (2019), MODIS Aqua (2019), SeaWiFS (2010), and MISR (2016). $PM_{2.5}$ (2019) surface mass concentration is derived from the multiple satellite measurements. Map lines delineate study areas and do not necessarily depict accepted national boundaries.

that offers extended atmospheric measurements from the orbit of Earth. The profiling and polarization features of this system provide distinct measurement capacities. Its primary purpose is to gather data on the distribution and characteristics of aerosols and clouds.

The progress in satellite remote sensing technology has facilitated the growth of research in aerosols, by utilizing the abundance of AOD data obtained from various satellites. The application potential of AOD data has further increased with the development of improved satellite retrieval methods (Lipponen et al., 2018). This development has stimulated fresh research endeavors aimed at investigating the connection between aerosols and various factors such as temperature and radiative feedback (Sorribas et al., 2019), properties of water clouds (Kumar et al., 2018), and the vertical distribution of aerosols (Liu et al., 2019). Previous studies have demonstrated the fusion of multisource AOD data to derive coherent information (Filonchyk et al., 2019; Singh et al., 2017). Nevertheless, with the growing availability of numerous aerosol measurements, it is imperative to undertake a comparative examination of AOD remote sensing.

Satellite-derived $PM_{2.5}$ has emerged as a key tool for air quality management (Dey et al., 2020; Gallagher et al., 2023). The direct observation of $PM_{2.5}$ concentrations from satellites is not feasible from the existing instruments yet, and therefore, it is necessary to rely on models or other sources such as ground-based measurements to make inferences. However, algorithms are devised to use satellites measurements of AOD and other related parameters of $PM_{2.5}$ in the atmosphere to derive concentrations of the latter. The satellite-based $PM_{2.5}$ derived from MODIS, MISR, and SeaWiFS AOD is shown in Fig. 21.1. The $PM_{2.5}$ data have a spatial resolution of 1 km, which is widely used for air quality monitoring across the globe (Hammer et al., 2020). We also observe high values of $PM_{2.5}$ in IGP, EC, and in the desert regions such as Sahara, Thar, and Takalamakan. Advanced satellite-based measurements are still needed on a regional scale to monitor air quality from a perspective of public health.

21.3 Satellite observations of trace gases

21.3.1 Measurements of NO_2

Nitrogen dioxide (NO_2) is regarded as a transient pollutant that significantly contributes to anthropogenic aerosols (Seinfeld & Pandis, 2016). NO_2 in the atmosphere is relatively short-lived, with a duration of less than a day (Beirle et al., 2020; Shah et al., 2020). This suggests that the high concentrations of NO_2 are mostly observed in proximity to the emission sources. The primary sources of emissions are derived from surface and anthropogenic activities, including but not limited to, traffic, energy generation, residential heating, and wastewater (Koziel et al., 2006; Liu et al., 2019). NO_2 is one of the two gases, along with NO (nitrogen monoxide or nitric oxide), collectively known as nitrogen oxides or NOx. The nitrogen oxides (NO and NO_2) are recognized as significant contributors to urban pollution, which adversely affects human health. Additionally, they play a crucial role in regulating ozone concentrations.

Monitoring of NO_2 pollution had big challenges due to the unavailability of satellite measurements in the past. Accurate data with fine resolution are necessary to assess the urban emissions. The GOME and SCIAMACHY satellites have consistently given measurements of NO_2 worldwide over the past 20 years (1995–2012, Richter et al., 2005). GOME was launched in 1995 on board the second European Remote Sensing Satellite (ERS-2), and it decommissioned in 2011. As its successor, SCIAMACHY was launched in 2002, which also measured NO_2 across the globe, and completed its mission in 2012. Then, GOME-2 was launched in 2006 on board European Organisation for the Exploitation of Meteorological Satellites (EUMETSAT) MetOp-A satellite, and is still in service. However, these data have a lower resolution, making it difficult to accurately detect small point sources, as illustrated in Fig. 21.2. When compared to GOME and SCIAMACHY, it is evident that GOME-2 is more capable of capturing the details of NO_2 distribution across the regions. The year 2005 witnessed a significant milestone in the field of satellite observation with the introduction of the Ozone Monitoring Instrument (OMI). This brought about notable enhancements in the monitoring of NO_2 pollution, with a spatial resolution of about 25 km.

As depicted in Fig. 21.2, the OMI instrument has demonstrated its capability to detect NO_2 emissions even at sites with limited spatial extent, but has been experiencing an instrument anomaly from the year 2014 onwards (e.g., Gopikrishnan and Kuttippurath, 2021). The debut of Sentinel 5-P with TROPOMI in 2017 marked a significant milestone in global air quality monitoring, as it has a superior ability for making fine scale measurements (3.5 km $\times$ 7 km). High amount of NO_2 in EC, IGP, Western Europe, and the United States is clearly depicted in the satellite observations, primarily due to the prevalence of industrial and traffic emissions in these regions. Similarly, TROPOMI effectively detects the areas of high NO_2 in South Africa, East USA, IGP, EC, and Central Africa. In addition, TROPOMI captures NO_2 pollution in cities and areas with small industrial units, and it is a significant advancement in the realm of

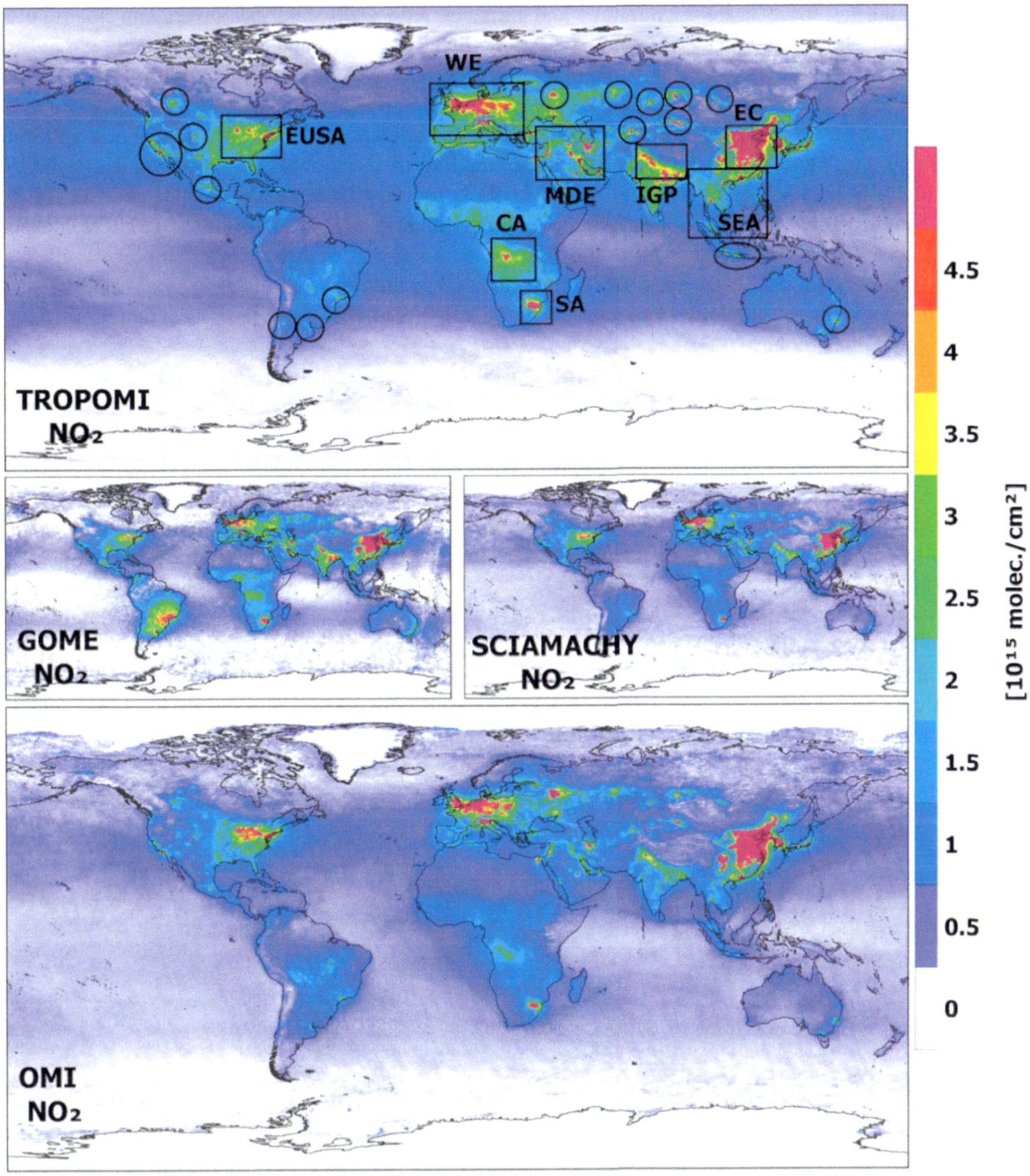

FIGURE 21.2 Distribution of NO_2 from TROPOMI (2019), OMI (2013), SCIAMACHY (2010), and GOME-2 (2019). Regions shown are East United States of America (EUSA), Western Europe (WE), Middle East (MDE), Indo-Gangatic Plain (IGP), East China (EC), Central Africa (CA), and South Africa (SA). Circles represent the high values of NO_2 on small spatial scales. Map lines delineate study areas and do not necessarily depict accepted national boundaries.

satellite-based NO_2 measurements (Sannigrahi et al., 2021; Zhang et al., 2021). In summary, the continuous utilization of TROPOMI measurements facilitate the monitoring of air quality across several spatial scales, and thus, aid in the formulation and execution of environmental rules and policies to mitigate air pollution.

21.3.2 Measurements of CO and NH_3

Carbon monoxide (CO) is a colorless and odorless gas, and it can pose a high risk to human health when inhaled in excessive quantities (Khalil & Rasmussen, 1990). CO is emitted as a byproduct during the combustion process, and the primary contributors to the emission of CO are automobiles and other forms of transportation or machinery that rely on the combustion of fossil fuels (Kumar et al., 2013; Pathak et al., 2023). Various household appliances, including unvented kerosene and gas space heaters, chimneys and furnaces with leaks, and gas stoves can potentially emit CO and thus, reduce indoor air quality (WHO, 2021). NH_3 is a chemically reactive and a readily soluble gaseous compound with alkaline properties. This has natural and manmade sources, but agriculture is the primary contributor (Behera, Sharma, & Aneja, 2013; Kuttippurath et al., 2020; Van Damme et al., 2018).

The application of Measurements of Pollution In The Troposphere (MOPITT) and Atmospheric Infrared Sounder (AIRS) measurements has significantly advanced our understanding of the global distribution and variability of atmospheric CO. Buchholz et al. (2021) and Joshi et al., 2023 have identified the global hotspots of CO for helping the implementation of environmental policies and technology across diverse industries (Fig. 21.3). These hotspots include IGP, EC, Eastern USA, Western Europe, Amazonia, Central Africa, and Southeast Asia. Similarly, AIRS effectively monitors atmospheric CO and aids in the identification of its sources and areas of high emissions. During the COVID-19 lockdown restrictions, these satellites have proven to be effective tools to record the changes in CO pollution throughout the globe (Pathak et al., 2023).

Since the early 2000s, many satellites have been deployed into space equipped with infrared sounders to quantify atmospheric concentrations of trace gases. Several satellite instruments have also been developed to measure NH_3, including the Tropospheric Emission Spectrometer (TES) (Shephard et al., 2011), Infrared Atmospheric Sounding Interferometer (IASI) (Clarisse et al., 2008; Van Damme et al., 2017), Cross-track Infrared Sounder (CrIS) (Shephard & Cady-Pereira, 2015), AIRS (Warner et al., 2017), and Greenhouse Gases Observing Satellite (GOSAT) (Someya et al., 2020). TES was launched into orbit in 2004 and concluded its mission in 2018. TES was responsible for introducing the first satellite-derived NH_3 product, but its capacity for measurements diminished in later stages. The IASI instrument offers Level-2 NH_3 product that have been available since 2008. These measurements provide a wider spatial coverage compared to TES. According to Warner et al. (2017), AIRS is one of the most extensive data records of NH_3 from a single satellite, for a period of 15 years (2002–16). The availability of NH_3 data from CrIS and GOSAT is currently limited (Someya et al., 2020).

AIRS, in conjunction with the Advanced Microwave Sounding Unit (AMSU), measures infrared and microwave radiation emitted by the Earth (Auman et al., 2003). Both instruments operate in coordination, allowing them to perform concurrent observations of the Earth's surface. The system employs over 2000 channels to scan different atmospheric areas, enabling the generation of a complete three-dimensional representation of atmospheric temperature, humidity,

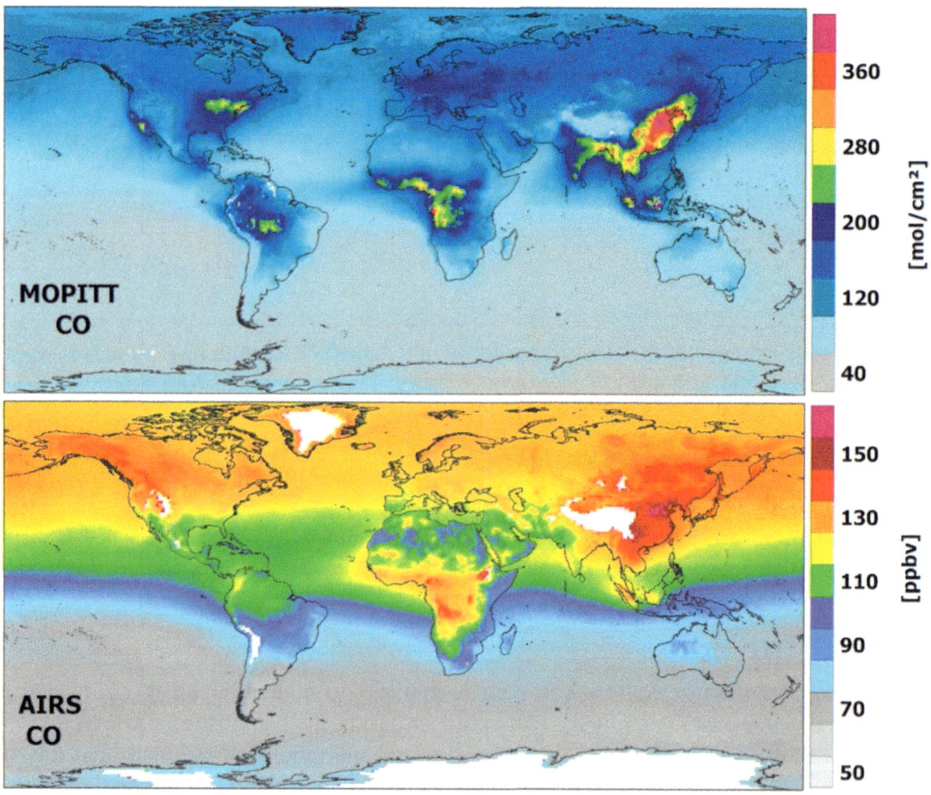

FIGURE 21.3 Distribution of CO derived from MOPITT and AIRS satellite measurements for the year 2019. Map lines delineate study areas and do not necessarily depict accepted national boundaries.

cloud characteristics (including quantities and altitudes), GHGs, and other meteorological processes. AIRS was not originally intended for NH_3 observations, but the instrument's sensitivity and stability enabled it to track NH_3 columns in the atmosphere. The significant rise in atmospheric NH_3, which was unforeseen, illustrates the rapid changes occurring in the atmosphere due to human activities, as detected by AIRS.

IASI is a remote sensing instrument that operates on the polar-orbiting MetOp-A/B/C satellites. These satellites were launched in 2006, 2012, and 2018, respectively (Clerbaux et al., 2009). The absorption properties of NH_3 are primarily measured within the spectral range of 800–1200 cm. The instrument offers NH_3 measurements twice a day, with flyover at 0930 and 2130 local solar time (LST). It possesses a swath width of 2400 km and a pixel size of 12 km in diameter when observed at nadir. The initial retrieval of NH_3 involved the utilization of lookup tables (LUTs) that were constructed using simulations derived from a forward radiative transfer model. The calculation of a Hyperspectral Range Index (HRI) was performed on each observation, and the HRI values were converted to NH_3 columns using LUTs afterwards. In a second iteration, the retrieval of NH_3 was enhanced through the utilization of an artificial neural network for IASI, referred to as ANNI-NH_3-v1. This approach, as described in Whitburn et al. (2016), involved converting the computed HRI values to column densities using the ANNI algorithm. The second iteration makes improvements in the algorithm through the implementation of distinct neural networks for land and ocean regions, as well as the streamlining of input parameters.

Both satellites can capture atmospheric NH_3 across the latitudes. Van Damme et al. (2018) used the IASI data to identify both point and industrial sources of atmospheric NH_3 on a global scale. It is observed that IASI effectively detects global atmospheric NH_3, with high values in IGP and EC (Fig. 21.4). In a study conducted by Kuttippurath et al. (2020) using IASI data showed that the intense agriculture activities in IGP produce high levels of atmospheric NH_3 there. IASI has made significant progress in monitoring the local (e.g., agriculture emissions) and remote (e.g., transport of emission of pollutants from biomass and agriculture waste burning) sources of atmospheric NH_3. Observations obtained from both AIRS and IASI greatly help in understanding the atmospheric distribution of NH_3 and its varied sources.

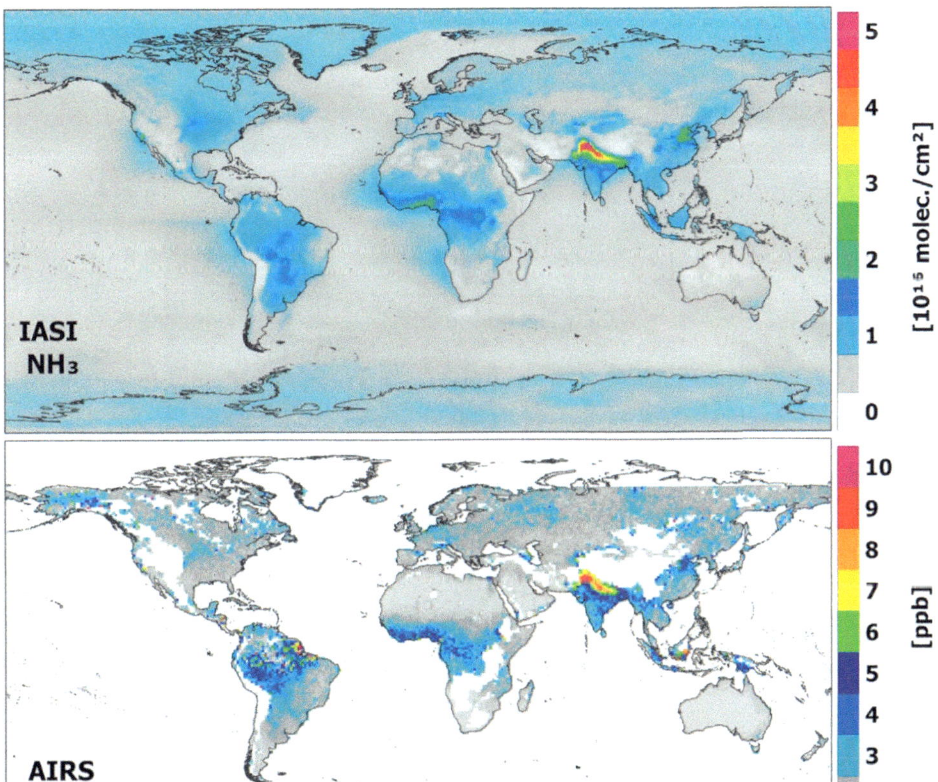

FIGURE 21.4 Distribution of atmospheric NH_3 derived from IASI (total column) and AIRS (mixing ratio) satellite observations for the year 2015. Map lines delineate study areas and do not necessarily depict accepted national boundaries.

21.3.3 Measurements of HCHO

According to the US Environmental Protection Agency (EPA, 2011), exposure to an average formaldehyde (HCHO) concentration of 1 $\mu g\ m^{-3}$ (about 0.7 ppb at STP) during one's lifetime can cause, up to 13 out of every million individuals, lung and nasopharyngeal cancer (Zhu et al., 2017). HCHO mostly has biogenic and pyrogenic sources that emits from biomass, including agricultural waste burning (Seinfeld & Pandis, 2016). Biogenic emissions are a significant contributor to the high levels of HCHO in India (Kuttippurath et al., 2022), which is particularly evident in the areas of forest and agriculture activities. For instance, the elevated levels of HCHO in the Western Ghats, Eastern Ghats, eastern India, and northeast India are primarily driven by biogenic sources (e.g., Kuttippurath et al., 2022). Additionally, there are occurrences of fire events, mainly during the periods of March to May and October to November in various parts of India such as the northeast India and IGP. In addition, stubble and biomass burning occur in several other locations, including Punjab and Haryana. Therefore the sources of pyrogenic emissions also contribute to the increased levels of HCHO in these areas. Furthermore, the northern IGP near the Himalaya exhibits abundant vegetation. These regions serve as a significant source of biogenic HCHO due to the presence of its precursors, such as isoprene. Therefore it is imperative to understand the distribution and variability of HCHO mainly in the regions where biomass burning is high such as Amazonia, Central Africa, Southeast Asia, IGP, and EC. On top of these, the high values of HCHO in the coastal regions, ports, and ship routes are also contributed by the ship emissions, as demonstrated in the OMI measurements (Gopikrishnan & Kuttippurath, 2021; Kuttippurath et al., 2022). The ground-based instruments are not sufficient to monitor HCHO variability in these regions. Therefore we consider two different satellite observations, namely OMI and TROPOMI, for the year 2013 and 2019, respectively, to analyze the distribution of HCHO, as depicted in Fig. 21.5. It is clearly shown that the HCHO values are very high, and both instruments capture higher concentrations in the above-mentioned regions compared to other parts of the world. Among the regions, Central Africa exhibits more HCHO than other regions in OMI measurements ($>0.9 \times 10^{16}$ molecules cm^{-2}), but similar distribution with higher values in TROPOMI in Central Africa, IGP, Southeast Asia, and EC, about $>18 \times 10^{-5}$ mol m^{-2}. Note that, the variability of HCHO in both satellite measurements are similar, but the spatial pattern in terms of distribution and regional variability is clearer and more dominant in the TROPOMI measurements.

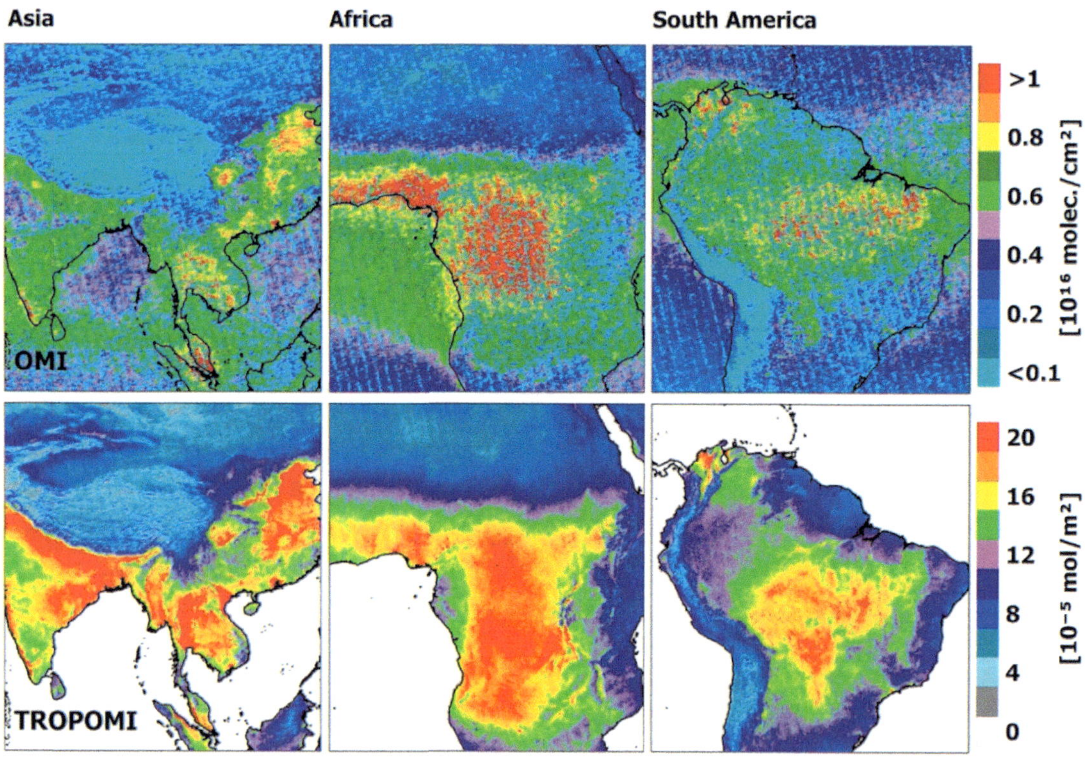

FIGURE 21.5 Distribution of HCHO derived from OMI (top) and TROPOMI (bottom) for the year 2013 and 2019, respectively, in prominent biomass burning regions in the world. Map lines delineate study areas and do not necessarily depict accepted national boundaries.

21.4 Application of satellite observations

21.4.1 Regional air quality monitoring

Rapid urbanization and economic development have led to the degradation of air quality in many countries. Prior to the year 2000, monitoring air quality was challenging due to the dearth of ground-based and satellite measurements. Recent advancements in EOS bridge the knowledge gap between air quality and its impact on human health and environment. Since the availability of remote sensing measurements, there have been multiple studies conducted in countries like India and China to understand the distribution and temporal changes in various air pollutants (Kuttippurath et al., 2020, 2022; Lin et al., 2018; Ramachandran et al., 2013). Both MODIS Terra and Aqua made a revolution in monitoring aerosols with a spatial coverage of 1 km (new Level 2 data), which can easily detect point sources, and these data can help even to understand the air quality in cities. On the other hand, most satellite measurements for trace gases such as MOPITT, AIRS, and IASI provide data at 1 degree spatial resolution, which challenges the small-scale (cities and rural areas) air quality monitoring. The TROPOMI sensor advances our understanding of gaseous pollutants on the regional scale. During the COVID-19 lockdown period, TROPOMI helped to monitor the city-scale changes in air pollutants. There is an improvement in air quality in China, India,and other countries, and satellite measurements played a pivotal role in identifying this progress in air quality (Gopikrishnan et al., 2021; Levelt et al., 2022; Miyazaki et al., 2020). In recent decades, satellite measurements of pollutants have acted as decisive tools for the trend detection, and evaluation of policies and environmental technologies to curb air pollution (Holloway et al., 2021).

We have also analyzed the effectiveness of satellite measurements for the regional monitoring of air quality, such as in India, China, USA, and Western Europe as a case study. Among these regions, all satellite measurements show high values of annual mean AOD in India (>0.32) followed by China, and small values are observed in the USA (<0.16) and Western Europe, as shown in Fig. 21.6. Also, there is a high spatial heterogeneity in the distributions of AOD. For example, MISR and SeaWiFS show relatively lower values (<0.5) of AOD in IGP of India, but MODIS measurements show very high values of about >0.64 there. Both MODIS Terra and Aqua observations show similar distribution patterns, but values of AOD in Aqua (0.64–0.8) in IGP is slightly lower than that of Terra (>0.8). Similar observation can also be made for EC, where MODIS measurements show higher values (>0.64) than the other two satellite measurements: MISR and SeaWiFS (<0.48). Among these AOD measurements, MODIS is well recognized and better captures the AOD distribution across the globe. In addition, MODIS measurements are still providing AOD data across the latitudes, which help in monitoring air quality, and plays a vital role in decision-making process to mitigate air quality and climate change.

Unlike other trace gases and particulates, distribution of NO_2 across the globe is highly variable owing to different sources. However, monitoring of NO_2 is very crucial due to its direct connection to anthropogenic activities. There are several regions across the world with poor air quality, where NO_2 plays a key role, as presented here for India, China, USA, and Western Europe. These regions also indicate their diverse sources, such as the vehicular emissions as the main NO_2 sources in the USA and Western Europe (Jiang et al., 2022; Kephart et al., 2023). On the other hand, industrial, vehicular, waste burning, and residential emissions are the major sources of NO_2 in India and China (Jiang et al., 2022; Pant & Harrison, 2012). It is also important to note that NO_2 has point sources, and high-resolution satellite measurements are needed to identify them, as illustrated in Fig. 21.7. In all four selected regions, GOME, SCIAMACHY, and OMI are unable to capture all-point sources or hotspots of NO_2. However, the areal extent of NO_2 across the regions is well captured by satellite measurements. GOME, SCIAMACHY, and OMI have relatively lower resolution of NO_2 measurements than TROPOMI. Therefore even small point sources are well detected in the regions of India, USA, and Western Europe by TROPOMI, and are widely used in the global air quality monitoring (Amritha, Varikoden, Patel, Kuttippurath, & Gopikrishnan, 2024).

21.4.2 Tracking rural air pollution

The deteriorating air quality in metropolitan areas has been extensively studied in recent years and air pollution is commonly perceived as an urban issue. Recent studies report that the rural air quality in different parts of the world is decreasing, which needs particular attention (Pathak & Kuttippurath, 2022; Ravishankara et al., 2007). Quantifying air pollutants in rural areas is a challenge due to the difficulty in accurately defining the rural boundaries, particularly in locations where urban sprawl is not well delineated (Ravishankara et al., 2007). Additionally, rural areas often have higher population density in countries like India, which further complicates the urban and rural separation and thus, the pollution in the latter region (Pathak & Kuttippurath, 2022). The higher resolution of TROPOMI enables the monitoring of air quality in rural regions with better accuracy. Here, we utilize TROPOMI measurements to assess the distribution

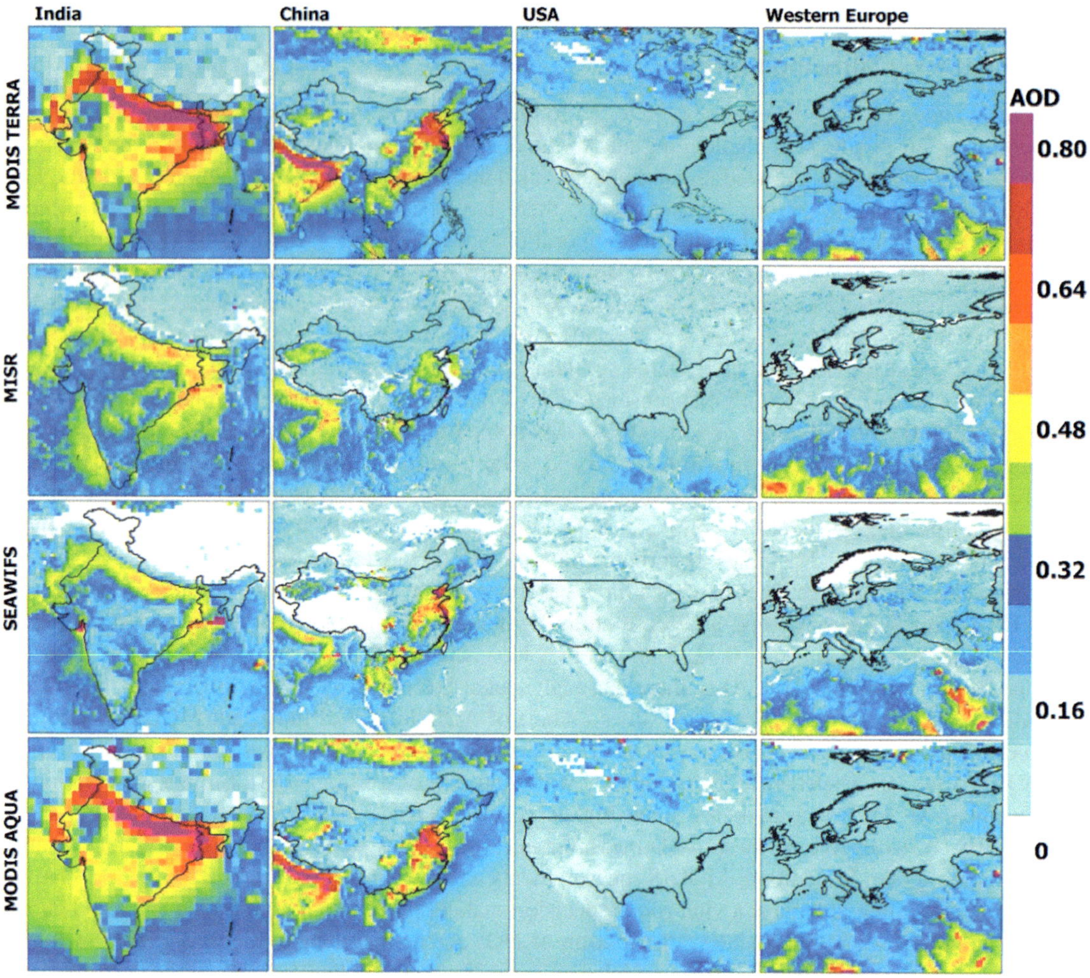

FIGURE 21.6 Distribution of AOD from MODIS Terra (2019), MODIS Aqua (2019), SeaWiFS (2010), and MISR (2016) over India, China, the USA, and Western Europe. Map lines delineate study areas and do not necessarily depict accepted national boundaries.

of CO, HCHO, NO_2, and SO_2 in rural India during the year 2019. We chose rural regions in India as a case study because of their large population, who are also exposed to poor air quality. Note that the area is segregated using the nighttime light data with values >1.15 nW cm^{-2} sr as urban land (Ravishankara et al., 2007), as shown in Fig. 21.8.

We observe high NO_2 in the rural areas, which are near industrial locations and cities, with values $>9\times10^{-5}$ mol m^{-2}. In India, the rural areas of IGP exhibit relatively higher NO_2 pollution compared to other regions. Similar results are found for $PM_{2.5}$, CO, and HCHO, where the rural areas of IGP have higher values (Fig. 21.8). Rural areas in the east, central, and near the west coast of India also have higher HCHO, about $18-21\times10^{-5}$ mol m^{-2}. These higher values in rural India can be attributed to biomass burning (or agricultural waste burning) and biogenic/pyrogenic emissions, as they are the major sources of HCHO, and TROPOMI well captures them. Similarly, the rural areas also have high SO_2 concentrations in the northernmost and eastern India, where the thermal power plants are located, which affect the nearby rural regions, and it is well detected by TROPOMI. In brief, due to the finer resolution of TROPOMI, even the air pollution in rural areas can be monitored with better accuracy than the previous satellites. For example, satellites with lower resolution may not provide accurate measurements of pollutants in the rural areas near urban centers due to the large size of each pixel that include both urban and rural lands. However, TROPOMI helps to better understand the rural pollution because of its higher resolution.

21.4.3 Air quality policy and public health assessment

Satellite data possess an unparalleled advantage in terms of spatial coverage that cannot be matched by any other technologies. Historically, this advantage was counteracted by the constraints in spatial resolution. However, with every new satellite, there have been notable improvements in terms of spatial monitoring. NASA, European Space Agency (ESA), and other

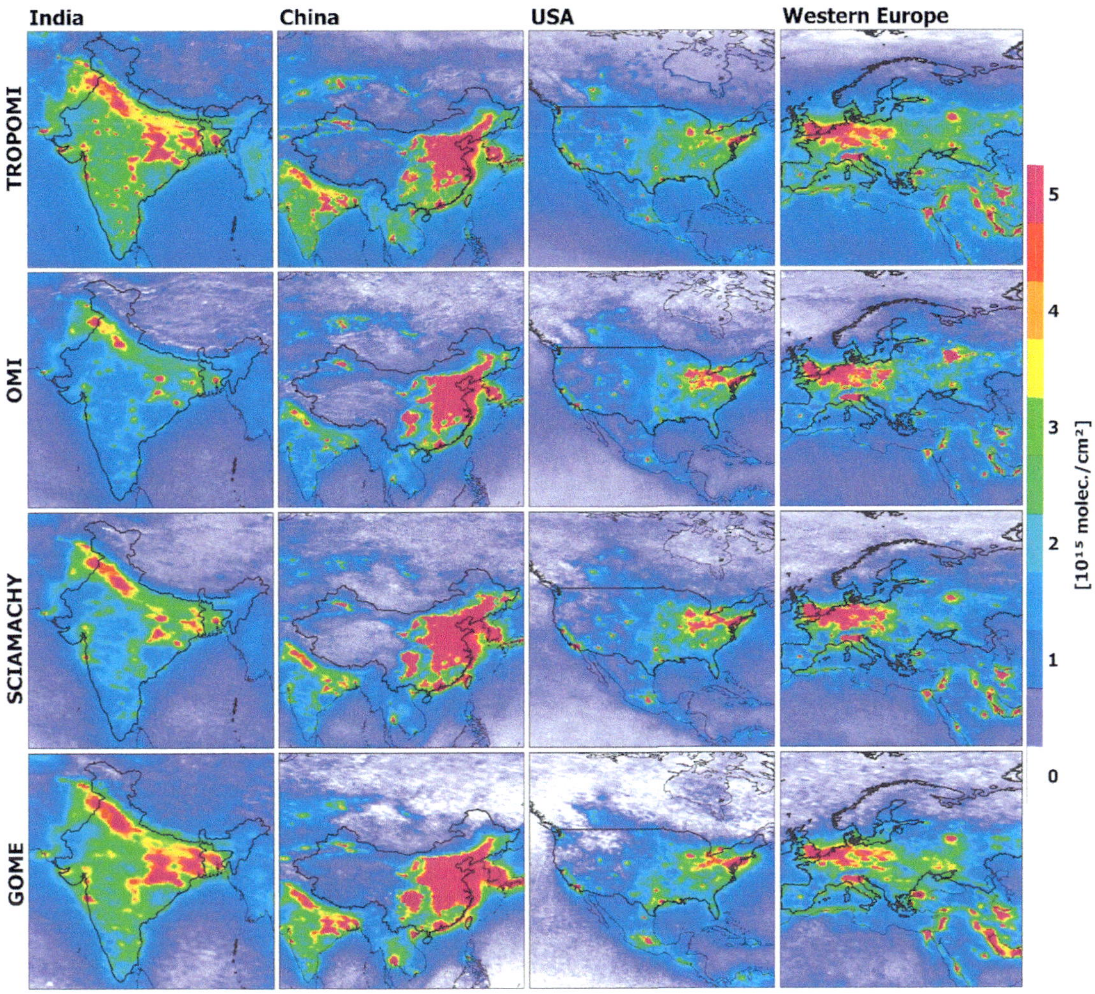

FIGURE 21.7 Distribution of NO_2 from TROPOMI (2019), OMI (2013), SCIAMACHY (2010), and GOME-2 (2019) over India, China, the USA, and Western Europe. Map lines delineate study areas and do not necessarily depict accepted national boundaries.

space agencies collaborate to launch an array of satellites that offer worldwide observations. These measurements are utilized to empower environmental and public health research, including activities such as monitoring air pollution, forecasting air quality, and assessing exposure. Satellite measurements pertaining to air quality have the potential to provide valuable information for policy-making. In the past 30 years, NASA and ESA have launched many satellite instruments that measure harmful trace gases that destroy environment, ecosystems, and human health (Burrows et al., 2011; Geddes et al., 2016; Goldberg et al., 2021; Martin, 2008). These measurements have advanced our fundamental comprehension of atmospheric chemistry and physics, and have enabled a deeper knowledge of the air quality challenges we currently encounter. Although these data have been widely utilized for scientific studies, there has been limited advancement in converting these data into outputs that are useful to draft air quality policy. The focus has primarily been on "fundamental" rather than "applied" science (Burrows et al., 2011). The accuracy and validation concerns, together with limitations regarding the spatial and temporal resolution, have resulted in a lack of attention and investigation toward these data for air quality management. This is further exacerbated by the historical lack of communication between the scientific community and decision makers.

Since the introduction of TROPOMI in 2017, and other groundbreaking satellite instruments for air quality monitoring, there has been a debate on using "applied" satellite data as an additional regulatory proof. Satellite sensors can measure important air contaminants throughout the world, with improved accuracy in terms of space and time. Satellites can watch the emission and movement of pollutants across a large region over time, as long as there are no clouds obstructing the view. This contrasts with ground monitors, which can only measure the concentration of pollutants at a specific location. It is important to note that column measurements cannot be directly converted into surface concentrations. Additionally, due to the limitations of satellite instruments in retrieving vertical distribution, there is a

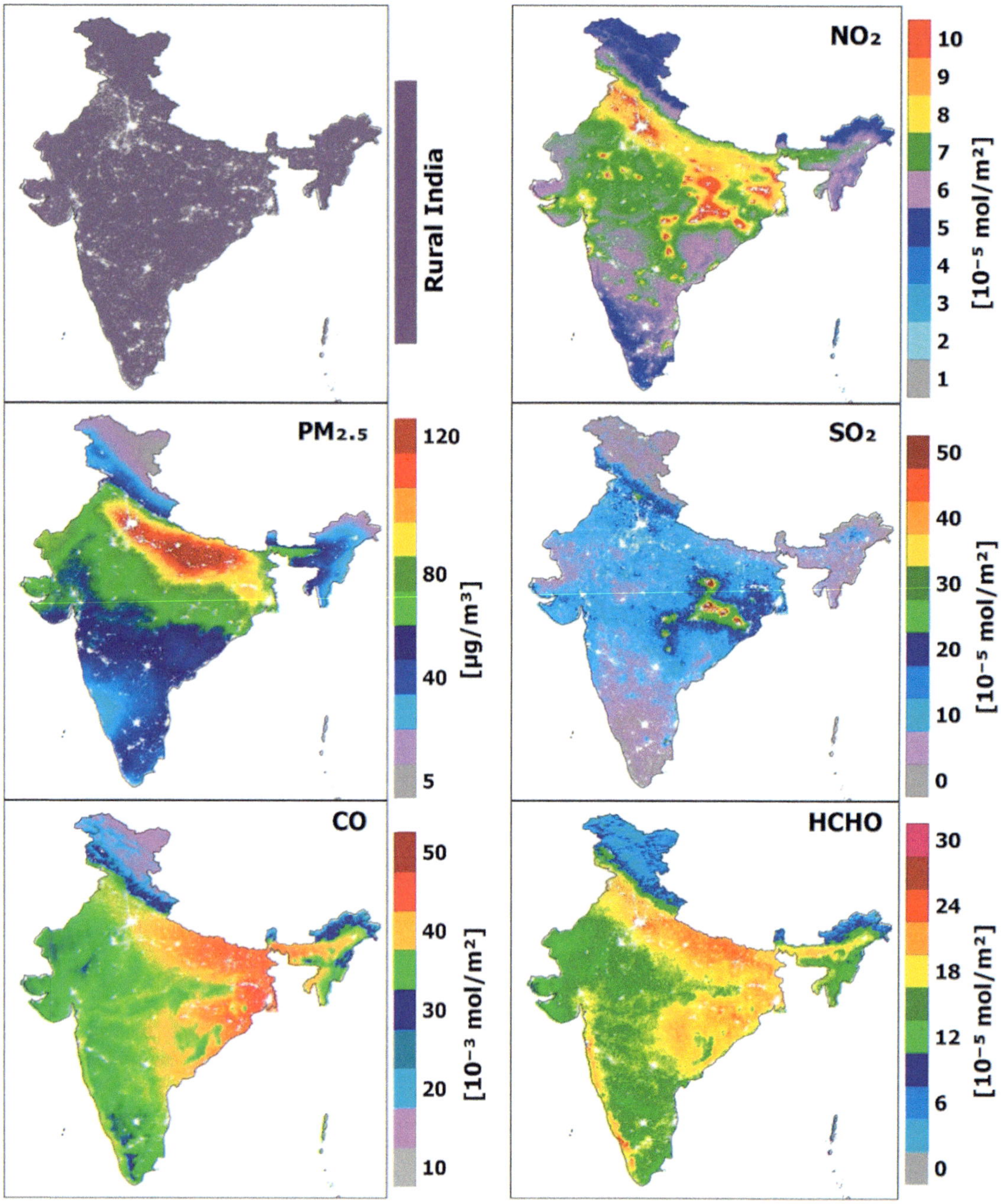

FIGURE 21.8 Top left: Illustration of demarcated rural India based on nighttime light data derived from visible infrared imaging radiometer suite (VIIRS) satellite. Remaining maps show the distribution of NO_2, SO_2, CO, HCHO, and $PM_{2.5}$ in rural India for the year 2019. Here, NO_2, SO_2, CO, and HCHO data are derived from TROPOMI and $PM_{2.5}$ is from gridded satellite product from NASA SEDAC. Map lines delineate study areas and do not necessarily depict accepted national boundaries.

significant level of uncertainty in the surface concentrations derived from satellite data (Van Donkelaar et al., 2010). Furthermore, the process of inverting these data requires the use of advanced chemical transport modeling (CTM), as highlighted by Martin (2008) and Van Donkelaar et al. (2010).

Modern satellites possess the capability to quantify the atmospheric concentrations of NO_2, SO_2, CO, HCHO, NH_3, CH_4, and aerosol properties, among other substances, at different spatial resolutions. Additionally, they can provide data with a temporal resolution of once per day, covering the entire globe. Each product's quality differs due to limitations of sensors, photo-chemical processes, and variations in vertical sensitivity. As a result, not all data products are currently acceptable for the end user needs (Martin, 2008). As an illustration, NO_2 has a high absorption capacity within the 425–450 nm range, which is not commonly observed in other substances. Consequently, it is very easy to accurately measure and obtain the amount of NO_2 in the atmosphere. It has a relatively short life time in the atmosphere, ranging from 2 to 24 hours (Beirle et al., 2011). As a result, the amount of NO_2 in the air is higher near its source and generally within the boundary layer. This makes it easier to determine the NO_2 sources, and shows a strong connection between surface concentrations and measurements taken by the satellites (Martin, 2008). Similarly, the MODIS satellite allows to capture particulate pollution at a finer resolution (1 km) and helps to monitor the air quality for better policy decisions. The quantification of individual species is important to make proper warning system and implementation of regulations that can help to curb air pollution. It is very difficult to monitor air pollution over large areas or in an entire city using ground-based measurements, as they are mostly confined to a particular place or randomly in few places of an urban region or city. Also, there is a huge disparity in the ground-based measurements because their network is denser in the developed countries than that in the developing or underdeveloped world. Therefore the advancement in satellite measurements is helping the developing world for their air quality monitoring programs.

The capacity to measure and record outdoor air pollution is rapidly growing, as more agencies and individuals are creating diverse technology for monitoring air quality. Nevertheless, there are still notable constraints in terms of resources and technology in numerous regions across the globe. Furthermore, a significant number of these stations operates in a state of relative seclusion, and thus, restrict the practical application of their technological and computational advancements. These differences are specifically noticeable across the monitoring disciplines, such as remote sensing by satellite technology, use of regulatory and nonregulatory air quality networks for station-based monitoring, computational air quality modeling, and implementation of low-cost sensor systems.

Effective collaboration across technological groups is crucial for optimizing the potential public health advantages of new technology-oriented measurements. Satellite observations were first utilized in environmental health research in the United States in 2009 (Hu, 2009). The study represents the first instance in which AOD was utilized to enhance the $PM_{2.5}$ exposure matrix and generate its comprehensive surface concentration in the USA. The purpose of this was to examine the relationship between $PM_{2.5}$ concentrations and the occurrence of coronary heart disease cases in the United States. The capacity to predict $PM_{2.5}$ concentrations using satellite data is expected to significantly accelerate the advancement of studies on the health effects of air pollution, especially in developing countries that may have poor air quality and limited ground monitoring networks.

A further constraint in using satellite remote sensing for health studies has been the incapacity to consistently observe air pollution at short time intervals. This is primarily attributed to the inadequacies in measurements, such as the absence of data on days with cloud cover or the inability to gather data at night. Additionally, the orbiting satellites have limitations in their ability to offer more than one measurement per day. Nevertheless, several newly deployed and forthcoming satellites present significant enhancements, such as increased spatiotemporal resolution (Holloway et al., 2021). These advancements will create new and captivating opportunities for applying these data to the public health research. The Tropospheric Emissions: Monitoring of POllution (TEMPO) and Geostationary Environment Monitoring Spectrometer (GEMS) missions offer hourly measurements of pollutants in several regions across the Northern Hemisphere. Additionally, the upcoming MultiAngle Imager for Aerosols (MAIA) mission will provide multiple measurements of particulate pollution on daily basis, including certain composition details, for specific regions in the globe. This future mission would further advance the temporal and spatial resolution of remotely observed air pollution for health and ecosystem-based studies.

21.5 Recent developments in air quality mapping: lessons learned from the COVID-19 lockdown

COVID-19 is an infectious disease that is caused by the severe acute respiratory syndrome coronavirus. The virus was initially detected in December 2019 in Wuhan, Hubei, China, and then turned out to be a global pandemic. This pandemic was a big issue in terms of human health. The rapid and worrisome rise in both the transmission and death rates of this pandemic made the world to come to a standstill to deal with its severe effects. In response, many countries implemented measures such as lockdown and travel bans, leading to significant disruption in society, energy consumption, and economic activities. This period of unrest has been commonly referred to as the anthropause. Although this lockdown has had significant impact on both social and economic realms of the world, it also had a beneficial effect on the environment, namely in terms of air quality. This is evident from the decrease in levels of $PM_{2.5}$, NO_2, and CO in major cities worldwide (Biswal et al., 2021; Cooper et al., 2022; Gopikrishnan & Kuttippurath, 2021; Kashyap et al., 2023; Kuttippurath et al., 2023; Patel et al., 2024; Pathak et al., 2023). The pandemic has exerted a discernible influence on the environment, as the changes in human activity and behavior have resulted in transient modifications in air pollution, GHG emissions, and water quality (Bhat et al., 2021).

Emission monitoring in real time over different regions of the globe is greatly aided by recent developments in sensors on board different satellites. The TROPOMI daily global data revealed remarkable reductions in tropospheric NO_2

in urban areas (Cooper et al., 2022; Gopikrishnan & Kuttippurath, 2021; Venter et al., 2020). These observations have been widely used by several studies to evaluate tropospheric NO_2 over polluted regions before and during the lockdown periods (Biswal et al., 2021; Cooper et al., 2022). Monitoring of air quality on city scale using TROPOMI is possible because of its finer spatial resolution of $3.5 \times 7\ km^2$ (Veefkind et al., 2012). From August 2019 onwards, resolution of the measurements has increased even to $3.5 \times 5.5\ km^2$, which made it easier to better map air quality.

To understand the city level changes in NO_2 (indicator of vehicular emissions) and SO_2 (represents industrial and coal-based thermal power plant emissions) pollution using TROPOMI measurements, we selected 3000 cities in the world, as shown in Fig. 21.9. We find that most cities worldwide exhibit a decrease in NO_2 pollution during the period of lockdown compared to prelockdown, about 5%–30% (Fig. 21.9). The decline is particularly noticeable in the cities where anthropogenic activities such as vehicular and industrial emissions are very high. For instance, New Delhi, Chandigarh, Kanpur, Jaipur, Mumbai, Ahmedabad, Chennai, Bengaluru, Patna, Hyderabad, Amritsar, Paris, Milan, Madrid, Berlin, Wuhan, Luxembourg, Brno, Chicago, Jambi, London, Amsterdam, Brussels, Beijing, Wuhan, Shanghai, Jakarta, Bangkok, Yangon, Hanoi show a reduction in NO_2 of about 5%–30% during lockdown (Fig. 21.9). Several other cities across the world also show a decline in NO_2 during lockdown, and they are clearly visible in Fig. 21.9. Similarly, a decrease of 5%–10% in SO_2 is observed during the lockdown compared to prelockdown period in major cities of India (Chennai, Delhi, Kolkata, Bangalore, and Mumbai). Also, the cites of the United States (New York, Charleston, Indianapolis, Jefferson City, St. Louis, and Washington) and Mexico (including Tampico, Aguascalientes, Guadalajara, Durango, and Culiacan) exhibit a reduction of about 5%–25%. A comparable decrease in SO_2 can be observed in various cities such as Muscat, Al Mukalla, Lahij, Aden, Abu Dhabi, and Al Kasab in the middle east, and Madrid, Braganca, Guarda, Arnhem, Ostersund, Hermansverk, Drammen, and Haarlem in Europe (Fig. 21.9). This decline is also evident in Russian cities Moscow, Skytyvkar, Smolensk, Klintsy, Tura, and Igarka, as well as in Chinese cities including Luoyang, Baotou, Zhengzhou, Jinan, Tiayuan, Fuzhou, and Chengdu. The decrease of pollution in cities (small spatial scale) is well detected by the high-resolution TROPOMI measurements.

In addition, to understand the spatial changes of NO_2 pollution within the cities, we have selected 16 cities in the world, as shown in Fig. 21.10. It is clearly observed by TROPOMI that most cities (e.g., Delhi, Kolkata, Mumbai, London, Paris, Beijing, and Shanghai) exhibit a reduction in NO_2 of about 5%–30%. In general, TROPOMI's spatial

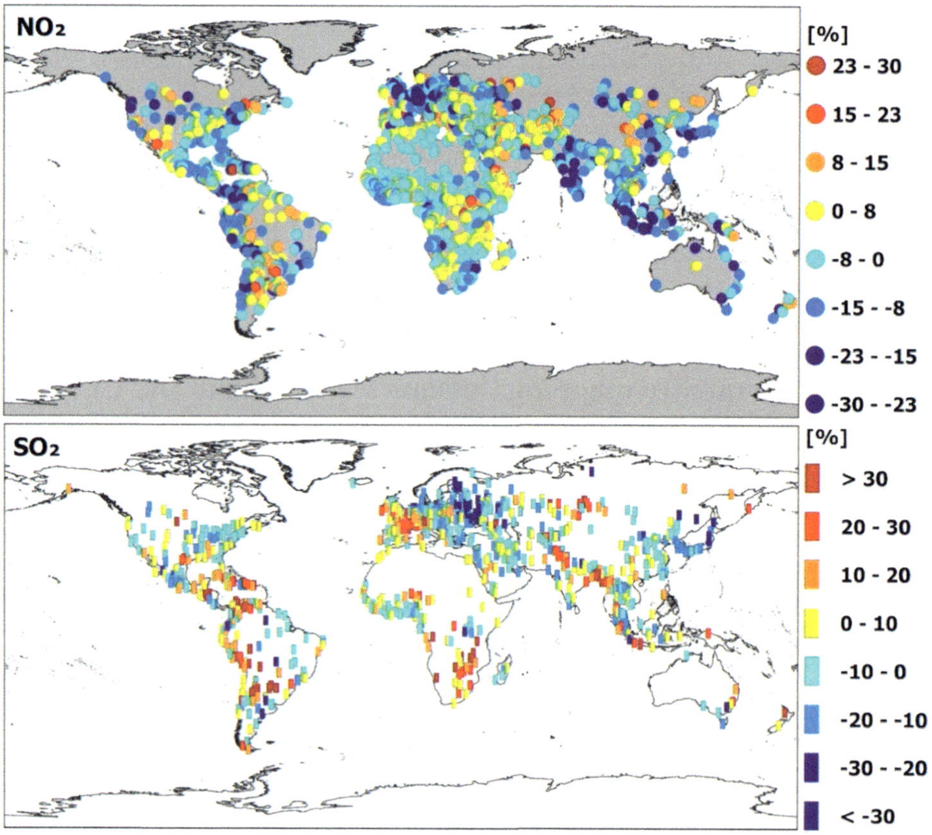

FIGURE 21.9 Changes in NO_2 and SO_2 concentrations during lockdown (April to May, 2020) compared to the prelockdown (April to May, 2017–19) for 3000 global cities as derived from the TROPOMI measurements. Map lines delineate study areas and do not necessarily depict accepted national boundaries.

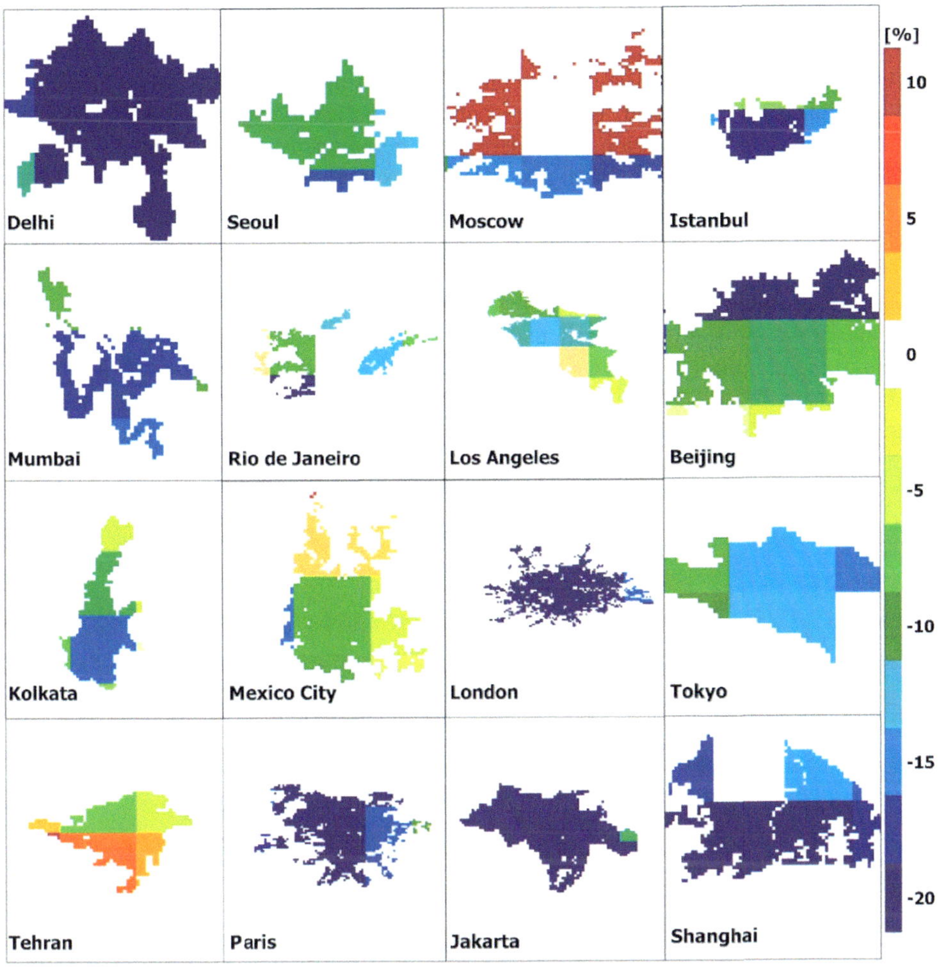

FIGURE 21.10 Changes in NO_2 concentrations during lockdown compared to prelockdown period derived from the TROPOMI measurements for selected cities in the world. Map lines delineate study areas and do not necessarily depict accepted national boundaries.

coverage within a metropolis exceeds 2 pixels, which enables the monitoring of spatial variability of pollution inside the city. This was not possible with previous satellite measurements due to their lower resolution. Therefore many studies use TROPOMI measurements during the COVID-19 lockdown period to detect the changes in urban pollution (e.g., Fioletov et al., 2022; Goldberg et al., 2020; Kuttippurath et al., 2024). However, still finer resolution satellite measurements are needed to fully understand the spatial heterogeneity of air pollution and its sources within the smaller cities.

21.6 Future air quality missions: geostationary satellites

Several satellite sensors, including OMI, GOME, GOME-2, SCIAMACHY, OMPS, and TROPOMI, detect solar radiance in the ultraviolet–visible (UV-Vis) spectral region when they are in the low earth orbit (Choi et al., 2024). GEMS is the first environmental instrument in Geostationary Earth Orbit (GEO). It is installed on the GEOstationary KOrea MultiPurpose SATellite-2B (GEO-KOMPSAT-2B), which was launched on February 19, 2020. GEMS is designed to measure the reflected radiance emitted by the earth's surface and atmosphere within the UV-Vis region, specifically in the wavelength range of 300–500 nm (Kim et al., 2020). The geostationary-air quality (Geo-AQ) constellation, consisting of GEMS in South Korea, TEMPO in the United States (Zoogman et al., 2017), and Sentinel-4 in Europe, are designed to monitor the transport of pollutants across the latitudes, including the Atlantic and Pacific Oceans in the Northern Hemisphere. This constellation aims at tracking both long- and short-range pollution movements. GEMS is positioned at an altitude of approximately 36,000 km above the Equator, at a longitude of 128.2 degrees E. Its main purpose is to observe the atmospheric environment in Asia, as its field of regard spans from 5 degrees S to 45 degrees N. It stretches from the longitude of India (75 degrees E) to the west, to the longitude of Japan (145 degrees E) to the east. GEMS observes aerosol, O_3, NO_2, SO_2, HCHO, clouds, and the UV index at a spatial resolution of 7×8 km^2. It operates exclusively during daylight hours at 1 hour intervals, which provide measurements eight times a day.

Similarly, TEMPO is a spectrometer that utilizes a grating to analyze light in the visible (540–740 nm) and ultraviolet (290–490 nm) ranges. The instrument is affixed to the Intelsat IS-40e telecommunication satellite, which was fabricated by Maxar and deployed on April 7, 2023, into a geostationary orbit above the United States, Canada, and Mexico (https://tempo.si.edu/). TEMPO is positioned to face Earth, ensuring continuous observations of the North America. This orientation enables the instrument's light-collecting mirror to perform a comprehensive sweep of the "field of regard" from East to West every hour. Its ultraviolet and visible light sensors measure the radiation that is reflected and scattered from Earth's surface and atmosphere back to the detectors. This allows the instrument to provide hourly measurements of O_3, NO_2, and others that are a part of the daily cycles of atmospheric chemistry at a spatial resolution of 4.5 km. The data obtained from TEMPO will enhance the regional air-quality information provided by the US Environmental Protection Agency (EPA), including its forecast and predictions.

These are the satellite instruments in the geostationary orbit that provide hourly measurements of all important atmospheric pollutants, and they help to monitor and forecast air quality (Hsu et al., 2024). We analyze the NO_2 pollution over Asia using GEMS, and USA using TEMPO. Note that TEMPO data used here are the Beta version and are not validated yet, and thus, we use the data only to demonstrate its spatial and temporal coverage to identify the hotspots (Fig. 21.11). It is clearly shown that GEMS capture high pollution in EC, South Korea, Japan, and IGP, as also

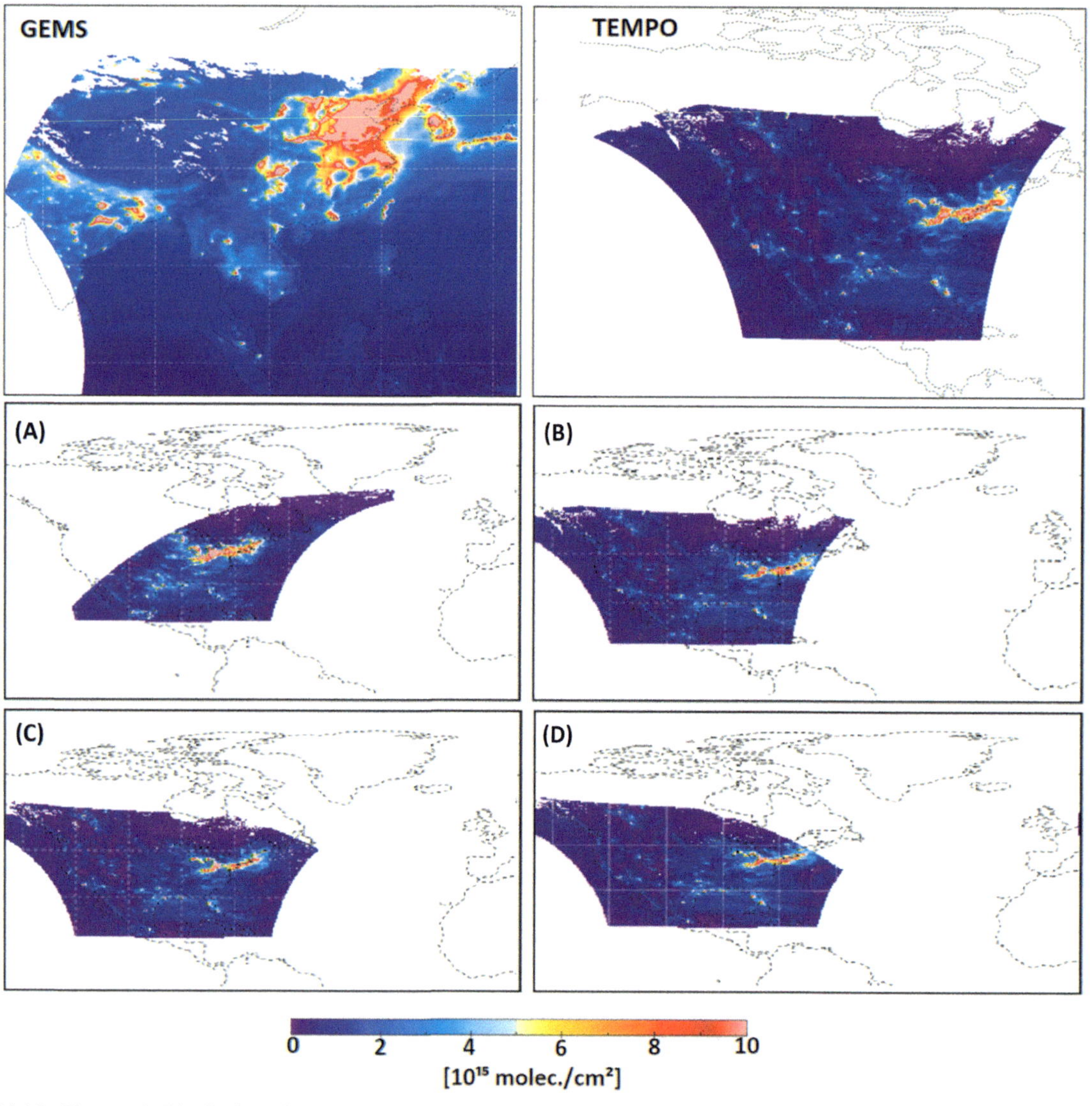

FIGURE 21.11 Top panel: Distribution of tropospheric column NO_2 derived from GEMS for the period April 2023 to March 2024 over the east Asia, and from TEMPO Level 3 Beta product for the time December 2023. Remaining row represents distribution of hourly NO_2 from TEMPO for the period December 30, 2023: (A) 14:11 PM to 15:11 PM, (B) 18:19 PM to 19:10 PM, (C) 19:11 PM to 20:11 PM, and (D) 20:10 PM to 21:10 PM. Map lines delineate study areas and do not necessarily depict accepted national boundaries.

found from TROPOMI measurements (Fig. 21.11). It indicates that the geostationary satellites well capture the NO_2 pollution, as in the case of low orbit satellites. The cities and industrial locations in India and China are also hotspots of NO_2 with high values of about $>0.8 \times 10^{16}$ molecules cm^{-2}, and these are well detected by TROPOMI (e.g., Zheng et al., 2021). Similarly, TEMPO identifies NO_2 pollution in the regions of high emissions such as the cities in the USA. East USA exhibits high levels of NO_2 pollution as found from TEMPO, which is also detected by TROPOMI. Advantage of both GEMS and TEMPO over TROPOMI is their hourly measurements (Fig. 21.11), which help to track the diurnal cycle of NO_2 and its emission sources, as demonstrated in Fig. 21.11. For instance, during 14:10–13:10 PM, NO_2 is higher in the regions of its "traditional" hotspots as compared to other time periods. As NO_2 has a short lifetime in the atmosphere, the hourly measurements from GEMS and TEMPO can assist in monitoring the transport of pollutants and aid in predicting air quality and its sources in specific areas.

21.7 Conclusions

Poor air quality is not only a regional problem, but a current global issue. To curb air pollution across the globe, monitoring air quality is essential. Observations of air quality using ground-based measurements are always the best option. However, due to the lack of monitoring stations in developing and underdeveloped countries, real-time monitoring of air pollution is very difficult. In addition, most monitoring stations are located in urban areas, and thus, the implementation of policies in rural areas is challenging. In recent decades, advanced satellite observations have made air quality monitoring relatively easy. The improvements in EOS in terms of their latitudinal coverage, and spatial, vertical, and temporal resolution, also help to identify emissions from various sources such as biomass burning, dust transport, industrial emissions, road transport, and stubble burning. For example, currently, fine resolution (3.5 km $\times$ 7 km) measurements are available from TROPOMI to examine gaseous air pollutants even in a city region. Similarly, fine-resolution MODIS level 2 AOD data (1 km) enhance our understanding of regional and point-source emissions of aerosols. The observations of vertical distribution of aerosols are also made possible by the CALIPSO satellite. Despite the relatively coarser resolution, NH_3 measurements from IASI, AIRS, and CrIS enabled large-scale global monitoring for several years, which would help to plan policy decisions on agricultural waste burning and the use of mineral fertilizers. On the other hand, there are also data available globally for $PM_{2.5}$ using satellite measurements. The recent TEMPO and GEMS missions facilitate hourly measurements of gaseous pollutants across a wide area in the Northern Hemisphere. This, together with the MAIA future mission, would significantly improve the real-time monitoring of air quality and the transport of pollution. In summary, this chapter discusses the time evolution of the improvement in satellite systems for monitoring key air pollutants. Yet, more advancement in EOS is needed to capture real-time synoptic views of atmospheric pollutants, and to make better policy decisions.

21.8 Data availability

MODIS data: https://modis.gsfc.nasa.gov/data/dataprod/mod04.php
MISR data: https://giovanni.gsfc.nasa.gov/giovanni/
OMI data: https://giovanni.gsfc.nasa.gov/giovanni/
TROPOMI data: Google Earth Engine
MOPITT data: https://giovanni.gsfc.nasa.gov/giovanni/
AIRS data: https://giovanni.gsfc.nasa.gov/giovanni/
IASI data: https://iasi.aeris-data.fr/nh3/
GOME data: https://www.temis.nl/
SCIAMACHY data: https://www.temis.nl/
SeaWiFS data: https://giovanni.gsfc.nasa.gov/giovanni/
GEMS data: https://nesc.nier.go.kr/en/html/cntnts/73/satellite/introduction.do
TEMPO data: https://search.earthdata.nasa.gov/search
Nighttime light data: https://www.earthdata.nasa.gov/learn/backgrounders/nighttime-lights

References

Alves, C., Evtyugina, M., Vicente, E., Vicente, A., Rienda, I. C., de la Campa, A. S., ... Duarte, I. (2023). PM2.5 chemical composition and health risks by inhalation near a chemical complex. *Journal of Environmental Sciences*, *124*, 860–874.

Amritha, S., Varikoden, H., Patel, V. K., Kuttippurath, J., & Gopikrishnan, G. S. (2024). Global, regional and city scale changes in atmospheric NO_2 with environmental laws and policies. *Sustainable Cities and Society*, *112*, 105617. doi:10.1016/j.scs.2024.105617.

Anenberg, S. C., Miller, J., Henze, D. K., Minjares, R., & Achakulwisut, P. (2019). The global burden of transportation tailpipe emissions on air pollution-related mortality in 2010 and 2015. *Environmental Research Letters*, *14*(9), 094012.

Auman, H. H., Chahine, M. T., Gautier, C., Goldberg, M. D., Kalnay, E., McMillin, L. M., ... Susskind, J. (2003). AIRS/AMSU/HSB on the Aqua mission: Design, science objectives, data products, and processing systems. *IEEE Transactions on Geoscience and Remote Sensing*, *41*, 253–264.

Behera, S. N., Sharma, M., Aneja, V. P., et al. (2013). Ammonia in the atmosphere: A review on emission sources, atmospheric chemistry and deposition on terrestrial bodies. *Environmental Science and Pollution Research*, *20*, 8092–8131.

Beirle, S., Boersma, K. F., Platt, U., Lawrence, M. G., & Wagner, T. (2011). Megacity emissions and lifetimes of nitrogen oxides probed from space. *Science (New York, N.Y.)*, *333*(6050), 1737–1739.

Bhat, S. A., Bashir, O., Bilal, M., Ishaq, A., Dar, M. U. D., & Kumar, R. (2021). Impact of COVID-related lockdowns on environmental and climate change scenarios. *Environmental Research*, *195*, 110839.

Biswal, A., Singh, V., Singh, S., Kesarkar, A. P., Ravindra, K., Sokhi, R. S., ... Mor, S. (2021). COVID-19 lockdown-induced changes in NO_2 levels across India observed by multi-satellite and surface observations. *Atmospheric Chemistry and Physics*, *21*, 5235–5251.

Buchholz, R., et al. (2021). Air pollution trends measured from Terra: CO and AOD over industrial, fire-prone, and background regions. *Remote Sensing of Environment*, *256*, 112275.

Burrows, J. P., Platt, U., & Borrell, P. (2011). *The remote sensing of tropospheric composition from space*. Springer Science & Business Media.

Choi, H., Liu, X., Jeong, U., Chong, H., Kim, J., Ahn, M. H., ... Lee, K. M. (2024). Geostationary Environment Monitoring Spectrometer (GEMS) polarization characteristics and correction algorithm. *Atmospheric Measurement Techniques*, *17*, 145–164.

Clarisse, L., Coheur, P. F., Prata, A. J., Hurtmans, D., Razavi, A., Phulpin, T., ... Clerbaux, C. (2008). Tracking and quantifying volcanic SO_2 with IASI, the September 2007 eruption at Jebel at Tair. *Atmospheric Chemistry and Physics*, *8*, 7723–7734.

Clerbaux, C., Boynard, A., Clarisse, L., George, M., Hadji-Lazaro, J., Herbin, H., ... Coheur, P. F. (2009). Monitoring of atmospheric composition using the thermal infrared IASI/MetOp sounder. *Atmospheric Chemistry and Physics*, *9*, 6041–6054.

Cooper, M. J., Martin, R. V., Hammer, M. S., Levelt, P. F., Veefkind, P., Lamsal, L. N., et al. (2022). Global fine-scale changes in ambient NO_2 during COVID-19 lockdowns. *Nature*, *601*(7893)), 380–387.

Dey, S., Purohit, B., Balyan, P., Dixit, K., Bali, K., Kumar, A., ... Shukla, V. K. (2020). A satellite-based high-resolution (1-km) ambient PM2.5 database for India over two decades (2000–2019): Applications for air quality management. *Remote Sensing*, *12*(23), 3872.

Diao, M., Holloway, T., Choi, S., O'Neill, S. M., Al-Hamdan, M. Z., et al. (2019). Methods, availability, and applications of PM2.5 exposure estimates derived from ground measurements, satellite, and atmospheric models. *Journal of the Air and Waste Management Association (1995)*, *69* (12), 1391–1414.

Dutta, A., Alexander, D., Karrison, T., et al. (2021). Household air pollution, ultrasound measurement, fetal biometric parameters and intrauterine growth restriction. *Environmental Health*, *20*, 1–9.

Filonchyk, M., Yan, H., Zhang, Z., Yang, S., Li, W., & Li, Y. (2019). Combined use of satellite and surface observations to study aerosol optical depth in different regions of China. *Scientific Reports*, *9*(1), 6174.

Fioletov, V., McLinden, C. A., Griffin, D., Krotkov, N., Liu, F., & Eskes, H. (2022). Quantifying urban, industrial, and background changes in NO_2 during the COVID-19 lockdown period based on TROPOMI satellite observations. *Atmospheric Chemistry and Physics*, *22*, 4201–4236.

Gallagher, C. L., Holloway, T., Tessum, C. W., Jackson, C. M., & Heck, C. (2023). Combining satellite-derived PM2.5 data and a reduced-form air quality model to support air quality analysis in US cities. *GeoHealth*, *7*, e2023GH000788.

Geddes, J. A., Martin, R. V., Boys, B. L., & van Donkelaar, A. (2016). Long-term trends worldwide in ambient NO_2 concentrations inferred from satellite observations. *Environmental Health Perspectives*, *124*(3), 281–289.

Goldberg, D. L., Anenberg, S. C., Griffin, D., McLinden, C. A., Lu, Z., & Streets, D. G. (2020). Disentangling the impact of the COVID-19 lockdowns on urban NO_2 from natural variability. *Geophysical Research Letters*, *47*, e2020GL089269.

Goldberg, D. L., Anenberg, S. C., Kerr, G. H., Mohegh, A., Lu, Z., & Streets, D. G. (2021). TROPOMI NO_2 in the United States: A detailed look at the annual averages, weekly cycles, effects of temperature, and correlation with surface NO_2 concentrations. *Earth's Future*, *9*(4), e2020EF001665.

Gopikrishnan, G. S., & Kuttippurath, J. (2021). A decade of satellite observations reveal significant increase in atmospheric formaldehyde from shipping in Indian Ocean. *Atmospheric Environment*, *246*, 118095.

Gopikrishnan, G. S., Kuttippurath, J., Raj, S., Singh, A., & Abbhishek, K. (2022). Air quality during the COVID–19 lockdown and unlock periods in India analyzed using satellite and ground-based measurements. *Environmental Processes*, *9*, 28. Available from https://doi.org/10.1007/s40710-022-00585-9.

Hammer, M. S., van Donkelaar, A., Li, C., Lyapustin, A., Sayer, A. M., Hsu, N. C., ... Martin, R. V. (2020). Global estimates and long-term trends of fine particulate matter concentrations (1998–2018). *Environmental Science and Technology*, *54*(13), 7879–7890.

Holloway, T., Miller, D., Anenberg, S., Diao, M., Duncan, B., Fiore, A. M., ... Zondlo, M. A. (2021). Satellite monitoring for air quality and health. *Annual Review of Biomedical Data Science*, *4*(1), 417–447.

Hsu, C. H., Henze, D. K., Mizzi, A. P., González Abad, G., He, J., Harkins, C., et al. (2024). An observing system simulation experiment analysis of how well geostationary satellite trace-gas observations constrain NOx emissions in the US. *Journal of Geophysical Research: Atmospheres*, *129*, e2023JD039323.

Hu, Z. (2009). Spatial analysis of MODIS aerosol optical depth, PM 2.5, and chronic coronary heart disease. *International Journal of Health Geographics*, *8*, 1–10.

Ignatov, A., & Stowe, L. (2002). Aerosol retrievals from individual AVHRR channels: II. Quality control, probability distribution, functions, information content and consistency checks of retrievals. *Journal of the Atmospheric Sciences*, *59*, 335–362.

Jiang, Z., Zhu, R., Miyazaki, K., McDonald, B. C., Klimont, Z., Zheng, B., et al. (2022). Decadal variabilities in tropospheric nitrogen oxides over United States, Europe, and China. *Journal of Geophysical Research: Atmospheres*, *127*, e2021JD035872.

Joshi, A., Pathak, M., Kuttippurath, J., & Patel, V. K. (2023). Adoption of cleaner technologies and reduction in fire events in the hotspots lead to global decline in carbon monoxide. *Chemosphere*, *336*, 139259.

Kacenelenbogen, M., Redemann, J., Vaughan, M. A., Omar, A. H., Russell, P. B., Burton, S., et al. (2014). An evaluation of CALIOP/CALIPSO's aerosol-above-cloud detection and retrieval capability over North America. *Journal of Geophysical Research*, *119*(1), 230–244.

Kashyap, R., Kuttippurath, J., & Patel, V. K. (2023). Improved air quality leads to enhanced vegetation growth during the COVID–19 lockdown in India. *Applied Geography*, *151*, 102869. Available from https://doi.org/10.1016/j.apgeog.2022.102869.

Kaufman, Y. J., Smirnov, A., Holben, B. N., & Dubovik, O. (2001). Baseline maritime aerosol: Methodology to derive the optical thickness and scattering properties. *Geophysical Research Letters*, *28*, 3251–3254.

Kephart, J. L., Gouveia, N., Rodriguez, D. A., Indvik, K., Alfaro, T., Texcalac-Sangrador, J. L., et al. (2023). Ambient nitrogen dioxide in 47, 187 neighbourhoods across 326 cities in eight Latin American countries: Population exposures and associations with urban features. *The Lancet Planetary Health*, *7*(12), e976–e984.

Khalil, M. A. K., & Rasmussen, R. A. (1990). The global cycle of carbon monoxide: Trends and mass balance. *Chemosphere*, *20*(1–2), 227–242.

Kibirige, G. W., Yang, M. C., Liu, C. L., & Chen, M. C. (2023). Using satellite data on remote transportation of air pollutants for PM2.5 prediction in northern Taiwan. *PLoS ONE*, *18*(3), e0282471.

Kim, J., Jeong, U., Ahn, M. H., Kim, J. H., Park, R. J., Lee, H., ... Choi, Y. (2020). New era of air quality monitoring from space geostationary environment monitoring spectrometer (GEMS). *Bulletin of the American Meteorological Society*, *101*, E1–E22. Available from https://doi.org/10.1175/BAMS-D-18-0013.1.

Kokhanovsky, A. A., et al. (2007). Aerosol remote sensing over land: A comparison of satellite retrievals using different algorithms and instruments. *Atmospheric Research*, *85*, 372–294.

Koziel, J. A., Aneja, V. P., & Baek, B. (2006). Gas-to-particle conversion process between ammonia, acid gases, and fine particles in the atmosphere. *Agricultural and Biosystems Engineering Publications, report available at Iowa State University Digital Repository 2006*. Available from http://lib.dr.iastate.edu/abe_.

Kumar, M., Parmar, K. S., Kumar, D. B., Mhawish, A., Broday, D. M., Mall, R. K., & Banerjee, T. (2018). Long-term aerosol climatology over Indo-Gangetic Plain: Trend, prediction and potential source fields. *Atmospheric Environment*, *180*, 37–50.

Kumar, R., Naja, M., Pfister, G. G., Barth, M. C., & Brasseur, G. P. (2013). Source attribution of carbon monoxide in India and surrounding regions during wintertime. *Journal of Geophysical Research*, *118*, 1981–1995.

Kuttippurath, J., Abbhishek, K., Gopikrishnan, G. S., & Pathak, M. (2022). Investigation of long–term trends and major sources of atmospheric HCHO over India. *Environmental Challenges*, *7*, 100477.

Kuttippurath, J., Kumar, P., Nair, P. J., & Chakraborty, A. (2018). Accuracy of satellite total column ozone measurements in polar vortex conditions: Comparison with ground-based observations in 1979–2013. *Remote Sensing of Environment*, *209*, 648–659.

Kuttippurath, J., Patel, V. K., Gopikrishnan, G. P., & Varikoden, H. (2023). Changes in air quality, meteorology and energy consumption during the COVID-19 lockdown and unlock periods in India. *Air*, *1*, 125–138. Available from https://doi.org/10.3390/air1020010.

Kuttippurath, J., Patel, V. K., Kashyap, R., Singh, A., & Clerbaux, C. (2024). Anomalous increase in global atmospheric ammonia during COVID-19 lockdown: Need policies to curb agricultural emissions. *Journal of Cleaner Production*, *434*, 140424. Available from https://doi.org/10.1016/j.jclepro.2023.140424.

Kuttippurath, J., Patel, V. K., Pathak, M., & Singh, A. (2022). Improvements in SO_2 pollution in India: Role of technology and environmental regulations. *Environmental Science and Pollution Research*, *29*, 78637–78649.

Kuttippurath, J., & Raj, S. (2021). Two decades of aerosol observations by AATSR, MISR, MODIS and MERRA-2 over India and Indian Ocean. *Remote Sensing of Environment*, *257*, 112363.

Kuttippurath, J., Singh, A., Dash, S. P., Mallick, N., Clerbaux, C., Van Damme, M., ... Varikoden, H. (2020). Record high levels of atmospheric ammonia over India: Spatial and temporal analyses. *Science of the Total Environment*, *740*, 139986.

Levelt, P. F., Stein Zweers, D. C., Aben, I., Bauwens, M., Borsdorff, T., De Smedt, I., ... Verhoelst, T. (2022). Air quality impacts of COVID-19 lockdown measures detected from space using high spatial resolution observations of multiple trace gases from Sentinel-5P/TROPOMI. *Atmospheric Chemistry and Physics*, *22*, 10319–10351.

Lin, H., Tao, J., Qian, Z., Ruan, Z., Xu, Y., Hang, J., ... Ma, W. (2018). Shipping pollution emission associated with increased cardiovascular mortality: A time series study in Guangzhou, China. *Environmental Pollution*, *241*, 862–868.

Lipponen, A., Mielonen, T., Pitkänen, M. R. A., Levy, R. C., Sawyer, V. R., Romakkaniemi, S., ... Arola, A. (2018). Bayesian aerosol retrieval algorithm for MODIS AOD retrieval over land. *Atmospheric Measurement Techniques*, *11*, 1529–1547.

Liu, D., et al. (2019). Analysis of global three-dimensional aerosol structure with spectral radiance matching. *Atmospheric Measurement Techniques*, *12*(12), 6541–6556.

Martin, R. V. (2008). Satellite remote sensing of surface air quality. *Atmospheric Environment*, *42*(34), 7823–7843.

Martin, R. V., Brauer, M., van Donkelaar, A., Shaddick, G., Narain, U., & Dey, S. (2019). No one knows which city has the highest concentration of fine particulate matter. *Atmospheric Environment: X*, *3*, 100040.

Miranda, M. L., Edwards, S. E., Keating, M. H., & Paul, C. J. (2021). Making the environmental justice grade: The relative burden of air pollution exposure in the United States. *International Journal of Environmental Research and Public Health*, *8*(6), 1755–1771.

Miyazaki, K., Bowman, K., Sekiya, T., Jiang, Z., Chen, X., Eskes, H., et al. (2020). Air quality response in China linked to the 2019 novel coronavirus (COVID-19) lockdown. *Geophysical Research Letters*, *47*, e2020GL089252.

Pant, P., & Harrison, R. M. (2012). Critical review of receptor modelling for particulate matter: A case study of India. *Atmospheric Environment*, *49*, 1–12.

Patel, V. K., Kuttippurath, J., & Kashyap, R. (2024). Increased global cropland greening as a response to the unusual reduction in atmospheric $PM_{2.5}$ concentrations during the COVID-19 lockdown period. *Chemosphere*, *358*, 142147.

Pathak, M., & Kuttippurath, J. (2022). Air quality trends in rural India: analysis of NO_2 pollution using satellite measurements. *Environmental Science: Processes and Impacts*, *24*(12), 2437–2449.

Pathak, M., Patel, V. K., & Kuttippurath, J. (2023). Spatial heterogeneity in global atmospheric CO during the COVID–19 lockdown: Implications for global and regional air quality policies. *Environmental Pollution*, *335*, 122269.

Ramachandran, A., Jain, N. K., Sharma, S. A., & Pallipad, J. (2013). Recent trends in tropospheric NO2 over India observed by SCIAMACHY: Identification of hot spots. *Atmospheric Pollution Research*, *4*, 354–361.

Ravishankara, A. R., David, L. M., Pierce, A., Jeffrey, R., & Venkataraman, C. (2007). Outdoor air pollution in India is not only an urban problem. *Proceedings of the National Academy of Sciences*, *117*(46), 28640–28644.

Richter, A., et al. (2005). Increase in tropospheric nitrogen dioxide over China observed from space. *Nature*, *437*(7055), 129–132.

Rosenfeld, D., et al. (2014). Global observations of aerosol-cloud-precipitation-climate interactions. *Reviews of Geophysics*, *52*, 750–808.

Sanatan, B., Vinoj, V., David, L. M., & Sen, I. S. (2024). Long-range transport of aerosols and regional sources using MODIS and NASA MERRA reanalysis over South Asia. *Earth Systems and Environment*. Available from https://doi.org/10.1007/s41748-024-00478-x.

Sannigrahi, S., Kumar, P., Molter, A., Zhang, Q., Basu, B., Basu, A. S., & Pilla, F. (2021). Examining the status of improved air quality in world cities due to COVID-19 led temporary reduction in anthropogenic emissions. *Environmental Research*, *196*, 110927.

Satheesh, S. K., Vinoj, V., & Moorthy, K. K. (2006). Vertical distribution of aerosols over an urban continental site in India inferred using a micro pulse lidar. *Geophysical Research Letters*, *33*, L20816.

Seinfeld, J. H., & Pandis, S. N. (2016). *Atmospheric chemistry and physics: from air pollution to climate change* (3rd ed.). New York: John Wiley & Sons, Inc.

Shaddick, G., Thomas, M. L., Green, A., Brauer, M., Donkelaar, A., et al. (2018). Data integration model for air quality: A hierarchical approach to the global estimation of exposures to ambient air pollution. *Journal of the Royal Statistical Society C*, *67*(1), 231–253.

Shah, V., Jacob, D. J., Li, K., Silvern, R. F., Zhai, S., Liu, M., … Zhang, Q. (2020). Effect of changing NOx lifetime on the seasonality and long-term trends of satellite-observed tropospheric NO_2 columns over China. *Atmospheric Chemistry and Physics*, *20*, 1483–1495.

Shephard, M. W., & Cady-Pereira, K. E. (2015). Cross-track Infrared Sounder (CrIS) satellite observations of tropospheric ammonia. *Atmospheric Measurement Techniques*, *8*, 1323–1336.

Shephard, M. W., Cady-Pereira, K. E., Luo, M., Henze, D. K., Pinder, R. W., Walker, J. T., … Clarisse, L. (2011). TES ammonia retrieval strategy and global observations of the spatial and seasonal variability of ammonia. *Atmospheric Chemistry and Physics.*, *11*, 10743–10763.

de Sherbinin, A., Levy, M. A., Zell, E., Weber, S., & Jaiteh, M. (2014). Using satellite data to develop environmental indicators. *Environmental Research Letters*, *9*, 084013.

Singh, N., Murari, V., Kumar, M., Barman, S. C., & Banerjee, T. (2017). Fine particulates over South Asia: Review and meta-analysis of PM2.5 source apportionment through receptor model. *Environmental Pollution*, *223*, 121–136.

Smirnov, A., Holben, B. N., Eck, T. F., Slutsker, I., Chatenet, B., & Pinker, R. T. (2002). Diurnal variability of aerosol optical depth observed at AERONET (Aerosol Robotic Network) sites. *Geophysical Research Letters*, *29*, 2115. Available from https://doi.org/10.1029/2002GL016305.

Someya, Y., Imasu, R., Shiomi, K., & Saitoh, N. (2020). Atmospheric ammonia retrieval from the TANSO-FTS/GOSAT thermal infrared sounder. *Atmospheric Measurement Techniques*, *13*, 309–321.

Sorribas, M., Andrews, E., Ogren, J. A., del Águila, A., Fraile, R., Sheridan, P., & Yela, M. (2019). Climatological study for understanding the aerosol radiative effects at Southwest Atlantic coast of Europe. *Atmospheric Environment*, *205*, 52–66.

Suresh Babu, S., et al. (2013). Trends in aerosol optical depth over Indian region: Potential causes and impact indicators. *Journal of Geophysical Research: Atmospheres*, *118*, 11794–11806.

Technical support document EPA's 2011 national-scale air toxics assessment, 2011 NATA TSD; United States Environmental Protection Agency: United States, 2015; https://www.epa.gov/sites/production/files/2015-12/documents/2011-nata-tsd.pdf (accessed on May 3, 2024).

Van Damme, M., Clarisse, L., Whitburn, S., et al. (2018). Industrial and agricultural ammonia point sources exposed. *Nature*, *564*, 99–103.

Van Damme, M., Whitburn, S., Clarisse, L., Clerbaux, C., Hurtmans, D., & Coheur, P. F. (2017). Version 2 of the IASI NH_3 neural network retrieval algorithm: near-real-time and reanalysed datasets. *Atmospheric Measurement Techniques*, *10*, 4905–4914.

Van Donkelaar, A., Martin, R. V., Brauer, M., Kahn, R., Levy, R., Verduzco, C., & Villeneuve, P. J. (2010). Global estimates of ambient fine particulate matter concentrations from satellite-based aerosol optical depth: development and application. *Environmental Health Perspectives*, *118*(6), 847–855.

Veefkind, J. P., Aben, I., McMullan, K., Förster, H., De Vries, J., Otter, G., et al. (2012). TROPOMI on the ESA Sentinel-5 Precursor: A GMES mission for global observations of the atmospheric composition for climate, air quality and ozone layer applications. *Remote Sensing of Environment*, *120*, 70–83.

Venter, Z. S., Aunan, K., Chowdhury, S., & Lelieveld, J. (2020). COVID-19 lockdowns cause global air pollution declines. *Proceedings of the National Academy of Sciences*, *117*(32), 18984–18990.

Vijayaraghavan, K., Hilary, E. S., & Christian, S. (2008). Practical aspects of using satellite data in air quality modeling. *Environmental Science and Technology*, *42*(22), 8187–8192.

Warner, J. X., Dickerson, R. R., Wei, Z., Strow, L. L., Wang, Y., & Liang, Q. (2017). Increased atmospheric ammonia over the world's major agricultural areas detected from space. *Geophysical Research Letters*, *44*, 2875–2884.

Weagle, C. L., Snider, G., Li, C., Van Donkelaar, A., Philip, S., Bissonnette, P., ... Martin, R. V. (2018). Global sources of fine particulate matter: Interpretation of PM2.5 chemical composition observed by SPARTAN using a global chemical transport model. *Environmental Science and Technology*, *52*, 11670–11681.

Wei, J., Li, Z., Peng, Y., & Sun, L. (2019). MODIS Collection 6.1 aerosol optical depth products over land and ocean: validation and comparison. *Atmospheric Environment*, *201*, 428–440.

WHO. (2021). WHO global air quality guidelines: Particulate matter (PM2.5 and PM10), ozone, nitrogen dioxide, sulfur dioxide and carbon monoxide. WHO Doc., 290. www.who.int/publications/i/.

WHO. (2022). Ambient air pollution: Policy and progress. https://www.who.int/teams/.

Whitburn, S., Van Damme, M., Clarisse, L., Bauduin, S., Heald, C., Hadji-Lazaro, J., ... Coheur, P.-F. (2016). A flexible and robust neural network IASI-NH_3 retrieval algorithm. *Journal of Geophysical Research: Atmospheres*, *121*, 6581–6599. Available from https://doi.org/10.1002/2016JD024828.

Winker, D. M., et al. (2010). The CALIPSO mission. *Bulletin of the American Meteorological Society*, *91*, 1211–1230.

Zhang, Y., et al. (2021). Satellite remote sensing of atmospheric particulate matter mass concentration: Advances, challenges, and perspectives. *Fundamental Research*, *1*(3), 240–258.

Zhu, L., Jacob, D. J., Keutsch, F. N., Mickley, L. J., Scheffe, R., Strum, M., et al. (2017). Formaldehyde (HCHO) as a hazardous air pollutant: Mapping surface air concentrations from satellite and inferring cancer risks in the United States. *Environmental Science and Technology*, *51*(10), 5650–5657.

Zheng, Z. W. Z., Chen, Y., Yang, Z., & Marinello, F. (2021). Analysis of temporal and spatial variation characteristics of NO2 pollutants in Guangdong-Hong Kong-Macao Greater Bay Area based on Sentinel-5P satellite data. *China Environmental Science*, *41*, 63–72.

Zoogman, P., Liu, X., Suleiman, R. M., Pennington, W. F., Flittner, D. E., et al. (2017). Tropospheric emissions: Monitoring of pollution (TEMPO). *Journal of Quantitative Spectroscopy and Radiative Transfer*, *186*, 17–39.

Chapter 22

Assessment of nitrogen dioxide concentrations in nonattainment cities during COVID-19 lockdown using GIS technique

Aniket Raut[1], Prakhar Misra[2], Nirmal Kumar[1], Hitesh Supe[1] and Ram Avtar[3]

[1]*Graduate School of Environmental Science, Hokkaido University, Sapporo, Japan,* [2]*Department of Civil Engineering, Indian Institute of Technology, Roorkee, Uttarakhand, India,* [3]*Faculty of Environmental Earth Science, Hokkaido University, Sapporo, Japan*

22.1 Introduction

Air pollution has a widespread impact globally, leading to 4.9 million premature deaths and a loss of 147 million healthy life years, as reported by the World Health Organization (Grisales-Romero et al., 2021). Over the years, rapid urbanization, industrialization, increased energy consumption, and a significant surge in the number of vehicles, particularly in developing nations like China and India (Panda, Mallik, Nath, Das, & Ramasamy, 2021), have exacerbated the problem. To address health concerns caused by anthropogenic factors and phenomena affecting the Earth's surface, there is a growing need for a heightened focus on air quality and the assessment of chemical concentrations (Halder et al., 2023). China and India, in particular, have the highest levels of fine particulate matter (PM2.5) and bear approximately half of this staggering burden (Yang et al., 2018). This fact showcases the multitude of challenges these countries face in combating air pollution. Middle-income countries like India experience severe health risks associated with air pollution (Gordon et al., 2014). Nitrogen dioxide (NO_2) also stands out as a significant contributor among diverse air pollutants, with its origins rooted in various sources. Primarily, NO_2 emanates from activities like fossil fuel combustion, power plant operations, vehicle emissions, biomass burning, soil bacterial activity, and even lighting (Kelly, 2010). This compound emerges as a major contributor to air pollution, underscoring the critical need for targeted efforts in mitigating its release and addressing its adverse environmental impacts. It only contributes to acid rain and a reddish-brown haze but also plays a crucial role as a pivotal criteria pollutant, influencing the tropospheric chemistry by acting as a precursor to surface ozone and subsequent aerosols (Brusseau, Pepper, & Gerba, 2019). Addressing their emissions is essential for combating air quality issues and their associated environmental impacts. Additionally, NOx poses health hazards, contributing to heart and lung diseases as well as unforeseen fatalities worldwide. NO_2 irritates the nose and throat, increases susceptibility to respiratory infections, and has carcinogenic effects (Misra et al., 2021).

The rapid industrial and various developmental activities, vehicular emissions, agricultural activities and crop residue burning, forest fire and shifting cultivation activities, population pressure, etc., increase air pollution in India. The National Clean Air Program (NCAP), started in January 2019 by the Ministry of Environment, Forest, and Climate Change, Government of India, aims to develop suitable management strategies to improve air quality. NCAP intend to achieve a 20%–30% reduction in PM10 and PM2.5 by 2024 compared to the 2017 baseline. The nonattainment areas are geographic regions that do not consistently meet the national ambient air quality standards (NAAQS) for PM10 or NO_2 over a 5-year period and are considered to have unhealthy levels of air pollution. The Central Pollution Control Board of India (CPCB) declared the list of nonattainment cities based on the pollution thresholds, which has identified 131 such cities, primarily across 21 states and 2 union territories in India (Fig. 22.1). Maharashtra stands out with the

Sustainable Development Perspectives in Earth Observation. DOI: https://doi.org/10.1016/B978-0-443-14072-3.00004-6

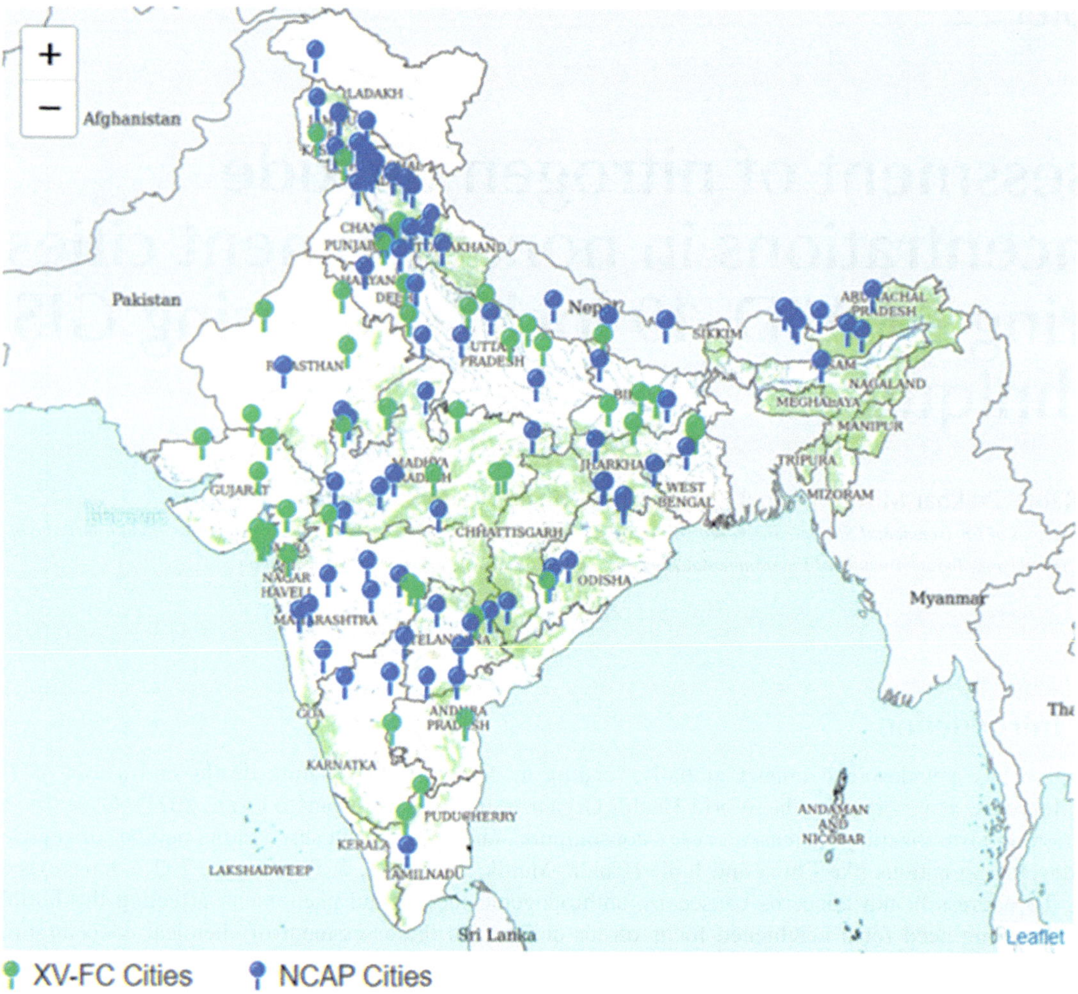

FIGURE 22.1 Monitoring dashboard of the PRANA (Portal for Regulation of Air-pollution in Nonattainment cities) portal for monitoring of implementation of National Clean Air Program (NCAP) in India. Map lines delineate study areas and do not necessarily depict accepted national boundaries.

highest count (19) of nonattainment cities in India, with pollution concentrations exceeding the NAAQS (https://prana.cpcb.gov.in/#/NCAPTracker accessed on June 8, 2024).

Most countries were forced to implement lockdowns in response to the COVID-19 pandemic, halting all activities as a precaution to curb the virus's spread (Mor et al., 2021). In India, a nationwide lockdown was enforced from March 25 to May 31, 2020, during which human activities were brought to a standstill (Avtar et al., 2021). Subsequently, the unlock phase began, involving the gradual relaxation of restrictions lasting from June 1, 2020, to March 31, 2022. A notable reduction in the concentrations of almost all pollutants was observed during the lockdown phase, owing to restrictions on vehicular, industrial, and anthropogenic activities (Mor et al., 2021). This led to a decrease in the air quality index (AQI) by 30%−50% and 40%−60% for NOx, accompanied by a substantial reduction in deaths due to air pollution (Banerji & Mitra, 2022; KS & Venkatasubramanian, 2021; Singh et al., 2022). Ruidas and Pal (2022) showed the dramatic changes in air pollution, in 10 cities of Maharashtra State, India, particularly PM2.5 concentration levels in prelockdown, lockdown, and various unlocking phases from 2018 to 2020. They used the CPCB hourly data of different lockdown phases for inverse distance weighted interpolation and reported a highest 74% reduction in the lockdown phase (March to June 2020) compared to the prelockdown period (2018 and 2019). Sahoo, Mangla, Pathak, Salãmao, and Sarkar (2021) examined the changes in multiple air quality parameters (CPCB data) during the COVID-19 pandemic (January to July 2020) and aimed to establish the relationships of environmental and demographic variables with COVID-19 cases in the state of Maharashtra, the worst-hit state in India. This study also indicated a significant decrease in multiple air quality parameters (PM2.5, PM10, NO_X, and CO) and the combined AQI as an impact of the lockdown, while no improvement in SO_2 and O_3 concentration. Saini, Darla, Lataye, Motghare, and Shingare (2022) analyzed the change in the concentration of three standard pollutants, namely PM10, SO_2, and NO_2, during the lockdown in Nagpur, India, and reported a significant improvement in AQI in the lockdown phase compared to pre- and postlockdown phases.

Although the ground-based monitoring stations provide the air quality parameters at a fine spatial (local scale) and temporal (e.g., hourly) resolution, it has limitations in assessing data over a large area. The satellite-based remote sensing monitoring systems, such as Environmental Satellite and Greenhouse Gases Observing Satellite, collect data on atmospheric constituents over a large area with coarse spatial (several kilometers) and temporal (daily to monthly) scales. The TROPOspheric Monitoring Instrument (TOPOMI) sensor is a space-borne, nadir-viewing imaging spectrometer covering wavelengths between the ultraviolet and shortwave infrared. It provides a variety of air pollutant datasets (Bodah et al., 2022; ESA, 2022). The sensor onboard TROPOMI senses ultraviolet, near-infrared, visible, and shortwave infrared wavelengths to monitor NO_2, O_3, CH_4, formaldehyde, CO, and SO_2 in the atmosphere. Several studies have used the TROPOMI data for air quality monitoring and linking with various natural and anthropogenic activities (Dekker et al., 2019; Pathak & Kuttippurath, 2022; Saw, Dey, Kaushal, & Lal, 2021). Recently, Das, Behera, and Abhilash (2024) studied the temporal variation of CO and NO_2 over India's capital Delhi and the surrounding region (NCR), which experienced a severe reduction in air quality due to crop residue burning in the postmonsoon season in the neighboring states annually.

Utilizing the capacity of TROPOMI data, several studies assessed the change in air quality parameters due to COVID-related lockdown phases in India. Biswal et al. (2021) assessed the spatial changes in NO_2 levels over different regions of India during the COVID-19 lockdown using the TROPOMI and Ozone Monitoring Instrument (OMI) data. Prakash, Goswami, Khan, and Nautiyal (2021) used TROPOMI NO_2 and Moderate Resolution Imaging Spectroradiometer (MODIS; MCD19A2) aerosol optical depth to show that the COVID-19 lockdown led to improved air quality in major Indian cities, with reduced concentrations of pollutants, especially NO_2 and SO_2. An assessment of NO_2 over eight districts of India was discussed by Ashwini, Saw, and Singh (2021), which depicted the spatial and temporal variations during and before the lockdown period. The findings provide insights into the changes in NO_2 levels and the environmental impact of the COVID-19 lockdown in tier-1 cities of India. Sarkar and Mondal (2022) used the TROPOMI data and studied the spatiotemporal changes in tropospheric NO_2 hotspots due to emission switch-off conditions during the COVID-related lockdown phases in India. They observed a higher NO_2 concentration in the major industrial and urban areas, mostly clustered in the eastern region (West Bengal, Odisha, and Chhattisgarh), and a significant improvement in air quality during the COVID lockdown phases. Balamadeswaran, Karthik, Ramakrishnan, and Bharath (2022) examined the fluctuations in NO_2 levels in India during the COVID-19 lockdown, employing open-source data from TROPOMI and ground-based sources, with a focus on major cities such as Mumbai and Delhi. Several other studies conducted air quality assessment during lockdown and unlock phases and highlighted the efficacy of TROPOMI data (Bauwens et al., 2020; Lal et al., 2020; Pei, Han, Ma, Su, & Gong, 2020; Venter, Aunan, Chowdhury, & Lelieveld, 2020).

The current study employs Sentinel-5P/TROPOMI observations to categorize NO_2 concentrations during lockdown and unlock phases in the Indian state of Maharashtra. The study applies the Mann-Kendall test (M-K test) to detect trends in air quality parameters over the study area. Moreover, considering the evident reduction in pollution levels across various regions and incorporating insights from the NCAP and nonattainment cities, this research offers a distinctive contribution by pinpointing and classifying regions with potential hazards.

22.2 Materials and methodology

22.2.1 Study area

The study was conducted in the state of Maharashtra, India, which has 18 nonattainment areas, according to the CPCB of India. Geographically, the study is located between 15.6°N to 22°N and 72.64°E to 80.9°E covering 307,713 km^2 area (Fig. 22.2). Maharashtra is the second-largest state in India in terms of land area, consisting of 36 districts and serves as the financial capital of India (Govt. of MHA, 2022). Home to 112.4 million people, the state accounts for 9.29% of India's total population. Its major cities boast high population densities, with 45% of the state's residents residing in metropolitan areas. Fig. 22.2 highlights the 18 nonattainment areas in the state of Maharashtra. Table 22.1 provides information about population density, the number of thermal power plants and their capacities, as well as the number of medium-scale medium enterprises (MSME) industries in the study area.

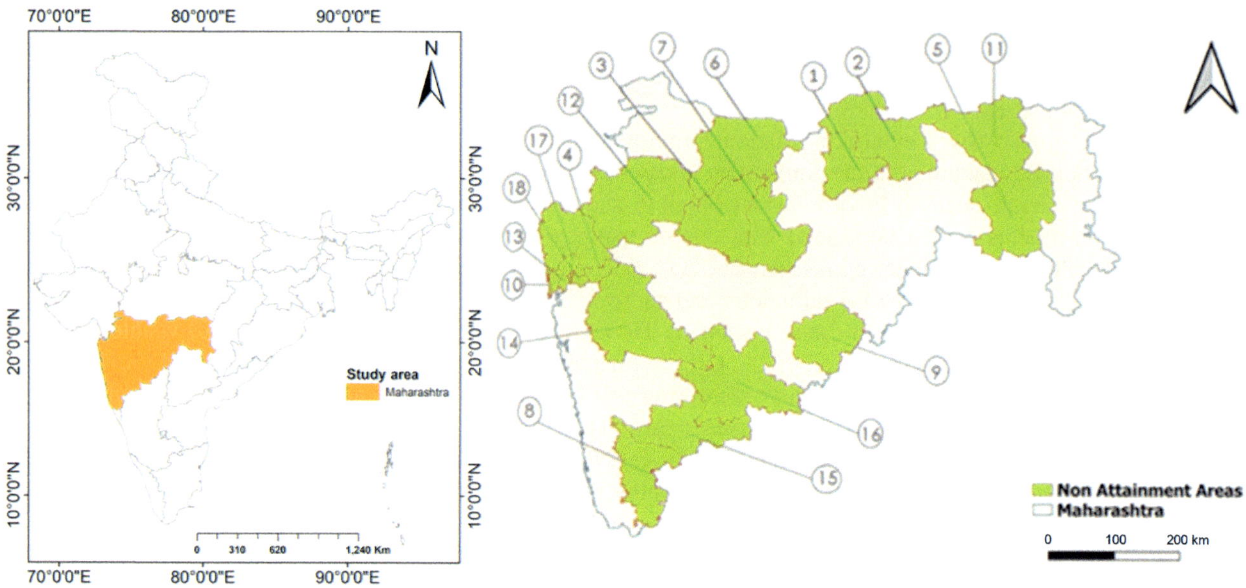

FIGURE 22.2 The study area map shows the location map and 18 nonattainment areas of Maharashtra, India. Map lines delineate study areas and do not necessarily depict accepted national boundaries.

TABLE 22.1 List of nonattainment cities with their respective population density, number of medium-scale medium enterprise (MSME) industries, and power plants with their capacities.

Sr. no.	City name	Population density (people per km^2)	MSME industries	Power plants (capacity)
1	Akola	320	82	Paras Power Plant (500 MW)
2	Amravati	237	39	Thermal Power Project—Rattan India (1350 MW)
3	Aurangabad	366	415	–
4	Badlapur	12,566	89	–
5	Chandrapur	293	42	Wardha Warora Power Plant EMCO (600 MW) Wardha Warora Power Plant (540 MW) Dhariawal Power Station (600 MW) Chandrapur Super Thermal Power Station (2920 MW) Ghugus Power Station (120 MW) Awarpur Ultratech Thermal Power Station (709 MW) Maratha Cement Captive Power Plant (70 MW)
6	Jalgaon	360	98	Bhusawal Thermal Power Station (1210 MW)
7	Jalna	254	103	–
8	Kolhapur	360	344	–
9	Latur	343	94	–
10	Mumbai	25,357	1514	Trombay Thermal Power Station (930 MW)
11	Nagpur	470	41	Khaperkheda Power Station (840 MW) Koradi Thermal Power Station (2400 MW) Mauda Power Station Phase II (2320 MW) MIHAN Power Station (246 MW) Butibori Power Plant (600 MW)

(Continued)

TABLE 22.1 (Continued)

Sr. no.	City name	Population density (people per km^2)	MSME industries	Power plants (capacity)
				Bela Power Station (270 MW)
12	Nashik	393	426	Eklahare Power Station (630 MW) Nashik Rattan India Power Plant (1350 MW)
13	Navi Mumbai	10,311	1805	–
14	Pune	603	168	–
15	Sangli	329	888	–
16	Solapur	290	41	Solapur Power Station (1320 MW)
17	Thane	1157	53	Dahanu Power Station (250 MW)
18	Ulhasnagar	18,000	104	–

22.2.2 Datasets

This study used the publicly available TROPOMI data for air quality assessment. The TROPOMI NO_2 data were accessed and processed using the Google Earth Engine (GEE) platform to determine concentration levels. GEE is a readily accessible and user-friendly cloud computing platform that provides multiple data catalogs, including datasets for monitoring air pollution. It offers interactive data access and algorithm development capabilities designed to handle extensive datasets for analysis and decision-making (Das et al., 2021). Along with GEE, open-access satellite observations and the Geographic Information System (GIS) were used in the study. The Sentinel-5P TROPOMI datasets are available on GEE from July 2018 onwards. Sentinel-5P has two dataset versions, near-real-time (NRTI) and offline (OFFL) (Van Geffen et al., 2022). In this study, we focus on OFFL, which has delayed availability but contains data from the entire orbit and is easier to work with for retrospective evaluation.

Population density data were acquired from Gridded Population of the World, Version 4, which contains estimates of the number of persons per square kilometer, based on counts with national census and population registers for the years 2000, 2005, 2010, 2015, and 2020 on 30 arc-second grid cells. The GIS boundary layer of Maharashtra state was used to subset various data layers.

22.2.3 Mann-Kendall test

The M-K test is one of the most commonly used statistical methods, which helps analyze the data statistics that are retrieved over time to determine if there is a steady increase or decrease or trends (Emami, Masiol, & Hopke, 2018). This test comprises two key interpretations: (1) null hypothesis (Ho), which means when the value appears to be zero (0); and (2) alternate hypothesis (Ha), which means when the value appears to be positive, negative, or nonzero. These two hypotheses determine whether there is a monotonic trend in the data or not (Jaiswal, Samuel, & Kadabgaon, 2018). Some important variables need to be considered to determine if the trend is increasing or decreasing. They are as follows: Kendall's tau: the number varies between −1 and 1. A negative value suggests a decreasing trend, while a positive value indicates an increasing trend. The P-value plays a vital role in determining and providing proof through significance level that the null hypothesis is true or not. If the P-value is less than .05, then the trend is significant and the alternate hypothesis is true. On the other hand, if the P-value goes above .05, it suggests that there is no trend in the data. Sen's slope (Theil-Sen slope): this parameter estimates the strength of the trend derived from the M-K test. Basically, it estimates the overall slope of the time series, corresponding to the median of all slopes calculated between each pair of data points in the time series (Aswad, Yousif, & Ibrahim, 2020). These are the important variables in the M-K trend test that are considered in this study to determine if the trend is upward or downward and its significance.

22.2.4 Methodology

Fig. 22.3 illustrates the methodology and step-by-step process carried out in this study. As described in the flowchart of methodology, GEE was used to extract Sentinel-5P/TROPOMI data. Additionally, the population density was extracted from the WorldPop dataset to determine the population density of the nonattainment areas. The time series analysis of these datasets was performed using GEE while considering the lockdown period from March 25, 2020 to May 31, 2020. To facilitate comparison and assess the percentage of days, we also incorporated NO_2 data from the previous year (2019) into our study. All 18 nonattainment cities were mapped using GIS to show the impact of power plants on these urban areas. To determine the trend of NO_2 concentration, the M-K test was carried out. In addition, XLSTAT software was used to indicate and signify the upward or downward trend in the study area. Considering all the results, this study was concluded by highlighting the critical areas of nonattainment cities that need to be monitored and underscoring some important key parameters for future studies.

The NO_2 concentration was analyzed and tracked using a time series approach over a two-time frame to reveal the effects of the lockdown. The lockdown time was scheduled as the first-time frame, starting on March 25 and ending on May 31, 2020. The second frame was scheduled in the previous year (2019) when no lockdown restrictions were in place. In this section, we look at the time series of NO_2 concentration for the lockdown period and then compare the mean of both years and see the overall percentage change to justify the levels of impact on each nonattainment city.

22.3 Results

22.3.1 Time series analysis of NO_2

Fig. 22.4 shows the results of six selected nonattainment cities among the 18 nonattainment areas of Maharashtra, underscoring the staggering consequences of COVID-19. Akola and Jalgaon showed a downward slope of NO_2 concentration, which was expected during the lockdown period due to the complete shutdown of anthropogenic activities. Although there was a lockdown, the mean slope of the graph showed a rising trend in Mumbai, Navi Mumbai, Chandrapur, and Nagpur, and Nagpur displayed an enormous rise in the amount of NO_2 concentration. As illustrated in Fig. 22.5(A–L), all other cities also demonstrated an upward inclination. Additionally, the mean NO_2 concentration of each city was examined together with its respective slope percentage (%); the slope percentage was measured from the start to the end of the lockdown (Table 22.2). The negative percentage in the table indicates a slope that was inclined downward, and the positive slope indicates a slope that was inclined upward. Comparing the above-mentioned figure,

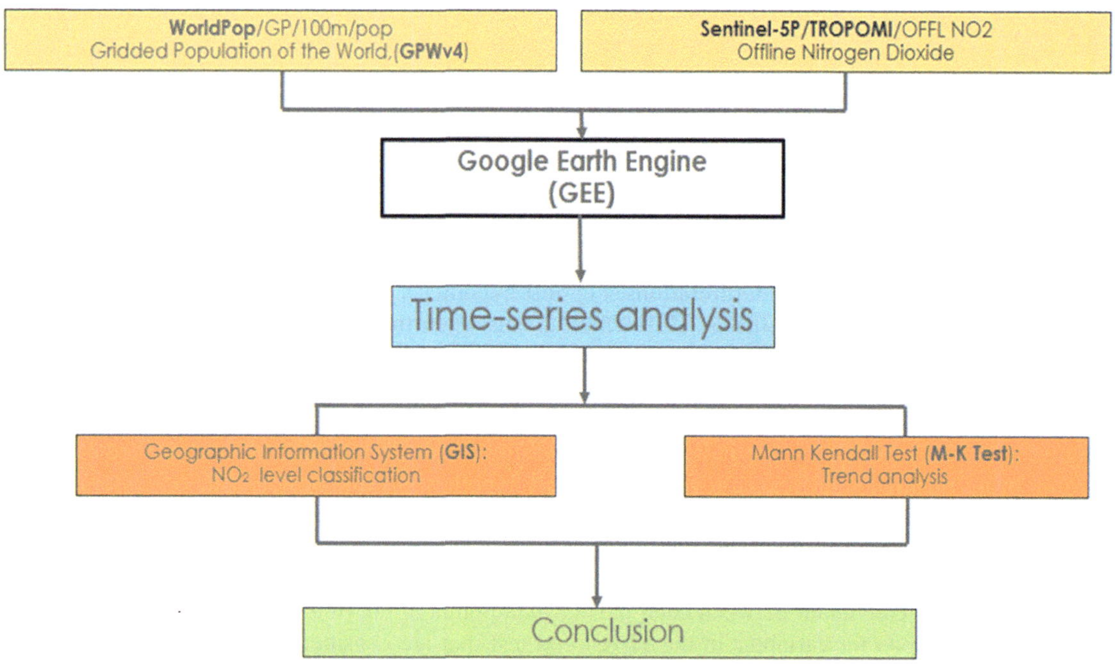

FIGURE 22.3 Flowchart of the study employed for assessing NO_2 concentration in Maharashtra.

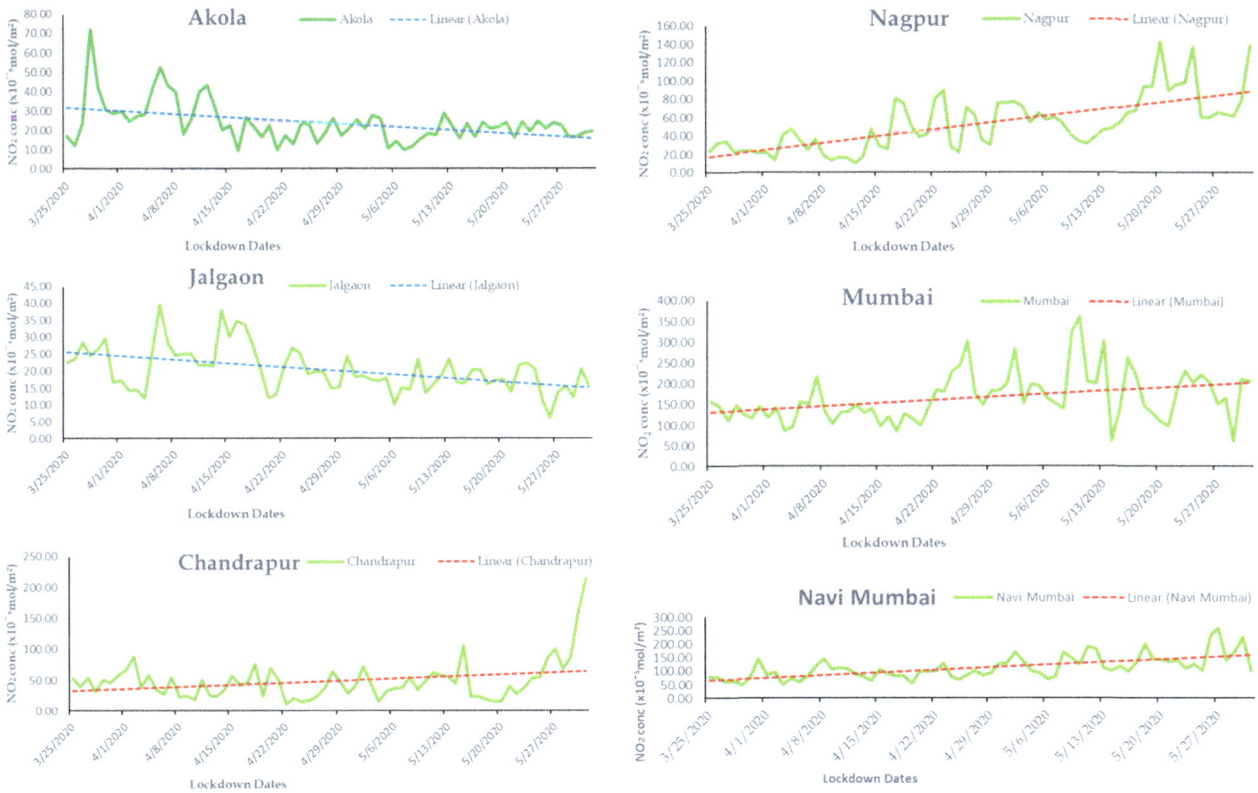

FIGURE 22.4 Time series plot of NO_2 concentration during lockdown for Akola, Jalgaon, Chandrapur, Nagpur, Mumbai, and Navi Mumbai.

we can find that cities like Nagpur have experienced the highest percentage increase (337%) since the lockdown began. At the same time, the lowest percentage increase was observed in Akola (−45%).

The mean NO_2 concentration for each of the 18 nonattainment cities during the lockdown period is shown in Fig. 22.6. Mumbai showed the highest mean NO_2 concentration at 166.553×10^{-5} mol m^{-2}, followed by Navi Mumbai at 112.8×10^{-5} mol m^{-2} and Thane at 93.122×10^{-5} mol m^{-2}. Additionally, out of 18 nonattainment cities, Solapur showed the lowest mean NO_2 concentration at 17.986×10^{-5} mol m^{-2}.

We have compared the NO_2 concentrations for the two time frames of 2020 and 2019 during the lockdown period (Table 22.3). In comparison to the previous year (i.e., 2019), we observed that cities like Badlapur, Chandrapur, Jalgaon, Nagpur, Pune, Sangli, and Solapur all exhibited higher percentage changes. Pune showed the largest percentage at 45.73% followed by Chandrapur at 45.25% and Sangli at 44.98%. Mumbai and Amravati experienced the least amount of change, both below 10%.

22.3.2 Trend test

The M-K trend test was performed using the data collected through the Sentinel-5P, and the results are displayed in Table 22.4 together with Kendall's tau and Sen's slope. The lockdown phase was the focus of this trend test, which was used to check the significance of trends. The trends in this section had been verified, and the *P*-value in this test suggested that the trend was significant, with the exception of Chandrapur and Solapur. Akola and Jalgaon depicted negative values for Kendall's tau and Sen's slope, which indicated that the trend of NO_2 concentration was downward in those areas.

22.3.3 Mapping unhealthy regions

The location of the power plants in Akola and Chandrapur is shown on the NO_2 map. The map was created using the aggregate value of the data obtained from Sentinel-5P. The regions were categorized based on air quality standards for NO_2 (WHO, 2022), and the Akola and Chandrapur regions were split into five classes: good, moderate, unhealthy (sensitive group), unhealthy, and very unhealthy (Fig. 22.7). The Akola city center, particularly its western portion, falls within the unhealthy to

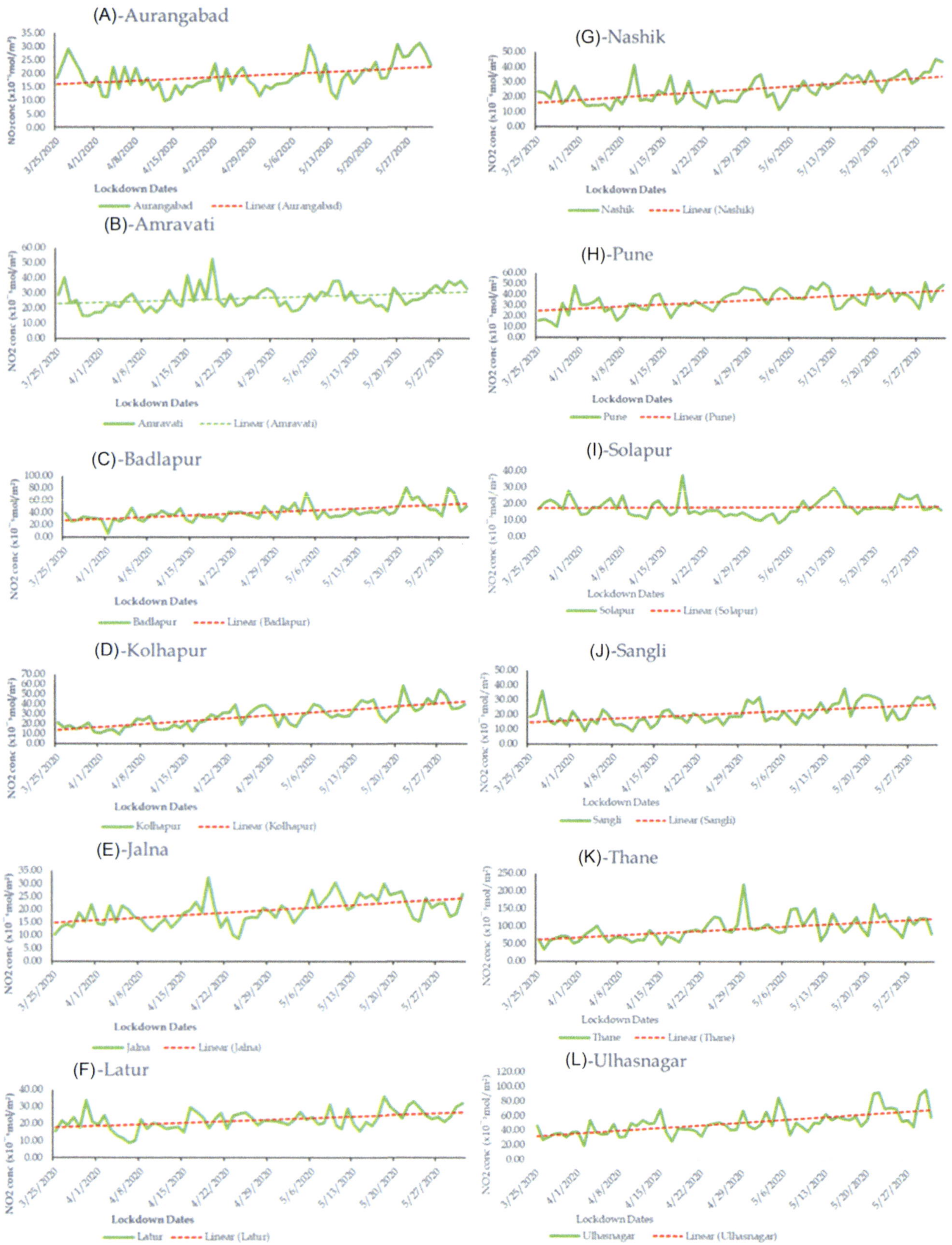

FIGURE 22.5 (A–L) Time series analysis of nonattainment cities during the lockdown period.

TABLE 22.2 Minimum, maximum, mean, and slope percentages of NO_2 concentration levels during the lockdown (March 25, 2020 to May 31, 2020).

Nonattainment city	Minimum ($\times 10^{-5}$ mol m^{-2})	Maximum ($\times 10^{-5}$ mol m^{-2})	Mean ($\times 10^{-5}$ mol m^{-2})	Slope percentage (%)
Akola	9.742	72.076	23.794	−45
Amravati	15.000	53.000	27.059	35
Aurangabad	9.914	31.260	19.135	44
Badlapur	6.186	81.915	40.828	93
Chandrapur	12.319	213.009	48.435	50
Jalgaon	6.221	39.611	20.352	−39
Jalna	8.893	32.369	19.554	61
Kolhapur	9.821	59.123	28.305	242
Latur	9.056	36.566	22.435	50
Mumbai	61.900	360.000	166.553	43
Nagpur	10.458	142.339	52.805	337
Nashik	11.243	46.231	25.220	100
Navi Mumbai	48.221	257.910	112.800	167
Pune	10.042	52.153	34.544	73
Sangli	8.856	38.204	21.007	81
Solapur	8.438	37.268	17.986	12
Thane	33.945	220.063	93.122	117
Ulhasnagar	19.855	96.852	50.609	106

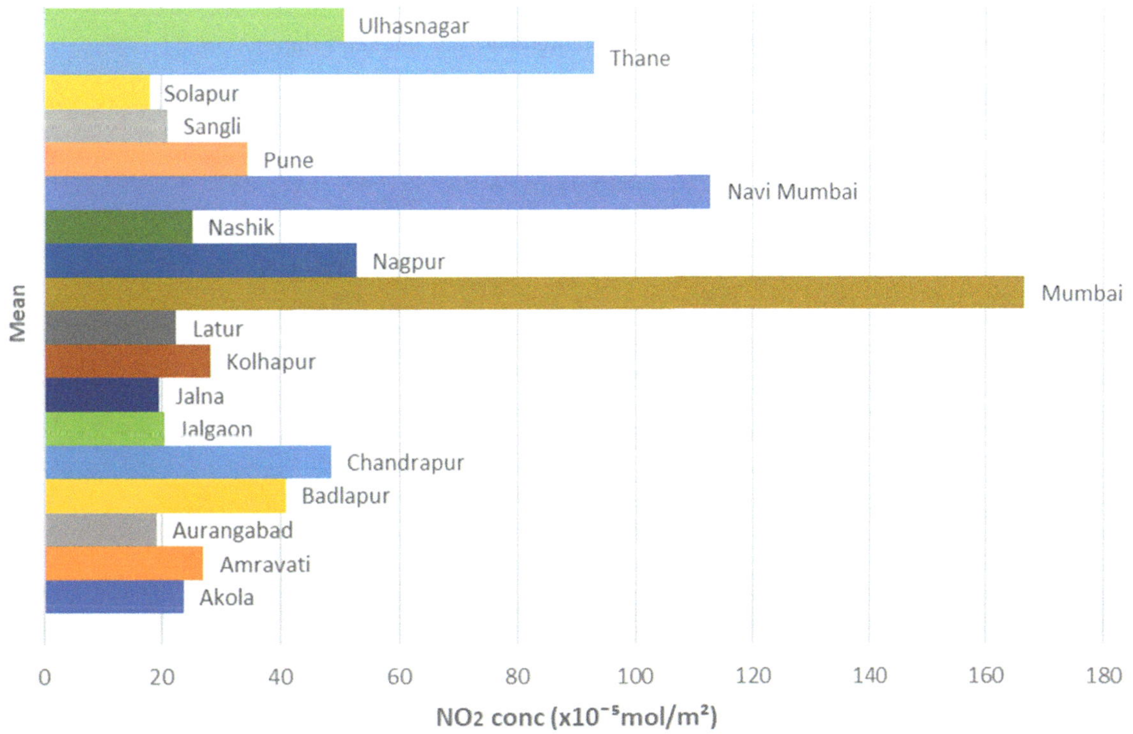

FIGURE 22.6 Mean NO_2 concentrations of each nonattainment city during lockdown.

TABLE 22.3 Comparison of two time frames from 2020 and 2019 and their respective percentage changes in values of NO_2 concentration.

Nonattainment city	2020 ($\times 10^{-5}$ mol m^{-2})	2019 ($\times 10^{-5}$ mol m^{-2})	Percentage change (%) in 2020 w.r.t. 2019
Akola	23.794	30.330	21.55
Amravati	27.059	29.737	9.00
Aurangabad	19.135	31.140	38.55
Badlapur	40.828	70.299	41.92
Chandrapur	48.435	88.478	45.25
Jalgaon	20.352	36.734	44.59
Jalna	19.554	26.922	27.36
Kolhapur	28.305	40.371	29.88
Latur	22.435	26.112	14.08
Mumbai	166.553	178.188	6.52
Nagpur	52.805	93.000	43.22
Nashik	25.22	34.979	27.89
Navi Mumbai	112.8	153.279	26.40
Pune	34.544	63.653	45.73
Sangli	21.007	38.185	44.98
Solapur	17.986	31.939	43.68
Thane	93.122	108.591	14.24
Ulhasnagar	50.609	79.529	36.36

TABLE 22.4 Mann-Kendall trend test results together with Kendall's tau and Sen's slope.

Nonattainment city	Kendall's tau	*P*-value	Sen's slope
Akola	−0.252	.002	−0.146
Amravati	0.285	.001	0.147
Aurangabad	0.262	.002	0.108
Badlapur	0.467	<.0001	0.329
Chandrapur	0.080	.338	0.150
Jalgaon	−0.337	<.0001	−0.160
Jalna	0.395	<.0001	0.150
Kolhapur	0.572	<.0001	0.430
Latur	0.312	.000	0.124
Mumbai	0.269	.001	1.059
Nagpur	0.489	<.0001	0.935
Nashik	0.473	<.0001	0.301
Navi Mumbai	0.463	<.0001	1.192
Pune	0.406	<.0001	0.301

(*Continued*)

TABLE 22.4 (Continued)

Nonattainment city	Kendall's tau	*P*-value	Sen's slope
Sangli	0.382	<.0001	0.188
Solapur	0.078	.349	0.029
Thane	0.450	<.0001	0.916
Ulhasnagar	0.507	<.0001	0.479

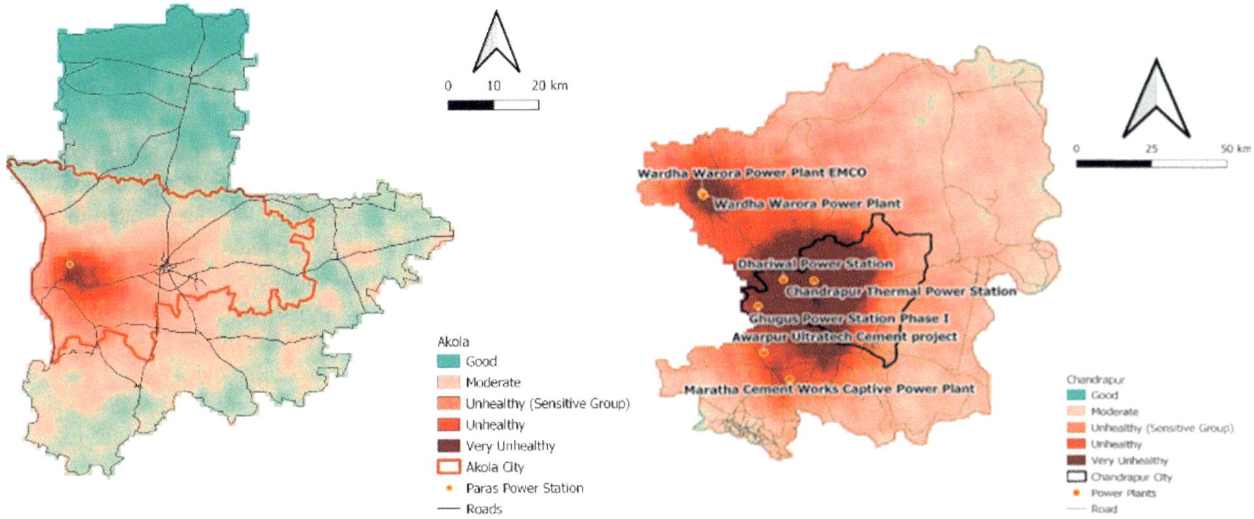

FIGURE 22.7 Area categorization based on air quality standards set for NO_2 concentration level maps of Akola and Chandrapur. Map lines delineate study areas and do not necessarily depict accepted national boundaries.

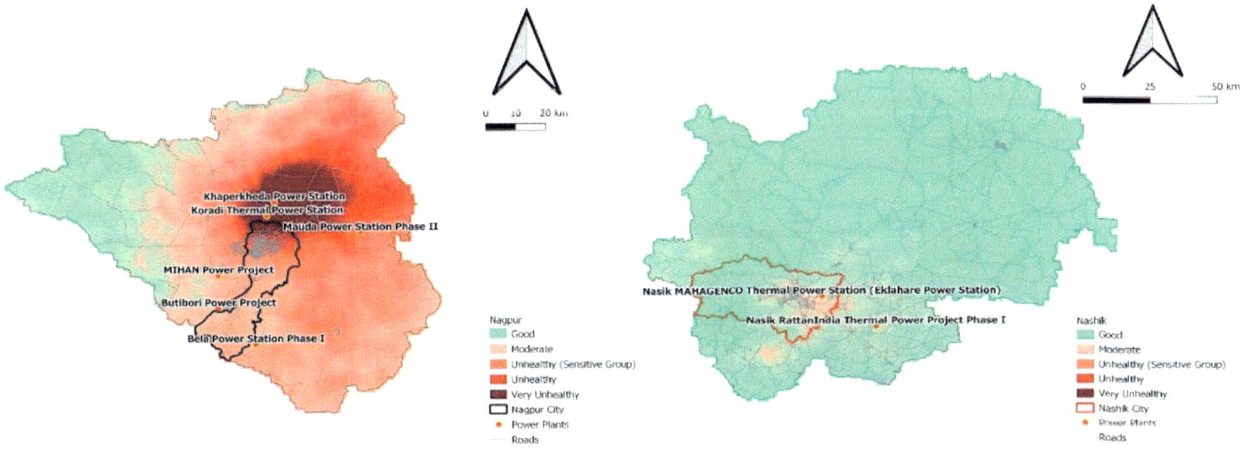

FIGURE 22.8 The area categorization is based on air quality standards set for the NO_2 concentration level map of Nagpur and Nashik. Map lines delineate study areas and do not necessarily depict accepted national boundaries.

very unhealthy category on the categorized map, depending on air quality standards. In contrast, the northern portion of the area was found to have relatively better air quality. Likewise, the north-eastern region of Chandrapur was also assessed as ranging from moderate to unhealthy (sensitive group), whereas the western region was classified as unhealthy to very unhealthy. Consequently, Akola and Chandrapur received suggestions to put into effect certain sustainable actions in the western part of the city. The classification for the Nagpur and Nashik city zones based on NO_2 air quality standards is presented in Fig. 22.8.

The findings demonstrated that the western portion of Nagpur was potentially rated in the good category, whereas the north- and south-eastern regions were classified under the moderate to very unhealthy zones. Furthermore, the city component of the red boundary was labeled as moderate to unhealthy (sensitive group) in relation to the region's southwestern portion, while other areas were placed in the good category. Mumbai city is renowned as the most densely populated and contaminated region in Maharashtra (citation). Based on air quality standards for NO_2 concentration, the eastern and south-eastern portions of Mumbai were classified as moderate to very unhealthy category, whereas the north-western portion was in a good category (Fig. 22.9). Mumbai city also showed the moderate to very unhealthy category. Given its significance as an economic hub, it is imperative to consider implementing more sustainable measures to address the fast-rising NO_2 concentration in this region. Likewise, the middle and southern portions of Navi Mumbai exhibit the moderate to very unhealthy category. Similarly, for other cities, the classification based on air quality standards for NO_2 concentration level maps is shown in Fig. 22.10(A–J).

22.4 Discussion

A typical situation developed in the months following the outbreak of COVID-19, forcing the implementation of lockdown in an effort to restrict the disease spread. Despite the difficulties brought on by this crisis, it provided a rare opportunity for researchers to analyze the differential role of climate and human disturbances on various land surface and atmospheric processes, including the emission of pollutants in the atmosphere due to anthropogenic activities. The emission reduction due to the complete and partial lockdown of industries and vehicular activities leads to reduced air pollutants in and around the industrial and urban areas. Discussion of the results and their perspective on nonattainment cities is essential, as NO_2 is an extremely toxic pollutant that is harmful to human health. In order to analyze the findings of the study, a simplified description of the nonattainment cities was explained here based on three parameters: (1) population density; (2) power plants; and (3) MSME and industries (Fig. 22.11). Cities categorized as nonattainment areas based on population density exhibited a population density greater than 1000 km^2. Meanwhile, the classification for power plants category considers the cities that either host power plants or are situated in close proximity to them. The nonattainment cities under the third category are based on the MSME sectors. Over the past decade, the MSME sector in India, particularly in Maharashtra, has experienced significant growth. Likewise, industries such as automobile manufacturing, construction, and infrastructure are overgrowing due to high demands. In accordance with the data about MSME shown in Table 22.1, there are many MSME industries close to cities like Aurangabad (415), Pune (168), Kolhapur (344), Latur (94), Jalna (103), Sangli (88), Badlapur (89), Navi Mumbai (1805), Ulhasnagar (104), Nashik (426), Akola (82), Jalgaon (98), and Mumbai (1514) that have a high number of MSME industries around them. The lockdown measures had a profound financial impact on these industries due to the complete shutdown of activities. The regions where these industries are significant contributors to pollution are visually represented through the Venn diagram. Notably, cities such as Pune, Sangli, Aurangabad, Jalna, and Latur, which rely primarily on MSME and other related industries, demonstrated

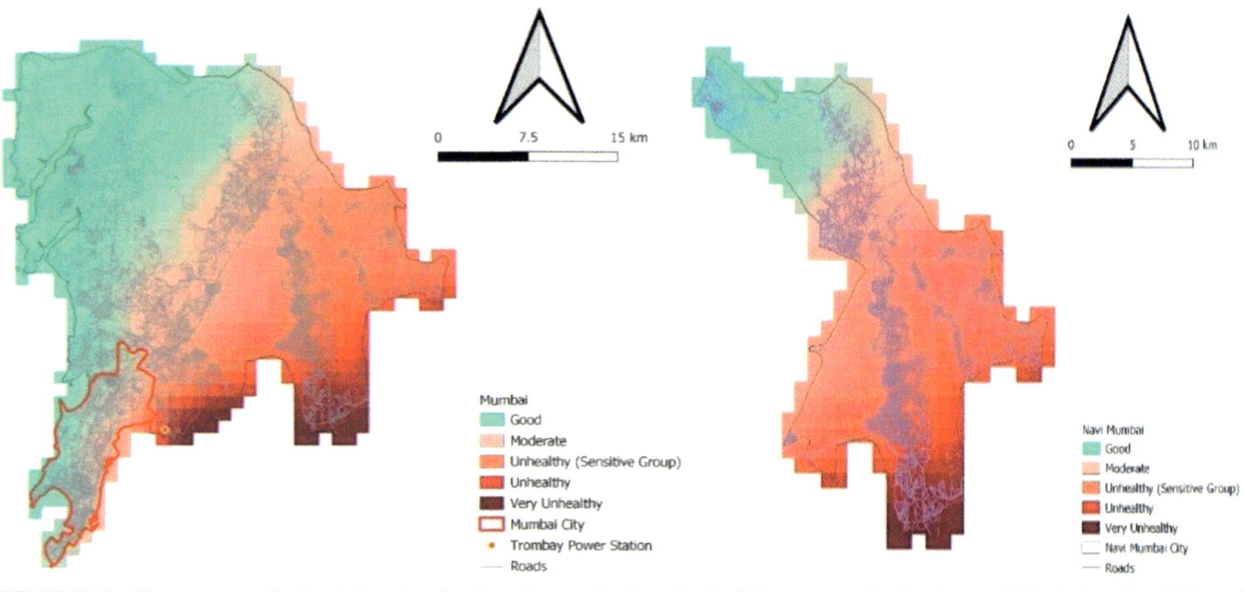

FIGURE 22.9 The area categorization is based on the air quality standards set for the NO_2 concentration level map of Mumbai and Navi Mumbai. Map lines delineate study areas and do not necessarily depict accepted national boundaries.

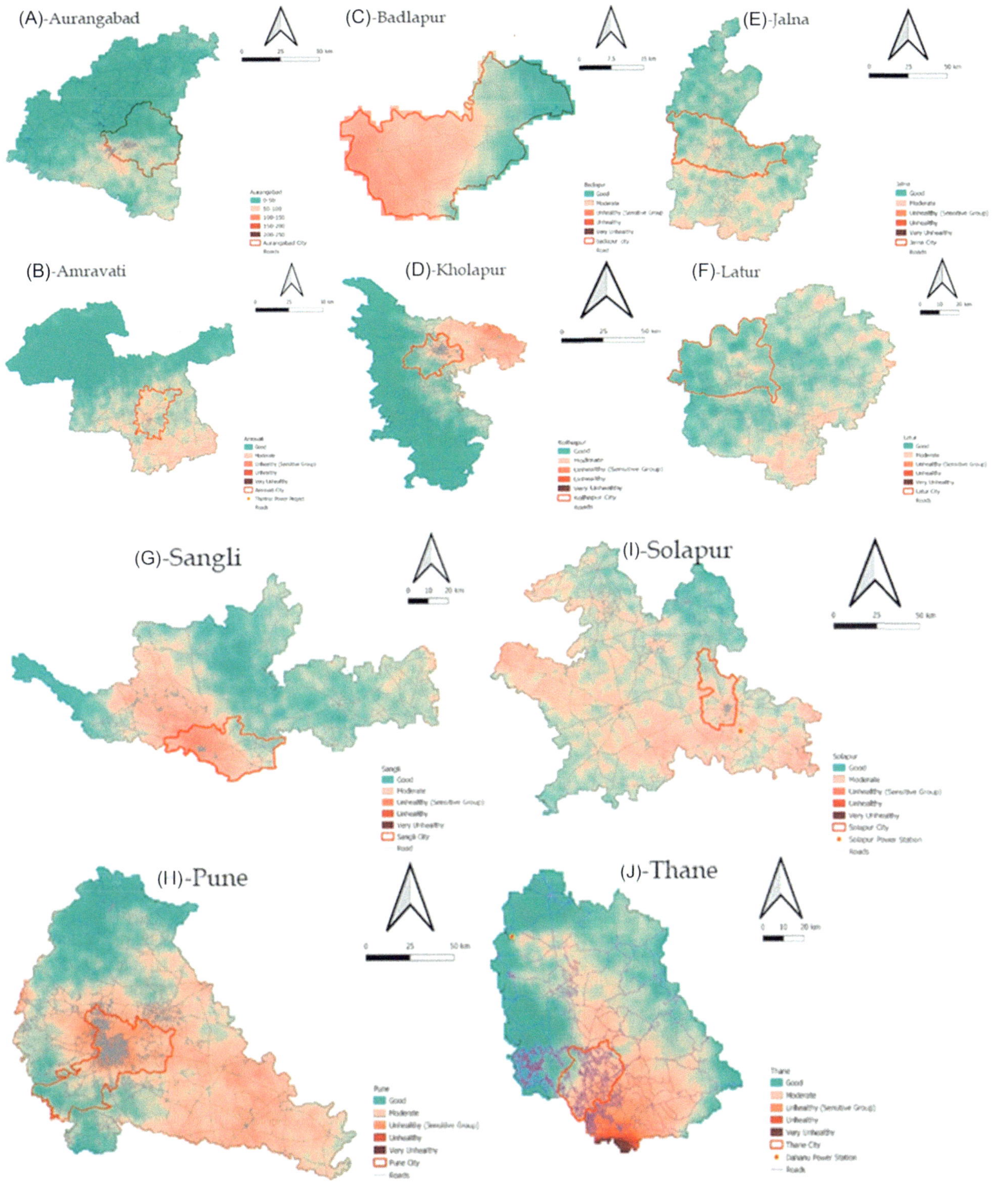

FIGURE 22.10 (A–J) NO_2 concentration level map of nonattainment cities. Map lines delineate study areas and do not necessarily depict accepted national boundaries.

a higher percentage change (Table 22.3). Among the cities mentioned, Badlapur, Navi Mumbai, and Ulhasnagar are the cities with higher population densities (greater than 1000 persons per km^2) and relatively lower change. This suggests that human indoor activities might be a significant contributor to the observed patterns, as indicated by previous research (Hu & Zhao, 2020; Jarvis, Adamkiewicz, Heroux, Rapp, & Kelly, 2010; Salonen, Salthammer, & Morawska, 2019). In contrast, cities like

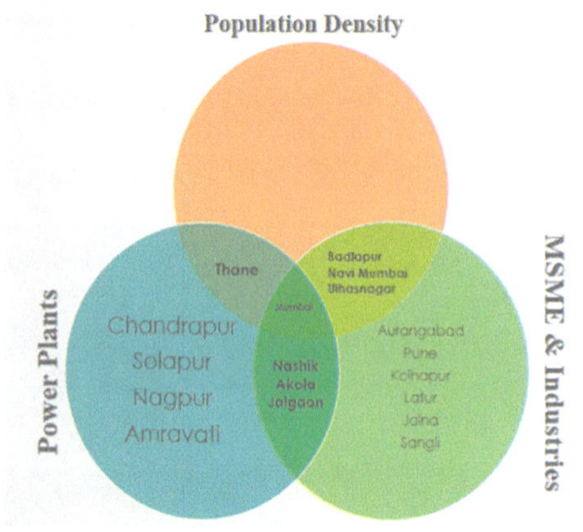

FIGURE 22.11 A Venn diagram of a simplified explanation of the nonattainment cities, divided based on three parameters.

Nashik, Akola, and Jalgaon, which are surrounded by power plants, exhibit comparatively different trends. Cities hosting major power plants with higher capacities, such as Chandrapur, Nagpur, Amravati, and Solapur, have consistently displayed an upward trend throughout the lockdown period. Among these, Kolhapur has experienced the most significant percentage increase after Nagpur. Amravati has the lowest percentage change among the previously mentioned cities while hosting a high-capacity power plant of 1350 MW. Thane City, in contrast, has comparatively little variation in percentage change despite being recognized for its higher population density and the presence of MSME sectors, notably in the Maharashtra Industrial Development Corporation zones where numerous small enterprises are located.

Mumbai is the most densely populated city in Maharashtra, one of the most densely populated cities, and also the financial capital of the country. Furthermore, the presence of the Trombay power station on its border contributes to its unique situation. Mumbai city, despite its high population density, has demonstrated the lowest percentage change during lockdown compared to the previous year. This outcome highlights the severity of pollution in the city and indicates the fact that the range of concentrations remained higher, as seen from the time series.

22.5 Conclusion

This study was conducted to examine the levels of NO_2 pollution during the COVID-related lockdown in nonattainment cities in Maharashtra state, India, using satellite remote-based observations. The air pollutant concentration levels during and before the lockdown phases for the nonattainment cities have yielded intriguing findings. Notably, it is apparent from the results that, except for Akola and Jalgaon, the NO_2 concentration experienced an upward trend during the period of lockdown. The M-K trend test has indicated that except for Chandrapur and Solapur, all other cities experience a statistically significant trend in NO_2 change. The comparison of time series data between the two periods of 2019 and 2020 has provided the percentage changes in NO_2 concentration. This analysis underscores the importance of considering various factors, such as the location of power plants, MSME, industries, and population density, in understanding the observed trends in NO_2 concentrations. The impact put down by these parameters emerges as the key determinant for the observed trend. Based on satellite-based air pollutants concentration data, this study serves as a valuable tool for identifying areas of high pollutant concentrations across various categories, thus aiding in assessing the health risks associated with the cities. The study advocates periodic monitoring of air pollutants and their sources, accompanied by the implementation of sustainable action plans to address the persistent pollution challenges. This study holds great significance as it highlights specific areas that are already recognized as highly polluted but have not demonstrated a downward trend in pollution levels, even during lockdown-like situations. Therefore this study suggests that the effectiveness of stringent restrictions cannot be the sole solution for pollution control. In a similar sense to NO_2 pollutants, it underscores the need for comprehensive monitoring of other pollutants such as SO_2, CO, and O_3 in such cities. This, in turn, will help to understand the root causes, strategize solutions, and contribute to initiatives like the NCAP and the Sustainable Development Goals.

Acknowledgments

The authors would like to thank Hokkaido University and JASSO for their financial support. The authors also express thanks to the Copernicus Sentinel-2 data hub and the United States Geological Survey (USGS) for providing satellite data.

References

Ashwini, K., Saw, G. K., & Singh, A. (2021). Phase-wise spatial and temporal variations of nitrogen dioxide during and pre-COVID-19 lockdown period in tier-1 cities of India. *Spatial Information Research*, *29*(6), 887–895.

Aswad, F. K., Yousif, A. A., & Ibrahim, S. A. (2020). Trend analysis using Mann-Kendall and Sen's slope estimator test for annual and monthly rainfall for SINJAR district, IRAQ. *Journal of Duhok University*, *23*(2), 501–508.

Avtar, R., Singh, D., Umarhadi, D. A., Yunus, A. P., Misra, P., Desai, P. N. ,, & Phanindra, K. B. V. N. (2021). Impact of COVID-19 lockdown on the fisheries sector: A case study from three harbors in Western India. *Remote Sensing*, *13*(2), 183.

Balamadeswaran, P., Karthik, J., Ramakrishnan, R., & Bharath, K. M. (2022). Impact of COVID-19 outbreak on tropospheric NO_2 pollution assessed using satellite-ground perspectives observations in India. *Modeling Earth Systems and Environment*, *8*(2), 1645–1655.

Banerji, S., & Mitra, D. (2022). Assessment of air quality in Kolkata before and after COVID-19 lockdown. *Geocarto International*, *37*(21), 6351–6374.

Bauwens, M., Compernolle, S., Stavrakou, T., Müller, J. F., Van Gent, J., Eskes, H., ... Zehner, C. (2020). Impact of coronavirus outbreak on NO_2 pollution assessed using TROPOMI and OMI observations. *Geophysical Research Letters*, *47*(11), e2020GL087978.

Biswal, A., Singh, V., Singh, S., Kesarkar, A. P., Ravindra, K., Sokhi, R. S., ... Mor, S., et al. (2021). COVID-19 lockdown-induced changes in NO_2 levels across India observed by multi-satellite and surface observations. *Atmospheric Chemistry and Physics*. Available from https://doi.org/10.5194/acp-21-5235-2021.

Bodah, B. W., Neckel, A., Stolfo Maculan, L., Milanes, C. B., Korcelski, C., Ramírez, O., ... Oliveira, M. L. S. (2022). Sentinel-5P TROPOMI satellite application for NO_2 and CO studies aiming at environmental valuation. *Journal of Cleaner Production*, *357*, 131960.

Brusseau, M. L., Pepper, I., & Gerba, C. (2019). *Atmospheric Pollution. Environmental and Pollution Science* (Third). Elsevier.

Das, P., Behera, M. D., & Abhilash, P. C. (2024). A rapid assessment of stubble burning and air pollutants from satellite observations. *Tropical Ecology*, *65*(1), 152–157. Available from https://doi.org/10.1007/s42965-022-00291-5.

Das, P., Mudi, S., Behera, M. D., Barik, S. K., Mishra, D. R., & Roy, P. S. (2021). Automated mapping for long-term analysis of shifting cultivation in Northeast India. *Remote Sensing*, *13*(6), 1066. Available from https://doi.org/10.3390/rs13061066.

Dekker, I. N., Houweling, S., Pandey, S., Krol, M., Röckmann, T., Borsdorff, T., ... Aben, I. (2019). What caused the extreme CO concentrations during the 2017 high-pollution episode in India? *Atmospheric Chemistry and Physics*, *19*(6), 3433–3445. Available from https://doi.org/10.5194/acp-19-3433-2019.

Emami, F., Masiol, M., & Hopke, P. K. (2018). Air pollution at Rochester, NY: Long-term trends and multivariate analysis of upwind SO_2 source impacts. *Science of the Total Environment*, *612*, 1506–1515.

ESA (2022). *Products and algorithms—Sentinel-5P technical guide—Sentinel online—Sentinel online*. Available fromhttps://sentinel.esa.int/web/sentinel/missions/sentinel-5p.

Gordon, S. B., Bruce, N. G., Grigg, J., Hibberd, P. L., Kurmi, O. P., Lam, K. B. H., ... Martin, W. J. (2014). Respiratory risks from household air pollution in low- and middle-income countries. *The Lancet Respiratory medicine*, *2*(10), 823–860.

Gov. of MHA. (2022). *Demography* | Mumbai Suburban District, Government of Maharashtra. Available from https://moef.gov.in/wp-content/uploads/2019/05/NCAP_Report.pdf.

Grisales-Romero, H., Piñeros-Jiménez, J. G., Nieto, E., Porras-Cataño, S., Montealegre, N., González, D., & Ospina, D. (2021). Local attributable burden disease to $PM_{2.5}$ ambient air pollution in Medellín, Colombia, 2010–2016. *F1000Research*. Available from https://doi.org/10.12688/f1000research.52025.2.

Halder, B., Ahmadianfar, I., Heddam, S., Mussa, Z. H., Goliatt, L., Tan, M. L., ... Yaseen, Z. M. (2023). Machine learning-based country-level annual air pollutants exploration using Sentinel-5P and Google Earth Engine. *Scientific Reports*.

Hu, Y., & Zhao, B. (2020). Relationship between indoor and outdoor NO_2: A review. *Building and Environment*, *180*, 106909.

Jaiswal, A., Samuel, C., & Kadabgaon, V. M. (2018). Statistical trend analysis and forecast modelling of air pollutants. *Global Journal of Environmental Science and Management*, *4*(4), 427–438.

Jarvis, D. J., Adamkiewicz, G., Heroux, M. E., Rapp, R., & Kelly, F. J. (2010). *Nitrogen dioxide. WHO guidelines for indoor air quality: Selected pollutants*. World Health Organization.

Kelly, F. J. (2010). WHO guidelines for indoor air quality: selected pollutants. *In WHO Guidelines for Indoor Air Quality: Nitrogen Dioxide Chapter*.

KS, V., & Venkatasubramanian, P. (2021). Air pollutants during COVID-19 lockdown period in India. *Journal of Disaster Research*, *16*(1), 88–96.

Lal, P., Kumar, A., Kumar, S., Kumari, S., Saikia, P., Dayanandan, A., ... Khan, M. L. (2020). The dark cloud with a silver lining: Assessing the impact of the SARS COVID-19 pandemic on the global environment. *Science of the Total Environment*, *732*, 139297.

Misra, P., Takigawa, M., Khatri, P., Dhaka, S. K., Dimri, A. P., Yamaji, K., ... Hayashida, S. (2021). Nitrogen oxides concentration and emission change detection during COVID-19 restrictions in North India. *Scientific Reports*, *11*(1), 9800.

Mor, S., Kumar, S., Singh, T., Dogra, S., Pandey, V., & Ravindra, K. (2021). Impact of COVID-19 lockdown on air quality in Chandigarh, India: Understanding the emission sources during controlled anthropogenic activities. *Chemosphere*, *263*, 127978.

Panda, S., Mallik, C., Nath, J., Das, T., & Ramasamy, B. (2021). A study on variation of atmospheric pollutants over Bhubaneswar during imposition of nationwide lockdown in India for the COVID-19 pandemic. *Air Quality, Atmosphere & Health*. Available from https://doi.org/10.1007/s11869-020-00916-5.

Pathak, M., & Kuttippurath, J. (2022). Air quality trends in rural India: analysis of NO_2 pollution using satellite measurements. *Environmental Science: Processes & Impacts*, *24*(12), 2437–2449.

Pei, Z., Han, G., Ma, X., Su, H., & Gong, W. (2020). Response of major air pollutants to COVID-19 lockdowns in China. *Science of the Total Environment*, *743*, 140879.

Prakash, S., Goswami, M., Khan, Y. I., & Nautiyal, S. (2021). Environmental impact of COVID-19 led lockdown: A satellite data-based assessment of air quality in Indian megacities. *Urban Climate*, *38*, 100900.

Ruidas, D., & Pal, S. C. (2022). Potential hotspot modelling and monitoring of PM2. 5 concentrations for sustainable environmental health in Maharashtra, India. *Sustainable Water Resources Management*, *8*(4), 98.

Sahoo, P. K., Mangla, S., Pathak, A. K., Salãmao, G. N., & Sarkar, D. (2021). Pre-to-post lockdown impact on air quality and the role of environmental factors in spreading the COVID-19 cases – A study from a worst-hit state of India. *International Journal of Biometeorology*, *65*, 205–222.

Saini, D., Darla, U. R., Lataye, D. H., Motghare, V. M., & Shingare, A. A. (2022). Effect on ambient air quality in Nagpur due to lockdown to contain the spread of COVID-19 pandemic in the year 2020: A case study. *Sādhanā*, *47*(2), 98.

Salonen, H., Salthammer, T., & Morawska, L. (2019). Human exposure to NO_2 in school and office indoor environments. *Environment International*, *130*, 104887.

Sarkar, S., & Mondal, D. (2022). Spatiotemporal changes in tropospheric nitrogen dioxide hotspot due to emission switch-off condition in the view of lockdown emergency in India. *Air Quality, Atmosphere & Health*, *15*(12), 2123–2135.

Saw, G. K., Dey, S., Kaushal, H., & Lal, K. (2021). Tracking NO_2 emission from thermal power plants in North India using TROPOMI data. *Atmospheric Environment*, *259*118514. Available from https://doi.org/10.1016/j.atmosenv.2021.118514.

Singh, H., Meraj, G., Singh, S., Shrivastava, V., Sharma, V., Farooq, M., ... Kumar, P. (2022). Status of air pollution during COVID-19-induced lockdown in Delhi, India. *Atmosphere*, *13*(12), 2090.

Van Geffen, J., Eskes, H., Compernolle, S., Pinardi, G., Verhoelst, T., Lambert, J. C., ... Veefkind, J. P. (2022). Sentinel-5P TROPOMI NO 2 retrievals: Impact of version v2. 2 improvements and comparisons with OMI and ground-based data. *Atmospheric Measurement Techniques*, *15*(7), 2037–2060.

Venter, Z. S., Aunan, K., Chowdhury, S., & Lelieveld, J. (2020). COVID-19 lockdowns cause global air pollution declines. *Proceedings of the National Academy of Sciences*, *117*(32), 18984–18990.

World Health Organization (WHO). (2022). WHO global air quality guidelines: Particulate matter (PM2.5 and PM10), ozone, nitrogen dioxide, sulfur dioxide and carbon monoxide. Available from https://apps.who.int/iris/handle/10665/345329. Accessed on January 15, 2022.

Yang, D., Ye, C., Wang, X., Lu, D., Xu, J., & Yang, H. (2018). Global distribution and evolvement of urbanization and $PM_{2.5}$ (1998–2015). *Atmospheric Environment*. Available from https://doi.org/10.1016/j.atmosenv.2018.03.053.

Chapter 23

Forecasting CO_2 emissions over India using deep learning models

Pringale Kumar Das[1], Mukunda Dev Behera[1], Buddolla Jagadish[1] and Pulakesh Das[2]

[1]Centre for Ocean, River, Atmosphere and Land Sciences (CORAL), Indian Institute of Technology Kharagpur, Kharagpur, West Bengal, India,

[2]School of Forest Resources, University of Maine, Orono, ME, United States

23.1 Introduction

Greenhouse gas (GHG) emissions from the start of the Industrial Revolution have been a major cause of global warming (Kweku et al., 2018). Carbon dioxide (CO_2) is one of the primary components responsible for the increased GHG in the atmosphere (Anderson, Hawkins, & Jones, 2016). The increased CO_2 emissions are adversely affecting our planet in various ways, for example, causing atmosphere warming and altering the seasonal climate, health, and well-being of all living organisms, influencing carbon sequestration in ecosystems, etc. (Waheed, Chang, Sarwar, & Chen, 2018). Such alterations modify various dependent events and their feedback mechanisms, such as precipitation, snow melting, river discharge, and water balance. These changes trigger droughts, heat waves, extreme rainfall, and flooding, which affect agricultural activities and lead to food insecurity. Therefore the increased GHG emissions have a multidimensional impact and affect the socio-economy of people globally. Given such a complex situation and the serious repercussions of increased GHG emissions, various intergovernmental and governmental authorities have taken several measures to control GHG emissions and reduce the impacts of climate change. Several major conventions were organized to discuss the pathways for sustainable economic growth while reducing GHG emissions and limiting global warming, such as the Kyoto Protocol (1997), the Paris Agreement (2015), and, more recently, the Conference of the Parties-28 (COP-28) in 2023. The Paris Agreement (2015) was one of the landmark treaties adopted by 196 parties at the UN Climate Change Conference, which aims to keep global temperature rise below 2°C above preindustrial levels and to pursue efforts to limit it to below 1.5°C by 2100. During such conventions, pledges and future action plan decisions are taken, including reducing emissions from industry, mining, and agriculture and committing to increased use of renewable and green energy. Such commitments require careful analysis and predictions of CO_2 emission and periodic monitoring to assess the progress and adopt course corrections (den Elzen, Hof, & Roelfsema, 2011). Knowing the emission sources and impending emission predictions will help develop suitable policies and mitigate climate change more efficiently.

CO_2 is a major GHG source that is leading to global warming. Accurate CO_2 emission monitoring will help assess the effectiveness of devised policies and offer suitable data to policymakers for effective mitigation measures. CO_2 emissions in India are primarily driven by the energy sector, industrial activities, transportation, thermal power plants running on coal, manufacturing, and vehicular emissions (Jayanthakumaran, Verma, & Liu, 2012). Burning coal is one of the major factors contributing to annual CO_2 emissions (Pata & Kumar, 2021) because of the combustion of coal, each carbon atom bonds with two oxygen atoms to form CO_2 (Adánez-Rubio et al., 2013). Burning oil and flaring are other included as the main causes of CO_2 emission, which is formed by combining unburnt carbon with atmospheric oxygen (Gavenas, Rosendahl, & Skjerpen, 2015). Moreover, the burning of natural gas produces CO_2, but it is 30% less than the emissions from burning oil and 45% less from burning coal (Burnham et al., 2012). Although natural gas is one of the major producers of CO_2, it does not contribute much to global warming. The cement production industry also plays a sizable role in adding to the total annual CO_2 emission, where CO_2 is produced during clinker production, which is an intermediate product of cement manufacturing and fossil fuel combustion (Andrew, 2018). In addition, different landscapes in India exhibit diverse and unique sources of carbon emissions and sinks. Northeast India is primarily

Sustainable Development Perspectives in Earth Observation. DOI: https://doi.org/10.1016/B978-0-443-14072-3.00006-X

forest-dominated, followed by agricultural land (FSI, 2021). The periodic deforestation for shifting cultivation causes significant forest cover loss and causes carbon emissions in the premonsoon dry season due to the burning of dead plant materials (Behera, Gupta, Barik, Das, & Panda, 2018; Das et al., 2021). The stubble burning in the post-monsoon season increases the atmospheric pollutants in various parts of the country, including New Delhi and national capital region of India (Das, Behera, & Abhilash, 2024). Therefore assessing the CO_2 changes in and around the major cities and the entire country would be important for decision-makers and policymakers in sustainable management and planning.

Various methods have been applied to analyze the time series data and predict the upcoming changes. Kumari and Singh (2023) employed multiple models, including the autoregressive integrated moving average (ARIMA) model for CO_2 emission prediction in India. Hosseini, Saifoddin, Shirmohammadi, and Aslani (2019) applied multiple polynomial regression (MPR) and multiple linear regression to forecast CO_2 emissions in Iran using time series data. They predicted CO_2 emissions by 2030 under two different scenarios, namely the business as usual (BAU) and the sixth development plan (SDP), and reported that the BAU scenario would not be able to meet its commitment to the Paris Agreement, which can be achieved by implementing SDP. Ding, Wu, and Lian (2020) developed a time series model to explore the dynamic relationship between carbon emission and enterprise business growth in Taiwan. Various time series methods, such as the augmented Dickey-Fuller unit root test, cointegration test, and vector autoregression (VAR) models were used to analyze 805 companies listed on the Taiwan Stock Exchange. The yearly increase in carbon emissions and total carbon emissions from 2012 to 2017 were assessed, which explained a dynamic relationship between carbon emissions and business growth. Akyol and Uçar (2021) studied time series data from 1990 to 2017 to forecast Turkey's carbon footprint by 2030. They used the Waikato Environment for Knowledge Analysis data mining software for time series forecasting, integrating multiple independent variables such as gross domestic product (GDP), population, energy consumption, and energy production.

The machine learning model has been one of the most effective data analysis and prediction tools. In this approach, the machine "learns" from several observations, which then predicts future outcomes with a certain degree of error. In comparison, the deep learning models are more complex in nature and better predict the observations. Several artificial intelligence (AI)-based models have been developed for time series analysis and prediction. Zhang, Li, Long, and Ling (2018) applied long short-term memory (LSTM) and wavelet transform for time series analysis of vehicular emissions in Beijing, China. Their study classified vehicular emissions into carbon monoxide (CO), nitric oxide (NO), and hydrocarbons (HC). They applied wavelet transformation to decompose the highly variable vehicle emission data time series into many smaller subseries with lesser variability, followed by the LSTM model to forecast these subseries data. Comparing with the observed data from May to December 2017, they reported that the developed approach improved prediction compared to conventional models like ARIMA and wavelet-ARIMA models. Jena, Managi, and Majhi (2021) applied a multilayer artificial neutral network (MLANN) model to forecast CO_2 emissions at a global level (17 major emitting countries). Urban population ratio, trade openness, and GDP were used as independent variables or predictors. They reported a higher prediction accuracy (96% on average) using the MLANN. CO_2 emission forecasting for China, India, Saudi Arabia, Indonesia, and Iran predicted further increases in their CO_2 emissions. Moreover, low-emission countries like South Korea, Turkey, Brazil, and South Africa would follow a similar trend of increasing CO_2 emissions. In contrast, countries like Japan, Italy, France, the United Kingdom, Australia, the United States, and Canada would reduce their CO_2 emissions.

Several studies were conducted for national CO_2 emission modeling and prediction in India. Nalini, Uma, Sijikumar, Tiwari, and Ramachandran (2018) investigated the distribution of CO_2 concentrations over India and adjacent oceanic regions from 2009 to 2012 using satellite observation (Greenhouse Gases Observing Satellite; GOSAT) and Atmospheric Infrared Sounder (Aqua), Carbon Tracker (CT) model simulations. Their study included the ground measurements from Sinhagad and Cape Rama stations. They reported a peak in CO_2 in the premonsoon season and a decline in postmonsoon, with the Indo-Gangetic Plain and northern India exhibiting higher concentrations. Kunchala et al. (2022) investigated the spatio-temporal variability of column-averaged dry-air mole fraction of CO_2 (XCO_2) over India using OCO-2 satellite data and Atmospheric Chemistry-Transport Model simulations from September 2014 to December 2018. The highest XCO_2 level was recorded during the pre-monsoon season due to increased respiration and reduced photosynthesis, while reduced concentration in the monsoon season due to enhanced photosynthesis. The XCO_2 showed a clear increasing trend and seasonality, with a ~ 12 ppm rise from 2015 to 2018, consistent with global trends. Although many studies assessed the CO_2 flux in India, limited studies applied deep learning models, such as LSTM, for time series analysis. Moreover, there is a lack of studies projecting CO_2 emissions using various machine learning or deep learning models. The current study aims to predict CO_2 flux and emission patterns for India and five of India's major cities. The objectives of this study have been divided into two parts. The first part deals with the CO_2 fluxes from various sources in five Indian cities, and the second part deals with total CO_2 emissions from the entire country.

23.2 Study area

The study was conducted in two different scales: (1) Indian landmass and (2) five major Indian cities. Indian landmass is located between 6.76°N to 37.08°N and 68.18°E to 97.42°E. The geographical area of the country is 3,287,260 km^2. Although most of the Indian landmass belongs to the equatorial climate zone, according to the Köppen-Geiger climate classification, India is divided into four major climate zones: (1) equatorial humid winter wet and dry, (2) warm temperate humid winter wet and dry, (3) arid and semiarid, and (4) montane (Das, Behera, Roy, & Barik, 2024). The major Indian land surface feature includes the Himalayan Mountain in the northeast with Indo-Gangetic plains in the north; the Thar desert in the West; Malwa Plateau, Deccan Plateau, and Chota Nagpur Plateau in central India; Western Ghats and Eastern Ghats along the west and east coastlines, respectively; and two major offshore islands, Andaman and Nicobar Islands in the Bay of Bengal and Lakshadweep islands in the Arabian Sea. Around 50% of the total geographical area in India is cropland, and the forest cover occupies around 21.7% of the total geographical area of India (FSI, 2021). Agriculture land expansion is one of the major causes of forest loss in India, as recorded in the past few decades (Reddy et al., 2016). India is the most populated country in the world, with nearly 1.45 billion people (https://www.worldometers.info/world-population/india-population [accessed on August 25, 2024]). GHG emissions by India are the third largest in the world, and the main source of this emission is coal burning. India emitted 2.8 Gt of CO_2 equivalent in 2016, wherein 79% was CO_2, 14% methane, and 5% nitrous oxide. The country emits ~7% of global emissions (Garg, Shukla, Kankal, & Mahapatra, 2017). As of September 2021, India generates 39.8% of its electricity from renewable energy sources and 60.2% using fossil fuels, of which 51% is generated from coal.

In addition to the total geographic area of India, the study analyzed the emissions from five major Indian cities: New Delhi (the national capital of India), Mumbai, Bangalore, Chennai, and Kolkata. The location of the five cities is shown in Fig. 23.1. The emission profiles of these five cities are characterized by high vehicular density, industrial activities, and substantial power demand. Each city faces unique challenges: Delhi's emissions are exacerbated by energy generation like power plants and vehicular pollution (Arora, Reddy, Sharma, Kilaparthi, & Gupta, 2023) from a disproportionately large number of vehicles in comparison to any other Indian city. Moreover, stubble burning in nearby states in the post-monsoon season causes a significant spike in the atmospheric GHG over New Delhi every year (Das et al., 2024). Kolkata's emissions are influenced by the large-scale use of adulterated and spurious fuel in outdated vehicles combined with rampant air conditioner usage due to high temperature and humidity (Paul, 2009). The GHG emissions in Mumbai are heightened by its extensive use of electricity, accounting for 71% of its total

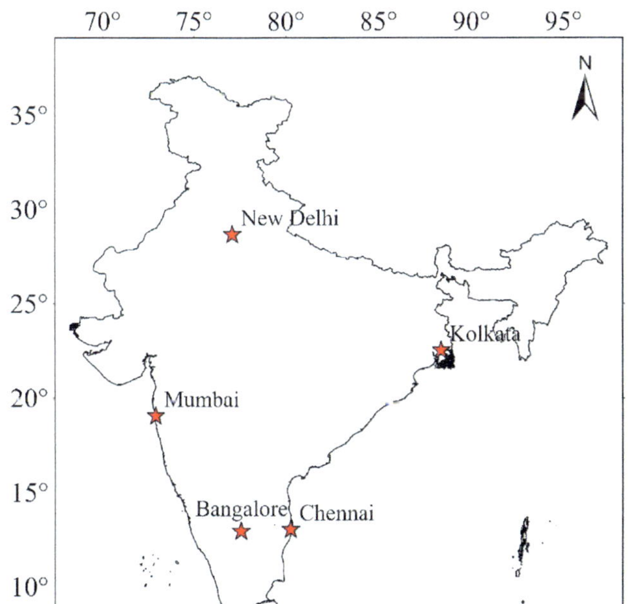

FIGURE 23.1 Location map of the five major Indian cities considered in the current study. Map lines delineate study areas and do not necessarily depict accepted national boundaries.

GHG emissions (Lewis, 2021) due to its large population and being the "financial capital" of India. Chennai has the highest per capita CO_2 emission at 4.79 tonnes (Ramachandra, Aithal, & Sreejith, 2015) and is facing emission issues due to large construction projects due to rapid urbanization; and Bangalore's emissions are influenced by its burgeoning tech industry, leading to an increase in industrial and commercial activities (Thakur, 2017). Despite these challenges, these cities possess potential carbon sinks, including urban green spaces, tree plantation initiatives, coastal mangroves, and wetlands, which are crucial in mitigating CO_2 emissions. Forests like Aarey near Mumbai, Sundarbans near Kolkata, and vast tree covers in Bangalore and New Delhi act as the carbon sink (Murthy et al., 2013). In addition, the large water bodies are great carbon sinks, such as the Arabian Sea for Mumbai and the Bay of Bengal for Chennai and Kolkata (Heinze et al., 2015).

23.3 Data and methodology

23.3.1 Objective I

The GOSAT, nicknamed "Ibuki," was launched in January 2009. GOSAT measures the atmospheric CO_2 and CH_4, and the latest retrievals are validated using XCO_2 from the upward-looking Fourier transform spectrometer (FTS) network (TCCON—total carbon column observing network). The single-shot precision of GOSAT measurements is 2.17 ppm. The Level 4A data product provides monthly averaged source/sink CO_2 and CH_4 fluxes in 64 and 43 regions, respectively. The uncertainties in CO_2 and CH_4 fluxes are estimated from the FTS short-wave infrared (SWIR) Level 2 column-averaged mixing ratios, ground-based observational data, and other meteorological data, such as wind direction and velocity. The column average XCO_2 (concentration: $\sim$1000 ppm) data are available for 15 pressure layers. The gridded CO_2 data from June 2009 to October 2019 were used in this study. Data were extracted for CO_2 flux from terrestrial biosphere exchanges, fossil fuel burning, biomass burning, air-sea interaction, and an optimized form of CO_2 flux, which accounted for all the aforementioned sources of CO_2 flux. The flux data from all the available sources were extracted over five metro cities.

The time series models were developed to predict CO_2 flux from terrestrial biosphere exchange, fossil fuel burning, and the optimized CO_2 flux. The input datasets were split into training and test sets in a 70:30 ratio, respectively. The training set was used to train the models, while the model performance was validated using the test set. The models were developed using two deep learning techniques, namely recurrent neural network (RNN) and LSTM. The loss functions were used to evaluate the prediction accuracy from each model, wherein the mean squared error (MSE) and mean absolute error (MAE) were derived.

23.3.2 Recurrent neural network (RNN)

RNN is a network with loops that uses previous information in a series of data spread over time. An RNN can be thought of as multiple copies of the same network, each passing a message to a successor. RNN is the artificial architecture of neural networks, and its chain-like nature reveals that RNNs are intimately related to sequences and lists (Amidi & Amidi, 2018). RNN forms connections between nodes and a directed or undirected graph along a temporal sequence. This allows it to exhibit temporal dynamic behavior. Derived from feedforward neural networks, RNNs can use their internal state (memory) to process variable-length sequences of inputs. RNNs are made recursively using the same set of weights in topological order by traversing a differentiable graph-like structure. RNNs can also be trained by reverse mode of automatic differentiation. Automatic differentiation applies chain rule to a set of elementary mathematical operations performed by a computer program and automatically computes derivatives of arbitrary orders to working precision (Cho et al., 2014). The problem with simple RNNs is computationally expensive in nature. During the training of a simple RNN using backpropagation, the long-term gradients, which are backpropagated, can tend to zero or infinity because of the computations involved in the process that use finite-precision numbers. This problem is partially tackled by using the *tanh* function as an activation function in the RNN model (23.1), which is given as follows:

$$g(z) = \frac{e^z - e^{-z}}{e^z + e^{-z}} \tag{23.1}$$

The *tanh* function squeezes the values between -1 and 1 for given input negative and positive values, respectively, and keeps the gradient in the linear region of the activation function; its second derivative sustains for a longer period before reducing to zero.

23.3.3 Long short-term memory (LSTM)

LSTM is a special case of RNN that has feedback connections. It can process single data points as well as entire sequences of data. A common LSTM unit is composed of a cell, an input gate, an output gate, and a forget gate. The cell remembers values over arbitrary time intervals, and the three gates regulate the flow of information into and out of the cell (Amidi & Amidi, 2018). Information can be read, written, or stored by opening and closing these gates. The gates store previous information in an analog form, which is differentiable and useful for backpropagation, and implement element-wise multiplication by a sigmoid activation function. For an input value, the sigmoid function value ranges between 0 and 1, corresponding to high negative and high positive values. It is used by the LSTM cell unit to forget or update any information. The sigmoid function determines the percentage of memory stored in a range of 0–1. A multiplication that results in close to 1 is stored, and one that results in close to 0 is forgotten. This helps the network to understand which data are important. A sigmoid function (23.2) is given as follows:

$$S(x) = \frac{1}{1 + e^{-x}} \tag{23.2}$$

The *tanh* functions used in the input and the output gates have a similar role to those in a normal RNN cell unit. These functions are nonlinear and calibrate the values flowing through them between 1 and −1. It helps store information for a longer period as its second derivative can survive longer.

The importance of information, that is, whether it is to be ignored or stored, is decided by the forget gate. For a particular timestep t, the information from hidden state h_{t-1} and present input x_t is passed through a sigmoid function with weight matrix W_f and bias b_f. It decides if the part of the older information is important by returning a value between 0 and 1, which the cell unit would later use for point-by-point multiplication. For a particular timestep t, the input gate firstly passes the current state x_t and previously hidden state h_{t-1} through a sigmoid function with weight matrix W_i and bias b_i to transform the values between 0 and 1, hence judging the importance of the information. In parallel to this operation, the input gate also passes the same set of information through a *tanh* function with weight matrix W_C and bias b_C, which transforms the values between 1 and −1. The output values from both these operations undergo point-by-point multiplication. The cell state stores information from the forget gate and the input gate. The forget vector f_t is multiplied by the previous cell state C_{t-1}, and if the resultant is 0, then older values are dropped. After this operation output from the input vector, i_t is added point by point to update the cell state and create a new cell state C_t.

The value of the next hidden state is determined by the output gate, which contains information about the previous inputs. Initially, the values from the previous hidden state and current state are passed through a sigmoid function, followed by passing the new cell state derived from the present cell state through a *tanh* function. Both these inputs are multiplied with each other on a point-by-point basis. The resultant value determines which information the hidden state should carry and use for prediction. Finally, the new hidden state and the new cell state are carried forward to the next cell unit. LSTMs were developed to deal with the vanishing gradient problem that can be encountered when training traditional RNNs (Greff, Srivastava, Koutník, Steunebrink, & Schmidhuber, 2016). Relative insensitivity to gap length is an advantage of LSTM over RNNs, hidden Markov models, and other sequence learning methods in numerous applications (Smagulova & James, 2020).

Both Simple RNN and LSTM models use Adam optimizers. Adam stands for adaptive moment estimation, which is an optimization algorithm for gradient descent. The method consumes less memory and, hence, is more efficient. This algorithm is a combination of momentum gradient descent and RMS Prop gradient descent. In momentum gradient descent, an exponentially weighted average of gradient descent is considered for speeding up the gradient descent algorithm by making the algorithm converge on minima faster. At a particular timestep t, with weights w, aggregate of gradients m, learning rate α, and moving average parameter β, momentum gradient descent can be given as follows:

$$w_{t+1} = w_t - \alpha m_t \tag{23.3}$$

$$m_t = \beta m_{t-1} + (1-\beta)\left[\frac{\partial L}{\partial w_t}\right] \tag{23.4}$$

RMS Prop stands for root mean squared prop. It is an adaptive learning algorithm, which is an improvement on the adaptive gradient algorithm. RMS Prop uses an exponential moving average. At a particular timestep t, the sum of the square of past gradients v, with a small positive constant ϵ, RMS Prop gradient descent is given as follows:

$$w_{t+1} = w_t - \frac{\alpha_t}{(v_t - \epsilon)^{1/2}} \times \left[\frac{\partial L}{\partial w_t}\right] \tag{23.5}$$

$$v_t = \beta v_{t-1} + (1 - \beta) \times \left[\frac{\partial L}{\partial w_t}\right]^2 \tag{23.6}$$

Adam optimizer takes the positive features of momentum and RMS Prop and enhances them to give a better-optimized gradient descent. The rate of gradient descent is controlled to minimize the oscillations to reach the global minima while taking large steps to pass the local minima along the way.

All the models were run in three settings, which were 10 cell units with 5000 epochs, 30 cell units with 12,000 epochs, and 70 cell units with 20,000 epochs. The model is supposed to fit the data better with more cell units and epochs; however, it can lead to higher variance and overfitting. The training errors are plotted for RNN and LSTM models for all three settings. The predictions generated by the RNN and LSTM models were compared for the test data to check the model's performance. Based on the training error and validation accuracy, appropriate models with optimal settings were decided, which were used to predict CO_2 fluxes for different cities up to 2030.

23.3.4 Objective II

The second phase of this study used annual CO_2 emissions from various sources in India from 1960 to 2020. The input data were classified into three categories: (1) CO_2 emission from fossil fuel and other pollutants, (2) socioeconomic factors, and (3) energy produced from renewable sources. CO_2 emissions (in tonnes) from fossil fuels and other pollutants included CO_2 emissions from coal, gas, oil, cement, flaring, and industries. The CO_2 emissions data were accessed from the Global Carbon Project (Ritchie, Roser, & Rosado, 2020). The socioeconomic data included GDP based on the 2010 constant value of the US dollar and population. The GDP and population data were accessed from the World Bank data catalog. Energy (TWh) produced from renewable sources, such as solar energy, wind energy, hydro energy, and geo biomass energy, was used. The data source for renewable energy sources is BP's Statistical Review of World Energy 2021.

Various regression models were developed to predict the annual CO_2 emissions. The relation between different variables was studied using a Pearson correlation coefficient chart. The data skewness was checked, and the skewed data were removed using log(x + 1) transformation. The data were scaled between 0 and 1 using the mean (μ) and the standard deviation (σ) before modeling (Eq. 23.7). The dataset was split into training and testing sets in the ratio 70:30, respectively. K-fold (k = 3) cross-validation was applied in model training.

$$z = \frac{x - \mu}{\sigma} \tag{23.7}$$

Four models are developed: (1) linear regression, (2) ridge regression, (3) least absolute shrinkage and selection operator (LASSO) regression, and (4) multilayer perceptron (MLP) regression. The learning rate, α, for the ridge and LASSO regression models was tuned using the R^2-α curve to achieve the highest accuracy. The cost functions for linear, ridge, and LASSO regressions are given below, where λ is the regularization parameter, β is a true parameter, n is the number of samples, $\bar{y}$ is the predicted value, and y is the actual value of the data.

$$J_{\text{Linear}} = \frac{1}{2n}\sum_{i=1}^{n} (\bar{y} - y)^2 \tag{23.8}$$

$$J_{\text{Ridge}} = \frac{1}{2n}\sum_{i=1}^{n} (\bar{y} - y)^2 + \lambda\sum_{i=1}^{n} \beta^2 \tag{23.9}$$

$$J_{\text{LASSO}} = \frac{1}{2n}\sum_{i=1}^{n} (\bar{y} - y)^2 + \lambda\sum_{i=1}^{n} \beta \tag{23.10}$$

The MLP network has two hidden layers, and the number of nodes in each layer is decided by the grid search cross-validation process, which regulates the modeling accuracy.

After model tuning, the future trend in CO_2 emission was predicted (extrapolated) by the various models based on the past observed trend.

The data processing flow diagrams for objectives (i) and (ii) are shown in Fig. 23.2.

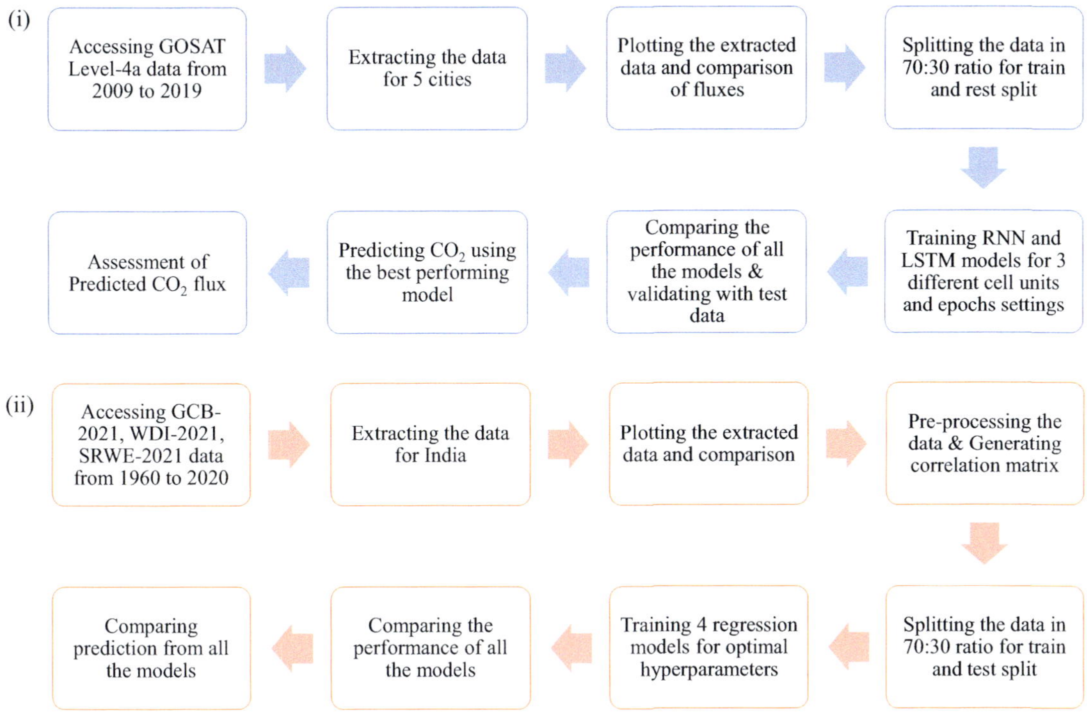

FIGURE 23.2 Data processing flow diagram.

23.4 Results

23.4.1 Objective I: CO_2 emissions from five major Indian cities

The optimized daily CO_2 flux recorded over five Indian cities from June 2009 to October 2019 is shown in Fig. 23.3(i). The lowest daily average CO_2 flux in New Delhi was recorded as 4.3 g m^{-2} day^{-1} in August 2009, which increased to 7 g m^{-2} day^{-1} in August 2019. While the highest daily average CO_2 flux increased from 5.09 g m^{-2} day^{-1} in March 2010 to 7.94 g m^{-2} day^{-1} in March 2019. Moreover, the seasonal change in CO_2 flux showed a continuous increasing trend in New Delhi. The daily average CO_2 flux in Kolkata was relatively lower than in New Delhi, which ranged between 1.85 and 2.4 g m^{-2} day^{-1} at the beginning of the study period (in 2009) and increased to 3.29–3.72 g m^{-2} day^{-1} at the end of the study period (in 2019). However, a similar continuous increasing trend in CO_2 flux was observed in Kolkata, like New Delhi. A high seasonal change in daily average CO_2 flux was recorded in Bangalore, ranging from a low negative value of −0.07 g m^{-2} day^{-1} in October 2009 to 1.19 g m^{-2} day^{-1} in May 2010, which gradually increased to 0.78 and 1.83 g m^{-2} day^{-1} in December 2018 and April 2019, respectively. In comparison, the seasonal daily average CO_2 flux in Chennai and Mumbai showed less variation, with a minor increasing trend from 2009 to 2019. Moreover, the optimized surface CO_2 flux was lower in Chennai, Mumbai, and Bangalore, moderate in Kolkata, and maximum in New Delhi.

The surface CO_2 flux from the terrestrial biosphere exchange and fossil fuel burning from June 2009 to October 2019 are shown in Fig. 23.3(ii) and (iii). It demonstrates that the burning of CO_2 emission fossil fuels primarily regulates the optimized CO_2 flux with a significantly lower contribution from the terrestrial biosphere, except for Bangalore. The lowest CO_2 flux from fossil fuel burning was recorded in Mumbai, followed by Bangalore and Chennai, moderate in Kolkata, and the maximum in New Delhi. A steady positive trend is observed in CO_2 flux due to fossil fuel burning from 2009 to 2019. In contrast, the terrestrial biosphere exchange varied between ± 0.2 g m^{-2} day^{-1} during the study period for all cities except Bangalore, wherein the CO_2 exchange varied between −0.7 g m^{-2} day^{-1} in post-monsoon and 0.3 g m^{-2} day^{-1} in pre-monsoon and monsoon. No significant change was observed during the study period. The sharp seasonal changes in CO_2 sequestration regulate the seasonal optimized CO_2 flux change in Bangalore.

23.4.1.1 *Development of recurrent neural network and long short-term memory models*

The RNN and LSTM models trained with variable cell units and epochs were compared to select the best predictive model. The training errors (MSE and MAE) during model development with various cell units and epochs are shown

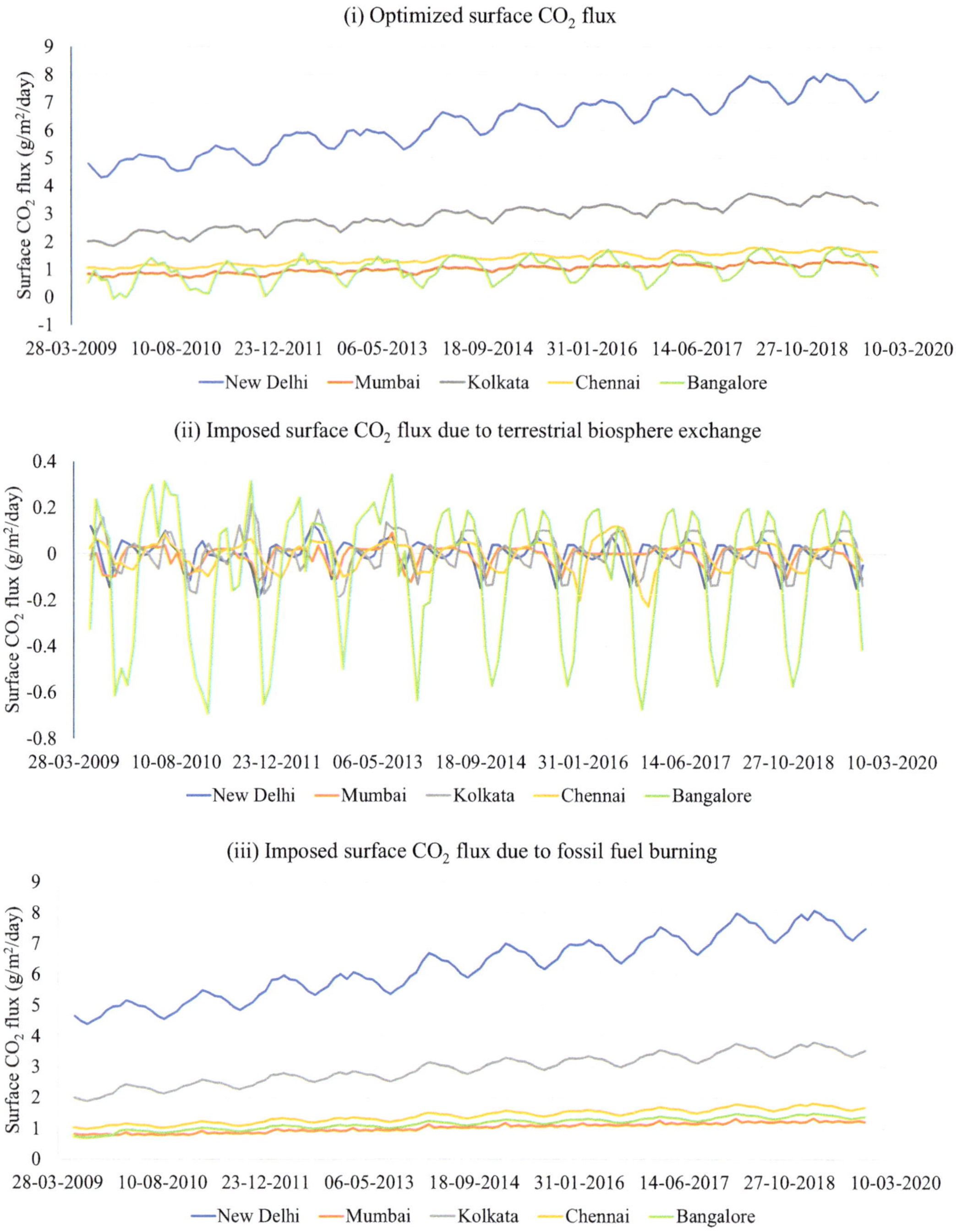

FIGURE 23.3 Daily average (i) optimized CO_2 flux, (ii) imposed surface CO_2 flux due to terrestrial biosphere exchange, and (iii) imposed surface CO_2 flux due to fossil fuel burning recorded from June 2009 to October 2019 in five Indian cities.

in Fig. 23.4. It can be observed that MSE and MAE values of the RNN models run with 10 cell units for 5000 epochs were higher than those of the LSTM models with similar hyperparameters in the initial epochs but converged toward the end. MSE and MAE values of RNN and LSTM models run with 30 cell units and 12,000 epochs mostly remain the same throughout the training with low training error at the start for the LSTM model; however, small variations can be seen in the MAE values in RNN modeling. Similarly, the MSE and MAE values of the RNN and LSTM models run for 70 cell units and 20,000 epochs were similar throughout the training period, though the LSTM model showed lower MAE after 9000 epochs than the RNN model. In comparing loss functions for all the cities, it was found that LSTM models were generally better predictors than RNN models.

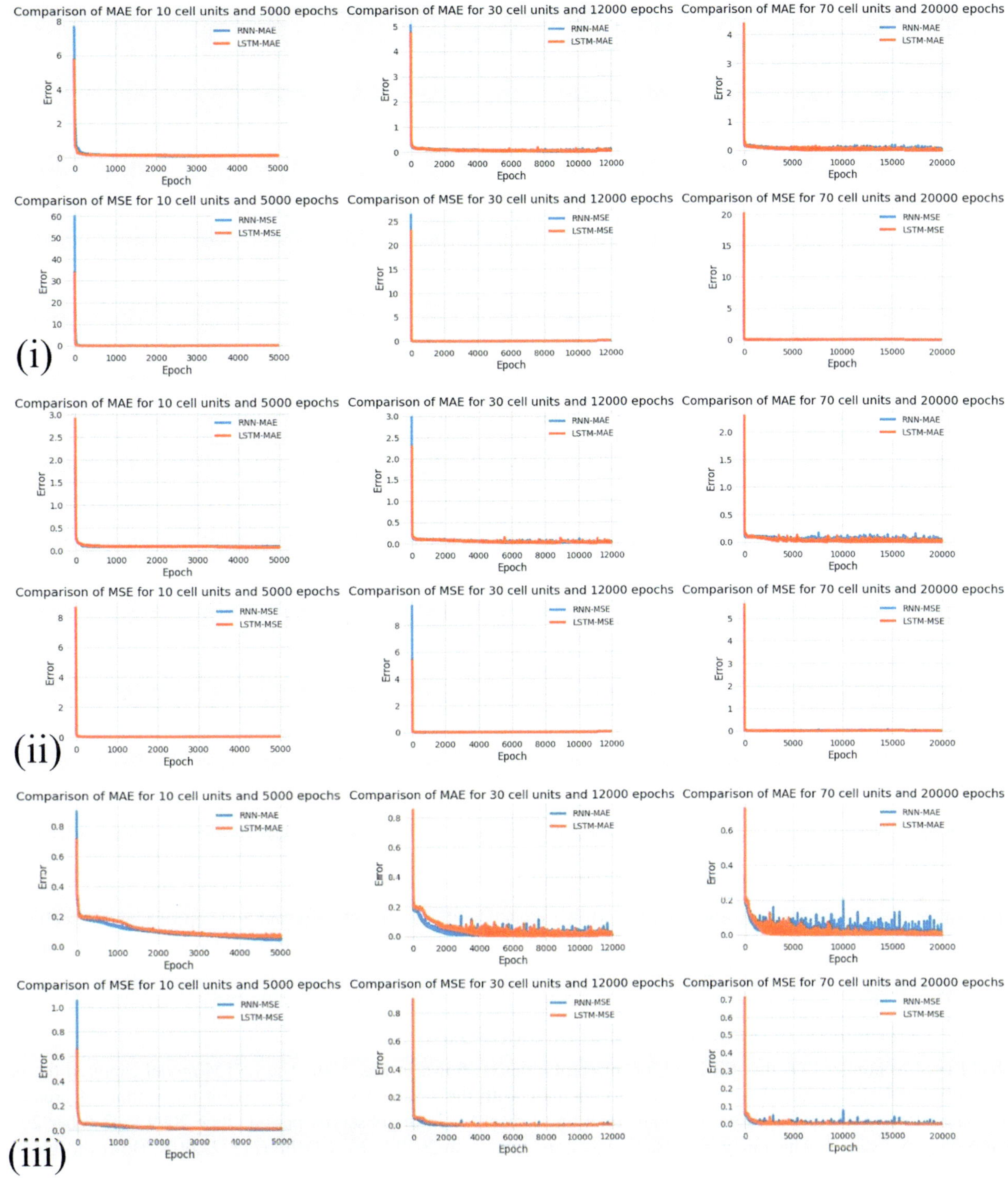

FIGURE 23.4 Comparison of training error of various time series models for (i) New Delhi, (ii) Kolkata, (iii) Bangalore, (iv) Mumbai, and (v) Chennai. The first, second, and third columns indicate model training with 10 cell units and 5000 epochs, 30 cells and 12,000 epochs, and 70 cells and 20,000 epochs, respectively. The first and second rows indicate errors (MAE and MSE).

The comparison of validating each model prediction with test data for different cities showed the highest divergence for models with 10 cell units and 5000 epochs, followed by the model with 30 cell units and 12,000 epochs, and the least for the model with 70 cell units and 20,000 epochs for RNN and LSTM. The best model was observed for Bangalore, which captured the seasonal variations and magnitude, and the worst performance was observed for

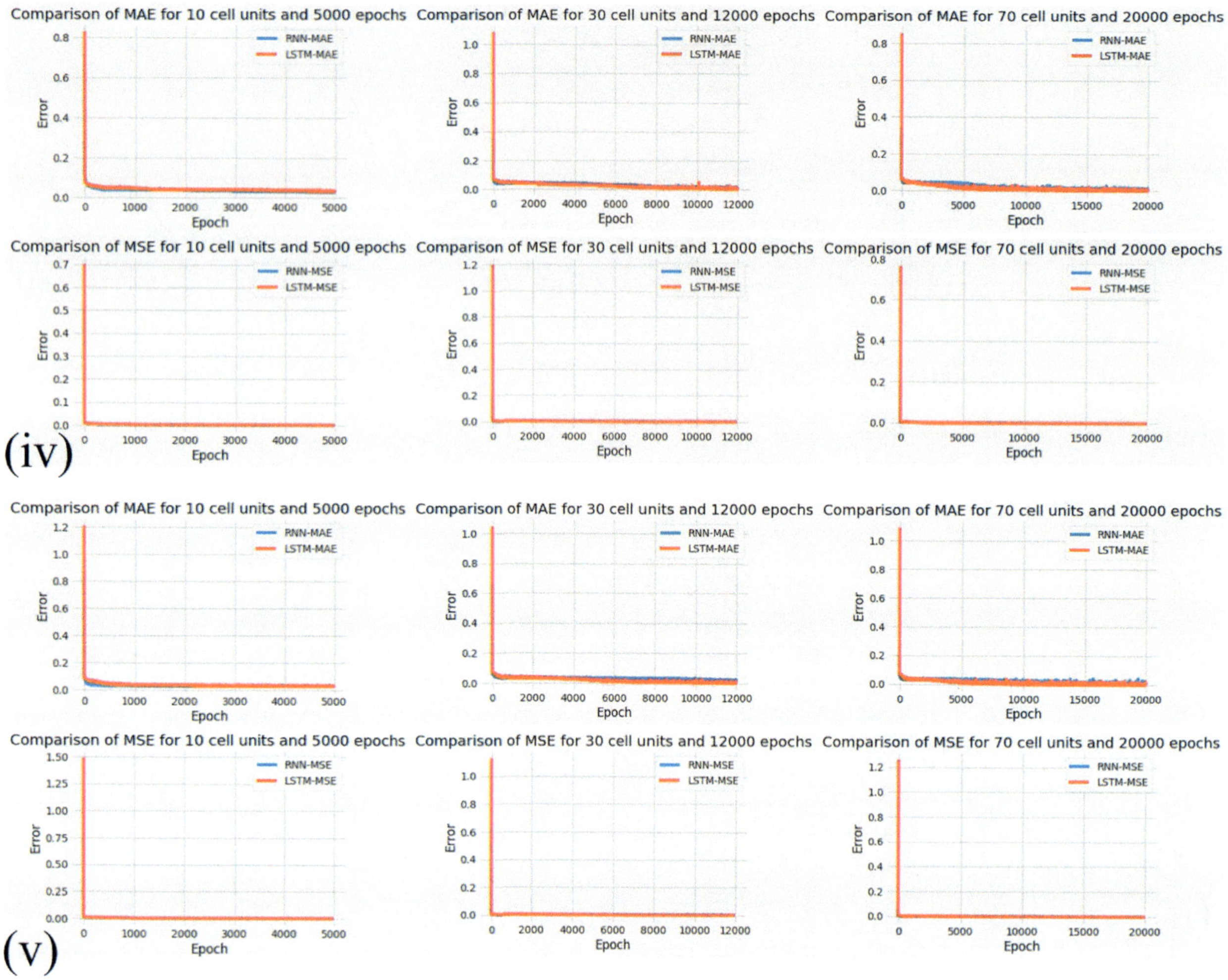

FIGURE 23.4 (Continued)

Mumbai. In most of the cases in RNN and LSTM, the models with the 10 cell units and 5000 epochs, and 30 cell units with 12,000 epochs could not capture the seasonality and magnitude. It is observed that neither the seasonality nor the values of the predictions are like that of the test data (Fig. 23.5).

23.4.1.2 Predictions of optimized carbon dioxide fluxes

LSTM used to predict CO_2 flux till 2030 for different cities is shown in Fig. 23.6. The LSTM model captured the past CO_2 flux trend and predicted a continuous flux increase with the seasonal variations, except for a minor decrease in Bangalore toward the end of 2030. The maximum value recorded in the observed period (June 2009 to October 2019) was 8.05 g m^{-2} day^{-1}, which may increase to 9.87 g m^{-2} day^{-1} in 2030 in New Delhi (Fig. 23.6(i)). However, the predicted change from 2026 to 2030 indicated a lower rate of change, as observed in the rest of the study period. A similar CO_2 flux was predicted for Kolkata, wherein the rate of projected change diminished from 2023 to 2028 and again increased after 2028 (Fig. 23.6(ii)). The maximum daily average CO_2 flux may change from 3.8 g m^{-2} day^{-1} in 2019 to 4.2 g m^{-2} day^{-1} in 2030. On the contrary, the predicted CO_2 flux for Bangalore showed a gradual reduction from 2024 to 2030. The current observed maximum CO_2 flux of 1.83 g m^{-2} day^{-1} in 2019 may reduce to 1.39 g m^{-2} day^{-1} in 2030. LSTM predicted CO_2 fluxes in Mumbai indicated sharp seasonal variations from 2020 to 2030 compared to 2009 to 2019. The observed seasonal variations indicated a gradual increase and decrease in month-wise values, which is predicted a sharp decline around November and December. Unlike other cities, the CO_2 flux predicted for Chennai indicated a monotonic increase till 2030.

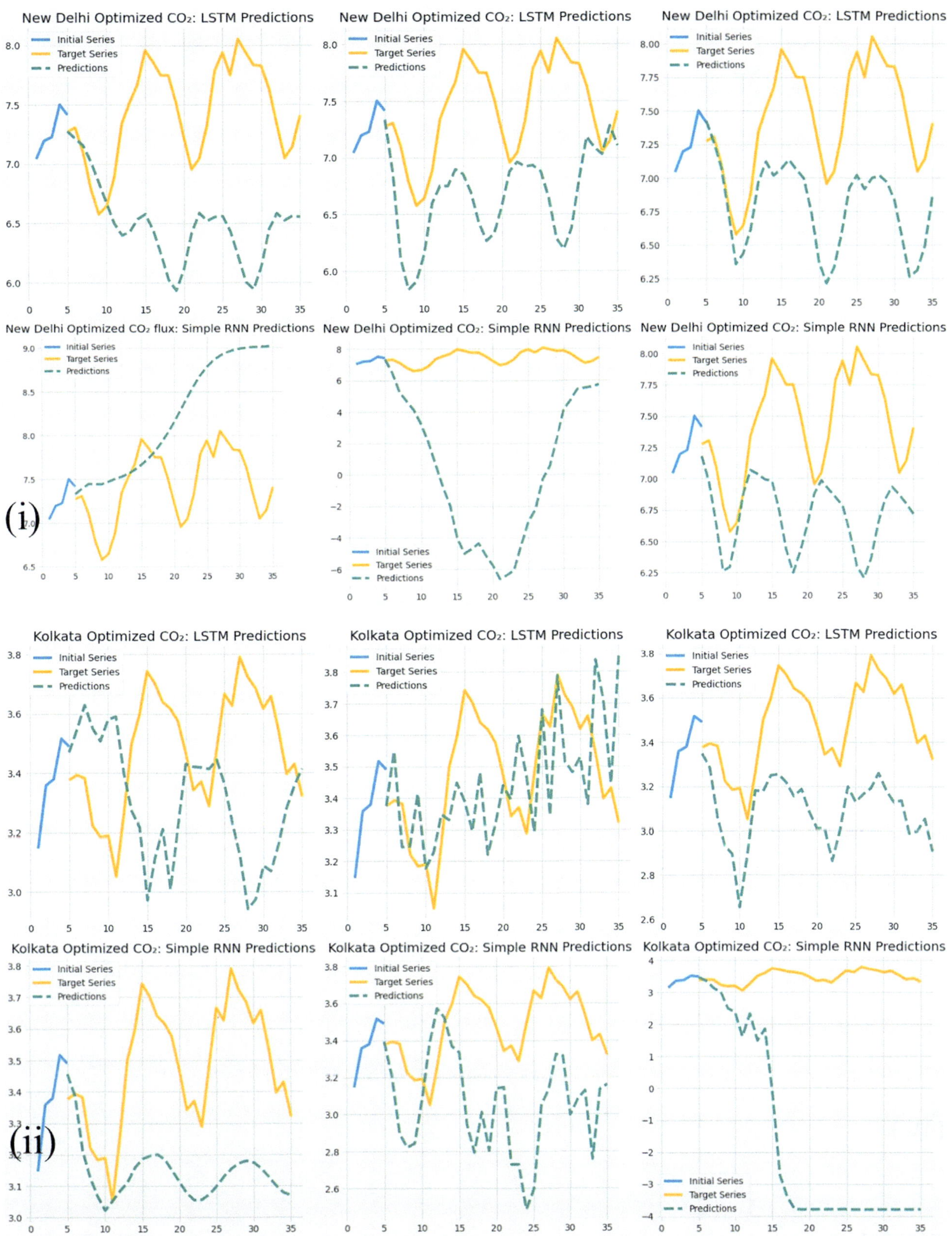

FIGURE 23.5 Comparison of validation (with test data) between RNN and LSTM for (i) New Delhi, (ii) Kolkata, (iii) Bangalore, (iv) Mumbai, and (v) Chennai. The first, second, and third columns indicate model training with 10 cell units and 5000 epochs, 30 cells and 12,000 epochs, and 70 cells and 20,000 epochs, respectively. The first and second rows indicate observed (initial and target) and predicted values by LSTM and RNN models, respectively.

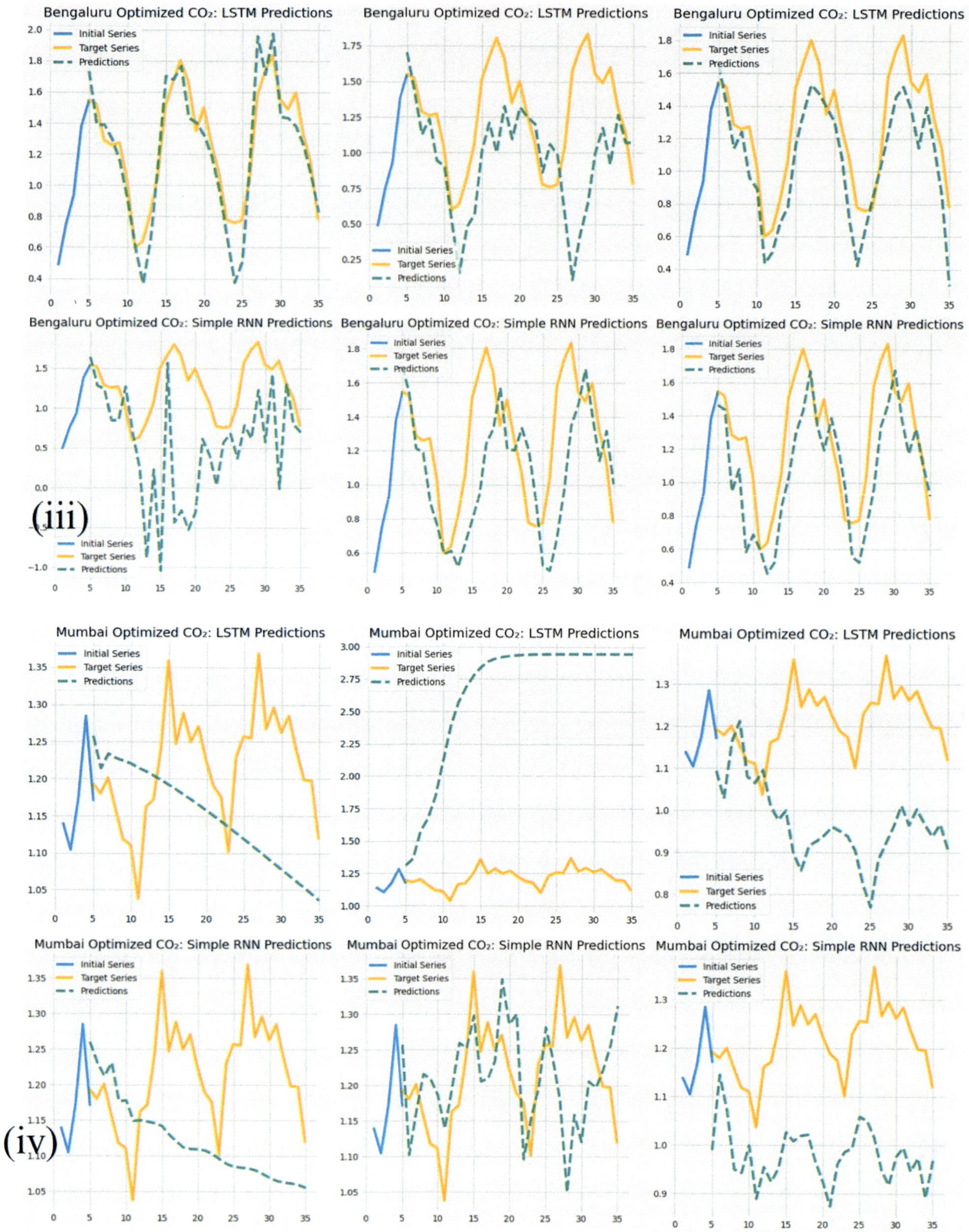

FIGURE 23.5 (Continued)

FIGURE 23.5 (Continued)

23.4.2 Objective II: CO_2 emissions from different sources in India

The annual CO_2 emissions from various sources in India from 1960 to 2020 are shown in Fig. 23.7(i). It demonstrated that the CO_2 emissions from coal and oil were the major contributors to total annual CO_2 emissions in India. Emissions from gas and cement industries were moderate, and significantly low from flaring and other industries. A sharp increase in CO_2 emissions from coal was evident from 2005 to 2019, which experienced a decline in 2020. On the other hand, the emissions from oil industries steadily increased till 2019 and experienced a decline in 2020. However, no prominent decline in emissions from the gas and cement industries was recorded in 2020. Fig. 23.7(ii) shows the amount of energy (in terawatt hours, i.e., trillion-watt hours: TWh) generated from various renewable sources in India from 1960 to 2020. It was observed that the energy generated from hydropower is far more than the energy generated from any other renewable sources. Wind energy and geo biomass energy began to grow significantly after 2002, while solar energy has shown rapid growth since 2015.

The correlation (coefficient of determination) among various parameters with CO_2 emissions is shown in Fig. 23.8. The highest correlation coefficient (0.98) between CO_2 emissions and other variables was observed for coal, oil, and energy from hydropower, followed by a strong correlation (0.97 and 0.96) with gas, GDP, and population. A similar high correlation (0.85–0.92) was observed between the CO_2 emissions and cement industries, flaring, and other industries (industry), and Geo (geo-thermal energy). In comparison, a lower correlation (<0.8) was observed between CO_2 emissions and solar and wind energy emissions.

23.4.2.1 Hyperparameter tuning and prediction

The performance of the models developed for national-scale CO_2 emission prediction using linear, LASSO, ridge, and MLP regression are shown in Table 23.1. The hyperparameters in the ridge and LASSO regression were tuned using

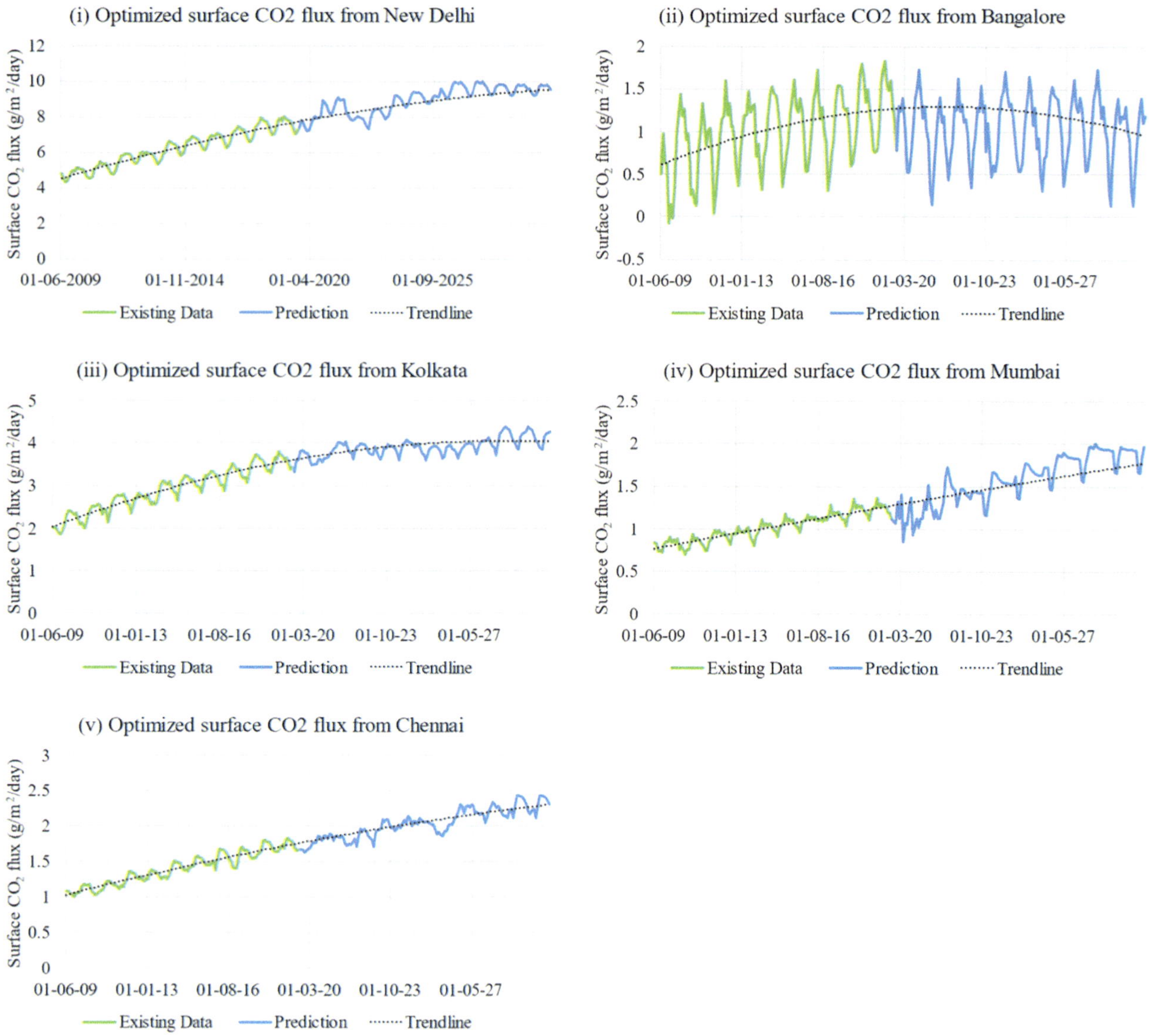

FIGURE 23.6 Observed (2009–2019) and LSTM predicted (2020–2030) optimized CO_2 flux for (i) New Delhi, (ii) Kolkata, (iii) Bangalore, (iv) Mumbai, and (v) Chennai.

the learning rate (α) based on the $R^2-\alpha$ curve. Fig. 23.9 shows the R^2- curve for LASSO and ridge regression, which indicated that the models performed well for $\alpha = 0.03$. The MLP used in the study had two layers, and the number of nodes for each layer was decided to maximize the model performance. The R^2 scores for various regressors showed the best performance by ridge regression and MLP, followed by LASSO and linear regression.

Fig. 23.10 demonstrates the predictions from all the trained models. The linear regression model predicted a sharp increase from 2020 to 2021, followed by a monotonic increase in CO_2 emissions, reaching 3.6 Gt of CO_2 by 2030. However, LASSO, ridge, and MLP predicted a gradual increase in emissions at the beginning (2021). Like linear regression, LASSO regression predicted a monotonic increase, and the annual emission may reach 4.6 Gt of CO_2 in 2030. On the contrary, ridge and MLP regression, the best-performing models, showed similar predictions. Although the rate of change in predicted emissions by ridge regression and MLP was initially low (till 2025), a sharp increase was observed thereafter. The predicted emissions may reach 5.5 Gt of CO_2 emission by 2030.

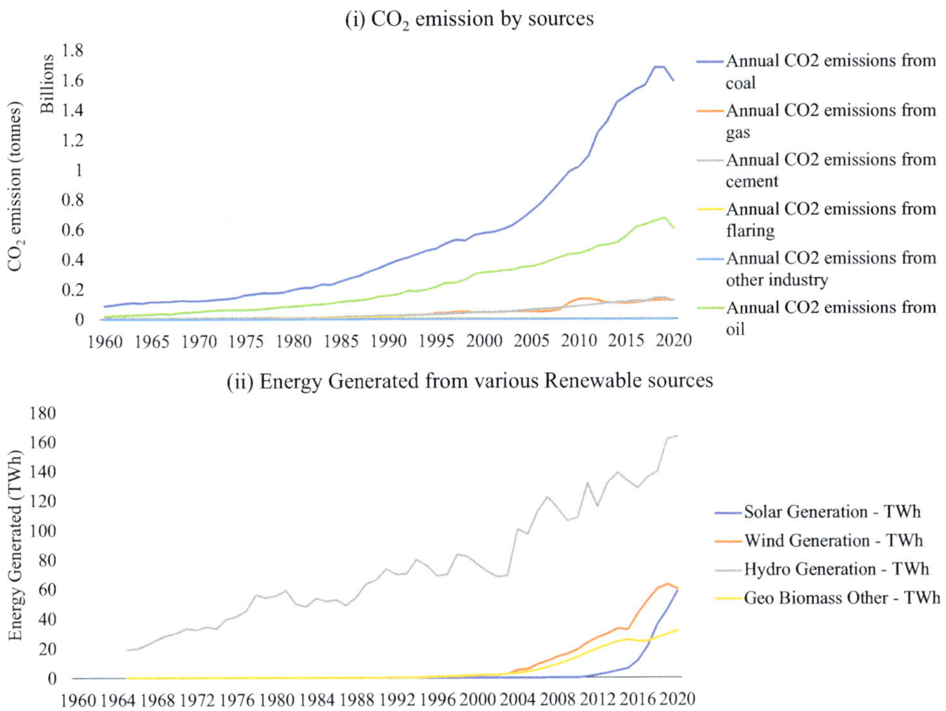

FIGURE 23.7 (i) CO_2 emission from various sources and (ii) energy (terawatt hours, i.e., trillion-watt hours: TWh) generated from various renewable energy sources in India.

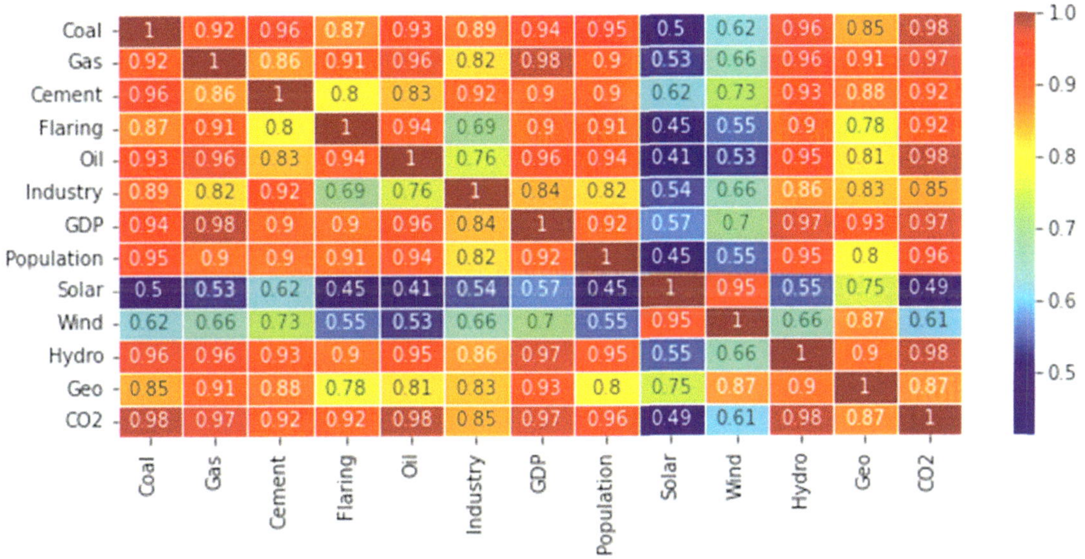

FIGURE 23.8 Correlation matrix among the CO_2 emissions and their various sources.

TABLE 23.1 Performance of various regression models used to predict CO_2 emissions in India.

	Linear	LASSO	Ridge	MLP
R^2 score	0.94	0.97	0.99	0.99

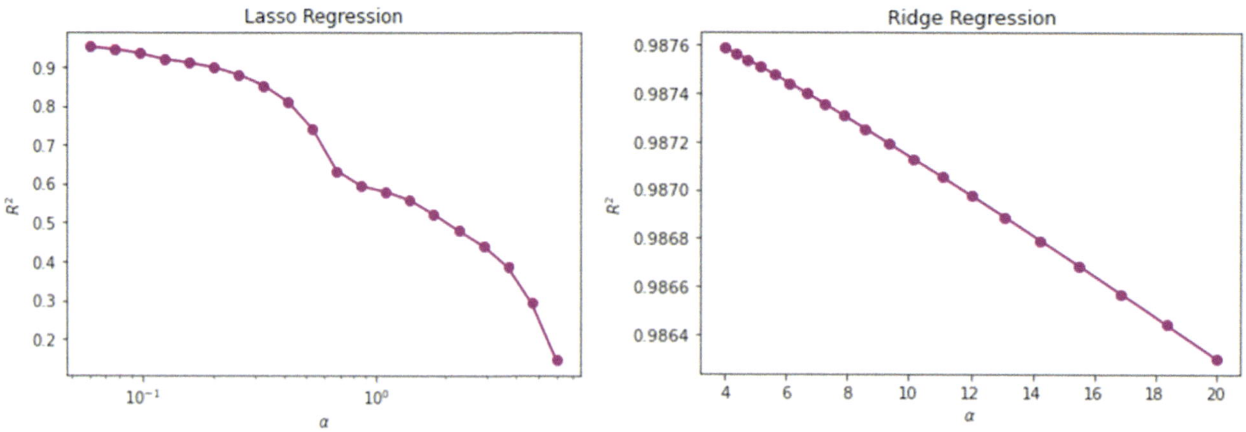

FIGURE 23.9 $R^2-\alpha$ curve for LASSO and ridge regression.

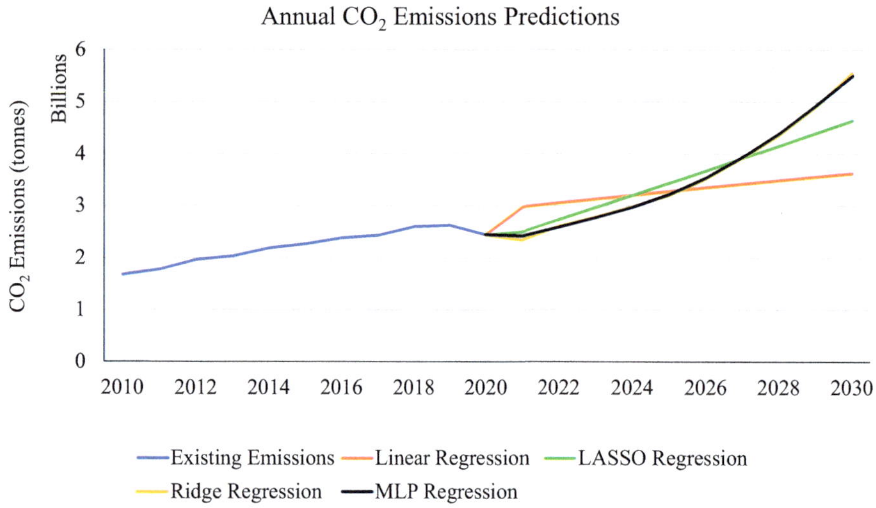

FIGURE 23.10 Predicted CO_2 emission using various regression models.

23.5 Discussion

With a significantly higher population density, Indian cities emerged as hotspots for CO_2 emission, and thus examining the urban CO_2 emission is important for developing suitable action plans (Kumar & Nagendra, 2016). The CO_2 emissions in five major Indian cities indicated a continuous increase from 2009 to 2019, wherein the maximum change was observed in New Delhi, followed by Kolkata. The change was mostly regulated by CO_2 flux due to increased fossil fuel burning, while the contribution from the terrestrial biosphere exchange was significantly low and did not experience observable change during this period. The overall CO_2 flux was observed much higher in New Delhi, followed by Kolkata, than in Chennai, Bangalore, and Mumbai. Kumar and Nagendra (2016) studied CO_2 emission in Chennai and reported a lower per-capita CO_2 flux than the national average. They also reported a higher CO_2 flux in summer (pre-monsoon) than in winter (post-monsoon). The seasonal changes in the terrestrial biosphere exchange showed higher emission and sequestration in Bangalore than in other cities. The LSTM model indicated a lower error and performed better than RNN. The LSTM model captured the seasonal trend better than the RNN model. Kumari and Singh (2023) used CO_2 emissions from 1980 to 2019 and applied three statistical models: (1) ARIMA, (2) seasonal autoregressive-integrated moving average with exogenous factors model, and (3) the Holt-Winters model, two machine learning models (linear regression and random forest model), and a deep learning-based (LSTM) model to predict the CO_2 emission for the next 10 years. They observed LSTM as the most accurate prediction model and reported a continuous increase in CO_2 emissions. The predicted CO_2 emission indicated an increase for all cities, excluding Bangalore. This could be attributed to the recent reduction in CO_2 emissions toward the end of the observation (2019). A monotonous increase is predicted for Mumbai and Chennai, while the rate of predicted change in Kolkata and New Delhi is reduced slightly.

Total annual CO_2 emission for India is affected most by the use of coal, natural gas, oil, GDP, and hydropower. A continuous increase in CO_2 emission was observed at the national scale, and a decline was recorded in 2020. Garg et al. (2017) studied the CO_2 emission in India from 1990 to 2013 and reported a ~5% compound annual growth rate. A marked decrease in emissions from all the sources was observed in 2020, which could be attributed to reduced anthropological activity during the COVID-19 lockdown (Behera et al., 2021; Venter, Aunan, Chowdhury, & Lelieveld, 2020). Mandal, Das, Behera, and Parida (2022) used the GOSAT to assess the CO_2 flux in India from 2010 to 2012. They observed a significant change in CO_2 concentration in the annual data, with the highest carbon emission occurring in the summer and the maximum carbon sequestration occurring during the post-monsoon season. Their study showed carbon emission from most of the forests in India during summer and monsoon, excluding a few forested regions in northeast India and the Western Himalayan region. The linear and LASSO regression indicated a monotonic increase with a variable gradient. In comparison, the ridge and MLP regression predicted a lower rate of change at the beginning and a comparatively higher rate after 2025.

GDP and population growth in India are accompanied by increased CO_2 emissions, as with a better economy and more people, there is more energy consumption and, hence, more avenues of CO_2 emission (Udemba, Güngör, Bekun, & Kirikkaleli, 2021). This could be seen from the correlation chart, which indicated a high positive relation with CO_2 emission. Hydropower, despite being a renewable energy source, shows a high correlation with total CO_2 emission. Earlier studies reported that the construction set-up, and running of a hydropower plant in the initial years led to the destruction of neighboring ecology and environment and the alteration of water temperature and chemistry, leading to a resultant increase in CO_2 emission. However, such renewable energy installations are more effective in reducing the net CO_2 emissions of a country in the long run (Bilgili et al., 2021). Solar power shows the least correlation with annual CO_2 emissions with a coefficient of 0.49, which means it is not a major contributor to total emissions. The reason for this is that every kilowatt-hour of energy produced by solar energy produces only 40 g of CO_2, which is quite minimal when compared to other sources (Hsu et al., 2012). Wind power is also seen to have a low correlation with the amount of CO_2 emission owing to the fact that studies have shown wind power produces only 5g to 26 g of CO_2 for every kWh of energy produced, which is even lower than the emission figures for solar power. However, the correlation coefficient for wind power is higher than solar as it is costlier to set up and can only be set up at very few locations where sufficient wind flows year-round (Wang & Sun, 2012). The energy produced from geothermal and biomass energies showed an intermediate level of correlation in comparison to all other energy sources because geothermal energy produces 99% less CO_2 than other fossil fuel sources (Ayuningtyas, Purba, & Adityatama, 2021), while biomass energy releases around 251 g of CO_2 equivalent per kWh of energy produced (Poeschl, Ward, & Owende, 2012).

The predicted increase in CO_2 emission will only add to intensifying global warming, which would be accompanied by undesirable effects like an increase in temperature and water scarcity, among many (Lal, 2003). Hence, it becomes important for policymakers to take prompt action, such as investing more in green energy and employing sustainable practices to prevent adverse climatic conditions. Moreover, forest protection and reduction in shifting cultivation can improve CO_2 sequestration. Various governmental and nongovernmental initiatives are taken to reduce people's dependency on shifting cultivation practices in various parts of northeast India (Behera et al., 2018). In urban areas, actions such as mass rapid transit system, fixed bus speed, hikes in parking fees, fuel efficiency, and stringent emission norms may reduce the per-capita energy demand, fossil fuel burning, and CO_2 emission significantly compared to the BAU (Aggarwal & Jain, 2016).

23.6 Conclusion

The increased CO_2 emissions due to fossil fuel burning and various anthropogenic activities are leading to global warming. Such changes modify climate change, damaging human health and triggering food insecurity. Moreover, urban areas have emerged as hotspots for CO_2 emissions due to higher population density and energy demand. Prediction models are developed at city and national scales to provide suitable data for policymakers to design suitable action plants to deal with increased emissions. The satellite data-driven CO_2 fluxes from various sources showed a higher contribution from fossil fuel burning and a comparatively lower contribution from biospheric exchange in urban areas, except Bangalore. The RNN and LSTM deep learning time series data analysis models were employed to develop prediction models for urban areas. The LSTM indicated a superior performance over RNN and predicted a continuous increase in CO_2 emissions in all the cities, except for a minor decrease in Bangalore. The national-scale CO_2 emission was predicted using regression models employing multiple covariants such as fossil fuel data, socioeconomic data, and renewable energy use data. The results revealed that CO_2 flux will continue to increase, and the annual CO_2 emission for India will become double that of the present values at present usage trends. The study advocates the use of various other indicators and the latest AI techniques for better insights into emission scenarios situations so that effective steps can be taken to combat any undesirable effects.

References

Adánez-Rubio, I., Abad, A., Gayán, P., De Diego, L. F., García-Labiano, F., & Adánez, J. (2013). Performance of CLOU process in the combustion of different types of coal with CO_2 capture. *International Journal of Greenhouse Gas Control*, *12*, 430–440.

Aggarwal, P., & Jain, S. (2016). Energy demand and CO_2 emissions from urban on-road transport in Delhi: Current and future projections under various policy measures. *Journal of Cleaner Production*, *128*, 48–61. Available from https://doi.org/10.1016/j.jclepro.2014.12.012.

Akyol, M., & Uçar, E. (2021). Carbon footprint forecasting using time series data mining methods: The case of Turkey. *Environmental Science and Pollution Research*, 1–11.

Anderson, T. R., Hawkins, E., & Jones, P. D. (2016). CO_2, the greenhouse effect and global warming: From the pioneering work of Arrhenius and Callendar to today's Earth System Models. *Endeavour*, *40*(3), 178–187.

Andrew, R. M. (2018). Global CO_2 emissions from cement production. *Earth System Science Data*, *10*(1), 195–217.

Arora, T., Reddy, C. S., Sharma, R., Kilaparthi, S. D., & Gupta, L. (2023). Greenhouse gas emissions of Delhi, India: A trend analysis of sources and sinks for 2017–2021. *Urban Climate*, *51*, 101634.

Amidi, A., & Amidi, S. (2018). Vip cheatsheet: Recurrent neural networks. https://storage.googleapis.com/kaggle-forum-message-attachments/904222/16155/recurrent%20neural%20network.pdf?GoogleAccessId = web-data@kaggle-161607.iam.gserviceaccount.com&Expires = 1731929218&Signature = nEkw7aML4%2FpjsDfWdYFQ5enUZNhWGJKey%2Fw9ZelJBS0Asd4JG1ApmIBUFb5tdcHNsgf5HJLe2C7ZQAs6QDqUF8HfbcSjbB17yxc%2BYEuf3QvgcNeX%2BMYwLwNNCBvLS7hwI0Joe8gGwZrKKQlCee%2FFQ1B9R2HhASTcmgxH%2BIem%2Bnhms5SCe0Ti4o8MCJD84952lAGi9VJxNMHahHylX3kLBKbYNoI%2FwFmVQMJGaIgjZ%2BkWt1cLWzfaN7ItVA%2Bhpi9pM%2B5SMEPmlviEeCfn7RGCPF%2F3fvp3ewPaqGtFFyOvQJaDihRqzuvVXHhfSb364VyJVw%2BYGWWfEd9YONBHHVFfEw%3D%3D accessed on 16.06.2024.

Ayuningtyas, R.R., Purba, D., & Adityatama, D.W. (2021). Geothermal role as a renewable energy in energy mix indonesia during and post-COVID-19 pandemic. Proceedings of the 46th Workshop on Geothermal Reservoir Engineering Stanford University, Stanford, California, February, 2021. Accessed from https://pangea.stanford.edu/ERE/db/GeoConf/papers/SGW/2021/Ayuningtyas2.pdf on 16.06.2024

Behera, M. D., Gupta, A. K., Barik, S. K., Das, P., & Panda, R. M. (2018). Use of satellite remote sensing as a monitoring tool for land and water resources development activities in an Indian tropical site. *Environmental Monitoring and Assessment*, *190*, 1–21. Available from https://doi.org/10.1007/s10661-018-6770-8.

Behera, M. D., Mudi, S., Shome, P., Das, P. K., Kumar, S., Joshi, A., ... Sirivella, Z. (2021). COVID-19 slowdown induced improvement in air quality in India: Rapid assessment using Sentinel-5P TROPOMI data. *Geocarto International*, 1–21.

Bilgili, F., Lorente, D. B., Kuşkaya, S., Ünlü, F., Gençoğlu, P., & Rosha, P. (2021). The role of hydropower energy in the level of CO_2 emissions: An application of continuous wavelet transform. *Renewable Energy*, *178*, 283–294.

Burnham, A., Han, J., Clark, C. E., Wang, M., Dunn, J. B., & Palou-Rivera, I. (2012). Life-cycle greenhouse gas emissions of shale gas, natural gas, coal, and petroleum. *Environmental Science & Technology*, *46*(2), 619–627.

Cho, K., Van Merriënboer, B., Gulcehre, C., Bahdanau, D., Bougares, F., Schwenk, H., & Bengio, Y. (2014). Learning phrase representations using RNN encoder-decoder for statistical machine translation. In *Proceedings of the 2014 Conference on Empirical Methods in Natural Language Processing (EMNLP)*, (pp. 1724–1734). Doha, Qatar: Association for Computational Linguistics.

Das, P., Behera, M. D., & Abhilash, P. C. (2024). A rapid assessment of stubble burning and air pollutants from satellite observations. *Tropical Ecology*, *65*(1), 152–157. Available from https://doi.org/10.1007/s42965-022-00291-5.

Das, P., Behera, M. D., Roy, P. S., & Barik, S. K. (2024). Predicting tipping points of vegetation resilience as a response to precipitation: Implications for understanding impacts of climate change in India. *Biodiversity and Conservation*, 1–18. Available from https://doi.org/10.1007/s10531-024-02804-1.

Das, P., Mudi, S., Behera, M. D., Barik, S. K., Mishra, D. R., & Roy, P. S. (2021). Automated mapping for long-term analysis of shifting cultivation in Northeast India. *Remote Sensing*, *13*(6), 1066. Available from https://doi.org/10.3390/rs13061066.

Ding, Y. J., Wu, P. C., & Lian, Y. H. (2020). Time series analysis for the dynamic relationship between an enterprise's business growth and carbon emission in Taiwan. *Sustainability*, *12*(14), 5560.

den Elzen, M. G., Hof, A. F., & Roelfsema, M. (2011). The emissions gap between the Copenhagen pledges and the 2 C climate goal: Options for closing and risks that could widen the gap. *Global Environmental Change*, *21*(2), 733–743.

FSI. (2021). *India State of Forest Report 2021*. Forest Survey of India, Dehradun.

Garg, A., Shukla, P. R., Kankal, B., & Mahapatra, D. (2017). CO_2 emission in India: Trends and management at sectoral, sub-regional and plant levels. *Carbon Management*, *8*(2), 111–123.

Gavenas, E., Rosendahl, K. E., & Skjerpen, T. (2015). CO_2 emissions from Norwegian oil and gas extraction. *Energy*, *90*, 1956–1966.

Greff, K., Srivastava, R. K., Koutník, J., Steunebrink, B. R., & Schmidhuber, J. (2016). LSTM: A search space odyssey. *IEEE Transactions on Neural Networks and Learning Systems*, *28*(10), 2222–2232.

Heinze, C., Meyer, S., Goris, N., Anderson, L., Steinfeldt, R., Chang, N., ... Bakker, D. C. (2015). The ocean carbon sink–impacts, vulnerabilities and challenges. *Earth System Dynamics*, *6*(1), 327–358.

Hosseini, S. M., Saifoddin, A., Shirmohammadi, R., & Aslani, A. (2019). Forecasting of CO_2 emissions in Iran based on time series and regression analysis. *Energy Reports*, *5*, 619–631.

Hsu, D. D., O'Donoughue, P., Fthenakis, V., Heath, G. A., Kim, H. C., Sawyer, P., … Turney, D. E. (2012). Life cycle greenhouse gas emissions of crystalline silicon photovoltaic electricity generation: Systematic review and harmonization. *Journal of Industrial Ecology*, *16*, S122–S135.

Jayanthakumaran, K., Verma, R., & Liu, Y. (2012). CO_2 emissions, energy consumption, trade and income: A comparative analysis of China and India. *Energy Policy*, *42*, 450–460.

Jena, P. R., Managi, S., & Majhi, B. (2021). Forecasting the CO_2 emissions at the global level: A multilayer artificial neural network modelling. *Energies*, *14*(19), 6336.

Kumar, M. K., & Nagendra, S. S. (2016). Quantification of anthropogenic CO_2 emissions in a tropical urban environment. *Atmospheric Environment*, *125*, 272–282. Available from https://doi.org/10.1016/j.atmosenv.2015.11.024.

Kumari, S., & Singh, S. K. (2023). Machine learning-based time series models for effective CO_2 emission prediction in India. *Environmental Science and Pollution Research*, *30*(55), 116601–116616. Available from https://doi.org/10.1007/s11356-022-21723-8.

Kunchala, R. K., Patra, P. K., Kumar, K. N., Chandra, N., Attada, R., & Karumuri, R. K. (2022). Spatio-temporal variability of XCO_2 over Indian region inferred from Orbiting Carbon Observatory (OCO-2) satellite and Chemistry Transport Model. *Atmospheric Research*, *269*, 106044.

Kweku, D. W., Bismark, O., Maxwell, A., Desmond, K. A., Danso, K. B., Oti-Mensah, E. A., … Adormaa, B. B. (2018). Greenhouse effect: Greenhouse gases and their impact on global warming. *Journal of Scientific Research and Reports*, *17*(6), 1–9.

Lal, M. (2003). Global climate change: India's monsoon and its variability. *Journal of Environmental Studies and Policy*, *6*(1), 1–34.

Lewis, C. (2021). *Power usage leads to 71% of greenhouse gas emissions in Mumbai*. The Times of India.

Mandal, S., Das, P., Behera, M. D., & Parida, B. R. (2022). Himalayan carbon flux demonstrates higher aseasonality GOSAT-derived spatio-temporal profile and comparison through wind and water vapour. In *Handbook of Himalayan Ecosystems and Sustainability* (Volume 1, pp. 183–201). CRC Press.

Murthy, I. K., Gupta, M., Tomar, S., Munsi, M., Tiwari, R., Hegde, G. T., & Ravindranath, N. H. (2013). Carbon sequestration potential of agroforestry systems in India. *Journal of Earth Science & Climatic Change*, *4*(1), 1–7.

Nalini, K., Uma, K. N., Sijikumar, S., Tiwari, Y. K., & Ramachandran, R. (2018). Satellite- and ground-based measurements of CO_2 over the Indian region: Its seasonal dependencies, spatial variability, and model estimates. *International Journal of Remote Sensing*, *39*(22), 7881–7900.

Pata, U. K., & Kumar, A. (2021). The influence of hydropower and coal consumption on greenhouse gas emissions: A comparison between China and India. *Water*, *13*(10), 1387.

Paul, P. (2009). *High carbon footprints in Kolkata, a major cause for worry*. Deccan Herald.

Poeschl, M., Ward, S., & Owende, P. (2012). Environmental impacts of biogas deployment–Part I: Life cycle inventory for evaluation of production process emissions to air. *Journal of Cleaner Production*, *24*, 168–183.

Ramachandra, T. V., Aithal, B. H., & Sreejith, K. (2015). GHG footprint of major cities in India. *Renewable and Sustainable energy reviews*, *44*, 473–495.

Reddy, C. S., Jha, C. S., Dadhwal, V. K., Hari Krishna, P., Vazeed Pasha, S., Satish, K. V., … Diwakar, P. G. (2016). Quantification and monitoring of deforestation in India over eight decades (1930–2013). *Biodiversity and conservation*, *25*, 93–116. Available from https://doi.org/10.1007/s10531-015-1033-2.

Ritchie, H., Roser, M., & Rosado, P. (2020). CO_2 and greenhouse gas emissions. *Our World in Data*.

Smagulova, K., & James, A. P. (2020). *Overview of long short-term memory neural networks. Deep learning classifiers with memristive networks* (pp. 139–153). Cham: Springer.

Thakur, A. (2017). Study of ambient air quality trends and analysis of contributing factors in Bengaluru, India. *Oriental Journal of Chemistry*, *33*(2), 1051–1056.

Udemba, E. N., Güngör, H., Bekun, F. V., & Kirikkaleli, D. (2021). Economic performance of India amidst high CO_2 emissions. *Sustainable Production and Consumption*, *27*, 52–60.

Venter, Z. S., Aunan, K., Chowdhury, S., & Lelieveld, J. (2020). COVID-19 lockdowns cause global air pollution declines. *Proceedings of the National Academy of Sciences*, *117*(32), 18984–18990.

Waheed, R., Chang, D., Sarwar, S., & Chen, W. (2018). Forest, agriculture, renewable energy, and CO_2 emission. *Journal of Cleaner Production*, *172*, 4231–4238.

Wang, Y., & Sun, T. (2012). Life cycle assessment of CO_2 emissions from wind power plants: Methodology and case studies. *Renewable Energy*, *43*, 30–36.

Zhang, Q., Li, F., Long, F., & Ling, Q. (2018). Vehicle emission forecasting based on wavelet transform and long short-term memory network. *IEEE Access*, *6*, 56984–56994.

Chapter 24

Role of ocean and climate in terrestrial ecosystems

Swadhin K. Behera[1], Roma Varghese[2,3], J.V. Ratnam[1], Takeshi Doi[1] and Mukunda Dev Behera[2]
[1]*Applications Laboratory, Research Institute for Value Added Information Generation, Japan Agency for Marine-Earth Science and Technology, Yokohama, Japan,* [2]*Centre for Ocean, River, Atmosphere and Land Sciences (CORAL), Indian Institute of Technology Kharagpur, Kharagpur, West Bengal, India,* [3]*Indian Institute of Tropical Meteorology, Pune, Maharashtra, India*

24.1 Introduction

Among various components of the Earth's climate system, the oceans play a pivotal role in regulating climate variations that resonate through terrestrial ecosystems. The dynamic interplay between oceanic processes and atmospheric conditions generates a complex form of interaction that basically regulates Earth's climate, with far-reaching consequences for the diverse landscapes and life forms that inhabit it. The oceans act as colossal heat reservoirs, absorbing, storing, and releasing heat over extended periods, thus mitigating rapid temperature fluctuations on land. However, the oceans' influence extends beyond the temperature modulation of the planet. Through their intricate currents, oceans distribute heat in a particular way around the planet, for example along the western boundaries of the oceans, regulating not only temperature but also precipitation patterns over the landmass. This distribution, in turn, shapes the climatic conditions that terrestrial ecosystems rely upon for their sustenance (Fig. 24.1). On the other hand, the rising air over the tropical oceans sinks over the subtropical regions, giving rise to some of the vast deserts on the planet. Hence, from the lush rainforests to the arid deserts, the oceans' role in moderating moisture and heat profoundly influences the distribution of habitats and species.

Understanding the intricate connections between oceans and terrestrial ecosystems is essential not only for unraveling the complex behavior of our climate but also for predicting and adapting to the changes that climate variations can bring. Also, ecosystems play a crucial role not only in the global carbon cycle but also in the climate, and not many studies have examined the changes in ecosystems associated with climate drivers. Particularly, few studies have analyzed how remote oceanic conditions affect land vegetation growth through atmospheric teleconnection, and the causes of the recent changes in the global ecosystem need further investigation. As humanity grapples with the challenges of a changing climate, from rising sea levels to altered precipitation patterns, comprehending the oceans' role in driving these shifts is crucial for devising strategies to mitigate their impacts on terrestrial environments, biodiversity, agriculture, and even our societies.

The climate factors that affect the terrestrial ecosystem most are precipitation, temperature, and radiation. Several studies have suggested ecosystems are temperature-limited at northern high latitudes, water-limited in arid and semiarid regions, and radiation-limited in the Amazon and eastern and southern Asia (e.g., Zhao, Dai, & Dong, 2018). On seasonal to interannual time scales, these climatic factors are driven by local as well as remote climate variations. For example, it is well documented that regional climates in parts of the world are affected by phenomena like El Niño/ Southern Oscillation (ENSO), which emerges from the tropical Pacific Ocean, triggering cascades of atmospheric changes. Similarly, the Indian Ocean Dipole (IOD) is a climate phenomenon characterized by SST anomalies in the equatorial Indian Ocean (Saji, Goswami, Vinayachandran, & Yamagata, 1999). The positive phase, known as a positive IOD, is marked by cold SST anomalies in the eastern Indian Ocean (Fig. 24.2A), and the negative phase, or negative IOD, is characterized by warm SST anomalies in the eastern Indian Ocean. The warm phase of ENSO, El Niño, is characterized by warm SST anomalies in the eastern Pacific Ocean (Fig. 24.2B). ENSO and IOD significantly influence regional climate patterns, impacting rainfall and temperature distributions in many parts of the world. For example, a positive IOD typically leads to increased rainfall in the western Indian Ocean, including Kenya and Somalia, while

Sustainable Development Perspectives in Earth Observation. DOI: https://doi.org/10.1016/B978-0-443-14072-3.00002-2

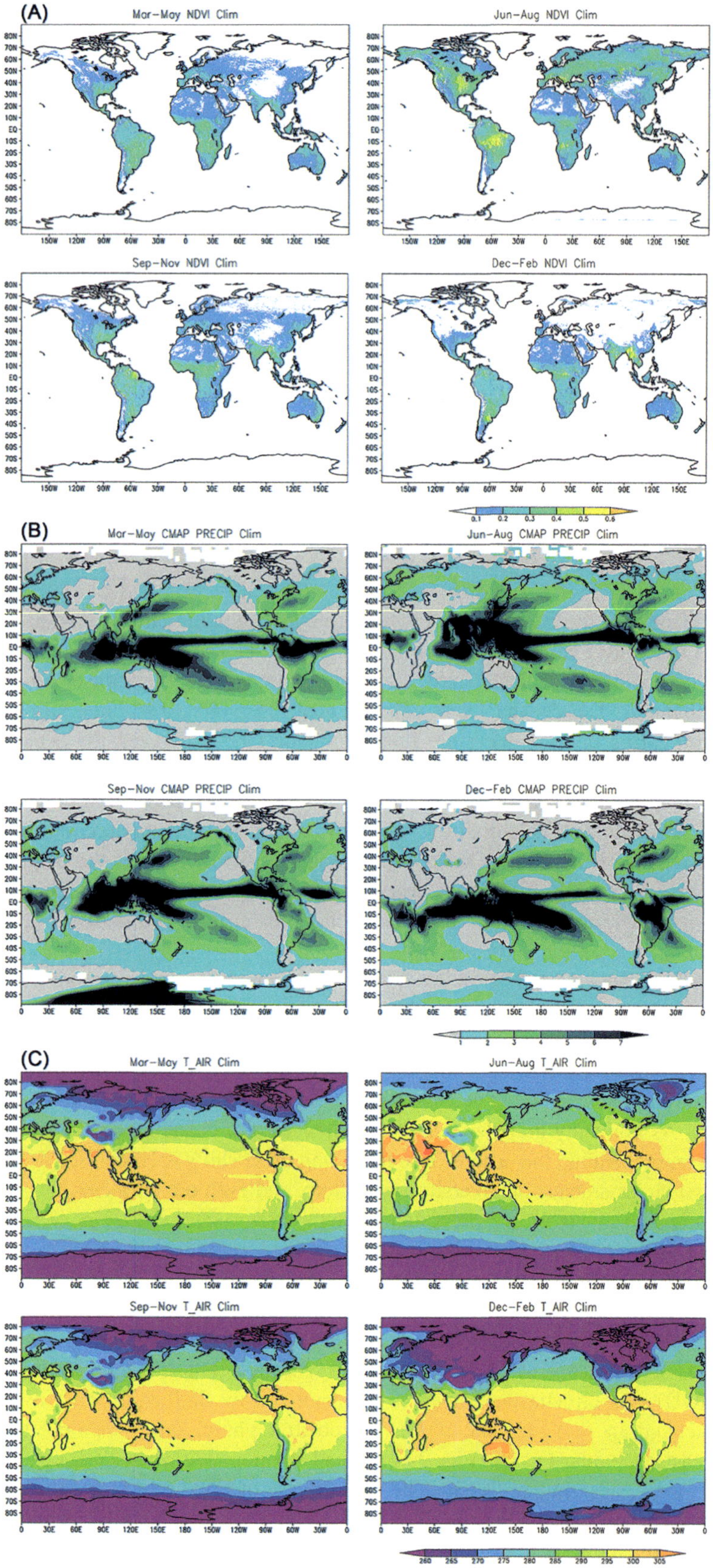

FIGURE 24.1 Seasonal march of (A) NDVI, (B) rainfall, and (C) surface air temperature for four seasons of boreal spring (Mar–May), summer (Jun–Aug), autumn (Sep–Nov) and winter (Dec–Feb). Units for rainfall is mm/day and temperature is K. Map lines delineate study areas and do not necessarily depict accepted national boundaries.

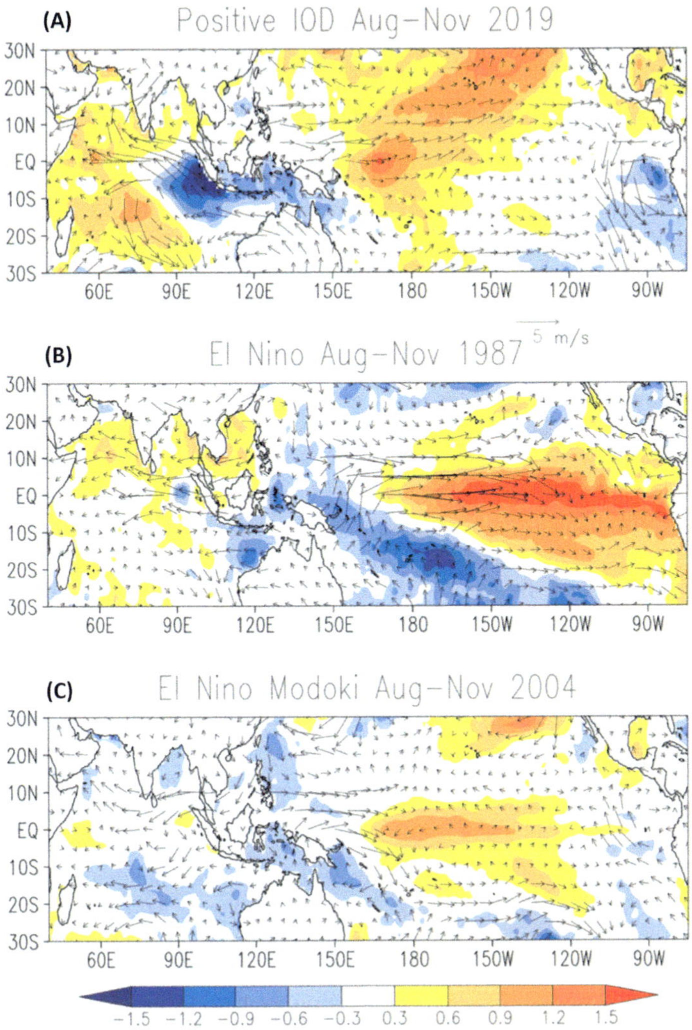

FIGURE 24.2 Anomalies of SST (shaded) and surface wind for (A) 2019 pure pIOD, (B) 1987 pure El Niño, and (C) 2004 pure El Niño Modoki. SST anomalies are in °C. Map lines delineate study areas and do not necessarily depict accepted national boundaries.

contributing to drought conditions in the eastern Indian Ocean region, including Australia and Indonesia. Conversely, a negative IOD tends to bring about opposite effects, with increased rainfall in the western Indian Ocean and drier conditions in the eastern part. Both phases affect the monsoon rainfall over India (Behera & Ratnam, 2018; Behera, Krishnan, & Yamagata, 1999). Similarly, an El Niño (La Niña) causes draughts (floods) in Indonesia and floods in Peru (draught). Another mode of ocean-atmosphere variations, known as the ENSO Modoki (Ashok, Behera, Rao, Weng, & Yamagata, 2007), exists in the tropical Pacific Ocean. In El Niño Modoki/La Niña Modoki events, the warming/cooling occurs in the central Pacific (Fig. 24.2C), while the eastern and western parts remain cooler/warmer. This unique pattern results in different climate impacts compared to traditional El Niño or La Niña events. ENSO Modoki can influence global weather patterns, affecting precipitation and temperature distribution in various regions. They can have significant impacts on agriculture, fisheries, and water resources, making them an important focus for climate research and prediction efforts. Therefore, observation and prediction of these climate phenomena are crucial for developing necessary mitigation measures.

These climate phenomena, like ENSO, IOD, and ENSO Modoki, have significant impacts on terrestrial vegetation activity through atmospheric teleconnections. In one of the early studies, Nagai, Ichii, and Morimoto (2007), using the satellite-observed Normalized Difference Vegetation Index (NDVI), suggested that a large portion of the tropical rainforest variations are explained by ENSO. In another study, Lyon and Barnston (2005) found that major peaks in the spatial extent of drought and excessively wet conditions are generally associated with extreme phases of El Niño and La Niña, respectively. A remarkably robust linear relationship was found between the spatial extent of drought in the tropics and El Niño strength, while such a relationship was comparatively weaker between La Niña and excessive wetness. Globally, ENSO is shown to be a leading climatic driver of the interannual variability of NDVI, especially over southern and eastern Africa, eastern Australia, northeastern Asia, and northern South America (e.g., Zhao et al., 2018). Regionally, El Niño events were found to be associated with the negative NDVI anomaly in central India (Bothale & Katpatal, 2014) and northeastern Brazil (Erasmi, Schucknecht, Barbosa, & Matschullat, 2014). The impact of ENSO on vegetation dynamics in Africa is indicated through its influences on rainfall, solar radiation, and temperature (Philippon, Martiny, Camberlin, Hoffman, & Gond, 2014). In general, warm ENSO events are associated with decreased greenness in Australia, Southeast Asia, northeastern South America, and southern Africa, but increased greenness in eastern Africa, central Asia, and northern North America (Miralles, van den Berg, & Gash, 2014; Woodward, Lomas, & Quaife, 2008). van Leeuwen, Orr, Marsh, and Herrmann (2006) indicated ENSO impacts on vegetation productivity and phenology in southeastern and northeastern Argentina, central and southern Chile, and Paraguay.

In a recent study, Arango, Lintner, Carvalho, and Lyon (2021) found similar relations while investigating the spatial extents of tropical droughts during El Niño in current and future climates using multiple observational and reanalysis products and ensembles of simulations from the Coupled Model Intercomparison Project Phase 5 (CMIP5). CMIP5 models forced with prescribed sea surface temperatures compare well to observations in capturing the increasing tropical drought area with increasing ENSO warm phase (El Niño) strength. The results from the ensemble of RCP8.5 simulations suggest lower El Niño strength-tropical drought area sensitivity at the end-of-the-century. Some of this lower sensitivity is attributed to the atmosphere-ocean coupling. While exploring such links with global warming, Dai (2013) highlighted the increasing intensity and frequency of drought events across the globe. Particularly, he found a global trend toward more intense and persistent droughts in various regions of the world over the last few decade. This increase in drought severity is particularly evident in regions that are already prone to dry conditions.

In the context of the Indian Ocean sector, the impact of ocean and climate factors on Indian vegetation is a complex interaction that involves various oceanic and atmospheric processes. The Indian subcontinent's climate and vegetation are influenced by multiple factors, including the Indian Ocean, monsoon patterns, ENSO, the Indian Ocean Dipole (IOD), and other sea surface temperature patterns (Varghese, Behera, & Behera, 2023). With rich natural vegetation, it is vital to understand the involvement of different climatic variables in the vegetation anomalies across India. Most of the previous studies were focused on the effects of rainfall and temperature on various structural and functional characteristics of Indian vegetation (Asoka & Mishra, 2015; Banger et al., 2015; Nayak, Patel, & Dadhwal, 2013; Sharma & Goyal, 2018). However, compared to these meteorological parameters, the slowly evolving sea surface temperature (SST) anomalies could have better predictive skills for the physiological traits of vegetation. But the connection between SST variations and Indian vegetation has not been thoroughly studied. In addition to the teleconnections arising from ENSO and the IOD, other patterns of SST anomalies in the tropical Indian Ocean (TIO) may also contribute to the vegetation anomalies in India. In this context, a recent study by Varghese et al. (2023) demonstrated a strong negative SST-solar-induced chlorophyll fluorescence (SIF) relationship. They found that the SIF anomalies in the northern and central Indian regions decrease (increase) when the SST of the tropical Indian Ocean is higher (lower) than normal. This SST-SIF link was found to be most pronounced during June in their study and is attributed to the atmospheric teleconnections during the first phase of the Indian summer monsoon.

In summary, oceanic and climate factors, particularly those related to the Indian Ocean and monsoon dynamics, play a vital role in shaping vegetation patterns and ecosystems in India. While several studies have examined the ENSO influence on terrestrial vegetation, the IOD and ENSO Modoki influences remain unclear. Therefore, it is vitally important to understand these regional SST patterns in addition to those related to ENSO for developing effective prediction systems, managing ecosystems, sustaining agriculture, and preparing for potential impacts from climate variation and change. In this investigation into the oceans' impact on climate variations and their consequences for terrestrial ecosystems, we intricately examine the mechanisms by which oceans shape climate dynamics. We then explore the resulting effects on terrestrial ecosystems and discuss strategies to enhance resilience in the context of a continuously evolving climate. In the following sections, we reviewed the current understanding of some of the large-scale climate phenomena as well as regional SST patterns and their role in vegetation and its predictability.

24.2 Data

Monthly gridded station precipitation and temperature data were obtained from the NOAA NCEP CPC CAMS climate anomaly monitoring system (Ropelewski, Janowiak, & Halpert, 1985). SST data were collected from the high-resolution product of the NOAA 0.25 degrees daily Optimum Interpolation Sea Surface Temperature version 2 (Huang et al., 2021). Monthly soil moisture data were derived from the Famine Early Warning Systems Network (FEWS NET) land data assimilation system (McNally & NASA/GSFC/HSL, 2018). The Normalized Difference Vegetation Index (NDVI), a well-known vegetation index, is computed as NDVI = (Ch2 − Ch1)/(Ch2 + Ch1), where Ch1 and Ch2 are the surface reflectance from AVHRR channels 1 and 2, respectively (e.g., Tarpley, Schneider, & Money, 1984). Arithmetically, NDVI ranges from −1.0 to 1.0, but actual values at the Earth's surface range from −0.1 to 0.7. The remotely sensed value of the NDVI allows the 'greenness' of vegetation to be analyzed at regional to global scales. NDVI data provide spatially continuous, long-term information on vegetation and are thus useful for studying seasonal vegetation cycles (e.g., Gutman, 1999). Monthly anomalies were derived by removing the monthly seasonal cycle from all the data sets, and then seasonal mean anomalies were constructed by taking the seasonal averages of the monthly anomalies for the respective seasons.

The climate indices of IOD, ENSO, and ENSO Modoki were derived from the SST anomalies. The ENSO index, Nino3, was derived by taking SST anomalies averaged over 5°S–5°N and 150°W–90°W. The IOD Index, DMI, was derived from NOAA NCEP Global SST analysis by taking the difference between SST anomalies over the western equatorial Indian Ocean (60°E–80°E, 10°S–10°N) and the eastern equatorial Indian Ocean (90°E–110°E, 10°S–Equator) as defined in Saji et al. (1999). The ENSO Modoki index, EMI, was derived by taking a difference of central Pacific SST anomalies (area average of 165°E–140°W, 10°S–10°N) from half of the eastern Pacific (area average of 110°W–70°W, 15°S–5°N) and half of the western Pacific (area average of 125°E–145°E, 10°S–20°N) following Ashok et al. (2007).

24.3 Climate teleconnections to terrestrial vegetation anomalies

The seasonal variations in the global vegetation cover are depicted in the seasonal plots of NDVI (Fig. 24.1A). The spatio-temporal transition of the vegetation cover is compared with the simulated seasonal transition of precipitation and surface air temperature (Fig. 24.1B and C), as the phenological features of vegetation are closely linked to climatological seasonal cycles. It may be noted here that the seasonal march of NDVI in different parts of the world generally follows the corresponding seasonal march in temperature and precipitation. Several studies have already pointed out that plants in mid- and high-latitudes basically start foliation in response to the rise in temperature from spring to summer and then gradually decay as temperatures fall in autumn and winter. Similarly, vegetation in tropical areas follows the seasonal cycle of precipitation more closely. In contrast, the dry regions do not show a marked seasonal cycle, nor does the leaf mass of the canopy of evergreen forests clearly change with the seasons.

Tropical modes of climate variability are well known for their substantial impact on terrestrial vegetation at different spatial and temporal scales, as discussed earlier in this article. These climate modes affect rainfall, temperature, and soil moisture directly through changes in atmospheric circulation. For example, ENSO affects the rainfall of Indonesia, South America, and Africa through changes in the atmospheric Walker circulation generated by the east–west movement of air masses along the equator. The Pacific Walker circulation is generated by the warm air that rises in the western Pacific, travels eastward in the upper troposphere, descends in the central and eastern Pacific, and returns westward near the surface. Similar vertical cells exist in the Indian and Atlantic Oceans. The anomalous changes in the Walker circulations of these ocean basins, associated with tropical modes of climate variations, alter the rainfall, as well as temperature and soil moisture, in those basins, thereby affecting the vegetation. Besides, these tropical climate modes affect the rainfall, temperature, and soil moisture of faraway regions through atmospheric teleconnections, a common process in which the atmospheric variability in a region affected by the rainfall variations gets projected onto mid-latitude wave guides to carry the signals around the globe. There are other modes of atmospheric teleconnections as well, when a series of atmospheric high-pressure and low pressure anomalies tied to each other carry the signal from the tropics to the high latitudes.

Here, we tried to understand the impacts of ENSO, IOD, and ENSO Modoki by considering typical years of their unique developments in the tropical Pacific and Indian Oceans. For this purpose, we have chosen the years of 2019, 1987, and 2004 as typical examples of positive IOD, El Niño, and El Niño Modoki years, respectively. Though we are trying to understand their unique impact on terrestrial vegetation (based on NDVI), we did not have many years of these climate phenomena in the short observational record (Fig. 24.2) of climate and NDVI data to pick more years for statistical analyses. So, in most of this article, we will discuss the vegetation variability in the context of these three typical IOD, ENSO, and ENSO Modoki years.

24.3.1 The Indian Ocean dipole impact on vegetation

The positive IOD (pIOD), characterized by an anomalously warm western tropical Indian Ocean and anomalously cold southeastern, is generally observed to develop in the boreal spring (March–May) and end by winter (December–February), though there exist some degrees of event-to-event variability (Fig. 24.3). Consecutive pIODs have also been observed in recent decades (Fig. 24.3), perhaps related to the changes in the tropical Indian Ocean related to global warming. Conversely, a negative phase (nIOD) occurs when anomalies exhibit opposite signs, evident in years when the index is negative (as depicted in Fig. 24.3) in contrast to pIOD conditions. However, the amplitude of the pIOD is generally larger than that of the nIOD (Behera, Doi, & Ratnam, 2021; Hong, Li, LinHo, & Kug, 2008). This intrinsic mode of ocean-atmospheric interaction within the tropical Indian Ocean is observed to influence anomalies in land photosynthesis across countries on the Indian Ocean rim (Wang et al., 2021).

The 2019 pIOD is one of the most extreme IOD events of the century (Behera et al., 2021) that began in the boreal spring and was already very strong by boreal summer, while it peaked in boreal fall. It remained quite strong until late boreal winter, and the warm SST anomalies of its western pole remained in the western Indian Ocean even after its decay. Those finally disappeared at the beginning of 2020. The peak phase is demonstrated in the anomalies of SST and horizontal wind averaged for August–November in 2019 (Fig. 24.2A). The strong surface southeasterly winds support evaporative cooling and coastal upwelling off the Sumatra-Java Islands, explaining the cold SST anomaly in the eastern tropical Indian Ocean. This led to reduced (enhanced) rainfall in the eastern tropical Indian Ocean, as discussed later in this section.

The anomalies of NDVI are analyzed to understand the IOD impacts on terrestrial vegetation growth. The anomalies are investigated during each phase of the pIOD in 2019 to track their spatial and temporal variability associated with the event (Figs. 24.4A, 24.5A, and 24.6A). As noted in those figures, enhanced NDVI anomalies were predominantly observed in the global terrestrial ecosystems, particularly in low and mid-latitudes. Large spatio-temporal variability is also observed across the globe in NDVI anomalies related to IOD. For example, the strong positive anomalies of NDVI detected across Russia, Europe, central South Africa, and America during the early pIOD phase started deteriorating in the following seasons. On the contrary, robust positive NDVI anomalies started appearing in the monsoon regions of Asia and Africa, while weak negative NDVI anomalies became evident in the polar terrestrial ecosystems during the peak phase of pIOD. Peak responses of the NDVI anomalies are visible, with a seasonal lag mostly in the Asian-African ecosystems (Figs. 24.4A, 24.5A, and 24.6A) as the positive anomalies persisted from boreal summer to winter in these areas. While some regional diversities are observed, most of these identified vegetation growth anomalies were either linked to relatively wet and/or warm temperatures associated with the pIOD, as demonstrated in the anomalies of precipitation and air temperature (Figs. 24.7A, 24.8A, and 24.9A), as well as soil moisture (not shown). The tropical teleconnection to mid- and high-latitude temperatures and rainfall anomalies associated with IOD and ENSO are discussed in several studies. For example, Behera, Ratnam, Masumoto, and Yamagata (2013) found that the extreme events in Eurasia are associated with a circumglobal teleconnection determined by the location of the tropospheric heating, which in turn is associated with the rainfall anomalies associated with IOD.

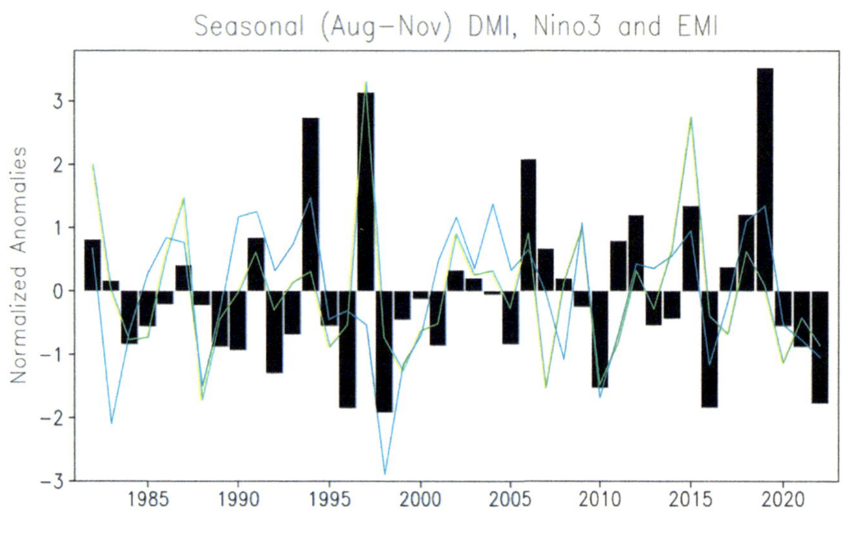

FIGURE 24.3 August–November averaged indices of IOD (bar), El Niño (green line), and El Niño Modoki (cyan line).

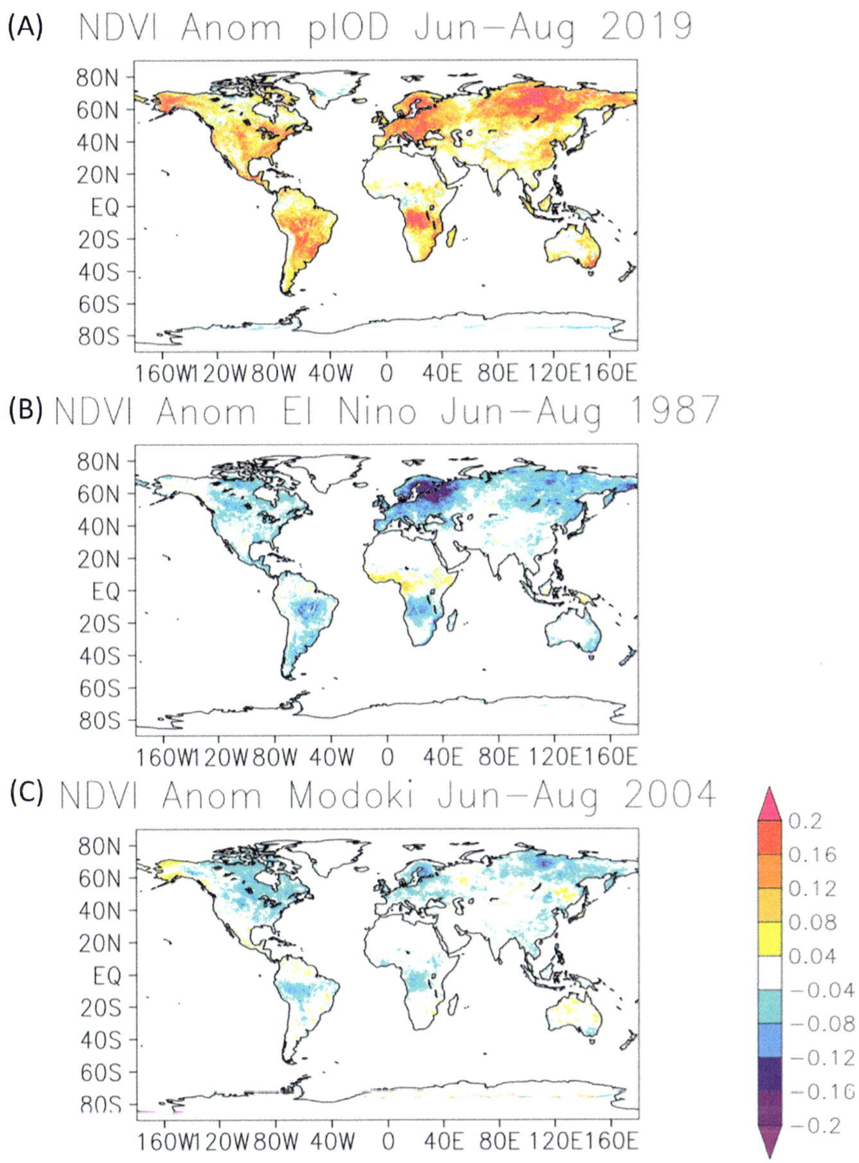

FIGURE 24.4 Same as Fig. 24.2 but for NDVI anomalies during boreal summer (Jun–Aug) season. Map lines delineate study areas and do not necessarily depict accepted national boundaries.

The detected NDVI anomalies in high latitudes during the peak phases of pIOD were consistent with the influence of air temperature on vegetation growth. In general, air temperature plays a crucial role in influencing plant growth and development. However, the relationship between air temperature and vegetation growth is complex and can vary depending on several factors, including the type of vegetation, local climate conditions, and specific physiological characteristics of the plants. For example, temperature affects the rate of photosynthesis, a fundamental process for plant growth. Most plants have an optimal temperature range for photosynthesis, and variations outside this range can impact their ability to convert sunlight into energy. Warmer temperatures generally increase the rate of photosynthesis, up to a certain point. The air temperature also influences the length and timing of the growing season. In colder climates, warmer temperatures may extend the growing season, allowing plants to grow for a longer period of time. In warmer climates, on the contrary, extremely high temperatures can limit plant growth. In our study, we found negative NDVI anomalies in northern high latitudes during the DJF season (Fig. 24.6A), despite elevated anomalies of air temperature on the Eurasian continent. This could be because of the critical role solar radiation plays in the vegetation activity of high-latitude ecosystems associated with the seasonal cycle of solar insolation. At the same time, we find positive NDVI anomalies over Europe and Central Africa associated with positive precipitation anomalies. Positive anomalies of both precipitation and soil moisture were maximum during the SON season, i.e., the

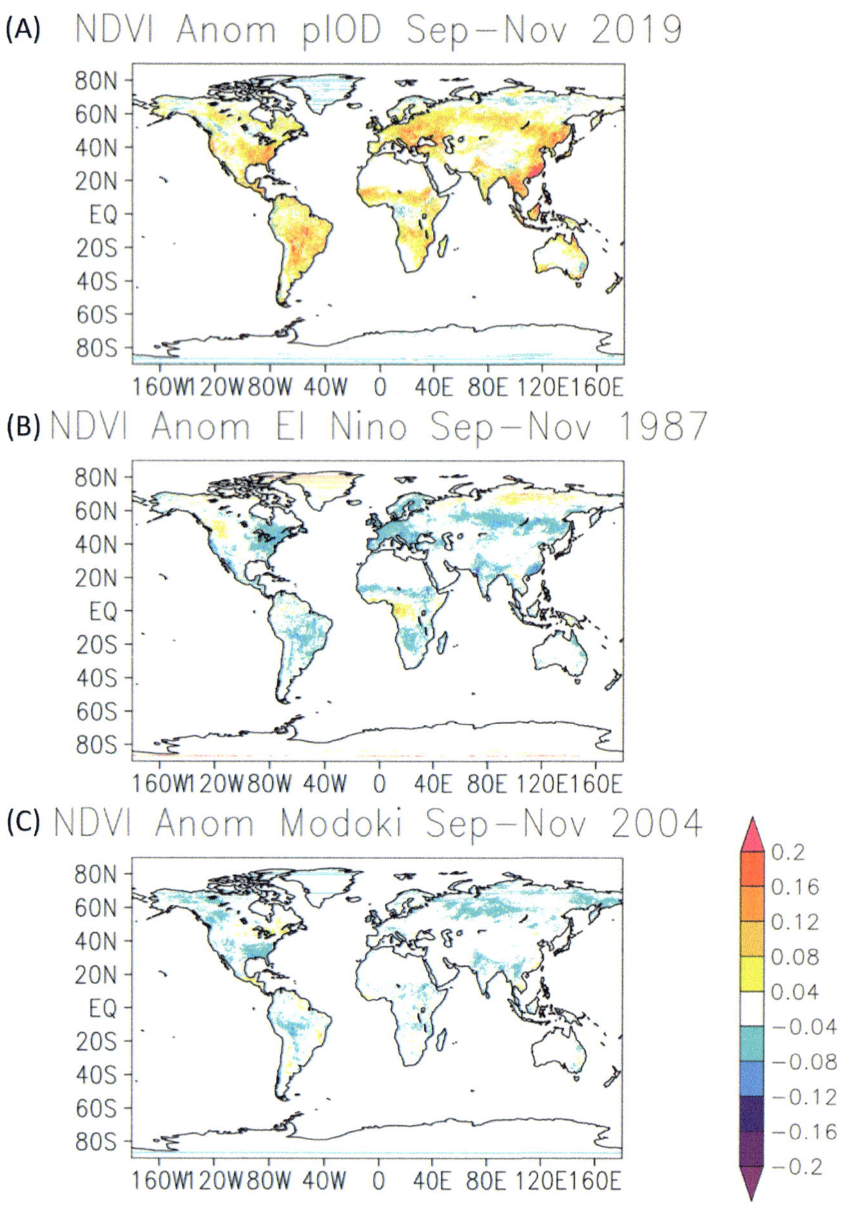

FIGURE 24.5 Same as Fig. 24.4 but fore boreal autumn (Sep–Nov) season. Map lines delineate study areas and do not necessarily depict accepted national boundaries.

peak phase of the pIOD. While enhanced precipitation anomalies began to deteriorate in the following season, soil moisture anomalies remained relatively stable. However, arid soil conditions are also seen in a few places, suggesting the contribution of soil moisture anomalies to the terrestrial vegetation variability in these cold deserts. Also, the observed seasonal lag in the impact of IOD on vegetation is in the response of growth anomalies, indicative of the persistence of the IOD impact through the decay season.

24.3.2 The El Niño/southern oscillation impact on vegetation

ENSO stands out as the primary driver of interannual climate variability on Earth. The alternating warm (El Niño) and cold (La Niña) phases of the ENSO cycle exhibit a close and intricate connection with hydrologic fluctuations worldwide, operating at diverse spatial and temporal scales. El Niño, characterized by abnormally warm surface waters in the eastern equatorial Pacific (Fig. 24.2B), exerts a significant adverse impact on global vegetation. Given its status as the most influential factor driving anomalies in global vegetation, the effects of El Niño on various vegetation biophysical properties have been extensively monitored in recent years. To comprehend the spatio-temporal characteristics of

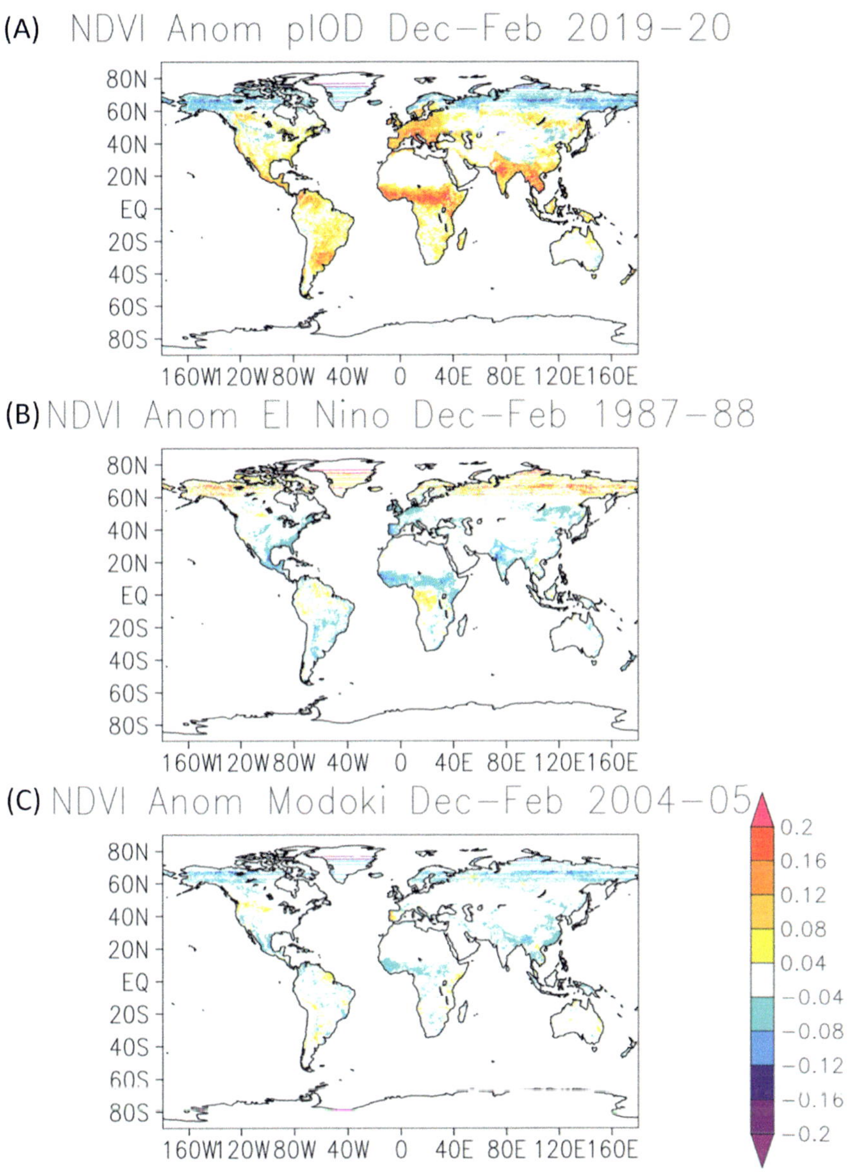

FIGURE 24.6 Same as Fig. 24.4 but for boreal winter (Dec–Feb) season. Map lines delineate study areas and do not necessarily depict accepted national boundaries.

terrestrial vegetation variability linked to El Niño events, we conducted a detailed examination of NDVI (normalized difference vegetation index) anomalies triggered by one of the pure El Niño events of 1987.

The response of global NDVI anomalies to the El Niño signal observed in our analyses can be broadly categorized into three categories. The first category involves strong negative NDVI anomalies observed in America, Eurasia, and eastern Australia during the boreal summer. Compared to other seasons, the El Niño influence on vegetation variability across these regions was maximum in the boreal summer (Fig. 24.4B), and surface air temperature anomalies (Fig. 24.7B) appear to have a dominant role in mediating El Niño teleconnections during this period. During the subsequent seasons, the initially observed negative NDVI anomalies weakened, while strong positive anomalies of NDVI began to appear in the terrestrial ecosystems of the northern high latitudes (Figs. 24.5B and 24.6B). The detected NDVI anomalies were mostly associated with negative air temperature anomalies in those regions. But the absence of clouds, as indicated by no positive precipitation anomalies, could have extended the plant-growing season in those higher latitudes, giving rise to higher-than-normal terrestrial vegetation.

The second category involves the delayed response of NDVI anomalies in the South Asian subcontinent to the El Niño signal, persisting from boreal autumn to boreal winter. An extremely dry hydroclimate situation persisted in those

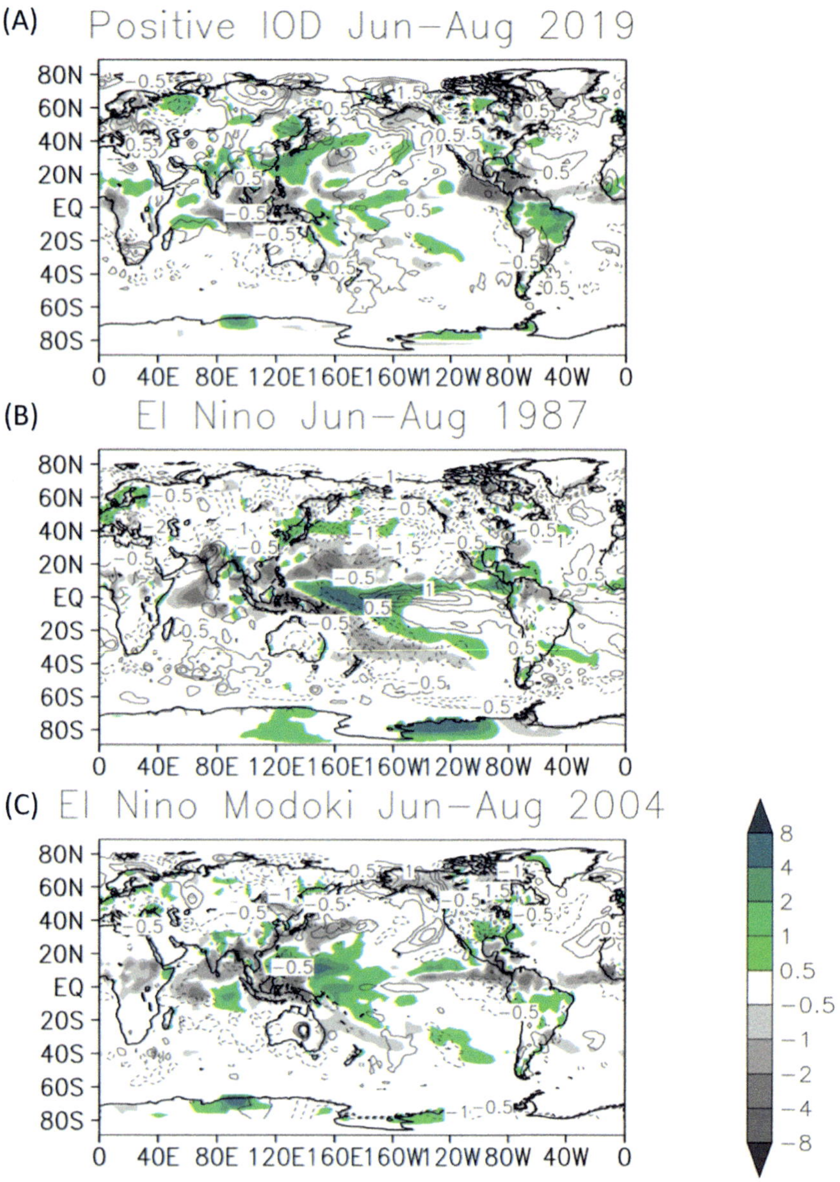

FIGURE 24.7 Same as Fig. 24.4 but for anomalies of rainfall (shaded) and surface air temperature (contour). Units for rainfall anomalies is mm/day and surface air temperature anomalies is K. Map lines delineate study areas and do not necessarily depict accepted national boundaries.

regions during the boreal summer, depicted by negative anomalies in precipitation (Fig. 24.7B and C) and soil moisture (not shown), as well as positive anomalies in air temperature. The peak NDVI anomalies observed during the SON and DJF climate seasons across India and Pakistan are a response to this preexisting climate condition induced by El Niño signals. Significant peaks in the spatial extent of drought in tropical regions are primarily linked to extreme phases of El Niño. Lyon and Barnston (2005) observed a robust linear relationship between drought extent in the tropics and the strength of El Niño, with a weaker connection for La Niña and excessive wetness in the opposite phase, revealing a doubling of both conditions between strong and weak ENSO events, especially in severe categories.

The last category pertains to the unique pattern of NDVI anomalies observed over Africa concerning the El Niño phase transitions. During the boreal summer, anomalies across the African continent showed a dipole pattern, with positive values in the equatorial belt and negative values further south (Fig. 24.4B). In the following seasons, observed positive NDVI anomalies gradually shifted southwards, while negative NDVI anomalies in the south moved northwards (Figs. 24.5B and 24.6B). By boreal winter, we can observe almost the same dipole pattern of NDVI anomalies with the opposite polarity (Fig. 24.6B) in equatorial and northern subtropical regions. The associated hydroclimatic conditions

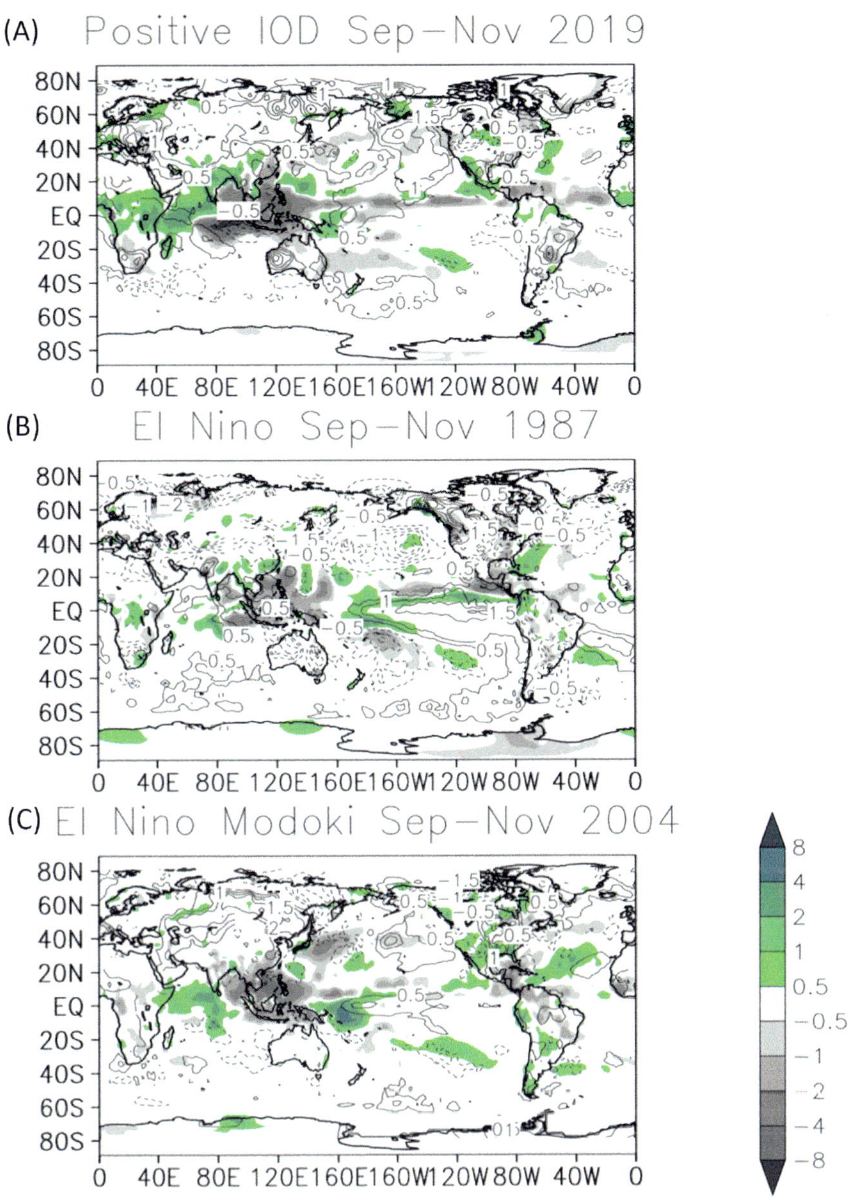

FIGURE 24.8 Same as Fig. 24.7 but for boreal autumn season. Map lines delineate study areas and do not necessarily depict accepted national boundaries.

during these transitions are more complex to explain the NDVI anomalies. The switch in the wetter and drier than normal conditions (Figs. 24.7B, 24.8B, and 24.9B) between the tropical and subtropical regions only to some extent explains the corresponding switch in the dipolar NDVI anomalies.

24.3.3 The El Niño/southern oscillation Modoki impact on vegetation

The ecological implications of El Niño are widely investigated; however, there is not much evidence on how the global vegetation anomalies respond to El Niño Modoki, a recently acknowledged face of coupled ocean-atmospheric phenomena in the tropical Pacific. The El Niño Modoki, though sometimes considered a central Pacific type El Niño, is characterized by a tripole pattern of SST anomalies in the tropical Pacific, with warm anomalies flanked by cold anomalies on the eastern and western sides, and its dynamics are entirely different from a canonical eastern Pacific type El Niño event. Therefore, El Niño Modoki is linked to a double-cell anomalous Walker circulation with rising motion in the central parts and sinking motion on both sides of the Pacific basin (Weng, Ashok, Behera, Rao, & Yamagata, 2008).

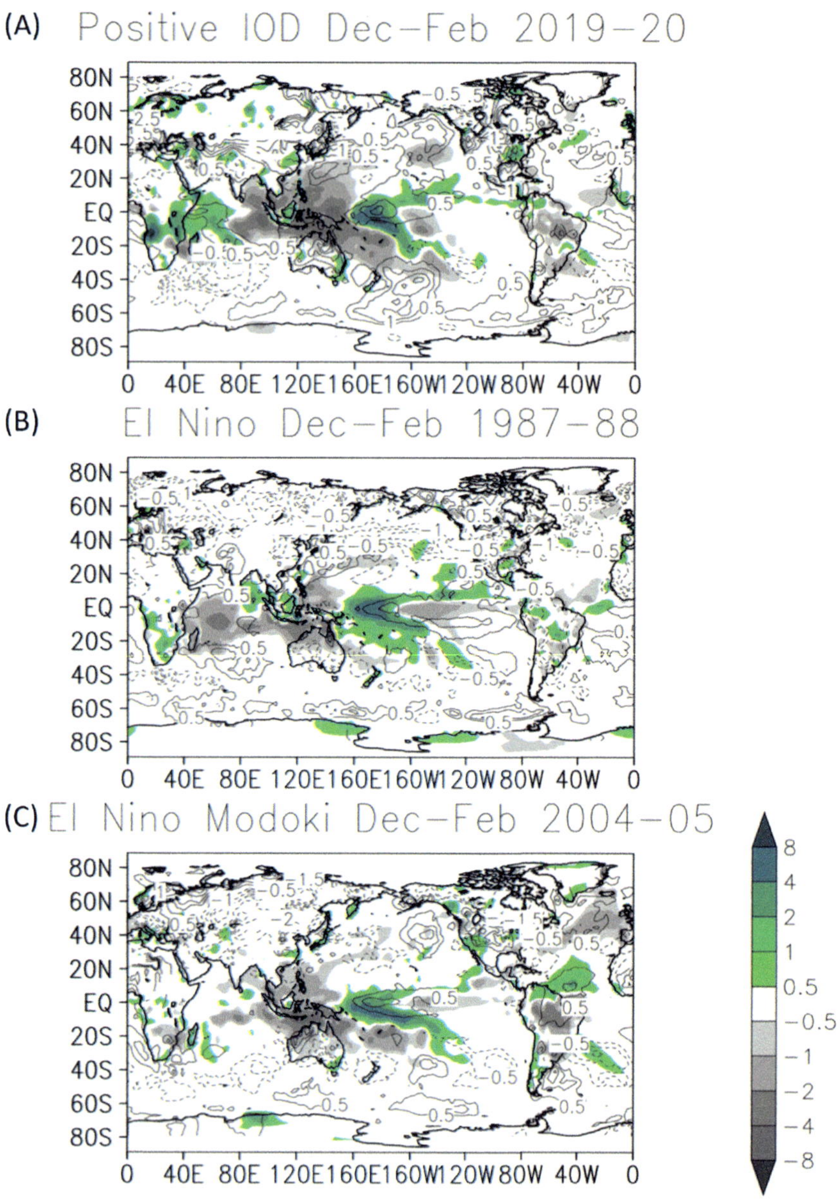

FIGURE 24.9 Same as Fig. 24.7 but for boreal winter season. Map lines delineate study areas and do not necessarily depict accepted national boundaries.

The characteristic SST pattern associated with El Niño Modoki normally persists from boreal summer through the following winter; however, peak phases can occur in any one of these seasons (Ashok et al., 2007).

Here we have considered the pure El Niño Modoki of 2004 for the analysis, and we found that El Niño Modoki also exerts a negative influence on global NDVI anomalies like El Niño. However, the magnitude of its impact appears to be less compared to that of a canonical El Niño event on a global scale. Nevertheless, the teleconnections seem to be prominent for the vegetation variability at the regional scale. Particularly during the boreal summer of the Modoki event, negative NDVI anomalies were pronounced in most parts of Europe, eastern Russia, southern Africa, and most parts of Brazil and Argentina (Fig. 24.4C). Meanwhile, positive NDVI anomalies were observed over Australia, Alaska, and northeast China. Like the El Niño teleconnections, surface air temperature anomalies acted as the mediators of El Niño Modoki signals in the NDVI anomalies during the boreal summer (Fig. 24.7C).

The observed NDVI anomalies in monsoon regions of Asia are different from other parts of the globe. Reduced NDVI anomalies appeared in eastern China and the southeast Asian mainland during the boreal summer and winter. In South Asia, El Niño Modoki imprints were observed in the negative NDVI anomalies over Himalayan regions after the boreal summer. The negative anomalies of NDVI were generally associated with lower-than-normal temperatures and precipitation, explaining the temperature and water stress in the high-altitude regions. In general, the observed NDVI anomalies are weaker for boreal autumn and winter seasons in all regions of the world.

24.3.4 Expected future changes and impacts

Climate models generally suggest that the IOD may intensify under global warming scenarios (Cai, Santoso, & Wang, 2014). This could result in more frequent and stronger pIOD events, characterized by warmer sea surface temperatures in the western Indian Ocean and cooler temperatures in the eastern Indian Ocean, as has already been observed in recent decades (Cai, Yang, & Wu, 2021). As discussed earlier in reference to the 2019 pure pIOD event, the pIOD events are associated with increased rainfall in some regions, like eastern Africa and parts of India, and drought in others, like Indonesia and Australia. These changes in precipitation patterns can affect vegetation growth, influencing ecosystems and agricultural productivity in some of the mentioned regions. In addition, vegetation in the high latitudes will also be affected by the pIOD teleconnections and associated changes in surface air temperature.

While projections for ENSO under future climate scenarios are uncertain, some models suggest an increase in the frequency and intensity of extreme ENSO events. This may include stronger El Niño events and more frequent La Niña events. El Niño tends to bring drier conditions to regions like, e.g., Australia, Southeast Asia, and parts of South America, potentially leading to drought stress and impacting vegetation. La Niña, on the other hand, is associated with increased rainfall in some areas, influencing vegetation growth.

The future behavior of ENSO Modoki is also notwell understood, and projections vary. Some studies suggest changes in the frequency and intensity of Modoki events. ENSO Modoki events can influence temperature and precipitation patterns and affect the vegetation of several parts of the world in boreal summer and the high-altitude regions of Himalaya during boreal spring and winter seasons. It is important to note that the impacts of climate variation on vegetation will depend on regional responses to changes in temperature, precipitation, and extreme weather events. Nevertheless, the interaction among these climate phenomena can lead to complex and sometimes nonlinear effects on ecosystems.

24.4 Summary

The Earth's climate system, primarily driven by oceanic processes, intricately regulates climate variations impacting terrestrial ecosystems. Oceans acting as vast heat reservoirs not only modulate global temperatures but also influence precipitation patterns through intricate currents. Understanding these connections is crucial for predicting and adapting to climate changes, which significantly affect biodiversity, agriculture, and societies. Climate phenomena such as ENSO, IOD, and ENSO Modoki have profound impacts on terrestrial vegetation, shaping global and regional ecosystems through atmospheric teleconnections. One important finding of this study is that tropical modes of climate variability like ENSO, IOD, and ENSO Modoki have a substantial influence on terrestrial vegetation at mid- and high-latitudes. The NDVI anomalies in those higher latitudes associated with pIOD are found to be opposite of what is seen in the cases of El Niño and El Niño Modoki. The response of vegetation indices to climate signals is maximum during boreal summer in America, Europe, Russia, Australia, and central and southern Africa. While in the monsoon regions of Asia, there is a seasonal lag in the response of NDVI anomalies to the tropical climate modes; NDVI anomalies were predominantly observed during SON and DJF. This observation made from individual event years is also supported by a correlation analysis made with individual climate indices of IOD, ENSO, and ENSO Modoki (Figs. 24.10 and 24.11). Particularly, the link to climate phenomena in some of the regions is much better represented in the correlation analyses; for example, IOD and ENSO Modoki impacts on the NDVI of Australia, compared to what is seen in the case of individual events. However, there is a possibility that the observed correlations may be influenced by concurrent events, such as the simultaneous occurrences of IOD, ENSO, ENSO Modoki, and various other climate phenomena.

Compared to the imprints of pIOD and El Niño, the implication of El Niño Modoki on terrestrial vegetation variability is less prominent at the planetary level. However, the negative influence of El Niño Modoki on NDVI anomalies across east and southeast Asian ecosystems is stronger than a canonical El Niño event, especially during the boreal summer and boreal winter. El Niño Modoki imprints in the global NDVI anomalies were mediated through anomalies of surface air temperature and precipitation. The influence of these phenomena on vegetation dynamics underscores the importance of observing and predicting them for effective climate mitigation and adaptation measures. Particularly, the strong impacts seen in high-latitude regions need a comprehensive study to understand the tropical ocean teleconnections in polar ecosystem functioning.

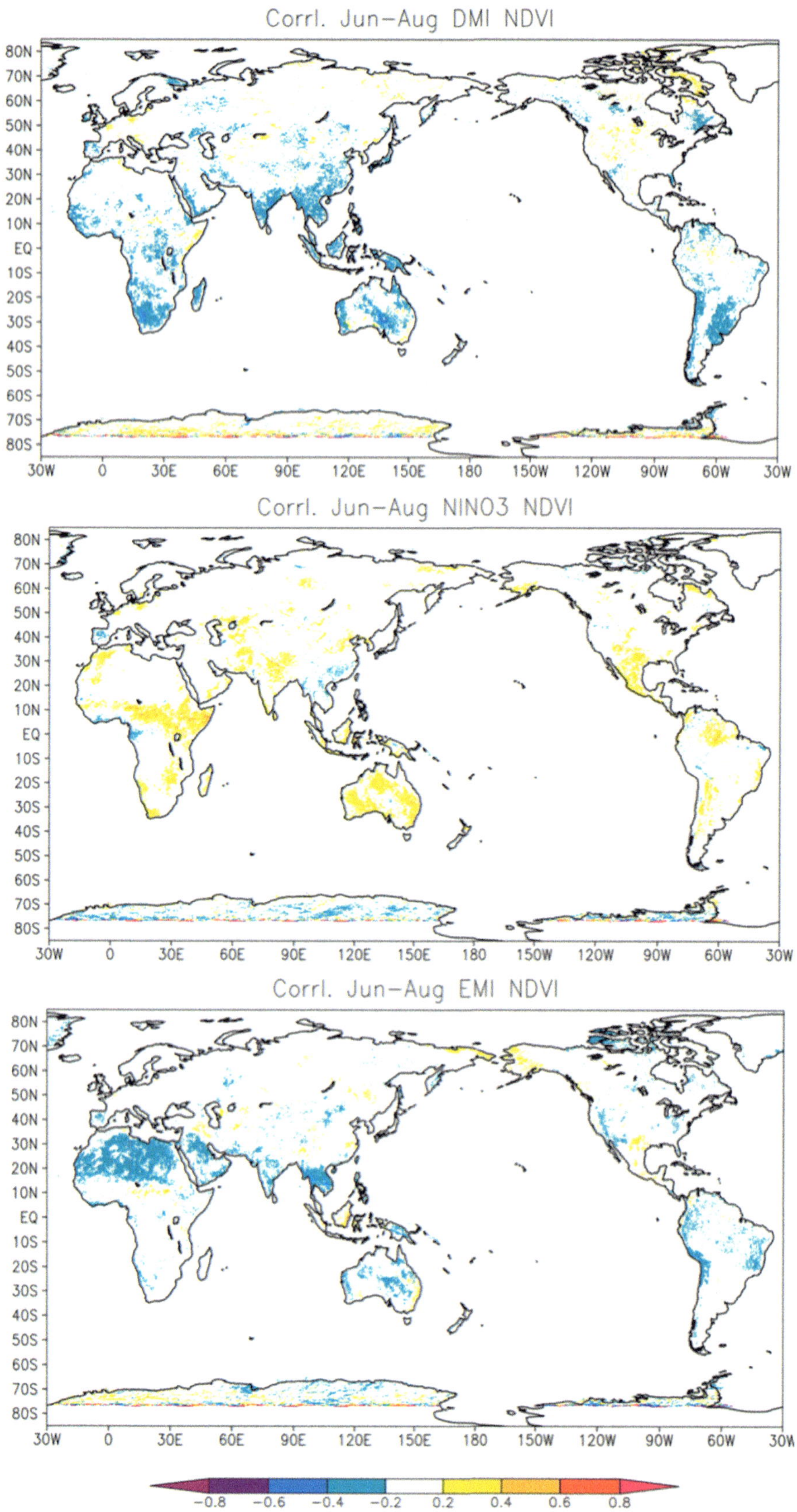

FIGURE 24.10 NDVI anomaly correlation coefficients with indices of IOD (top), ENSO (middle) and ENSO Modoki (bottom) for boreal summer season. Values below 95% significant level using a 2-tailed *t*-test are not shaded. Map lines delineate study areas and do not necessarily depict accepted national boundaries.

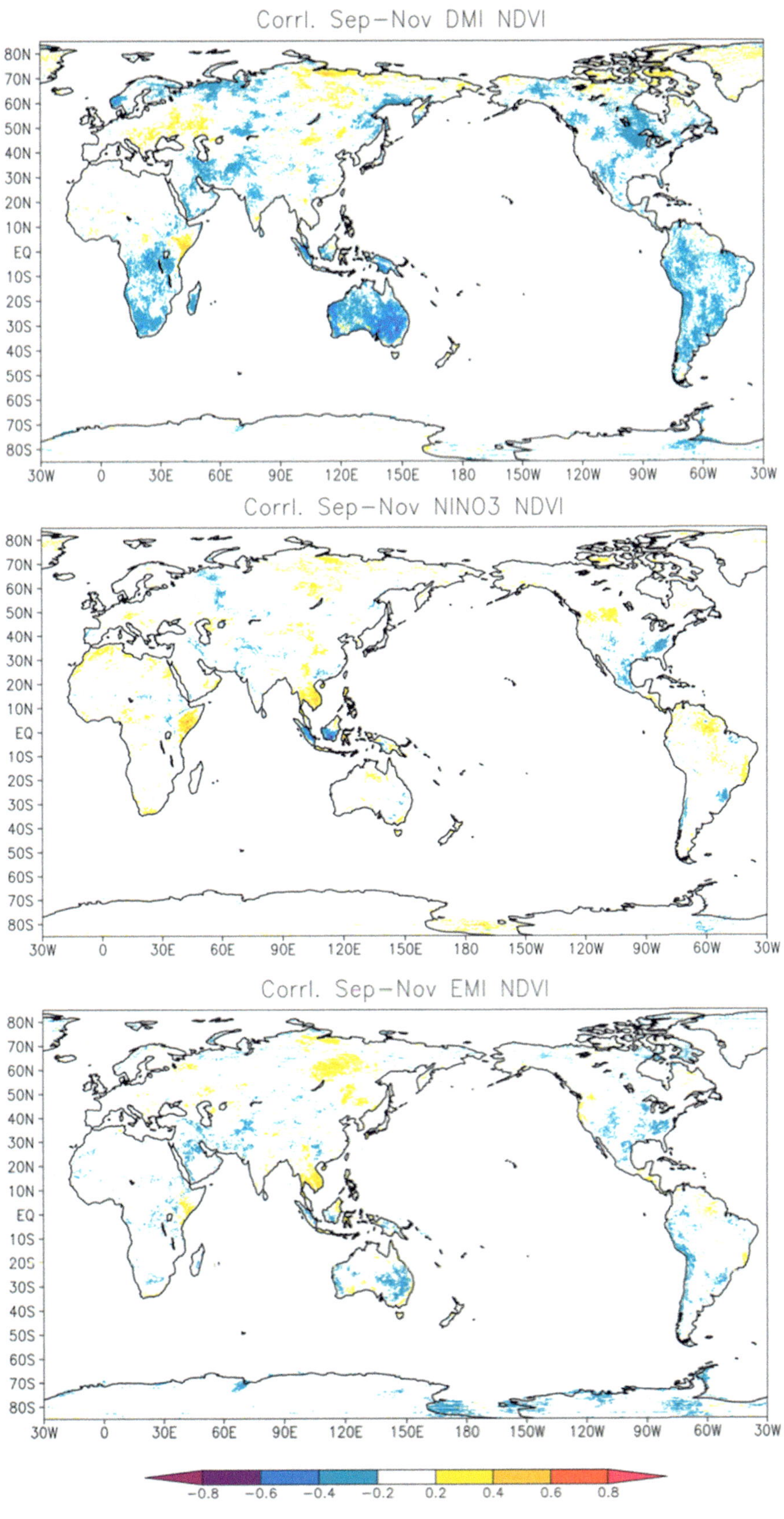

FIGURE 24.11 Same as Fig. 24.10 but for boreal spring season. Map lines delineate study areas and do not necessarily depict accepted national boundaries.

References

Arango, J. D. P., Lintner, B. R., Carvalho, L. M. V., & Lyon, B. (2021). Spatial extents of tropical droughts during El Niño in current and future climate in observations, reanalysis, and CMIP5 models. *Geophysical Research Letters*, *48*, e2021GL093701. Available from https://doi.org/10.1029/2021GL093701.

Ashok, K., Behera, S. K., Rao, S. A., Weng, H., & Yamagata, T. (2007). El Niño Modoki and its possible teleconnection. *Journal of Geophysical Research*, *112*, C11007. Available from https://doi.org/10.1029/2006JC003798.

Asoka, A., & Mishra, V. (2015). Prediction of vegetation anomalies to improve food security and water management in India. *Geophysical Research Letters*, *42*, 5290–5298.

Banger, K., et al. (2015). Terrestrial net primary productivity in India during 1901–2010: Contributions from multiple environmental changes. *Climatic Change*, *132*, 575–588.

Behera, S. K., Krishnan, R., & Yamagata, T. (1999). Unusual ocean-atmosphere conditions in the tropical Indian Ocean during 1994. *Geophysical Research Letters*, *26*, 3001–3004.

Behera, S., Ratnam, J. V., Masumoto, Y., & Yamagata, T. (2013). Origin of extreme summers in Europe: the Indo-Pacific connection. *Climate Dynamics*, *41*, 663–676.

Behera, S. K., & Ratnam, J. V. (2018). Quasi-asymmetric response of the Indian summer monsoon rainfall to opposite phases of the IOD. *Science Reports*, *8*, 123. Available from https://doi.org/10.1038/s41598-017-18396-6.

Behera, S. K., Doi, T., & Ratnam, J. V. (2021). Air-sea interactions in tropical Indian Ocean: The IOD. In S. K. Behera (Ed.), *Tropical and extratropical air-sea interactions* (pp. 115–139). Elsevier. Available from https://doi.org/10.1016/B978-0-12-818156-0.00001-0.

Bothale, R. V., & Katpatal, Y. B. (2014). Response of rainfall and vegetation to ENSO events during 2001-2011 in Upper Wardha Watershed, Maharashtra, India. *Journal of Hydraulic Engineering*, *19*(3), 583–592.

Cai, W., Santoso, A., Wang, G., et al. (2014). Increased frequency of extreme Indian Ocean dipole events due to greenhouse warming. *Nature*, *510*, 254–258. Available from https://doi.org/10.1038/nature13327.

Cai, W., Yang, K., Wu, L., et al. (2021). Opposite response of strong and moderate positive Indian Ocean dipole to global warming. *Nature Climate Change*, *11*, 27–32. Available from https://doi.org/10.1038/s41558-020-00943-1.

Dai, A. (2013). Increasing drought under global warming in observations and models. *Nature Climate Change*, *3*, 52–58.

Erasmi, S., Schucknecht, A., Barbosa, M. P., & Matschullat, J. (2014). Vegetation greenness in Northeastern Brazil and its relation to ENSO warm events. *Remote Sensing*, *6*(4), 3041–3058. Available from https://doi.org/10.3390/rs6043041.

Gutman, G. (1999). On the use of long-term global data of land reflectance's and vegetation indices derived from the advanced very high resolution radiometer. *Journal of Geographical Research*, *104*, 6241–6255.

Hong, C., Li, T., Ho, L., & Kug, J. (2008). Asymmetry of the Indian Ocean dipole. Part I: Observational analysis. *Journal of Climate*, *21*, 4834–4848. Available from https://doi.org/10.1175/2008JCLI2222.1.

Huang, B., Liu, C., Banzon, V., Freeman, E., Graham, G., Hankins, B., . . . Zhang, H. (2021). Improvements of the daily optimum interpolation sea surface temperature (DOISST) version 2.1. *Journal of Climate*, *34*, 2923–2939. Available from https://doi.org/10.1175/JCLI-D-20-0166.1.

Lyon, B., & Barnston, A. G. (2005). ENSO and the spatial extent of interannual precipitation extremes in tropical land areas. *Journal of Climate*, *18*, 5095–5109.

McNally, A., & NASA/GSFC/HSL. (2018). FLDAS Noah Land Surface Model L4 Global Monthly 0.1 x 0.1 degree (MERRA-2 and CHIRPS), Greenbelt, MD, USA, Goddard Earth Sciences Data and Information Services Center (GES DISC). Available from https://doi.org/10.5067/5NHC22T9375G.

Miralles, D. G., van den Berg, M. J., Gash, J. H., et al. (2014). El Niño-La Niña cycle and recent trends in continental evaporation. *Nature Climate Change*, *4*, 122–126.

Nagai, S., Ichii, K., & Morimoto, H. (2007). Interannual variations in vegetation activities and climate variability caused by ENSO in tropical rainforest. *International Journal of Remote Sensing*, *28*, 1285–1297.

Nayak, R. K., Patel, N. R., & Dadhwal, V. K. (2013). Inter-annual variability and climate control of terrestrial net primary productivity over India. *International Journal of Climatology*, *33*, 132–142.

Philippon, N., Martiny, N., Camberlin, P., Hoffman, M., & Gond, V. (2014). Timing and patterns of the ENSO signal in Africa over the last 30 years: Insights from normalized difference vegetation index data. *Journal of Climate*, *27*, 2509–2532.

Ropelewski, C. F., Janowiak, J. E., & Halpert, M. S. (1985). The analysis and display of real time surface climate data. *Monthly Weather Review*, *113*, 1101–1106.

Saji, N. H., Goswami, B. N., Vinayachandran, P. N., & Yamagata, T. (1999). A dipole mode in the tropical Indian Ocean. *Nature*, *401*, 360–363.

Sharma, A., & Goyal, M. K. (2018). Assessment of ecosystem resilience to hydroclimatic disturbances in India. *Global Change Biology*, *24*, e432–e441.

Tarpley, J. D., Schneider, S. R., & Money, R. L. (1984). Global vegetation indices from the NOAA-7 meteorological satellite. *Journal of Applied Meteorology and Climatology*, *23*, 491–494.

van Leeuwen, W. J., Orr, B. J., Marsh, S. E., & Herrmann, S. M. (2006). Multi-sensor NDVI data continuity: Uncertainties and implications for vegetation monitoring applications. *Remote Sensing of Environment*, *100*, 67–81. Available from https://doi.org/10.1016/j.rse.2005.10.002.

Varghese, R., Behera, S. K., & Behera, M. D. (2023). Significant inverse influence of tropical Indian Ocean SST on SIF of Indian vegetation during the summer monsoon onset phase. *Remote Sensing*, *15*(7), 1756. Available from https://doi.org/10.3390/rs15071756.

Wang, J., Wang, M., Kim, J.-S., Joiner, J., Zeng, N., Jiang, F., et al. (2021). Modulation of land photosynthesis by the Indian Ocean dipole: Satellite-based observations and CMIP6 future projections. *Earth's Future*, *9*, e2020EF001942. Available from https://doi.org/10.1029/2020EF001942.

Weng, H., Ashok, K., Behera, S. K., Rao, S. A., & Yamagata, T. (2008). Impacts of recent El Niño Modoki on dry/wet conditions in the Pacific Rim during boreal summer. *Climate Dynamics*, *29*, 113–129.

Woodward, F. M., Lomas, T., & Quaife. (2008). Global responses of terrestrial productivity to contemporary climatic oscillations. *Philosophical Transactions of the Royal Society B*, *363*, 2779–2785.

Zhao, L., Dai, A., & Dong, B. (2018). Changes in global vegetation activity and its driving factors during 1982–2013. *Agricultural and Forest Meteorology*, *249*, 198–209.

Chapter 25

Significant rise in aerosol concentration in the past two decades over the Kalinga Nagar industrial area, Odisha, India

S. Kumar[1], Pulakesh Das[2] and Mukunda Dev Behera[1]

[1]*Centre for Ocean, River, Atmosphere and Land Sciences (CORAL), Indian Institute of Technology Kharagpur, Kharagpur, West Bengal, India,* [2]*School of Forest Resources, University of Maine, Orono, ME, United States*

25.1 Introduction

The iron and steel industries are major sources of greenhouse gases and particulate matter (PM) or aerosol in the atmosphere, causing high air pollution (Yoro & Daramola, 2020). There are multiple other sources that cause air pollution, such as thermal power production, industry emissions, burning fossil fuels, agricultural activities, stubble burning, forest fires, transportation, and construction activities (Guttikunda, Goel, & Pant, 2014). The increasing population with ever-increasing energy demand, resource utilization, industrialization, increasing number of vehicles, etc., causes an increased rate of air pollutant release into the atmosphere in urban and industrial areas (Sharma et al., 2019). On the other hand, mining areas, mineral and crude oil processing units, waste treatments, and various production industries, such as cement and fertilizer, produce several pollutants. The emission of toxic pollutants and their dispersion in the atmosphere is the major reason behind air pollution (Leelőssy et al., 2014). The hazardous air pollutants cause multiple health issues, including premature mortality, globally, with a more significant impact in low-income and middle-income countries (Brauer et al., 2019; Cohen et al., 2017). According to the World Health Organization (WHO), each year, air pollution is responsible for nearly 7 million deaths around the globe. In developing countries like India, air pollution is an unavoidable concern that cannot be removed entirely, but stern actions can be taken to mitigate the risk (Balakrishnan et al., 2019).

The aerosol optical depth (AOD) determines the extinction of solar irradiance in the atmosphere due to the presence of suspended atmospheric particles, including liquid, dust, smoke, pollutants, and haze. AOD is a dimensionless quantity that measures aerosol concentration in a column of air. The AOD is measured through the aerosol optical thickness (AOT), wherein a lower value indicates a clear atmosphere and a higher value indicates a higher aerosol concentration or hazy conditions. In addition to the regional weather and events, the AOD in an area depends on local events such as agricultural crop residue burning, forest fires and dust storms. Studies have indicated increased air pollutants in Delhi and the national capital region (Delhi-NCR) during postmonsoon due to stubble burning (Das et al., 2022). The aerosol concentration regulates the formation of clouds, wherein the higher AOT reduces the visibility and alters Earth's radiation budget (Kuttippurath & Raj, 2021). The measurement of AOD varies with the operating wavelength, which is denoted by the Ångström Exponent that relates AOD changes with wavelength. AOD is a proxy measure of PM, which is commonly measured at different particle sizes, such as 1, 2.5, and 10 μm (i.e., PM_1, $PM_{2.5}$, and PM_{10}).

Periodic monitoring of AOD and its sources is important to assess air quality, which is important for developing various management practices and policy-making. Various in situ monitoring networks have been established in India to monitor air quality. For example, the System of Air Quality and Weather Forecasting and Research established multiple monitoring stations in the cities and industrial areas for robust landscape-level air quality monitoring (https://safar.tropmet.res.in/MONITORING%20SYSTEM-10-3-Details accessed on September 2, 2024). The Central Pollution Control Board (CPCB), India, operates more than 800 monitoring stations in more than 340 cities/towns for ambient air quality monitoring, including sulfur dioxide (SO_2), nitrogen oxide (NO_2), respirable suspended particulate matter (PM_{10}), and

Sustainable Development Perspectives in Earth Observation. DOI: https://doi.org/10.1016/B978-0-443-14072-3.00023-X

fine particulate matter ($PM_{2.5}$), under the National Air Quality Monitoring Program (NAMP) (https://cpcb.nic.in/about-namp/ accessed on September 2, 2024). Sharma, Gupta, Gargava, and Kota (2024) studied the air quality in India from 1987 to 2019 using the NAMP data and reported a significant rise in the available in situ data, including the data deficit regions such as the Himalayan and Northeast India. Although many in situ measurement stations have been installed for air quality monitoring, the number of available data stations in India is significantly low than developed countries (Brauer et al., 2019).

In comparison to the footprint-based observations by the in situ observatories, remote sensing technology can potentially play a significant role in monitoring AOD at a large scale with limited resources in terms of labor and cost, which provides useful data to decision-makers for controlling and managing air pollution (Lal et al., 2020). Various sensors, such as Moderate Resolution Imaging Spectroradiometer (MODIS), TROPOspheric Monitoring Instrument (TROPOMI), and Greenhouse Gases Observing Satellite (GOSAT), mounted on satellites can efficiently measure the atmospheric constituents present in the atmosphere (Bauwens et al., 2020; Guo et al., 2017; Li et al., 2007). Satellite-derived products are being used to measure traces of gases and aerosols, forecast air quality, and identify the sources of pollution and their change dynamics. Moreover, satellite data are essential for monitoring air quality where ground stations or monitoring stations are absent, mostly in remote areas and underdeveloping countries (Brauer et al., 2019). However, satellite data-driven air quality monitoring depends on data accuracy and spatial and temporal resolution of the satellite data.

The terrestrial ecosystem plays an important role in reducing atmospheric carbon and regulating air pollution, while deforestation and various disturbances reduce the efficacy of forest carbon assimilation (Das et al., 2022; Deb Burman & Das, 2024; Singh et al., 2024). Moreover, the climate influences the carbon sequestration of the terrestrial ecosystem at the regional scale (Varghese, Behera, & Behera, 2023). Mandal, Das, Behera, and Parida (2022) used GOSAT satellite-driven data to assess CO_2 in India from 2010 to 2012. They reported the highest carbon sequestration by the terrestrial ecosystems in the postmonsoon season and lower in monsoon and summer, mostly regulated by agricultural activities, seasonal growth, and anthropogenic activities. Behera et al. (2022) studied the changes in NO_2, SO_2, formaldehyde (HCHO), methane (CH_4), carbon monoxide (CO), aerosol (ultraviolet aerosol index, UVAI), and ozone (O_3) to assess the impact of human disturbances considering the COVID lockdown, when the human disturbances were least in the past few decades. Their study indicated a significant reduction in air pollution during the lockdown phases compared to the concentration before and after the COVID lockdown.

Several studies were conducted to measure AOD in India using ground and satellite-based observations. Gupta, Ratnam, Madhavan, and Narayanamurthy (2022) took several satellite-driven data as MODIS and Multiangle Imaging Spectral Radiometer (MISR) onboard in Terra satellite, and Cloud-Aerosol Lidar with Orthogonal Polarization (CALIOP) onboard in Cloud-Aerosol Lidar and Infrared Pathfinder Satellite Observation (CALIPSO) satellite, and compared with the ground-based AEronet RObotic NETwork (AERONET) observations from 2001 to 2020 for more than 80 stations globally. They reported moderate to high correlation (0.3 to above 0.7) among the various satellite data, except CALIOP (over a few sites), underestimation by MISR, and overestimation by CALIOP in the high aerosol concentration regions. Ramachandran, Kedia, and Srivastava (2012) employed MODIS data (Level 2 data with 10 km × 10 km resolution at daily temporal scale) for AOD change assessment from 2000 to 2010 and reported a general increasing trend in India in most Indian cities. Kuttippurath and Raj (2021) studied aerosols over India and North Indian Ocean (NIO) from 2000 to 2019, including multiple datasets such as MODIS and MISR (Terra satellite), Advanced Along Track Scanning Radiometer (AATSR) [Envisat satellite], and Modern-Era Retrospective analysis for Research and Applications version 2 (MERRA-2) reanalysis data. They reported the highest AOD in the Indo-Gangetic Plain (IGP) and the least in the Himalayan region over the Indian landmass. Mukherjee and Vinoj (2020) used a ground-based instrument (MICROTOPS-II sun photometer) to measure AOD in five narrow-band wavelengths in the visible near-infrared range (from 380 to 870 nm) from October 2015 to October 2016 near Bhubaneswar City, India. They compared ground-measured AOD with satellite sensor-based observations from MODIS and MISR and reported a low bias in the satellite data.

The aerosol concentration assessment conducted by Dutta and Chatterjee (2022) from 2005 to 2019 identified Odisha as one of the vulnerable states in India with a positive trend in AOD. Mhawish et al (2021) assessed the aerosol characteristics using satellite data over South Asia from 2000 to 2019. Their study indicated a predominance of smaller particles in Odisha, which is dominated by the mining industries. Limited studies were conducted to assess the AOD in the eastern region of India, including the industrial areas in Odisha state. This study used MODIS AOD data (MOD19A2) and MERRA-2 reanalysis data to understand the spatio-temporal variations in AOD over Kalinga Nagar, one of the industrial cities of Jajpur district, Odisha. The study aims to assess the efficacy of satellite data for air quality monitoring to support future air quality regulations formulated by various government agencies. This research will help monitor urban air quality with satellite sensors, a field of applied research with policy and environmental health implications.

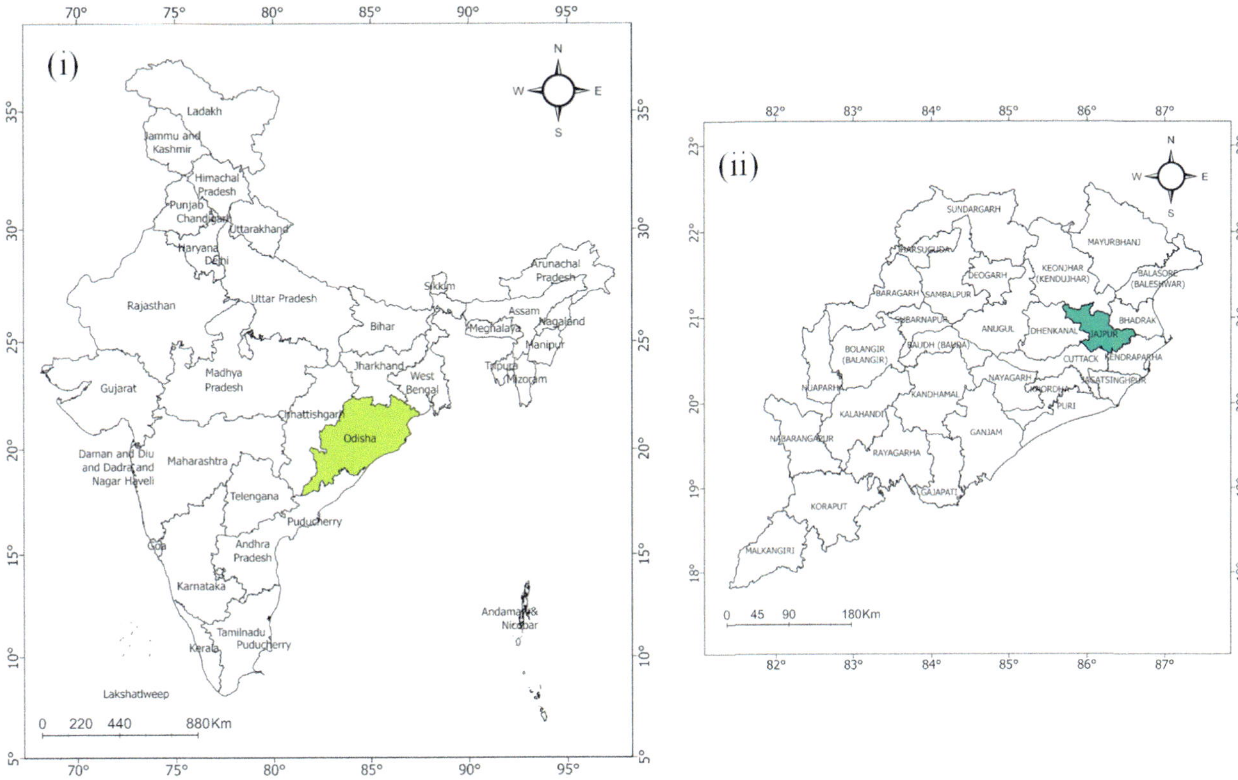

FIGURE 25.1 Location map of the study area: (i) Odisha state and (ii) Jajpur district. Map lines delineate study areas and do not necessarily depict accepted national boundaries.

25.2 Study area

The study was conducted in Kalinga Nagar, an industrial city in Jajpur district, Odisha, India. It is one of the major manufacturing hubs in Odisha state, which has multiple heavy industries. Industries have rapidly grown in the past decades and are recognized as a National Manufacturing and Investment Zone by the Government of India. The Jajpur district is located in the eastern coastal region and is surrounded by Keonjhar, Kendrapara, Cuttack, Dhenkanal, and Bhadrak (Fig. 25.1). The district is situated between 20°35'N and 21°10'N and between 85°41'E and 86°38'E. The district has more than 10 mines rich in huge chromite, iron ore, and quartzite deposits, associated with pyroxenite and nickel ore. The region is dominated by the tropical climate, with the peak summer temperature exceeding 40°C and the winter temperature reaching below 12°C, with an average annual precipitation exceeding 1000 mm. Forests are the major land cover feature, followed by agricultural land, wherein rice is grown dominantly. Although more than 35 stations are installed for air quality monitoring in Odisha, no air quality monitoring station is available in the Jajpur district, which is one of the important industrial areas (https://ospcboard.odisha.gov.in/programs-projects/ekamra/ accessed on September 2, 2024).

25.3 Materials and methodology

25.3.1 Data used

In this study, the MODIS (MOD19A2) product onboard the Terra satellite was used to analyze the spatio-temporal variation in AOD from 2001 to 2021 (Table 25.1). MODIS AOD data in the 550-nm range is susceptible to dust particles in the water droplets and captures cloud droplets as aerosol particles. The MERRA-2 AOT at 550-nm data was used to validate the results of the AOD change analysis. Various factors, such as wet deposition, dry deposition, relative humidity, and AOT from the MERRA-2 reanalysis data, were analyzed to understand the significance of these variables concerning air pollution (Table 25.1). We examined the dispersion pattern of aerosols caused by wind blows in the study area using wind data from the CPCB. The satellite data were retrieved using the Google Earth Engine platform, which is a webGIS platform that provides access to a large set of publicly available geospatial data and allows online processing (Behera et al., 2022). The wind data were accessed from the Talchar CPCB station, located 100 km from Jajpur city.

TABLE 25.1 Data used in this study.

S. no	Data	Spatial resolution	Temporal resolution
1	MOD19A2	1 km	Daily
2	MERRA-2 dry deposition	0.625° × 0.5°	Monthly
3	MERRA-2 wet deposition	0.625° × 0.5°	Monthly
4	MERRA-2 relative humidity	0.625° × 0.5°	Monthly

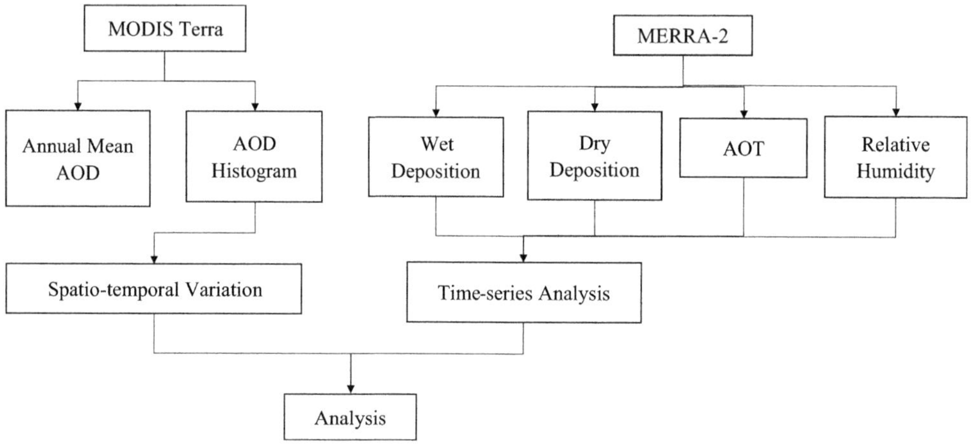

FIGURE 25.2 Methodology flow chart.

25.3.2 Methodology

MOD19A2 data were mainly used to study the atmospheric aerosol concentration present in the atmosphere at various time scales from 2001 to 2021. The annual mean AOD was calculated using long-term daily AOD data for the Jajpur district of Odisha. A histogram showing the AOD data distribution was generated to analyze the annual AOD range. Long-term monthly variations in AOD, along with a histogram showing the data distribution range, were generated using the MOD19A2 dataset to monitor aerosols in the study area and to examine the variations with respect to meteorological forcing in the atmosphere. Wet deposition, dry deposition, AOT, and relative humidity from MERRA-2 were used to examine the intraannual variations. Time series were plotted for these variables, and correlations were derived with AOD data. MERRA-2 AOT was used to represent the monthly data and the significance of monthly variations.

The wind rose diagram was generated using the wind data accessed from the Talchar CPCB station to understand the role of winds in the AOD distribution. It describes the transport direction/trajectory of aerosols or dust emitted from industries. The overall data processing flowchart is shown in Fig. 25.2.

25.4 Results

25.4.1 Interannual and monthly variations in aerosol optical depth

Annual mean AOD from 2001 to 2021 were derived for Jajpur district using MODIS data. A lower AOD was observed in the northern part of Jajpur compared to a higher value in the southern parts, which is dominated by the industries (Fig. 25.3). The annual mean AOD varied between 0.3 and 0.5 from 2001 to 2006. A minor increase was observed, ranging between 0.5 and 0.7 from 2007 to 2014. A further increase was observed from 2015 to 2019 in annual mean

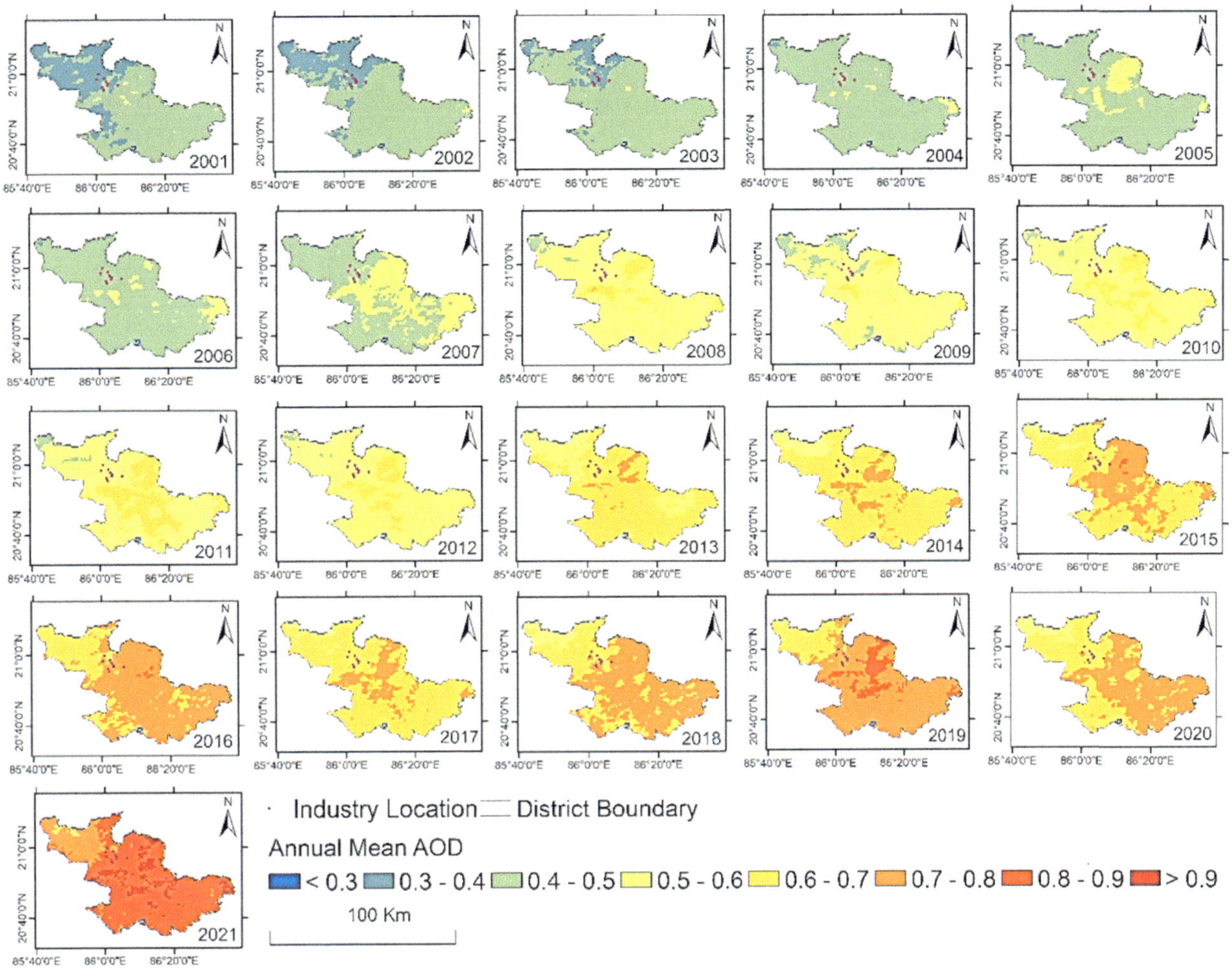

FIGURE 25.3 Annual mean aerosol optical depth variations in the study area from 2001 to 2021. Map lines delineate study areas and do not necessarily depict accepted national boundaries.

AOD, which reached 0.8. A significant reduction was observed in 2020 when the mean AOD varied between 0.6 and 0.8, which again increased to >0.9 in 2021. The spatio-temporal AOD maps from 2001 to 2021 showed a monotonic increase in the study area, wherein the mean values close to 0.4 in 2001 increased to 0.9 in 2021.

A histogram of mean AOD over the past 21 years was generated to analyze the variations in the AOD range (Fig. 25.4). It was observed that the majority (mode) of AOD ranged between 0.4 and 0.5 from 2001 to 2007. After 2007, the AOD mode in the study area shifted toward a higher AOD range, that is, 0.6–0.8, from 2008 to 2019. A small shift toward a lower AOD range was observed in 2020; however, it shifted toward higher AOD values, ~0.8–0.9 in 2021. The temporal annual mean AOD data indicated a major shift in 2008 and 2021 compared to other years.

The monthly variation in mean AOD over Jajpur from 2001 to 2021 was studied. It was observed that the AOD in January was 0.5–0.7, which reduced to 0.4–0.6 in February (Figs. 25.5 and 25.6). It increased to 0.6–0.7 in March and remained constant until July. The maximum AOD value was observed in July, followed by higher values before and after July, that is, June, May, August, and September. Moderate AOD values were observed in March, April, December, and January. A low AOD value was observed in October and November, with the lowest value in February.

The monthly mean AOD value range in the study area was analyzed. It shows the distribution of AOD values in frequency and range over the study area. In January, the highest frequency (mode) was near 0.6, which decreased to 0.5 in February. Again, the monthly mean AOD in March increased to nearly 0.6, which slowly increased till June. A very

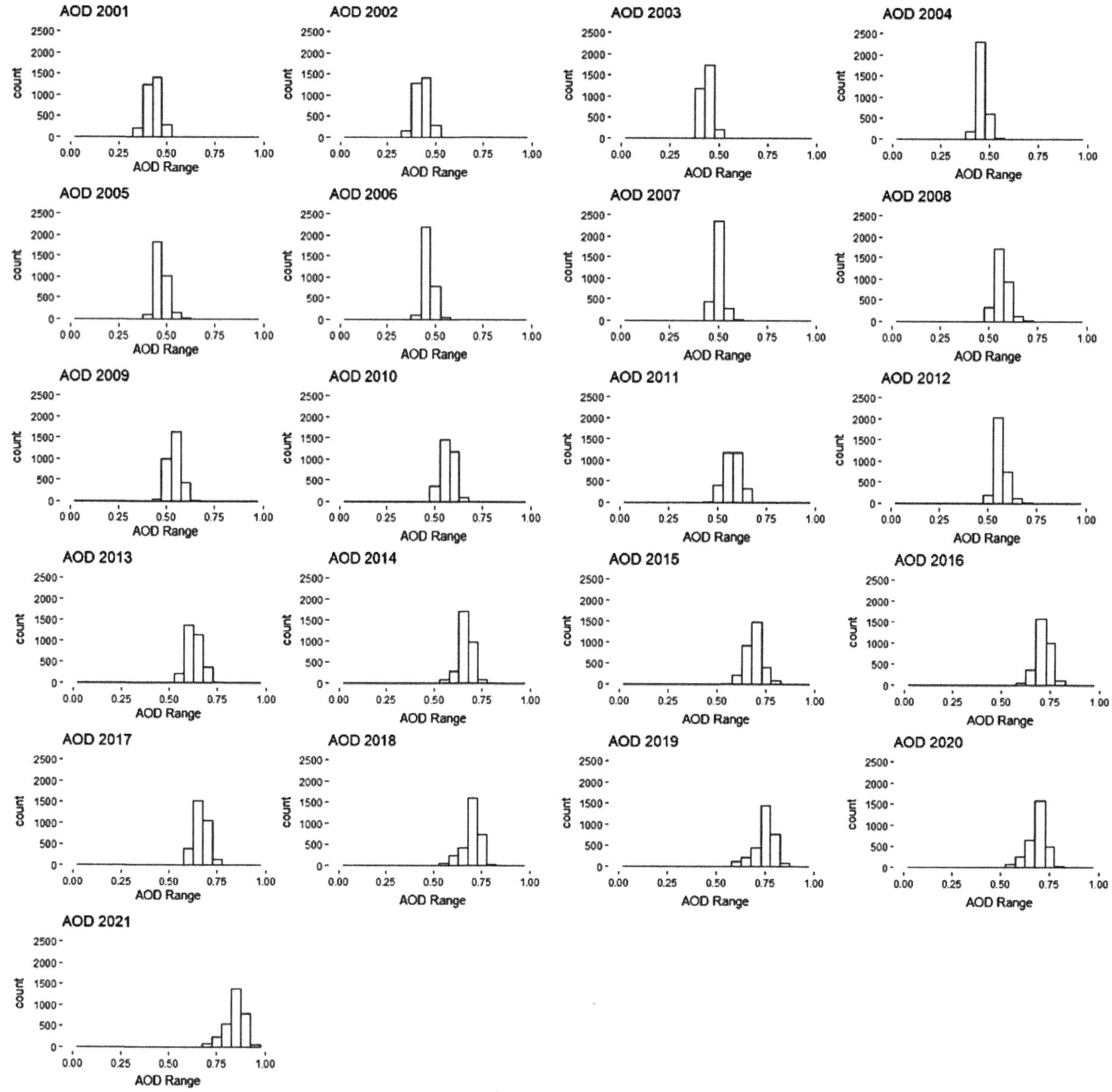

FIGURE 25.4 Histogram of annual mean aerosol optical depth showing the data distribution variations from 2001 to 2021.

high AOD range (0.7–0.9) is observed in July and then gradually decreases from August to November. Again, the mean AOD mode in December and January was relatively higher. The highest range in AOD was observed during the monsoon period, followed by premonsoon and postmonsoon in the study area.

25.4.2 Wind speed and direction analysis using CPCB station data

Winds blow the aerosol particles and play an important role in predicting their transport direction (Fig. 25.7). The wind rose diagram generated using the Talchar CPCB station indicated a prominent wind direction and speed over the study area. It indicated a prominent wind direction constantly blowing toward the South and Southeastern regions in the study, which clarifies the higher AOD values in the southern parts of Jajpur. The transport of pollutants from industries was observed to move toward the South, which caused more aerosol accumulation or higher concentration in the southern regions of Jajpur district.

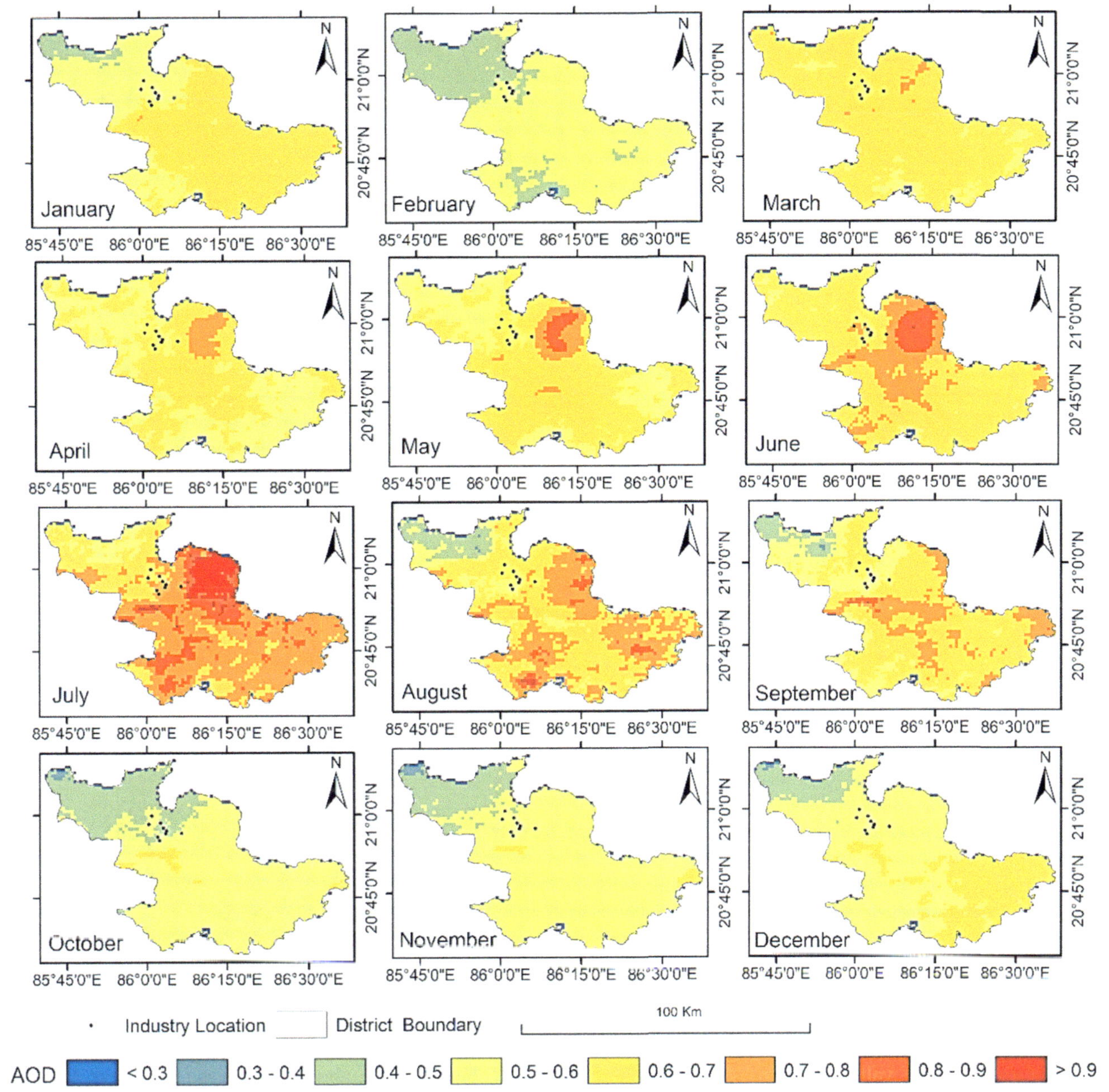

FIGURE 25.5 Mean monthly variations in aerosol optical depth in Jajpur district from 2001 to 2021. Map lines delineate study areas and do not necessarily depict accepted national boundaries.

25.5 Time series analysis using MERRA-2

Dry and wet deposition of aerosols with respect to relative humidity was analyzed using a time series analysis (Fig. 25.8). It was observed that dry deposition was maximum in peak summer (May) at 0.03–0.04 g m^{-2} $month^{-1}$. It starts to decrease to 0.01 g m^{-2} $month^{-1}$ due to high relative humidity and precipitation in the monsoon. The dry deposition was the minimum (0–0.02 g m^{-2} $month^{-1}$) in postmonsoon (September–October). In comparison, the maximum wet deposition was observed in monsoons (July), >0.35 g m^{-2} $month^{-1}$. Wet deposition and relative humidity showed a similar trend, having a high value in monsoon and lower values in winter months. The lowest wet deposition was observed in the winter months (December and January).

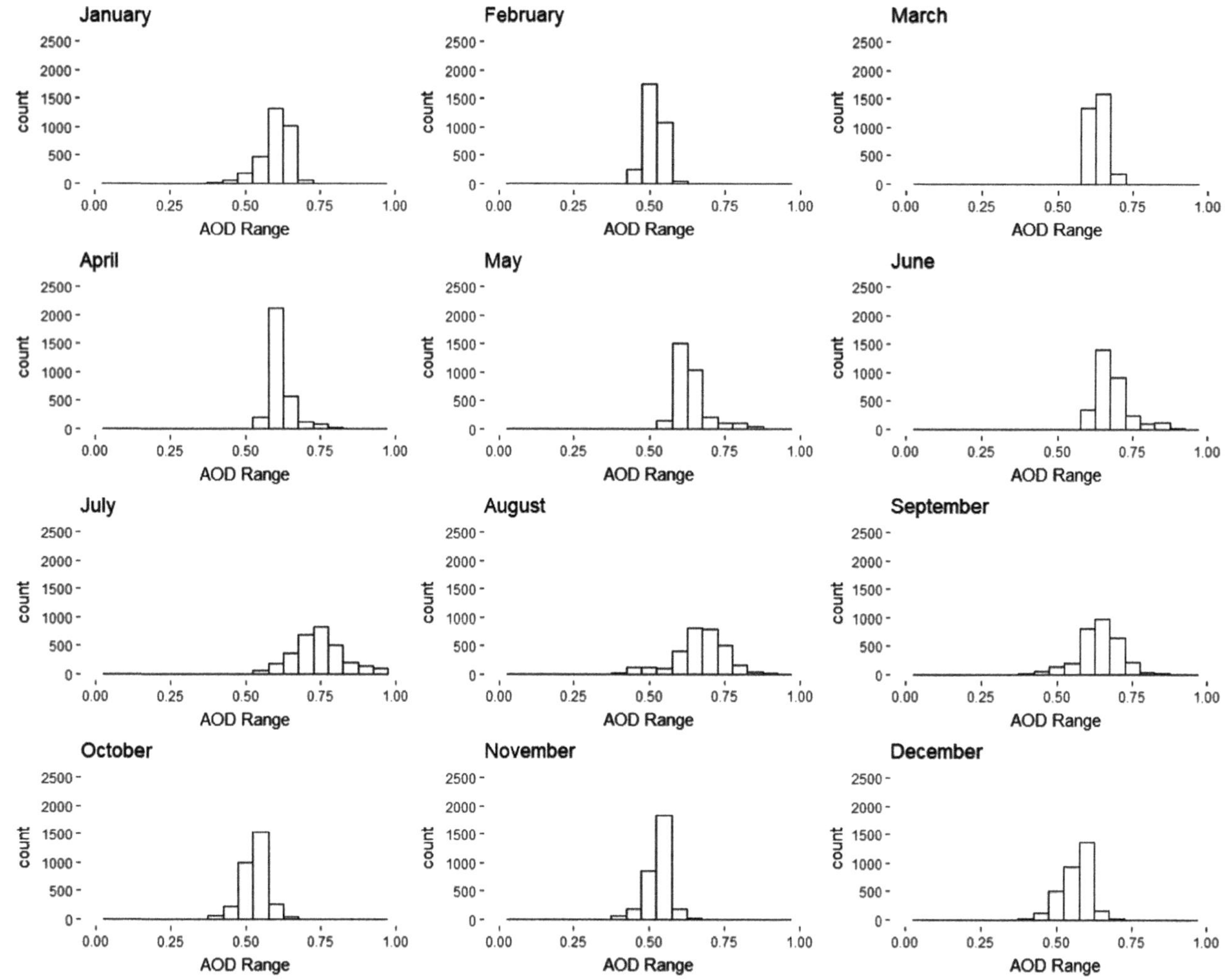

FIGURE 25.6 Histogram showing the mean (from 2001 to 2021) monthly aerosol optical depth range in the study area.

Variations in AOT and relative humidity were examined using MERRA-2 reanalysis data over the study area (Fig. 25.9). The AOT value was above 0.1 from March to June. The peak AOT (above 0.16) was recorded in May, and then it started to decrease with an increase in precipitation and relative humidity. The lowest AOT was observed in September (below 0.04), with lower values in winter (≤0.04). However, the maximum relative humidity was observed in July, and a minor decreasing trend followed thereafter. The monthly MODIS AOD and MERRA-2 AOT were compared (Fig. 25.9). The monthly variations in AOD derived from MODIS indicated a shift from the AOT data recorded by MERRA-2 (Fig. 25.9). The AOD recorded in MODIS indicated the maximum in June, while the maximum AOT in MERRA-2 was recorded in July.

25.6 Discussion

MODIS AOD showed higher aerosol concentration in June, which could be attributed to the presence of cloud particles. MODIS AOD was found more sensitive than MERRA-2 AOT to the particle sizes present in clouds during monsoon months. On the other hand, MERRA-2 AOT data show the maximum value in May and then the lowest in monsoon and postmonsoon. The seasonal AOD data indicated higher values in summer (June through September) and lower in winter (October to February). Past studies reported contrasting results in seasonal variations. Mukherjee and Vinoj (2020) studied AOD using a ground-based instrument and satellite data from October 2015 to October 2016 near

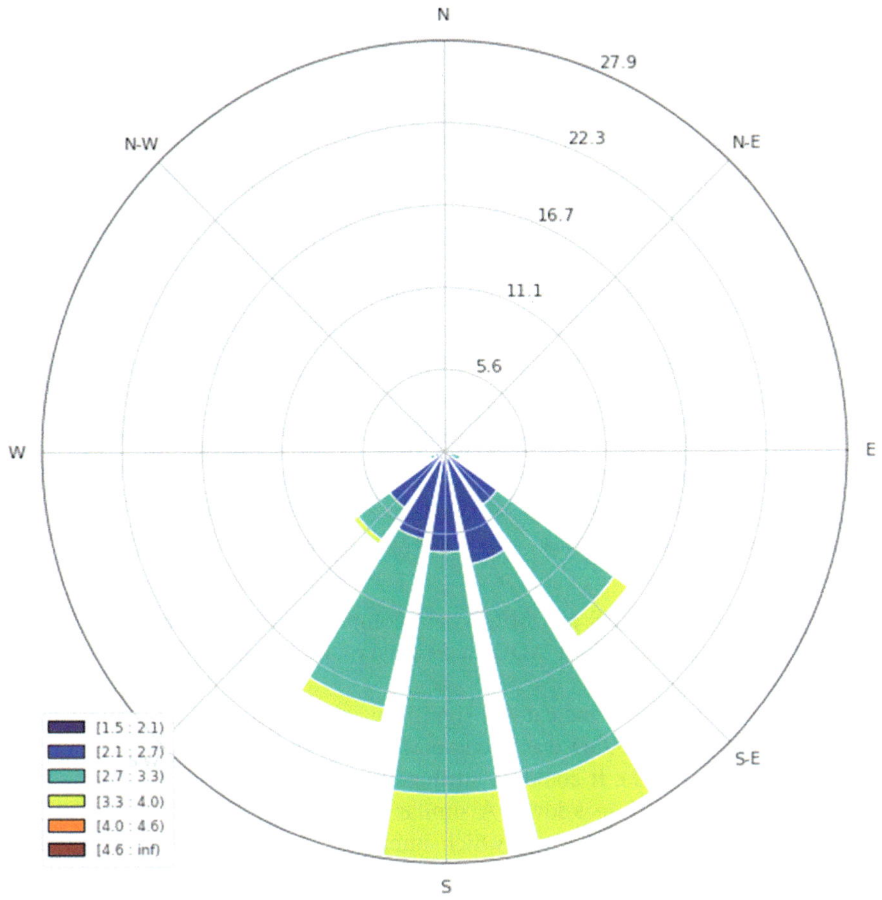

FIGURE 25.7 Wind rose generated using CPCB station data showing prominent wind directions in the study area.

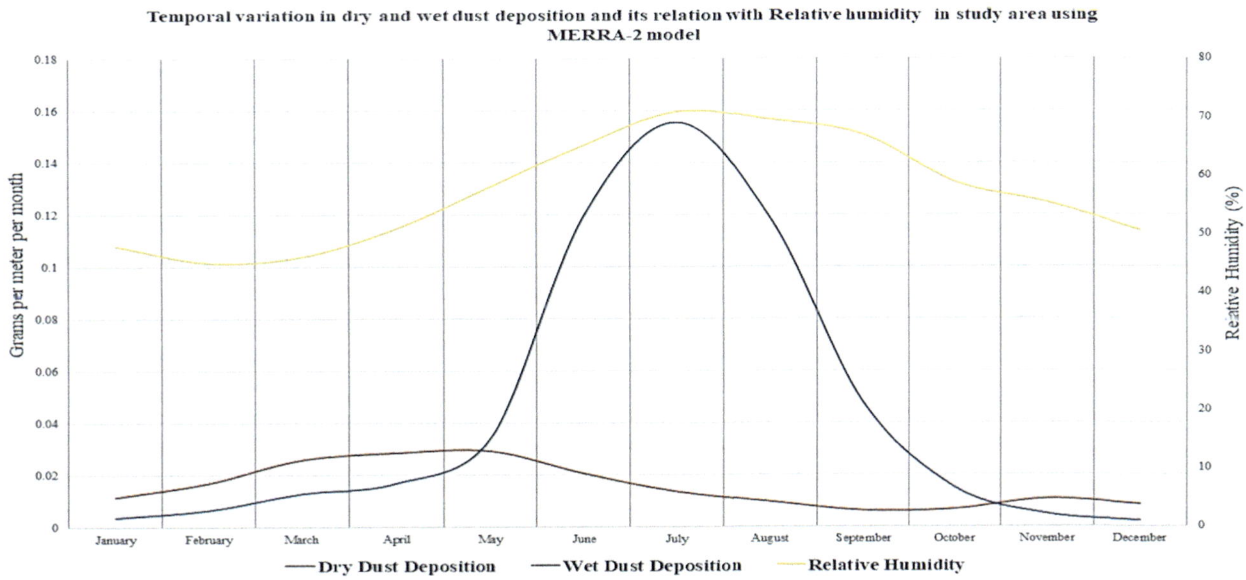

FIGURE 25.8 Variations in wet and dry depositions with respect to the relative humidity in the study area using MERRA-2 reanalysis data.

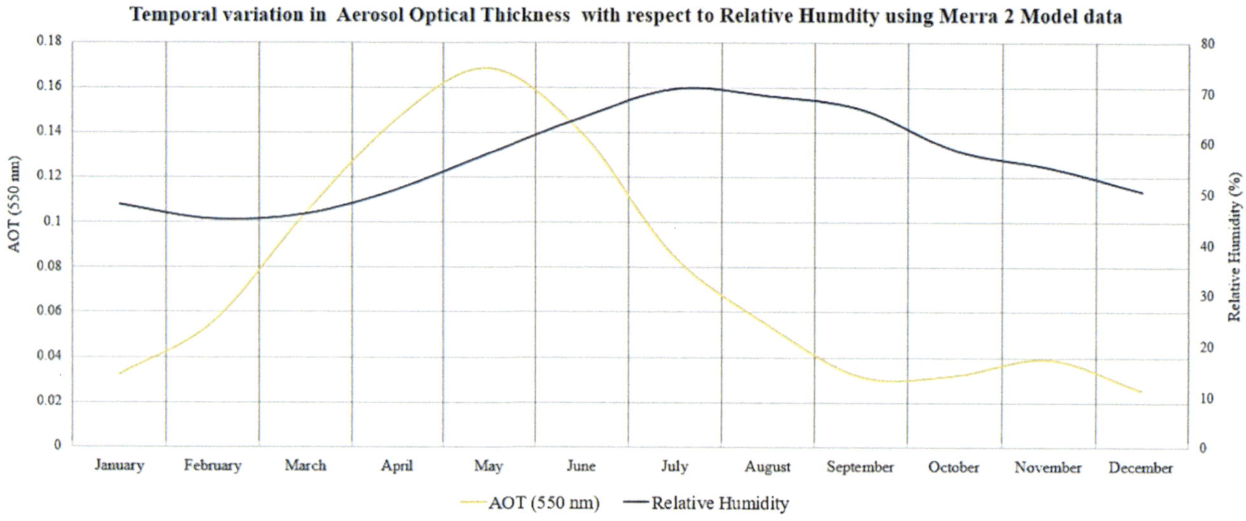

FIGURE 25.9 Variations in aerosol optical thickness with respect to relative humidity using MERRA-2 model data in the study area.

Bhubaneswar City, India. They reported a low bias in MODIS and MISR data compared to the ground-based assessment. They observed strong seasonal variations, with the highest AOD in February, the lowest in July, and a mean annual AOD of 0.8 ± 0.3.

The examination of aerosol concentration in the study area indicated a significant change from ~0.4 in 2001 to 0.9 in 2021, with a continuous increase. A sudden decrease in AOD was observed in 2020, ranging between 0.6 and 0.8, compared to the past few years and post-2020 (i.e., in 2021). It could be attributed to less anthropogenic activities and partial restrictions on industrial work during the COVID-19 lockdown. A similar change was reported in past studies, which observed reduced air pollutants due to the COVID lockdown, which improved air quality nationally (Behera et al., 2022). Pathakoti et al. (2021) used TROPOMI tropospheric columnar NO_2 (TCN) and studied the impact of the COVID lockdown, comparing the 2020 data from 2015 to 2019. Their study reported a significant drop (~18% reduction in TCN in 2020 compared to previous years) at the national scale, including Odisha. In contrast, several studies indicated a negative aerosol change in Odisha. For example, Rani and Kumar (2022) used MODIS AOD data for phase-I lockdown (March 25 to April 15, 2020) using MODIS AOD data from 2001 to 2020. Their study indicated an overall 9% reduction in AOD during the phase-I lockdown compared to the past two decades at the national scale, while an increase in AOD in several states, including Odisha. They attributed this to the forest fire events in these regions and smoke dispersal. Ranjan, Patra, and Gorai (2020) employed MODIS AOD data to study the effect of the COVID lockdown on aerosol concentration in urban and mining regions in India. They used the AOD data from 2000 to 2019 as a reference and observed a ~45% reduction in AOD over India. On the contrary, they observed a positive anomaly, that is, an increase in AOD in the coal mining areas. They reported a high AOD concentration similar to the prelockdown phase during the lockdown phase in parts of India, including Odisha. Several other studies also reported a similar trend in AOD change at the national scale and in Odisha, with overall reduction and increase, respectively (Gouda, Gogeri, & ThippaReddy, 2022).

The continuous increase in AOD in the study area could be attributed to industrial expansions and urbanization. Several past studies conducted on air quality reported similar outcomes. Mhawish et al (2021) reported an increase in AOD from 2000 to 2019 due to industrial emissions and anthropogenic activities, which exceeded 80% for the Odisha state. Sharma et al. (2024) conducted a national-scale assessment of air quality (based on PM_{10}, SO_2, and NO_2), including urban and rural residential areas and industrial areas from 1987 to 2019, using the monitoring stations operated under the NAMP, India. They observed a spike with more than 120% increase in PM_{10} concentrations from 2006 to 2009. The highest PM_{10} concentration is observed in the IGP, whereas the lowest concentration was recorded in Northeast India. Ramachandran et al. (2012) employed MODIS data for AOD assessment in India from 2000 to 2010. They attributed the aerosol increase in Hyderabad and Bengaluru to urbanization, while fossil fuel and biomass burning in New Delhi, and biomass burning and forest fires in Northeast India. Kuttippurath and Raj (2021) employed multiple datasets and studied aerosols over India and NIO from 2000 to 2019. They observed the highest trend in AOD increase in the eastern part of IGP, including coastal Odisha and West Bengal, and attributed it to population rise, urbanization, and industrial activities.

Industrial process upgradation, energy efficiency, agricultural stubble burning control, and fuel conversion are important aspects of reducing pollutants that create industrial air pollution. Several policies have been implemented to tackle pollution; for example, using compressed natural gas instead of petrol and diesel in Delhi has significantly reduced emissions and improved air quality. In addition, reducing deforestation and forest fire events may help reduce the GHG emission from the atmosphere (Das et al., 2022). The National Clean Air Program was initiated by the Government of India in 2019 to combat air pollution. The program aims to reduce $PM_{2.5}$ and PM_{10} by more than 20% in 2024 with respect to the baseline of 2017 (Dutta & Chatterjee, 2022). Robust monitoring of the aerosol concentration is one of the prerequisites. The approach applied in the current study and generated outputs provide relevant data to the decision-makers and policy-makers.

25.7 Conclusions

This study concludes that AOD levels rose to >0.9 in 2021 compared to the past 10 years. The rise in AOD was prominent after 2007 (>0.8 from 0.5 to 0.6), which corresponds to the establishment of the industrial cluster. However, a small decrease in AOD was observed during the 2020 lockdown period due to restricted emissions from industries. Monthly observations of AOD showed that the AOD level increased (0.5−0.8) from January to July and then dropped until November (0.4−0.5). A very high AOD (0.8−0.9) was observed in May, June, and July. MERRA-2 data confirm that there is high wet deposition during the monsoon period and the highest dry deposition observed during summer. It is also linked with the rise and fall of relative humidity in the study area. MERRA-2 AOT data clarifies why the high AOD values were observed during the monsoon period. MODIS AOD data in the 550-nm range is susceptible to dust particles in the water droplets and captures cloud droplets as aerosol particles. It was observed that dry dust deposition increased during summer months, followed by a sudden decrease during monsoon months and a gradual increase further. In comparison, wet dust deposition increased with an increase in relative humidity in the months of monsoon season, as relative humidity was higher. The study outcome would help prescribing suitable measure to mitigate the industrial pollution in and around the Kalinga Nagar industrial area, Odisha, India.

References

Balakrishnan, K., Dey, S., Gupta, T., Dhaliwal, R. S., Brauer, M., Cohen, A. J., . . . Dandona, L. (2019). The impact of air pollution on deaths, disease burden, and life expectancy across the states of India: The Global Burden of Disease Study 2017. *The Lancet Planetary Health*, *3*(1), e26−e39. Available from https://doi.org/10.1016/S2542-5196(18)30261-4.

Bauwens, M., Compernolle, S., Stavrakou, T., Müller, J.-F., Gent, J., Eskes, H., . . . Zehner, C. (2020). Impact of coronavirus outbreak on NO_2 pollution assessed using TROPOMI and OMI observations. *Geophysical Research Letters*, *47*(11). Available from https://doi.org/10.1029/2020GL087978.

Behera, M. D., Mudi, S., Shome, P., Das, P. K., Kumar, S., Joshi, A., & Sirivella, Z. (2022). COVID-19 slowdown induced improvement in air quality in India: Rapid assessment using Sentinel-5P TROPOMI data. *Geocarto International*, *37*(25). Available from https://doi.org/10.1080/10106049.2021.1993351.

Brauer, M., Guttikunda, S. K., Nishad, K. A., Dey, S., Tripathi, S. N., Weagle, C., & Martin, R. V. (2019). Examination of monitoring approaches for ambient air pollution: A case study for India. *Atmospheric Environment*, *216*, 116940. Available from https://doi.org/10.1016/j.atmosenv.2019.116940.

Cohen, A. J., Brauer, M., Burnett, R., Anderson, H. R., Frostad, J., Estep, K., . . . Forouzanfar, M. H. (2017). Estimates and 25-year trends of the global burden of disease attributable to ambient air pollution: An analysis of data from the Global Burden of Diseases Study 2015. *The Lancet*, *389*(10082), 1907−1918. Available from https://doi.org/10.1016/S0140-6736(17)30505-6.

Das, P., Behera, M. D., Barik, S. K., Mudi, S., Jagadish, B., Sarkar, S., & Chauhan, P. S. (2022). Shifting cultivation induced burn area dynamics using ensemble approach in Northeast India. *Trees, Forests and People*, *7*, 100183. Available from https://doi.org/10.1016/j.tfp.2021.100183.

Deb Burman, P. K., & Das, P. (2024). A data-driven approach to assess the impact of climate change on a tropical mangrove in India. *Journal of Geophysical Research: Biogeosciences*, *129*(8). Available from https://doi.org/10.1029/2023JG007911, e2023JG007911.

Dutta, M., & Chatterjee, A. (2022). A deep insight into state-level aerosol pollution in India: Long-term (2005−2019) characteristics, source apportionment, and future projection (2023). *Atmospheric Environment*, *289*, 119312. Available from https://doi.org/10.1016/j.atmosenv.2022.119312.

Gouda, K. C., Gogeri, I., & ThippaReddy, A. S. (2022). Assessment of aerosol optical depth over Indian subcontinent during COVID-19 lockdown (March−May 2020). *Environmental Monitoring and Assessment*, *194*(3), 195. Available from https://doi.org/10.1007/s10661-022-09855-3.

Guo, M., Li, J., Xu, J., Wang, X., He, H., & Wu, L. (2017). CO_2 emissions from the 2010 Russian wildfires using GOSAT data. *Environmental Pollution*, *226*, 60−68. Available from https://doi.org/10.1016/j.envpol.2017.04.014.

Gupta, G., Ratnam, M. V., Madhavan, B. L., & Narayanamurthy, C. S. (2022). Long-term trends in aerosol optical depth obtained across the globe using multi-satellite measurements. *Atmospheric Environment*, *273*, 118953. Available from https://doi.org/10.1016/j.atmosenv.2022.118953.

Guttikunda, S. K., Goel, R., & Pant, P. (2014). Nature of air pollution, emission sources, and management in the Indian cities. *Atmospheric Environment*, *95*, 501−510. Available from https://doi.org/10.1016/j.atmosenv.2014.07.006.

Kuttippurath, J., & Raj, S. (2021). Two decades of aerosol observations by AATSR, MISR, MODIS and MERRA-2 over India and Indian Ocean. *Remote Sensing of Environment*, *257*, 112363. Available from https://doi.org/10.1016/j.rse.2021.112363.

Lal, P., Kumar, A., Kumar, S., Kumari, S., Saikia, P., Dayanandan, A., ... Khan, M. L. (2020). The dark cloud with a silver lining: Assessing the impact of the SARS COVID-19 pandemic on the global environment. *Science of The Total Environment*, *732*, 139297. Available from https://doi.org/10.1016/j.scitotenv.2020.139297.

Leelőssy, Á., Molnár, F., Izsák, F., Havasi, Á., Lagzi, I., & Mészáros, R. (2014). Dispersion modeling of air pollutants in the atmosphere: A review. *Open Geosciences*, *6*(3). Available from https://doi.org/10.2478/s13533-012-0188-6.

Li, L., Wang, Y., Zhang, Q., Yu, T., Zhao, Y., & Jin, J. (2007). Spatial distribution of aerosol pollution based on MODIS data over Beijing, China. *Journal of Environmental Sciences*, *19*(8), 955–960. Available from https://doi.org/10.1016/S1001-0742(07)60157-0.

Mandal, S., Das, P., Behera, M. D., & Parida, B. R. (2022). *Himalayan carbon flux demonstrates higher aseasonality GOSAT-derived spatio-temporal profile and comparison through wind and water vapour, . Handbook of Himalayan ecosystems and sustainability* (Volume 1, pp. 183–201). CRC Press. Available from https://doi.org/10.1201/9781003268383-13.

Mhawish, A., Sorek-Hamer, M., Chatfield, R., Banerjee, T., Bilal, M., Kumar, M., & Kalashnikova, O. (2021). Aerosol characteristics from earth observation systems: A comprehensive investigation over South Asia (2000–2019). *Remote Sensing of Environment*, *259*, 112410. Available from https://doi.org/10.1016/j.rse.2021.112410.

Mukherjee, T., & Vinoj, V. (2020). Atmospheric aerosol optical depth and its variability over an urban location in Eastern India. *Natural Hazards*, *102*(2), 591–605. Available from https://doi.org/10.1007/s11069-019-03636-x.

Pathakoti, M., Muppalla, A., Hazra, S., D. Venkata, M., A. Lakshmi, K., K. Sagar, V., & Vijayasundaram, U. (2021). Measurement report: An assessment of the impact of a nationwide lockdown on air pollution–a remote sensing perspective over India. *Atmospheric Chemistry and Physics*, *21*(11). Available from https://doi.org/10.5194/acp-21-9047-2021.

Ramachandran, S., Kedia, S., & Srivastava, R. (2012). Aerosol optical depth trends over different regions of India. *Atmospheric Environment*, *49*, 338–347. Available from https://doi.org/10.1016/j.atmosenv.2011.11.017.

Rani, S., & Kumar, R. (2022). Spatial distribution of aerosol optical depth over India during COVID-19 lockdown phase-1. *Spatial Information Research*, *30*(3), 417–426. Available from https://doi.org/10.1007/s41324-022-00442-9.

Ranjan, A. K., Patra, A. K., & Gorai, A. K. (2020). Effect of lockdown due to SARS COVID-19 on aerosol optical depth (AOD) over urban and mining regions in India. *Science of the Total Environment*, *745*, 141024. Available from https://doi.org/10.1016/j.scitotenv.2020.141024.

Sharma, G., Gupta, M., Gargava, P., & Kota, S. H. (2024). Mapping air quality trends across 336 cities in India: Insights from three decades of monitoring (1987–2019). *Environment International*, 108979. Available from https://doi.org/10.1016/j.envint.2024.108979.

Sharma, R., Kumar, R., Sharma, D. K., Son, L. H., Priyadarshini, I., Pham, B. T., & Rai, S. (2019). Inferring air pollution from air quality index by different geographical areas: Case study in India. *Air Quality, Atmosphere & Health*, *12*. Available from https://doi.org/10.1007/s11869-019-00749-x.

Singh, R. K., Biradar, C. M., Behera, M. D., Prakash, A. J., Das, P., Mohanta, M. R., & Rizvi, J. (2024). Optimising carbon fixation through agroforestry: Estimation of aboveground biomass using multi-sensor data synergy and machine learning. *Ecological Informatics*, *79*, 102408. Available from https://doi.org/10.1016/j.ecoinf.2023.102408.

Varghese, R., Behera, S. K., & Behera, M. D. (2023). Significant inverse influence of tropical Indian Ocean SST on SIF of Indian vegetation during the summer monsoon onset phase. *Remote Sensing*, *15*(7), 1756. Available from https://doi.org/10.3390/rs15071756.

Yoro, K. O., & Daramola,, M. O. (2020). *CO_2 emission sources, greenhouse gases, and the global warming effect.. Advances in carbon capture* (pp. 3–28). Elsevier. Available from https://doi.org/10.1016/B978-0-12-819657-1.00001-3.

Index

Note: Page numbers followed by "*f*" and "*t*" refer to figures and tables, respectively.

D

E

9780443140723